Human Physiology

Human Physiology

Stuart Ira Fox

**Los Angeles City College
and California State at Northridge**

wcb

Wm. C. Brown Publishers
Dubuque, Iowa

Cover Photo

© William B. Westwood

Book Team

Edward T. Jaffe
Senior Editor
Judith A. Clayton
Senior Developmental Editor
Laura Beaudoin
Senior Production Editor
Mark E. Christianson
Designer
Shirley M. Charley
Visual Research Editor
Vicki Krug
Permissions Editor

wcb
group

Wm. C. Brown
Chairman of the Board
Mark C. Falb
President and Chief Executive Officer

wcb

Wm. C. Brown Publishers, College Division

Lawrence E. Cremer
President
James L. Romig
Vice-President, Product Development
David Wm. Smith
Vice-President, Marketing
David A. Corona
Vice-President, Production and Design
E. F. Jogerst
Vice-President, Cost Analyst
Marcia H. Stout
Marketing Manager
Linda M. Galarowicz
Director of Marketing Research
William A. Moss
Production Editorial Manager
Marilyn A. Phelps
Manager of Design
Mary M. Heller
Visual Research Manager

To Ellen and Laura, for their patience and support; to my parents, Bess and Sam; to Eileen and to the memory of my father-in-law, Harold, in the hope that scientific advances quickly outdate some of the information in this text

Contents

vii

Tables

Preface

Human physiology courses at the introductory level are unique. Undergraduates majoring in English, the humanities, or the social sciences enroll in large numbers out of sheer interest, but most students who take a human physiology course enroll because it is required of their major. These include students who are majoring in health areas such as pre-nursing, radiation technology, nuclear medicine technology, nutrition, dental hygiene, medical technology; students who are pre-medical, pre-dental, and pre-veterinarian majors; and others who are majoring in areas such as physical education, hospital administration, education, early childhood development, home economics, and many others. The degree of academic preparation in the sciences and the level of motivation vary tremendously among students in these courses, as does the degree of rigor required—some students only want to learn whatever is required to pass, while for others a relatively detailed mastery of the subject is absolutely necessary for success in their career. For these reasons, courses in introductory human physiology are uniquely difficult to teach.

I have taught human physiology at a university and at a community college for many years. During this time I have longed for a textbook that could make my task easier by presenting physiological concepts in a contemporary, comprehensive fashion that would be interesting and understandable to all of the diverse students who take this class. While there were many excellent texts available, all had major shortcomings for use at this level; some were too abbreviated and too simple to be useful to students in pre-health professions, while others were either too detailed or too advanced for many other students. I therefore have attempted to write a textbook that is sufficiently contemporary and detailed for the needs of pre-professional students and at the the same time, interesting and understandable to students who lack a science background. Features of the text that aided in accomplishing this difficult task are described briefly here.

Content

All areas of human physiology are presented in a detailed and contemporary fashion. Basic concepts are emphasized, and the practical applications of these concepts—in health and disease and in exercise physiology—are presented in sufficient detail to stimulate interest and improve the transfer of knowledge from the classroom to the professional arena. This textbook, in other words, is written so that students can learn, at both the identification and conceptual levels, the subjects that must be mastered in a human physiology course. Background information in biology and chemistry is presented, so that all students, regardless of their prior academic preparation, can understand the physiological concepts presented.

Writing Style

The concepts in *Human Physiology* are presented in a simple and straightforward style. Specialized jargon and complex sentences have been intentionally avoided. Since it is assumed that most students do not have a prior background in the subject matter, technical terms are defined and explained when they are introduced. I have endeavored to make the writing style so clear that even the most potentially difficult concepts can be understood by the average student in an introductory human physiology course.

Organization

I tell my students that basic concepts are those that provide a foundation for further study; the term *basic* does not mean "simple." In fact, basic concepts are often more difficult to grasp than more specialized concepts. For these reasons, the first few chapters of this textbook are both comprehensive and contemporary so that those students who do not have as good a background in biology and chemistry as others do will be able to understand the physiological concepts that are presented later.

The importance of the nervous system in human physiology is recognized by the fact that four chapters are devoted to this subject, following those chapters on basic biological and chemical concepts. The middle portion of the book concentrates on "blood and guts" physiology—that is, the physiology of the cardiovascular, pulmonary, renal, and digestive systems. These subjects are described with the care and detail required by students who intend to enter the health professions. A separate chapter on the immune system is presented in a manner that is more complete than that provided by other textbooks at this level; I believe this is justified by the obvious clinical relevance of this topic and the rapid advances of research in this area. The endocrine system is treated in a modern and comprehensive fashion in the last three chapters, ending at the beginning—with a chapter on reproduction.

Practical Applications

Each student—regardless of major—has a personal interest in health, and most students have an intense curiosity about the physiological mechanisms involved during exercise and in disease processes. Students who intend to enter the health professions are additionally motivated to learn the practical applications of physiology to their proposed career. Practical applications thus help motivate students to learn fundamental concepts, and these applications provide a means of making concepts more accessible to students in an introductory course. I have therefore introduced applications of concepts to health, disease processes, and exercise physiology whenever appropriate throughout this textbook.

Pedagogical Aids

The following devices are provided within *Human Physiology* to help students study and learn the material presented:

1. Outline of the concepts discussed, with page references, at the beginning of each chapter
2. Behavioral objectives at the beginning of each chapter
3. Boldface and italics to highlight important concepts and scientific terms within each chapter
4. Study activities within each chapter
5. Outline summary at the end of each chapter
6. Self-study quiz at the end of each chapter
7. Answers to self-study quizzes in Appendix A
8. Supplementary readings in Appendix B
9. Glossary at the end of the book

Ancillary Material

The following material is available from Wm. C. Brown Publishers to supplement the use of this textbook in the teaching of human physiology:

1. *Instructor's Guide,* by Stuart I. Fox. The instructor's guide provides suggested answers to study activity questions, answers to self-study quizzes, and additional objective-type test questions with answers. The instructor's guide also provides suggestions for ancillary material such as laboratory exercises, films, slides, and other support media.
2. *Laboratory Guide to Human Physiology: Concepts and Clinical Applications, Third Edition,* by Stuart I. Fox. While this laboratory guide is self-contained and can be used with any textbook, it is particularly well adapted for use in courses that employ this book as a lecture text.
3. Overhead transparencies of some of the figures in this textbook.

Acknowledgments

Human Physiology could not have been written without the support and encouragement of the chairman of the Life Science Department at Los Angeles City College, Professor Robert J. Lyon. For these favors, and for his illuminating example of brilliant scholarship and humanism, I wish to express my deepest gratitude. I also would like to thank my colleagues of the Life Science Department at Los Angeles City College—particularly Professors Alice Logrip, Chen-Hau Poon, Lester Schneider, and James Sandoval—for their support and advice during the writing of this textbook. Sandy Sanford, California State University, Northridge, helped me greatly in the preparation of the Glossary.

My editor, Ed Jaffe, demonstrated an amazing degree of patience and perseverance during the production of *Human Physiology* and in so doing, contributed enormously to the final completion of the project. I would also like to thank the reviewers of the book, whose suggestions, criticisms and praise have been invaluable. These reviewers are: T. Daniel Kimbrough, Virginia Commonwealth University; James M. Rankin, Michigan State University; Joe William Crim, University of Georgia; Richard W. Heninger, Brigham Young University; Charles J. Imig, University of Iowa; Donovan J. Nielsen, Mankato State University; Maureen Diggins, Augustana College; Robert Catlett, University of Colorado. Despite the high degree of professional competence of these reviewers, I am sure I have made some mistakes in *Human Physiology.* For these, I am entirely to blame. I urge all readers of this book—professors and students—to inform me of these errors, and to send me any criticisms or suggestions you may have. I look forward to working with you and changing this textbook as the field of physiology itself grows and changes.

Human Physiology

Introduction: *Tissues, Organs, and Control Systems*

Objectives

By studying this chapter, you should be able to:

1. Describe, in a general way, the topics studied in physiology and the importance of physiology in modern medicine

2. Describe the four primary tissues in terms of their distinguishing characteristics, subtypes, and locations in the body

3. Describe how the primary tissues are organized into organs, using the skin as an example

4. Define homeostasis and describe the importance of this concept in the study of physiology and medicine

5. Describe negative feedback, and give examples of the way negative feedback loops help maintain homeostasis

Physiology is the study of biological function—of how the body works, from cell to tissue, tissue to organ, organ to system, and of how the organism as a whole accomplishes particular tasks essential for life. In the study of physiology, the emphasis is on *mechanisms*—with questions that begin with the word *how* and answers that involve cause-and-effect sequences. These sequences can be woven into larger and larger stories that include descriptions of the structures involved (anatomy) and that overlap other sciences, particularly biochemistry. In order to understand the physiology of the kidneys, for example, you must know their structure, understand the principles of filtration and transport mechanisms, and learn how the kidneys help to maintain fluid, electrolyte, and pH balance. Further, to completely understand the physiology of the kidneys, you must appreciate how kidney function affects the cardiovascular system and how the kidneys are in turn regulated by the nervous and endocrine systems.

Figure 1.1 Three skeletal muscle fibers showing the characteristic cross-striations.

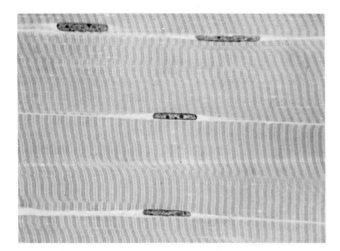

Figure 1.2 Human cardiac muscle. Notice the striated appearance and dark-staining intercalated discs.

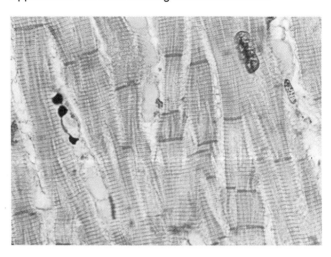

The separate facts of these cause-and-effect stories are empirical, and their causal relationships are based on experimental evidence. Stories that seem logical are not necessarily true—they are only as valid as the data on which they are based, and they can change as new techniques are developed and further experiments are performed.

One standard technique for investigating the function of an organ is to observe what happens when it is surgically removed from an experimental animal or when its function is biochemically altered in a specific way. This study is often aided by "experiments of nature"—diseases—that involve specific damage to the normal function of an organ. The study of disease processes has thus aided the understanding of normal function, and the study of normal physiology has provided the scientific basis of modern medicine. This relationship is recognized by the Nobel prize committee who award prizes in "Physiology or Medicine."

Although physiology is the study of function, it is difficult to properly conceptualize the function of the body without some knowledge of its anatomy, particularly at a microscopic level. The anatomy of individual organs will be discussed together with their function in later chapters. All organs, however, have a common "fabric," which is composed of only four *primary tissues*. An understanding of this common fabric can thus aid the study of many areas of physiology.

The Primary Tissues

Cells that have similar functions are grouped into categories called *tissues*. The entire body is composed of only four types of tissues. These **primary tissues** include (1) muscle, (2) nerve, (3) epithelial, and (4) connective tissues. Groupings of these four primary tissues into anatomical and functional units are known as *organs*. Organs, in turn, may be grouped together by common functions into *systems*.

Muscles

Muscle tissue is specialized for contraction. There are three types of muscles: (1) *skeletal muscles,* (2) *cardiac muscle,* and (3) *smooth muscle.* Skeletal muscle is often called *voluntary muscle* because we have conscious control of its contraction without special training. Both skeletal and cardiac muscle are *striated*—they have striations, or stripes, that extend across the width of the muscle cell, and for this reason they have similar mechanisms of contraction (see chapter 8). Smooth muscle lacks these cross-striations and has a different mechanism of contraction (see chapter 9).

Figure 1.3 Smooth muscle in a human uterus showing a longitudinal section of smooth muscle. Arrows show nuclei *(N)*.

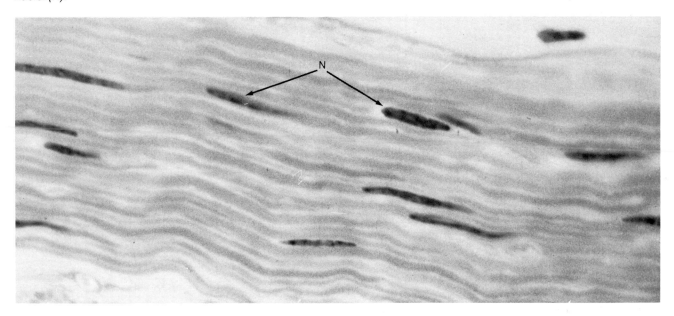

Skeletal Muscle Skeletal muscles are usually attached by means of tendons to bones at both ends so that contraction produces movements of the skeleton. The same type of muscle, however, is also located in the tongue, oral cavity, and esophagus—providing voluntary control of speech and swallowing—and in the anal sphincter (allowing voluntary control of defecation). The diaphragm—a domelike sheet of muscle that forms the floor of the thoracic (chest) cavity—is also composed of skeletal muscle.

Since skeletal muscle cells are long and thin they are called **fibers** or **myofibers** (myo = muscle). Despite their specialized structure and function, each myofiber contains structures common to all cells (nuclei, mitochondria, and other organelles—see chapter 3).

In some skeletal muscles, the muscle fibers extend in parallel from one end of the muscle to the other. Contraction of the entire muscle is therefore due to summation of the separate myofiber contractions. The parallel arrangement of myofibers allows each cell to be controlled individually—one can thus contract fewer or more muscle fibers and vary the strength of the whole muscle's contraction. The ability to vary (or "grade") the strength of skeletal muscle contraction is obviously needed for proper control of skeletal movements.

Cardiac Muscle Though striated, cardiac muscle has a very different appearance than skeletal muscle. Cardiac muscle is found only in the heart, where the **myocardial cells** are short, branched, and intimately interconnected to form a continuous fabric. Special areas of contact between adjacent cells stain darkly to show *intercalated discs* that are characteristic of heart muscle (see figure 1.2).

The intercalated discs couple myocardial cells together, both mechanically and electrically. Unlike skeletal muscle, therefore, the heart cannot produce a graded contraction by varying the number of cells stimulated to contract. Because of the way it is constructed, stimulation of one myocardial cell results in stimulation of all other cells in the mass and a whole-hearted contraction.

Smooth Muscle As implied by the name, smooth-muscle cells do not have the cross-striations characteristic of skeletal and cardiac muscle. Smooth muscle (see figure 1.3) is found in the digestive tract (from the lower portion of the esophagus to the anal sphincter), blood vessels, bronchioles (small air passages in the lungs), and in the

Figure 1.4 A neuron within the spinal cord, showing the cell body with nucleus, dendrites, and axon.

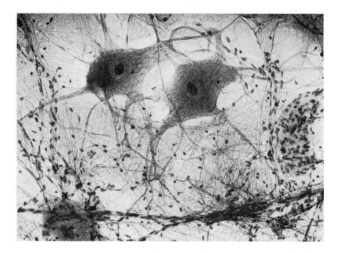

urinary and reproductive systems. Circular arrangements of smooth muscles in these organs produce constriction of the *lumen* (cavity) when the muscle cells contract. The digestive tract also contains longitudinally arranged layers of smooth muscle. Rhythmic contraction of circular and longitudinal muscles pushes food from one end of the digestive tract to the other.

Nerve Tissue

Nerve tissue is specialized for generation and conduction of electrical events. By means of these electrical events and by means of chemical secretions that result from this electrical activity, nerve tissue provides the major regulatory system of the body.

Nerve cells are called **neurons.** Each neuron consists of three parts: (1) a *cell body,* which contains the nucleus and which serves as the metabolic center of the cell; (2) numerous, highly branched cytoplasmic extensions from the cell body called *dendrites* (literally, "branches"), which receive input from other neurons or from receptor cells; and (3) a single cytoplasmic process, which may be as long as a few feet in length, called the *axon,* or *nerve fiber,* which is specialized for conduction of nerve impulses over long distances.

Figure 1.5 Formation of exocrine and endocrine glands from epithelial membranes.

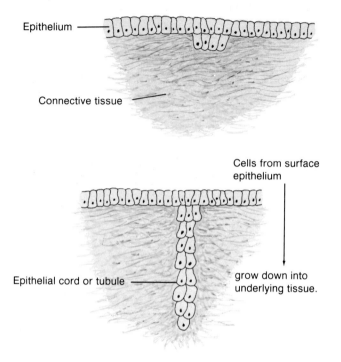

Epithelium

Connective tissue

Cells from surface epithelium

Epithelial cord or tubule

grow down into underlying tissue.

If **exocrine** gland forms,

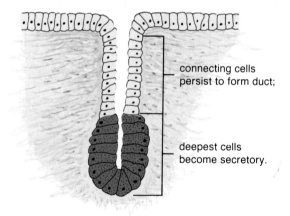

connecting cells persist to form duct;

deepest cells become secretory.

If **endocrine** gland forms,

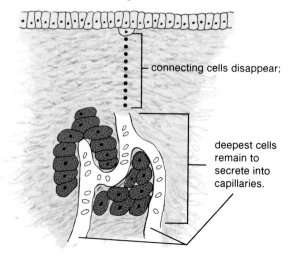

connecting cells disappear;

deepest cells remain to secrete into capillaries.

Figure 1.6 Simple squamous *(a)*, simple cuboidal *(b)*, and simple columnar *(c)* epithelial membranes. Tissue beneath each membrane is connective tissue.

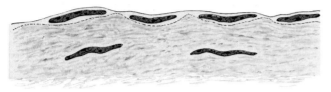

(a)

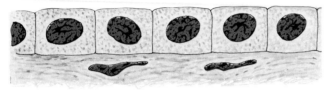

(b)

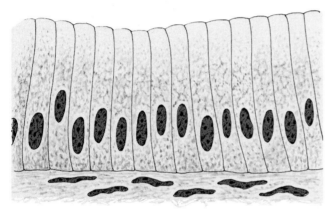

(c)

Epithelial Tissue

Epithelial tissue consists of cells that form **membranes** that cover and line the body surfaces, and of **glands** that are derived from these membranes. There are two main types of glands. *Exocrine glands* (exo = outside) secrete chemicals through a duct that leads to the outside of the membrane (and thus to the outside of the body). *Endocrine glands* (endo = within) secrete chemicals called *hormones* into the blood.

Epithelial membranes are classified according to their number of layers and according to the shape of the cells in the upper layer. Epithelial cells that are flattened in shape are *squamous;* those that are taller than they are wide are *columnar,* and those that are as wide as they are tall are *cuboidal.* Those epithelial membranes that are only one cell layer in thickness are known as *simple* membranes, while those that are composed of a number of layers are *stratified* membranes.

A simple squamous membrane forms the lining of all blood vessels, where it is known as an *endothelium.* Most small ducts of exocrine glands are lined by a simple cuboidal epithelium. A simple columnar epithelium lines the stomach and intestine.

The *epidermis* of the skin is a stratified squamous epithelial membrane. Since the epidermis is dry and exposed to the potentially desiccating effects of the air, the surface is covered with dead cells that are filled with a water-resistant protein known as *keratin.* This protective layer is constantly flaked off the surface of the skin and therefore must be constantly replaced by division of cells in deeper layers of the epidermis.

The constant loss and renewal of cells is characteristic of epithelial membranes. The entire epidermis is completely replaced about every two weeks; the stomach lining is renewed every two to three days. Examination of cells that are lost (or "exfoliated") from the surface of the female genital tract is a common procedure in gynecology (as in the Pap smear).

In order to form a strong membrane that is effective as a barrier at the body surfaces, epithelial cells are very closely packed and joined together by structures collectively called *junctional complexes.* There is no room for blood vessels between adjacent epithelial cells; the epithelium must receive nourishment from the tissue beneath, which has large intercellular spaces that can accommodate blood vessels and nerves. This underlying tissue is called *connective tissue.*

Figure 1.7 Stratified squamous epithelial membranes.
(a) shows the non-cornified membrane of the esophagus.
(b) is a photomicrograph of the thick skin of the palm,
showing a thick cornified layer.

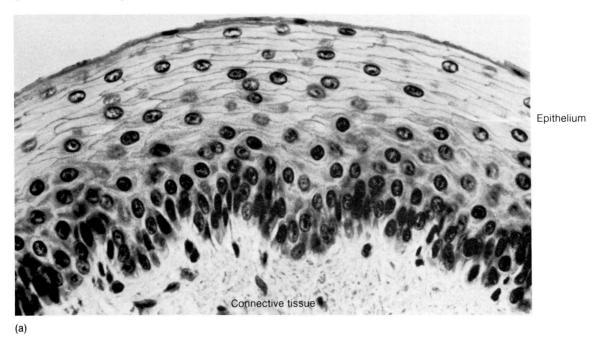

Epithelium

Connective tissue

(a)

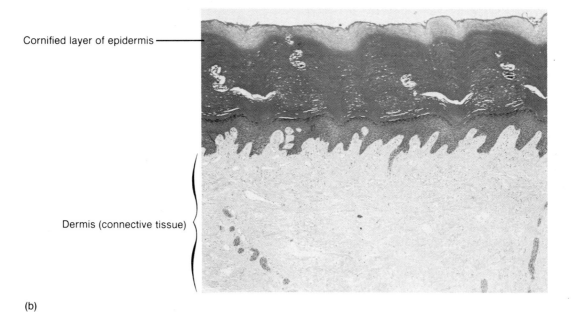

Cornified layer of epidermis —

Dermis (connective tissue)

(b)

Connective Tissue

Connective tissue has large amounts of extracellular material in the spaces between adjacent cells. The *dermis* of skin is an example of *loose connective tissue,* in which the extracellular spaces are filled with protein-rich fluid and scattered fibrous proteins such as *collagen.*

Tendons, which connect muscles to bone, and ligaments, which connect bones together at joints, are examples of *dense regular connective tissue.* This tissue—as implied by its name—contains a dense arrangement of collagen fibers that are parallel to each other. *Dense, irregular connective tissue* contains a meshwork of collagen fibers. This connective tissue forms tough capsules and sheaths around organs.

Figure 1.8   A section of skin showing the loose
connective tissue dermis beneath the cornified epidermis.
Loose connective tissue contains scattered collagen
fibers in a matrix of protein-rich fluid. The intercellular
spaces also contain cells and blood vessels.

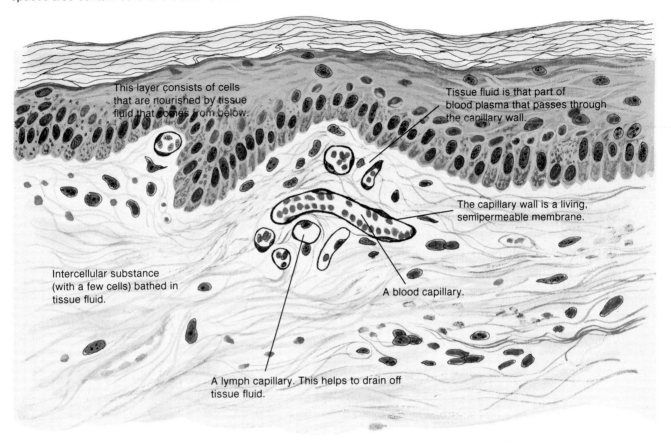

This layer consists of cells
that are nourished by tissue
fluid that comes from below.

Tissue fluid is that part of
blood plasma that passes through
the capillary wall.

The capillary wall is a living,
semipermeable membrane.

Intercellular substance
(with a few cells) bathed in
tissue fluid.

A blood capillary.

A lymph capillary. This helps to drain off
tissue fluid.

Figure 1.9   Dense irregular connective tissue. Note the
tightly packed, irregularly arranged collagen proteins.

Figure 1.10   Photomicrograph of a tendon showing
dense, regular arrangement of collagen fibers.

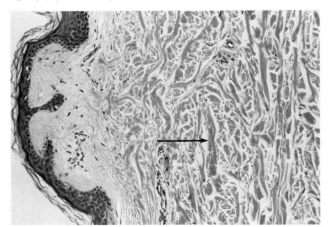

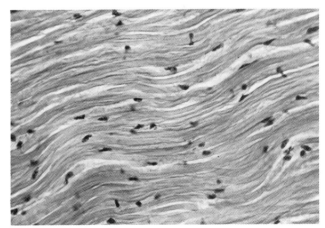

Figure 1.11   Diagram illustrating how a bone grows in width. Cells within the outer connective tissue covering of the bone (the periosteum) add new bone lamellae around blood vessels within the periosteum. This produces new haversian systems with a central canal containing the blood vessel and lined by the same connective tissue (which is now an endosteum).

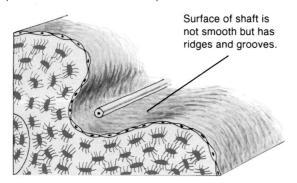

Surface of shaft is not smooth but has ridges and grooves.

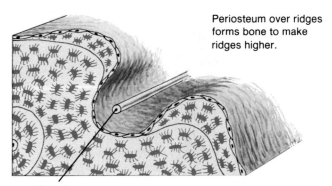

Periosteum over ridges forms bone to make ridges higher.

Vessel in groove.

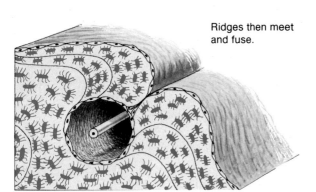

Ridges then meet and fuse.

This makes groove a tunnel.

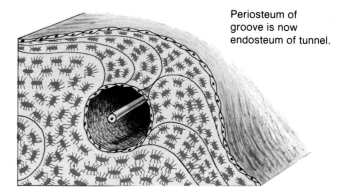

Periosteum of groove is now endosteum of tunnel.

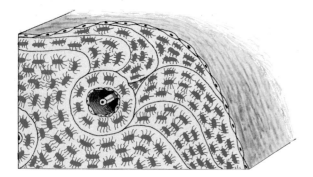

By depositing layers of bone inside tunnel,

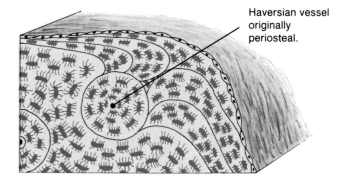

Haversian vessel originally periosteal.

Endosteal cells make it a haversian system.

Figure 1.12 Cross section of a tooth showing pulp, dentin, and enamel. The cells that form dentin (odontoblasts) are located in the pulp, and their processes extend into the dentin-forming tubules. When these processes die, the tubules fill with air and appear dark. The odontoblasts in the pulp can form new reparative dentin to seal off the pulp from the dead tracts.

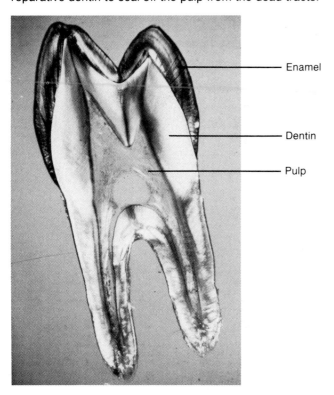

Enamel

Dentin

Pulp

Blood, cartilage, and the calcified ("hard") tissues—bone and the hard layers in teeth (enamel, dentin, and cementum)—are collectively called *special connective tissues*. Blood is usually classified as connective tissue because about half its volume is composed of extracellular fluid known as plasma. The hard, calcified tissues are formed by precipitation of calcium phosphate (as hydroxyapatite crystals) in the extracellular spaces of this connective tissue.

Bone is produced as concentric layers, or *lamellae*, of calcified material laid around blood vessels. The cells that form bone—called *osteoblasts*—become trapped within cavities (called *lacunae*) surrounded by their calcified products. The trapped cells, which are now called *osteocytes*, remain alive because they are nourished by "lifelines" of cytoplasm that extend from the cell to the blood vessel in *canaliculi* (little canals). The blood vessel lies within a central canal surrounded by concentric rings of bone lamellae with their trapped osteocytes; this entire unit is called a *Haversian system* (see figure 1.11).

Dentin is similar in composition to bone, but the cells that form this calcified tissue are located within the tooth in the pulp cavity (a connective tissue area that also contains blood vessels, and nerve endings that mediate pain). The cells that form dentin—odontoblasts—send cytoplasmic extensions called *dentinal tubules* into the dentin. Tooth dentin, like bone, therefore contains living cytoplasm. Both bone and tooth dentin, as a result, are able to respond to stresses by growth and resorption (loss of calcified material). The cells that form the outer enamel of a tooth, in contrast, are lost as the tooth erupts. Enamel is highly calcified material (harder than bone or dentin) that cannot be regenerated; artificial "fillings" are therefore required to patch holes.

Organs

An organ is a structure composed of at least two, and usually all four, types of the primary tissues. Each primary tissue contributes to the structure and function of each organ. The largest organ of the body, in terms of its surface area, is the skin. The numerous functions of the skin serve to illustrate how primary tissues cooperate in the service of organ physiology.

The Skin

The connective tissue dermis of the skin is covered by the epidermis. The cornified epidermis protects the skin against water loss and against invasion by disease-causing organisms. Invaginations of epithelium into the underlying connective tissue creates exocrine glands of the skin. These include hair follicles (which secrete the proteins that compose hair), sweat glands, and sebaceous glands. Secretions of sweat glands cool the body by evaporation and produce odors that, at least in lower animals, serve as sexual attractants. Sebaceous glands secrete oily sebum into hair follicles, where it is transported to the surface of the skin (unless these ducts are blocked, in which case "blackheads" are produced). Sebum lubricates the cornified surface of the skin, helping to prevent it from drying and cracking (as in chapped lips; there are no sebaceous glands in the lips—one must therefore moisten the lips periodically with the tongue).

Figure 1.13 A diagram showing the structure of skin. Notice that all four types of primary tissues are present.

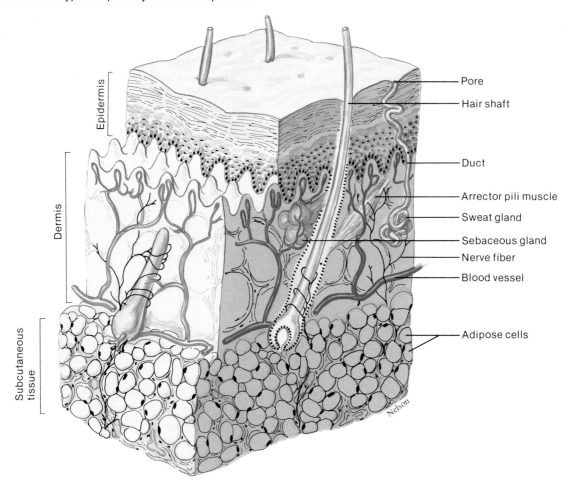

Epidermis

Dermis

Subcutaneous tissue

Pore

Hair shaft

Duct

Arrector pili muscle

Sweat gland

Sebaceous gland

Nerve fiber

Blood vessel

Adipose cells

Nelson

The skin is nourished by blood vessels within the dermis. In addition to blood vessels, the dermis contains wandering white blood cells and other types of cells that protect against invading disease-causing organisms, as well as containing nerve fibers and fat cells. Most of the fat cells, however, are grouped together to form the *hypodermis* (a layer beneath the dermis). Although fat cells are a type of connective tissue, masses of fat deposits throughout the body—such as subcutaneous fat—are referred to as **adipose tissue.**

Sensory nerve endings within the dermis mediate the cutaneous sensations of touch, pressure, heat, cold, and pain. Some of these sensory stimuli directly affect the sensory nerve endings. Others act via sensory structures derived from epithelial cells. These include Pacinian corpuscles in the dermis that monitor sensations of pressure (see figure 1.14). Motor nerve fibers in the skin stimulate effector organs, such as secretion of exocrine glands and contraction of arrector pili muscles that attach to hair follicles and surrounding connective tissue (producing goose bumps). The degree of constriction or dilation of cutaneous blood vessels, and therefore the rate of blood flow, is also regulated by nerve fibers.

The epidermis itself is a dynamic structure that can respond to environmental stimuli. The rate of its cell division, and consequently the thickness of the cornified layer, increases under the stimulus of constant abrasion. This produces calluses. The skin can also protect itself against the dangers of ultraviolet light by increasing its production of *melanin* pigment, which absorbs ultraviolet light while producing a tan. The skin is also an endocrine gland that produces and secretes vitamin D (derived from cholesterol under the influence of ultraviolet light) which functions as a hormone.

Figure 1.14 Examples of sensory organs derived from epithelial tissue. *(a)* is a taste bud in the tongue. *(b)* is a receptor for deep pressure, the Pacinian corpuscle. The latter consists of epithelial cells and connective tissue proteins that form concentric layers around the ending of a sensory nerve.

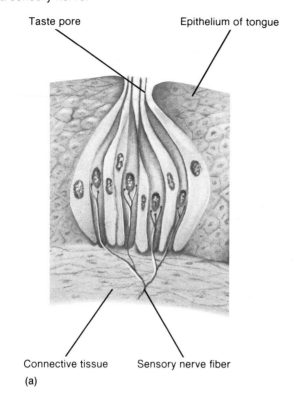

Taste pore   Epithelium of tongue

Connective tissue   Sensory nerve fiber

(a)

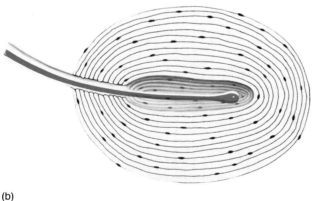

(b)

The architecture of most organs is similar to that of the skin. Most are covered by an epithelium immediately over a connective tissue layer. This connective tissue contains blood vessels, nerve endings, scattered cells for fighting infection, and may contain glandular cells with ducts (if there are exocrine glands) derived from epithelial tissue. If the organ has a hollow area *(lumen)*—as in the digestive tract or in blood vessels—the lumen is also lined with an epithelium immediately over a connective tissue layer. The presence and type of muscle, glandular, and nervous tissue varies in different organs.

Organ Systems

Organs that are located in different regions of the body and that perform related functions are grouped together into **systems.** These include the nervous system, endocrine system, cardiovascular system, respiratory system, urogenital system, digestive system, and immune system. By means of numerous regulatory mechanisms, these systems work together to maintain the life and health of the entire organism.

1. List the four primary tissues. Describe the distinguishing characteristics of each and some of the structures that each form.
2. Describe the roles of each primary tissue in the structure and function of the skin.

Homeostasis

Over a century ago the French physiologist Claude Bernard observed that the *melieu interior* (internal environment) remains remarkably constant despite changing conditions in the external environment. In a book entitled *The Wisdom of the Body* (published in 1932), Walter Cannon coined the term **homeostasis** to describe this internal constancy. Cannon further suggested that mechanisms of physiological regulation exist for one purpose—maintenance of internal constancy.

Although Cannon may have overstated his case somewhat (mechanisms for growth and reproduction do not help maintain homeostasis), the concept of homeostasis has been of inestimable value in understanding physiological control mechanisms. When particular regulatory functions are not working properly, homeostasis is not maintained and a person is sick. The concept of homeostasis therefore also provides a major foundation of modern medical diagnostic procedures. When a particular

Figure 1.15 Examples of three different organs. *(a)* is a photomicrograph of the inner layer of the heart (the endocardium). *(b)* is a photomicrograph of the cornea of the eye (the stroma is composed of dense regular connective tissue). *(c)* is a diagram of a small brochus in the lungs.

Endothelium

Connective tissue

Cardiac muscle

(a)

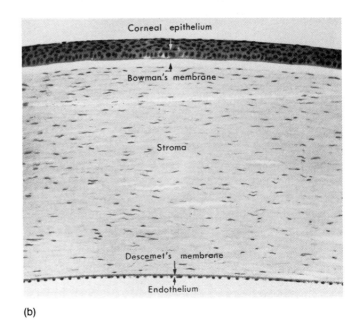

Corneal epithelium

Bowman's membrane

Stroma

Descemet's membrane

Endothelium

(b)

Epithelium

Alveolus

Mixed glands

Muscle

Alveolus

Cartilage

Cartilage

Muscle   Artery   Nerve   Vein   Adipose tissue

(c)

Figure 1.16 A rise in some factor of the internal environment (↑X) is detected by a sensor. The sensor activates an effector (generally, nerves or hormones), which causes a decrease in this factor (↓X). In this way the factor returns to its initial level and homeostasis is maintained. Completion of the negative feedback loop is shown by a dotted arrow and a negative sign.

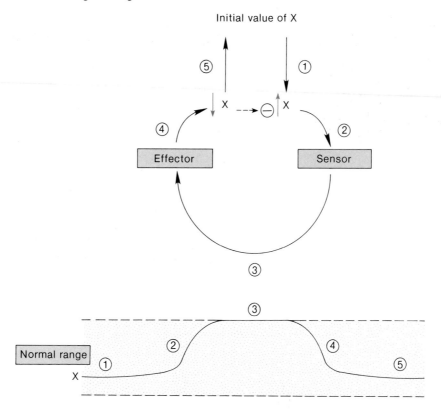

Table 1.1 Approximate normal ranges for measurements of some blood values. Deviations of these values from the normal range indicate that homeostasis is not maintained.

Measurement	Normal Range
Arterial pH	7.35–7.43
Body water	830–865 g/L
Bicarbonate	21.3–28.5 mEq/L
Sodium	136–151 mEq/L
Calcium	4.6–5.2 mEq/L
Oxygen content	17.2–22.0 ml/100ml
Urea	12–35 mg/100 ml
Amino acids	3.3–5.1 mg/100ml
Protein	6.5–8.0 g/100ml
Total lipids	350–850 mg/100ml
Glucose	75–110 mg/100ml

measurement of the internal environment deviates significantly from the normal homeostatic range (see table 1.1), the defective mechanisms involved in the illness may be detected.

Negative Feedback

In order for internal constancy to be maintained, the body must have *sensors* that are able to detect deviations from the normal range of particular internal conditions and *effectors* that are able to reduce, and ultimately reverse, these deviations. A detectable change in the internal environment thus initiates mechanisms that produce changes in the reverse direction. Internal constancy is maintained by such **negative feedback loops.**

An increase in a value (indicated as X in figure 1.16) above its normal range is detected by a sensor; the sensor in turn activates an effector that causes X to decrease. This negative feedback loop prevents X from rising too far above

Figure 1.17   A negative feedback loop in which a decrease in some factor of the internal environment is compensated for by the actions of an effector. Compare this diagram with that shown in figure 1.16.

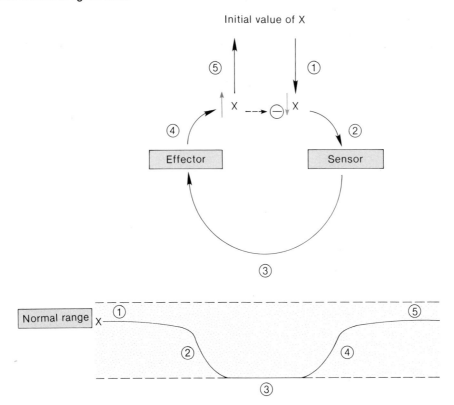

its normal range. It is important to realize, however, that even when X is in its normal range there is some activity in the effector. Even under normal, "resting" conditions, to be more specific, there is always some spontaneous nerve activity and some hormone secretion. If increased effector activity causes a decrease in X (shown in figure 1.16), decreased effector activity will cause an increase in X. Changes in either direction from the normal range, therefore, can be compensated by reverse changes in effector activity (see figure 1.17).

Insulin, for example, is a hormone that lowers the blood-sugar (glucose) concentration by promoting the uptake of blood glucose into tissue cells. Insulin secretion is increased as blood glucose concentrations rise after meals,

so that blood glucose is restored to pre-meal levels. As the blood-glucose concentration falls, the secretion of insulin gradually diminishes. During fasting, when blood glucose begins to fall below the normal range, insulin secretion is likewise reduced. This decreases the entry of glucose into tissue cells and helps to maintain constancy of the blood glucose concentration.

Homeostasis is best conceived as a state of **dynamic constancy,** rather than simply a state of absolute constancy. The various aspects of the internal environment fluctuate above and below an average value, or *set point,* within a normal range. Values are therefore at their set point only in passing, and regulatory mechanisms are constantly active to greater or lesser degrees in ongoing negative feedback loops.

Figure 1.18   Deviations of blood glucose concentration
from the normal range regulate insulin secretion. Insulin,
in turn, helps to regulate blood glucose (completing the
negative feedback loop) and to maintain homeostasis. In
this system, the Islets of Langerhans (clusters of
endocrine cells in the pancreas) are both sensors and
effector organs—they sense deviations from the normal
range and secrete insulin appropriately.

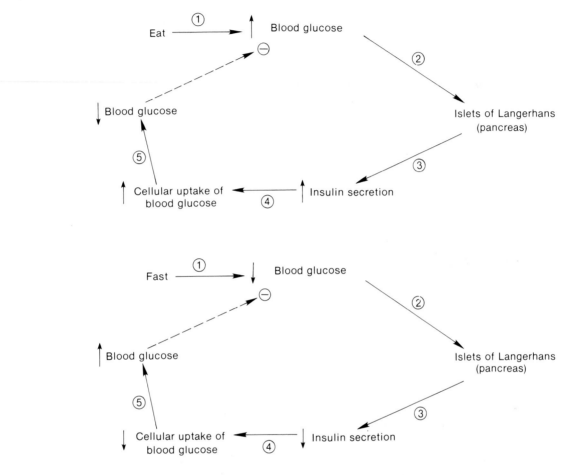

Figure 1.19   Negative feedback loops (indicated by
negative signs) maintain a state of dynamic constancy
within the internal environment.

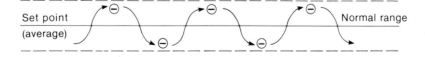

Figure 1.20 A simplified scheme by which body temperature is maintained within the normal range (with a set point of 37°C) by two antagonistic mechanisms— shivering and sweating. Shivering is induced when the body temperature falls too low, and gradually subsides as the temperature rises. Sweating occurs when the body temperature is too high, and diminishes as the temperature falls. Most aspects of the internal environment are regulated by the antagonistic actions of different effector mechanisms.

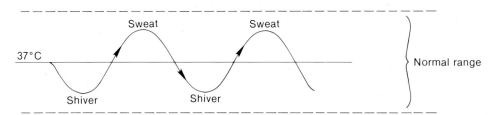

Antagonistic Effectors   Most factors in the internal environment are controlled by more than one effector, which often have antagonistic effects. Control by antagonistic effectors is sometimes described as "push-pull." This affords a finer degree of control than could be achieved by simply switching one effector off and on. Normal body temperature, for example, is maintained about a set point of 37° C by the antagonistic effects of sweating, shivering, and other mechanisms (see figure 1.20).

The blood concentrations of glucose, calcium, and other substances are regulated by negative feedback loops that involve hormones that promote opposite effects. While insulin, for example, lowers blood glucose (as previously described), other hormones raise the blood-sugar concentration. The heart rate, similarly, is controlled by nerve fibers that produce opposite effects—stimulation of one group of nerve fibers increases heart rate, while stimulation of another group slows the beat. These and numerous other examples of effector actions, together with the negative feedback control of these effectors, will be discussed in subsequent chapters.

Positive Feedback

Constancy of the internal environment is maintained by activation of effectors that produce changes that compensate for those that activated the effector. A thermostat, for example, maintains a constant temperature by increasing heat production when it is cold and decreasing heat production when it is warm. This is negative feedback. The opposite occurs during **positive feedback**—in this case, the action of effectors *amplifies* those changes that activated the effectors. A thermostat that works by positive feedback, for example, would increase heat production as the temperature increased.

It is clear that homeostasis must ultimately be maintained by negative rather than by positive feedback mechanisms. The effectiveness of some negative feedback loops, however, is increased by positive feedback mechanisms that amplify the action of negative feedback mechanisms. Blood clotting, for example, occurs as a result of sequential activation of clotting factors; activation of one clotting factor results in activation of many in a positive feedback manner. In this way, a single change is amplified to produce a blood clot. Formation of the clot, however, can prevent further blood loss and thus represents the completion of a negative feedback loop.

1. Define homeostasis and describe the importance of this concept in physiology and medicine.
2. Describe the meaning of the phrase "negative feedback loop" and its physiological significance.
3. Draw a flowchart (using arrows to represent cause-and-effect sequences) of the negative feedback control of blood glucose by the hormone insulin. Try to do this first without looking back at the chapter.

Summary

Tissues and Organs
I. Physiology is the study of how cells, tissues, organs, and systems function.
   A. In the study of physiology, cause-and-effect sequences are emphasized.

B. Knowledge of physiological mechanisms is deduced from data obtained experimentally.
II. The body is composed of four primary tissues.
   A. Muscle tissue is specialized for contraction.

1. There are three types of muscle tissues—skeletal, cardiac, and smooth muscle.
   a) Skeletal and cardiac muscles are striated.

b) Skeletal muscle is under voluntary control; cardiac and smooth muscles are generally involuntary.

B. Nerve tissue comprises a regulatory system that produces and conducts electrical events and secretes regulatory chemicals.

C. Epithelial tissue includes cellular membranes that cover and line the body surfaces, and also includes glands that are derived from these membranes.
  1. Epithelial tissue contains cells closely joined together by junctional complexes.
  2. Glands are divided into two categories: exocrine and endocrine.
    a) Exocrine glands secrete chemicals into ducts that carry these secretions to the outside of the body.
    b) Endocrine glands secrete directly into the blood; their secretions are called hormones.

D. Connective tissue is characterized by large spaces that contain abundant extracellular material.

  1. Connective tissue is subdivided into loose dense regular and dense irregular connective tissue.
  2. Special connective tissue includes cartilage, bone, and blood.

III. Organs are composed of at least two, and usually all four types of primary tissues.
  A. The skin is an organ that contains all four types of primary tissue.
    1. The epidermis of the skin is an epithelial membrane.
    2. Sweat and sebaceous glands are exocrine glands derived from the epidermis.
    3. The dermis is a type of loose connective tissue.
    4. Blood vessels and nerve fibers travel in the dermis.
    5. The epidermis is also an endocrine gland because it produces vitamin D that can function as a hormone.
    6. Smooth muscle in the skin can produce goose bumps.
  B. Organs that perform related functions are grouped together into systems.

Homeostasis

I. Homeostasis refers to the dynamic constancy of the internal environment.
  A. Diverse regulatory mechanisms exist to maintain internal values within normal limits.
  B. Deviations from homeostasis indicate disease; the nature of such deviations can be used to diagnose the cause of many illnesses.

II. Homeostasis is maintained by effectors that are controlled by negative feedback loops.
  A. Deviations from the normal range are detected by sensors that activate effector organs.
  B. Actions of the effectors reduce the deviation by causing reverse changes.
  C. Effector action is stimulated when an internal condition changes in one direction and inhibited when the change is in the opposite direction.
  D. Many effectors have antagonistic actions that provide a finer control of the internal environment.

Self-Study Quiz

Match the following:
1. Glands are derived from
2. Cells are joined closely together in
3. Cells are separated by large extracellular spaces in
4. Blood vessels and nerves are usually located within

(a) nerve tissue
(b) connective tissue
(c) muscular tissue
(d) epithelial tissue

Multiple Choice
5. The wall of a blood vessel is composed of:
  (a) epithelial tissue
  (b) muscle tissue
  (c) connective tissue
  (d) all of these
6. Sweat is secreted by exocrine glands. This means that:
  (a) it is produced by epithelial cells
  (b) it is a hormone
  (c) it is secreted into a duct
  (d) it is produced outside the body

7. Which of the following statements about homeostasis is TRUE?
  (a) the internal environment is maintained absolutely constant
  (b) negative feedback mechanisms act to correct deviations from a normal range within the internal environment
  (c) homeostasis is maintained by switching effector actions on and off
  (d) all of these

8. In a negative feedback loop the effector organ produces changes that are:
  (a) similar in direction to that of the initial stimulus
  (b) opposite in direction to that of the initial stimulus
  (c) unrelated to the initial stimulus
9. A hormone called *parathyroid hormone* acts to help raise the blood-calcium concentration. According to the principles of negative feedback, an effective stimulus for parathyroid hormone secretion would be:
  (a) a fall in blood calcium
  (b) a rise in blood calcium

2 Chemical Composition of the Body

Objectives

By studying this chapter, you should be able to:

1. Understand the meaning of the terms *atom, ion, atomic number,* and *atomic weight*

2. Understand the nature and different types of chemical bonds

3. Know the meaning of the pH scale and the definitions of the terms *acid* and *base*

4. Know the meaning of the terms *polar* and *nonpolar* compounds, and be able to relate these to water solubility

5. Describe the different types of carbohydrates and give examples of each type

6. Describe the mechanisms and significance of dehydration synthesis and hydrolysis reactions

7. Describe the different types of lipids and their common characteristics

8. Understand the structure and functions of proteins

9. Describe the structure of DNA and the law of complementary base pairing

10. Describe the structure of RNA and name the different types of RNA

Water is the major solvent in the body and contributes 65 to 75 percent of the total weight of an average adult. Thirty to forty percent of this amount is within the body cells (in the *intracellular compartment*); the remainder is in the *extracellular compartment,* including the blood and tissue fluids. Dissolved in this water are a large amount of organic molecules (carbon-containing molecules such as carbohydrates, lipids, proteins, and nucleic acids) and inorganic molecules and ions (atoms with a net charge). Before describing the structure and function of organic molecules within the body, some basic chemical concepts, terminology, and symbols will be introduced.

Table 2.1 Atoms commonly present in organic molecules.

Atom	Symbol	Atomic Number	Atomic Weight	Orbital 1	Orbital 2	Orbital 3	Number of Chemical Bonds
Hydrogen	H	1	1.01	1	0	0	1
Carbon	C	6	12.01	2	4	0	4
Nitrogen	N	7	14.01	2	5	0	3
Oxygen	O	8	16.00	2	6	0	2
Phosphorous	P	15	30.97	2	8	5	5
Sulfur	S	16	32.06	2	8	6	2

Atoms

Although atoms are much too small to be seen, they do have mass and occupy space. Most of the mass of an atom is located in the center of its volume, in a structure called the *nucleus*. The nucleus contains two types of particles—*protons*, which have a positive charge, and *neutrons*, which are noncharged. The nucleus therefore has a positive charge. The sum of the number of protons and neutrons in an atom is shown as its *mass number*, or *atomic weight*.

Orbiting the positively charged nucleus are negatively charged particles called *electrons*. The number of electrons in an atom is equal to the number of its protons; each atom, therefore, has a net charge of zero. The number of protons (or the number of electrons) in an atom is called its *atomic number* (see table 2.1).

The exact location of an electron at a given time cannot be predicted. One can only predict that, with a given probability, the electron will be within a given volume of space. The *orbital* of an electron describes the outer boundary of the volume of space within which the electron will be 90 percent of the time. This orbital is drawn as a circle around the nucleus, much like the orbit of a satellite (like the moon) around a planet. Since these orbitals roughly describe the outer boundaries of an atom they are called *shells*.

The first orbital, or shell, can only contain two electrons. If an atom contains more than two electrons (as do all atoms except hydrogen and helium), the additional electrons must occupy orbitals that surround the first. Each of these outer orbitals can contain a maximum of eight electrons. Each orbital is filled, in turn, from the most inner (the first) to the last. Carbon—with six electrons—therefore has two electrons in the first orbital and four in the second orbital.

Chemical Bonds

Atoms are joined together by chemical bonds to form **molecules.** These bonds can be strong or weak, depending upon the amount of energy required to break them. The strongest bonds are those formed when atoms share electrons. These are called **covalent bonds.** Bonds become progressively weaker as the sharing of electrons becomes less equal.

The number of bonds that each atom can have with other atoms is determined by the number of electrons in its outer orbital. A hydrogen atom with only one electron, for example, requires the sharing of only one more electron to complete the first orbital (which can contain a maximum of only two electrons—see figure 2.2). Carbon, in contrast, must obtain four more electrons (and have four chemical bonds with other atoms) to complete its maximum of eight electrons in the second orbital (see figure 2.3).

Nonpolar and Polar Covalent Bonds

Covalent bonds formed between identical atoms—as in oxygen gas (O_2) and hydrogen gas (H_2)—are particularly strong because their electrons are equally shared. Since the electrons are equally distributed between the two atoms, these molecules are said to be **nonpolar.** When a covalent bond is formed between two different atoms, however, the electron may be "pulled" more towards one atom than the other. The end of the molecule towards which the electron is pulled will therefore be electrically negative in comparison to the other end. The molecule is thus said to be **polar** (has a positive and negative "pole"). Atoms such as oxygen, nitrogen, and phosphorous are very *electronegative* because they have a strong affinity for electrons.

Figure 2.1 Diagrams of the hydrogen and carbon atoms. The electron orbitals on the left are represented by dots indicating probable positions of the electrons. The orbitals on the right are represented by concentric circles.

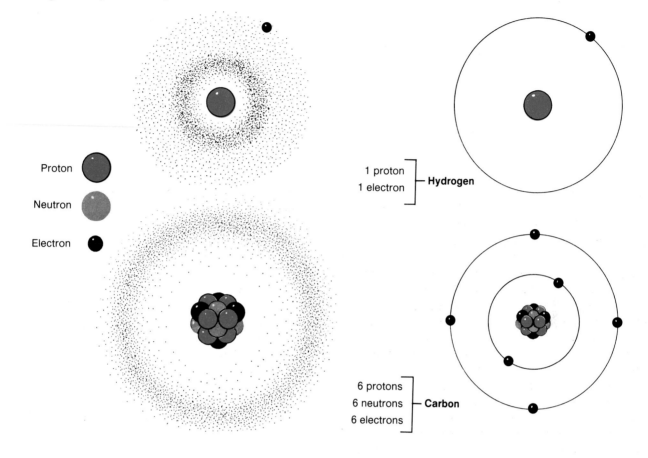

Figure 2.2 The hydrogen molecule, showing the covalent bond between hydrogen atoms formed by the equal sharing of electrons.

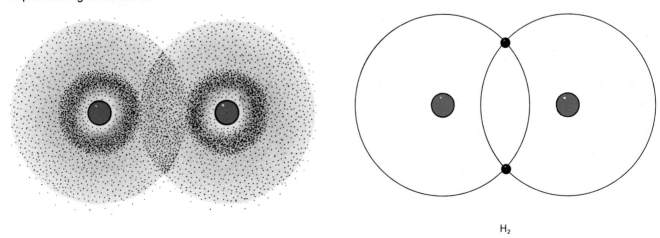

Figure 2.3 The molecules *methane* and *ammonia* represented in three different ways. Notice that a bond between two atoms consists of a pair of shared electrons (one electron from the outer orbital of each atom).

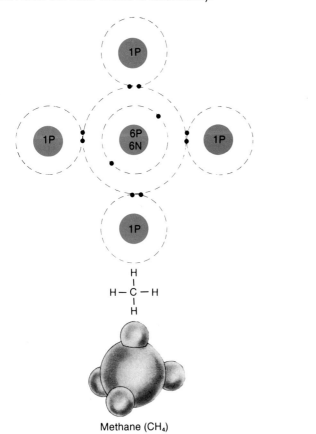

Methane (CH₄)

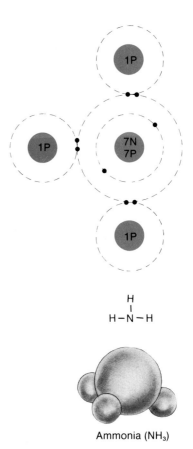

Ammonia (NH₃)

Figure 2.4 A model of a water molecule showing its polar nature. Notice that the oxygen side of the molecule is negative while the hydrogen side is positive. Polar covalent bonds are weaker than nonpolar covalent bonds. Some water molecules, as a result, ionize to form OH⁻ (hydroxy ion) and H⁺ (hydrogen ion).

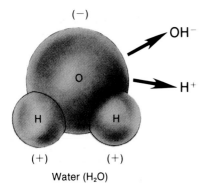

Water (H₂O)

Water—the most ubiquitous molecule and major solvent in living organisms on earth—is a polar molecule. This is because oxygen is much more electronegative than hydrogen and tends to "pull" electrons to its side of the molecule (like bedcovers pulled to one side of the bed). Hydrogen is left with its "feet cold"—that is, with a slight positive charge—while the side of the molecule containing oxygen has a slight negative charge. The polar nature of water is very important (as will be described later) in its function as a solvent.

Ionic Bonds

In covalent bonds, atoms are joined together by the sharing of electrons—equally in nonpolar covalent bonds; unequally in polar bonds. In an **ionic bond** the electrons are not shared. Instead, they are transferred from one atom to another. Common table salt, sodium chloride (NaCl), can serve as an example. Sodium, with an atomic number of 11, has only one electron in its outer orbital (it needs

Figure 2.5 Ionization of sodium and chlorine to produce sodium and chlorine ions. The positive sodium and the negative chloride ions attract each other to produce the ionic compound sodium chloride (NaCl).

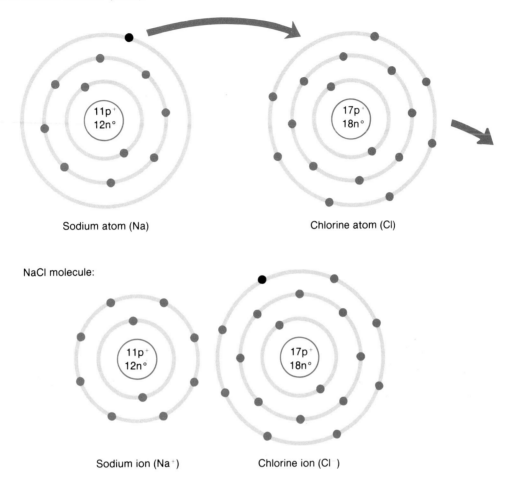

Sodium atom (Na)   Chlorine atom (Cl)

NaCl molecule:

Sodium ion (Na^+)   Chlorine ion (Cl^-)

eight to complete this shell). Chlorine, conversely, is only one electron short of completing its outer orbital of eight electrons. The single outer electron in a sodium atom is strongly attracted to the chlorine. This creates a *chloride ion* (Cl^-) with one extra electron and a *sodium ion* (Na^+) that has one electron less than its number of protons. These positive and negative ions attract each other to produce an *ionic compound*—NaCl.

Ionic bonds are significantly weaker than polar covalent bonds and therefore break more easily when dissolved in water. Sodium chloride, for example, almost completely ionizes when dissolved in water. Each of the ions released (Na^+ and Cl^-) attracts polar water molecules; the negative ends of water molecules are attracted to the

Na^+, and the positive ends of water molecules are attracted to the Cl^- (see figure 2.6). The water molecules that surround these ions in turn attract other molecules of water to form *hydration spheres* around each ion.

Formation of hydration spheres makes an ion or molecule soluble in water. Glucose, amino acids, and other organic molecules are water soluble because hydration spheres can form around atoms of oxygen, nitrogen, and phosphorous that are joined by polar covalent bonds to carbon atoms. Molecules composed primarily of hydrocarbons with nonpolar covalent bonds (such as fat), which therefore contain few charges, are not soluble in water. This is because there are very few locations where hydration spheres can be formed. Such water-insoluble nonpolar molecules are said to be **hydrophobic** (water-fearing).

Figure 2.6   The negatively charged oxygen-ends of water molecules *(large circles)* are attracted to the positively charged Na⁺, while the positively charged hydrogen-ends of water molecules *(small circles)* are attracted to the negatively charged Cl⁻. Other water molecules are attracted to this first concentric layer of water, forming hydration spheres around the sodium and chloride ions.

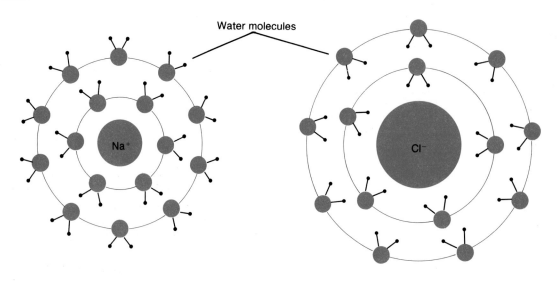

Water molecules

Figure 2.7   The oxygen atoms of water molecules are weakly joined together by attraction of the electronegative oxygen for the positively charged hydrogen. These weak bonds are called **hydrogen bonds.**

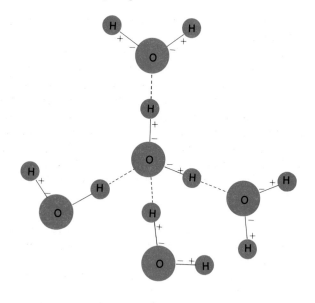

Hydrogen Bonds

Hydrogen bonds are weaker than ionic bonds. These bonds are formed when an atom of hydrogen shares its single electron with *two* other atoms, usually oxygen or nitrogen, which have a strong tendency to pull electrons toward themselves (are strongly electronegative). The hydrogen is therefore left with a slight positive charge that gives it a weak attraction for the two electronegative atoms.

Water molecules are bonded to each other by weak hydrogen bonds. These bonds are responsible for many of the physical properties of water, such as its *surface tension* and its ability to be pulled as a column through narrow channels—a process called *capillary action.*

pH

Since the electrons in the bonds of water molecules are pulled towards the oxygen atoms, the hydrogen atoms in some of these molecules may completely lose their electrons. When this happens these hydrogen atoms may dissociate (break bonds with) the oxygen atoms to become free protons. (A hydrogen atom consists of only one proton in the nucleus, surrounded by one electron.) The free proton, in other words, is a hydrogen atom that is missing an electron. This is represented by the symbol H⁺. Any atom that is missing electrons, or that has gained electrons (and therefore has a net charge), is called an **ion.**

Table 2.2 Common acids and bases.

Acid	Symbol	Base	Symbol
Hydrochloric acid	HCl	Sodium hydroxide	NaOH
Phosphoric acid	H_3PO_4	Potassium hydroxide	KOH
Nitric acid	HNO_3	Calcium hydroxide	$Ca(OH)_2$
Sulfuric acid	H_2SO_4	Ammonium hydroxide	NH_4OH
Carbonic acid	H_2CO_3		

Table 2.3 The pH scale.

	H^+ Concentration (molar)	pH	OH^- Concentration (molar)
	1.0	0	10^{-14}
	0.1	1	10^{-13}
	0.01	2	10^{-12}
Acids	0.001	3	10^{-11}
	0.0001	4	10^{-10}
	10^{-5}	5	10^{-9}
	10^{-6}	6	10^{-8}
Neutral	10^{-7}	7	10^{-7}
	10^{-8}	8	10^{-6}
	10^{-9}	9	10^{-5}
	10^{-10}	10	0.0001
Bases	10^{-11}	11	0.001
	10^{-12}	12	0.01
	10^{-13}	13	0.1
	10^{-14}	14	1.0

Table 2.4 The pH of various body fluids.

Body Fluid	pH
Gastric juice	1.9–2.6
Urine	5.7
Sweat	4–6.8
Bile	6.0
Saliva	6.4
Breast milk	7.0
Feces	7.15
Semen	7.19
Cerebrospinal fluid	7.35
Arterial blood	7.40
Pancreatic juice	7.5–8.8

An **acid** is defined as a molecule that can liberate H^+, and thus can raise the H^+ concentration above that of pure water. A **base** is defined as a molecule that lowers the H^+ concentration of a solution (by combining with free H^+). Most bases liberate OH^-, which lowers the H^+ concentration by combining with free H^+ to form water. Examples of common acids and bases are shown in table 2.2.

The H^+ concentration of a solution is traditionally indicated in pH units. The pH is inversely equal to the logarithm (to the base 10) of the molar H^+ concentration:

$$pH = \frac{1}{\log [H^+]}$$

In pure water, the H^+ concentration is 10^{-7} molar; this neutral solution thus has a pH of 7. The pH number is easier to write than the molar concentration, but it is admittedly confusing because it is *inversely related* to the H^+ concentration. A solution with a high H^+ concentration has a low pH number; one with a low H^+ concentration has a high pH number. A strong acid with a pH of 2, for example, has a H^+ concentration of 10^{-2} molar, while a solution that has only 10^{-10} molar H^+ has a pH of 10. Acidic solutions therefore have a pH less than 7, while basic solutions have a pH between 7 and 14 (see table 2.3).

Dissociation of hydrogen from a water molecule thus results in the production of two ions—H^+ (hydrogen ion) and OH^- (hydroxyl ion—which is negatively charged because it has "stolen" the electron from hydrogen). Since positive ions tend to move towards the negative pole in a battery (called the *cathode*), they are known as *cations*. Conversely, since negative ions move towards the positive pole (the anode), they are called *anions*.

When water molecules dissociate, or *ionize,* the concentration of H^+ is of course equal to the concentration of OH^- produced. It should be noted that only a relatively small proportion of water molecules ionize in this way, yielding a H^+ concentration of 10^{-7} molar (the term *molar* is a unit of concentration described in chapter 6). A solution with this H^+ concentration—that produced by pure water—is said to be *neutral.* Any solution that contains a higher H^+ concentration is described as *acidic,* and any solution with a lower H^+ concentration is called *basic.*

1. List the types of chemical bonds in order of decreasing strength.
2. Define a covalent bond and distinguish between polar and nonpolar covalent bonds.
3. Explain why ionic compounds and molecules with many polar covalent bonds are water soluble, while molecules composed mainly of nonpolar covalent bonds cannot be dissolved in water.
4. Write the definition of an acid, and explain in words the relationship between the pH of a solution and its H^+ concentration.

Figure 2.8 Two carbons joined by a single covalent bond *(above)* or by a double covalent bond *(below)*. In both cases each carbon shares four pairs of electrons (has four bonds) to complete the eight electrons required to fill its outer orbital.

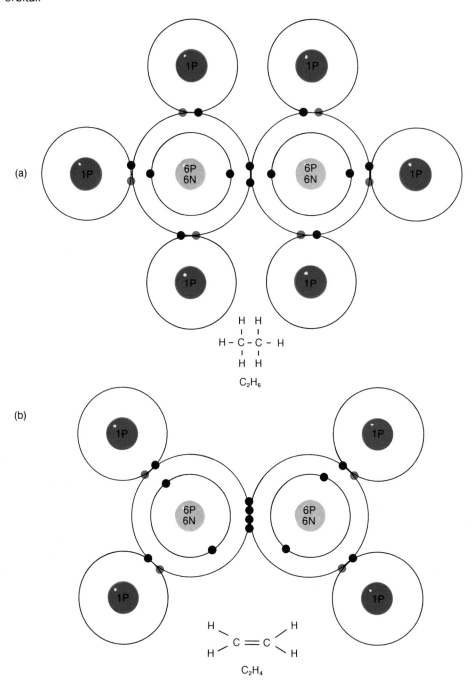

Figure 2.9 Hydrocarbons that are *(a)* linear; *(b)* cyclic; and *(c)* aromatic rings.

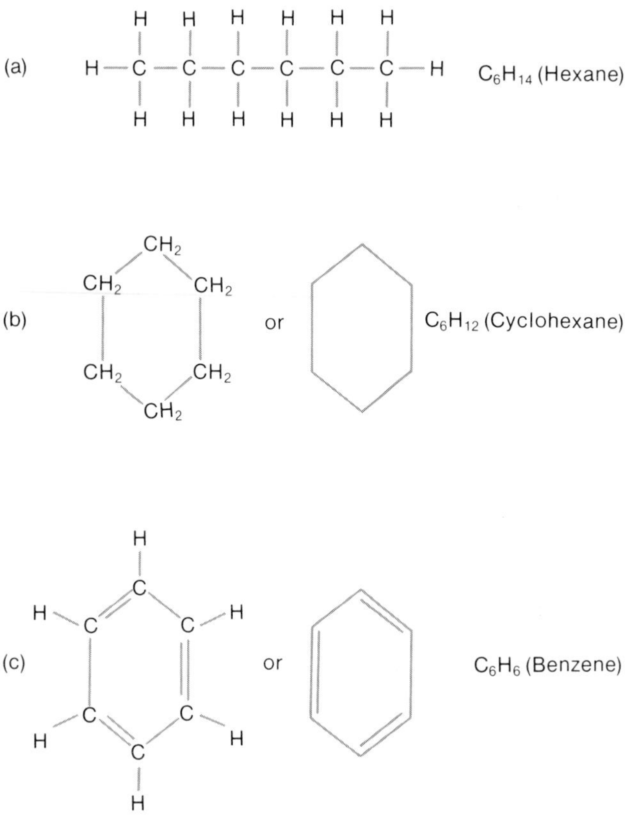

Figure 2.10 Various functional groups of organic molecules.

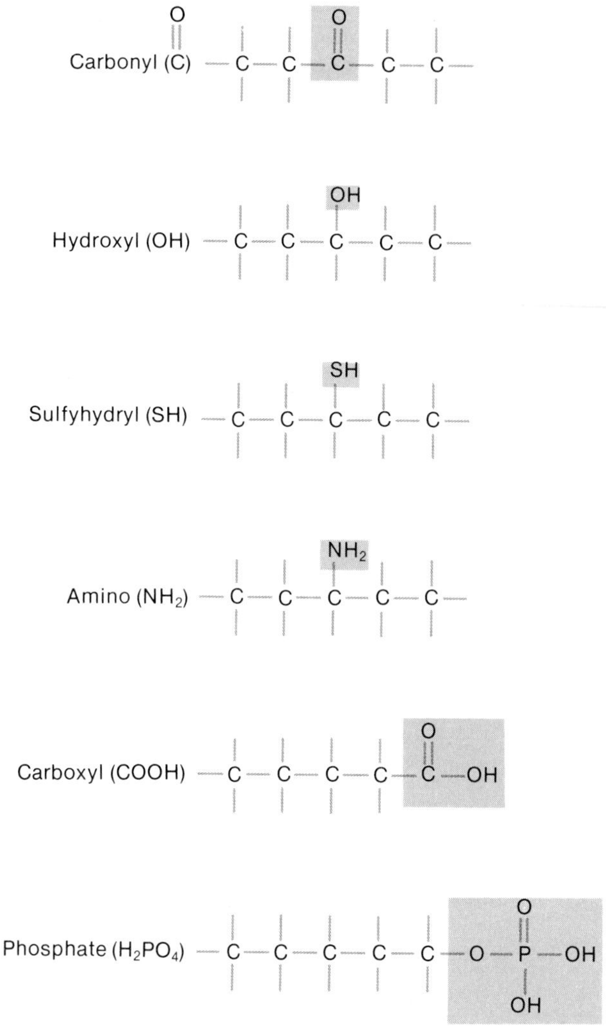

Organic Molecules

Organic molecules are those that contain the atom *carbon*. Since carbon has four electrons in its outer orbital, it must share four additional electrons by covalent bonding with other atoms to fill its outer orbital with eight electrons. The unique bonding requirements of carbon enable it to join with other carbon atoms—forming carbon chains or rings—while it can also bond to hydrogen and other atoms.

Most organic molecules in the body contain hydrocarbon chains and rings, together with other atoms bonded to carbon. Two adjacent carbons in a hydrocarbon chain or ring may share one or two pairs of electrons. If the two carbons share one pair of electrons, they are said to have a *single covalent bond;* this leaves each carbon free to bond to as many as three other atoms. If the two carbons share two pairs of electrons—have a *double covalent bond*— each carbon can only bond to a maximum of two additional atoms (see figure 2.8).

The ends of some hydrocarbons are joined together to form rings. In the shorthand structural formulas of these molecules the carbon atoms are not shown, but are understood to be located at the corners of the ring. Some of these cyclic hydrocarbons have a double bond between two adjacent carbons. Benzene and related carbons are shown as a six-sided ring with alternating double bonds. Such compounds are called *aromatic*. Since all of the carbons in an aromatic ring are equivalent, double bonds can be shown between any two adjacent carbons in the ring (see figure 2.9).

The hydrocarbon chain, or ring, of many organic molecules provides a relatively inactive molecular "backbone," to which is attached more reactive groups of atoms. These *functional groups* of the molecule usually contain atoms of oxygen, nitrogen, phosphorous, or sulfur and are, in large part, responsible for the unique chemical properties of the molecule.

Figure 2.11 Categories of organic molecules based on functional groups.

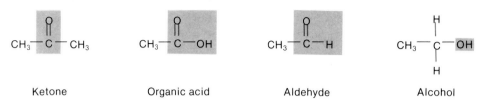

Ketone   Organic acid   Aldehyde   Alcohol

Classes of organic molecules can be named according to their functional groups. *Ketones,* for example, have a carbonyl group within the carbon chain. An organic molecule is an *alcohol* if it has a hydroxyl group at one end of the chain. All *organic acids* (citric acid, acetic acid, and so on) have a carboxyl group on one of its ends (see figure 2.11).

Optical Isomers   Two molecules may have exactly the same atoms arranged in exactly the same sequence, yet may differ with respect to the spatial orientation of key functional groups. Such molecules are called *stereoisomers* of each other. They are also known as *optical isomers* because they rotate polarized light to either the right or left, depending on the orientation of their functional groups. The isomers that rotate light to the right (and have functional groups respresented on the right side of the molecule) are called **D-isomers.** The *D* stands for *dextro*—right-handed. Molecules that rotate light to the left, and which are represented by structures showing functional groups on the left side, are called **L-isomers** (for *levo,* or left-handed).

The two optical isomers are mirror images of each other—they cannot be superimposed. These subtle differences in structure are extremely important biologically, because enzymes—which interact with such molecules in a stereo-specific way in chemical reactions—cannot combine with the "wrong" stereoisomer. The enzymes of all cells (human and others) can only combine with L-amino acids and D-sugars, for example. The opposite stereoisomers (D-amino acids and L-sugars) cannot be used by the body.

Carbohydrates

Carbohydrates are organic molecules that contain carbon, hydrogen, and oxygen in the ratio described by their name—*carbo* (carbon) and *hydrate* (water—H_2O). The general formula of a carbohydrate is thus CH_2O; the molecule contains twice the number of hydrogen atoms as it contains carbon or oxygen atoms. The category known as carbohydrates includes simple sugars, or **monosaccharides** (mono = one; sacchar = sugar) and longer molecules that contain a number of these monosaccharides joined together.

Monosaccharides may contain three carbons (triose sugars), five carbons (pentose sugars), or six carbons (hexose sugars). Notice that the suffix *-ose* denotes a sugar molecule. A hexose sugar, for example, can be represented by the formula $C_6H_{12}O_6$. This notation is adequate for some purposes, but it does not distinguish between related hexose sugars, which are *structural isomers* of each other. Glucose, fructose, and galactose, for example, have the same ratio of atoms arranged in slightly different ways (see figure 2.13).

Two monosaccharides can be joined covalently to form a **disaccharide,** or double sugar. Common disaccharides include table sugar, or *sucrose,* (glucose and fructose), milk sugar, or *lactose,* (glucose and galactose), and malt sugar, or *maltose,* (glucose and glucose). Long chains formed by monosaccharides joined together are known as **polysaccharides.** *Starch,* for example, is a polysaccharide found in many plants, which consists of thousands of repeating glucose subunits joined together. Animal starch, or **glycogen,** found in liver and muscles, likewise consists of repeating glucose molecules, but differs from plant starch in that it is more highly branched (see figure 2.14).

Figure 2.12 Stereoisomers of glyceraldehyde and glucose. D-isomers are right-handed, L-isomers are left-handed.

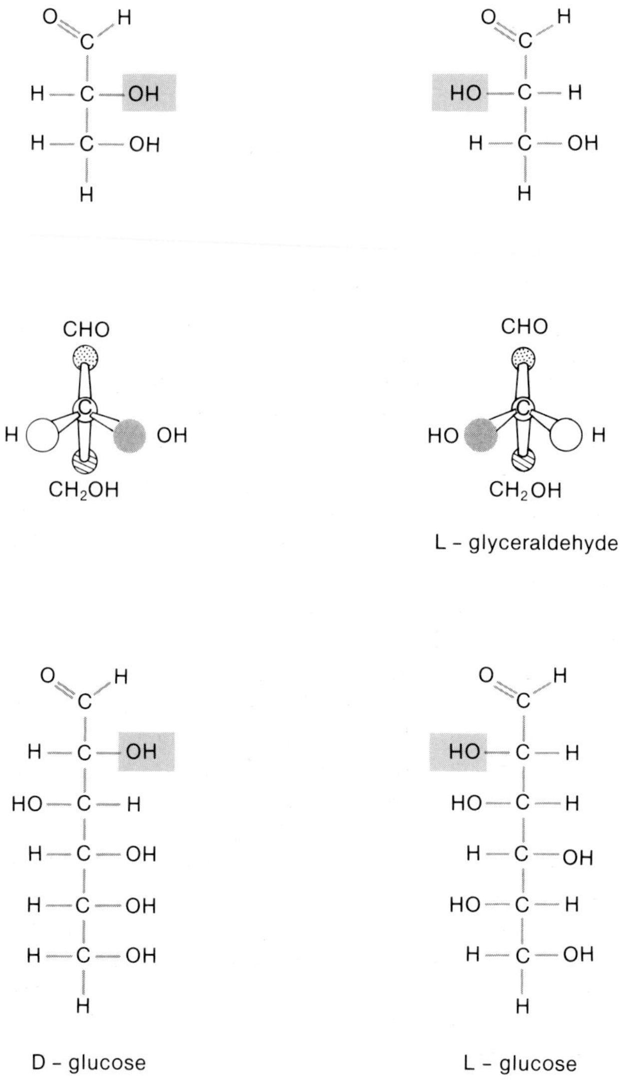

D – glucose

L – glucose

L – glyceraldehyde

Figure 2.13 Structural formulas of three hexose sugars—*(a)* glucose, *(b)* galactose and *(c)* fructose. All three have the same ratio of atoms—$C_6H_{12}O_6$.

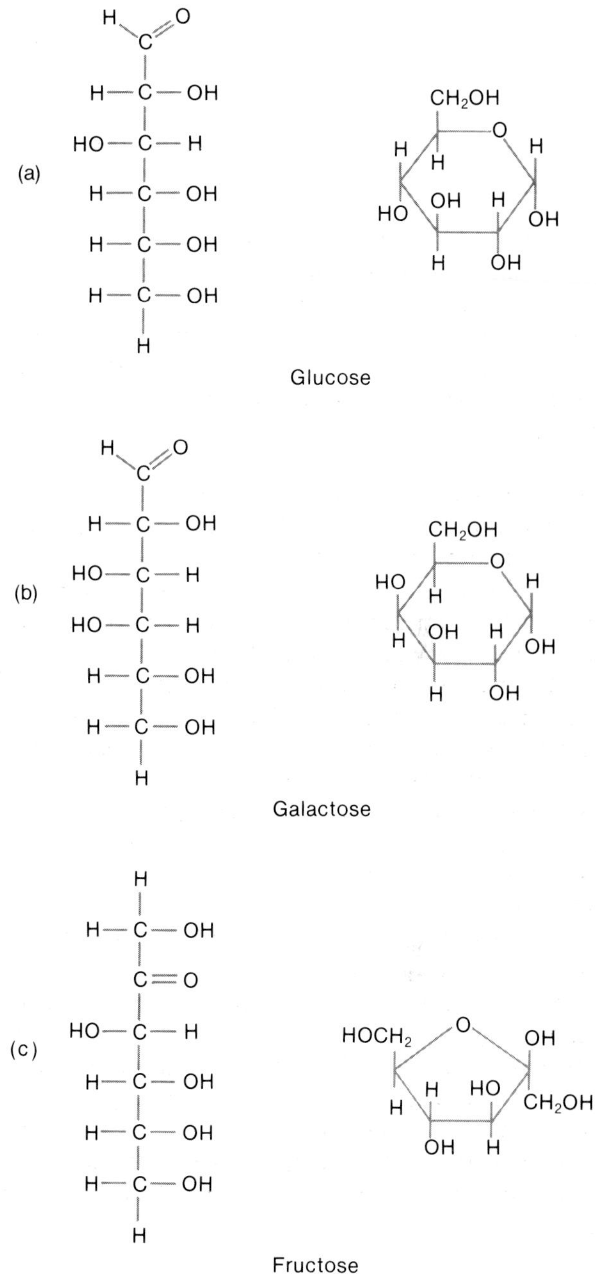

(a)

Glucose

(b)

Galactose

(c)

Fructose

Figure 2.14 Glycogen is a polysaccharide composed of glucose subunits joined together to form a large, highly branched molecule.

(a)   (b)

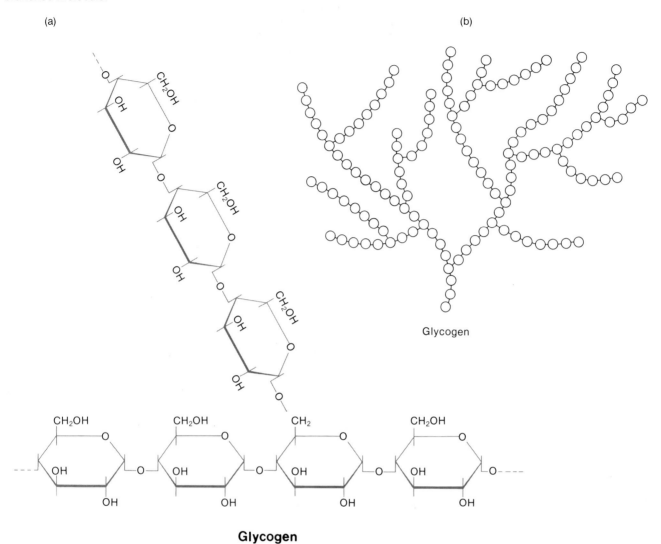

Glycogen

Glycogen

Dehydration Synthesis and Hydrolysis

In the formation of disaccharides and polysaccharides, the separate subunits (monosaccharides) are bonded together covalently by means of a type of reaction called **dehydration synthesis** or **condensation.** In these reactions, which require the participation of specific enzymes (as described in chapter 4), a hydrogen is removed from one monosaccharide while a hydroxyl group (OH) is removed from another. This results in the formation of water (H_2O) as a covalent bond is formed between the two monosaccharides. Dehydration synthesis reactions are illustrated in figure 2.15.

When a person eats disaccharides and polysaccharides, or when the stored glycogen in liver and muscles is used by tissue cells, the covalent bonds that join monosaccharides together must be broken. These *digestion* reactions occur by means of **hydrolysis.** Hydrolysis is the reverse of dehydration synthesis. A water molecule is split, as implied by the word *hydrolysis* (hydro = water; lysis = break), and the resulting hydrogen is added to one of the free glucose molecules as hydroxyl is added to the other (see figure 2.16).

Figure 2.15 Dehydration synthesis of two disaccharides, (a) maltose and (b) sucrose. Notice that as the disaccharides are formed a molecule of water is produced.

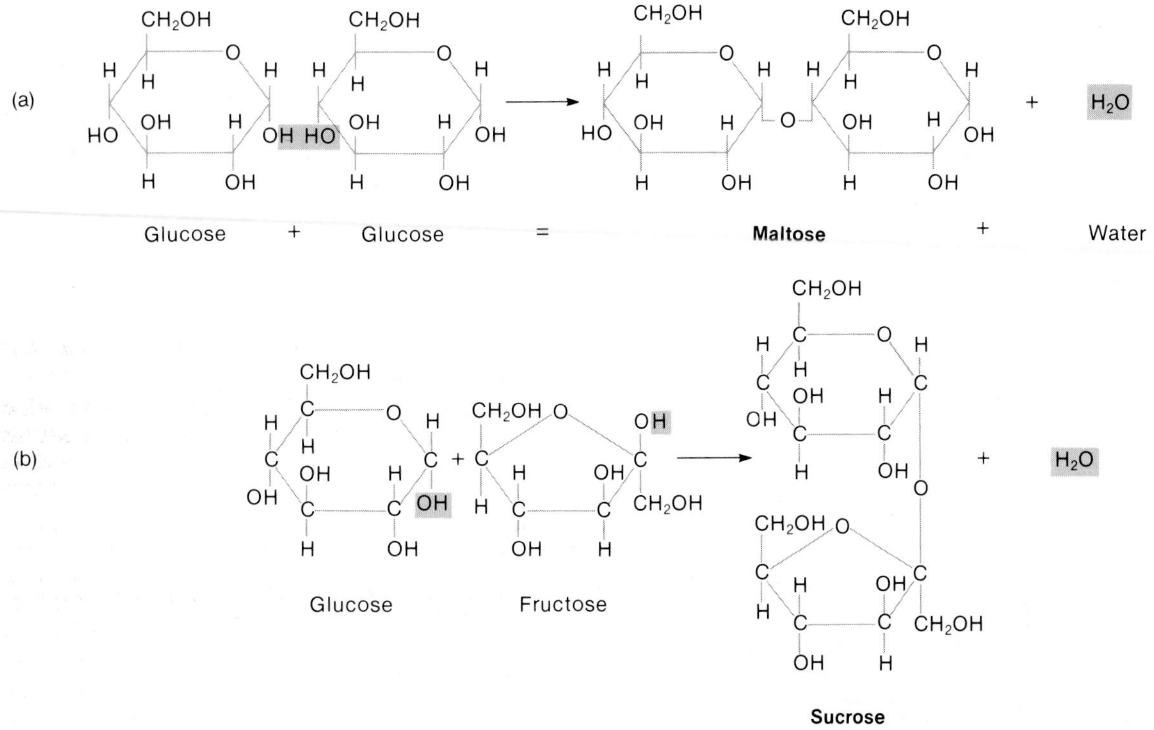

(a)

Glucose   +   Glucose   =   **Maltose**   +   Water

(b)

Glucose   Fructose   **Sucrose**

Figure 2.16 The hydrolysis of starch into disaccharides (maltose) and into monosaccharides (glucose). Notice that as the covalent bond between subunits breaks a molecule of water is split. In this way hydrogen and hydroxyl from water is added to the ends of the released subunits.

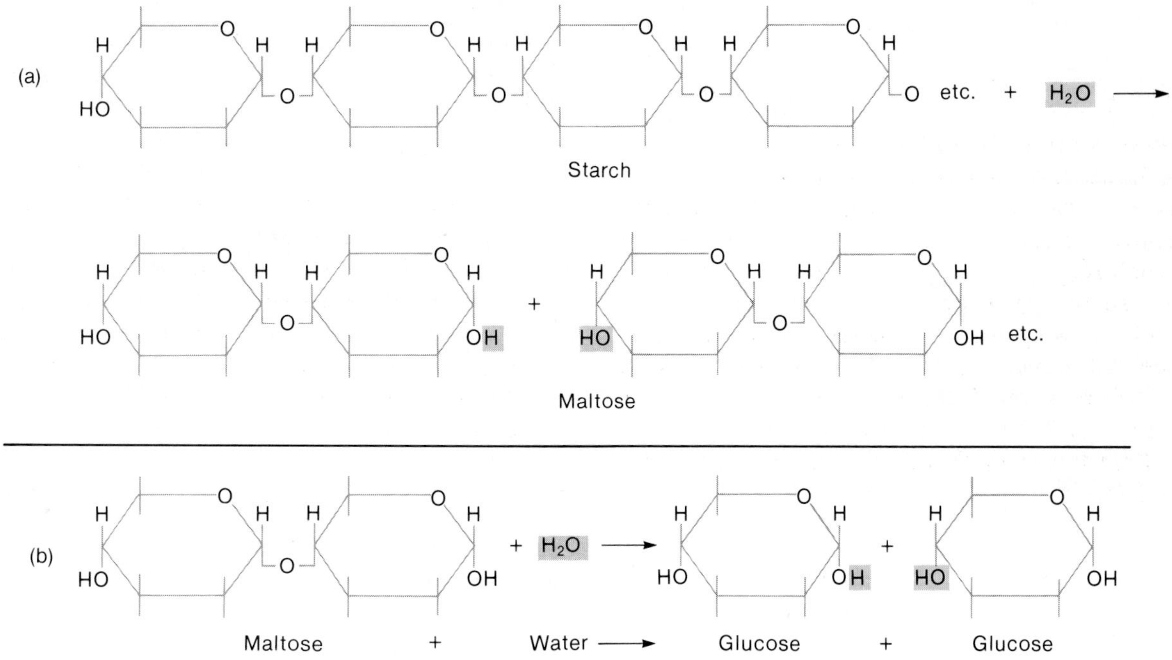

(a)

Starch

Maltose

(b)

Maltose   +   Water ⟶   Glucose   +   Glucose

Figure 2.17   Structural formulas for *(a)* saturated and *(b)* unsaturated fatty acids.

(a)

Palmitic acid

a saturated fatty acid

(b)

Linolenic acid

an unsaturated fatty acid

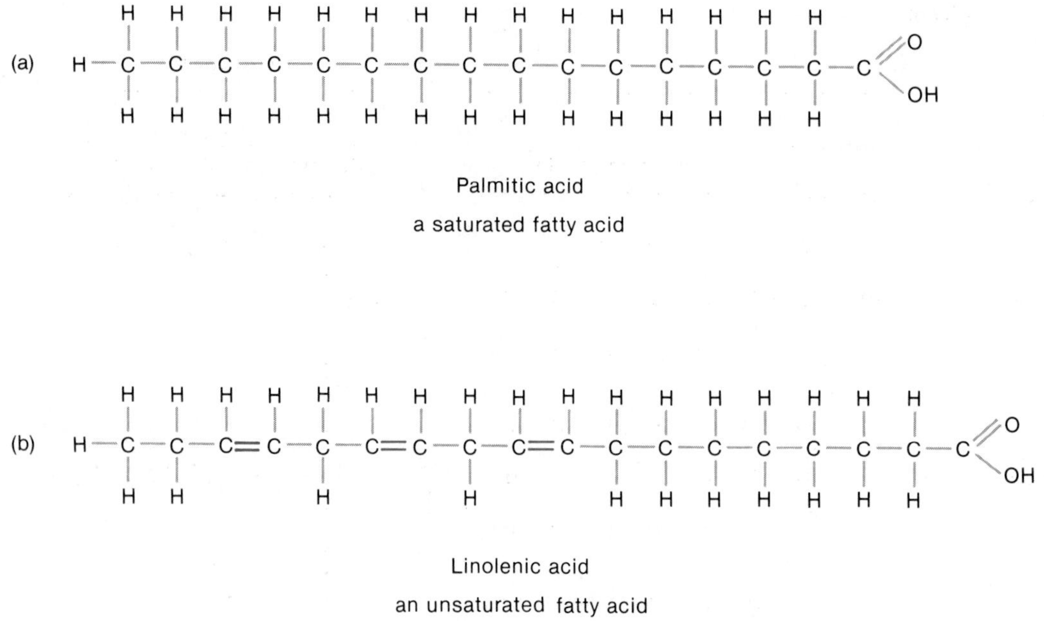

When a potato is eaten, therefore, the starch in it is hydrolyzed into separate glucose molecules within the intestine. This glucose is absorbed into the blood, which carries it to the tissues. Some of these tissues may use glucose for energy (as described in chapter 5). The liver and muscles can store excess glucose in the form of glycogen by dehydration synthesis reactions within these cells. In times of fasting or prolonged exercise, when more glucose needs to be added to the blood, this stored glycogen in the liver can be hydrolyzed to release additional glucose molecules.

It should be noted that dehydration synthesis and hydrolysis reactions do not occur spontaneously in the body—they require the action of specific enzymes. Similar reactions, in the presence of other enzymes, build and break down proteins, lipids, and nucleic acids. It can thus be stated that all digestion reactions occur by hydrolysis, and all reactions that build larger molecules composed of subunits occur by means of dehydration synthesis.

1. Define carbohydrates and give examples of monosaccharides, disaccharides, and polysaccharides.
2. Define dehydration synthesis and hydrolysis, and give examples of how these reactions are used in the metabolism of carbohydrates.
3. Draw a flow diagram (using arrows) to show the hydrolysis and dehydration synthesis reactions involved in the conversion of potato starch to liver glycogen.

Lipids

Unlike that of carbohydrates, the category of organic molecules known as lipids includes molecules with a great diversity of chemical structures. These diverse molecules are all placed in the same category (lipids) by virtue of a common physical property—they are all *insoluble in polar solvents* such as water. This is because lipids consist primarily of hydrocarbon chains and rings, which are nonpolar and thus hydrophobic. While insoluble in water, lipids can be dissolved in nonpolar "organic solvents" such as ether, chloroform, benzene, and related compounds.

Triglycerides

The subcategory of lipids known as triglycerides, or "neutral fats," includes fat and oil. These compounds are formed by dehydration synthesis of one molecule of *glycerol* (a three-carbon-long alcohol) with three molecules of *fatty acids*. Each fatty acid molecule contains a carboxylic acid group (COOH) at one end of a nonpolar hydrocarbon chain. If the carbons in this chain are joined by single covalent bonds so that each carbon is free to bond with two hydrogen atoms, the fatty acid is said to be *saturated*. If there are a number of double covalent bonds within the hydrocarbon chain, each carbon can bond with only one hydrogen (remember—one carbon can have only four bonds). Such a fatty acid is said to be *unsaturated*. Triglycerides that contain saturated fatty acids are called **saturated fats;** those that contain unsaturated fatty acids are **unsaturated fats.**

Figure 2.18 Dehydration synthesis of a triglyceride molecule from glycerol and three fatty acids. A molecule of water is produced as an ester bond forms between each fatty acid and glycerol.

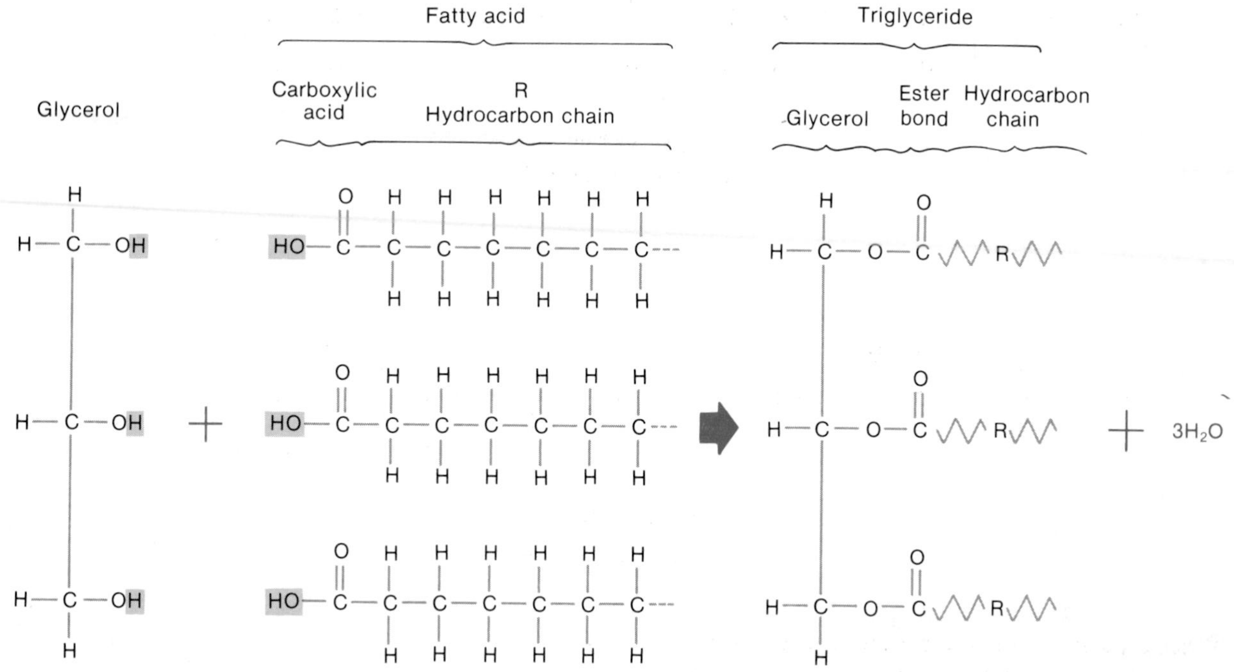

Within the adipose cells of the body, triglycerides are formed as the carboxylic acid ends of fatty acids condense with glycerol. In these dehydration synthesis reactions, the hydrogen from the carboxylic acid ends combine with hydroxyl (OH) groups of glycerol to form water (see figure 2.18). In the process, each fatty acid is joined to each carbon in glycerol through a mutual bond with an oxygen atom. These bonds are called *ester bonds,* and the fatty acids are said to be *esterified.* Since the hydrogen from the carboxyl ends of fatty acids is used to form water, fatty acids that are esterified to glycerol cannot ionize further. They can no longer release H^+ and function as acids. Thus, triglycerides are *neutral fats.*

Ketone Bodies   Hydrolysis of triglycerides occurs when specific enzymes that split water and the ester bonds are active. The hydroxyl group from the water is added back to the glycerol, while hydrogen is added to the carboxylic acid ends of the fatty acids. The *free fatty acids* liberated into the blood from adipose tissue can thus ionize and release H^+—that is, they can function as acids. Most of these fatty acids are usually used as a source of energy by various tissues, including skeletal muscles and the liver.

Those fatty acids released into the blood that are not immediately required as an energy source may be converted by enzymes in the liver into derivatives known as *ketone bodies.* These four-carbon-long derivatives—including acids such as acetoacetic acid and β-hydroxybutyric acid, as well as acetone (the active ingredient in nail polish remover)—are secreted by the liver into the blood. People who are experiencing rapid breakdown of fat, such as dieters and those with uncontrolled diabetes mellitus (a disease that will be described in a later chapter), may thus have elevated blood concentrations of ketone bodies. This condition is called **ketosis.**

Since the kidneys filter ketone bodies into the urine, dieters can monitor the rate of their fat loss by checking their urinary concentration of ketone bodies. Although ketone bodies are acidic, ketosis is not necessarily dangerous because the blood contains buffers (primarily bicarbonate) that help prevent pH changes. In uncontrolled diabetes mellitus, however, the level of ketone bodies may rise so high that the buffers cannot prevent a fall in blood pH. When this occurs the person is said to have **ketoacidosis.**

Figure 2.19 The structure of lecithin, a typical phospholipid *(above)*, and its more simplified representation *(below)*.

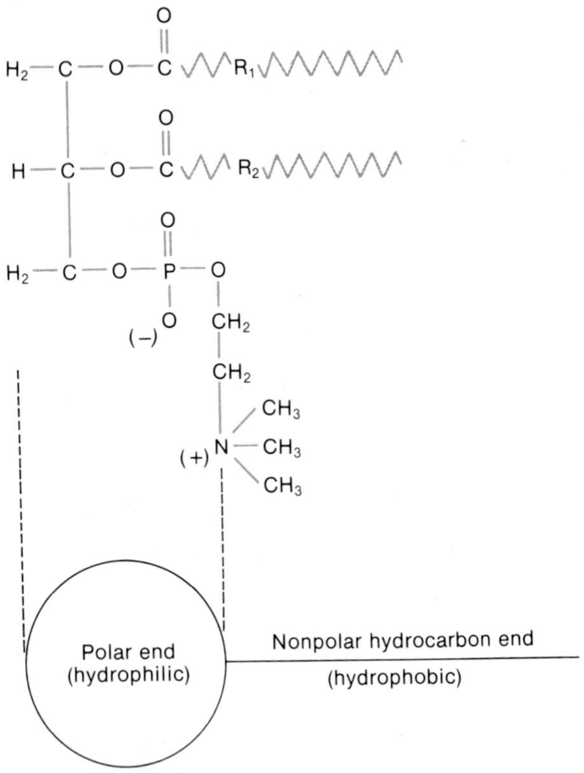

Lecithin

Figure 2.20 Formation of micelle structure by phospholipids such as lecithin. The straight lines represent the hydrophobic fatty acid parts of the molecule, while the colored circles represent the polar phosphate part of the molecule. The detailed structure of lecithin is shown in one part of the micelle.

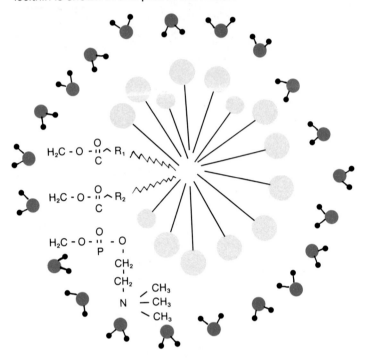

Phospholipids

Phospholipids are a diverse subclass of lipids that contain phosphate. The most common type of phospholipids contain glycerol attached to two fatty acids. The third carbon in the glycerol is attached to a phosphate group, which is in turn bonded to other molecules such as the nitrogen-containing choline molecule. This type of phospholipid is known as *lecithin* (see figure 2.19).

When phospholipids are added to water, the phosphate group and its attached molecule (such as choline) can partially ionize, and by this means can attract water molecules and form hydration spheres. This polar end of the phospholipid contrasts with the rest of the molecule, which is nonpolar and thus hydrophobic. Figure 2.19 shows a simple way of representing this dual nature of a phospholipid; it is shown as a "lollipop," with the circular part representing the polar end and the stick representing the nonpolar end.

Since the nonpolar ends of phospholipids are hydrophobic they tend to aggregate together when mixed with water, much like the circling of a wagon train, so that they are "protected" by the polar groups in the outer layer. Such aggregates are called **micelles** (see figure 2.20). Similar associations of phospholipids form the major structure of cell membranes, which are surrounded by water both outside and inside the cell (this will be discussed in chapter 3). Phospholipids in the lungs serve an additional function—hydration spheres formed around the micelles disrupt hydrogen bonding between water molecules and reduce its surface tension. Phospholipids in the lungs thus act as "surface-active agents," or *surfactants*. This prevents collapse of the lungs, as will be described in chapter 14.

Steroids

Steroids have a quite different structure than that of triglycerides and phospholipids, yet they are still in the lipid category because of the fact that they are nonpolar and

Figure 2.21 Cholesterol and some steroid hormones derived from cholesterol.

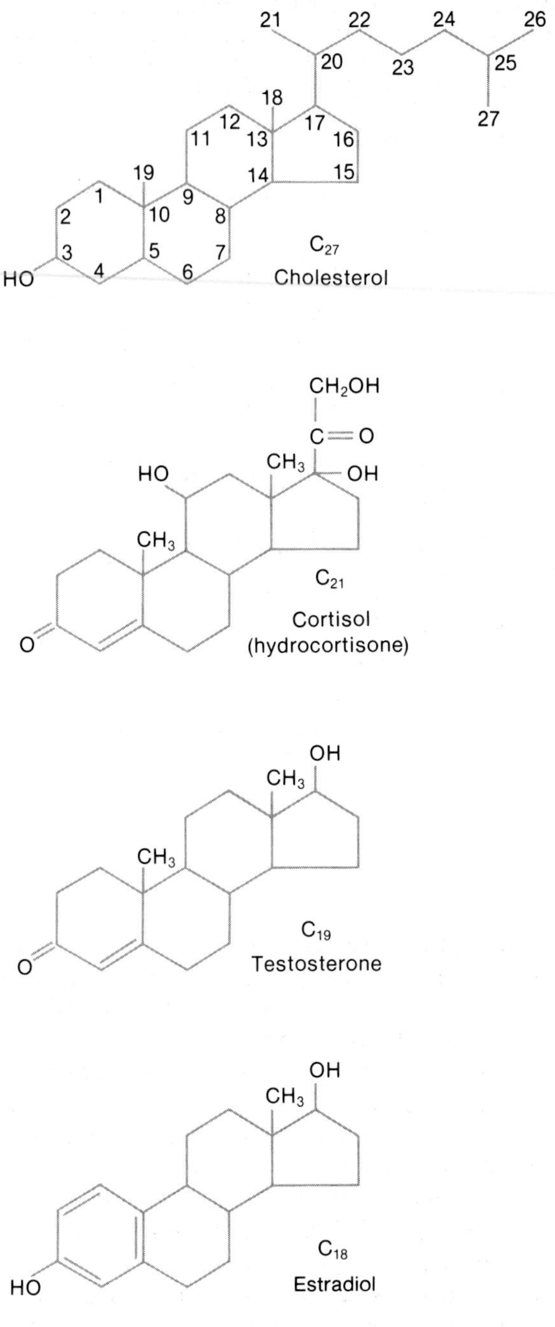

C$_{27}$
Cholesterol

C$_{21}$
Cortisol
(hydrocortisone)

C$_{19}$
Testosterone

C$_{18}$
Estradiol

Figure 2.22 Structural formulas of some prostaglandins.

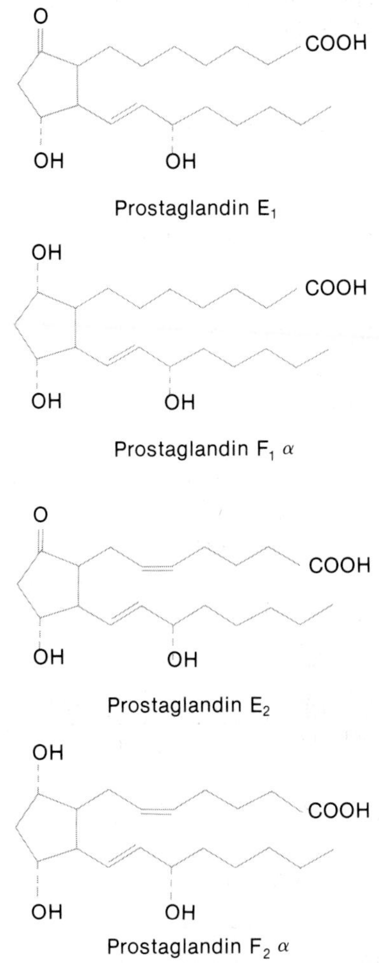

Prostaglandin E$_1$

Prostaglandin F$_1$ α

Prostaglandin E$_2$

Prostaglandin F$_2$ α

(as will be described in a later chapter). Steroid hormones are secreted by three endocrine glands: testes, ovaries, and the adrenal cortex. The testes and ovaries secrete *sex steroids* (testosterone and other male sex hormones from the testes; estradiol and progesterone from the ovaries). The adrenal cortex secretes *corticosteroids* such as cortisol, cortisone, and aldosterone (these will be discussed in later chapters).

Prostaglandins

While not usually considered a separate subclass of lipids, prostaglandins are a special type of cyclic fatty acids that have a variety of regulatory functions in different tissues. Although their name is derived from the fact that they were initially discovered as a prostate secretion in semen, they have since been shown to be produced by almost all tissues in the body. Prostaglandins are implicated in the regulation of blood-vessel diameter, ovulation, uterine contraction during labor, inflammation reactions, blood clotting, and many other functions. Some of the different types of prostaglandins are shown in figure 2.22.

insoluble in water. All steroids have the same basic structure—three rings of six carbons each and one ring containing five carbons (see figure 2.21). Different steroids, however, have different functional groups attached to this basic structure and can have different numbers and positions of double covalent bonds between adjacent carbons.

Cholesterol is an extremely important molecule in the body because it functions as the parent compound from which steroids are derived. Cholesterol itself is of clinical significance because of its association with arterial disease

Figure 2.23   Representative amino acids, showing different types of functional ("R") groups.

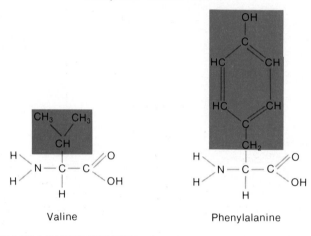

Functional group

Amino group   Carboxylic acid
                   group

Nonpolar amino acids

Valine   Phenylalanine

Polar amino acids

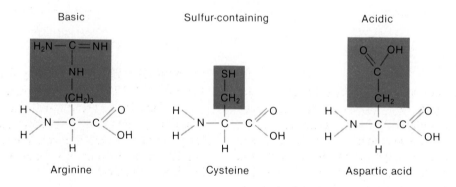

Basic   Sulfur-containing   Acidic

Arginine   Cysteine   Aspartic acid

---

1. Define lipids, and describe the structure and functions of triglycerides, phospholipids, steroids, and prostaglandins.
2. Suppose you were measuring the pH of a solution of dairy cream (a triglyceride) as it was digested by an enzyme. Describe and explain the pH changes that would occur.
3. Describe the meaning of the terms *ketosis* and *ketoacidosis* and the conditions under which these occur.

Proteins

Proteins are very big molecules *(macromolecules)* composed of long chains of subunits known as **amino acids.** As their name implies, each amino acid contains an *amine group* (NH_2) on one end of the molecule and a *carboxylic acid group* (COOH) on the other end. There are approximately twenty different amino acids in the body. The unique structure and chemical properties of each amino acid are due to its *functional group,* which is abbreviated as *R* in the general formula shown in figure 2.23. The *R* symbol actually stands for the word *residue,* but it can be thought of as indicating the "rest of the molecule."

Figure 2.24 Amphoteric nature of amino acids (see text for description).

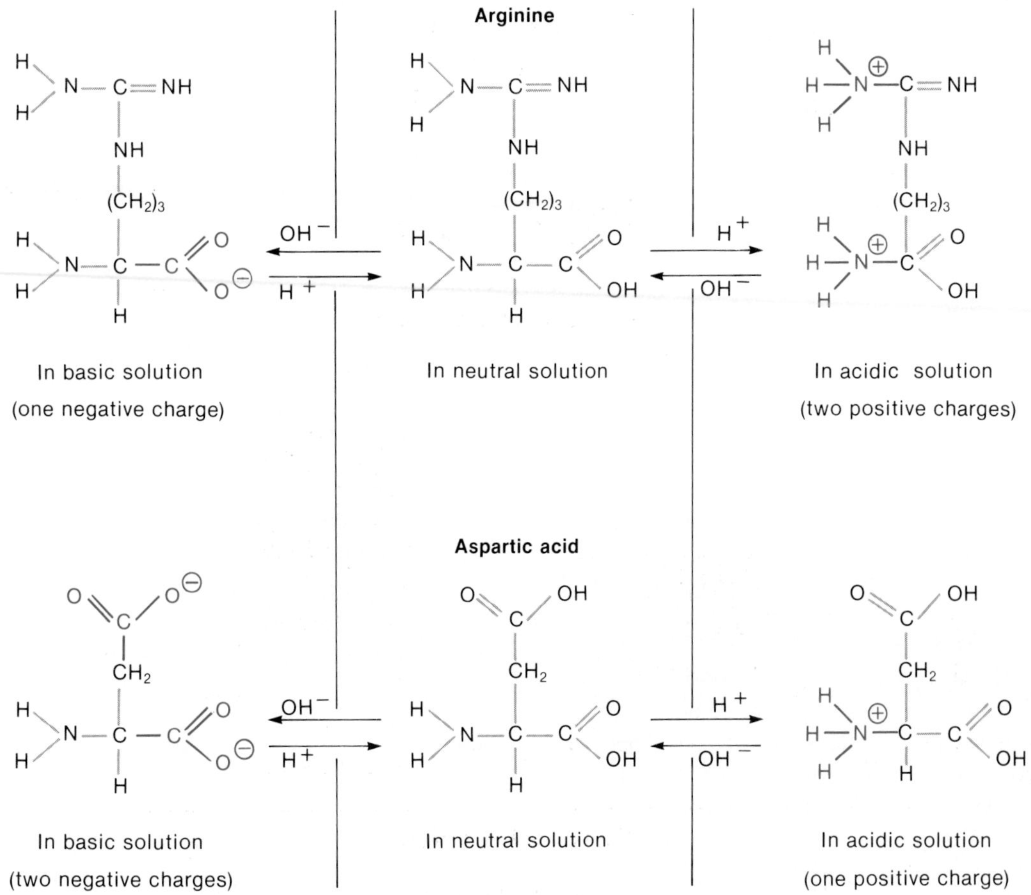

Arginine

In basic solution
(one negative charge)

In neutral solution

In acidic solution
(two positive charges)

Aspartic acid

In basic solution
(two negative charges)

In neutral solution

In acidic solution
(one positive charge)

Amino Acids

The functional groups of some amino acids contain additional amino groups, while others contain additional carboxyl groups. In acidic solutions, the amine groups can take up an extra H^+, giving them a net positive charge; in basic solutions, the carboxyl groups can lose a H^+ and have a negative charge. The *net charge* of each amino acid therefore depends on the nature of its *R* group and the pH of the solution. Since amino acids can have a net positive or negative charge, depending on the pH of the solution, they are said to be *amphoteric* (see figure 2.24).

When amino acids are joined together by dehydration synthesis, the hydrogen from the amino end of one amino acid combines with the hydroxyl group of the carboxylic acid end of another amino acid. This produces water as a covalent bond is formed between the two amino acids (see figure 2.25). The bond between adjacent amino acids is called a **peptide bond,** and the compound formed is called a *peptide*. When many amino acids are joined in this way a chain of amino acids, or **polypeptide,** is produced.

The length of polypeptide chains can vary greatly. A hormone called *thyrotrophin-releasing hormone* (TRH), for example, is only three amino acids long, while myosin (a muscle protein involved in contraction) contains about forty-five hundred amino acids. When the length of polypeptide chains becomes very long (greater than about a hundred amino acids) the molecule is called a **protein.**

Protein Structure

If one wants to describe the structure of a protein, one can do so at four different levels. First, one can describe the sequence of amino acids in the polypeptide chains that make up the protein. This is called the **primary structure** of the protein. The billions of copies of each type of protein in a cell all have the same amino acid sequences, despite the fact that an almost infinite variety of sequences could be produced using twenty different amino acids. The information needed to build proteins of given amino acid sequences is contained within the genetic code of the cells, as described in chapter 3.

Figure 2.25 Formation of peptide bonds by dehydration synthesis reaction between amino acids.

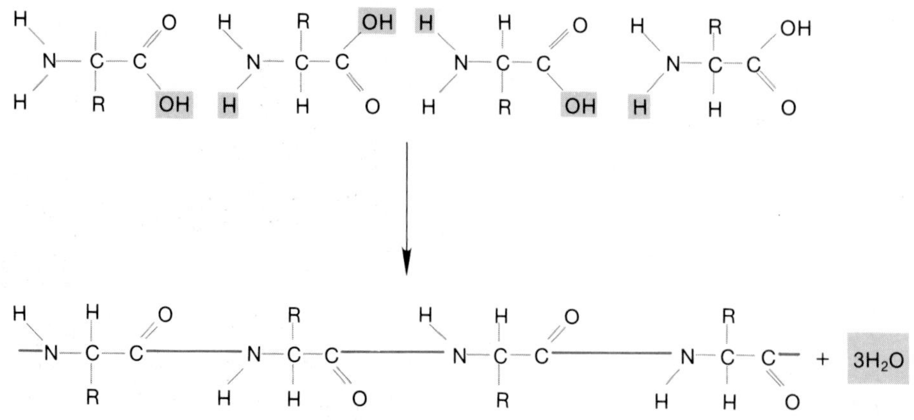

Weak interactions (such as hydrogen bonds) between functional groups in different positions in the polypeptide chain can cause this chain to twist into a *helix*. The extent and location of the helical structure is different in each type of polypeptide chain because of differences in amino acid composition. A description of the helical structure of a protein is termed its **secondary structure** (see figure 2.26).

Most polypeptide chains bend and fold on themselves to produce complex three-dimensional shapes. These are called the **tertiary structures** of the proteins. Each type of protein has its own characteristic tertiary structure. If one were to gently disrupt this structure (as by—figuratively—holding the ends and pulling it straight), the original tertiary structure would be produced again. This illustrates that the specific folding and bending are determined by chemical interactions within the protein itself, not by external "molding" forces.

The tertiary structure of a protein is formed and stabilized primarily by weak chemical interactions (such as with hydrogen bonds) between amino acids located in different regions of the polypeptide chain. The tertiary structure of some proteins, however, is made more stable by covalent bonds between sulfur atoms (called *disulfide bonds* and abbreviated S-S) in the functional groups of the amino acids known as cysteines (see figure 2.27). These covalent bonds are the exception; since most of the tertiary structure of a protein is stabilized by weaker bonds, this structure can be easily disrupted by changes in temperature or in pH. Changes in the tertiary structure induced by such treatment is called *denaturation.*

Denatured proteins retain their primary structure but have altered chemical properties. Egg-albumin proteins, for example, are soluble in their natural state in which they form a yellow-colored viscous fluid. When denatured by cooking, they change shape, cross-bond with each other, and by this means form an insoluble white precipitate—egg white. Changes in the amino acid composition of a protein—due to inherited genetic defects—may produce changes in the tertiary structures of all copies of that protein within the affected person, and by this means produce different genetically inherited diseases.

Some proteins (such as hemoglobin and insulin) are composed of a number of polypeptide chains covalently bonded together. This is the **quaternary structure** of these proteins. Insulin, for example, is composed of two polypeptide chains—one that is twenty-one amino acids long and another that is thirty amino acids long—bonded together. Hemoglobin (the protein in red blood cells that carries oxygen) is composed of four separate polypeptide chains.

Conjugated Proteins Many proteins in the body are normally found combined, or "conjugated," with other molecules. Proteins can be conjugated with carbohydrates, lipids, and organic pigment molecules (those that have a color).

Proteins conjugated with carbohydrates are called *glycoproteins*. Examples of such gylcoproteins include all protein hormones (such as growth hormone), proteins located in the outer surface of cell membranes (including glycoproteins that function as cellular receptors for hormones), and cell-surface antigens. Examples of the latter include the glycoproteins that determine red blood cell type (such as type A, type B, and so on).

Figure 2.26   A polypeptide chain, showing *(a)* its primary structure and *(b)* secondary structure.

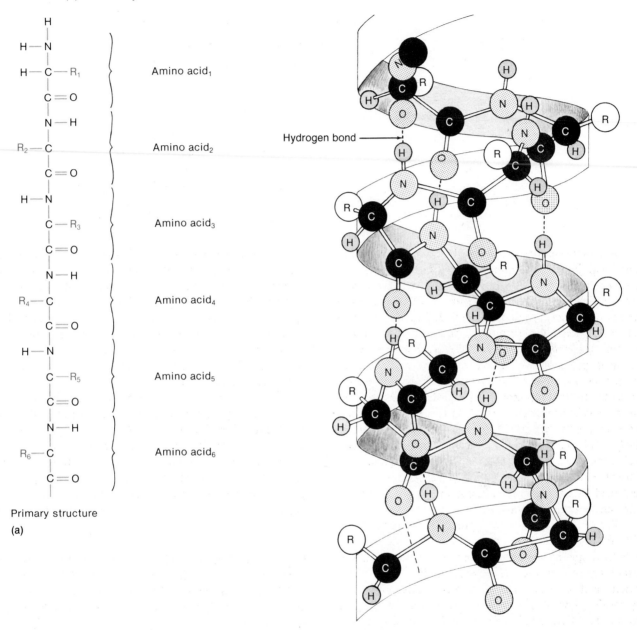

Primary structure

(a)

Hydrogen bond

α-Helix
secondary structure

(b)

Figure 2.27   The tertiary structure of a protein.
(a) *Interactions between functional ("R") groups of amino acids result in *(b)* the formation of complex three-dimensional shapes of proteins.

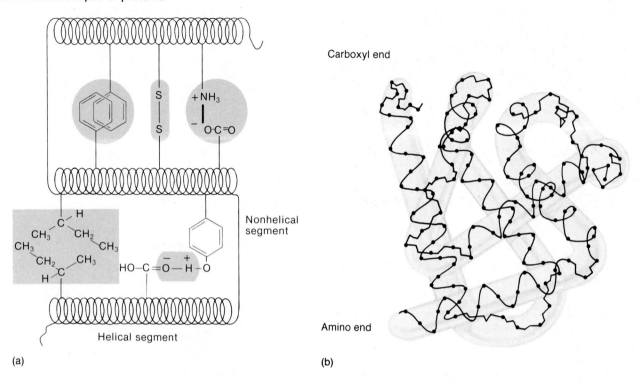

(a)

(b)

Table 2.5   Composition of selected proteins in the body.

Protein	Number of Polypeptide Chains	Nonprotein Component	Function
Hemoglobin	4	Heme pigment	Carries oxygen in the blood
Myoglobin	2	Heme pigment	Stores oxygen in muscle
Insulin	2	None	Hormone regulating metabolism
Luteinizing hormone	1	Carbohydrate	Hormone that stimulates gonads
Fibrinogen	1	Carbohydrate	Involved in blood clotting
Mucin	1	Carbohydrate	Forms mucus
Blood group proteins	1	Carbohydrate	Produces blood types
Lipoproteins	1	Lipids	Transports lipids in blood

Proteins conjugated with lipids are called *lipoproteins*. These are found in cell membranes and in plasma (the fluid portion of the blood). Since lipids are insoluble in water (the major constituent of plasma), lipids must travel in the blood attached to protein *carriers*. Certain plasma proteins therefore act as "shuttles" for transporting triglycerides, cholesterol, and steroid hormones through the blood.

Proteins conjugated to pigment molecules are sometimes called *chromoproteins* (chromo = color). These include hemoglobin (in blood) and myoglobin (in muscle), which bind oxygen, and the cytochromes in tissue cells, which are needed for oxygen utilization in the production of cellular energy (see chapter 5).

Functions of Proteins

Because of their tremendous structural diversity, proteins can serve a wider variety of functions than any other type of molecule in the body. Many proteins, for example, contribute significantly to the structure of different tissues and

Figure 2.28 Photomicrograph of collagen fibers.

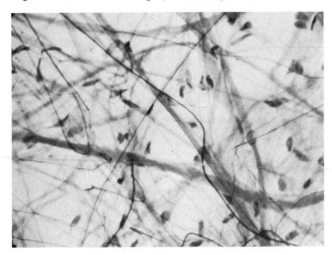

Figure 2.29 General structure of a nucleotide and the formation of sugar-phosphate bonds between nucleotides to form a polymer.

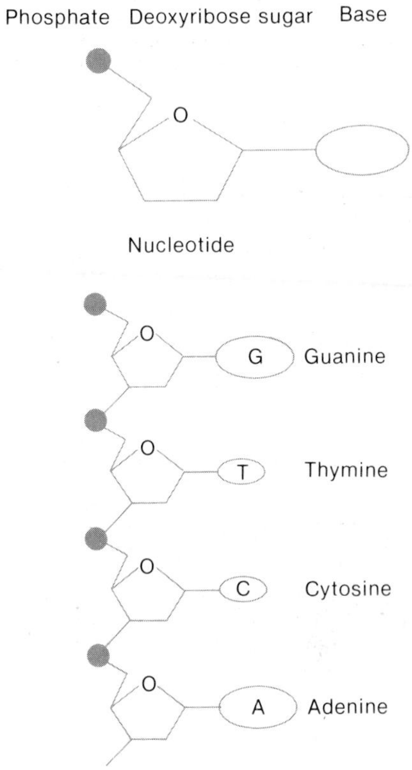

Phosphate   Deoxyribose sugar   Base

Nucleotide

G Guanine

T Thymine

C Cytosine

A Adenine

in this way play a passive role in the functions of these tissues. Examples of such *structural proteins* include collagen and keratin. *Collagen* is a fibrous protein (unlike most, which are globular in shape) that provides tensile strength to connective tissues such as tendons, ligaments, and others. Keratin is found in the outer layer of dead cells in the epidermis, where it serves to prevent water loss through the skin.

Many proteins serve a more active role in the body where specialized structure and function are required. *Enzymes* and *antibodies,* for examples, are proteins—no other type of molecule could provide the vast array of different structures needed for these functions. Proteins in cell membranes serve as receptors and transporters of specific molecules. These proteins provide specific shapes and chemical properties that allow them to interact in a specific way with other molecules. In addition, proteins serve a wide variety of other functions that will be discussed in subsequent chapters.

Nucleic Acids

The nucleic acids include the macromolecules of **DNA (deoxyribonucleic acid)** and **RNA (ribonucleic acid),** which play a central role in genetic regulation, and the subunits from which these molecules are formed. These subunits are known as **nucleotides.** In addition to their function as the "building blocks" of DNA and RNA, nucleotides are involved in energy transformations within the cell and in various regulatory processes.

Although nucleotides are used as subunits in the formation of the long polynucleotide chains of DNA and RNA, the nucleotides are themselves composed of subunits. Each nucleotide is composed of three parts—a five-

carbon sugar, a phosphate group bonded to one end of the sugar, and a *nitrogenous base* that is bonded to the other end of the sugar. The nucleotide bases are cyclic nitrogen-containing molecules with either one ring of carbons (the *pyrimidines*) or two rings (the *purines*).

Deoxyribonucleic Acid

The structure of DNA provides the information of the genetic code. One might therefore expect DNA to have an extremely complex structure. Actually, however, DNA—though the largest molecule in the cell—has a structure simpler than that of most proteins. This simplicity of structure deceived some of the early investigators into believing that the protein content of chromosomes, rather than their DNA content, provided the basis for the genetic code.

Sugar molecules in the nucleotides of DNA are a type of pentose (five-carbon) sugar called *deoxyribose* (hence the name for this nucleic acid). Each deoxyribose sugar can be covalently bonded to one of four possible bases. These bases include the two purines (adenine and guanine) and the two pyrimidines (cytosine and thymine). There are thus four different types of nucleotides that can be used to produce the long DNA chains.

Figure 2.30   The four nitrogenous bases in deoxyribonucleic acid (DNA). Notice that hydrogen bonds can form between guanine and cytosine, and between thymine and adenine.

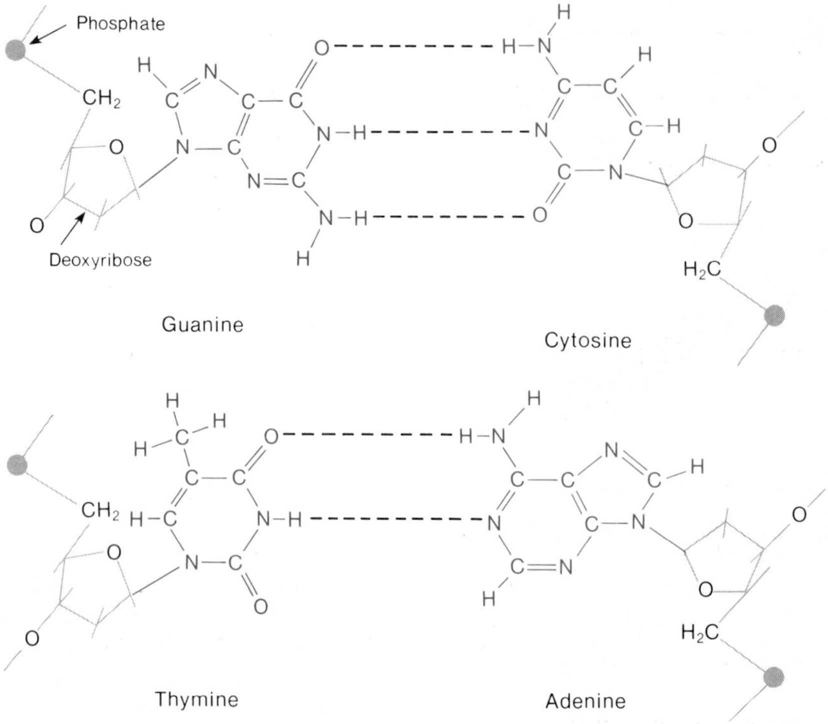

Guanine   Cytosine

Thymine   Adenine

When nucleotides combine to form a chain, the phosphate group of one condenses with the deoxyribose sugar of another nucleotide. This forms a sugar-phosphate chain as water is removed in dehydration synthesis. Since the nitrogenous bases are attached to the sugar molecules, the sugar-phosphate chain looks like a "backbone" from which the bases project. Each of these bases can form hydrogen bonds with other bases, which are in turn joined to a different chain of nucleotides. Such hydrogen bonding between bases thus produces a *double-stranded* DNA molecule; the two strands are like a staircase, with the paired bases as steps.

Actually, the two chains of DNA twist about each other to form a **double helix**—the molecule is much like a spiral staircase. The number of purine bases in DNA, it has been shown, is equal to the number of pyrimidine bases. The reason for this is explained by the **law of complementary base pairing;** adenine can only pair with thymine (through two hydrogen bonds), while guanine can only pair with cytosine (through three hydrogen bonds). Knowing this rule, one can predict the base sequence of one DNA strand if one knows the sequence of bases in the complementary strand.

While one can predict which base is opposite a base at a given position in DNA, one cannot predict the bases above or below that level. Although there are only four

Figure 2.31   Hydrogen bonds between bases in two DNA strands produce a double-stranded DNA molecule.

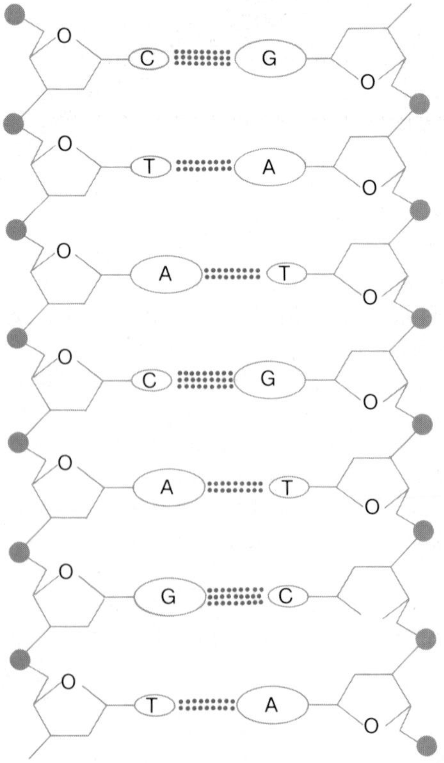

Figure 2.32 The double helix structure of DNA.

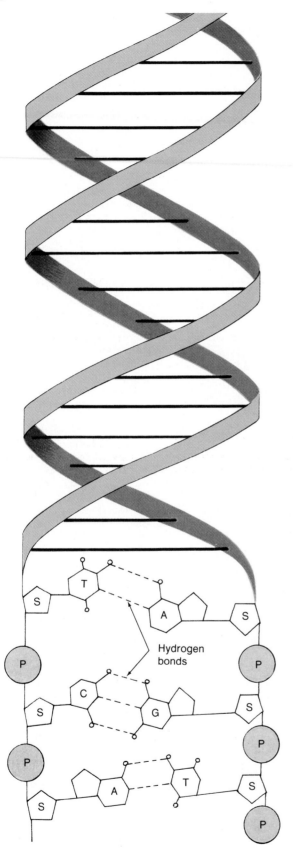

Hydrogen bonds

Figure 2.33 Differences between the nucleotides and sugars in DNA and RNA.

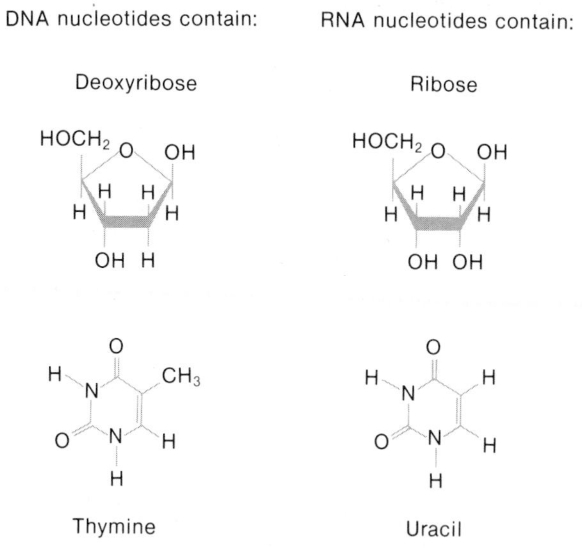

DNA nucleotides contain: RNA nucleotides contain:

Deoxyribose   Ribose

Thymine   Uracil

bases, the number of possible base sequences along a stretch of several thousand nucleotides (the length of a gene) is almost infinite. Despite this infinite possible variety, almost all of the billions of copies of a gene in a person will be identical. The mechanisms by which this is achieved will be discussed in chapter 3.

Ribonucleic Acid

The genetic information contained in DNA is able to direct the activities of the cell through production of another type of nucleic acid—RNA (ribonucleic acid). Like DNA, RNA consists of long chains of nucleotides joined together by sugar-phosphate bonds. Nucleotides in RNA, however, differ from those in DNA in three ways: (1) ribonucleotides contain the sugar *ribose* (instead of deoxyribose); (2) the base *uracil* is found in place of thymine; and (3) RNA is composed of a single polynucleotide strand (it is not double stranded like DNA).

There are three types of RNA molecules that function in the cytoplasm of cells. These are *messenger RNA (mRNA), transfer RNA (tRNA),* and *ribosomal RNA (rRNA).* All three types are made within the cell nucleus, using the information contained within DNA as a guide. The synthesis of RNA and its role in gene expression are discussed in chapter 3.

1. Write the general formula for an amino acid and describe how amino acids differ from each other.
2. Describe the primary, secondary, tertiary, and quaternary structure of proteins.
3. Write the DNA bases that would pair with the following base sequence: adenine, cytosine, thymine, guanine.
4. Compare the structure of DNA with that of RNA.

Summary

Atoms and Chemical Bonds

I. Covalent bonds are formed by the sharing of electrons between atoms.
   A. Covalent bonds are the strongest bonds and are usually not broken without the actions of enzymes.
   B. There are two types of covalent bonds.
      1. Nonpolar covalent bonds are formed when atoms share electrons equally.
      2. Polar covalent bonds are formed when atoms—such as oxygen or nitrogen—pull electrons toward themselves and do not share equally.
   C. Molecules with many nonpolar bonds are not soluble in water; those with many polar bonds are water soluble.
      1. Examples of nonpolar bonds are those between two carbons and between a carbon and a hydrogen.
      2. Polar covalent bonds include those between an oxygen and a hydrogen atom or those between nitrogen and hydrogen.

II. Ionic bonds in inorganic salts are weaker than covalent bonds.
   A. In sodium chloride, for example, the chloride takes an electron from sodium.
      1. Chloride is therefore negatively charged while sodium is positively charged.
      2. The sodium and chloride attract each other to form the ionic bond.
   B. When salt is dissolved in water the NaCl ionizes to produce Na^+ and Cl^-, each of which attracts water molecules.
      1. Elements with a net charge, due to loss of gain of electrons, are called ions.
      2. Ions form hydration spheres and are water soluble.

III. Hydrogen bonds, such as those between water molecules, are the weakest bonds.
   A. Hydrogen bonds are formed when two electronegative atoms (such as oxygen or nitrogen) bond with one hydrogen atom.

B. Hydrogen bonds are responsible for some of the physical properties of water and in part for the shape of proteins.

IV. The pH is a measure of H^+ concentration.
   A. Hydrogen that is linked by ionic or polar covalent bonds to other atoms can dissociate and enter solution as H^+.
      1. Compounds that release free H^+ to solution are called *acids*.
      2. Compounds that remove H^+ from solution are called *bases*—these usually have an OH^- (hydroxyl) group that can form water.
   B. Pure water has a H^+ concentration of 10^{-7} molar and a pH of 7.
      1. Acidic solutions have a higher H^+ concentration and a lower pH because the pH is inversely proportional to the log of the H^+ concentration.
      2. Basic solutions have a lower H^+ concentration and a pH between 7 and 14.

Carbohydrates and Lipids

I. Carbohydrates are organic molecules that contain carbon, hydrogen, and oxygen in the ratio of 1:2:1.
   A. The simplest carbohydrates are monosaccharides such as glucose, fructose, and galactose.
      1. Monosaccharides can be joined together to form double sugars or disaccharides such as sucrose (glucose plus fructose).
      2. Polysaccharides such as glycogen are formed by combining many monosaccharides (such as glucose) together.
   B. Bonds between monosaccharide subunits are formed by dehydration synthesis and broken by hydrolysis.
      1. In dehydration synthesis, a water molecule is formed as two subunits become bonded together.

2. In hydrolysis, a water molecule is broken into hydrogen and OH, as the bonds between the subunits are broken.
   C. Liver and skeletal muscles store carbohydrates in the form of glycogen.
      1. Glycogen can be hydrolyzed into glucose.
      2. Glucose is used for energy, particularly by the brain.

II. Lipids are organic molecules that are nonpolar and therefore not water soluble.
   A. Triglycerides—or neutral fats—are the principal type of energy storage molecule in the body.
      1. Triglycerides are formed from one glycerol (an alcohol) and three fatty acids.
      2. Some fatty acids are converted by the liver into four-carbon-long derivatives called *ketone bodies*.
   B. Phospholipids are nonpolar on one end (the part containing the fatty acids) and polar on the end containing the phosphate group.
      1. The cell membrane is composed primarily of phospholipids.
      2. Phospholipids are also surface-active agents—they lower the surface tension of water.
   C. Steroids and prostaglandins are regulator molecules.
      1. Steroid hormones, derived from cholesterol, are secreted from the adrenal cortex and gonads.
      2. Prostaglandins are produced by most tissues, where they act as local regulators.

Proteins and Nucleic Acids

I. Proteins are macromolecules formed from condensation of amino acids.
   A. Amino acids contain an amine group, an acidic group, and a "functional group."
      1. There are about twenty different amino acids, with an equal number of different functional groups.

2. Dehydration synthesis reactions join amino acids together, by peptide bonds, to make polypeptide chains.

B. The amino acid sequence of a protein is termed its primary structure.
1. The primary structure of a particular type of protein is always the same—it is specified by the genetic code.
2. The unique primary structure of each protein affects its three-dimensional structure.

C. Chemical bonding between amino acids affects the three-dimensional structure of proteins.
1. Interactions between amino acids cause some regions of the polypeptide chain to form a helix—this is known as the secondary structure of the protein.

2. Bonding between functional groups in different amino acids cause the polypeptide chain to fold and bend in characteristic ways—this is the tertiary structure of the protein.

II. Deoxyribonucleic acid (DNA) is composed of a long chain of nucleotides.
A. Each DNA nucleotide consists of a sugar (deoxyribose) bonded to a phosphate and a nitrogenous base.
1. The bases are adenine, guanine, cytosine, and thymine.
2. Nucleotides join together by bonds between sugar and phosphate, with the bases projecting from the sugar-phosphate "backbone."

B. DNA consists of two polynucleotide chains joined together by hydrogen bonds between complementary bases.
1. Adenine and thymine bond together.
2. Guanine and cytosine bond together.

III. RNA is produced in the nucleus as a complementary copy of a region of one strand of DNA.
A. RNA nucleotides are the same as DNA nucleotides except for the fact that they contain ribose sugar and the base uracil instead of thymine.
B. RNA is single stranded, compared to DNA, which is double stranded.
C. There are three types of RNA that enter the cytoplasm—messenger RNA (mRNA), transfer RNA (tRNA), and ribosomal RNA (rRNA).

Self-Study Quiz

1. Which of the following statements about atoms is TRUE?
   (a) they have more protons than electrons
   (b) they have more electrons than protons
   (c) they are electrically neutral
   (d) they have as many neutrons as they have electrons

2. The bond between oxygen and hydrogen in a water molecule is a:
   (a) hydrogen bond
   (b) polar covalent bond
   (c) nonpolar covalent bond
   (d) ionic bond

3. Which of the following is a nonpolar covalent bond?
   (a) the bond between two carbons
   (b) the bond between sodium and chloride
   (c) the bond between two water molecules
   (d) the bond between nitrogen and hydrogen

4. Solution A has a pH of 2, while solution B has a pH of 10. Which of the following statements about these solutions is TRUE?
   (a) solution A has a higher H^+ concentration than solution B
   (b) solution B is basic
   (c) solution A is acidic
   (d) all of these are true

5. Glucose is a:
   (a) disaccharide
   (b) polysaccharide
   (c) monosaccharide
   (d) phospholipid

6. Digestion reactions occur by means of:
   (a) dehydration synthesis
   (b) hydrolysis

7. Carbohydrates are stored in liver and muscles in the form of:
   (a) glucose
   (b) triglycerides
   (c) glycogen
   (d) cholesterol

8. Lecithin is a:
   (a) carbohydrate
   (b) protein
   (c) steroid
   (d) phospholipid

9. Which of the following lipids have regulatory roles in the body?
   (a) steroids
   (b) prostaglandins
   (c) triglycerides
   (d) both a and b
   (e) both b and c

10. The tertiary structure of a protein is *directly* determined by:
    (a) the genes
    (b) the primary structure of the protein
    (c) enzymes that "mold" the shape of the protein
    (d) the position of peptide bonds

11. If four bases in one DNA strand are A (adenine), G (guanine), C (cytosine), and T (thymine), the complementary bases in the opposite strand will be:
    (a) T,C,G,A
    (b) C,G,A,T
    (c) A,G,C,T
    (d) U,C,G,A

12. Which of the following statements about RNA is TRUE?
    (a) it is made in the nucleus
    (b) it contains the base uracil
    (c) it is double stranded
    (d) both (a) and (b) are true
    (e) both (b) and (c) are true

3

Cell Structure
and Genetic Control

Objectives

By studying this chapter, you should
be able to:

1. Describe the structure of the cell
   membrane and the cytoskeleton
2. Describe endocytosis and
   exocytosis
3. Describe the structure of cilia
   and flagella
4. Understand and explain the
   semiconservative replication of
   DNA
5. Describe the phases of the cell
   cycle
6. Understand and explain genetic
   transcription and the post-
   transcriptional modifications that
   change pre-mRNA into mRNA

7. Describe the functions of mRNA,
   tRNA, and rRNA
8. Understand and explain the
   mechanisms of genetic
   translation of the RNA code into
   the synthesis of proteins
9. Describe the functions of the
   rough endoplasmic reticulum and
   Golgi apparatus in post-
   translational modifications of
   secretory proteins
10. Describe the sequences of
    events that occur in the
    synthesis, packaging, and
    exocytosis of secretory proteins
11. Describe the functions of
    lysosomes and mitochondria

*T*he cell is so small, and so simple in appearance when viewed with the ordinary (light) microscope, that it is difficult to conceive that each cell is a living entity unto itself. Equally amazing is the fact that the physiology of our organs and systems derive from the complex functions of the cells of which they are composed. Complexity of function demands complexity of structure, even at the subcellular level. This complexity of structure requires, in part, separation of the different functioning units by membranes.

Figure 3.1 The fluid-mosaic model of the cell membrane. The membrane consists of a double layer of phospholipids, with the phosphates *(open circles)* oriented outward and the hydrophobic hydrocarbons *(wavy lines)* oriented towards the center. Proteins may completely (P₁) or partially (P₂) span the membrane. Carbohydrates *(closed circles)* are attached to the outer surface.

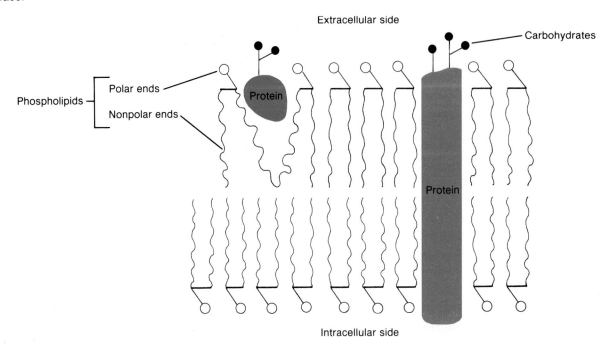

Since both the intracellular and extracellular environments (or "compartments") are aqueous (aqua = water), a barrier must exist to prevent loss of cellular enzymes, nucleotides, and other polar organic molecules. This function is provided by the *cell membrane,* also called the *plasma membrane,* which is composed primarily of lipids. In addition to "trapping" needed molecules within the cell, the plasma membrane is itself a complex, active structure that permits selective interaction between the cell and its environment. Specialized functions within the cell are performed by specialized structures called *organelles* ("little organs"). The part of the intracellular compartment that is not contained within organelles is known as *cytoplasm.*

Cell Membrane

The cell membrane, and indeed all of the membranes surrounding organelles within the cell, are composed primarily of phospholipids. Phospholipids, as described in chapter 2, are polar on the end that contains the phosphate group, while nonpolar (and hydrophobic) throughout the rest of the molecule. Since there is an aqueous environment on each side of the membrane, the hydrophobic ends of the molecules "huddle together" in the

Figure 3.2 Formation of endocytotic vesicles at the cell membrane of a smooth muscle cell. The technique known as "freeze-fracture" reveals craters (caveola) at the inner surface of the membrane.

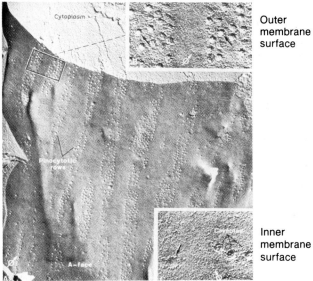

Outer membrane surface

Inner membrane surface

Figure 3.3 Electron micrograph of the membranes of two adjacent cells showing different stages in the formation of endocytotic vesicles.

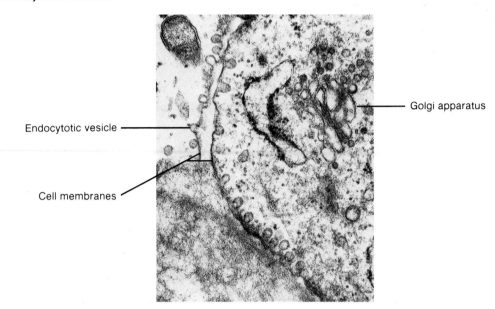

Golgi apparatus

Endocytotic vesicle

Cell membranes

center of the membrane, leaving the polar ends exposed to water at both surfaces. This is believed to result in the formation of a continuous double layer of phospholipids in the cell membrane (although other possible models have been proposed).

The hydrophobic core of the membrane restricts passage of water and water-soluble molecules and ions. The specialized functions and selective transport properties of the membrane are believed to be due to its protein content. Proteins are found partially submerged on each side of the membrane; some proteins appear to span the membrane completely from one side to the other. Since the membrane is not solid—phospholipids and proteins are free to move laterally—the proteins within the phospholipid "sea" are not uniformly distributed. This view of membrane structure has been aptly termed the **fluid-mosaic model.**

The cell membrane is a mosaic in that different amounts and kinds of proteins are found in different areas of the membrane. These proteins can serve many functions, including (1) structural support; (2) transport of molecules across the membrane; (3) enzymatic control of chemical reactions at the cell surfaces; (4) receptors for hormones and other regulatory molecules that arrive at the outer surface of the membrane; and (5) cellular "markers" that identify the blood and tissue type.

Endocytosis and Exocytosis

Regions of the cell membrane can bulge inwards towards the cytoplasm, creating structures that look like craters when seen in a surface view (see figure 3.2). Continued invagination of this membrane creates a pouch that encloses part of the extracellular environment. This pouch ultimately closes and pinches off membrane from the cell to form a membrane-enclosed *vesicle.* This process is called **endocytosis** (endo = within; cyto = cell).

There are different types of endocytosis. When the endocytotic vesicle contains only extracellular water, for example, the process is termed *pinocytosis* (pino = drink). When large bodies of organic matter—such as bacteria or degenerated cells—are engulfed, the process is called *phagocytosis* (phago = eat). Liver cells and some cells of the immune system are active phagocytes. These cells extend their cytoplasm to form pseudopods (pseudo = false; pod = foot) that surround their "victim" and then join together. Fusion of the inner membrane of the pseudopods creates a "food vacuole" within the cell that contains the ingested particles.

Figure 3.4 Scanning electron micrographs of phagocytosis showing the formation of pseudopods and the entrapment of the prey within a food vacuole.

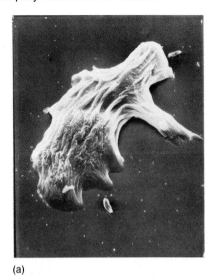
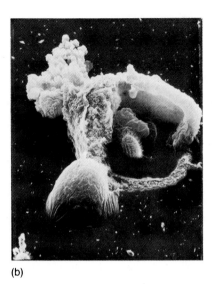

(a)   (b)

Figure 3.5 Immunofluorescence photograph of microtubules forming the cytoskeleton of a cell. Microtubules are visualized with the aid of antibodies against tubulin, the major protein component of the microtubules.

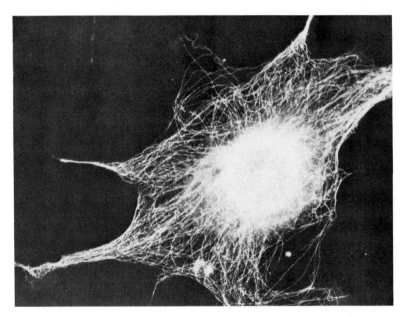

Proteins produced by the cell, and destined for "export" (secretion), are also packaged within vesicles. In the process of **exocytosis** (exo = outside), these secretory vesicles fuse with the cell membrane and release their contents to the extracellular environment. This process—the reverse of endocytosis—adds new membrane material that replaces that which was lost from the cell membrane in endocytosis. The source of the secretory vesicles will be discussed together with proteins synthesis later in this chapter.

Endocytosis and exocytosis account for only part of the two-way traffic between the intracellular and extracellular compartments. The bulk of this traffic is due to movement of molecules and ions *through* the cell membrane. These membrane transport processes will be discussed in detail in chapter 6.

Figure 3.6 Diagram of a proposed structure of the cytoskeleton.*

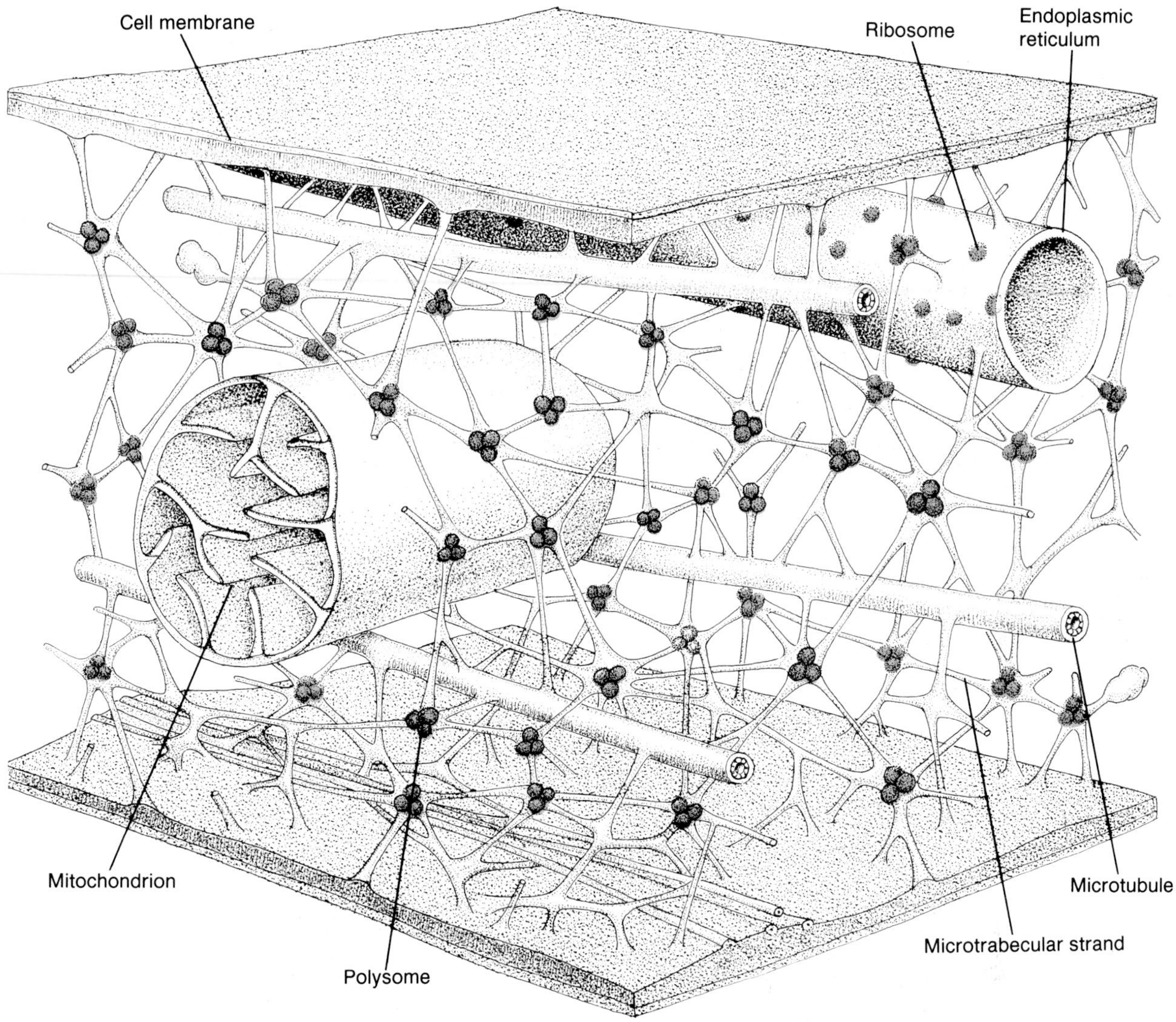

Cell membrane

Ribosome

Endoplasmic reticulum

Mitochondrion

Polysome

Microtubule

Microtrabecular strand

Membrane Cholesterol and Carbohydrates

Endocytosis, exocytosis, and indeed all cellular movements require flexibility of the cell membrane. This flexibility is derived in part from the fluid nature of the double phospholipid layers. The fluidity of the membrane depends upon a proper ratio of different phospholipids and upon a proper ratio of phospholipids to cholesterol (found in the cell membranes of all higher organisms). Flexibility is lost when there is an inherited defect in the ratios of these lipids. This may result, for example, in the inability of red blood cells to bend at the middle when squeezing through narrow blood channels (and thus result in occlusion of these small vessels).

In addition to containing lipids and proteins, the cell membrane also contains carbohydrates. These are attached primarily to the outer surface of the cell to form glycoproteins and glycolipids. The surface carbohydrates have many negative charges (they are polyanions) that affect the interaction of regulatory molecules with the cell surface and the interactions between cells—they help keep red blood cells apart, for example. Stripping carbohydrates from the outer red blood cell surface results in their more rapid destruction by phagocytic cells in the liver, spleen, and bone marrow.

Cilia, Flagella, and the Cytoskeleton

According to recent evidence, the cytoplasm of cells is not a homogenous solution of proteins, but is, rather, a highly organized structure in which protein fibers—in the form of *microfilaments* and *microtubules*—are arranged in a complex latticework. These interconnected microfilaments and microtubules are believed to provide structural organization for cytoplasmic enzymes and to provide support for various organelles (see figures 3.5 and 3.6).

Figure 3.7   Electron micrographs of cilia, showing *(a)* longitudinal and *(b)* cross sections. Notice the characteristic "9 + 2" arrangement of microtubules in the cross sections.

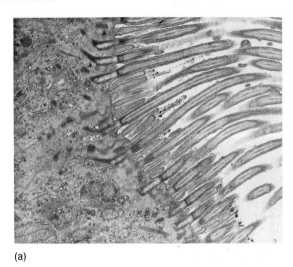

(a)

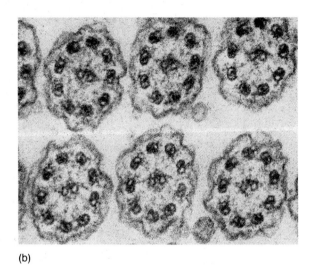

(b)

The interconnected microfilaments and microtubules are thus said to function as a **cytoskeleton.** The structure of this "skeleton" is not rigid; it has been shown experimentally to be capable of quite rapid reorganization. Contractile proteins—including actin and myosin, which are responsible for muscle contraction—may be able to shorten the length of some microfilaments. The cytoskeleton, in other words, may also represent the cellular "musculature." Microtubules, for example, form the *spindle apparatus* that pulls chromosomes away from each other in cell division (this will be described later), and form the central part of cilia and flagella.

Cilia and Flagella

Cilia are tiny hairlike structures that protrude from the cell and, like the coordinated action of oarsmen in a boat, stroke in unison. Cilia in the human body are found on the apical surface (the surface facing the lumen, or opening) of epithelial cells of the respiratory and female genital tracts. In the respiratory system, the cilia transport strands of mucus, which act as a filtering system. Particles in the inspired air become trapped in the mucus, which is then conveyed by ciliary action to a region (the pharynx) where the mucus can either be swallowed or expectorated. In the female genital tract, ciliary movements in the epithelium lining draw the egg (ovum) into the uterine or fallopian tube and move it towards the uterus.

Sperm are the only cells in the human body that have **flagella.** The flagellum (Latin for *whip*) is a single, whip-like structure that propels the sperm through its environment. Both cilia and flagella are composed of microtubules arranged in a characteristic way. Nine pairs of microtubules in the periphery of a cilium or flagellum surround a single pair of microtubules in the center. This arrangement is called a "nine-plus-two" structure.

1. Draw the fluid mosaic model of the cell membrane, and describe the structure of the membrane in words.
2. Using a flow chart (arrows), describe the sequence of events that occur in endocytosis and exocytosis.
3. Describe the structure and properties of the cytoskeleton.

Nuclear Structure and DNA Function

Most cells in the body have a single nucleus, although some—such as skeletal muscle cells—are multinucleate. The nucleus is surrounded by a *nuclear envelope* composed of an inner and outer membrane. These membranes (separated by a small space in most regions) fuse together in many areas to form openings, or *nuclear pores*. These pores prevent DNA from leaving the nucleus but apparently allow RNA to exit into the cytoplasm.

Figure 3.8 Electron micrograph of freeze-fractured nuclear envelope, showing nuclear pores.

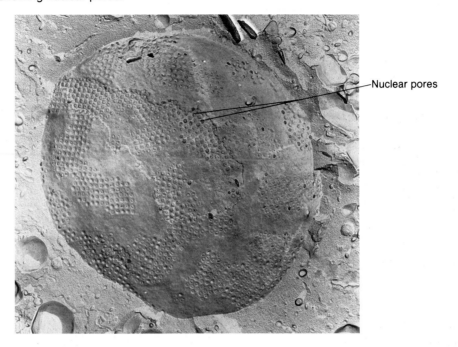

Nuclear pores

Figure 3.9 The nucleus of a liver cell showing nuclear envelope, heterochromatin, and nucleolus.

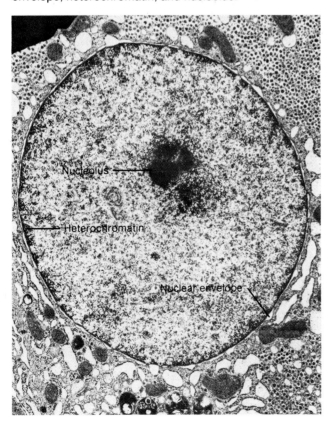

Nucleolus

Heterochromatin

Nuclear envelope

Many granulated threads in the nuclear fluid (or nucleoplasm) can be seen with the electron microscope. These threads are called *chromatin* and consist of a combination of DNA and protein. There are two forms of chromatin. Thin, extended chromatin—or *euchromatin*—appears to represent the active form of DNA in a nondividing cell. Regions of condensed, "blotchy"-appearing chromatin is believed to contain inactive DNA. These regions are called *heterochromatin.*

Within each nucleus, one or more large dark areas can be seen. These regions, which are not surrounded by membranes, are called *nucleoli.* The DNA within nucleoli contain genes that code for the production of ribosomal RNA (rRNA), an essential component of ribosomes. The function of ribosomes (in protein synthesis) will be discussed in a later section.

DNA Replication

When a cell is going to divide, each DNA molecule replicates itself, and each of the identical DNA copies thus produced is distributed to the two daughter cells. Replication of DNA requires the activation of a specific enzyme known as *DNA polymerase.* This enzyme moves along the DNA molecule, breaking the weak hydrogen bonds between complementary bases as it travels. As a result, the bases of each of the two DNA strands become free to bond to new complementary bases that are part of free nucleotides within the surrounding fluid.

Figure 3.10 Replication of DNA. Each new double helix is composed of one new and one old strand. The base sequences of each of the new molecules is identical to that of the parent DNA because of complementary base pairing.

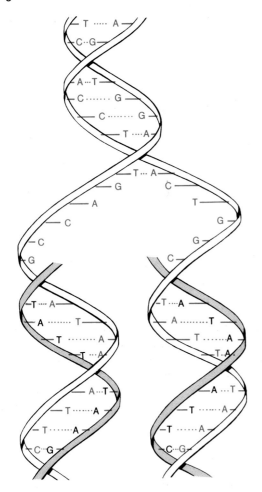

The bases of each original strand, because of the rules of complementary base pairing, will bond to the appropriate free nucleotides: adenine bases pair with thymine-containing nucleotides; guanine pairs with cytosine-containing nucleotides; and so on. In this way two new molecules of DNA, each containing two complementary strands, are formed. The DNA polymerase enzyme links the phosphate groups and dexoyribose sugar groups together to form a second polynucleotide chain in each DNA that is complementary to each of the first DNA strands. In this way two new double-helix DNA molecules are produced that contain the same DNA base sequence as the "parent" molecule.

When DNA replicates, therefore, each copy is composed of one new strand and one strand from the original DNA molecule. Replication is said to be **semiconservative** (half of the original DNA is "conserved" in each of the new DNA molecules). Through this mechanism, the sequence of bases in DNA—which is the basis of the genetic code—is preserved from one cell generation to the next.

The Cell Cycle and Mitosis
Unlike the life of an organism, which can be pictured as a linear progression from birth to death, the life of a cell follows a cyclical pattern. Each cell is produced as a part of its "parent" cell; when the daughter cell divides, it in turn becomes two new cells. In a sense, then, each cell is potentially immortal as long as its progeny can continue to divide. Some cells in the body divide frequently; the skin, for example, is renewed approximately every two weeks, and the stomach lining is renewed about every two or three days. Other cells, however—such as nerve cells and striated muscle cells in the adult—do not divide at all. All cells in the body, of course, live only as long as the person lives (some cells die sooner than others, but eventually all cells die when vital functions cease).

Certain types of cells can be removed from the body and grown in nutrient solutions (outside the body, or *in vitro*). Under these artificial conditions the potential longevity of different cell lines can be studied. For unknown reasons normal connective tissue cells (called fibroblasts) stop dividing *in vitro* after about forty to seventy population doublings. Cells that become transformed into cancer, however, apparently do not age and continue dividing indefinitely in culture. It is ironic that these potentially immortal cells commit suicide by killing their host.

The Cell Cycle   The non-dividing cell is in a part of its life cycle known as **interphase.** The chromosomes are in their extended form (as euchromatin) and actively direct the synthesis of RNA. Through their direction of RNA synthesis, the genes control the metabolism of the cell. During this time the cell may be growing, and so this part of interphase is known as the G_1 *phase* (G stands for *growth*). Although sometimes described as "resting," cells in the G_1 phase perform the physiological functions characteristic of the tissue in which they are found.

If a cell is going to divide, it replicates its DNA during a part of interphase known as the *S phase* (S stands for *synthesis*). The mechanisms that cause transformation of a cell from the G_1 to the S phase are not known.

Figure 3.11 The life cycle of a cell.

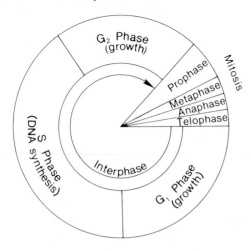

The cell life cycle

Figure 3.12 Photograph of homologous pairs of chromosomes from a human male cell at metaphase of mitosis (homologous chromosomes have been paired and numbered according to convention).

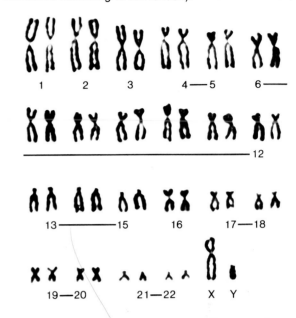

Some experiments suggest that two nucleotides—cyclic adensosine monophosphate (cAMP) and cyclic guanosine monophosphate (cGMP) may regulate this transformation. An increase in the intracellular concentration of cAMP and a decrease in cGMP can stimulate cell division. Reverse changes in the concentration of these nucleotides inhibits cell division. Besides helping to regulate cell division, these cyclic nucleotides have a wide variety of other regulatory functions (as described in subsequent chapters).

Once DNA has replicated, the chromatin condenses to form short, thick, rodlike structures. This is the more familiar form of chromosomes because they are easily seen in the ordinary (light) microscope. It should be remembered that this form of the chromatin represents a "packaged" state of DNA—not the extended, threadlike form that is active in directing the metabolism of the cell during the G_1 phase.

When the chromosomes become short and thick they can be photographed, matched, and counted. By this method, it can be seen that human cells contain twenty-three matched pairs of chromosomes.

The matched pairs of chromosomes are called **homologous chromosomes.** One member of each homologous pair is derived from a chromosome inherited from the father, while the other member is a copy of one of the chromosomes inherited from the mother. Homologous

chromosomes do not have identical DNA base sequences—one member of the pair may code for blue eyes, for example, while the other codes for brown eyes. There are twenty-two homologous pairs of *autosomal chromosomes* and one pair of *sex chromosomes,* described as X and Y. Females have two X chromosomes, while males have one X and one Y chromosome.

Mitosis Following the S phase of the cell cycle, each chromosome contains two strands called *chromatids,* which are joined together by a *centromere.* The two chromatids within a chromosome contain identical DNA base sequences because each is produced by the semiconservative replication of DNA. Each chromatid, therefore, contains a complete double-helix DNA molecule that is a copy of the single DNA molecule existing prior to replication. Each chromatid will become a separate chromosome once cell division has been completed.

Following a second resting phase (G_2), which is usually shorter than the G_1, the cell proceeds through the various stages of cell division or **mitosis** (the *M phase* of the cycle). In mitosis, the chromosomes line up single file along the equator of the cell as *spindle fibers* (composed of microtubules) are produced by structures called *centrioles* located at the cell poles. The spindle fibers attach to the centromeres of each chromosome.

Figure 3.13 The stages of mitosis.

(a) Interphase

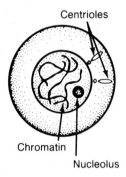
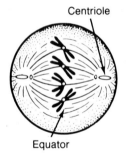

Chromosomes are in an extended form and seen as chromatin in the electron microscope.

The nucleolus is visible.

(b) Prophase

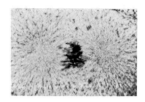

Chromosomes are seen and observed to consist of two chromatids joined together by a centromere.

Centrioles move apart towards opposite poles of the cell.

Spindle fibers are produced and extend from each centriole.

The nuclear membrane starts to disappear.

The nucleolus is no longer visible.

(c) Metaphase

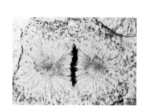
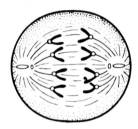

Chromosomes are lined up at the equator of the cell.

The spindle fibers from each centriole are attached to the centromeres of the chromosomes.

The nuclear membrane has disappeared.

(d) Anaphase

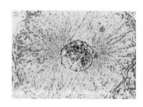
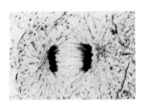

The centromeres split and the sister chromatids separate as each is pulled to an opposite pole.

(e) Telophase

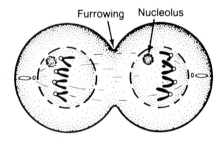
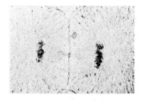

Chromosomes become longer, thinner, and less distinct.

New nuclear membranes form.

The nucleolus reappears.

Cell division (cytokinesis) is nearly complete.

Figure 3.14 The synthesis of single-stranded messenger RNA using one of the two DNA strands as a guide, or template. Each triplet of bases in the messenger RNA serves as a code word, or "codon," when it later serves as a guide for the synthesis of proteins.

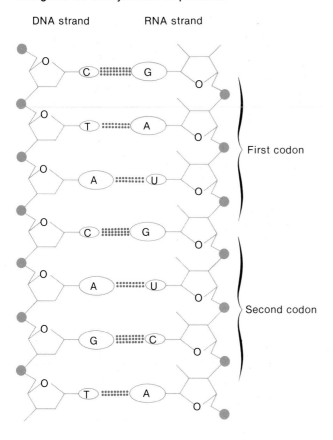

When the spindle fibers shorten, the centromeres split apart and the two chromatids in each chromosome are pulled to opposite poles. Each pole therefore gets one copy of each of the forty-six chromosomes. Division of the cytoplasm *(cytokinesis)* results in the production of two daughter cells that are genetically identical to each other and to the original parent cell. The stages of mitosis are summarized in figure 3.13.

Hypertrophy and Hyperplasia Individuals grow from a fertilized egg into an adult by an increase in cell number accompanied by an increase in cell size. Growth due to increased cell numbers is always due to mitotic cell division, and is called *hyperplasia*. Growth of an organ that is due to increased cell size occurs during the G_1 phase of the cycle and is termed *hypertrophy*. The production of a callous in the hands—where frequent abrasion results in increased skin thickness—is an example of growth by hyperplasia. An increase in skeletal muscle size as a result of exercise, in contrast, is generally believed to be produced by hypertrophy (see chapter 8).

Meiosis A special type of cell division occurs in the gonads (testes and ovaries) of sexually mature males and females. This type of cell division, known as meiosis, is used to produce gametes (sperm and ova). In the process of meiotic division, homologous chromosomes are separated into daughter cells before the duplicate chromatids are separated. The cells produced at the end of this process thus contain only one of each pair of homologous chromosomes. Sperm and ova therefore contain twenty-three, rather than forty-six chromosomes. The details of meiotic division will be discussed, together with other aspects of reproduction, in chapter 21.

Genetic Transcription—RNA Synthesis
When a cell completes mitotic division it again enters the G_1 phase and its chromosomes become extended into the form known as chromatin. This is the "working" form of DNA; the more familiar short, stubby form of chromosomes during cell division provide more efficient "packaging" of DNA, but the genes are inactive until the chromosomes unravel. When the combination of DNA and protein that comprises euchromatin is active, it directs the metabolism of the cell by indirectly regulating the synthesis of enzyme proteins.

One gene codes for one polypeptide chain. Each gene is a stretch of DNA that is several thousand nucleotide pairs long. The DNA in a human cell contains three to four billion base pairs—enough to code for at least three million proteins. Since the average human cell contains "only" 30,000 to 150,000 different types of proteins, it follows that only a fraction of the DNA in each cell is used to code for proteins. The remainder of the DNA may be inactive, may be redundant, or may serve to regulate those regions that do code for proteins.

In order for the genetic code to be translated into the synthesis of specific proteins, the DNA code must first be transcribed into an RNA code. This is accomplished by DNA-directed RNA synthesis, or **genetic transcription.**

In DNA-directed RNA synthesis, the enzyme *RNA polymerase* breaks the weak hydrogen bonds between paired DNA bases. This does not occur throughout the length of DNA, but only in the regions that are to be transcribed (there are base sequences that code for "start" and "stop"). Double-stranded DNA therefore separates in these regions so that the freed bases can pair with complementary RNA nucleotide bases. This pairing of bases, like that which occurs in DNA replication, follows the law of complementary base pairing: guanine bonds with cytosine (and vice versa), and adenine bonds with uracil (because uracil in RNA is equivalent to thymine in DNA—see chapter 2). Unlike DNA replication, however, only *one* of the two freed DNA strands serves as a guide for RNA synthesis.

Figure 3.15 The synthesis of ribosomal RNA on DNA in the nucleolus.

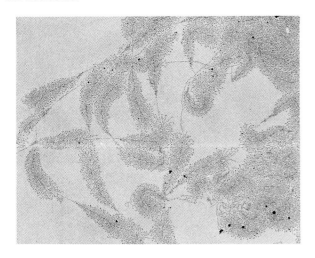

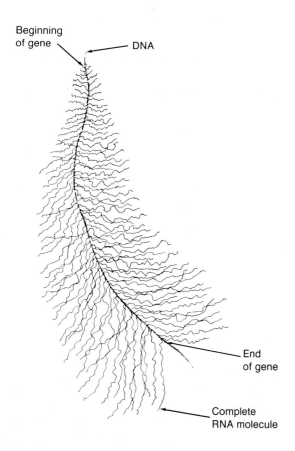

Beginning of gene

DNA

End of gene

Complete RNA molecule

At the end of transcription the single-stranded RNA detaches from DNA. While a gene is active, many RNA molecules can be produced in this way (see figure 3.15). Because of complementary base pairing, each of these RNA molecules is a complementary copy, analogous to a "blueprint," of the base sequence in the DNA that was transcribed.

Figure 3.16 Electron micrograph of polyribosomes. An RNA strand *(arrow)* joins the ribosomes together.

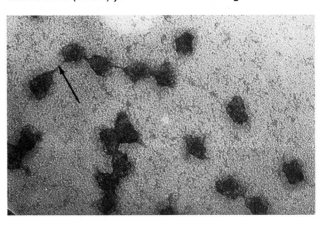

Types of RNA

There are four types of RNA produced within the nucleus by genetic transcription. These types include (1) **precursor messenger RNA (pre-mRNA),** which is altered within the nucleus to form mRNA; (2) **messenger RNA (mRNA),** which contains the code for synthesis of specific proteins; (3) **transfer RNA (tRNA),** which is needed for decoding the genetic message contained in mRNA (this mechanism will be described in the next section); and (4) **ribosomal RNA (rRNA),** which forms part of the structure of ribosomes (described in the next section). The genes that code for rRNA synthesis are located in the part of the nucleus called the nucleolus (as previously described). The genes that code for pre-mRNA and tRNA synthesis are located elsewhere in the nucleus.

In bacteria, where the molecular biology of the gene is best understood, a gene that codes for one type of protein produces an mRNA molecule that begins to direct protein synthesis as soon as it is transcribed. This is *not true* for the cells of higher organisms, including humans. In higher cells a pre-mRNA is produced that must be modified within the nucleus before it can enter the cytoplasm as mRNA and direct protein synthesis.

Precursor mRNA is much bigger than the mRNA that it forms. This large size of pre-mRNA is, surprisingly, not due to excess bases at the ends that must be trimmed; the excess bases are *within* the pre-mRNA. The genetic code for a particular protein, in other words, is split up by stretches of base pairs that do not contribute to the code. As a result, pre-mRNA must be cut and spliced to make mRNA. After cutting and splicing, the mRNA is further modified by the addition of characteristic bases at its ends: about two hundred adenine-containing nucleotides are added to one end; and a modified guanine-containing nucleotide is added to the other end of mRNA. These *post-transcriptional modifications* produce the form of mRNA that enters the cytoplasm.

Figure 3.17 The genetic code is first transcribed into base triplets (codons) in mRNA and then translated into a specific sequence of amino acids in a protein.

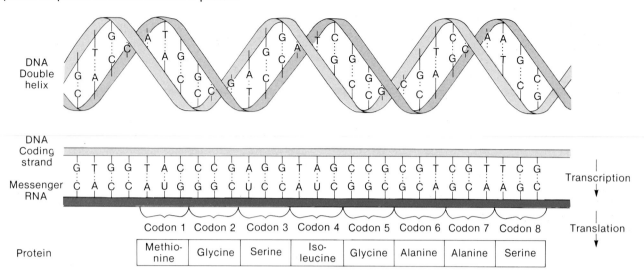

1. Using stick figures and two colors, draw a simplified picture showing the semiconservative replication of DNA. Describe this process in words.
2. Rearrange this random sequence in proper order: G_2, S, G_1, and M. Briefly describe the meaning of these symbols. Using brackets, show which of these is included within interphase.
3. List the phases of mitosis and briefly describe the events that occur in each phase.
4. Using stick figures and two colors, draw a simplified picture of genetic transcription (RNA synthesis). Describe this process in words.

Protein Synthesis

When mRNA enters the cytoplasm it attaches to **ribosomes,** which are seen in the electron microscope as numerous small particles. Ribosomes are composed of rRNA and protein, and each ribosome has two subunits, which are unequal in size. The mRNA passes through a number of ribosomes to form a "string-of-pearls" structure called a *polyribosome* (or *polysome* for short). Association of mRNA with ribosomes is needed for **genetic translation**—the production of specific proteins according to the code of mRNA nucleotide bases.

Table 3.1 Selected DNA base triplets and mRNA codons.

DNA Triplet	RNA Codon	Amino Acid
TAC	AUG	"Start"
ATC	UAG	"Stop"
AAA	UUU	Phenylalanine
AGA	UCU	Serine
ACA	UGU	Cysteine
GGG	CCC	Proline
GAA	CUU	Leucine
GCT	CGA	Arginine
TTT	AAA	Lysine
TGC	ACG	Tyrosine
CCC	GGG	Glycine
CTC	GAG	Aspartic acid

Each mRNA molecule contains several hundred or more nucleotides that are ordered into a specific base sequence, as determined by complementary base pairing with DNA during genetic transcription. Every three bases (base triplet) in mRNA is a code word—or **codon**—for a specific amino acid (see table 3.1). As mRNA moves through the ribosome, the sequence of codons is translated into a sequence of specific amino acids within a growing polypeptide chain.

Figure 3.18   The structure of transfer RNA (tRNA).

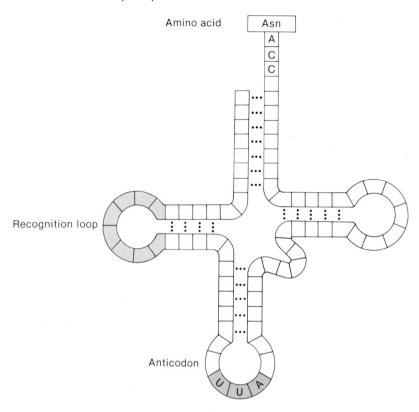

Transfer RNA   Translation of the codon "language" into the polypeptide "language" is accomplished by tRNA and particular enzymes. Each tRNA molecule, like molecules of mRNA and rRNA, is single stranded, and is produced in the nucleus by specific regions of DNA. Although tRNA is single stranded, it bends on itself to form three loop regions (see figure 3.18). One of these loops contains the **anticodon**—three nucleotide bases that are complementary to a specific codon in mRNA. Since there are about twenty different types of amino acids, there must be at least twenty different mRNA codons and twenty different types of tRNA (which differ in their anticodon loop).

Enzymes in the cell cytoplasm called *aminoacyl-tRNA synthetases* join specific amino acids to the ends of tRNA, so that a tRNA with a given anticodon is always bonded to one specific amino acid (see figure 3.18). These enzymes, it has been learned, don't actually "read" the anticodon. Rather, each tRNA with a different anticodon has a different "recognition loop" that is detected by these enzymes.

Formation of a Polypeptide   As mRNA moves through the ribosomes, complementary base pairing occurs between the codons of mRNA and the anticodons of tRNA that are joined to amino acids. By trial and error, the correct tRNA bonds to two adjacent codons and the amino acids join together by a peptide bond. The first amino acid detaches from its tRNA so that a dipeptide is linked by the second amino acid to the second tRNA. When a third tRNA bonds to the third codon, amino-acid number three forms a peptide bond with amino-acid number two, which then detaches from its tRNA. A tripeptide is then attached by the third amino acid to the third tRNA. The polypeptide chain thus grows as new amino acids are added to its growing tip (see figure 3.19).

Figure 3.19 Translation of messenger RNA (mRNA). As the anticodon of each new aminoacyl-tRNA bonds to a codon of the mRNA, new amino acids are joined to the growing tip of the polypeptide chain.

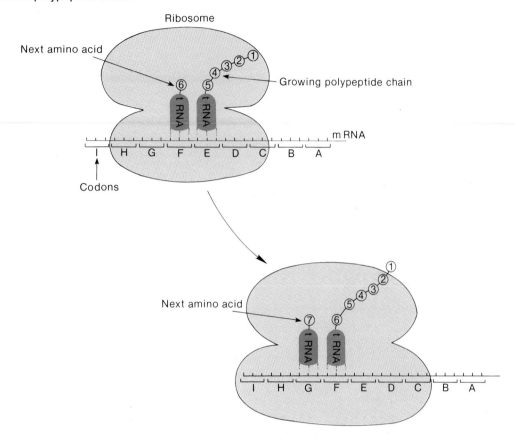

As the polypeptide chain grows in length, interactions between its amino acids cause the chain to twist into a helix (secondary structure) and to fold and bend on itself (tertiary structure). At the end of this process, the new protein detaches from the tRNA as the last amino acid is added. Many proteins are further modified after they are formed; these modifications occur in organelles called the endoplasmic reticulum and Golgi apparatus, which are described in the next section.

1. List the three types of RNA molecules and describe the role of each in protein synthesis.
2. Explain the meaning of the terms "genetic transcription" and "genetic translation." What serves as the "interpreter" in each of these processes?
3. Using table 3.1, indicate the amino acids coded by the *DNA* base sequence: AAATACCTCGCT

Endoplasmic Reticulum and Golgi Apparatus

Most cells contain a system of membranes known as the endoplasmic reticulum. There are two types: (1) a *smooth endoplasmic reticulum* and (2) *a rough*, or *granular, endoplasmic reticulum*. The smooth endoplasmic reticulum is used for a variety of purposes in different cells; it serves as a site for enzyme reactions in steroid hormone production, for example, and a site for storage of Ca^{++} in skeletal muscle cells. In cells that are active secretors of proteins (such as exocrine and endocrine gland cells), the endoplasmic reticulum is studded with ribosomes, producing a rough or granular appearance.

Proteins that are to be used within the cell are produced in polyribosomes that are free in the cytoplasm. If the protein is a secretory product of the cell, however, it is made by mRNA-ribosome complexes located in the rough endoplasmic reticulum. The membranes of this system enclose fluid-filled spaces (cisternae), which the newly formed protein may enter.

Figure 3.20 The rough endoplasmic reticulum. The small dark particles located on the membranes are ribosomes.

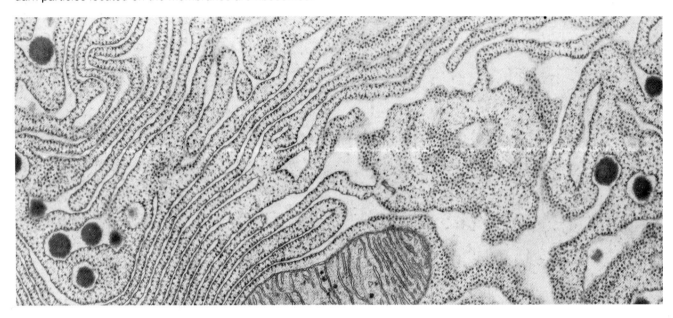

Figure 3.21 Protein that is destined for secretion begins with a "leader sequence" that enables it to be inserted into the endoplasmic reticulum. Once it has been inserted, the leader sequence is removed and carbohydrate is added to the protein.

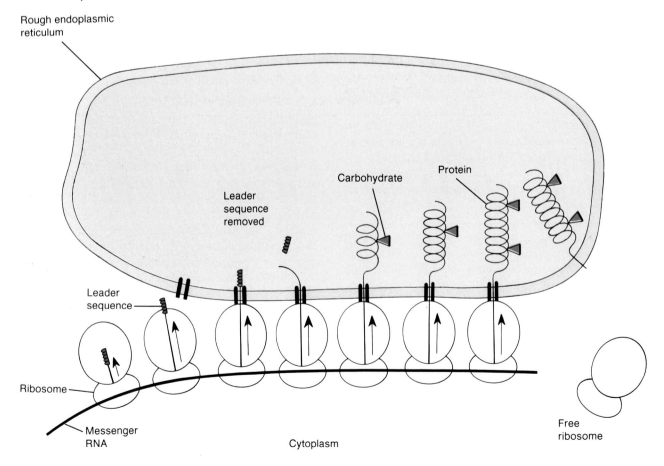

Figure 3.22 Electron micrograph of a Golgi apparatus. Notice the formation of vesicles at the ends of some of the flattened sacs.

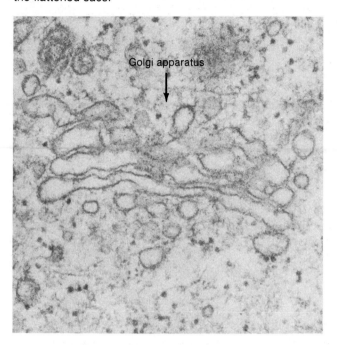

Golgi apparatus

When protein that is destined for secretion is produced, the first thirty or so amino acids are primarily hydrophobic. This *leader sequence* is attracted to the lipid component of the membranes of the endoplasmic reticulum. As the polypeptide chain elongates, it is "injected" into the cisterna (space) within the endoplasmic reticulum. The leader sequence is, in a sense, an "address" that directs secretory proteins into the endoplasmic reticulum. Once the protein is in the cisterna, the leader sequence is removed so the protein cannot reenter the cytoplasm.

The processing of the hormone insulin can serve as an example of the post-translational changes that occur within the endoplasmic reticulum. Insulin enters the cisterna as a single polypeptide composed of 109 amino acids. The first twenty-three amino acids serve as a leader sequence and are quickly removed. The remaining chain folds within the cisterna so that the first and last amino acids in the polypeptide are brought close together. The central region is then enzymatically removed, producing two chains—one that is twenty-one amino acids long, the other that is thirty amino acids long—which are subsequently joined together by disulfide bonds. This is the form of insulin that is normally secreted from the cell.

Golgi Apparatus

Secretory proteins do not remain trapped within the endoplasmic reticulum, but are instead transported to another organelle within the cell—the *Golgi apparatus*. This organelle serves two interrelated functions: (1) proteins are further modified within the Golgi apparatus (by addition of carbohydrates, for example) and (2) different types of proteins are separated according to their function and destination—proteins that are destined for secretion out of the cell are separated from those that will be incorporated into the plasma membrane; these, in turn, are separated from those that will be introduced into organelles called *lysosomes* (discussed in a later section); and (3) proteins are "packaged" within membrane-enclosed organelles that "bud" off the Golgi apparatus.

The Golgi apparatus consists of several flattened sacs that enclose spaces (cisternae) that communicate through hollow, membranous tubules. These cisternae, in turn, communicate with those of the endoplasmic reticulum. This latter communication may be due to a direct connection between the two organelles, or may occur by means of membranous vesicles that pass from the endoplasmic reticulum to the Golgi apparatus.

Proteins that are going to be secreted from the cell (as plasma proteins, digestive enzymes in the gastrointestinal tract, or protein hormones in the plasma) are combined with carbohydrates within the Golgi apparatus. Secretory proteins, therefore, are almost always *glycoproteins*. The carbohydrate portion of secreted proteins appears to provide protection against phagocytosis; those proteins that have been experimentally removed of their carbohydrates are more rapidly degraded into inactive products by the liver and other organs.

The Golgi apparatus sorts proteins in some unknown way. Some are placed in organelles called *lysosomes;* others are placed within vesicles that ultimately fuse with the plasma membrane. These vesicles may simply be used to add new membrane material to replace that which was lost in endocytosis, or may contain secretory products that are released from the cell in exocytosis. In either event, new membranes are formed as the sacs of the Golgi apparatus form buds that contain the secretory material.

The Golgi apparatus and the rough endoplasmic reticulum contain enzymes that modify the structure of proteins. These changes, combined with those that occur during genetic translation and transcription, combine to provide numerous possible sites for the regulation of genetic expression. A summary of the processes involved in genetic expression, and possible site for its regulation, is provided in table 3.2.

Table 3.2 Stages in genetic expression and possible sites for its regulation.

Stage of Gene Expression	Possible Sites for Regulation
RNA synthesis	Association of histone and non-histone proteins with the chromatin
Nuclear processing of RNA	Cutting and splicing of nuclear RNA
	Capping of messenger RNA with 7-methyl guanosine, and addition of about two hundred adenylate nucleotides
Translation of mRNA	Availability of specific tRNA molecules
Post-translational changes in protein structure	Cleavage of parent protein into biologically active fragments
	Chemical modification of amino acids
	Association of polypeptide chains together to form quaternary structure
	"Addressing" of protein for secretion by leader sequence and insertion into rough endoplasmic reticulum
	Addition of carbohydrate to secreted protein

Lysosomes and Mitochondria

After a phagocytic cell has engulfed the proteins, polysaccharides and lipids present in a particle of "food," these molecules are still kept in a compartment that is separated from the cytoplasm. This is because, after phagocytosis is complete, the food particle is surrounded by a piece of the cell membrane, which encloses it in a "food vacuole." These large molecules must first be digested into their smaller subunits (monosaccharides, amino acids, and so on) before they can cross the vacuole membrane and enter the cytoplasm.

The digestive enzymes of a cell are isolated from the cytoplasm and concentrated within membrane-bound organelles called **lysosomes.** A *primary lysosome* may fuse with a food vacuole to form a *secondary lysosome*. After the products of digestion have been absorbed across the membrane of the secondary lysosome, the organelle that contains the remaining undigested wastes is known as a *residual body*. Residual bodies may eliminate their wastes by exocytosis, or they may accumulate within the cell as the cell ages.

Figure 3.23 Electron micrograph showing primary lysosomes *(Lys₁)* and secondary lysosomes *(Lys₂)*. Mitochondria *(Mi)*, Golgi apparatus *(GA)*, and the nuclear envelope *(NE)* are also seen.

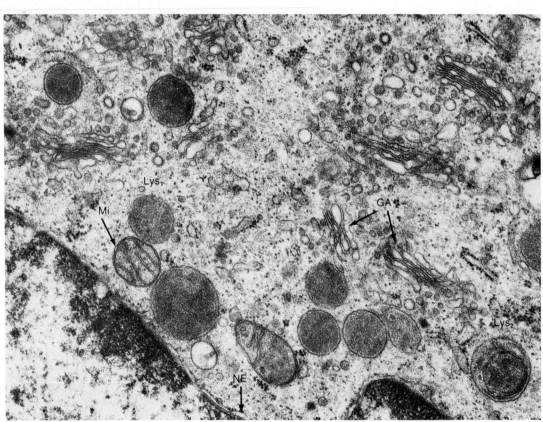

Figure 3.24 A mitochondrion. The outer membrane *(OM)* and the infoldings of the inner membrane—the cristae *(Cr)*—are clearly seen. The fluid in the center is the matrix *(Ma)*.

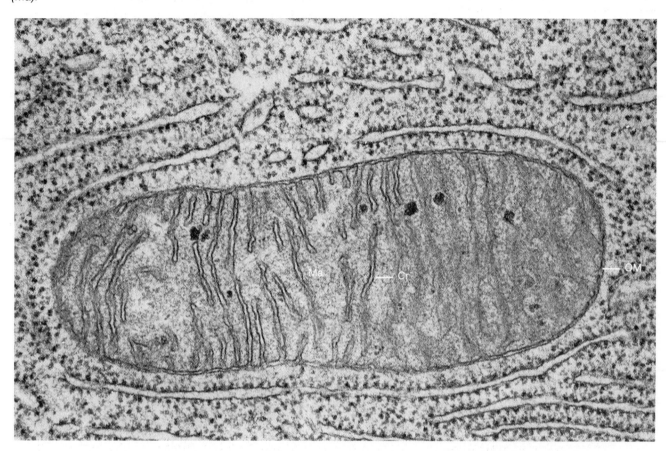

Autophagy Partly digested membranes of various or- ganelles, and other cellular debris, are often observed within secondary lysosomes. This is a result of *autophagy* (auto = self; phag = eat), a process that destroys worn- out organelles so that they can be continuously replaced by the cell.

Lysosomes are thus aptly referred to as the "digestive system" of the cell. They have also been called *suicide bags,* because if their membranes should break, the diges- tive enzymes would be released, destroying the cell. This happens normally as part of *programmed cell death* in which destruction of tissues is part of embryological de- velopment. This also occurs in some pathological (dis- ease) states.

Most—if not all—molecules in the cell have a limited life span. They are continuously destroyed and must be continuously replaced. Glycogen and some complex lipids in the brain, for example, are digested normally at a par- ticular rate by lysosomes. If a person, because of a genetic defect, does not have the proper amount of these lysoso- mal enzymes, the resulting abnormal accumulation of glycogen and lipids could destroy the tissues.

Mitochondria All cells in the body, with the exception of mature red blood cells, have a hundred to a few thou- sand organelles called *mitochondria*. Mitochondria serve as the sites for production of most of the cellular energy, as will be described in detail in chapter 5.

Mitochondria vary in appearance in different tissues, but all have the same basic structure. Each is surrounded by an *outer membrane* that is separated by a narrow space from an *inner membrane*. The inner membrane has many folds, or *cristae,* that extend into the central area (or *ma- trix*) of the mitochondrion. The cristae and the matrix provide different compartments in the mitochondrion and have different roles in the generation of cellular energy.

1. Construct a flow chart (using arrows) to show the events that occur between synthesis of a protein and its secretion from a cell.
2. Describe the post-translational changes in protein structure that occur in the endoplasmic reticulum and Golgi apparatus.
3. Describe the role of lysosomes in phagocytosis and autophagy.

Summary

Cell Membrane and Cytoskeleton

I. The cell membrane is composed primarily of a double layer of phospholipids.
   A. The hydrophobic ends of each layer face the center of the membrane and provide the main barrier to diffusion of water-soluble molecules and ions.
   B. The membrane also contains irregularly arranged proteins.
      1. These proteins are free to move within the phospholipids.
      2. This model of the cell membrane is known as the fluid-mosaic model.
II. In endocytosis, the cell membrane invaginates to form vesicles that carry extracellular material into the cell.
   A. Ingestion of large particles, such as bacteria and degenerating cells, is called phagocytosis ("cell eating").
   B. Exocytosis is the reverse process, in which secretory vesicles formed within the cell fuse with the cell membrane and release their products into the extracellular environment.
III. Protein in the cytoplasm of cells is not in the form of an amorphous gel, but is rather highly organized.
   A. Proteins form microfilaments and microtubules that provide an organized cytoskeleton.
   B. Ribosomes and other organelles are fixed in specific locations by the cytoskeleton.
   C. The cytoskeleton can be rearranged and is capable of contraction.
   D. Cilia, flagella, and the spindle fibers are composed of microtubules arranged in a characteristic way.
      1. They contain a nine-plus-two arrangement of microtubules.
      2. Their movements may be due to the same proteins (actin and myosin) that cause muscle contraction.

The Nucleus and DNA Function

I. DNA, associated with proteins as chromatin, is localized in the nucleus.
   A. Extended euchromatin is active in directing the activities of the cell; clumps of heterochromatin are inactive.
   B. When DNA is in the extended euchromatin form, it directs the synthesis of RNA, which in turn controls protein synthesis in the cell.
      1. RNA can leave the nucleus through nuclear pores.
      2. A region of DNA that codes (through its production of RNA) for the synthesis of a specific type of protein is called a gene.
      3. RNA and protein synthesis occurs during the G_1 phase of the cell cycle; cellular growth can occur during this time.
   C. When a cell is going to divide, it enters the S phase of the cell cycle.
II. During the S phase of the cycle, DNA replicates itself.
   A. In this process, the weak hydrogen bonds between complementary bases in the double helix break.
      1. The freed bases pair with other DNA bases within the nucleus.
      2. Because of complementary base pairing, the new DNA strand contains the same base sequence as the original complementary strand.
      3. Two new double helix DNA molecules are thus formed, each of which contains one old strand and one new strand; this is called semiconservative replication.
   B. After the S phase, a usually short resting phase occurs, known as the G_2.
III. During mitotic cell division (the M phase), the duplicate DNA is distributed to two new daughter cells.
   A. Extended chromatin first condenses to form the short, thick chromosomes seen during cell division.
      1. Each chromosome at this stage contains two duplicate DNA molecules—these are called chromatids.
      2. The chromatids are joined together by a centromere.
      3. There are forty-six chromosomes, representing twenty-two pairs of autosomal chromosomes and one pair of sex chromosomes; each chromosome in a pair is called a homologous chromosome.
   B. The forty-six chromosomes line up in single file along the equator of the cell during metaphase of mitosis.
      1. At this time, centrioles at the poles produce spindle fibers that attach to the centromeres.
      2. Contraction of the spindle fibers during anaphase pulls the chromatids apart and to opposite poles.
      3. Division of the cell therefore produces two daughter cells, with the same complement of chromosomes.
IV. When the cell is not going to divide, DNA directs the synthesis of RNA.
   A. In order to produce RNA, complementary bases within the double helix structure of DNA must break their hydrogen bonds.
      1. This occurs by means of the enzyme RNA polymerase.
      2. Only a limited region of DNA is involved in RNA production.
   B. One of the two strands of DNA is used as a guide in the formation of RNA.
      1. Complementary base pairing of RNA nucleotides with the bases of one of the DNA strands produces an RNA molecule that is a "blueprint" of this region of DNA.
      2. The RNA thus produced detaches from DNA as a single-stranded molecule.
   C. There are four types of RNA produced in the nucleus, using DNA as a guide.
      1. Pre-messenger RNA (pre-mRNA) is cut and spliced to form the smaller mRNA.
      2. Messenger RNA leaves the nucleus and contains the code for production of a specific type of protein.

3. Transfer RNA (tRNA) also enters the cytoplasm where it serves to deliver amino acids to the mRNA.
4. Ribosomal RNA (rRNA) forms part of ribosomes; these organelles associate with mRNA to form the unit that is active in protein synthesis.

Protein Synthesis and Secretion

I. Each triplet of bases in mRNA codes for a specific amino acid; these triplets are called codons.
   A. One loop of tRNA contains three bases that act as anticodons.
      1. There are over twenty different tRNA types, each with a different anticodon.
      2. Specific enzymes recognize each different type of tRNA and join it to a specific amino acid.
   B. Complementary base pairing of anticodons in tRNA with codons in mRNA bring specific amino acids together according to the code specified in mRNA.

1. Peptide bonds between these amino acids join them together to form a specific protein.
2. In this indirect way, therefore, genes code for the synthesis of specific proteins.

II. If a protein is to enter the cytoplasm, it is produced by free polyribosomes; if it is to be secreted, it is produced by ribosomes located on a rough endoplasmic reticulum.
   A. The secretory protein is directed, by a "leader sequence" of amino acids, into the cisternae of the endoplasmic reticulum.
   B. The protein is then transferred to a Golgi apparatus.
      1. Within the Golgi apparatus, the protein is further modified, by addition or deletion of polypeptides and by addition of carbohydrates.
      2. The Golgi apparatus packages secretory proteins within vesicles, which can fuse with the plasma membrane in exocytosis.

Lysosomes and Mitochondria

I. Lysosomes are organelles that contain digestive enzymes.
   A. Digestion of particles contained in food vacuoles occurs by fusion of these vacuoles with lysosomes.
   B. The digestive enzymes in lysosomes are also used to eliminate old organelles within the cell—this is called autophagy.
      1. All structures and molecules in the cell therefore have limited life spans.
      2. Lysosomes are also called "suicide bags" because breakage of the lysosomal membrane results in cellular death.

II. Mitochondria are the organelles used to produce most of the energy within the cell.
   A. Each mitochondrion contains an outer and inner membrane.
   B. The inner mitochondrial membrane is folded into structures called cristae.
   C. The inner fluid of the mitochondrion is called matrix.

Self-Study Quiz

1. According to the fluid-mosaic model of the cell membrane:
   (a) protein and phospholipids form a regular, repeating structure
   (b) the membrane is a rigid structure
   (c) phospholipids form a double layer, with the polar parts facing each other
   (d) proteins are free to move within a double layer of phospholipids

2. After the DNA has replicated itself, the duplicate strands are called:
   (a) homologous chromosomes
   (b) chromatids
   (c) centromeres
   (d) spindle fibers

3. Nerve and skeletal muscle cells in the adult, which do not divide, remain in:
   (a) the G_1 phase
   (b) the S phase
   (c) the G_2 phase
   (d) the M phase

4. The phase of mitosis in which the chromosomes line up at the equator of the cell is called:
   (a) interphase
   (b) prophase
   (c) metaphase
   (d) anaphase
   (e) telophase

5. The phase of mitosis in which the chromatids separate is called:
   (a) interphase
   (b) prophase
   (c) metaphase
   (d) anaphase
   (e) telophase

6. The RNA nucleotide base that pairs with adenine in DNA is:
   (a) thymine
   (b) uracil
   (c) guanine
   (d) cytosine

7. Which of the following statements about RNA is TRUE?
   (a) it is made in the nucleus
   (b) it is double stranded
   (c) it contains the sugar deoxyribose
   (d) it is a complementary copy of the entire DNA molecule

8. Which of the following statements about mRNA is FALSE?
   (a) it is produced as a larger pre-mRNA
   (b) it forms associations with ribosomes
   (c) its base triplets are called anticodons
   (d) it codes for the synthesis of specific proteins

9. The organelle that combines proteins with carbohydrates and packages them within vesicles for secretion is:
   (a) the Golgi apparatus
   (b) the rough endoplasmic reticulum
   (c) the smooth endoplasmic reticulum
   (d) the ribosomes

10. The organelle that contains digestive enzymes is:
   (a) the mitochondria
   (b) the lysosomes
   (c) the endoplasmic reticulum
   (d) the Golgi apparatus

Enzymes and Cellular Energy

Objectives

By studying this chapter, you should be able to:

1. Describe the principles of catalysis and the way enzymes function as catalysts

2. Describe the effects of pH, temperature, substrate concentration, enzyme concentration, and cofactor and coenzyme concentration on the rate of enzyme-catalyzed reactions

3. Describe the characteristics of metabolic pathways

4. Understand and explain allosteric inhibition of enzymes and how this relates to end-product inhibition of metabolic pathways

5. Explain the effects of end-product inhibition on branched metabolic pathways

6. Explain the mechanisms by which inborn errors of metabolism result in accumulation of certain molecules and depletion of others in a cell, and provide clinical examples of the effects of such metabolic disorders

7. Explain how the first and second laws of thermodynamics apply to the flow of energy in cells

8. Define the terms *exergonic reaction* and *endergonic reactions,* and provide examples of each

9. Describe the formation and function of ATP in cells

10. Describe cellular oxidation-reduction reactions and define the terms *oxidizing agent* and *reducing agent*

11. Describe, in a general way, the functions of NAD, FAD, and oxygen within the cell

*T*he ability of yeast cells to make alcohol from glucose (a process called *fermentation*) had been known since antiquity, yet no chemist by the mid-nineteenth century could duplicate the trick in the absence of living yeast. Also, yeast and other living cells could perform a vast array of chemical reactions at body temperature that could not be duplicated in the chemical laboratory without adding a substantial amount of heat energy. These observations led many mid-nineteenth century scientists to believe that chemical reactions in living cells were aided by a "vital force" that operated beyond the laws of the physical world. This "vitalist" concept was squashed along with the yeast cells when a pioneering biochemist, Eduard Buchner, demonstrated that juice obtained from yeast could ferment glucose to alcohol. The yeast juice was not alive—evidently some chemical or group of chemicals in the cells were responsible for fermentation. Buchner didn't know what these chemicals were, so he simply named them *enzymes* (which is German for *in yeast*).

Figure 4.1   Comparison of a non-catalyzed reaction with a catalyzed reaction. Upper figures compare the proportion of reactant molecules that have sufficient activation energy to participate in the reaction (shown as the shaded portion of the curves). This proportion is increased in the enzyme-catalyzed reaction because enzymes lower the activation energy required for the reaction (shown as a barrier on top of an energy "hill" in the bottom figures). Reactants that can overcome this barrier are able to participate in the reaction, as shown by arrows pointing to the bottom of the energy "hill."

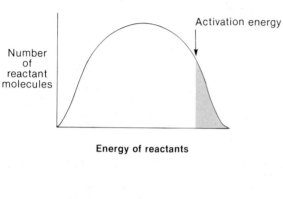

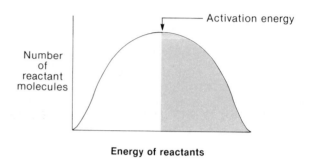

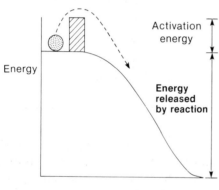

Non-catalyzed reaction

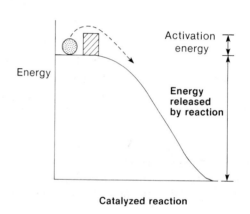

Catalyzed reaction

Enzymes as Catalysts

In more recent times, biochemists have demonstrated that *all enzymes are proteins,* and that enzymes act as *biological catalysts.* A catalyst is a chemical that (1) increases the rate of a reaction; (2) is not itself changed at the end of the reaction; and (3) does not change the nature of the reaction or its final result. The same reaction would have occurred to the same degree in the absence of the catalyst, but at a much slower rate.

In order for a given reaction to occur (for example, the breaking of a larger organic molecule into two smaller molecules), the reactants must have sufficient energy. The amount of energy required for a reaction to proceed is called the **energy of activation.** In a large population of molecules only a small fraction will possess sufficient energy for a reaction. Adding heat to the reaction will raise the energy level of all the reactant molecules, thus increasing the fraction of the population that have the activation energy. Heat would make the reaction go faster, but it would produce undesirable side effects in the cell. Catalysts make the reaction go faster at a lower temperature by *lowering the activation energy* so that a larger fraction of the population of molecules have sufficient energy to participate in the reaction (see figure 4.1).

Since a small fraction of the reactants will have the activation energy required for the reaction even in the absence of a catalyst, the reaction could theoretically occur spontaneously at a slow rate. This rate, however, would

Figure 4.2 Lock-and-key model of enzyme action (see text for description).

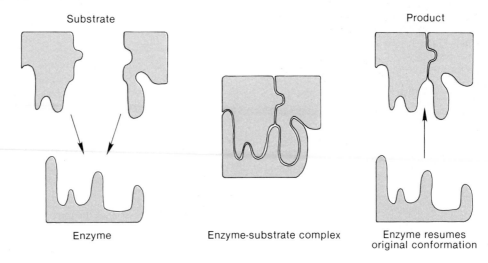

Table 4.1 Some examples of the names of enzymes, the reactions they catalyze, and their turnover numbers (molecules of substrate converted to product per minute under optimum conditions).

Enzyme	Reaction Catalyzed	Turnover Number
Catalase	$H_2O_2 \rightarrow H_2O + O_2$	600×10^6
Carbonic anhydrase	$H_2CO_3 \rightarrow H_2O + CO_2$	3.6×10^6
Amylase	starch + $H_2O \rightarrow$ maltose	1.0×10^5
Lactate dehydrogenase	lactic acid $\rightarrow$ pyruvic acid + H_2	6.0×10^4
Ribonuclease	RNA + $H_2O \rightarrow$ ribonucleotides	600

be much too slow for the needs of a cell. So, from a biological standpoint, the presence or absence of a specific enzyme catalyst acts as a switch—the reaction will occur if the enzyme is present and will not occur if the enzyme is absent.

Mechanism of Enzyme Action

The ability of enzymes to lower the activation energy of a reaction is a result of their structure. Enzymes are proteins—they are thus very large molecules with complex, highly ordered three-dimensional shapes produced by chemical interaction between their amino acids (see chapter 2). Each type of enzyme protein has characteristic ridges, grooves, and pockets that are lined with specific amino acids. The particular pockets that are active in catalyzing a reaction are called the **active sites** of the enzyme.

The reactant molecules—which are the *substrates* of the enzyme—have shapes that allow them to fit into the active sites. The fit may not be perfect at first, but a perfect fit may be induced as the substrate gradually slips into the active site. This induced fit, together with temporary bonds that form between the substrate and the amino acids that line the active sites of the enzyme, weaken existing bonds within the substrate molecules. This allows these bonds to be more easily broken. New bonds are more easily formed as substrates are brought close together in the proper orientation. The *enzyme-substrate complex,* formed temporarily in the course of the reaction, then dissociates to yield *products* and the free unaltered enzyme. This model of how enzymes work is known as the **lock-and-key** model of enzyme activity (see figure 4.2).

By this mechanism, enzymes lower the activation energy and greatly increase the rate of reactions. Hydrogen peroxide (H_2O_2), for example, will slowly decompose by itself if left uncapped for a few days. Addition of catalase, an enzyme found in many tissues such as blood and liver, will make hydrogen peroxide break down into water and oxygen gas a trillion times faster. The ability of enzymes to increase the rate of reactions can be measured by their *turnover number.* This is the number of substrate molecules that the enzyme can convert into products per minute (see table 4.1 for examples).

Table 4.2 Examples of the diagnostic value of some enzymes found in plasma.

Enzyme	Diseases Associated with Abnormal Plasma Enzyme Concentrations
Alkaline phosphatase	Obstructive jaundice, Paget's disease (osteitis deformans), carcinoma of bone
Acid phosphatase	Benign hypertrophy of prostate, cancer of prostate
Amylase	Pancreatitus, perforated peptic ulcer
Aldolase	Muscular dystrophy
Creatine kinase (or creatine phosphokinase-CPK)	Muscular dystrophy, myocardial infarction
Lactate dehydrogenase (LDH)	Myocardial infarction, liver disease, renal disease, pernicious anemia
Transaminases (GOT and GPT)	Myocardial infarction, hepatitis, muscular dystrophy

Identification of Enzymes

Although an international committee has established a uniform naming system for enzymes, the names that are in common use do not follow a completely consistent pattern. With the exception of the digestive enzymes that were discovered first (trypsin and pepsin), all enzyme names end with the suffix -*ase.* Classes of enzymes are named according to their job category. *Hydrolases,* for example, promote hydrolysis reactions. Some other enzyme categories include *phosphatases,* which catalyze the removal of phosphate groups, *synthetases,* which catalyze dehydration synthesis reactions, and *dehydrogenases,* which remove hydrogen atoms from their substrates. Atoms are rearranged within the substrate molecule to form different structural isomers (such as isomerization of glucose to fructose) by enzymes called *isomerases.*

The names of many enzymes specify both the substrate of the enzyme and the activity ("job category") of the enzyme—lactic acid dehydrogenase, for example, removes hydrogen from lactic acid and by this means converts it to pyruvic acid. Since enzymes are very specific as to their substrates and activity, the concentration of a specific enzyme in a sample (such as plasma) can be measured relatively easily. This is usually done by measuring the rate of conversion of the enzyme's specific substrate into product under specified conditions (described in the next section).

Table 4.3 Some clinical uses of isoenzyme measurements.

Enzyme	Isoenzyme Form Number	Disease Associated with Abnormal Elevation of Isoenzyme
Creatine phosphokinase (CPK)	1	Muscular dystrophy
	2	Myocardial infarction
Lactic acid dehydrogenase (LDH)	1	Myocardial infarction
	5	Liver disease (such as hepatitis)

When tissues become damaged due to disease, some of the dead cells disintegrate and release their enzymes into the blood. These enzymes are not normally active in the blood, but their presence is useful clinically in identifying certain diseases. Myocardial infarction (irreversible damage to a part of the heart), for example, results in characteristic increases in blood levels of the enzymes *creatine kinase* and *lactate dehydrogenase* (see table 4.2).

Isoenzymes   Enzymes that do exactly the same job (catalyze the same reaction) in different organs have the same name, since the name describes the activity of the enzyme. Different organs, however, may make slightly different "models" of the enzyme that differ in one or a few amino acids. These different models of the same enzyme are called *isoenzymes.* The differences in structure do not affect the active sites (which are the same because the isoenzymes catalyze the same reaction), but they change the chemistry sufficiently so that isoenzymes can be separated from each other and identified by standard biochemical tests. Abnormal elevations in specific isoenzyme forms (identified by numbers—see table 4.3) occur in certain disease states.

1. Draw a labeled picture of the lock-and-key model of enzyme action. Write a formula for this reaction.
2. Explain how catalysts in general (enzymes or inorganic catalysts) increase the rate of chemical reactions.

Figure 4.3 The effect of temperature on enzyme activity, as measured by the rate of the enzyme-catalyzed reaction under standardized conditions.

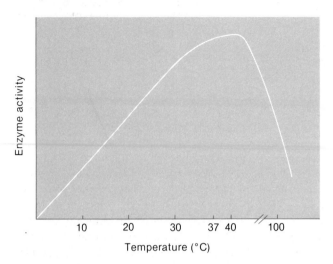

Figure 4.4 The effect of pH on activity of three digestive enzymes.

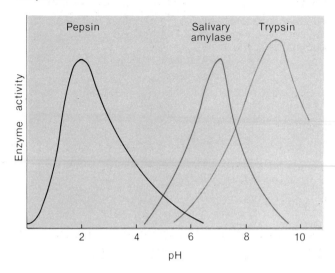

Control of Enzyme Activity

The activity of an enzyme, as measured by the rate at which its substrates are converted to products, is influenced by a variety of factors. These include (1) the temperature and pH of the solution; (2) the concentration of cofactors and coenzyme molecules, which are needed by many enzymes as "helpers" for catalytic activity; (3) the concentration of enzyme and substrate molecules in the solution; and (4) the stimulatory and inhibitory effects of some products of enzyme action on the activity of the enzymes that helped form these molecules.

Effects of Temperature and pH

An increase in temperature, as previously described, will increase the rate of non-enzyme-catalyzed reactions because a higher proportion of reactant molecules will have the activation energy required. A similar relationship between temperature and reaction rate occurs in enzyme-catalyzed reactions. At a temperature of 0°C the reaction rate is unmeasurably slow. As the temperature is raised above 0°C the reaction rate increases, but only up to a point. At a few degrees above body temperature (which is 37°C) the reaction rate reaches a plateau; further increases in temperature actually *decrease* the rate of the reaction. This is due to the fact that enzymes denature (see chapter 2) at high temperature and thus lose their catalytic ability.

A similar relationship occurs when the rate of an enzymatic reaction is measured at different pH values. Each enzyme characteristically has its peak activity in a very narrow pH range. This is the **pH optimum** for the enzyme. If the pH is changed from this optimum, the reaction rate decreases (see figure 4.4). This decreased enzyme activity is due to changes in the conformation of the enzyme and in the charges of the R groups of the amino acids that line the active sites.

The pH optimum of an enzyme usually reflects the pH of the body fluid in which the enzyme is found. The acidic pH optimum of the protein-digesting enzyme *pepsin,* for example, allows it to be active in the strong hydrochloric acid of gastric juice. Similarly, the neutral pH optimum of salivary amylase and the alkaline pH optimum of trypsin in pancreatic juice allow these enzymes to digest starch and protein, respectively, in other parts of the digestive tract.

While the pH of other body fluids shows less variation than the fluids of the digestive tract, significant differences exist between the pH optima of different enzymes found throughout the body (see table 4.4). Some of these

Figure 4.5 Roles of cofactors in enzyme function. In *(a)* the cofactor changes the conformation of the active site, allowing a better fit between the enzyme and its substrates. In *(b)* the cofactor participates in the temporary bonding between the active site and the substrates.

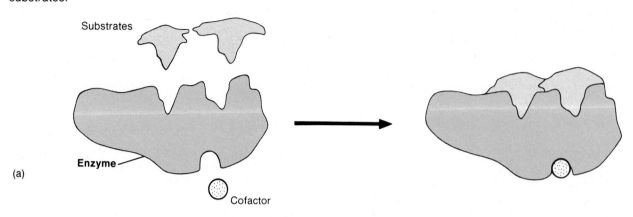

(a)

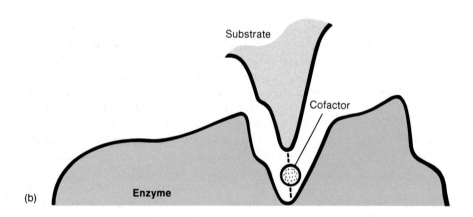

(b)

Table 4.4 The pH optima of selected enzymes.

Enzyme	Reaction Catalyzed	pH Optimum
Pepsin (stomach)	Digestion of protein	2.0
Acid phosphatase (prostate)	Removal of phosphate group	5.5
Salivary amylase (saliva)	Digestion of starch	6.8
Lipase (pancreatic juice)	Digestion of fat	7.0
Alkaline phosphatase (bone)	Removal of phosphate group	9.0
Trypsin (pancreatic juice)	Digestion of protein	9.5
Monoamine oxidase (nerve endings)	Removal of amine group from norepinephrine	9.8

differences can be exploited for diagnostic purposes. Disease of the prostate, for example, may be associated with elevated blood levels of a prostatic phosphatase with an acidic pH optimum (descriptively called acid phosphatase). Bone disease, on the other hand, may be associated with elevated blood levels of a phosphatase with a higher pH optimum—alkaline phosphatase.

Cofactors and Coenzymes
Many enzymes are completely inactive when they are isolated in a pure state. Evidently some of the ions and smaller organic molecules that were removed in the purification procedure are needed for enzyme activity. These are called *cofactors* and *coenzymes*.

Cofactors are metal ions such as Ca^{++}, Mg^{++}, Mn^{++}, Cu^{++}, and Zn^{++}. Some enzymes that have a cofactor requirement do not have a properly shaped active site in the absence of their cofactor. In these cases, the attachment of cofactors causes a conformational change in the protein that allows it to combine with its substrate. The cofactors of other enzymes participate in the temporary bond between the enzyme and its substrate when the enzyme-substrate complex is formed (see figure 4.5).

Coenzymes are organic molecules that are derived from water-soluble vitamins, such as niacin and riboflavin. The coenzymes participate in enzyme-catalyzed reactions by transporting hydrogen atoms and small molecules from one enzyme to another. Examples of cofactors and coenzymes will be given in the context of their roles in cellular metabolism in chapter 5.

Substrate Concentration and Reversible Reactions

If the pH is optimal, and if the proper cofactors and coenzymes are present, the rate of an enzymatic reaction will depend on the enzyme concentration and the amount of substrate available to be converted into product. When the enzyme concentration is constant, the rate of product formation will increase as the substrate concentration increases. Eventually, however, a point will be reached where additional increases in substrate concentration do not result in comparable increases in reaction rate. When the relationship between substrate concentration and reaction rate reaches a plateau, the enzyme is said to be *saturated.* If one thinks of enzymes as workers and substrates as jobs, there is 100 percent employment when the enzyme is saturated—further availability of jobs (substrate) cannot further increase employment (conversion of substrate to product). This is illustrated in figure 4.6.

Some enzymatic reactions within a cell are reversible, with both the forward and backward reactions catalyzed by the same enzyme. The enzyme *carbonic anhydrase,* for example, is named because it can catalyze the following reaction:

$$H_2CO_3 \longrightarrow H_2O + CO_2$$

The same enzyme, however, can also catalyze the reverse reaction:

$$H_2O + CO_2 \longrightarrow H_2CO_3$$

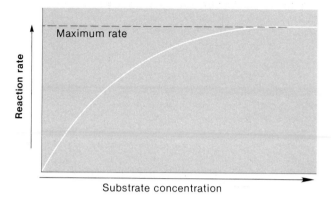

Figure 4.6 The effect of substrate concentration on the reaction rate of an enzyme-catalyzed reaction. When the reaction rate is maximum the enzyme is said to be "saturated."

The two reactions can be more conveniently illustrated by a single equation:

$$H_2O + CO_2 \rightleftharpoons H_2CO_3$$

The direction of the reversible reaction depends, in part, on the relative concentrations of the molecules to the left and right of the arrows. If the concentration of CO_2 is very high (as it is in the tissues), the reaction will be driven to the right. If the concentration of CO_2 is low and that of H_2CO_3 is high (as it is in the lungs), the reaction will be driven to the left. The ability of reversible reactions to be driven from the side of the equation where the concentration is higher to the side where the concentration is lower is known as the **law of mass action.**

The ability of reversible reactions to be driven in one direction or another by mass action also depends on the relative affinities (bonding strength) of the enzyme for the molecules on either side of the equation. Lactic acid dehydrogenase (LDH), for example, can catalyze the following reversible reaction:

$$H_2 + \text{Pyruvic acid} \rightleftharpoons \text{Lactic acid}$$

The LDH enzyme in liver has high affinities for both pyruvic acid and lactic acid, so the particular direction of the reaction depends on the effects of mass action. The LDH isoenzyme in some skeletal muscles, however, has a relatively low affinity for lactic acid as substrate, so the

Figure 4.7   A metabolic pathway, where the product of one enzyme becomes the substrate of the next in a multi-enzyme system.

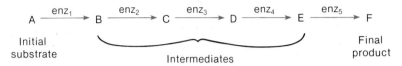

Figure 4.8   A branched metabolic pathway.

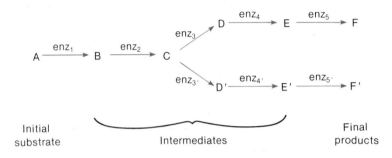

"back-reaction" occurs to only a very slight degree despite high levels of lactic acid. In order for lactic acid that is produced by these skeletal muscles to be converted into pyruvic acid, therefore, the lactic acid must leave the muscles and travel through the blood to the liver. The significance of this will be discussed in chapter 5.

While some enzymatic reactions are not directly reversible, the net effects of the reactions can be reversed by the action of different enzymes. The enzymes that convert glucose to pyruvic acid, for example, are different from those that reverse the pathway and produce glucose from pyruvic acid. Likewise, the formation and breakdown of glycogen (a polymer of glucose—see chapter 2) are catalyzed by different enzymes.

Metabolic Pathways

The many thousands of different types of enzymatic reactions within a cell do not occur independently of each other. They are, rather, all linked together by intricate webs of interrelationships, the total pattern of which constitutes cellular metabolism. A part of this web that begins with an *initial substrate,* progresses through a number of *intermediates,* and ends with a final product is known as a **metabolic pathway.**

The enzymes in a metabolic pathway cooperate in a manner analogous to workers on an assembly line where each contributes a small part to the final product. In this process, the product of one enzyme in the line becomes the substrate of the next enzyme, whose product in turn becomes the substrate of the next, and so on (see figure 4.7).

Few metabolic pathways are completely linear. Most are branched so that one intermediate at the branch point can serve as a substrate for two different enzymes. Two different products can thus be formed that serve as intermediates of two divergent pathways (see figure 4.8).

The rate at which an initial substrate is converted into final products by a metabolic pathway is subject to regulation. Some metabolic pathways are switched on or off by hormones, which affect the synthesis of a particular enzyme, or which activate previously synthesized but inactive enzymes (described in chapter 19). The rate of most enzymatic reactions, however, is limited by the amount of substrate available, rather than by the amount of enzyme. Most enzymes, in other words, are not normally saturated with substrate, so that they can convert all of the substrate presented to them quickly into products.

Figure 4.9 End-product inhibition in a branched metabolic pathway. Inhibition is shown by dotted arrow and negative sign.

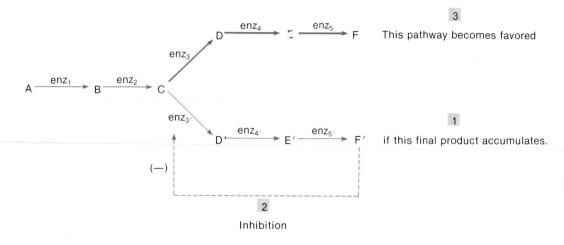

3 This pathway becomes favored

1 if this final product accumulates.

2 Inhibition

Figure 4.10 The effects of an inborn error of metabolism on a branched metabolic pathway.

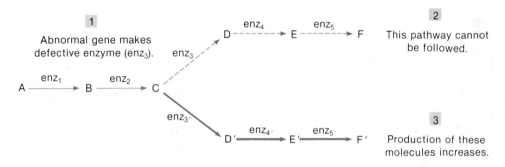

1 Abnormal gene makes defective enzyme (enz_3).

2 This pathway cannot be followed.

3 Production of these molecules increases.

End-Product Inhibition The activities of enzymes at the branch points of metabolic pathways are often regulated by a process called *end-product inhibition*. In this process, one of the final products of a divergent pathway inhibits the branch point enzyme that began the path towards production of the inhibitor. This prevents that final product from accumulating excessively, and results in a shift towards the final product of the alternate divergent pathway.

The mechanism by which a final product inhibits an earlier enzymatic step in the pathway is known as **allosteric inhibition.** The allosteric inhibitor combines with a part of the enzyme located away from the active site (allosteric means "other site"), causing the active site to change shape so that it can no longer combine properly with its substrate.

Inborn Errors of Metabolism Each enzyme in a metabolic pathway is coded by a different gene. An inherited defect in one of these genes may result in a disease known as an "inborn error of metabolism." In these diseases there is an *increased* amount of intermediates formed *prior* to the defective step and a *decrease* in the intermediates and final products formed *after* the defective enzymatic step. If the defective enzyme is active at a step after a branch point in a pathway, the intermediates and final products of the divergent pathway will increase as a result of the block in the alternate pathway. (see figure 4.10).

Table 4.5   Selected examples of inborn errors in the metabolism of amino acids, carbohydrates, and lipids.

Metabolic Defect	Disease	Abnormality	Clinical Result
Amino acid metabolism	Phenylketonuria (PKU)	Increase in phenylalanine	Mental retardation, epilepsy
	Albinism	Lack of melanin	Carcinoma of skin
	Maple-syrup disease	Increase in leucine, isoleucine, and valine	Degeneration of brain, early death
	Homocystinuria	Accumulation of homocystine	Mental retardation, eye problems
Carbohydrate metabolism	Lactose intolerance	Lactose not utilized	Diarrhea
	Glucose-6-phosphatase deficiency (Gierke's disease)	Accumulation of glycogen in liver	Liver enlargement, hypoglycemia
	Glycogen phosphorylase deficiency (McArdle's syndrome)	Accumulation of glycogen in muscle	Muscle fatigue and pain
Lipid metabolism	Gaucher's disease	Lipid accumulation (glucocerebroside)	Liver and spleen enlargement, brain degeneration
	Tay-Sach's disease	Lipid accumulation (ganglioside G_{M_2})	Brain degeneration, death by age 5
	Hypercholestremia	High blood cholesterol	Atherosclerosis of coronary and large arteries

Figure 4.11   Metabolic pathways for the degradation of the amino acid phenylalanine. Defective $enzyme_1$ produces phenylketonuria (PKU), defective $enzyme_2$ produces alcaptonuria, and defective $enzyme_3$ produces albinism (see text).

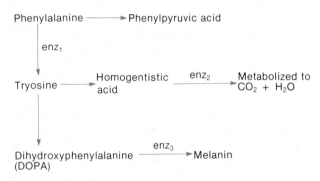

The branched metabolic pathway that begins with phenylalanine as the initial substrate is subject to a number of inborn errors of metabolism (see figure 4.11). When the enzyme that converts this amino acid to the amino acid *tyrosine* is defective, the final products of a divergent pathway accumulate and can be detected in the blood and urine. This disease—*phenylketonuria (PKU)*—can result in severe mental retardation and a shortened life span. Although no inborn error of metabolism is common, PKU occurs frequently enough, and is easy enough to detect, that all newborn babies are tested for this defect. If this

disease is detected early, brain damage can be prevented by placing the child on an artificial diet low in the amino acid phenylalanine.

One of the conversion products of phenylalanine is a molecule known as *DOPA,* which is an acronym for dihydroxyphenylalanine. DOPA is a precursor of the pigment molecule *melanin.* An inherited defect in the enzyme that catalyzes the formation of melanin from DOPA results in an albino. Besides PKU and albinism, there are a large number of other inborn errors of amino acid metabolism, as well as errors in carbohydrate and lipid metabolism (see table 4.5).

1. Draw graphs to represent the effects of pH, temperature, and substrate concentration on the rates of enzyme-catalyzed reactions. Describe these effects in words.
2. Draw a flow chart of a metabolic pathway (using arrows and letters such as *A, B, C,* etc.) with one branch point. Define end-product inhibition and illustrate its effects, using your diagram of a branched metabolic pathway.
3. Suppose, due to an inborn error of metabolism, that the enzyme that catalyzes the third reaction in your pathway (question no. 2) is defective. Describe the effects this would have on the concentrations of the intermediates in your pathway. Give two "real-life" examples of inborn errors of metabolism.

Figure 4.12 Simplified diagram of photosynthesis. Some of the sun's radiant energy is captured by plants and used to produce glucose from carbon dioxide and water. As the product of this endergonic reaction, glucose has a higher free energy content than the initial reactants.

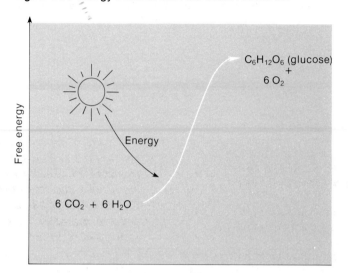

Figure 4.13 Since glucose contains more free energy than carbon dioxide and water, combustion of glucose is an exergonic reaction. The same amount of energy is released if the glucose is broken down stepwise within the cell.

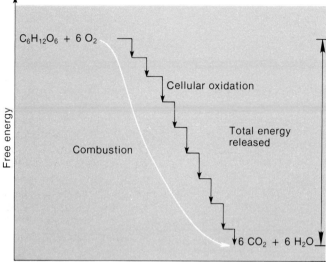

Bioenergetics

Bioenergetics refers to the flow of energy in living systems. Organisms maintain their highly ordered structure and life-sustaining activities through the constant expenditure of energy obtained ultimately from their environment. The energy flow in living systems, like that of nonliving systems, obeys the first and second laws of a branch of physics known as thermodynamics.

Energy, according to the **first law of thermodynamics,** can be transformed but it cannot be created nor destroyed. This is sometimes called the *law of conservation of energy*. As a result of energy transformations, according to the **second law of thermodynamics,** the universe and its parts (such as living systems) have increased amounts of *entropy* (or disorder). Only energy that is in an organized state—called *free energy*—can be used to do work. Thus, since entropy increases in every energy transformation, the amount of available or free energy decreases. As a result of the second law, systems tend to go from a state of higher free energy to a state of lower free energy.

Matter is a form of energy (from Einstein's $E = mc^2$). Atoms that are organized into complex organic molecules such as glucose thus have more free energy (less entropy) than six separate molecules each of carbon dioxide and water. Therefore, in order to convert carbon dioxide and water to glucose, energy must be added. Plants perform this feat using energy from the sun in *photosynthesis* (see figure 4.12).

Endergonic and Exergonic Reactions

Chemical reactions that require the input of energy are known as **endergonic reactions.** Since energy is added to make these reactions "go," the products of endergonic reactions must contain more free energy than the reactants. A portion of the energy added, in other words, is contained within the product molecules. This follows from the fact that energy cannot be destroyed (first law of thermodynamics) and from the fact that a more organized state of matter contains less entropy (more free energy) than a less organized state (second law).

The fact that glucose contains more free energy than carbon dioxide and water can be easily proven by combustion of glucose to CO_2 and H_2O. This reaction releases energy in the form of heat. Reactions that convert molecules with more free energy to molecules with less, and which therefore release energy as they proceed, are called **exergonic reactions.**

As illustrated in figure 4.13, the amount of energy released by an exergonic reaction is the same whether the energy is released in a single combustion reaction or in the many small, enzymatically controlled steps that occur in tissue cells. The energy that the body obtains from consumption of particular foods can therefore be measured as the amount of heat energy released when these foods are combusted.

Table 4.6 Composition and caloric value of 100-gram portions of some foods.

Food	Percent Protein	Percent Lipid	Percent Carbohydrate	Calories
Apples	0.3	0.6	15.0	58
Avocados	2.2	17.0	6.0	171
Grapefruit	0.6	0.1	9.8	39
Peanuts	26.2	48.7	20.6	582
Butter	0.6	81.0	0.7	716
Cheese, cheddar	25.0	32.2	2.1	398
Eggs	12.8	11.5	0.7	162
Milk, whole	3.2	3.7	4.6	64
Beef, sirloin	21.5	5.7	0	143
Chicken, fryers	20.6	5.6	0	138
Ham, smoked	16.9	35.0	0.3	389
Salami	17.8	49.7	0	524
Halibut	18.6	5.2	0	126
Trout	19.2	2.1	0	101

Source: From Diem, K., and Lentner, C., Eds., Documented Geigy Scientific Tables, 7th ed., J. R. Geigy S. A., Basle, Switzerland, 1970. With permission.

Figure 4.14 Model of coupling of exergonic and endergonic reactions. The driveshaft (representing energy of activation) turns the exergonic gear, which turns the endergonic gear. The reactants of the exergonic reaction have more free energy than the products of the endergonic reaction (the gear is larger) because the coupling is not 100 percent efficient—some energy is lost as heat.

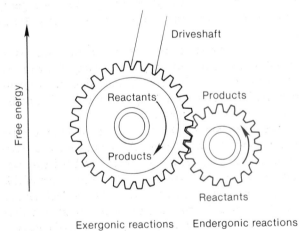

Figure 4.15 Formation and structure of adenosine triphosphate (ATP).

Adenosine diphosphate (ADP)

+

Inorganic phosphate (Pi)

Adenosine triphosphate (ATP)

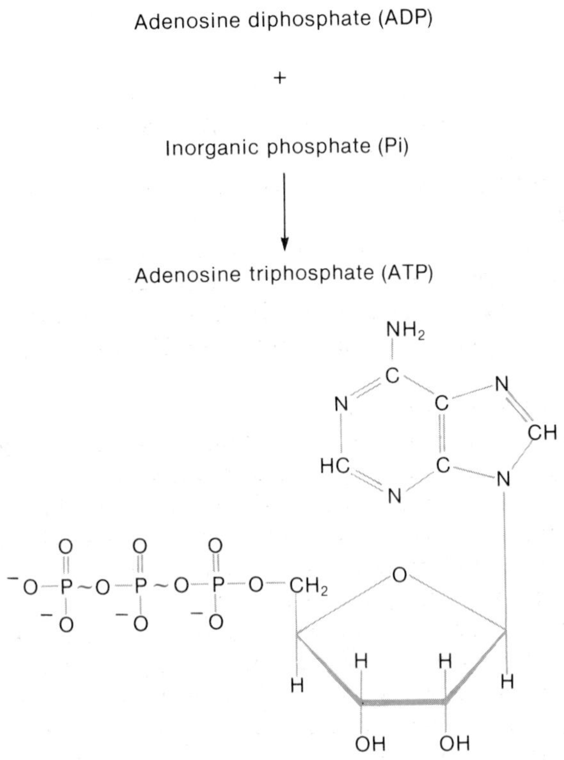

Figure 4.16 Model of ATP as the universal energy donor of the cell. Exergonic reactions are shown as gears with arrows going down (reactions produce decrease in free energy); endergonic reactions are shown as gears with arrows going up (reactions produce increase in free energy).

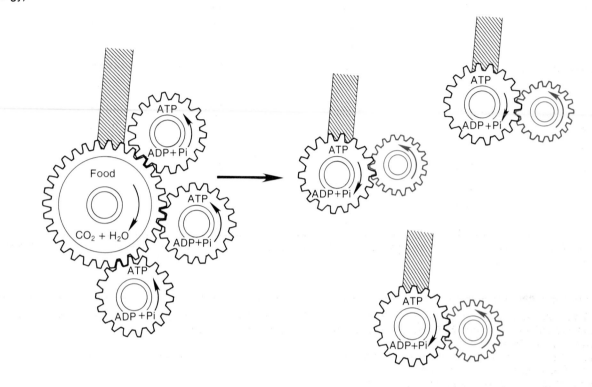

Heat is measured in units called *calories*. One calorie is defined as the amount of heat required to raise the temperature of one cubic centimeter (cc) of water from 14.5°C to 15.5°C. The caloric value of food is usually indicated in kilocalories (one kilocalorie equals 1000 calories). This is commonly expressed with a capital letter—Calories. The amount of energy, expressed in Calories, that is contained in some common foods is shown in table 4.6.

Coupled Reactions: ATP

In order to remain alive a cell must maintain its highly organized, low entropy state at the expense of free energy in its environment. The cell therefore contains many enzymes that catalyze exergonic reactions, using substrates that come ultimately from the environment. The energy released by these exergonic reactions is used to drive the energy-requiring processes (endergonic reactions) in the cell. Since the cell cannot use heat energy to drive energy-requiring processes, chemical bond energy that is released in the exergonic reactions must be directly transferred to chemical bond energy in the products of endergonic reactions. Energy-liberating reactions are thus *coupled* to

energy-requiring reactions. This is like two meshed gears: turning of one (the energy-releasing, exergonic gear) causes turning of the other (the endergonic gear—see figure 4.14).

The energy released by most exergonic reactions in the cell is used, either directly or indirectly, to drive *one* endergonic reaction: the formation of **adenosine triphosphate** (**ATP**) from adenosine diphosphate (ADP) and inorganic phosphate (abbreviated P_i).

The formation of ATP requires the input of a fairly large amount of energy. Since this energy must be conserved (first law of thermodynamics), the bond that was produced by joining P_i to ADP must contain a part of this energy. This bond is accordingly called a *high-energy bond*. When specific enzymes break this bond, and thus convert ATP to ADP and P_i, a large amount of energy is therefore released. Energy released from the breakdown of ATP is used to power the energy-requiring processes in all cells. As the **universal energy donor,** ATP serves to couple more efficiently the energy released by the breakdown of food molecules to the energy required by the diverse energy-requiring processes in the cell (see figure 4.16).

Figure 4.17 Structures of *(a)* the oxidized form of NAD (nicotinamide adenine dinucleotide), and *(b)* the reduced form of FAD (flavine adenine dinucleotide). Note the two additional hydrogen atoms (shown in color) that reduce FAD.

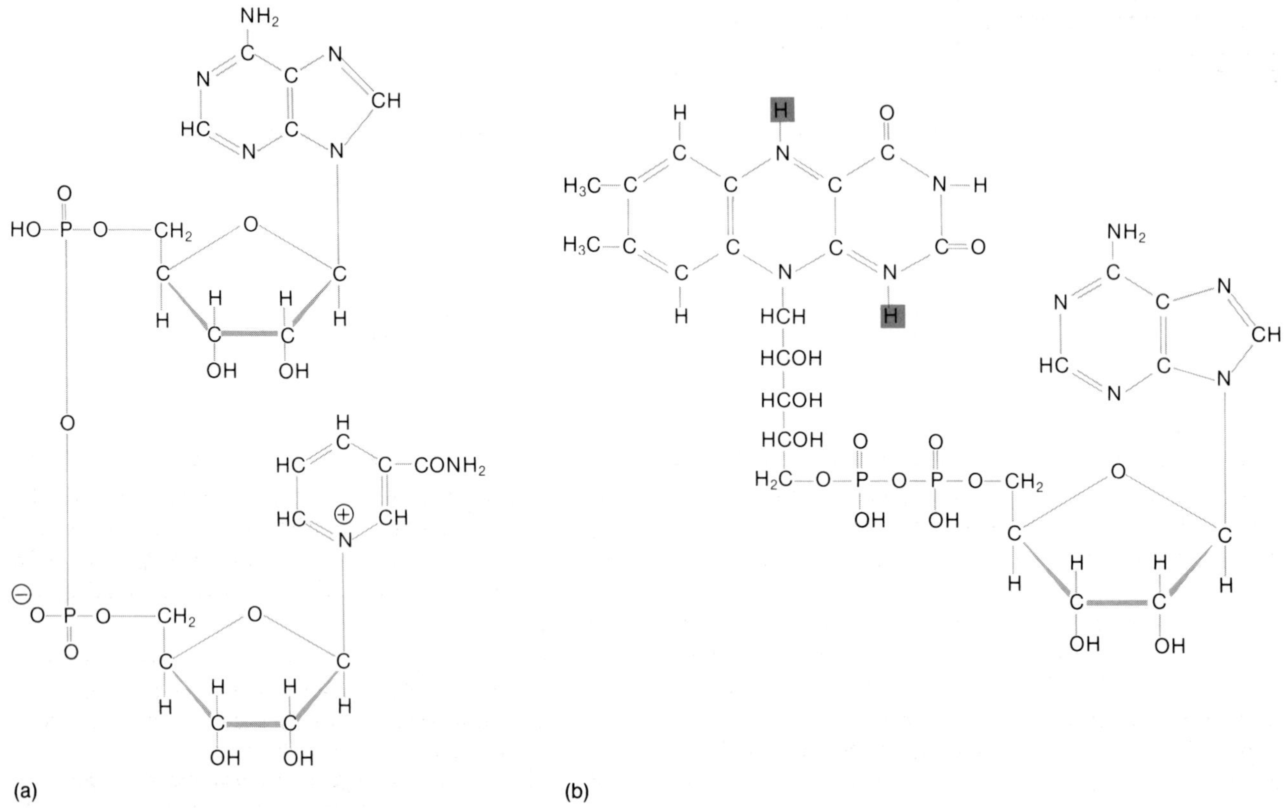

(a)   (b)

In summary, part of the energy released from the conversion of glucose to CO_2 and H_2O in the cell is trapped as high energy bonds in ATP; breakage of these bonds, in turn, releases energy that is directly used to drive the energy-requiring processes needed to sustain life. These endergonic processes include synthesis of large molecules (glycogen, protein, DNA, RNA, and so on), cell movement, and some types of membrane transport (discussed in chapter 6).

Coupled Reactions: Oxidation-Reduction

When an atom or molecule gains electrons it is said to become **reduced;** when it loses electrons it is said to become **oxidized.** These are always coupled reactions—an atom or molecule cannot become oxidized unless it donates electrons to another, which therefore becomes reduced. The molecule that donates electrons is thus a *reducing agent,* while the one that accepts electrons from another is an *oxidizing agent.* It should be noted that a molecule may function as an oxidizing agent in one reaction and a reducing agent in another reaction; it may gain electrons from one molecule and pass them on to another in a series of coupled oxidation-reduction reactions—like a bucket brigade.

Notice that the term *oxidation* does not imply that oxygen participates in the reaction. This term is derived from the fact that oxygen has a great tendency to accept electrons—that is, to act as a strong oxidizing agent. This property of oxygen is exploited by cells: oxygen acts as the final electron acceptor in a chain of oxidation-reduction reactions that provides energy for ATP production (this will be discussed in chapter 5).

Oxidation-reduction reactions in cells often involve the transfer of hydrogen atoms rather than of free electrons. Since a hydrogen atom contains one electron (and one proton in the nucleus—see chapter 2), a molecule that

Figure 4.18 NAD becomes reduced by addition of two electrons from hydrogen atoms removed from an organic molecule (X). In this reaction, NAD acts as an oxidizing agent. In another cellular location, NAD_{red} can donate these two electrons to a different organic molecule (Y). If Y can also bind to protons (H^+), it therefore receives two complete hydrogen atoms (H_2). The molecule Y is thus reduced by this reaction in which NAD serves as a reducing agent.

NAD is oxidizing
agent

NAD is reducing
agent

loses hydrogen becomes oxidized, while one that gains hydrogen becomes reduced. In many oxidation-reduction reactions, pairs of electrons—either separately or in the form of a pair of hydrogen atoms—are transferred from the reducing agent to the oxidizing agent.

Two molecules that serve important roles in the transfer of hydrogens are **nicotinamide adenine dinucleotide (NAD),** which is derived from the vitamin niacin (vitamin B_3), and **flavin adenine dinucleotide (FAD),** which is derived from the vitamin riboflavin (vitamin B_2). These molecules are sometimes called *hydrogen carriers* because they accept hydrogens (become reduced) in one cellular location and donate hydrogens to other molecules (becoming oxidized in the process) at a different cellular location.

Each FAD can accept two electrons and can bind two protons. The reduced form of FAD can therefore combine with the equivalent of two hydrogen atoms and can be written as $FADH_2$. Each NAD can also accept two electrons, but can only bind one proton. The reduced form of NAD may therefore be shown as $NADH + H^+$ (the H^+ represents a free proton). In order to represent these two hydrogen carriers in a consistent fashion, the oxidized forms will subsequently be shown as NAD_{ox} and FAD_{ox} and the reduced forms will be shown as NAD_{red} and FAD_{red}.

1. Describe, in terms of the first and second laws of thermodynamics, why the chemical bonds in glucose represent a source of potential energy.
2. Define the terms *exergonic reaction* and *endergonic reaction.* Use these terms to describe the formation and utilization of ATP within the cell.
3. Using the symbols $X-H_2$ and Y, draw a coupled oxidation-reduction reaction. Identify the molecule that is reduced and the one that is oxidized, and tell which one is the reducing agent and which is the oxidizing agent.
4. Describe, in the general terms used in this chapter, the functions of NAD and oxygen within the cell.

Summary

Enzymes and Metabolic Pathways

I. Enzymes are biological catalysts.
   A. Catalysts increase the rate of chemical reactions.
      1. A catalyst is not altered by the reaction.
      2. Catalysts do not change the final result of a reaction.
   B. Catalysts lower the activation energy of chemical reactions.
      1. The activation energy is the amount of energy needed by the reactant molecules to participate in a reaction.
      2. In the absence of a catalyst, only a small proportion of the reactants have the activation energy.
      3. By lowering the activation energy, enzymes allow a larger proportion of the reactants to participate in the reaction, thus increasing the reaction rate.

II. All enzymes are proteins.
   A. Protein enzymes have specific three-dimensional shapes, which are determined by the amino acid sequence and, ultimately, by the genes.
   B. The reactants in an enzyme-catalyzed reaction—called the substrates of the enzyme—fit into a specific pocket in the enzyme called the active site.
   C. By forming an enzyme-substrate complex, substrate molecules are brought into proper orientation and existing bonds are weakened; this allows new bonds to be more easily formed.

III. The activity of an enzyme is affected by a variety of factors.
   A. The rate of enzyme-catalyzed reactions increases with increasing temperature, up to a maximum.
      1. This is because increasing the temperature increases the energy in the total population of reactant molecules, thus increasing the proportion of reactants that have the activation energy.
      2. At a few centigrade degrees above body temperature, however, most enzymes start to denature, and the rate of the reactions at which they catalyze therefore decreases.

   B. Each enzyme has optimal activity at a characteristic pH—called the pH optimum for that enzyme.
      1. This is because the pH affects the shape and charges within the active site.
      2. The pH optima of different enzymes can be quite different—pepsin has a pH optimum of 2, for example, while trypsin is most active at a pH of 9.
   C. Many enzymes require metal ions in order to be active—these ions are therefore said to be cofactors for the enzymes.
   D. Many enzymes require smaller organic molecules for activity. These smaller organic molecules are called coenzymes.
      1. Many coenzymes are derived from water-soluble vitamins.
      2. Coenzymes transport hydrogen atoms and small substrate molecules from one enzyme to another.
   E. The rate of enzymatic reactions increases when either the substrate concentration or the enzyme concentration is increased.
      1. If the enzyme concentration is constant, the rate of the reaction increases as the substrate concentration is raised, up to a maximum rate.
      2. When the rate of the reaction does not increase upon further addition of substrate, the enzyme is said to be saturated.

IV. Metabolic pathways involve a number of enzyme-catalyzed reactions.
   A. A number of enzymes usually cooperate to convert an initial substrate to a final product by way of several intermediates.
   B. Metabolic pathways are produced by multi-enzyme systems in which the product of one enzyme becomes the substrate of the next.
   C. If an enzyme is defective due to an abnormal gene, the intermediates formed after the step catalyzed by the defective enzymes decrease, and the intermediates formed prior to the defective step accumulate.

      1. Diseases that result from defective enzymes are called inborn errors of metabolism.
      2. Accumulation of intermediates often results in damage to the organ that contains the defective enzyme.
   D. Many metabolic pathways are branched so that one intermediate can serve as the substrate for two different enzymes.
   E. The activity of a particular pathway can be regulated by end-product inhibition.
      1. In end-product inhibition, one of the products of the pathway inhibits the activity of a key enzyme.
         a. This is an example of allosteric inhibition, in which the product combines with its specific site on the enzyme, changing the conformation of the active site.

Bioenergetics

V. The flow of energy in the cell is called bioenergetics.
   A. According to the first law of thermodynamics, energy can neither be created nor destroyed, but only transformed from one form to another.
   B. According to the second law of thermodynamics, all energy transformation reactions result in an increase in entropy (disorder).
      1. As a result of the increase in entropy, there is a decrease in free (usable) energy.
      2. Atoms that are organized into large organic molecules thus contain more free energy than more disorganized, smaller molecules.
   C. In order to produce glucose from carbon dioxide and water, energy must be added as sunlight.
      1. This process is called photosynthesis.
      2. Reactions that require the input of energy to produce molecules with higher free energy than the reactants are called endergonic reactions.

D. The combustion of glucose to carbon dioxide and water releases energy in the form of heat.
   1. When a reaction releases energy and thus forms products that contain less free energy than the reactants it is called an exergonic reacton.
   2. The same total amount of energy is released when glucose is converted into carbon dioxide and water within cells, even though this process occurs in many small steps.
E. The exergonic reactions that convert food molecules into carbon dioxide and water in cells are coupled to endergonic reactions that form ATP.

   1. Some of the chemical-bond energy in glucose is therefore transferred to the "high-energy" bonds of adenosine triphosphate (ATP).
   2. The breakdown of ATP into adenosine diphosphate (ADP) and inorganic phosphate results in the liberation of energy.
   3. The energy liberated by the breakdown of ATP is used to power all of the energy-requiring processes of the cell—ATP is thus the "universal energy donor" of the cell.
VI. Oxidation-reduction reactions have several characteristics.
   A. A molecule is said to be oxidized when it loses electrons, and to be reduced when it gains electrons.

   B. A reducing agent is thus an electron donor, and an oxidizing agent is an electron acceptor.
   C. Although oxygen is the final electron acceptor in the cell, other molecules can act as oxidizing agents.
   D. A single molecule can be an electron acceptor in one reaction and an electron donor in another.
      1. NAD and FAD can become reduced by accepting electrons from hydrogen atoms removed from other molecules.
      2. NAD_{red} and FAD_{red}, in turn, donate these electrons to other molecules in other locations within the cells.
      3. Oxygen is the final electron acceptor (oxidizing agent); since protons (H^+) follow the electrons, oxygen becomes reduced to H_2O.

Self-Study Quiz

1. Which of the following statements about enzymes is TRUE?
   (a) all proteins are enzymes
   (b) all enzymes are proteins
   (c) enzymes are changed by the reactions they catalyze
   (d) the active sites of enzymes have little specificity for substrates
2. Which of the following statements about enzyme-catalyzed reactions is TRUE?
   (a) the rate of reaction is independent of temperature
   (b) the rate of all enzyme-catalyzed reactions is decreased when the pH is lowered from 7 to 2
   (c) the rate of reaction is independent of substrate concentration
   (d) under given conditions of substrate concentration, pH, and temperature, the rate of product formation varies directly with enzyme concentration, up to a maximum, at which point the rate can not be further increased

3. Which of the following statements about lactate dehydrogenase is TRUE?
   (a) it is a protein
   (b) it oxidizes lactic acid
   (c) it reduces another molecule (pyruvic acid)
   (d) all of these
4. In a metabolic pathway:
   (a) the product of one enzyme becomes the substrate of the next
   (b) the substrate of one enzyme becomes the product of the next
5. In an inborn error of metabolism:
   (a) a genetic change results in the production of a defective enzyme
   (b) intermediates produced before the defective step accumulate
   (c) alternate pathways are taken by intermediates at branch points located before the defective step
   (d) all of these

6. Which of the following represents an *endergonic* reaction?
   (a) $ADP + P_i \rightarrow ATP$
   (b) $ATP \rightarrow ADP + P_i$
   (c) $glucose + O_2 \rightarrow CO_2 + H_2O$
   (d) $CO_2 + H_2O \rightarrow glucose$
   (e) both (a) and (d)
   (f) both (b) and (c)
7. Which of the following statements about ATP is TRUE?
   (a) the bond joining ADP and the third phosphate is a high-energy bond
   (b) formation of ATP is coupled to energy-liberating reactions
   (c) conversion of ATP to ADP and P_i provides energy for biosynthesis, cell movement, and other cellular processes that require energy
   (d) ATP is the "universal energy donor" of cells
   (e) all of these
8. When oxygen is combined with two hydrogens to make water:
   (a) oxygen is reduced
   (b) the molecule that donated the hydrogens becomes oxidized
   (c) oxygen acts as a reducing agent
   (d) both (a) and (b)
   (e) both (a) and (c)

Cell Respiration and Metabolism

Objectives

By studying this chapter, you should be able to:

1. Describe glycolysis in terms of its products and functional significance

2. Explain how lactic acid is formed in anaerobic respiration, and describe the functional significance of this pathway

3. Explain the physiological roles of glycogen stored in skeletal muscles and liver

4. Define gluconeogenesis, and describe how this process contributes to the Cori cycle

5. Describe the Krebs cycle in general terms and explain its functional significance

6. Describe electron transport and oxidative phosphorylation

7. Compare anaerobic and aerobic respiration in terms of initial substrates, final products, cellular locations, and the total numbers of ATP molecules produced per glucose respired

8. Describe the role of oxygen in aerobic respiration

9. Explain, in terms of the metabolic pathways involved, how excess calories in the form of carbohydrates or protein can be converted to fat

10. Define the terms *lipolysis* and *β-oxidation,* and describe how these processes function in cellular energy production

11. Explain how ketone bodies are formed

12. Define the terms *transamination* and *deamination,* and explain the functional significance of these processes

13. Describe the changes in skeletal muscle metabolism during exercise

All of the reactions in the body that involve energy transformations are collectively termed **metabolism** (metab = change). Metabolism may be divided into two categories: *anabolism* and *catabolism.* Catabolic reactions are those that release energy, usually by the breakdown of larger organic molecules into smaller molecules; anabolic reactions include the synthesis of large, energy-storage molecules such as glycogen, fat, and protein—these reactions require the input of energy.

Figure 5.1   Blood glucose that enters tissue cells is rapidly converted to glucose-6-phosphate. This intermediate can be metabolized for energy in glycolysis, or can be converted to glycogen (1)—a process called glycogenesis. Glycogen represents a storage form of carbohydrates, which can be used as a source for new glucose-6-phosphate (2), in a process called glycogenolysis. The liver contains an enzyme that can remove the phosphate from glucose-6-phosphate; liver glycogen thus serves as a source for new blood glucose.

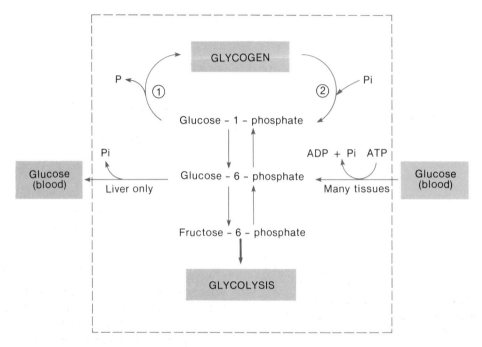

The catabolic reactions that break down glucose, amino acids, and fatty acids serve as the primary sources of energy for the cellular synthesis of ATP (adenosine triphosphate, discussed in chapter 4). These metabolic pathways are known collectively as *cellular respiration.* When oxygen serves as the final electron acceptor, these processes are called **aerobic cell respiration.** The final products of aerobic respiration are carbon dioxide, water, and energy (which is partially trapped in the chemical bonds of ATP). The overall equation for aerobic respiration, therefore, is identical to the equation that describes combustion:

$$\text{Glucose } (C_6H_{12}O_6) + O_2 \longrightarrow CO_2 + H_2O + \text{energy}$$

Notice that the term *respiration* refers to chemical reactions that liberate energy for the production of ATP. The oxygen used in aerobic respiration by tissue cells is obtained from the blood; the blood, in turn, becomes oxygenated in the lungs by the process of breathing. Breathing, which is also called *ventilation* or *external respiration,* is thus needed for, but different from, aerobic respiration.

Unlike combustion, conversion of glucose to carbon dioxide and water within cells occurs in small, enzymatically catalyzed steps. Oxygen is only used at the last step

(as described in a later section). Since a small amount of the chemical-bond energy of glucose is released at early steps in the metabolic pathway, some cells in the body can obtain energy for ATP production in the temporary absence of oxygen. This process is called **anaerobic respiration.**

Although several types of molecules can be utilized for energy in cell respiration, the metabolic fate of glucose will be emphasized in the following discussion. This is because (1) carbohydrates make up about 40 percent of the average American diet; (2) glucose is by far the major carbohydrate in blood; it is *the* blood sugar; (3) blood glucose is the major energy source of many tissues, including the brain; and (4) the metabolic pathways by which fat and protein are respired for energy and the pathway for respiration of glucose converge to common intermediates. Study of the cellular respiration of glucose therefore provides a framework for the study of fat and protein metabolism.

Glycogenesis and Glycogenolysis

Many tissues, particularly the liver, skeletal muscles, and cardiac muscle, store carbohydrates in the form of glycogen. The formation of glycogen (a polymer of glucose—see chapter 2) from glucose is called **glycogenesis.** In this

Figure 5.2 Energy expenditure and gain in glycolysis. Notice that there is a "net profit" of 2 ATP and 2 NAP$_{red}$ per glucose molecule in glycolysis. Molecules listed by number are (1) fructose-1, 6-diphosphate; (2) 1, 3-diphosphoglyceric acid; and (3) 3-phosphoglyceric acid.

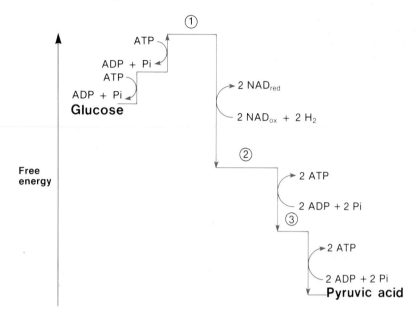

process, glucose is converted to glucose-6-phosphate utilizing the terminal phosphate group of ATP. Glucose-6-phosphate is then converted to its isomer, glucose-1-phosphate. Finally, the enzyme *glycogen synthetase* removes these phosphate groups as it polymerizes glucose to form glycogen.

The reverse reactions are similar; the enzyme *glycogen phosphorylase* catalyzes the breakdown of glycogen to glucose-1-phosphate molecules (the phosphates are derived from inorganic phosphate, not from ATP). Glucose-1-phosphate is then converted to its isomer, glucose-6-phosphate. The process by which glucose-6-phosphate molecules are released from stored glycogen is called **glycogenolysis.** In most tissues (such as skeletal muscles), glucose-6-phosphate can then be used to resynthesize glycogen or it can be respired for energy (as described in the next section). Glucose-6-phosphate *cannot* be released from the cell into the blood because organic molecules with phosphate groups cannot cross cell membranes.

Unlike skeletal muscles, the liver contains an enzyme—known as *glucose-6-phosphatase*—that can remove the phosphate groups and produce free glucose. This free glucose can be transported through the cell membrane and can therefore be secreted by the liver into the blood. Liver glycogen can thus supply blood glucose for use by other organs, including exercising skeletal muscles that have depleted much of their own stored glycogen.

Glycolysis and Anaerobic Respiration

Glycolysis (glyco = sugar; lysis = break) is the metabolic pathway by which glucose—a six-carbon (hexose) sugar—is converted into two molecules of pyruvic acid. Each pyruvic acid molecule contains three carbons, three oxygens, and four hydrogens. The number of carbon and oxygen atoms in glucose—$C_6H_{12}O_6$—can thus be accounted for in the two pyruvic acid molecules. Since the two pyruvic acids together account for only eight hydrogens, however, it is clear that four hydrogen atoms are removed from the intermediates in glycolysis. These hydrogens are used to reduce two molecules of NAD$_{ox}$ (producing two NADH plus two H$^+$, or two NAD$_{red}$—see chapter 4).

Glycolysis is exergonic, and a portion of the energy that is released is used to drive the endergonic reaction ADP + P$_i$→ATP. At the end of the glycolytic pathway there is a net gain of two ATP per glucose, as indicated in the overall equation for glycolysis:

$$\text{Glucose} + 2NAD_{ox} + 2ADP + 2P_i \longrightarrow 2\text{pyruvic}$$
$$\text{acid} + 2NAD_{red} + 2ATP$$

Although the overall process of glycolysis is exergonic, glucose must be "activated" at the beginning of the pathway before energy can be obtained. This activation requires the addition of two phosphate groups derived from two molecules of ATP. Energy from the reaction ATP→ADP + P$_i$ is therefore consumed at the beginning of glycolysis. This is shown as an "up-staircase" in figure 5.2. At later steps in glycolysis, however, four molecules

Figure 5.3 In glycolysis, one glucose molecule is converted into two pyruvic acid molecules in nine separate steps. Since two pyruvic acids are produced from one glucose, the products are multiplied by two in the pathway shown. In addition to two pyruvic acids, these products include two molecules of NAD_{red} and four molecules of ATP. Since two ATP molecules were used at the beginning of glycolysis, however, the net gain is 2 ATP per glucose.

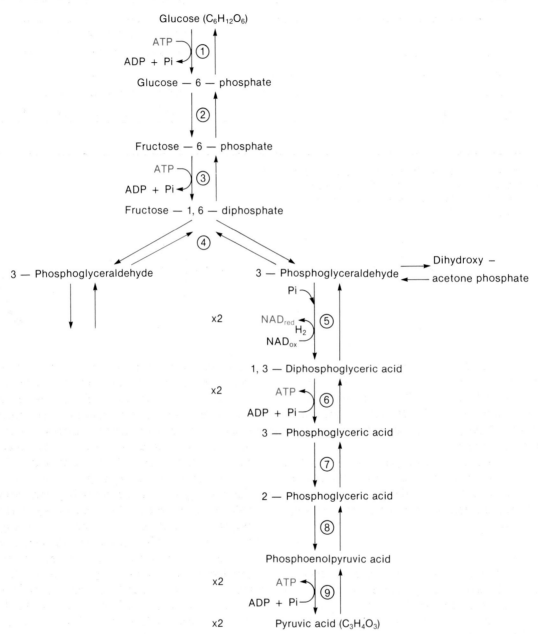

Table 5.1 Enzymes, cofactors, and coenzymes required for glycolysis.

Step	Enzyme	Coenzyme or Cofactor	Comments
1	Hexokinase	Mg^{++}	Liver enzyme catalyzes the phosphorylation of glucose, fructose, or mannose.
2	Hexose phosphate isomerase	——	Interconverts glucose and fructose (both are $C_6H_{12}O_6$).
3	Phosphofructokinase	Mg^{++}, ATP	Allosteric enzyme that is inhibited by high ATP.
4	Aldolase	——	Splits hexose sugar into two three-carbon compounds.
5	Phosphoglyceraldehyde dehydrogenase	NAD	Adds inorganic phosphate and oxidizes aldehyde to acid as NAD is reduced by removal of two hydrogens.
6	Phosphoglycerate kinase	Mg^{++}	Two molecules of ATP are formed at this step.
7	Phosphoglyceromutase	——	Phosphate group is transferred to different carbon.
8	Enolase	Mg^{++}, Mn^{++}	Catalyzes molecular rearrangement.
9	Pyruvate kinase	Mg^{++}, K^+	Two molecules of ATP are formed at this step.

Figure 5.4 The addition of two hydrogen atoms *(colored circles)* from reduced NAD to pyruvic acid produces lactic acid and oxidized NAD. This reaction is catalyzed by lactate dehydrogenase (LDH).

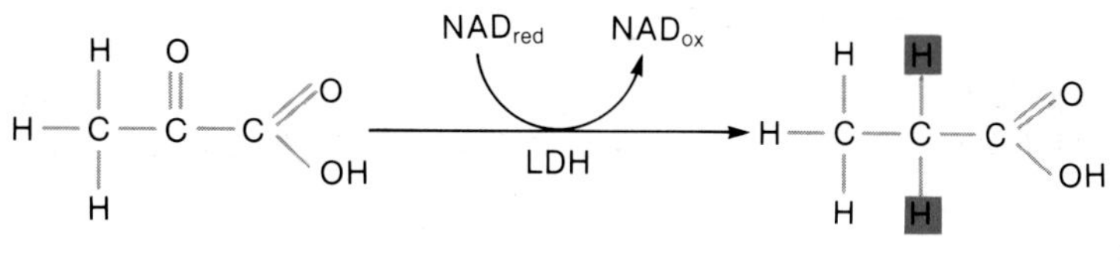

Pyruvic acid   Lactic acid

of ATP are produced (and two molecules of NAD are reduced) as energy is liberated (the "down-staircase" in figure 5.2). The two molecules of ATP used in the beginning therefore represent an energy investment; the net gain of two ATP and two NAD_{red} by the end of the pathway represents an energy "profit."

The overall equation for glycolysis obscures the fact that this is a metabolic pathway consisting of nine separate steps. The individual steps in this pathway are shown in figure 5.3, and the enzymes that catalyze these steps are listed in table 5.1

Anaerobic Respiration

In order for glycolysis to continue, there must be adequate amounts of NAD_{ox} available to become reduced by hydrogen atoms that are removed in the conversion of 3-phosphoglyceraldehyde to 1,3-diphosphoglyceric acid (see figure 5.3). The NAD_{red} that is produced in glycolysis, therefore, must become oxidized by donating its electrons to another molecule. When respiration is aerobic, the NAD_{red} that is produced in glycolysis can become oxidized by donating its electrons to special molecules within the mitochondria (this will be discussed in a later section).

These molecules comprise an electron transport system that ultimately donates two electrons (together with two protons—two H^+) to oxygen, forming water (H_2O).

When oxygen is not available in sufficient amounts this electron transport system "backs up"—that is, it remains in the reduced state and is unable to accept more electrons. In this case, the two electrons in NAD_{red} are instead donated as two hydrogen atoms to pyruvic acid. The NAD_{red} is converted to NAD_{ox}, and pyruvic acid is reduced to *lactic acid* (see figure 5.4). The metabolic pathway in which glucose is converted through pyruvic acid to lactic acid is called *anaerobic respiration*.

Anaerobic respiration yields a net gain of two ATP (produced by glycolysis) per glucose molecule. A cell can thus survive anaerobically as long as it can produce sufficient energy for its needs in this way, and as long as lactic acid concentrations do not become excessive. Some tissues are better adapted to anaerobic respiration than

Figure 5.5 Relative contributions of anaerobic and aerobic respiration to total energy in a well-trained person performing at maximal effort.

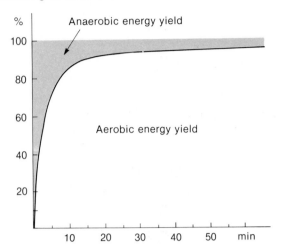

others—skeletal muscles survive better than the cardiac muscle, which in turn can survive under anaerobic conditions longer than can the brain.

Except for red blood cells, which can only respire anaerobically (thereby sparing the oxygen they carry), anaerobic respiration provides only temporary sustenance for tissues that have energy requirements that exceed their aerobic ability. Anaerobic respiration can only occur for a limited period of time (longer for skeletal muscles; shorter for the heart; shortest for the brain) when the *ratio of oxygen supply* (from the blood) *to NAD_red* falls below a critical level. Anaerobic respiration is, in a sense, an emergency procedure that provides some ATP until the emergency (oxygen deficiency) is overcome.

Physical Exercise

Skeletal muscles respire anaerobically for the first forty-five to ninety seconds of moderate-to-heavy exercise. This is because the energy requirements (need for ATP) increase faster than does the rate of oxygen supply by the cardiopulmonary system. If exercise is moderate and the person is in good physical condition, aerobic respiration contributes the major portion of the skeletal muscle energy requirements following the first two minutes of exercise (see figure 5.5).

The maximum rate of oxygen consumption (by aerobic respiration) in the body is called the **maximal oxygen uptake.** The maximal oxygen uptake in a given person is determined primarily by the person's age, size, and sex. It is about 15 to 20 percent higher for males than for females, and the highest at age twenty for both sexes. Some world-class athletes have maximal oxygen uptakes that are twice the average for their age and sex—this appears to be due largely to genetic factors, but training can increase this value by about 20 percent.

Lactic acid concentrations in the blood decrease after the first two minutes of light exercise; in moderate to severe exercise, however, lactic acid concentrations increase. The rate of this increase depends on the severity of the exercise and the physical condition of the person. Increased lactic acid concentrations, together with depletion of stored glycogen and ATP, contribute to muscle fatigue. The amount of exercise that can be performed before muscle fatigue occurs therefore depends on the person's maximal oxygen uptake.

When a person stops exercising, the rate of oxygen uptake does not immediately go back to pre-exercise levels; it returns slowly (the person continues to breathe heavily for some time afterwards). This extra oxygen is used to repay the **oxygen debt** incurred during exercise. The oxygen debt includes oxygen that was withdrawn from savings deposits (hemoglobin in blood and myoglobin in muscle—see chapter 8), extra oxygen required for metabolism by tissues warmed during exercise, and oxygen needed for metabolism of lactic acid produced during anaerobic respiration.

Lactic acid produced under anaerobic conditions during exercise is reconverted to pyruvic acid under aerobic conditions. Conversion of lactic acid to pyruvic acid results in the production of NAD_red, which can be oxidized to NAD_ox if oxygen is available. This occurs in some muscle fibers (the slow-twitch fibers—see chapter 8) and in the heart where pyruvic acid derived from lactic acid can be aerobically respired for energy. In the liver, however, some of the lactic acid produced by exercising skeletal muscles is converted into glucose.

Gluconeogenesis and the Cori Cycle

Some of the lactic acid produced by exercising skeletal muscles is delivered by the blood to the liver. Lactic acid dehydrogenase within liver cells can convert lactic acid to pyruvic acid (as NAD_ox is reduced to NAD_red). Unlike most other organs, the liver contains the enzymes needed

Figure 5.6 The Cori cycle. The direction of reversible reactions that occurs is favored in the Cori cycle as shown by heavy arrows; the sequence of steps is indicated by numbers 1 through 9.

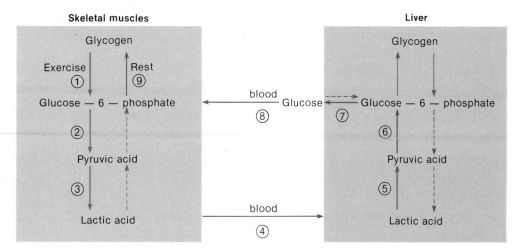

to convert pyruvic acid to glucose-6-phosphate; this is essentially the reverse of glycolysis. Glucose-6-phosphate in liver cells, as previously discussed, may be used as an intermediate for glycogen synthesis, or may be converted to free glucose, which is secreted into the blood.

The formation of glucose from pyruvic acid, and by this means, from non-carbohydrate molecules that can be converted to pyruvic acid (lactic acid, amino acids, and glycerol) is called **gluconeogenesis.** In starvation and prolonged exercise, when liver glycogen stores are depleted, formation of new glucose in this way becomes the only means for maintaining constant blood-sugar levels. Under these conditions, gluconeogenesis in the liver is the only way that adequate blood sugar levels can be maintained to prevent brain death. Conversion of amino acids to glucose prolongs survival during starvation (this will be discussed more in later sections).

Some of the lactic acid produced by skeletal muscles may, during exercise, be transformed through gluconeogenesis in the liver to blood glucose. This new glucose can serve as an energy source during exercise and can be used after exercise is over to replenish the depleted muscle glycogen. This two-way traffic between skeletal muscles and the liver is called the **Cori cycle** (see figure 5.6). Recent evidence also suggests that some muscle fibers may be able to convert lactic acid to glucose (and thus to glycogen) under aerobic conditions, after exercise has ceased.

Through the Cori cycle, gluconeogenesis in the liver (together perhaps with gluconeogenesis in the muscles) allows depleted skeletal muscle glycogen to be restored within forty-eight hours after exercise.

Myocardial Ischemia

Ischemia refers to inadequate blood flow to an organ in which the rate of oxygen delivery is insufficient to maintain aerobic respiration. Sufficiency of blood and oxygen supply is relative—a given level of blood flow may be sufficient at rest but may become insufficient when the metabolic demand of the organ increases. Inadequate blood flow through the heart, or *myocardial ischemia,* may occur if the coronary blood flow is occluded by a clot (coronary thrombosis) or artery spasm. The degree of ischemia, however, depends not only on the reduction of blood flow, but also on the metabolic demands of the tissue (which are increased in exercise and other stress situations).

People with ischemic heart disease often experience *angina pectoris*—severe pain in the chest and left arm area. The degree of ischemia, and of pain, can be decreased by various vasodilator drugs. These vasodilators—including nitrates such as nitroglycerin, and nitrites such as amyl nitrite—can increase blood flow to the ischemic tissues and also decrease the work of the heart by dilating peripheral blood vessels.

Figure 5.7   The formation of acetyl coenzyme A in
aerobic respiration.

Pyruvic acid   Coenzyme A   Acetyl coenzyme A

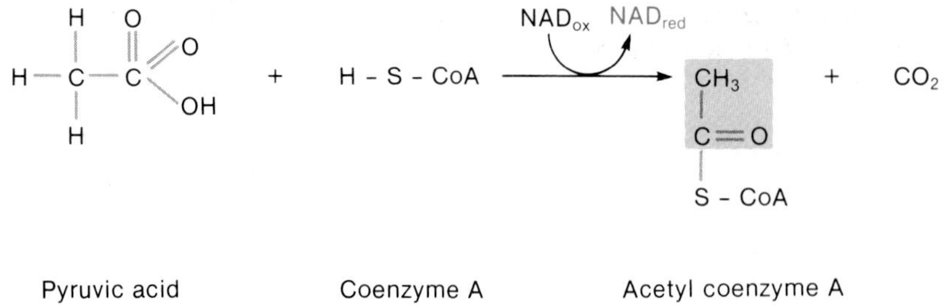

1. Write the overall equation for anaerobic respiration, beginning with the initial substrate (glucose) and ending with the final products. Include NAD and ATP in your equation.
2. Describe the oxidation-reduction reactions that occur in anaerobic respiration and why lactic acid is the final product of this pathway.
3. Explain the functional significance of anaerobic respiration in physical activity. What are the benefits and costs of anaerobic respiration?
4. A person with a coronary thrombosis may have increased blood lactic acid concentrations after exercise on a treadmill, but normal blood lactate levels at rest. Explain this observation.

Aerobic Respiration

Aerobic respiration is equivalent to combustion in terms of its final products (CO_2 and H_2O) and in terms of the total amount of energy liberated. In aerobic cell respiration, however, the energy is released in small, enzymatically controlled steps, and a portion of this energy (38 to 40 percent) is captured in the chemical bonds of ATP (the rest of the energy escapes as heat).

Aerobic respiration of glucose, like anaerobic respiration, begins with glycolysis. This results in the production of two molecules of pyruvic acid, two molecules of ATP, and two molecules of NAD_{red} per glucose. Unlike anaerobic respiration, however, the electrons in NAD_{red} are not donated to pyruvic acid and lactic acid is not formed.

The enzymes that catalyze glycolysis (and the conversion of pyruvic to lactic acid) are located in the cell cytoplasm. In aerobic respiration, pyruvic acid leaves the cytoplasm and enters the interior (the matrix—see chapter 3) of mitochondria. Once pyruvic acid is inside a mitochondrion, carbon dioxide is enzymatically removed from each three-carbon-long pyruvic acid to form a two-carbon organic acid—acetic acid. The enzyme that catalyzes this reaction combines the acetic acid with a coenzyme (derived from the vitamin pantothenic acid) called *coenzyme A*. The combination thus produced is called *acetyl coenzyme A* (abbreviated *acetyl CoA*).

Since glycolysis converts one glucose molecule to two molecules of pyruvic acid, two molecules of acetyl CoA and two molecules of CO_2 are derived from each glucose. The acetyl CoA molecules serve as substrates for mitochondrial enzymes in the aerobic pathway; carbon dioxide is a waste product that diffuses into the blood and is carried to the lungs for elimination in the expired breath. It should be noted that the oxygen in CO_2 is derived from pyruvic acid, not from oxygen gas.

The Krebs Cycle

Conversion of pyruvic acid to acetic acid, and the combination of this product with coenzyme A, is the first of a series of metabolic reactions that occur in mitochondria during aerobic respiration. Once acetyl CoA is formed, the acetic acid subunit (two carbons long) is combined

Figure 5.8 A simplified diagram of the Krebs cycle showing how the original four-carbon-long oxaloacetic acid is regenerated at the end of the cyclic pathway. Only the numbers of carbon atoms in the Krebs cycle intermediates are shown; the numbers of hydrogens and oxygens are not accounted for in this simplified scheme.

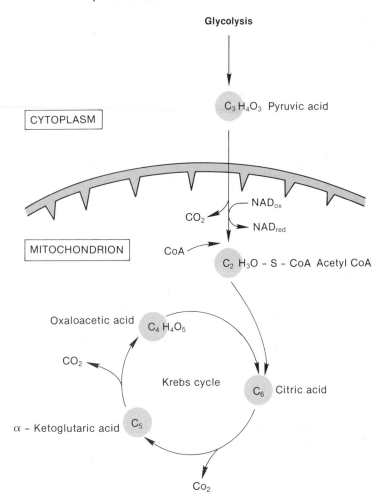

with oxaloacetic acid (four carbons long) to form a molecule of citric acid (six carbons long). Coenzyme A acts only as a transporter of acetic acid from one enzyme to another (similar to the transport of hydrogen by NAD). Formation of citric acid begins a cyclic metabolic pathway known as the **citric acid cycle,** or **TCA cycle** (for tricarboxylic acid; citric acid has three carboxylic acid groups). Most commonly, however, this cyclic pathway is named after its principal discoverer, Sir Hans Krebs, and is thus called the **Krebs cycle.**

Through a series of reactions involving the elimination of two carbons and four oxygens (as two CO_2 molecules) and the elimination of hydrogens, citric acid is eventually converted to oxaloacetic acid. This completes the cyclic metabolic pathway, and in the process the following occur: (1) one guanosine triphosphate (GTP) is produced at step 5 of figure 5.9; this is converted to ATP; (2) three molecules of NAD are reduced (steps 4, 5, and 8 of figure 5.9); and (3) one molecule of FAD is reduced (step 6). The reduced NAD and FAD produced by each "turn" of the Krebs cycle is far more significant, in terms of energy production, than the single GTP (converted to a single ATP) produced directly by the cycle.

Figure 5.9   The Krebs cycle.

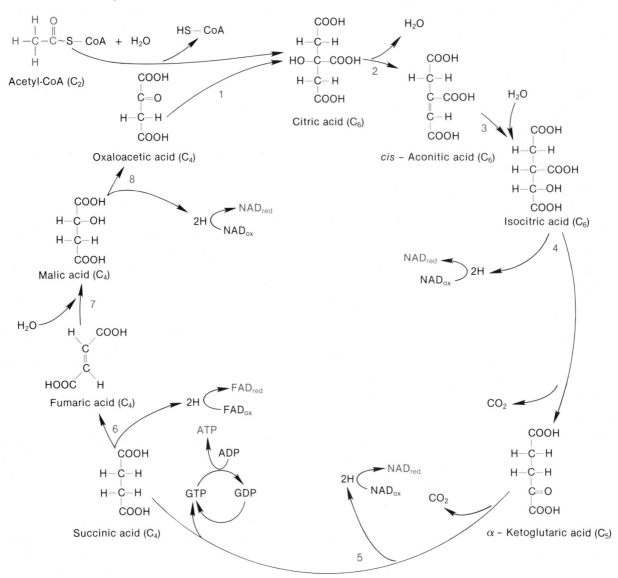

Electron Transport and Oxidative Phosphorylation

Built into the foldings, or cristae, of the inner mitochondrial membrane (see chapter 3) are a series of molecules that serve in **electron transport** during aerobic respiration. This electron transport chain consists of a flavoprotein (derived from the vitamin riboflavin), a molecule called coenzyme Q (derived from vitamin E), and a group of iron-containing pigments (related to hemoglobin) called cytochromes. The last of this group of molecules is an *ATPase enzyme* that can produce ATP from ADP and P_i, using energy derived from electron transport. This subunit protrudes from the cristae (under some conditions) like a lollipop (see figure 5.10).

In aerobic respiration, NAD_{red} and FAD_{red} become oxidized by transferring their pairs of electrons to the electron transport system of the cristae. In this way the

oxidized form of NAD and FAD is regenerated and able to continue to "shuttle" electrons from the Krebs cycle intermediates to the electron transport chain. The first molecule of the electron transport chain, in turn, becomes reduced when it accepts the electron pair from NAD_{red} or FAD_{red}. When the cytochromes receive a pair of electrons, two ferric ions (Fe^{+++}) become reduced to two ferrous ions (Fe^{++}). Notice that the gaining of an electron is indicated by reduction of the number of positive charges.

The electron transport chain thus acts as an oxidizing agent for NAD and FAD. Each element in the chain, however, also functions as a reducing agent; one reduced cytochrome transfers its electron pair to the next cytochrome in the chain. In this way, the iron ions in each cytochrome alternately become reduced (to ferrous iron) and oxidized (to ferric iron)—like a "ferrous" wheel. The

Figure 5.10 The ATPase at the end of the electron transport system is attached to the cristae by a stalk. This unit produces ATP, using energy from electron transport (a process called oxidative phosphorylation).

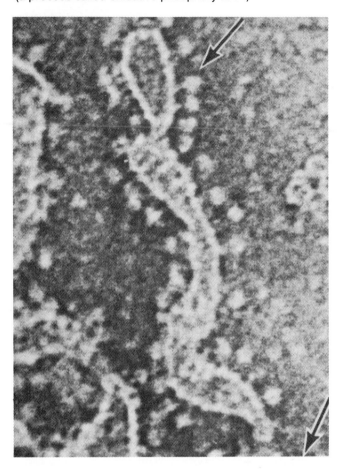

passage of electrons along the electron transport chain is an exergonic process; the reduced form of the last cytochrome contains less free energy than the first. Energy released by electron transport is used, by a process whose description is beyond the scope of this book, to "drive" the reaction in which ADP is "phosphorylated" (by the addition of P_i) to ATP. The production of ATP in this way is appropriately termed **oxidative phosphorylation.**

Function of Oxygen If the last cytochrome remained in a reduced state it would be unable to accept more electrons. Electron transport would then progress only to the next-to-last cytochrome. This process would continue until all of the elements of the electron transport chain remained in the reduced state. When this occurred, NAD_{red} and FAD_{red} could not become oxidized by donating their electrons to the chain, and through inhibition of Krebs cycle enzymes, no more NAD_{red} and FAD_{red} would be produced in the mitochondria (the Krebs cycles would stop). Only NAD_{red} made in glycolysis would be produced, and the only way this NAD_{red} could become oxidized would be to donate its electrons to pyruvic acid. Lactic acid would thus be formed, and respiration would be anaerobic.

Oxygen, from the air we breathe, allows electron transport to continue by functioning as the *final electron acceptor* of the electron transport chain. This oxidizes the last cytochrome so that electron transport and oxidative phosphorylation can continue. At the very last step of aerobic respiration, therefore, oxygen becomes reduced by the two electrons that were passed to the chain from NAD_{red} or FAD_{red}. This reduced oxygen binds two protons, and a molecule of water (H_2O) is formed.

ATP Balance Sheet

Each time the Krebs cycle turns, three molecules of NAD are reduced by electrons from three pairs of hydrogens removed from Krebs cycle intermediates. Each NAD_{red} donates a pair of electrons to the electron transport chain; transport of this pair of electrons to oxygen generates energy for production of three molecules of ATP (through oxidative phosphorylation). Electrons from FAD_{red} enter the electron transport chain "down the line" from where the first ATP is produced. Each pair of electrons from FAD_{red}, therefore, produces only two molecules of ATP from oxidative phosphorylation.

The three NAD_{red} produced per turn of the Krebs cycle therefore results in the production of nine ATP molecules. The single FAD_{red} per turn of the Krebs cycle results in the production of two ATP molecules. Together with the single ATP made directly by the Krebs cycle (not by oxidative phosphorylation—see step 5 of figure 5.9), each turn of the Krebs cycle therefore yields a total of twelve ATP molecules. Since one molecule of glucose produces two pyruvic acids, and thus two turns of the Krebs cycle, a total of twenty-four ATP molecules are produced by a single molecule of glucose through the Krebs cycle and oxidative phosphorylation.

The conversion of pyruvic acid to acetyl CoA also involves reduction of one NAD. Since two pyruvic acids are produced per glucose, and since each NAD_{red} yields three ATP molecules by oxidative phosphorylation, a total of twenty-four plus six, or thirty, ATP molecules are made in the mitochondrion from the steps that occur after pyruvic acid formation.

Two molecules of NAD_{red} are produced in the cytoplasm during glycolysis (conversion of glucose to pyruvic acid). These NAD_{red} cannot directly enter mitochondria; instead, they donate their electrons to other molecules that "shuttle" these electrons into the mitochondria. Depending on which shuttle is used, either two or three ATP molecules can be produced from each pair of these cytoplasmic electrons through oxidative phosphorylation. A total of thirty-four or thirty-six ATP molecules are thus produced (thirty plus four, or thirty plus six). Together with the two molecules of ATP made directly by glycolysis in the cytoplasm, a grand total of thirty-six or thirty-eight ATP molecules are thus produced per glucose that is aerobically respired. This represents an efficiency of 38 to 40 percent in the conversion of chemical bond energy of glucose to chemical bond energy in ATP.

Figure 5.11 Electron transport and oxidative phosphorylation. Each element in the electron transport chain alternately becomes reduced and then oxidized as it transports electrons to the next member of the chain. This process provides energy for the formation of ATP. At the end of the electron transport chain the electrons are donated to oxygen, which becomes reduced (by the addition of two hydrogen atoms) to water.

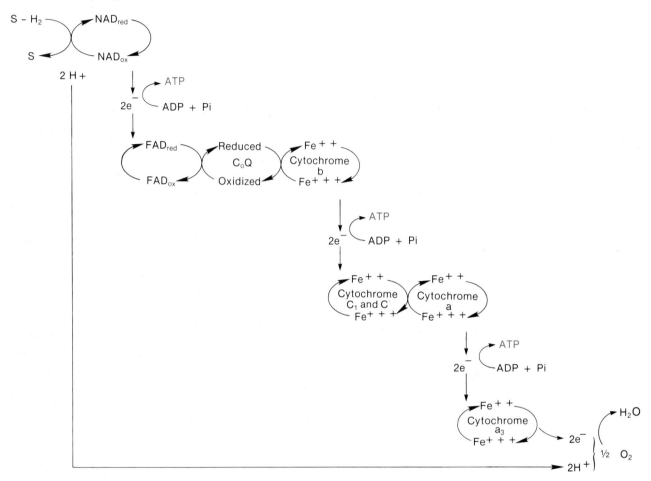

Table 5.2 Maximum ATP yield per glucose molecule respired anaerobically and aerobically.

Phase of Respiration	High-Energy Products	ATP from Oxidative Phosphorylation	ATP Subtotal
Glycolysis (glucose to pyruvic acid)	2 ATP	_____	2 (total if anaerobic)
	2 NAD$_{red}$	6	8 (if aerobic)
Pyruvic acid to acetyl CoA	1 NAD$_{red}$ (X2)	6	14
Krebs cycle	1 ATP (X2)	_____	16
	3 NAD$_{red}$ (X2)	18	34
	1 FAD$_{red}$ (X2)	4	38
		Total (aerobic)	38 ATP

Table 5.3 Calorie expenditure per minute for various activities (values are only approximate and subject to wide variation).

Rest (1 to 2 Cal/min)	Light Work 2.5 to 5 Cal/min	Moderately Heavy Work 5 to 7.4 Cal/min	Heavy Work 7.5 to 9.9 Cal/min	Very Heavy Work over 10 Cal/min
Sitting	Housework	Dancing	Football	Lumbering
	Painting	Farm work	Running	Climbing
	Driving	Bicycling		Wrestling
	Golf	Walking fast		
	Bowling	Tennis		
		Swimming		

Figure 5.12 Conversion of glucose into glycogen and fat due to allosteric inhibition of respiratory enzymes when cell has adequate amounts of ATP. Favored pathways are indicated by heavy arrows.

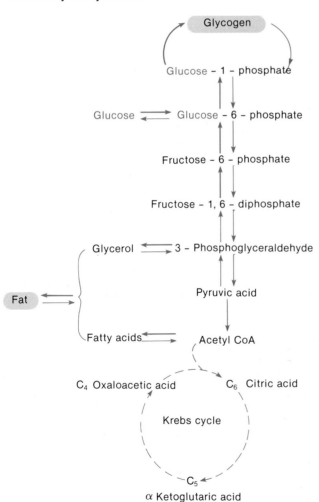

Control of Carbohydrate Metabolism

When oxygen is not available in sufficient amounts, the activity of the Krebs cycle enzymes decreases, while glycolysis (and anaerobic respiration) increases. This switching from aerobic to anaerobic respiration is accomplished largely by allosteric inhibition of enzymes (see chapter 4). Key glycolytic enzymes are inhibited by high ATP and stimulated by high ADP levels. When the Krebs cycle is inhibited (by lack of sufficient NAD_{ox}), therefore, the decreased ATP concentrations stimulate glycolysis. This increased rate of glycolysis under anaerobic conditions is called the *Pasteur effect*.

Cells do not store energy as ATP. When cellular ATP concentrations rise, because more energy (from food) is available than can be immediately used, high ATP concentrations inhibit glycolysis. Under conditions of high cellular ATP concentrations, when glycolysis is inhibited, glucose is instead converted into glycogen.

High ATP concentrations also inhibit Krebs cycle enzymes, preventing acetyl CoA from joining oxaloacetic acid and thereby forming citric acid. Inhibition of Krebs cycle enzymes results in increased availability of acetyl CoA, phosphoglyceraldehyde, and dihydroxyacetone phosphate (formed from phosphoglyceraldehyde—see figure 5.2) for alternate pathways. These intermediates are instead channeled into pathways leading to triglyceride (fat) production. When food intake occurs at a faster rate than energy consumption, therefore, the excess energy (calories) is stored in the form of glycogen and fat (see figure 5.12).

The rate of energy consumption by the body is increased by skeletal muscle activity in exercise (see table 5.3) and in shivering (when increased energy consumption increases heat production). The rate of cell respiration is also affected by hormones. The rate of aerobic respiration, and thus of oxygen consumption under basal conditions (at rest)—known as the *basal metabolic rate* or *BMR*)—is determined by thyroid hormone secretion.

Figure 5.13   Divergent metabolic pathways for acetyl coenzyme A.

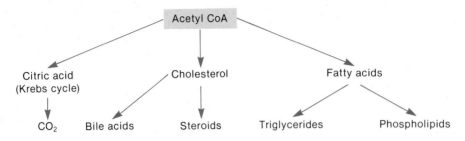

Table 5.4   Contribution of different organs to the total oxygen consumption measured under basal conditions (basal metabolic rate, or BMR).

Organ	Percent of Body Weight	Oxygen Consumed (ml/min)	Percent of Total BMR
Liver	2.1	66	26.4
Skeletal muscles	39.7	64	25.6
Brain	2.0	46	18.3
Heart	0.43	23	9.2
Kidneys	0.43	18	7.2

Increased secretion of epinephrine (adrenalin) under stress increases energy utilization, while increased secretion of insulin promotes formation of glycogen and fat. The hormonal control of metabolism is discussed in detail in chapter 20.

1. Aerobic and anaerobic respiration diverge after pyruvic acid is formed; pyruvic acid, in other words, is the branch point for these two pathways. Draw a figure illustrating glycolysis (glucose to pyruvic acid) and the two alternate fates of pyruvic acid in anaerobic and aerobic respiration.
2. Draw a simplified Krebs cycle, using C_2 for acetic acid, C_4 for oxaloacetic acid, C_6 for citric acid, and C_5 for α-ketoglutaric acid. Indicate the high-energy products that result from each turn of the Krebs cycle.
3. Show how electrons from NAD_{red} and FAD_{red} are passed by the cytochromes. Use Fe^{++} and Fe^{+++} to represent the reduced and oxidized forms of the cytochromes. Indicate the role of oxygen in the electron transport system.
4. Explain how glycogen formation and breakdown (glycogenesis and glycogenolysis, respectively) and the rate of glycolysis are regulated by cellular ATP concentrations.

Lipid Metabolism

It is common experience that ingestion of excessive calories in the form of carbohydrates (cake, ice cream, candy, and so on) increases fat production. The rise in blood glucose that follows carbohydrate-rich meals stimulates insulin secretion, and this hormone, in turn, promotes the entry of blood glucose into adipose cells. Increased availability of glucose within adipose cells, under conditions of high insulin concentrations, promotes the conversion of glucose to fat. Hormonal control of carbohydrate and fat metabolism is discussed in chapter 20.

Phosphoglyceraldehyde and dihydroxyacetone phosphate, produced by glycolysis from glucose, can be converted to glycerol. Not all of the glycolytic intermediates are converted to glycerol, however—some are used to form pyruvic acid, which in turn is converted to acetyl CoA. The two-carbon acetic acid subunits of acetyl CoA can then be used to produce a variety of lipids, including steroids such as cholesterol, ketone bodies, and fatty acids (see figure 5.13). In the formation of fatty acids, a number of acetic acid subunits are joined together to form the fatty acid chain; three of these chains can then condense with glycerol to form a triglyceride (neutral fat) molecule.

Formation of fat *(lipogenesis)* occurs in a number of tissues—primarily adipose tissue and the liver—when blood glucose is elevated. This represents the major form of energy storage in the body. In a non-obese 70-kilogram man, 80 to 85 percent of the body's energy is stored as fat; this amounts to about 140,000 Calories. Stored glycogen, in contrast, accounts for less than 2,000 Calories. Most of this amount (about 350 g) is stored in skeletal muscles and is available only for the muscle's use (for reasons previously discussed). The liver contains about eighty to ninety grams of glycogen, which can be converted to glucose and used by other organs. Protein accounts for about 15 to 20 percent of the stored calories in the body, but this is usually not used extensively as an energy source because that would involve loss of muscle mass.

Breakdown of Fat (Lipolysis)

When fat stored in adipose tissue is going to be used as an energy source, lipase enzymes hydrolyze triglycerides into *glycerol* and *free fatty acids.* These molecules (primarily the free fatty acids) serve as *blood-borne energy carriers* that can be used by the liver, skeletal muscles, and other organs for aerobic respiration.

A few organs can utilize glycerol for energy, by virtue of an enzyme that converts glycerol to β-phosphoglyceraldehyde. Free fatty acids, however, serve as the major energy source derived from triglycerides. Most fatty acids, as described in chapter 2, consist of a long hydrocarbon chain with a carboxylic acid group (COOH) at one end. In a process known as β-**oxidation**, enzymes remove two-carbon acetic acid molecules from the acid end of a fatty acid. This results in the formation of acetyl CoA, as the third carbon from the end becomes oxidized to produce a new acid group. The fatty acid chain is thus decreased in length by two carbons. The process of β-oxidation continues until the entire fatty acid molecule is converted into acetyl CoA.

A sixteen-carbon-long fatty acid, for example, yields eight acetyl CoA molecules. Each of these can enter a Krebs cycle and produce twelve ATP per turn of the cycle, so that eight times twelve, or ninety-six ATP are produced. In addition, every time an acetyl CoA is formed and the end-carbon of the fatty acid chain is oxidized, one NAD and one FAD are reduced. Oxidative phosphorylation produces three ATP per NAD_{red} and two ATP per FAD_{red}. For a sixteen-carbon-long fatty acid, these five ATP molecules would be formed seven times (producing five times seven, or thirty-five, ATP). Not counting the ATP that is used to start β-oxidation (see figure 5.14), this fatty acid could yield a grand total of 35 + 96, or 131, ATP molecules!

Figure 5.14 β-oxidation of a fatty acid. After the attachment of coenzyme A to the carboxylic acid group *(step 1)*, a pair of hydrogens is removed from the fatty acid and used to reduce one molecule of FAD *(step 2)*. When this electron pair is donated to the cytochrome chain, two ATP are produced. Addition of a hydroxyl group from water in *step 3*, followed by the oxidation of the β-carbon in *step 4*, results in the production of three ATP from the electron pair donated by reduced NAD. In *step 5*, the bond between the α and β carbons in the fatty acid is broken, releasing acetyl coenzyme A and a fatty acid chain that is two carbons shorter than the original. The addition of a new coenzyme A to the shorter fatty acid begins the process again at *step 2*, while acetyl CoA enters the Krebs cycle and generates twelve ATP.

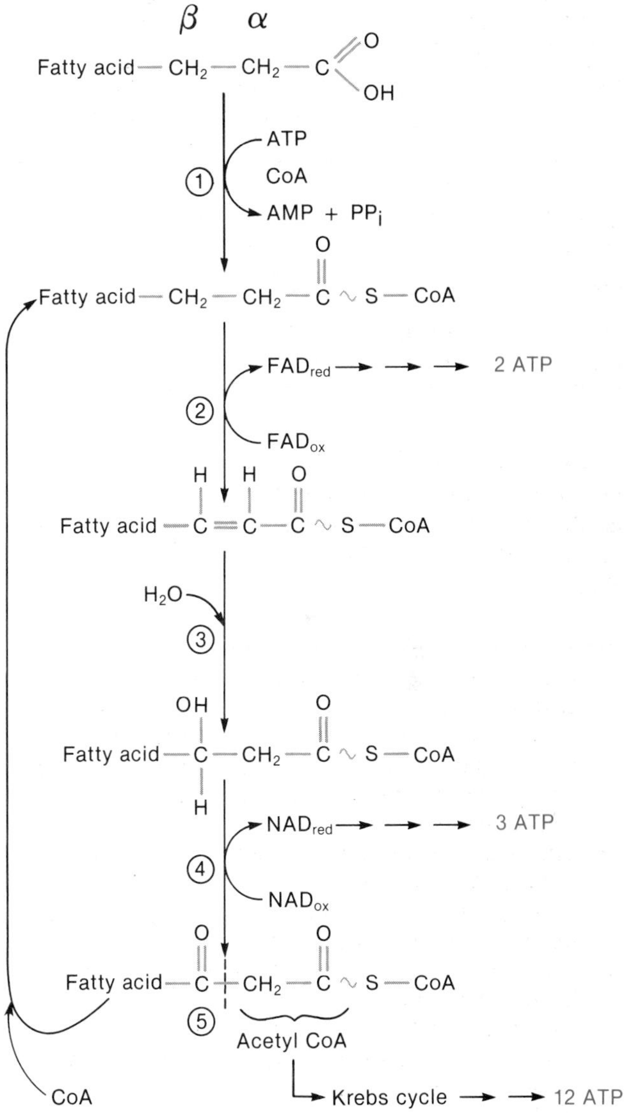

Table 5.5 The essential, semiessential, and nonessential amino acids. The semiessential amino acid can be produced by transamination, but not in sufficient quantities to classify it as a nonessential amino acid. The body cannot produce the essential amino acids, and so must obtain them in the diet.

Essential Amino Acids	Semiessential Amino Acid	Nonessential Amino Acids
Lysine	Arginine	Aspartic acid
Tryptophan		Glutamic acid
Phenylalanine		Proline
Threonine		Glycine
Valine		Serine
Methionine		Alanine
Leucine		Cysteine
Isoleucine		
Histidine (children)		

Figure 5.15 General formula for a transamination reaction. The amine group from amino acid$_1$ is transferred to a keto acid, forming a new amino acid and a new keto acid.

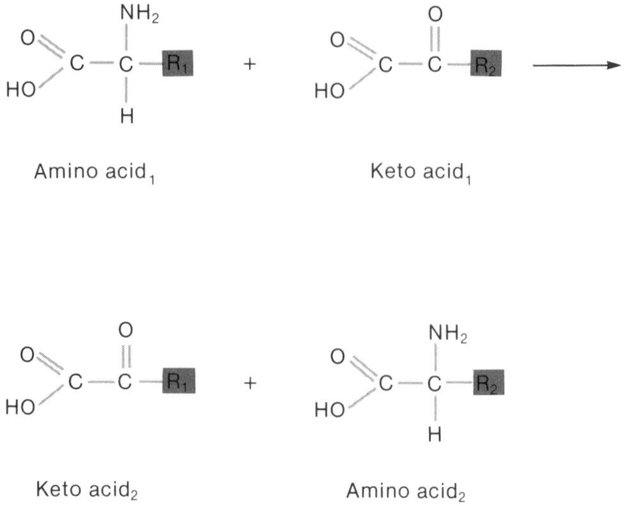

Amino acid$_1$ Keto acid$_1$

Keto acid$_2$ Amino acid$_2$

Ketone Bodies and Ketosis

There is a continuous turnover of triglycerides in adipose tissue, even when a person is not losing weight. New triglycerides are produced, while others are hydrolyzed into glycerol and fatty acids. This turnover insures that the blood normally contains a sufficient amount of fatty acids for aerobic respiration by skeletal muscles, the liver, and other organs. When the rate of lipolysis exceeds the rate of fatty acid utilization—as it does in starvation, dieting, and diabetes mellitus—the blood concentrations of fatty acids increase.

If the liver cells contain sufficient amounts of ATP, so that further production of ATP through the Krebs cycle is not needed, some of the acetyl CoA derived from fatty acids is channeled into an alternate pathway. This pathway involves the conversion of two molecules of acetyl CoA into four-carbon long acidic derivatives: *acetoacetic acid* and *β-hydroxybutyric acid*. Together with *acetone*, which is a three-carbon long derivative of acetoacetic acid, these products are known as **ketone bodies.**

The liver secretes increased amounts of ketone bodies into the blood—a condition called *ketosis*—when fatty acid levels in the blood rise during starvation or diabetes mellitus. The hormonal control of ketone body formation and its clinical significance are discussed in chapter 20.

Amino Acid Metabolism

Nitrogen is ingested primarily as amino acids, and is excreted mainly as urea in the urine. In childhood, the amount of nitrogen excreted is less than the amount ingested because amino acids are incorporated into proteins during growth. Growing children are thus said to be in a state of *positive nitrogen balance*. People who are starving or who are suffering from prolonged wasting diseases, in contrast, are in a state of *negative nitrogen balance;* they excrete more nitrogen than they ingest because they are breaking down their tissue proteins.

Healthy adults maintain a state of nitrogen balance, in which the amount of nitrogen excreted is equal to the amount ingested. This does not imply that the amino acids ingested are unnecessary; on the contrary, they are needed to replace the approximately four hundred grams of body protein that is "turned over" each day. When more amino acids are ingested than are needed to replace proteins, the excess amino acids are not "stored" as additional protein (one cannot build muscles by eating large amounts of protein). Rather, the amine groups can be removed, and the "carbon skeletons" of the organic acids that are left can be used for energy or converted to carbohydrates and fat.

Figure 5.16 Formation of the amino acids *aspartic acid* and *alanine* using glutamic acid as the amine donor in transamination. (GOT= glutamic acid oxaloacetic acid transaminase; GPT= glutamic acid pyruvic acid transaminase). Shaded areas show parts of molecules that are changed by transamination reactions.

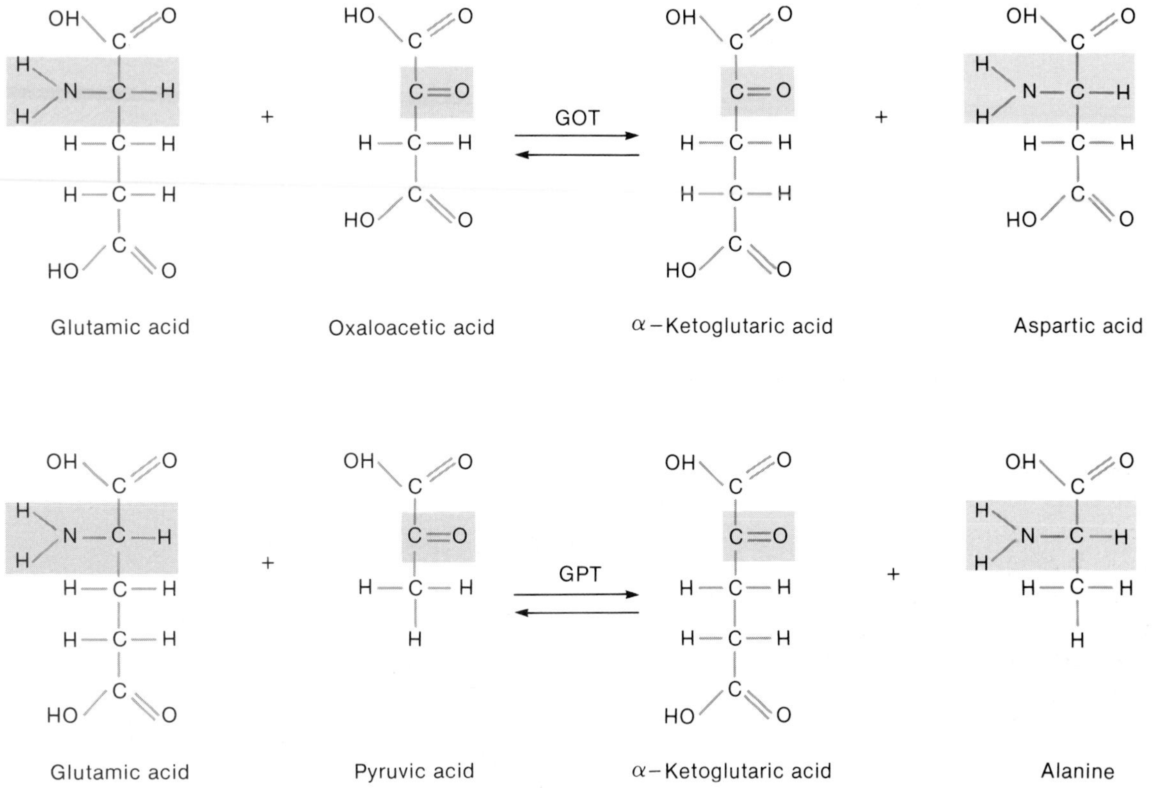

Glutamic acid   Oxaloacetic acid   α–Ketoglutaric acid   Aspartic acid

Glutamic acid   Pyruvic acid   α–Ketoglutaric acid   Alanine

Making New Amino Acids: Transamination

An adequate amount of all twenty amino acids are required to build proteins for growth and for replacement of proteins that are turned over. Fortunately, only eight (in adults) or nine (in children) amino acids must be obtained from the diet. These are the *essential amino acids.* The remaining amino acids are "nonessential" only in the sense that the body can produce them if it is given the essential amino acids and carbohydrates.

When the body has sufficient energy (ATP) so that pyruvic acid and the Krebs cycle acids do not have to complete the aerobic pathway, these *keto acids* (organic acids with a ketone group—see chapter 2) can be combined with amino groups to form new amino acids. In order to become amino acids, however, the keto acids must obtain an amine group. This amine group is usually "cannibalized" from another amino acid; a new amino acid is thus formed, while the one that was cannibalized is converted to a new keto acid (pyruvic acid or one of the Krebs cycle acids). This type of reaction, in which the amine group is moved "across" from one amino acid to form another, is called **transamination.**

Each transamination reaction is catalyzed by a specific enzyme (a transaminase) that requires vitamin B_6 (pyridoxine) as a coenzyme. The amine group from glutamic acid, for example, may be transferred to either pyruvic acid or to oxaloacetic acid. The former reaction is catalyzed by the enzyme *glutamate pyruvate transaminase* (GPT), while the latter reaction is catalyzed by *glutamate oxaloacetate transaminase* (GOT). Addition of an amine group to pyruvic acid produces the amino acid alanine; addition of an amine to oxaloacetic acid produces the amino acid known as aspartic acid.

Figure 5.17   Oxidative deamination. Glutamic acid is converted to α-ketoglutaric acid as it donates its amine group to a metabolic pathway that results in the formation of urea.

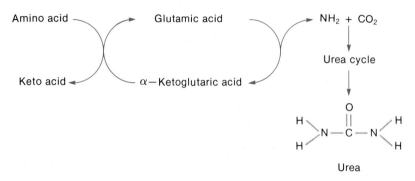

Figure 5.18   Pathways by which amino acids can be catabolized for energy. These pathways are indirect for some amino acids, which must be transaminated into other amino acids before being converted into keto acids by deamination.

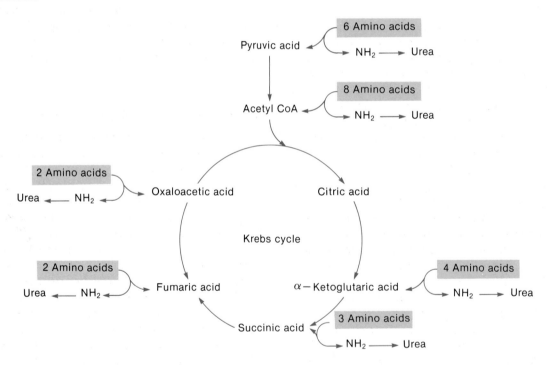

Figure 5.19 Simplified metabolic pathways showing how glycogen, fat, and protein can be interconverted.

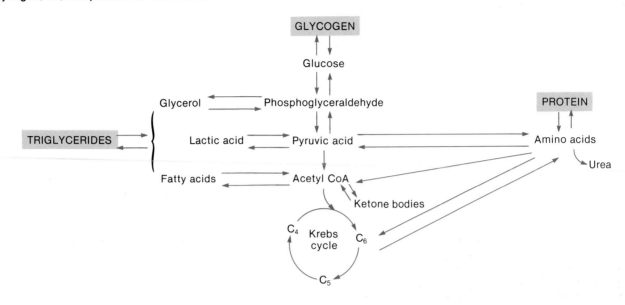

Oxidative Deamination

As shown in figure 5.17, glutamic acid can be formed through transamination by the combination of an amine group with α-ketoglutaric acid. Glutamic acid is also produced in the liver from ammonia that is generated by intestinal bacteria and carried to the liver in the hepatic portal vein. Since free ammonia is very toxic, its removal from the blood and incorporation into glutamic acid is an important function of the healthy liver (ammonia may be present in the blood when the liver is diseased).

If there are more amino acids than are needed for protein synthesis, the amine group from glutamic acid may be removed and excreted as *urea* in the urine (see figure 5.17). The metabolic pathway that removes amine groups from amino acids—leaving a keto acid and ammonia (used for urea production)—is known as **oxidative deamination.**

A number of amino acids can be converted into glutamic acid by transamination. Since glutamic acid can donate amine groups to urea (through deamination), it serves as a "funnel" through which other amino acids can be used to produce keto acids (pyruvic and Krebs cycle acids) and urea. Urea is excreted by the liver into the blood as a non-toxic waste product; the remaining keto acids can enter a variety of metabolic pathways.

Depending on the amino acid that is deaminated, the keto acid left over may be either pyruvic acid or one of the Krebs cycle acids. These can be respired aerobically for energy, converted via pyruvic acid to glucose (*gluconeogenesis*), or converted (via 3-phosphoglyceraldehyde and acetyl CoA) to fat. Interrelationships between carbohydrates, fat, and amino acids are illustrated in figure 5.19.

Uses of Different Energy Sources

The blood serves a common trough from which all the cells in the body are fed. If all cells used the same energy source—such as glucose—this source would quickly be depleted and cellular starvation would occur. This is normally prevented by the fact that the blood contains different energy sources: glucose and ketone bodies from the liver; fatty acids from adipose tissue; lactic acid and amino acids from muscles. Some organs preferentially use one energy source more than the others, so that each energy source is "spared" for organs with strict energy needs.

Under normal conditions, for example, the brain uses blood glucose as its major energy source. This glucose is supplied primarily by the liver, which secretes about 150 milligrams of glucose into the blood per minute—about 75 percent of this glucose is supplied by glycogenolysis (at rest) and 25 percent is produced in the liver by gluconeogenesis using amino acids, lactic acids, and glycerol as initial substrates. Many organs spare blood glucose by using fatty acids, ketone bodies, and lactic acid as energy sources (see table 5.6).

Table 5.6   Relative importance of different molecules in blood to the energy yield in various tissues.

Organ	Glucose	Fatty Acids	Ketone Bodies	Lactic Acid
Brain	+++	+	−	−
Skeletal muscles (resting)	+	+++	+	−
Liver	+	+++	++	+
Heart	+	+	+	+++

Table 5.7   Relative energy yield from different sources for skeletal muscles during different phases of exercise.

Phase	Free Fatty Acids	Glucose from Muscle Glycogen	Blood Glucose	Liver Glucose Output (Percent)	
				From Glycogenolysis	From Gluconeogenesis
Rest	+++	−	+	75 percent	25 percent
5–10 min	−	+++	+	Decreases	Increases
10–40 min	+	++	++	Decreases	Increases
40–90 min	++	+	+++	Decreases	Increases
By 4 hours	+++	−	++	55 percent	45 percent

Muscle Metabolism during Exercise

Skeletal muscles at rest use fatty acids as their major energy source. Metabolism changes at the beginning of moderate exercise, when glucose derived from stored muscle glycogen becomes the prime source of energy. At this time the energy contribution from blood glucose is still low.

After exercise has continued for ten to forty minutes, however, the metabolic pattern changes as muscle glycogen becomes depleted. The contribution of blood glucose to the energy requirements of exercising muscles increases to about 40 percent. This increased demand for blood glucose is met by the liver, which increases its rate of glycogenolysis two to five times over resting levels.

By forty minutes of exercise, the liver has added about eighteen grams of glucose to the blood—roughly 25 percent of its stored glycogen. Liver glycogenolysis continues beyond this time, but much of the burden is removed by increased use of fatty acids as an energy source by the exercising muscles. Since some consumption of blood glucose continues, however, the blood glucose concentration is slightly decreased by ninety minutes of exercise. True hypoglycemia (low blood glucose), however, is rare—which is fortunate, because the glucose requirements of the brain do not change with exercise.

As stored glycogen in the liver becomes depleted during prolonged exercise, the contribution of gluconeogenesis to liver glucose output increases from 25 percent (at rest) to 45 percent. By four hours of exercise, the liver contains only about 25 percent of its initial glycogen stores, and the muscles increasingly rely on free fatty acids as an energy source (see table 5.7).

During and immediately following exercise, lactic acid produced by skeletal muscles can be used as an energy source by the heart, and can be used by the liver as a substrate for gluconeogenesis (as previously described in relation to the Cori cycle). A recovery period of about forty-eight hours is needed to completely restore the muscle glycogen stores that were depleted in prolonged exercise.

---

1. Construct a flow chart (using arrows) to show the metabolic pathway by which glucose can be converted to fat. Indicate only the major intermediates and processes (do not include all of the steps in glycolysis).
2. Construct a flow chart to show how fat can be used for energy. Include lipolysis and β-oxidation, but do not include the detailed steps of the Krebs cycle.
3. Compare transamination with deamination, in terms of their products and functional significance.
4. List five blood-borne energy carriers. Indicate which of these can serve as substrates for gluconeogenesis.
5. Define glycogenolysis and gluconeogenesis, and describe how these processes help to maintain constant blood glucose concentrations during exercise.

Summary

Anaerobic Respiration

I. Anaerobic respiration provides energy for ATP production when the rate of oxygen delivery to a tissue is less than the rate needed for aerobic respiration.
   A. In glycolysis, glucose is converted into two molecules of pyruvic acid.
      1. Two molecules of ATP are consumed at the beginning of glycolysis, as glucose is converted to fructose 1, 6-diphosphate.
      2. Four ATP molecules are produced when fructose 1, 6-diphosphate is converted to two molecules of pyruvic acid.
      3. Glycolysis therefore yields a net gain of two ATP per glucose molecule.
   B. In the process of glycolysis, two molecules of NAD_{ox} are reduced as electrons from four hydrogens convert them into two molecules of NAD_{red}.
      1. The two pyruvic acids produced from glycolysis of one glucose contain only a total of eight hydrogens.
      2. In anaerobic respiration, the two NAD_{red} produced in glycolysis are oxidized by pyruvic acid, which accepts the hydrogen atoms.
      3. The combination of pyruvic acid with two hydrogen atoms produces lactic acid.

II. Different organs have varying abilities to respire anaerobically.
   A. The brain cannot respire anaerobically—after a few minutes without oxygen, the brain usually dies.
   B. Skeletal muscles respire anaerobically for the first minute or two of exercise and when exercise is extremely heavy.
   C. The heart can respire anaerobically, but only to a limited degree.
      1. When the heart respires anaerobically (due to myocardial ischemia), the lactic acid produced contributes to angina pectoris—pain in the chest and left pectoral region.
      2. When heart cells are deprived of oxygen for too long they die.
   D. Lactic acid produced by exercising skeletal muscles contributes to muscle fatigue.
      1. Lactic acid can be converted to pyruvic acid and can be used as an energy source by the heart and some skeletal muscle fibers.
      2. The liver can convert lactic acid to pyruvic acid and can convert this pyruvic acid to glucose; this process is called gluconeogenesis.
      3. Glucose produced by the liver from lactic acid secreted from muscles can be used to produce new muscle glycogen; this cycle (from muscle lactate to skeletal muscle glycogen) is called the Cori cycle.

Aerobic Respiration and Carbohydrate Metabolism

I. If respiration is aerobic, a pyruvic acid (formed in the cytoplasm from glycolysis) enters a mitochondrion.
   A. In the matrix of mitochondria, pyruvic acid is converted (by decarboxylation) to acetyl CoA.
   B. The acetic acid part of acetyl CoA joins with a four-carbon-long acid called oxaloacetic acid to form the six-carbon citric acid.
   C. Formation of citric acid starts a cyclic metabolic pathway called the citric acid cycle, or Krebs cycle.
      1. The pathway is cyclic because oxaloacetic acid is regenerated (by elimination of two molecules of CO_2) at the end of the pathway.
      2. As the cycle turns, one molecule of GTP (converted to ATP) is made, and three molecules of NAD and one molecule of FAD is reduced.

II. Reduced NAD and FAD, made by the Krebs cycle, by conversion of pyruvic acid to acetyl CoA, and by glycolysis, are oxidized by molecules located in the cristae of mitochondria.
   A. These molecules form an electron transport chain.
      1. Cytochromes containing iron are reduced to ferrous (Fe^{++}) and oxidized to ferric (Fe^{+++}) iron as they pass electrons from one molecule to the next.
      2. Finally, the last cytochrome passes a pair of electrons to an oxygen atom.
      3. The oxygen atom binds two protons to form a molecule of water (H_2O).
      4. Oxygen thus serves as the final electron acceptor of the electron transport chain.
   B. Since the reduced form of water contains less free energy than the reduced form of NAD and FAD, electron transport is an exergonic process.
      1. Energy released by electron transport is used to phosphorylate ADP to ATP.
      2. This process is called oxidative phosphorylation.
   C. Three ATP are made by oxidative phosphorylation from one pair of electrons donated to the electron transport chain by NAD_{red}; two ATP are made from the electron pair donated by FAD_{red}.
      1. Counting the one ATP made directly by each turn of the Krebs cycle and the ATP made by oxidative phosphorylation from electrons donated by NAD_{red} and FAD_{red}, twelve ATP are made per turn of the Krebs cycle.
      2. All together, thirty-four or thirty-six ATP are made in the mitochondria from the aerobic respiration of glucose.
      3. Together with the two ATP made in glycolysis, thirty-six or thirty-eight ATP are made from the aerobic respiration of one glucose molecule.

III. The rate of glycolysis and aerobic respiration is regulated by allosteric inhibition of ATP on glycolytic and Krebs cycle enzymes.
  A. When oxygen is in limited supply, low concentrations of ATP stimulate glycolytic enzymes—this is known as the Pasteur effect.
  B. When cellular ATP concentrations are high, glycolysis is inhibited and glucose is converted instead into glycogen.
  C. When cellular ATP concentrations are high, Krebs cycle enzymes are inhibited.
    1. Glycerol is therefore produced from phosphoglyceraldehyde and dihydroxyacetone phosphate.
    2. Acetyl CoA is used as "building blocks" in the formation of fatty acids.
    3. Combination of three fatty acids with glycerol produces a molecule of triglyceride.
  D. High cellular ATP concentrations thus result in the shunting of blood-borne energy carriers into the energy storage molecules of glycogen and fat.

Metabolism of Fat and Amino Acids

I. Triglycerides in adipose tissue are produced mainly from glycerol and fatty acids derived from glucose.
  A. When cellular ATP concentrations are high, acetyl CoA is used to produce fatty acids and phosphoglyceraldehyde is used to produce glycerol.
  B. Condensation of three fatty acids with one molecule of glycerol produces a molecule of triglyceride.
II. In lipolysis, triglycerides are hydrolyzed into glycerol and free fatty acids.
  A. Glycerol is converted into phosphoglyceraldehyde in some tissue and used for energy.
  B. Fatty acids are converted into acetyl CoA units by β-oxidation and are respired for energy via the Krebs cycle and oxidative phosphorylation.
  C. The acetyl CoA produced from some fatty acids is converted into ketone bodies within the liver.
III. Amino acids can be used as substrates for energy or for gluconeogenesis.
  A. Some amino acids can be used to form other amino acids by transamination.
    1. The amine group from one amino acid is transferred to a keto acid.
    2. The combination of the keto acid and amine group forms a new amino acid; the old amino acid, which donated its amine group, becomes another keto acid.
  B. Removal of the amine group from some amino acids results in the production of a keto acid and an amine group for incorporation into urea.
    1. Oxidative deamination thus results in urea production in the liver; urea is secreted into the blood and excreted in the urine.

    2. Transamination reactions result in glutamic acid production; the amine group from this amino acid is used to make urea in oxidative deamination.
    3. Deamination of various amino acids produces keto acids—pyruvic acid or Krebs cycle acids—that can be used for energy.
IV. Different blood-borne energy carriers are used by different tissues.
  A. The brain has an almost absolute requirement for blood glucose as its energy source.
  B. The heart respires lactic acid for energy.
  C. Liver and resting skeletal muscles use fatty acids as their major energy source.
    1. In the early phases of exercise, muscles use stored glycogen and blood glucose as a major energy source.
    2. As exercise progresses, the liver first uses its stored glycogen for energy, then relies increasingly on gluconeogenesis to add new glucose to the blood.
    3. In prolonged exercise, the muscles use increasing amounts of blood-borne free-fatty acids for energy.
  D. After exercise is over, some of the lactic acid produced by exercising muscles is converted, via gluconeogenesis in the liver, to new glucose, which is used to restore some of the depleted skeletal muscle glycogen (this is called the Cori cycle).

Self-Study Quiz

1. The net gain of ATP per glucose molecule in anaerobic respiration is _____ ; the net gain in aerobic respiration is _____ :
   - (a) 2;4
   - (b) 2;38
   - (c) 38;2
   - (d) 24;30
2. In anaerobic respiration, the oxidizing agent for NAD_{red} (the molecule that removes electrons from NAD_{red}) is:
   - (a) pyruvic acid
   - (b) lactic acid
   - (c) citric acid
   - (d) oxygen
3. When organs respire anaerobically, there is an increased blood concentration of:
   - (a) oxygen
   - (b) glucose
   - (c) lactic acid
   - (d) ATP
4. Conversion of lactic acid to pyruvic acid occurs:
   - (a) in anaerobic respiration
   - (b) in the heart, where lactic acid is aerobically respired
   - (c) in the liver, where lactic acid can be converted to glucose
   - (d) in both (a) and (b)
   - (e) in both (b) and (c)

5. The oxygen in the air we breathe:
   - (a) functions as the final electron acceptor of the electron transport chain
   - (b) combines with hydrogen to form water
   - (c) combines with carbon to form CO_2
   - (d) both (a) and (b)
   - (e) both (a) and (c)
6. In terms of the number of ATP molecules directly produced, the major energy-yielding process in the cell is:
   - (a) glycolysis
   - (b) the Krebs cycle
   - (c) oxidative phosphorylation
   - (d) gluconeogenesis
7. Ketone bodies are derived from:
   - (a) fatty acids
   - (b) glycerol
   - (c) glucose
   - (d) amino acids
8. Conversion of glycogen to glucose-6-phosphate occurs in the:
   - (a) liver
   - (b) skeletal muscles
   - (c) both of these organs

9. Conversion of glucose-6-phosphate to free glucose, which can be secreted into the blood, occurs in:
   - (a) the liver
   - (b) the skeletal muscles
   - (c) both of these organs
10. Formation of glucose from pyruvic acid that is derived from lactic acid, amino acids, or glycerol is called:
    - (a) glycogenesis
    - (b) glycogenolysis
    - (c) glycolysis
    - (d) gluconeogenesis
11. Which of the following organs has an almost absolute requirement for blood glucose as its energy source?
    - (a) liver
    - (b) brain
    - (c) skeletal muscles
    - (d) heart
12. When amino acids are used as an energy source:
    - (a) oxidative deamination occurs
    - (b) pyruvic acid or one of the Krebs cycle acids (keto acids) are formed
    - (c) urea is produced
    - (d) all of these

6

Membrane Transport and the Membrane Potential

Objectives

By studying this chapter, you should be able to:

1. Describe diffusion and explain the factors that affect diffusion rate

2. Describe the simple diffusion of polar and nonpolar molecules and ions through the cell membrane

3. Define facilitated diffusion and describe the characteristics of carrier-mediated transport

4. Define active transport and describe how it differs from facilitated diffusion

5. Define osmosis and describe the conditions under which it occurs

6. Explain the mechanisms that help to maintain a constant plasma osmolality

7. Describe the meaning of the terms *osmolality* and *osmotic pressure* and how these factors relate to osmosis

8. Explain the meaning of the phrase *membrane potential* and describe how this membrane potential is developed

9. Describe the relationship between the membrane potential and the potassium equilibrium potential

10. Describe the action of the Na^+/K^+ pump and explain its physiological significance

*T**he* cell membrane separates the intracellular environment (or "compartment") from the extracellular compartment. Proteins, nucleotides (see chapter 2), and other molecules needed for the structure and function of the cell cannot penetrate (or "permeate") the membrane. Other molecules and many ions can penetrate the membrane to varying degrees. The cell membrane is **selectively permeable**; it provides the two-way traffic in nutrients and wastes needed to sustain metabolism while it prevents the passage of other substances between the intracellular and extracellular compartments.

Figure 6.1 Diffusion occurs when there is a concentration difference (or concentration gradient) between two regions of a solution *(a)*, provided that the membrane separating these regions is permeable to the diffusing substance. Diffusion tends to equalize the concentration of these solutions *(b)* and thus to abolish the concentration differences.

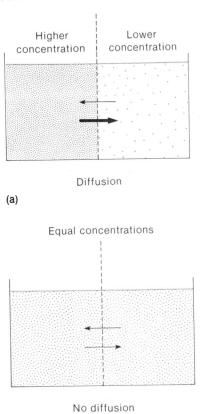

(a)

(b)

Figure 6.2 Gas exchange between the intracellular and extracellular compartments occurs by diffusion. The regions of higher concentration are represented by the larger symbols.

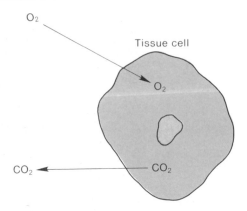

There are two major categories of membrane transport processes: *passive transport* and *active transport.* Transport is said to be passive when the net movement of molecules or ions through the cell membrane does not depend upon metabolic energy; passive transport can continue even after the cell has been killed. Passive transport includes *simple diffusion, facilitated diffusion,* and *osmosis.* Active transport, in contrast, requires the expenditure of energy by hydrolysis of ATP, and thus depends upon cellular metabolism.

Simple Diffusion

Molecules in a gas, and molecules and ions dissolved in a solution, are in a constant state of random motion as a result of their thermal (heat) energy. This random motion tends to make the gas or solution evenly mixed. Whenever a *concentration difference* (also called a *concentration gradient*) exists between two parts of a solution, therefore, random molecular motion tends to abolish the gradient and to make the molecules uniformly distributed. In terms of the second law of thermodynamics, as discussed in chapter 4, the concentration difference represents an unstable state of high organization (low entropy), which changes to produce a uniformly distributed solution with maximum disorganization (entropy).

As a result of random motion, molecules in the part of the solution with higher concentration will enter the area of lower concentration. Molecules will also randomly move in the opposite direction, but not as frequently. As a result, there will be a *net movement* from the region of higher to the region of lower concentration until the concentration difference is abolished. This net movement is called **diffusion.**

The oxygen concentration is relatively high, for example, in the extracellular fluid because oxygen is carried from the lungs to the body tissues by the blood. Since oxygen is converted to water in aerobic cell respiration, the oxygen concentration within the cells is lower than in the extracellular compartment. The concentration gradient for carbon dioxide is in the opposite direction because cells produce CO_2. *Gas exchange* thus occurs by diffusion between the tissue cells and their extracellular environment (see figure 6.2).

Movement of molecules by diffusion across a cell membrane is known as **passive transport.** The term *passive* is used because cellular energy is not needed—this transport could even occur if the cell was killed (as long as a concentration gradient remained). The *rate of diffusion,* measured by the number of diffusing molecules passing through the membrane per unit time, depends on (1) the magnitude of the concentration difference across the membrane (the "steepness" of the concentration gradient); (2) the permeability of the membrane to the diffusing substances; and (3) the surface area of the membrane through which the substances are diffusing.

Figure 6.3 Microvilli *(Mv)* in the small intestine, as seen with the transmission *(a)* and scanning *(b)* electron microscopes.

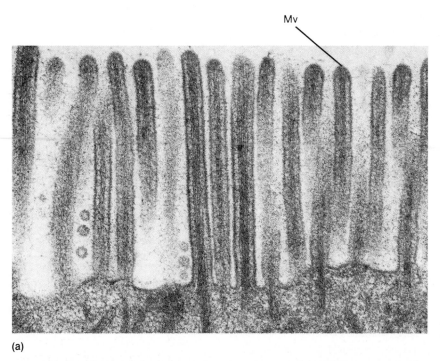

(a)

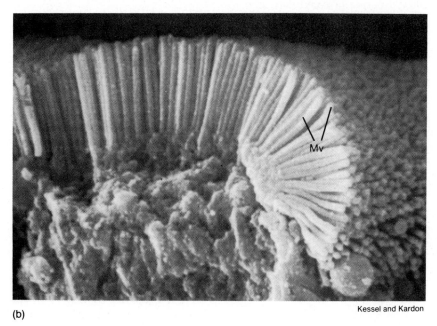

(b)

Kessel and Kardon

The permeability of the membrane to various molecules and ions is regulated by molecular changes within the membrane structure. Changes in the permeability of the membrane to certain ions, for example, allows diffusion of these ions and the production of electrical currents in nerve and muscle cells (as discussed in chapter 7).

Rapid passage of the products of digestion across epithelial membranes in the intestine is aided by structural adaptations that increase the surface area of these cells. Increased surface area of the apical membrane of these cells (the side facing the intestinal lumen) is increased by many tiny folds that form fingerlike projections of the membrane. These membrane folds are called *microvilli* (see figure 6.3). Similar microvilli are also found in the kidney tubule epithelium, which must reabsorb various solutes that were filtered out of the blood.

Figure 6.4 Water may penetrate the membrane through pores within integral proteins that span the thickness of the double phospholipid layers.

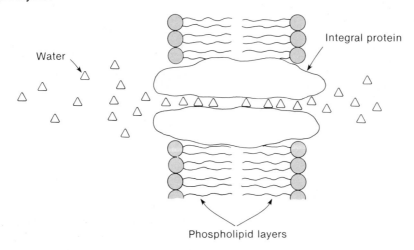

Molecules that are nonpolar and therefore lipid soluble, such as oxygen, carbon dioxide, and steroid hormones, can easily diffuse through the phospholipid layers of the cell membrane. Inorganic ions and water molecules, which are polar (see chapter 2), however, cannot pass through the hydrophobic lipid barrier. Despite this, the membrane does allow diffusion of some ions under special circumstances and is freely permeable to water.

Tiny passages, or *pores,* must therefore be present in the membrane. Although these pores are too small to be seen, even with an electron microscope, they must be present in large numbers to account for the rapid rate of water transport through cell membranes (a process called osmosis and discussed in the next section). Membrane pores are believed to be part of the *integral proteins* that span the thickness of the membrane (see figure 6.4).

There are two subcategories of passive transport. The first is **simple diffusion,** which includes diffusion through the phospholipid layers of the membrane and through pores in the membrane. Osmosis is thus a type of simple diffusion. The second subcategory of passive transport is called **facilitated diffusion,** in which molecules move from high to low concentration with the aid of membrane carriers (this will be discussed in a later section together with other examples of carrier-mediated transport). In contrast to these examples of passive transport, some membrane carriers move molecules from areas of lower to areas of higher concentration—this requires cellular energy (ATP), and is called active transport. Active transport will be discussed in a later section.

Figure 6.5 Model of osmosis, or the net movement of water *(closed circles)* from the solution of lesser solute *(open circles)* concentration to the solution of greater solute concentration.

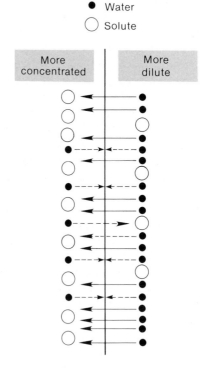

Figure 6.6 In *(a)*, a movable semipermeable membrane (permeable to water but not glucose) separates two solutions of different glucose concentration. As a result, water moves by osmosis into the solution of greater concentration until *(b)* the volume changes equalize the concentrations on both sides of the membrane.

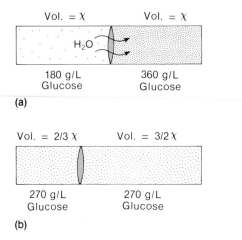

(a)

(b)

Osmosis

Osmosis refers to the simple diffusion of water (the solvent) across a membrane. As with the diffusion of solute molecules, diffusion of solvent occurs when water is more concentrated on one side of the membrane than on the other—that is, when one solution is more dilute than the other. The more-dilute solution has a higher concentration of water molecules because the less-dilute solution contains more solute molecules, which reduces the water concentration. The principles of osmosis are the same as those for diffusion of any molecule, but the terminology is backwards because the term *concentration* is usually used to refer to the density of solute rather than to solvent molecules.

Suppose a cylinder is divided into two equal compartments by a membrane partition that can freely move, and that one compartment initially contains 180 g/L (grams per liter) of glucose while the other compartment contains 360 g/L of glucose. If the membrane is permeable to glucose, glucose will diffuse from the 360 g/L compartment to the 180 g/L compartment until both compartments contain 270 g/L of glucose. If the membrane is not permeable to glucose but is permeable to water, the same result (270 g/L solutions on both sides of the membrane) can be achieved by diffusion of water (osmosis). As water diffused from the 180 g/L compartment to the 360 g/L compartment the former solution would become more concentrated as the latter became more dilute. This would be accompanied by changes in volume, as illustrated in figure 6.6.

Figure 6.7 If a semipermeable membrane separates pure water from a 180 g/L glucose solution, water tends to move by osmosis into the glucose solution, thus creating a hydrostatic pressure that pushes the membrane to the left and expands the volume of the glucose solution. The amount of pressure that must be applied to just counteract this volume change is equal to the osmotic pressure of the glucose solution.

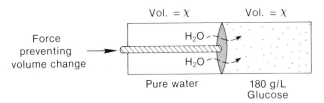

Molality, Osmolality, and Osmotic Pressure

Osmosis, and the movement of the membrane partition, could be prevented by an opposing force. If one compartment contained 180 g/L of glucose and the other compartment contained pure water, osmosis of water into the glucose solution could be prevented by pushing against the membrane with a force of 22.4 atmospheres pressures (see figure 6.7).

The force that would have to be exerted to prevent osmosis in this situation is the **osmotic pressure** of the solution. This backwards measurement indicates how strongly the solution "draws" water into it by osmosis. The greater the solute concentration of a solution the greater its tendency to draw water, and the greater its osmotic pressure. Pure water thus has an osmotic pressure of zero; a 360 g/L glucose solution has twice the osmotic pressure of a 180 g/L glucose solution.

Suppose that pure water was separated by a membrane partition from a solution that contained 342 g/L of sucrose. The pressure that would have to be applied to prevent osmosis—the osmotic pressure of the sucrose solution—would be 22.4 atmospheres pressure. This is the same as the osmotic pressure of the 180 g/L glucose solution because both solutions contain an equal number of solute molecules per liter.

Glucose is a monosaccharide with a molecular weight of 180 (the sum of its atomic weights). Sucrose is a disaccharide (see chapter 2) of glucose and fructose, which have molecular weights of 180 each. When glucose and fructose join together by dehydration synthesis to form sucrose, a molecule of water (molecular weight = 18) is split off. Sucrose therefore has a molecular weight of 342 (the sum of 180 + 180 − 18). Each sucrose molecule thus weighs 342/180 times as much as each glucose molecule; 342 grams of sucrose must thus contain the same number of molecules as 180 grams of glucose.

Figure 6.8 Diagram illustrating the difference between a one molar (1.0 M) and a one molal (1.0 m) glucose solution.

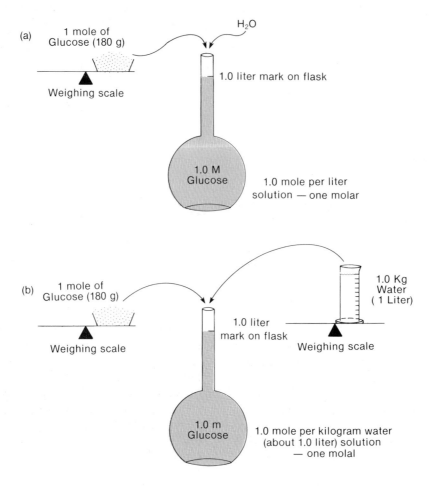

Notice that the molecular weight in grams of any compound must contain the same number of molecules as the gram molecular of any other compound. This unit of weight is called a *mole,* and it always contains 6.02×10^{23} molecules (*Avogadro's number*). One mole of solute dissolved in water to make one liter of solution is described as a one-**molar** solution (abbreviated 1.0M). While this unit of measurement is commonly used in chemistry, it is not completely desirable in discussions of osmosis because the exact amount of water needed to make one liter of solution is variable. More water, for example, is needed to make a 1.0M salt solution (where a mole of NaCl weighs 58.5 grams) than is needed to make a 1.0M glucose solution because 180 grams of glucose take up more volume than 58.5 grams of salt.

Since the ratio of solute to water molecules is of critical importance in osmosis, a more desirable measurement of concentration is **molality.** In a one-molal solution (abbreviated 1.0m), one mole of solute (180 grams of glucose, for example) is dissolved in one kilogram of water (equal to 1,000 ml at 4°C). A 1.0m NaCL solution and a 1.0m glucose solution both contain a mole of solute dissolved in exactly the same amount of water (see figure 6.8).

Figure 6.9 The osmolality (Osm) of a solution is equal to the sum of the molalities of each solute in the solution. If a semipermeable membrane separates two solutions with equal osmolalities, no osmosis will occur.

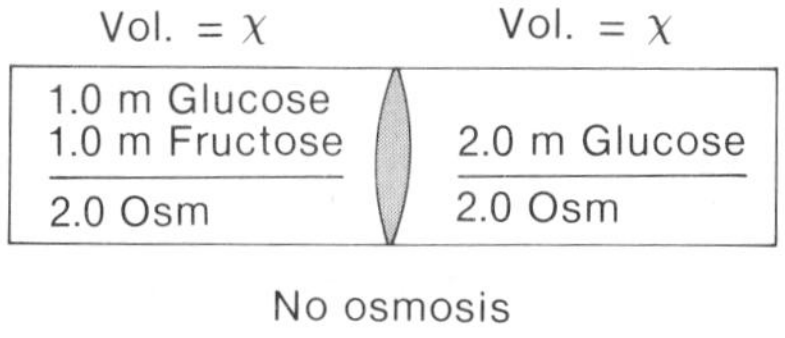

Vol. = χ   Vol. = χ

1.0 m Glucose 1.0 m Fructose	2.0 m Glucose
2.0 Osm	2.0 Osm

No osmosis

Osmolality   If 180 grams of glucose and 180 grams of fructose were dissolved in the same kilogram of water, the osmotic pressure of the solution would be the same as that of a 360 g/L glucose solution. Osmotic pressure depends on the ratio of solute to solvent, *not* on the chemical nature of the solute molecules. An expression is needed that gives the total molality of a solution—this is called **osmolality**. The 1.0m glucose plus 1.0m fructose solution thus has a total molality, or osmolality, of 2.0 (abbreviated 2.0 Osm). The 360 g/L glucose solution, which is 2.0m, has the same osmolality and therefore the same osmotic pressure.

Unlike glucose, fructose, and sucrose, electrolytes such as NaCl ionize when they dissolve in water. One molecule of NaCl dissolved in water yields two ions (Na^+ and Cl^-); one mole of NaCl ionizes to form one mole of Na^+ and one mole of Cl^-. A 1.0m NaCl solution thus has a total concentration of 2.0 Osm. If a 1.0m NaCl solution was separated by a membrane partition from a 1.0m glucose solution, and if the membrane was permeable to water but not to glucose, Na^+ or Cl^-, osmosis would occur from the glucose to the salt solution until each had a total concentration of 1.5 Osm (see figure 6.10).

Measurement of Osmolality. Plasma, cerebrospinal fluid, and other biological fluids contain many organic molecules and electrolytes. The osmolality of such complex fluids can only be estimated by calculations; fortunately, however, there is a relatively simple procedure for directly measuring osmolality. This method is based on the fact that the freezing point of a solution, like its osmotic pressure, is affected by the total concentration of the solution and not by the chemical nature of the solute.

Figure 6.10 If a semipermeable membrane (permeable to water but not to glucose, Na^+ or Cl^-) separates a 1.0 m glucose solution from a 1.0 m NaCl solution *(a)*, water will move by osmosis into the NaCl solution. This is because NaCl can ionize to yield one molal Na^+ plus one molal Cl^-. After osmosis *(b)*, the total concentration or osmolality of the two solutions are equal.

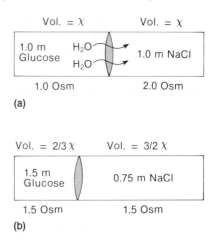

(a)

(b)

One mole of solute depresses the freezing point of water by $-1.86°C$. A 1.0m glucose solution thus freezes at a temperature of $-1.86°C$, while a 1.0m NaCl solution freezes at a temperature of $2 \times -1.86 = -3.72°C$ (because of ionization). The *freezing point depression* is thus a measure of the osmolality. Since plasma freezes at about $-0.56°C$, its osmolality is equal to 0.56/1.86, or 0.3 Osm. This is more commonly indicated as three hundred milliosmolal (or 300 mOsm).

Tonicity

A 0.3m glucose solution thus has the same osmolality and osmotic pressure as plasma. The same is true of a 0.15m NaCl solution, which ionizes to produce a total concentration of 300 mOsm. Both of these solutions are used clinically as intravenous infusions, labeled *5.0 percent dextrose* (5 g/100 ml glucose or 0.30m) and *normal saline* at 0.9 g percent (0.9 g/100 ml NaCl or 0.15m). All solutions that have the same osmolality and osmotic pressure as plasma are said to be **isotonic** (iso = same; tonic = strength).

Figure 6.11 Photomicrograph of normal and crenated red blood cells.

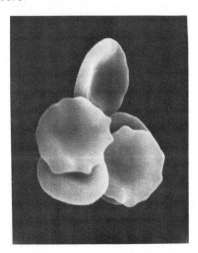

Since isotonic solutions, by definition, have the same osmotic pressure as plasma, red blood cells placed in these solutions would neither gain nor lose water. If red blood cells are placed in a solution that has a lower osmotic pressure and osmolality than plasma, water will move from this more dilute solution into the cells. Solutions with a lower osmotic pressure than plasma are called **hypotonic solutions** (hypo = beneath). When red blood cells are placed in a **hypertonic solution** (such as sea water), which has a higher osmolality and osmotic pressure than plasma, osmosis of water out of the cells will cause them to shrink. In this process, called *crenation,* the cell surface becomes scalloped in appearance (see figure 6.11).

Regulation of Blood Osmolality

The osmolality of the blood plasma is normally maintained within very narrow limits by a variety of regulatory mechanisms. When a person becomes dehydrated, for example, the blood becomes more concentrated as the total blood volume is reduced. The increased blood osmolality and osmotic pressure stimulates *osmoreceptors,* which are neurons located in a part of the brain called the hypothalamus.

As a result of increased osmoreceptor stimulation, the person becomes thirsty and—if water is available—drinks. Along with increased water intake, a person who is dehydrated excretes a lower volume of urine. This occurs as a result of the following sequence of events: (1) increased plasma osmolality stimulates osmoreceptors in the hypothalamus; (2) the osmoreceptors stimulate the posterior pituitary gland, via a nerve tract, to secrete **antidiuretic hormone (ADH);** (3) ADH acts on the kidneys to promote water retention; so that (4) a lower volume of urine is excreted.

A person who is dehydrated therefore drinks more and urinates less. The effects of increased osmoreceptor stimulation thus act to increase blood volume as the plasma osmolality is decreased back to the normal range. This negative feedback loop is illustrated in figure 6.12.

As the person continues to drink water, this water is absorbed across the membranes of the intestine and makes the blood more dilute as its volume increases. When the plasma osmolality decreases below the normal range, the desire to drink is extinguished. At the same time, osmoreceptor stimulation of ADH secretion by the posterior pituitary is reduced. As ADH secretion decreases, less water is retained by the kidneys, which excrete a larger volume of more dilute urine. The excess water is thus eliminated in the urine as the blood osmolality is increased back towards the normal range.

1. Define diffusion and describe three factors that influence diffusion rate.
2. Define osmosis, and describe the conditions required for it to occur.
3. Illustrate the following hypothetical experiment: a 150 mM NaCl solution is separated by a semipermeable membrane from a 200 mM glucose solution. Below this figure, indicate the following:
   (a) The milliosmolality of each solution
   (b) Which solution has the higher and which the lower osmotic pressure
   (c) Which direction (if any) water will move through the membrane by osmosis
4. Explain why hospitals use 5 percent dextrose and normal saline as intravenous infusions.
5. A person who is dehydrated drinks more and urinates less. Explain these observations by drawing a negative feedback loop (try not to look at figure 6.12).

Figure 6.12 An increase in plasma osmolality (increased concentration and osmotic pressure), due to dehydration, stimulates thirst and increased ADH secretion. These effects cause the person to drink more and urinate less. The blood volume, as a result, is increased while the plasma osmolality is decreased. These effects help bring the blood volume and osmolality back to the normal range and complete the negative feedback loop (indicated by −).

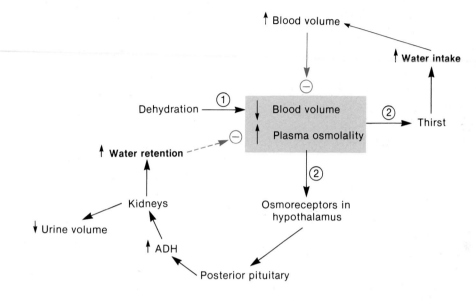

Carrier-Mediated Transport

In order to sustain metabolism, cells must be able to take up glucose, amino acids, and other organic molecules from the extracellular environment. In muscles and adipose tissue the rate of glucose uptake determines the rate of cell metabolism. Yet glucose is too large to fit through the membrane pores (the molecular weight of glucose is 180; ten times the molecular weight of water), and too polar to pass through the lipid barrier. Despite this, glucose penetrates the cell membrane at a rate that is ten times that predicted from its lipid solubility.

The transport of glucose, amino acids, and some other molecules is mediated by protein **carriers** within the membrane. Although such carriers cannot be directly observed, their presence has been inferred by the observation that such transport has characteristics in common with enzyme activity. These characteristics include (1) *specificity;* (2) *competition;* and (3) *saturation.* Because of their similarity to enzymes, the transport carriers are sometimes called *permeases.*

Like enzyme proteins, carrier proteins interact with only specific molecules. Glucose carriers, for example, can only interact with D-glucose (not with L-glucose—see chapter 2). As a further example of specificity, particular carriers for amino acids transport some types of amino acids but not others. Two amino acids that are transported by the same carrier compete with each other, so that the rate of transport of each is lower than it would be if each amino acid were present separately.

Table 6.1   Inherited defects of transport carriers in the kidney and intestine.

Disease	Defect	Clinical Significance
Cystinuria	Excessive urinary excretion of cystine, lysine, arginine, and ornithine	Calculi (stones) in urinary tract
Phosphaturia	Excessive urinary excretion of phosphate	Rickets; treated with large doses of vitamin D
Renal glycosuria	Kidney tubules have lower than normal T_m for glucose	None known
Glucose malabsorption	Dietary glucose not absorbed from intestine	Sometimes fatal; must use fructose as only dietary carbohydrate
Hartnup disease	Delayed intestinal absorption of tryptophan and related molecules	Cerebellum dysfunction; photosensitive dermatitis

Figure 6.13   Carrier-mediated transport displays the characteristics of saturation (illustrated by the transport maximum) and competition. Molecules X and Y compete for the same carrier, so that when they are present together the rate of transport of each is less than when either is present separately.

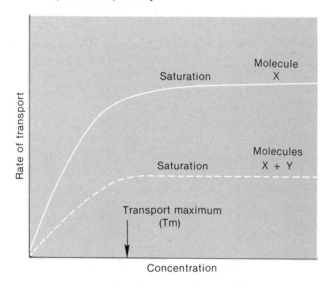

As the concentration of a transported molecule is increased, its rate of transport will also be increased—but only up to a maximum rate. Beyond this rate—called the *transport maximum* (or T_m)—further increases in concentration do not further increase transport rate. This indicates that the carriers have become saturated (see figure 6.13).

As an example of saturation, imagine a bus stop that is serviced once per hour by a bus that can hold a maximum of forty people (its "transport maximum"). If ten people wait at the bus stop, ten will be transported per hour. If twenty people wait at the bus stop, twenty will be transported per hour. This linear relationship will hold up to a maximum of forty people—if eighty people are at the bus stop, the transport rate will still be forty per hour.

The kidneys transport a number of molecules from the blood filtrate (which will become urine) back into the blood. Glucose, for example, is normally completely reabsorbed so that urine is normally glucose free. If the glucose concentration of the blood and filtrate is too high (a condition called *hyperglycemia*), however, the transport maximum can be exceeded. In this case, glucose will be found in the urine (a condition called *glycosuria*). This may result from eating too many sweets, or from inadequate action of the hormone *insulin* (in *diabetes mellitus*). Other molecules may be abnormally present in the urine as a result of a genetic defect in the production of transport carriers (see table 6.1).

Facilitated Diffusion

The membrane transport of glucose in most tissues (with the exception of the gut and kidney) is accomplished by **facilitated diffusion.** Facilitated diffusion, like simple diffusion, involves the transport of molecules across a cell membrane from the side of higher concentration to the side of lower concentration. Both simple and facilitated diffusion are examples of passive transport that does not require the expenditure of cell energy.

Unlike simple diffusion, however, the facilitated diffusion of glucose displays the properties of carrier-mediated transport: specificity, competition, and saturation.

Figure 6.14 Model of facilitated diffusion*.

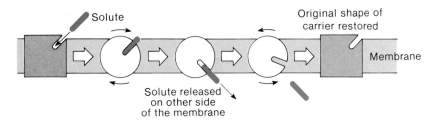

Solute

Original shape of
carrier restored

Membrane

Solute released
on other side
of the membrane

The diffusion of glucose through a cell membrane must therefore be mediated by carrier proteins. These may be conceptualized as specific revolving doors built into the membrane. The direction of the facilitated diffusion of glucose depends on the concentration gradient; the revolving doors can revolve both ways, but more glucose is moved in the direction of lower concentration.

Active Transport

There are some aspects of cell transport that cannot be explained by simple or facilitated diffusion. The epithelial lining of the intestine and of the kidney tubules, for example, move glucose from the side of lower to the side of higher concentration (from the lumen to the blood). Similarly, all cells extrude Ca^{++} into the extracellular environment, and by this means maintain an intracellular Ca^{++} concentration that is one thousand to ten thousand times lower than the extracellular Ca^{++} concentration.

Movement of molecules and ions against their concentration gradient, from lower to higher concentrations, requires the expenditure of cellular energy obtained from ATP. This type of transport is thus termed **active transport.** If a cell is poisoned with cyanide (which inhibits oxidative phosphorylation—see chapter 5), active transport is inhibited. This is the distinguishing feature of active transport—it is *energy dependent.* This contrasts with passive transport, which can continue even when metabolic poisons (such as cyanide) kill the cell by preventing formation of ATP.

Active transport, like facilitated diffusion, is carrier mediated. These carriers appear to be integral proteins that span the thickness of the membrane. According to one theory of active transport, the following events may occur: (1) the molecule or ion to be transported bonds to a specific "recognition site" on one side of the carrier protein; (2) this bonding stimulates breakdown of ATP, which in turn results in phosphorylation of the transport protein; (3) as a result of phosphorylation, the carrier protein

Figure 6.15 Model of active transport, showing hingelike motion of integral proteins subunits (see text for description).

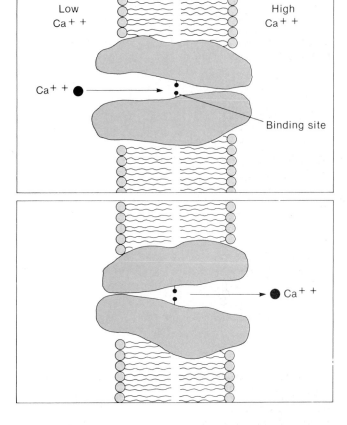

Low
Ca^{++}

High
Ca^{++}

Ca^{++}

Binding site

Ca^{++}

undergoes a conformation change, like that which occurs in enzyme proteins as a result of allosteric effects (see chapter 4); and (4) the carrier protein undergoes a hingelike motion, which releases the transported molecule or ion on the other side of the membrane. This model of active transport is illustrated in figure 6.15.

Figure 6.16   The Na^+/K^+ pump actively exchanges intracellular Na^+ for K^+. The carrier is itself an ATPase that breaks down ATP for energy. Dotted lines indicate direction of passive transport (diffusion); solid arrows indicate direction of active transport.

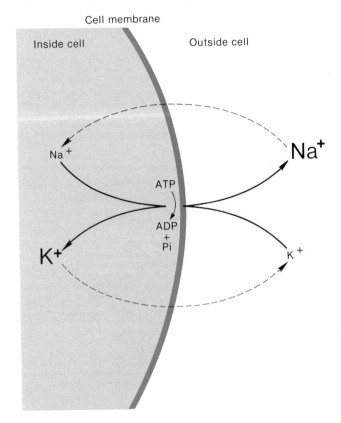

The Na^+/K^+ Pump   Active transport carriers are often referred to as "pumps." While some of these carriers transport only one molecule or ion at a time, others exchange one molecule or ion for another. The most important of these latter type of carriers is the **Na^+/K^+ pump.** This carrier protein, which is also an ATPase enzyme that converts ATP to ADP and P_i, actively extrudes three Na^+ ions from the cell as it transports two K^+ into the cell. This transport is energy dependent because Na^+ is more highly concentrated outside the cell and K^+ is more highly concentrated within the cell. Both ions, in other words, are moved against their concentration gradients.

All cells have numerous Na^+/K^+ pumps that are constantly active. This represents an enormous expenditure of energy used to maintain a steep gradient of Na^+ and K^+ across the cell membrane. This serves three known functions: (1) the steep Na^+ gradient is used to provide energy for the "co-transport" of other molecules (described in the next section); (2) activity of the Na^+/K^+ pumps can be adjusted (primarily by thyroid hormones) to regulate the resting calorie expenditure and basal metabolic rate; and (3) the Na^+ and K^+ gradients across the cell membranes of nerve and muscle cells are used to produce electrical impulses (described in chapter 7).

Co-transport   In co-transport, the energy needed for the "uphill" movement of a molecule or ion is obtained from the "downhill" transport of Na^+ into the cell. The active extrusion of Ca^{++} from some cells, for example, is coupled to the passive diffusion of Na^+ into the cell. Cellular energy, obtained from ATP, is not used to move Ca^{++} directly out of the cell, but energy is constantly required to maintain the steep Na^+ gradient. The inward diffusion of Na^+ due to this concentration gradient, in turn, is used to power the active extrusion of Ca^{++}.

A similar mechanism is used to actively transport glucose across the membranes of epithelial cells that line the gut and the kidney tubules. In this case, the inward diffusion of Na^+ is coupled to the cellular uptake of glucose (as well as amino acids) against their concentration gradients. Active transport in the kidney tubules accounts for as much as 6 percent of the body's energy expenditure at rest.

1. Draw a figure illustrating the simple diffusion of water and the facilitated diffusion of glucose through a membrane. List the similarities of these two forms of membrane transport.
2. List the three characteristics of facilitated diffusion that distinguish it from simple diffusion. Illustrate (using a graph) two of the distinguishing features of carrier-mediated transport.
3. Describe active transport, including co-transport in your description. In what way does active transport differ from facilitated diffusion?
4. Describe the functional significance of the Na^+/K^+ pump.

Figure 6.17 Co-transport of glucose. Glucose accumulates inside the cell against its concentration gradient, using energy derived from the passive transport (diffusion) of Na$^+$ into the cell. Metabolic energy is indirectly required because the steep electrochemical gradient for Na$^+$ is maintained by the Na$^+$/K$^+$ pump. Dotted arrows indicate direction of diffusion; solid arrows show direction of active transport.

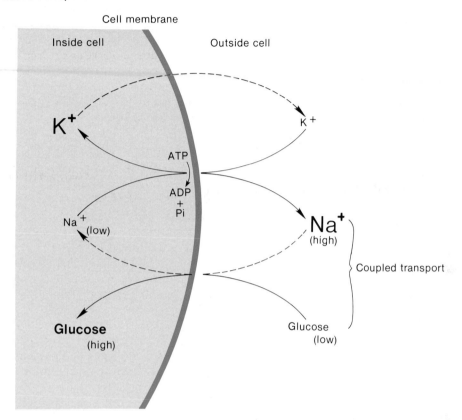

The Membrane Potential

Cellular proteins, and the phosphate groups of ATP and other nucleotides (see chapter 2), are negatively charged within the cell cytoplasm. These anions (negative ions) are "fixed" within the cell by the fact that they cannot penetrate the cell membrane. Since these negatively charged organic molecules cannot leave the cell, they attract positively charged inorganic ions (cations) from the extracellular fluid that are small enough to diffuse through the membrane pores. The distribution of small, inorganic cations (mainly K$^+$, Na$^+$, and Ca^{++}) between the intracellular and extracellular compartments, in other words, is influenced by the negatively charged fixed ions within the cell.

Since the cell membrane is much more permeable to K$^+$ than to any other cation, K$^+$ accumulates within the cell more than the others as a result of electrical attraction for the fixed anions. Instead of being evenly distributed between the intracellular and extracellular compartments, therefore, K$^+$ becomes more highly concentrated within the cell. In the human body, the intracellular K$^+$ concentration is 155 mEq/L compared to an extracellular concentration of 4 mEq/L (mEq = milliequivalents, which is the millimolar concentration multiplied by the valence of the ion—in this case, by 1).

Figure 6.18 Proteins, organic phosphates, and other organic anions that cannot leave the cell create a "fixed" negative charge on the inside of the membrane. This attracts positively charged inorganic ions (cations), which therefore accumulate within the cell at a higher concentration than in the extracellular fluid. The amount of cations that accumulate within the cell is limited by the fact that a concentration gradient builds up, which favors the diffusion of the cations out of the cell.

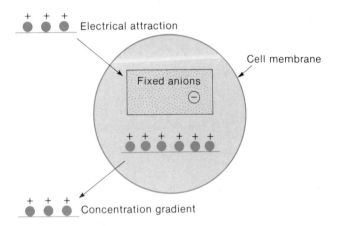

Figure 6.19 If K$^+$ was the only ion able to diffuse through the cell membrane, it would distribute itself between the intracellular and extracellular compartment until an equilibrium would be established. At equilibrium the K$^+$ concentration within the cell would be higher than outside the cell due to the attraction of K$^+$ for the fixed anions. Not enough K$^+$ would accumulate within the cell to neutralize these anions, however, so the inside of the cell would be 90 millivolts negative compared to the outside of the cell. This membrane voltage is the equilibrium potential (E$_K$) for potassium.

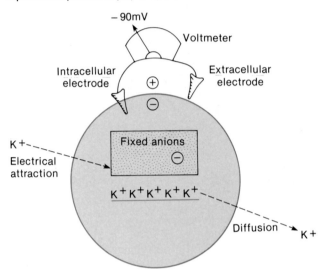

Even without the action of the Na$^+$/K$^+$ pump, therefore, K$^+$ would be more highly concentrated within the cell than in the extracellular fluid. The magnitude of the intracellular K$^+$ concentration depends on the amount of fixed anions, and on the K$^+$ concentrations that are normally found in the extracellular compartment (plasma and tissue fluid). An increase in the extracellular K$^+$ concentration above normal, for example, would cause a corresponding increase in the intracellular K$^+$ concentration.

If (1) the extracellular K$^+$ concentration is normal, and (2) K$^+$ is the only inorganic cation able to diffuse through the membrane, then an *equilibrium* will be established in which the intracellular and extracellular K$^+$ concentrations remain stable. There is no net movement of K$^+$ at this equilibrium because if a certain amount of K$^+$ was to move into the cell (by electrical attraction), an identical amount of K$^+$ would diffuse out of the cell (down its concentration gradient). At this equilibrium, the intracellular K$^+$ concentration would be such that only part of the fixed anions would be "neutralized."

At equilibrium, therefore, the inside of the cell membrane would have a higher concentration of negative charge than the outside of the membrane. The membrane, in other words, separates charges because the fixed anions cannot leave the cell and K$^+$ cannot accumulate sufficiently to neutralize these negative charges. The magnitude of the charge separation—or **potential difference**

between the two sides of the membrane—under these conditions is equal to ninety millivolts (90 mV). This is shown with a negative sign (as -90 mV) to indicate that the inside of the cell is the negative pole.

The potential difference of -90 mV, which would be developed if K$^+$ was the only diffusible cation, is sometimes called the K$^+$ **equilibrium potential** (abbreviated E$_K$). It should be noted, however, that this description is only a "first approximation" to the truth. Potassium *is* the most diffusible cation, but it is not the only cation that can cross the membrane. The cell membrane is slightly permeable to Na$^+$, and the concentration of Na$^+$ in the extracellular compartment (145 mEq/L) is much higher than it is in the cell (12 mEq/L). This concentration gradient, together with the electrical attraction of the fixed cellular anions, causes Na$^+$ to move into the cell at a rate permitted by the permeability of the cell membrane.

If the cell membrane were very permeable to Na$^+$, so much Na$^+$ would enter the cell that the membrane potential would actually reverse polarity and become positively charged on the inside (as discussed in chapter 7).

Figure 6.20 Because some Na⁺ leaks into the cell by diffusion, the actual resting membrane potential is less than the K⁺ equilibrium potential. As a result, some K⁺ diffuses out of the cell (shown by dotted lines).

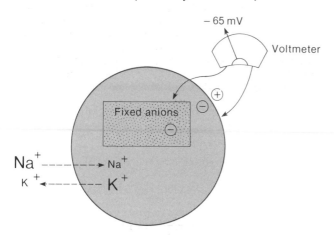

Figure 6.21 The concentrations of Na⁺ and K⁺ both inside and outside the cell do not change as a result of diffusion *(dotted arrows)* because of active transport *(solid arrows)* by the Na⁺/K⁺ pump. Since the pump transports three Na⁺ for every two K⁺, the pump itself helps to create a charge separation (a potential difference or voltage) across the membrane.

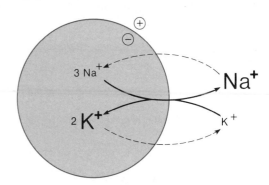

In a normal resting cell, however, the membrane permeability to Na^+ is so low that only a "trickle" of Na^+ can enter the cell. This small rate of Na^+ entry is sufficient, however, to reduce the *membrane potential* from E_K (−90 mV) to −85 to −65 mV, depending on the cell.

The Na^+/K^+ Pump   Since the membrane potential is reduced below E_k, as a result of Na^+ entry, some K^+ leaks out of the cell. The cell is *not* at equilibrium with respect to K^+ and Na^+ concentrations. Despite this, the concentrations of K^+ and Na^+ *are* maintained constant; this is due to the constant expenditure of energy in active transport by the Na^+/K^+ pump. The Na^+/K^+ pump "repairs the leak" and thus helps maintain the membrane potential. Actually, the Na^+/K^+ pump does more than simply repair the leak—since it transports three Na^+ ions out for every two K^+ ions that it moves in, it helps generate a potential difference across the membrane. As a result of all of these activities, a real cell has (1) a relatively constant intracellular concentration of K^+ and Na^+; and (2) a constant membrane potential (in the absence of stimulation in nerves or muscles) of −65 to −85 mV due to action of Na^+/K^+ pumps.

Hyperkalemia   Although changes in the extracellular concentration of many ions can affect the membrane potential, this potential is particularly sensitive to changes in plasma potassium. Since maintenance of a particular membrane potential is critical for the generation of electrical events in nerves and muscles (including the heart),

the body has a variety of mechanisms that serve to maintain plasma K^+ concentrations within very narrow limits. These mechanisms act primarily through the kidneys, which can excrete K^+ in the urine or reabsorb it into the blood. Excretion of K^+ is stimulated by hormones of the adrenal cortex (particularly aldosterone); if the adrenal glands of an experimental animal are removed, the animal will die as a result of accumulation of K^+ in the blood. This condition is called **hyperkalemia.**

When hyperkalemia occurs, the diffusion gradient that favors the extrusion of K^+ from the cell is reduced. As a result, more K^+ can enter the cell and neutralize more of the fixed negative charges. This reduces the membrane potential (brings it closer to zero) and thus alters the function of many organs, particularly the heart. For these reasons, the blood electrolyte concentrations are monitored very carefully in patients with heart or kidney disease.

1. Describe the "fixed anions," and explain their functional significance.
2. Define the term *membrane potential* and describe how it is measured.
3. Describe, in the hypothetical case in which K^+ is the only diffusible cation, how an equilibrium is established between K^+ concentrations and the membrane potential.
4. Explain why the true resting membrane potential of a cell is less than the equilibrium potential for K^+ (the E_K).
5. Explain the role of the Na^+/K^+ pump in the generation and maintenance of the membrane potential.

Summary

Simple Diffusion and Osmosis

I. Diffusion is the net movement of molecules or ions from regions of high to regions of low concentration.
   A. This is a type of passive transport—energy is provided by the thermal energy of the molecules, not by cellular metabolism.
   B. Diffusion stops when the concentration is equal on both sides of the membrane.

II. The rate of diffusion is dependent on a variety of factors.
   A. The rate of diffusion depends on the concentration difference on the two sides of the membrane.
   B. The rate depends on the permeability of the cell membrane to the diffusing substance.
   C. The rate of diffusion through a membrane is also directly proportional to the surface area of the membrane, which can be increased by such adaptations as microvilli.

III. Simple diffusion is the type of passive transport in which molecules, such as water, and inorganic ions, such as Na^+ and K^+, move through membrane pores.
   A. These polar substances cannot pass through the lipid barrier.
   B. Lipids such as steroid hormones can pass by simple diffusion directly through the phospholipid layers of the membrane.

IV. Osmosis is the simple diffusion of solvent (water) through a membrane that is more permeable to the solvent than it is to the solute.
   A. Water moves from the solution that is more dilute to the solution that has a higher solute concentration.
   B. Osmosis depends on a difference in total solute concentration, not on the chemical nature of the solute.
      1. The concentration of total solute (in moles) per kilogram (liter) of water is measured in osmolality units.
      2. The solution with the higher osmolality has the higher osmotic pressure.
      3. Water moves by osmosis from the solution of lower osmolality and osmotic pressure to the solution of higher osmolality and osmotic pressure.
   C. Solutions that have the same osmolality and osmotic pressure as plasma (such as 0.9 percent NaCl and 5 percent glucose) are said to be isotonic.
      1. Solutions with a lower osmotic pressure are hypotonic; those with a higher osmotic pressure are hypertonic.
      2. Cells in a hypotonic solution gain water and swell; those in a hypertonic solution lose water and shrink (crenate).
   D. The osmolality and osmotic pressure of the plasma is maintained within a normal range by the action of osmoreceptors in the hypothalamus.
      1. Increased osmolality stimulates osmoreceptors; decreased osmolality inhibits electrical activity of osmoreceptors.
      2. Stimulation of osmoreceptors causes thirst and the secretion of antidiuretic hormone (ADH) from the pituitary.
      3. ADH stimulates water retention by the kidneys, which serves to maintain a normal blood volume and osmolality.

Carrier-Mediated Transport

I. The passage of glucose, amino acids, and other substances through the cell membrane is mediated by carrier proteins in the cell membrane.
   A. Carrier-mediated transport exhibits the properties of specificity, competition, and saturation.
   B. The transport rate of molecules such as glucose reaches a maximum when the carriers are saturated—this maximum rate is called the transport maximum, or T_m.

II. Transport of molecules such as glucose from the side of higher to the side of lower concentration by means of membrane carriers is called facilitated diffusion.
   A. Like simple diffusion, this is passive transport—cellular energy is not required.
   B. Unlike simple diffusion, facilitated diffusion is specific, exhibits competition, and can be saturated.

III. Active transport of molecules and ions across a membrane requires the expenditure of cellular energy (ATP).
   A. In active transport, carriers move molecules or ions from the side of lower to the side of higher concentration.
   B. One example of active transport is the action of the Na^+/K^+ pump.
      1. Sodium is more concentrated on the outside of the cell, while potassium is more concentrated on the inside of the cell.
      2. The Na^+/K^+ pump helps to maintain these concentration differences by transporting Na^+ out of the cell and K^+ into the cell.
      3. Three Na^+ are transported out for every two K^+ ions that are transported into the cell.

The Membrane Potential

I. The cytoplasm of the cell contains negatively charged organic ions (anions) that cannot leave the cell—they are "fixed anions."
   A. These fixed anions attract K^+, which is the inorganic cation that can most easily pass through the cell membrane.
   B. As a result of this electrical attraction, the concentration of K^+ within the cell is greater than the concentration of K^+ in the extracellular fluid.
   C. If K^+ was the only diffusible ion, the concentrations of K^+ on the inside and outside of the cell would reach an equilibrium.
      1. At this point, the rate of K^+ entry (due to electrical attraction) would equal the rate of K^+ exit (due to diffusion).
      2. At this equilibrium, there would still be a higher concentration of negative charges within the cell (due to the fixed anions) than outside the cell.

3. At this equilibrium, the inside of the cell would be ninety millivolts (90 mV) negative compared to the outside of the cell—this is called the K^+ equilibrium potential (E_K).

D. The true membrane potential is less than E_K—it is usually -65 mV to -85 mV.

   1. This is because some Na^+ can also enter the cell.

2. Na^+ is more highly concentrated outside than inside the cell, and the inside of the cell is negative—these forces attract Na^+ into the cell.

3. The rate of Na^+ entry is generally slow because the membrane is usually not very permeable to Na^+.

II. The slow rate of Na^+ entry is accompanied by a slow rate of K^+ exit from the cell.

A. These changes are corrected by the active transport Na^+/K^+ pump, which maintains constant concentrations and a constant resting membrane potential.

B. There are numerous Na^+/K^+ pumps in all cells of the body that require a constant expenditure of energy.

C. The Na^+/K^+ pump itself contributes to the membrane potential because it pumps more Na^+ out than it pumps K^+ in (by a ratio of three to two).

Self-Study Quiz

1. Movement of water across a cell membrane occurs by:
   (a) active transport
   (b) facilitated diffusion
   (c) simple diffusion (osmosis)
   (d) all of these

2. Which of the following statements about the facilitated diffusion of glucose is TRUE?
   (a) there is a net movement from the region of low to the region of high concentration
   (b) carrier proteins in the cell membrane are required for this transport
   (c) this transport requires energy obtained from ATP
   (d) this is an example of co-transport

3. If a poison such as cyanide stops production of ATP, which of the following transport processes would cease?
   (a) movement of Na^+ out of a cell
   (b) osmosis
   (c) movement of K^+ out of a cell
   (d) all of these

4. Red blood cells crenate in:
   (a) a hypotonic solution
   (b) an isotonic solution
   (c) a hypertonic solution

5. Plasma has an osmolality of about 300 mOsm. Isotonic saline has an osmolality of:
   (a) 150 mOsm
   (b) 300 mOsm
   (c) 600 mOsm
   (d) none of these

6. A 0.5m NaCl solution and a 1.0m glucose solution:
   (a) have the same osmolality
   (b) have the same osmotic pressure
   (c) are isotonic to each other
   (d) all of these

7. The diffusible ion that is most important in the establishment of the membrane potential is:
   (a) K^+
   (b) Na^+
   (c) Ca^{++}
   (d) Cl^-

8. An increase in blood osmolality:
   (a) can occur as a result of dehydration
   (b) causes a decrease in blood osmotic pressure
   (c) is accompanied by a decrease in ADH secretion
   (d) all of these

9. In hyperkalemia, the membrane potential:
   (a) increases
   (b) decreases
   (c) is not changed

10. Which of the following statements about the Na^+/K^+ pump is TRUE?
   (a) Na^+ is actively transported into the cell
   (b) K^+ is actively transported out of the cell
   (c) an equal number of Na^+ and K^+ ions are transported with each cycle of the pump
   (d) the pumps are constantly active in all cells

7

The Nervous System:
Organization, Electrical Activity, and Synaptic Transmission

Objectives

By studying this chapter, you should be able to:

1. Describe the parts of a neuron and their functional significance

2. Describe the anatomical divisions of the nervous system

3. Explain how a myelin sheath is formed, and describe saltatory conduction

4. Describe the different types of neuroglial cells and their functions

5. Define depolarization, repolarization, and hyperpolarization

6. Describe the regulation of voltage-regulated ion gates and their significance in the production of action potentials

7. Explain how action potentials are regenerated along an axon

8. Describe the absolute and relative refractory periods, and explain their functional significance

9. Describe the events that occur between electrical excitation of an axon terminal and release of neurotransmitter

10. Describe how ACh and other neurotransmitters stimulate a postsynaptic cell

11. Describe the mechanisms of presynaptic and postsynaptic inhibition

12. Describe how ACh and catecholamine neurotransmitters are inactivated

13. Describe post-tetanic potentiation

*N*erve cells, or **neurons,** can be divided into three basic regions. The cell nucleus is contained within the *cell body,* or *perikaryon* (peri = around; karyon = nucleus), which is the "nutritional center" of the neuron where macromolecules are produced. Thin, highly branched cytoplasmic extensions called *dendrites* (dendr = tree) extend from the cell body and serve as a receptive area that transmits electrical impulses towards the cell body. The *axon, or nerve fiber,* which can be a fraction of a millimeter or three feet in length, transmits electrical impulses from the cell body to another cell.

Each neuron has one axon, which can produce a number of *collateral branches* that terminate at other neurons, muscle cells, or glands. The functional connection between an axon of a neuron and another cell is called a **synapse.** Neurons produce electrical impulses that are conducted from the origin of the axon—in the *axon hillock*—to its termination at the synapse.

The perikaryon contains granular, densely staining material known as *Nissl bodies,* which are not found in the dendrites or axon. The Nissl bodies are granular (rough) endoplasmic reticulum—an organelle involved in protein synthesis. Synthesis of proteins, therefore—as well as of other important organic molecules—occurs in the perikaryon. If an axon is severed, the part still attached to the perikaryon survives while the severed segment degenerates. Movement of molecules from the perikaryon through to the end of the axon can be demonstrated by experiments in which the axon is constricted—this results in outward bulging of the axon on the perikaryon side of the constriction (see figure 7.2).

Proteins and other molecules are transported through the axon at faster rates than could be achieved by simple diffusion. This rapid movement is produced by two different mechanisms: *axoplasmic flow* and *axonal transport.* Axoplasmic flow is the slower of the two, and results from rhythmic waves of contraction that push the axoplasm (cytoplasm of the axon) from the axon hillock to the nerve endings. Axonal transport is more rapid and more selective, and may occur in a retrograde as well as a forward direction. Indeed, such retrograde transport may be responsible for movement of herpes virus, rabies virus, and tetanus toxin from nerve terminals into cell bodies.

Types of Neurons

Neurons may be classified into three types, based on the number of their cytoplasmic extensions. *Pseudounipolar neurons* have only one "pole" or extension, but they are "pseudo" rather than truly unipolar because this single extension divides to produce a "T." Sensory neurons are pseudounipolar—one end of the process formed by the T receives sensory stimuli and produces nerve impulses; the other end of the T delivers these impulses to synapses within the spinal cord or brain.

Figure 7.1   The parts of a neuron.

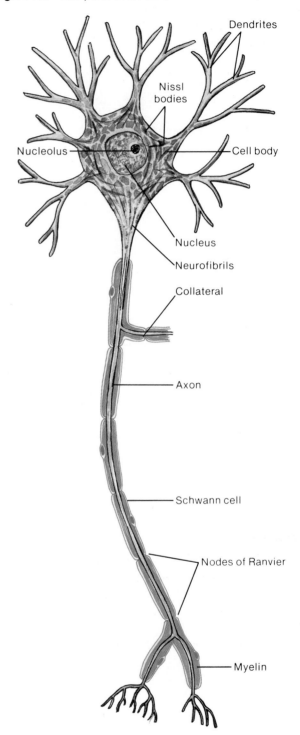

Figure 7.2 Effects of constriction experiment, demonstrating axoplasmic flow.

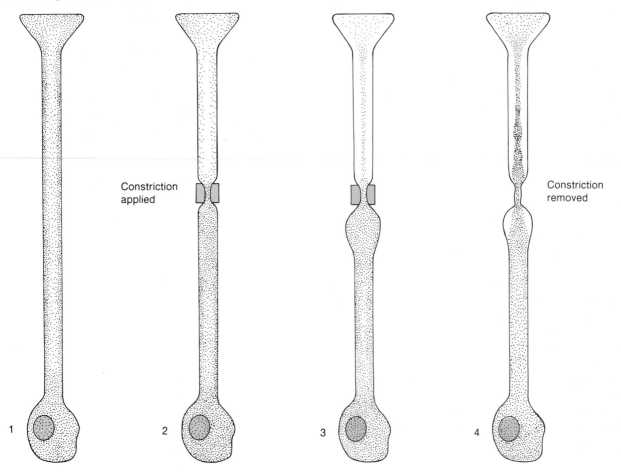

Constriction applied

Constriction removed

1   2   3   4

Table 7.1 Comparison of axoplasmic flow with axonal transport.

Axoplasmic Flow	Axonal Transport
Transport rate comparatively slow (1–2mm/day)	Transport rate comparatively fast (200–400mm/day)
Molecules transported only from cell body	Molecules transported from cell body to axon endings and in reverse direction
Bulk movement of proteins in axoplasm, including microfilaments and tubules	Transport of specific proteins, mainly of membrane proteins and acetylcholinesterase
Transport accompanied by peristaltic waves of axon membrane	Transport dependent on cagelike microtubule structure within axon, and on actin and Ca^{++}

Figure 7.3   Three different types of neurons.

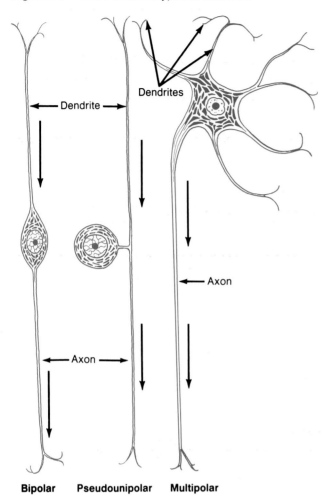

Bipolar   Pseudounipolar   Multipolar

Figure 7.4   Major anatomical divisions of the nervous system.

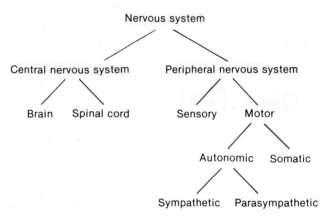

Bipolar neurons have two extensions—one dendrite (which can branch extensively) and one axon—which are located at opposite sides of the cell body. Bipolar neurons are found in the retina of the eye and in the cochlear, vestibular, and olfactory nerves. *Multipolar neurons* are the most common type of neurons within the brain and spinal cord. These neurons have many dendrites and so can receive synaptic input from the axons of many other neurons (some of these neurons receive input from as many as one thousand other neurons).

Organization of the Nervous System

The nervous system is divided into main parts: the **central nervous system (CNS)** and the **peripheral nervous system.** The central nervous system consists of the brain and the spinal cord; the peripheral nervous system consists of axons and entire neurons that extend out of the CNS. Neurons whose cell bodies, dendrites, and axons are located entirely within the CNS are known as *association neurons,*

interneurons, or *internuncial neurons.* Relatively well-defined groupings of cell bodies within the CNS are called *nuclei.* Collections of axons that conduct impulses up, down, or across the CNS are called *tracts.*

The peripheral nervous system consists of cranial and spinal *nerves,* with their associated *ganglia.* A nerve is a collection of axons, packaged within the same connective tissue sheath, within the peripheral nervous system. In an anatomical dissection, a nerve looks like a white string. Functionally, however, it is more like a telephone cable that contains many separate wires (axons) that conduct information (nerve impulses) in two directions. Nerve fibers of **sensory neurons** conduct impulses from sense organs into the CNS (these are thus called *afferent fibers*); axons that transmit impulses from the CNS to effector organs (muscles and glands) are *efferent fibers,* which are part of **motor neurons.**

There are two subgroupings of motor neurons—*somatic motor* (which innervate skeletal muscles) and *autonomic motor* (which innervate involuntary effector organs—smooth muscle, cardiac muscle, and glands). There are two subdivisions of autonomic nerves: *sympathetic* and *parasympathetic.* These will be discussed in detail in chapter 9.

The term *ganglia* usually refers to collections of nerve cell bodies located outside the CNS. There are two types of ganglia in the peripheral nervous system: the *dorsal root ganglia* and the *autonomic ganglia.* The dorsal root ganglia contain the cell bodies of sensory neurons, which are located outside the dorsal surface of the spinal cord. The autonomic ganglia contain the cell bodies of sympathetic and parasympathetic neurons (see figure 7.5 for a schematic view; these will be discussed in detail in chapters 8 and 9).

Table 7.2 Anatomical terms used in describing the nervous system.

Term	Definition
Central nervous system (CNS)	Brain and spinal cord.
Peripheral nervous system (PNS)	Nerves and ganglia.
Sensory nerve fiber	Nerve that transmits impulses from sensory receptor into CNS. (An afferent fiber.)
Motor nerve fiber	Nerve that transmits impulses from CNS to an effector organ (e.g., muscle). (Efferent.)
Nerve	Cablelike collection of nerve fibers. May be "mixed" (contain both sensory and motor fibers).
Somatic motor nerve	Nerve that stimulates contraction of skeletal muscles.
Autonomic motor nerve	Nerve that stimulates contraction (or inhibits contraction) of smooth muscle and cardiac muscle, and stimulates secretion of glands.
Ganglion	Collection of neuron cell bodies located outside the CNS.
Nucleus	Groupings of neuron cell bodies within the CNS.
Tract	Collections of nerve fibers that interconnect regions of the CNS.

Figure 7.5 The relationship between sensory and motor fibers of the peripheral nervous system (PNS) and the central nervous system (CNS).

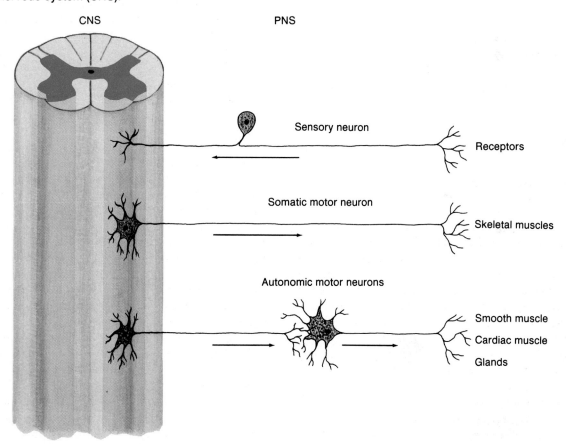

Table 7.3   Types of neuroglial cells and their functions.

Neuroglia	Functions
Schwann cells	Surrounds axons of all peripheral nerve fibers, forming neurilemmal sheath, or sheath of Schwann. Wraps around many peripheral fibers to form myelin sheaths.
Oligodendrocytes	Forms neurilemmal sheath around axons in CNS. Formation of myelin sheaths around central axons produces "white matter" of CNS.
Astrocytes	Perivascular foot process covers capillaries within brain and contributes to the blood-brain barrier.
Microglia	Amoeboid cells within CNS that are phagocytic.
Ependyma	Forms epithelial lining of brain cavities (ventricles) and central canal of spinal cord. Covers tufts of capillaries to form chorid plexus—structures that produce cerebrospinal fluid.

Neuroglia

Unlike other organs that are "packaged" in connective tissue (which is derived from mesoderm—the middle embryonic tissue layer), the nervous system is held together by *neuroglia* (literally, "nerve glue") that is derived from the same embryonic tissue layer (ectoderm) as the nerve cells. Neuroglial (or, more simply, glial) cells far outnumber neurons in the nervous system. Unlike neurons in the CNS, neuroglial cells can divide throughout life. Most of the brain tumors in adults, in fact, are composed of glial cells. Although glial cells do not conduct electrical impulses, they do modify the electrical activity of neurons, and they are believed to have many other important functions in the nervous system (see table 7.3).

There are five types of neuroglial cells: (1) *Schwann cells,* which form the myelin sheaths around peripheral axons (see the next section); (2) *oligodendrocytes,* which form myelin sheaths around central axons; (3) *astrocytes,* which help regulate the passage of molecules from the blood to the brain, and thus contribute to the "blood-brain barrier," (4) *microglia,* which are phagocytic cells that migrate through the CNS and remove foreign and degenerated material by phagocytosis; and (5) *ependyma,* which form a lining epithelium around the brain cavities (ventricles) and around the central canal of the spinal cord.

Neurilemmal and Myelin Sheaths

Some types of neuroglial cells form sheaths around nerve fibers. In the peripheral nervous system, this function is served by *Schwann cells;* in the central nervous system, *oligodendrocytes* serve a similar function. When these cells simply surround an axon, they form a living sheath called a *neurilemma* (or *sheath of Schwann* in peripheral axons). If this is the only covering around an axon, the nerve fiber is said to be *unmyelinated.*

Axons that are less than two micrometers in diameter are usually unmyelinated. Axons that are larger in diameter are generally surrounded by a **myelin sheath** in addition to their neurilemmal sheath. Myelin sheaths are formed by Schwann cells (in the peripheral nervous system) and oligodendrocytes (in the CNS). These cells wrap themselves around an axon, so that successive layers of their cell membranes insulate the axon the way electrical tape insulates a wire.

Unlike electrical tape, which is usually wound at an angle so that the entire surface of the wire is covered, the myelin sheath is discontinuous. The gaps between adjacent Schwann (or oligodendrocyte) membrane wrappings produce gaps in the insulation; these are known as the *nodes of Ranvier.* As each of these glial cells wraps itself

Figure 7.6   Electron micrograph of unmyelinated *(left)* and myelinated axons *(right)*.

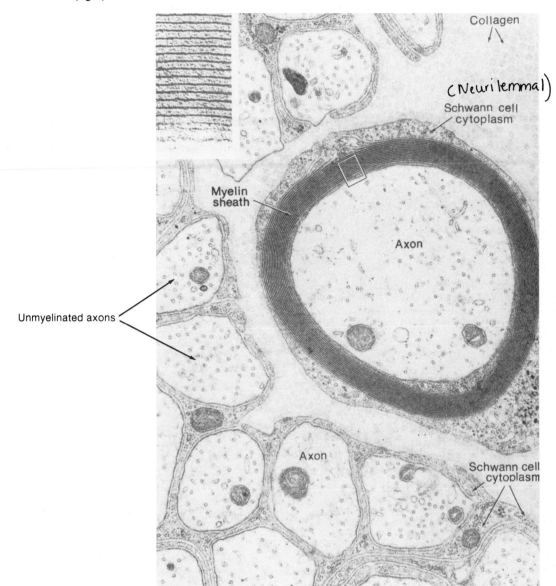

Figure 7.7   Formation of a myelin sheath in a peripheral axon. The myelin sheath is formed by successive wrappings of Schwann cell membranes, leaving most of the Schwann cell cytoplasm outside the myelin. The neurilemmal sheath of Schwann cells is thus located outside the myelin sheath.

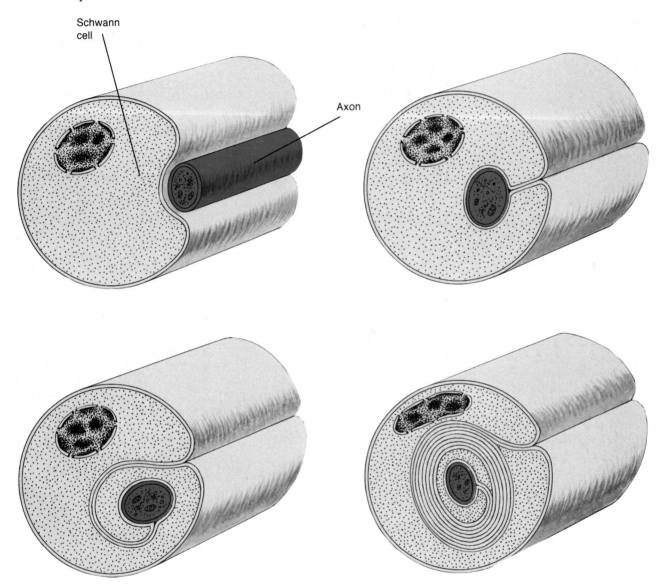

Schwann cell

Axon

Figure 7.8 Electron micrograph of a nerve fiber *(Nf)* surrounded by a myelin sheath *(My)* produced from the membranes of Schwann cells *(SC)*. In the center of the picture, the myelin sheath is interrupted to form a node of Ranvier; this is indicated as a pore *(Pr)* between the myelin layers on both sides *(X)*. Endoplasmic reticulum *(ER)* and mitochondria *(M)* are seen within the nerve fiber, and a basement membrane *(BM)* and connective tissue *(CT)* are shown outside the sheath of Schwann cells.

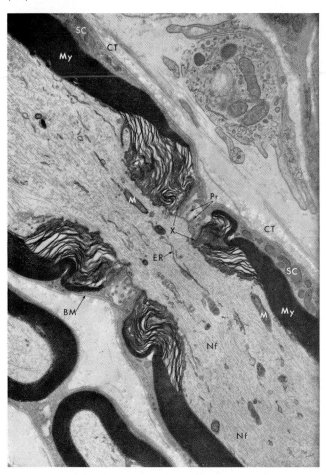

around the axon, the cytoplasm becomes squeezed towards the outermost regions (much as toothpaste that is rolled from the bottom of a tube is squeezed towards the top). The neurilemmal sheath of Schwann cells or oligodendrocytes is thus outside the myelin sheath in myelinated nerve fibers.

The lipid membranes of the myelin sheath give a white appearance to areas of the CNS that contain closely packed axons. Since there is no myelin sheath around dendrites and cell bodies, areas of the CNS that are rich in cell bodies have a grey appearance. The *grey matter* in the brain is located near the surface, with *white matter* underneath. This relationship between grey and white matter is reversed in the spinal cord (see figure 7.9).

Astrocytes and the Blood-Brain Barrier

Unlike other organs, the brain does not obtain molecules from blood plasma by a nonspecific filtering process through capillary walls. There is a *blood-brain barrier* that very selectively restricts the passage of certain molecules into the brain. This blood-brain barrier presents difficulty in the chemotherapy of brain diseases because drugs that could enter other organs may not be able to enter the brain. In the treatment of Parkinson's disease, for example, patients who need a chemical called dopamine in the brain (see chapter 8) must be given a precursor molecule—L-dopa—because dopamine cannot cross the blood-brain barrier as effectively as L-dopa.

Capillaries in the brain, unlike those of most other organs, do not have pores between adjacent endothelial cells (the cells that compose the walls of capillaries). Molecules in the blood within brain capillaries must thus be moved through the endothelial cells—by active transport, endocytosis, and exocytosis—to the brain. Glial cells called *astrocytes* contribute to the blood-brain barrier by means of numerous extensions called *perivascular feet,* which

Figure 7.9 Arrangement of grey and white matter in the brain and spinal cord.

White matter

Grey matter

Pedigo

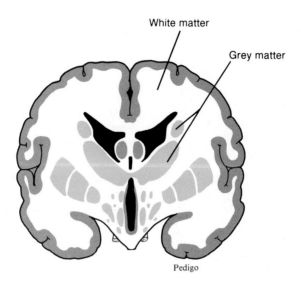

(a)

Spinal cord

White matter

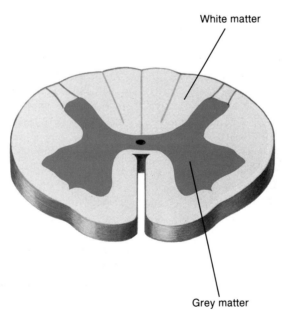

Grey matter

(b)

Figure 7.10 Diagram showing the perivascular feet of astrocytes (a type of neuroglial cell) that cover most of the surface area of brain capillaries.

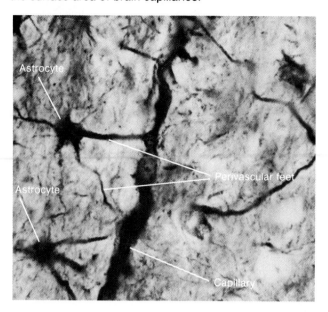

surround most of the outer surface of brain capillaries (see figure 7.10).

Neuroglia and Regeneration of a Cut Axon

When an axon in a peripheral nerve is cut, the distal region of the axon that was severed from the cell body degenerates, and is phagocytosed by Schwann cells. The Schwann cells then form a *regeneration tube,* as the part of the axon that is connected to the cell body begins to grow and exhibit amoeboid movement. The Schwann cells of the regeneration tube are believed to secrete chemicals that attract the growing axon tip, and the tube itself helps to guide the regenerating axon to its proper destination. The CNS lacks Schwann cells, and central axons are generally believed to have a much more limited ability to regenerate than peripheral axons.

In addition to the role of Schwann cells in nerve regeneration, and of astrocytes in the blood-brain barrier, neurons and neuroglial cells are believed to interact with each other in many complex and poorly understood ways. Neuroglial cells may aid the development of the nervous system, for example, by secreting a polypeptide called *nerve growth factor.* Glial cells have been shown to take up neurotransmitter chemicals released by axons at synapses, and to "buffer" the ionic environment of axons by taking up K^+ released during nerve-impulse production. The physiological significance of these and other glial cell functions are currently under investigation.

1. Construct a chart or outline of the anatomic divisions of the nervous system, including the different types of neurons in each division.
2. List the different types of neuroglial cells found in the CNS and in the peripheral nervous system, and give a brief description of their functions.
3. Briefly describe the function of dendrites, cell bodies, and axons.
4. Describe a neurilemmal sheath in an unmyelinated nerve fiber. Describe the neurilemmal and myelin sheath of a myelinated nerve fiber and how the myelin sheath is formed.

Electrical Activity in Nerve Fibers

While all cells in the body maintain a potential difference across their cell membrane (see chapter 6), only nerve and muscle cells can alter this membrane potential in response to appropriate stimuli. These alterations are achieved by varying the membrane permeability to different ions. Tissues that can respond to stimuli in this manner are said to be *irritable.*

An increase in membrane permeability to a specific ion results in diffusion of that ion down its concentration gradient (either into or out of the cell). These *ion currents* occur across only limited patches of membrane (located fractions of a millimeter apart), where specific ion channels are located. Changes in the potential difference across the membrane at these points can be measured by the voltage developed between two electrodes: one placed inside the cell, the other placed outside the cell membrane

Figure 7.11 The difference in potential (in millivolts) between an intracellular and extracellular recording electrode is displayed on an oscilloscope screen. The resting membrane potential (rmp) of the axon may be reduced (depolarization) or increased (hyperpolarization).

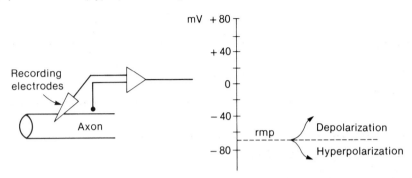

at the region being recorded. The voltage between these two *recording electrodes* can be visualized by connecting these electrodes to an oscilloscope (see figure 7.11).

In an oscilloscope, electrons from a cathode ray "gun" are sprayed across a fluorescent screen, producing a line of light. Changes in the potential difference between the two recording electrodes produces a deflection of this line. The oscilloscope can be calibrated in such a way that an upward deflection of this line indicates that the inside of the membrane has become less negative (or more positive) compared to the outside of the membrane. A downwards deflection of the line, conversely, indicates that the inside of the cell has become more negative.

If both recording electrodes were placed outside of the cell, the potential difference between the two would be zero (there is no charge separation). When one of the two electrodes penetrates the cell membrane, the oscilloscope shows that the intracellular electrode is electrically negative with respect to the extracellular electrode. If appropriate stimulation causes positive charges to flow into the cell, the line would deflect upwards—this change is called **depolarization** (or hypopolarization) because the potential difference between the two recording electrodes is reduced. If the inside of the membrane becomes more negative as a result of stimulation, the line on the oscilloscope would deflect downwards; this is called **hyperpolarization.**

Ion "Gating" in Nerve Fibers
As discussed in chapter 6, the intracellular potassium concentration is much higher than the concentration of potassium in the extracellular fluid. The sodium concentration, conversely, is about twenty times higher in extracellular fluid than it is within the cell. These concentration differences are a result of (1) electrical attraction by negatively charged proteins and organic phosphates that are "fixed" within the cell; (2) the greater permeability of the

resting cell membrane to K^+ than to Na^+; and (3) active transport by the Na^+/K^+ pump. As a result of this ion distribution, the inside of a neuron is about sixty-five millivolts negative (-65 mV) compared to the extracellular fluid.

The permeability of the membrane to K^+, Na^+, and other ions is regulated by proteins within the membrane that function as *gates*. At the **resting membrane potential** of an axon (equal to about -65 mV), the Na^+ gates are almost—but not completely—closed. The K^+ gates, however, are slightly open, so that the membrane is more permeable to K^+ than to Na^+.

Depolarization of a small region of an axon can be experimentally induced by a pair of stimulating electrodes. If a pair of recording electrodes are placed in the same region (one electrode within the axon and one outside), an upward deflection of the oscilloscope line will be observed as a result of this depolarization. If a certain level of depolarization is achieved (from -65 mV to -55 mV) by this artificial stimulation, a sudden and very rapid change in the membrane potential will be observed. This is due to the fact that depolarization to a threshold level causes the *opening of Na^+ gates*. Sodium, as a result, diffuses into the cell, down its concentration gradient.

A fraction of a second after the Na^+ gates open, they close again. At about this time, the depolarization stimulus causes *opening of the K^+ gates*. Membrane permeability to K^+, in other words, greatly increases, and K^+ diffuses out of the cell down its concentration gradient. The Na^+ and K^+ gates in an axon membrane are thus *voltage regulated*—they are closed (in the case of Na^+) or partially closed (for K^+) at the resting membrane potential (-65mV), but open for a brief time when the membrane is depolarized to a threshold value (about -55 mV).

Table 7.4 Representative values for intracellular and extracellular ion concentrations (in millimoles/L) and equilibrium potentials (in millivolts).

		Inside Cell	Outside Cell	Equilibrium Potential
K⁺	Squid axon	410	22	−74
	Frog muscle	125	3	−98
	Human erythrocyte	136	5	−86
Na⁺	Squid axon	49	460	+56
	Frog muscle	15	110	+50
	Human erythrocyte	19	155	+55

Figure 7.12 Depolarization of an axon has two effects: (1) Na⁺ gates open and Na⁺ diffuses into the cell and (2) after a brief period, K⁺ gates open and K⁺ diffuses out of the cell. Inward diffusion of Na⁺ causes further depolarization—this causes further opening of Na⁺ gates in a positive feedback ⊕ fashion. Opening of K⁺ gates and outward diffusion of K⁺ make the inside of the cell more negative, and thus has a negative feedback effect ⊖ on the initial depolarization.

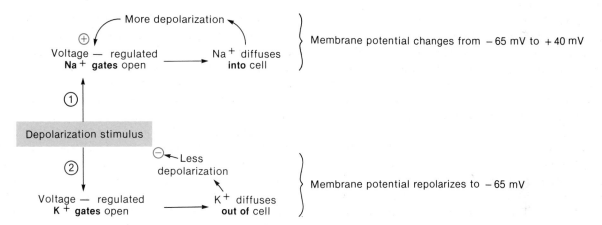

Action Potentials

When the axon membrane has been depolarized to a threshold level—by stimulating electrodes, for example—the Na⁺ gates open as previously described. This permits Na⁺ to enter the axon by diffusion, which further depolarizes the membrane (makes the inside less negative or more positive). Since the Na⁺ gates of the axon are voltage regulated, this depolarization causes a further increase in Na⁺ permeability. A positive feedback loop is thus created, as increased Na⁺ entry into the cell causes more depolarization, and more depolarization causes increased Na⁺ diffusion into the cell.

The effect of depolarization of the axon membrane on K⁺ permeability has the opposite effect. Depolarization—with a slight time delay—causes an increase in K⁺ permeability (opens the K⁺ gates). Potassium, as a result, moves out of the cell and makes the inside of the axon membrane more negative. This process is called **repolarization** and helps to bring the membrane potential back towards the resting level of −65 mV.

Figure 7.13   An action potential *(upper figure)* is produced by an increase in sodium conductance followed, with a short time delay, by an increase in potassium conductance *(bottom figure)*. This drives the membrane potential first towards the sodium equilibrium potential and then towards the potassium equilibrium potential.

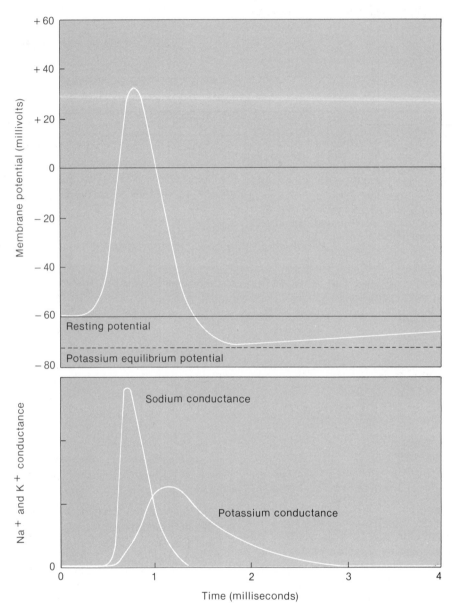

The stimulus of depolarization thus has two effects on the axon: (1) Na^+ gates open and Na^+ diffuses into the cell—this makes the inside of the cell positively charged compared to the outside of the cell (up to about $+40\,mV$); and (2) K^+ gates open and K^+ diffuses out of the cell, helping to reestablish the resting membrane potential of $-65\,mV$. Depolarization thus causes a transient increase in Na^+ permeability (and depolarization) followed by a transient increase in K^+ permeability (and repolarization). The effects of depolarization on the passage of Na^+ and K^+ ions through the axon membrane—and on the membrane potential of the axon—are shown in figure 7.13.

All-or-None Law

Once a region of axon membrane has been depolarized to a threshold value, the positive feedback effect of depolarization on Na^+ permeability, and of Na^+ permeability on depolarization, causes the membrane potential to shoot towards about $+40\,mV$. It does not normally become more positive because of the fact that the Na^+ gates quickly close and the K^+ gates open. The length of time that the Na^+ and K^+ gates stay open is independent of the strength of the depolarization stimulus.

The amplitude of this potential change—which is called an **action potential** or **nerve impulse**—is thus fixed. The amplitude of the action potential, in other words, is **all or none.** When depolarization is below a threshold value, the voltage-regulated gates are closed; when depolarization reaches threshold, a maximum potential change (the action potential) is produced. Since the change from $-65\,mV$ to $+40\,mV$, and back to $-65\,mV$, lasts only about three msec (milleseconds—thousandths of a second), the image of the action potential in an oscilloscope looks like a spike. Action potentials are therefore sometimes called *spike potentials.*

Both the rising (depolarization) and falling (repolarization) phases of an action potential are, in summary, produced by gating the *diffusion* of ions down their opposing concentration gradients. Since the gates are open for a fixed period of time, the duration of each action potential is about the same (3 msec in an axon). Likewise, since the concentration gradient for Na^+ is relatively constant, the amplitude of each action potential is about the same in all axons at all times (from $-65\,mV$ to $+40\,mV$, or about 100 mV total amplitude).

Figure 7.14 Recordings from a single sensory fiber of a sciatic nerve of a frog stimulated by varying degrees of stretch of the gastrocnemius muscle. Note that increasing amounts of stretch (induced by increasing weights attached to the muscle) result in increased frequency of action potentials.

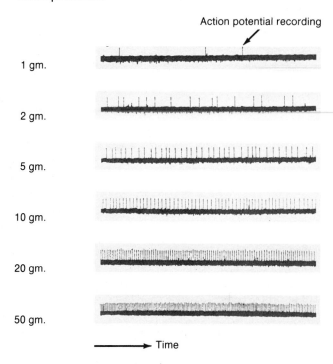

Coding for Stimulus Intensity

If one depolarization stimulus is greater than another, therefore, the greater stimulus strength is not coded by a greater amplitude of action potentials. The code for stimulus strength in the nervous system is not *AM* (amplitude modulated). When a greater stimulus strength is applied to a neuron, identical action potentials are produced more frequently (more are produced per minute). Therefore the code for stimulus strength in the nervous system is frequency-modulated *(FM)*. This is shown in figure 7.14.

When an entire collection of axons (in a nerve) is stimulated, different axons will be stimulated at different stimulus intensities. A low-intensity stimulus will only activate those few fibers with low thresholds, while high-intensity stimuli can activate fibers with higher thresholds. As the intensity of stimulation increases, more and more fibers will become activated. This process—called *recruitment*—represents another mechanism by which the nervous system can code for stimulus strength.

Figure 7.15 The absolute and relative refractory periods. While a segment of axon is producing an action potential, the membrane is absolutely or relatively resistant (refractory) to further stimulation.

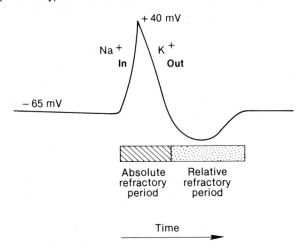

Refractory Periods

If a stimulus of a given intensity is maintained at one point of an axon (depolarizing it to threshold), action potentials will be produced at that point at a given frequency (number per minute). As the stimulus strength is increased, the frequency of action potentials produced at that point will increase accordingly. As action potentials are produced with increasing frequency, the lag time between successive action potentials will decrease—but only up to a minimum time interval. The interval between successive action potentials will never get so short that one action potential is produced before the preceeding one has finished.

During the short time that a patch of axon membrane is producing an action potential, it is incapable of responding—or *refractory*—to further stimulation. If a second stimulus is applied, for example, while the Na⁺ gates are open in response to a first stimulus, the second stimulus cannot have any effect—the gates are already open. During the time that the Na⁺ gates are open, therefore, the membrane is in an **absolute refractory period,** and cannot respond to any subsequent stimulus. If a second stimulus is applied while the K⁺ gates are open (and the membrane is in the process of repolarizing), the membrane is in a **relative refractory period.** During this time only a very strong stimulus can depolarize the membrane and produce a second action potential (see figure 7.15).

As a result of the fact that the cell membrane is refractory during the time it is producing an action potential, each action potential remains a separate, all-or-none event. In this way, as a continuously applied stimulus increases in intensity, its strength can be coded strictly by the frequency of action potentials it produces at each point of an axon membrane.

One might think that as an axon produces a large number of action potentials, the relative concentrations of Na⁺ and K⁺ would be changed in the extracellular and intracellular compartments. This is not true. In a typical mammalian axon that is one micrometer in diameter, for example, only one intracellular K⁺ ion in three thousand would be exchanged for a Na⁺. Since a typical neuron has about one million Na⁺/K⁺ pumps, able to transport nearly 200 million ions per second, these small changes can be quickly corrected.

Cable Properties of Neurons

If a pair of stimulating electrodes produce a depolarization that is too weak to cause opening of voltage-regulated Na⁺ gates—that is, if the depolarization is below threshold (about −55 mV)—the change in membrane potential will be *localized* to within one to two millimeters of the point of stimulation. For example, if the stimulus causes depolarization from −65 mV to −60 mV at one point, and the recording electrodes are placed only three millimeters away from the stimulus, the membrane potential recorded will remain at −65 mV (the resting potential). The axon is thus a very poor conductor compared to metal wires.

The ability of a neuron to transmit charges through its cytoplasm is known as its *cable properties*. These cable properties are quite poor because there is a high internal resistance to the spread of charges, and because many charges leak out of the axon through its membrane. If an axon had to conduct only through its cable properties, therefore, no axon could be more than a millimeter in length. The fact that some axons are a meter or more in length suggests that conduction of nerve impulses does not rely only on the cable properties of the axon.

Figure 7.16 Conduction of nerve impulse (action potential) in unmyelinated nerve fiber (axon). Each action potential "injects" positive charges that spread to adjacent regions. The region that has previously produced an action potential is refractory. The previously unstimulated region is partially depolarized. As a result, its voltage-regulated Na⁺ gates open, repeating the process.

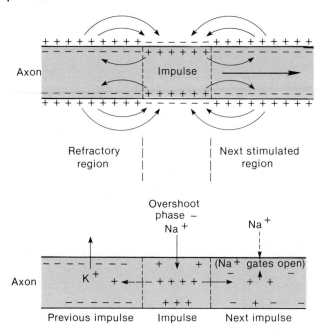

Conduction of Nerve Impulses

When stimulating electrodes artificially depolarize one point of an axon membrane to a threshold value, voltage-regulated gates open and an action potential is produced at that small region of axon membrane. For about the first millisecond of the action potential, when the membrane voltage changes from -65 mV to $+40$ mV, a current of Na⁺ enters the cell by diffusion due to opening of the Na⁺ gates. Each action potential thus "injects" positive charges in the form of Na⁺ ions into the axoplasm.

These positively charged Na⁺ ions are conducted, by the cable properties of the axon, to an adjacent region that still has a membrane potential of -65 mV. Within the limits of the cable properties of the axon (one to two millimeters), this helps to depolarize the adjacent region of axon membrane. When this adjacent region of membrane reaches a threshold level of depolarization, it too produces an action potential as its voltage-regulated gates open. Each action potential thus acts as a stimulus for the production of another action potential at the next region of membrane that contains voltage-regulated gates.

Conduction in an Unmyelinated Nerve Fiber

In an unmyelinated axon, which is not insulated by successive wrappings of Schwann cell (or oligodendrocyte) membranes, new action potentials are produced only tenths of a micrometer apart along the axon membrane. This is because voltage-regulated gates are located only a fraction of a micrometer apart in the axon membrane, and because the cablelike spread of depolarization induced by Na⁺ influx during one action potential can open voltage-regulated gates in the next patch of membrane (see figure 7.16).

Notice that action potentials are not really *conducted*, although it is convenient to use that word. Each action potential is a separate, complete event (flow of Na⁺ and K⁺ across the membrane to produce an all-or-none change in membrane potential) that is repeated—or *regenerated*—along the axon's length. The action potential that is produced at the end of the axon is thus a completely new event that was produced by depolarization from the previous action potential. The last action potential has the same amplitude as the first, which was produced at the beginning of the axon (at the axon hillock). Action potentials are thus said to be **conducted without decrement** (without decreasing in amplitude).

Since action potentials or nerve impulses are produced at every fraction of a micrometer of axon membrane in an unmyelinated fiber, the conduction rate is relatively slow. This conduction rate is greater when the axon increases in diameter, because a thicker fiber conducts depolarization (by cable properties) faster than a thinner fiber. The conduction rate is also faster in a myelinated axon.

Figure 7.17 Conduction of the nerve impulse in a myelinated nerve fiber. Since the myelin sheath prevents inward Na⁺ current, action potentials can only be produced at the interruptions in the myelin sheath, or Nodes of Ranvier. This "leaping" of the action potential from node to node is known as *saltatory conduction.*

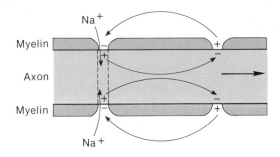

Table 7.5 Examples of conduction velocities and functions of mammalian nerves of different diameters.

Diameter (μm)	Conduction Velocity (m/sec)	Examples of Functions Served
12–22	70–120	Somatic motor fibers
5–13	30–70	Sensory: muscle position, pressure
3–8	15–40	Sensory: touch, pressure
1–5	12–30	Sensory: pain, temperature
1–3	3–15	Autonomic fibers to ganglia
0.3–1.3	0.7–2.2	Autonomic fibers to muscles

1. Define the resting membrane potential, depolarization and repolarization. Draw a figure that illustrates these terms.
2. Describe how the rate of Na⁺ and K⁺ diffusion is regulated. Describe the effect that increased Na⁺ and K⁺ conductance have on the resting membrane potential.
3. Define an action potential. Draw a figure of an action potential, and label which parts of this figure are produced by increased Na⁺ conductance and which parts are produced by increased K⁺ conductance.
4. Define the all-or-none law of action potentials. Describe the effect of increased stimulus strength on action potential production.
5. Define the absolute and relative refractory periods, and describe the functional significance of refractory periods in neuron physiology.
6. Describe the conduction of action potentials in unmyelinated nerve fibers. Describe saltatory conduction in myelinated fibers.

Conduction in a Myelinated Nerve Fiber

The myelin sheath, which is composed of successive wrappings of Schwann (or oligodendrocyte) cell membranes, provides insulation for the axon. Inward diffusion of Na⁺ and outward diffusion of K⁺ cannot occur across the layers of myelin sheath, even when threshold levels of depolarization are produced. In a myelinated fiber, action potentials can only be produced at the breaks in the myelin sheath—that is, at the nodes of Ranvier.

Since action potentials can only be produced at the nodes of Ranvier, and since the cable properties of the axon can only conduct depolarization a very short distance (1 to 2 mm), it follows that the nodes must be no further than about one millimeter apart. Action potentials thus seem to "leap" from node to node; this is called **saltatory conduction** (salt = leap).

Saltatory conduction in a myelinated nerve fiber allows a *faster rate of conduction* than is possible in an unmyelinated fiber. Conduction rate, as previously discussed, is also directly related to fiber diameter. For a fiber of a given diameter, the presence of a myelin sheath (and thus saltatory conduction) increases conduction rate by a factor of about twenty-five. Conduction rate in the human nervous system varies from 1.0 m/sec (for thin, unmyelinated fibers that mediate slow visceral responses) to greater than 100 m/sec (225 miles per hr) for thick myelinated fibers involved in quick stretch reflexes of skeletal muscles (see table 7.5).

Synaptic Transmission

Synapses are the functional connections between a neuron and a second cell. In the CNS, this other cell is a neuron. In the peripheral nervous system, the other cell is an *effector cell*—either a muscle or a gland. Although the physiology of neuron-neuron synapses and neuron-muscle synapses is similar, the latter synapses are often distinguished by the names *myoneural* or *neuromuscular junctions.*

Figure 7.18   Different types of synapses: axodendritic (*a*); axoaxonic (*b*); dendrodendritic (*c*); and axosomatic (*d*).

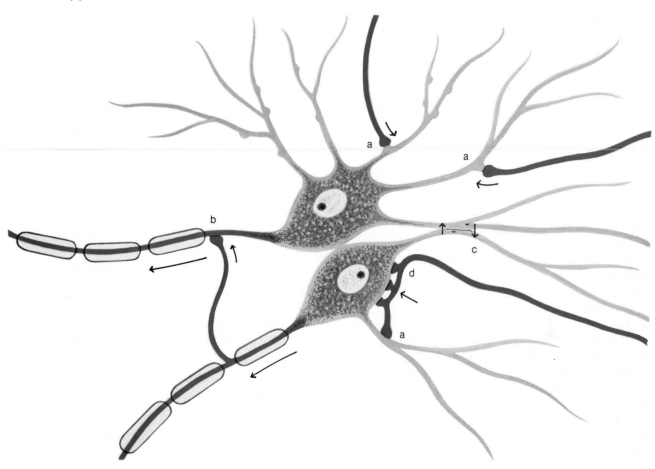

Neuron-neuron synapses usually involve a connection between the axon of one neuron and the dendrites, cell body, or axon of a second neuron. These are called, respectively, *axodendritic, axosomatic,* and *axoaxonic* synapses (see figure 7.17). In almost all synapses, transmission is in *one direction only*—from the axon of the first (or presynaptic) cell to the second (or postsynaptic) cell. *Dendrodendritic* synapses do not fit this classic pattern; in these synapses, two dendrites from different neurons make "reciprocal innervations"—some of these synapses conduct in one direction while others conduct in the opposite direction.

In the early part of the twentieth century, most physiologists believed that synaptic transmission was *electrical*—that is, that action potentials were conducted continuously from one cell to the next. This belief was reasonable in view of the facts that nerve endings appeared to touch the postsynaptic cells and that the delay in synaptic transmission was extremely short (about 0.5 msec). It was known, however, that the actions of autonomic nerves could be duplicated by certain chemicals. This lead to the suspicion that synaptic transmission might be *chemical*—that the presynaptic endings might release chemical **neurotransmitters** that stimulated action potentials in the postsynaptic cell.

In 1921, a physiologist named Otto Loewi published the results of an experiment that suggested that synaptic transmission was chemical, at least at the junction between a branch of the parasympathetic vagus nerve and the heart. He had isolated the heart of a frog and—while stimulating the branch of the vagus that innervated the heart—perfused this heart with a salt solution balanced to the extracellular solution of the frog (Ringer's solution). Stimulation of this nerve slowed the heart beat, as expected. More importantly, application of this Ringer's solution to the heart of a second frog caused the second heart to slow its rate of beat.

Loewi concluded that the nerve endings of the vagus must have released a chemical—which he called *vagus-stoff*—that inhibited the heart rate. This chemical was subsequently identified as **acetylcholine** (abbreviated **ACh**). In the decades following Loewi's discovery many other examples of chemical synapses were discovered and the theory of electrical synaptic transmission fell into disrepute. More recent evidence, ironically, has shown that electrical synapses do exist (though they are the exception) in the nervous system, within smooth muscles, and between cardiac cells in the heart.

Electrical Synapses: Gap Junctions

In order for two cells to be electrically coupled they must be approximately equal in size, and must be joined by areas of contact with low electrical resistance. In this way, impulses can be regenerated from one cell to the next without interruption—and without Frankenstein-like sparks between cells.

Adjacent cells that are electrically coupled are joined together by **gap junctions.** In gap junctions, the membranes of the two cells are separated by only two nanometers (10^{-9} meter). When gap junctions are observed in the electron microscope from a surface view, hexagonal arrays of particles are seen that are believed to be channels through which ions and molecules may pass from one cell to the next (see figure 7.19).

Gap junctions are present in smooth and cardiac muscle, where they allow excitation and rhythmic contractions of large masses of muscle cells. Gap junctions have also been observed in various regions of the brain. Although their functional significance here is unknown, it has been speculated that they may allow two-way transmission of impulses, (in contrast to chemical synapses, which are always one-way). Gap junctions have also been observed between glial cells, which do not produce electrical impulses; perhaps these act as channels for the passage of informational molecules between cells. It is interesting in this regard that many embryonic tissues have gap junctions, and that these gap junctions disappear as the tissue becomes more specialized.

Figure 7.19 Gap junctions are shown in electron micrographs in *(a)* and *(b)*. The photograph in *(a)* shows that cell membranes of two cells are fused together in the gap junction; a surface view of a gap junction is seen in *(b)*. The information presented in these and other electron micrographs is interpreted by the illustration shown in *(c)*.

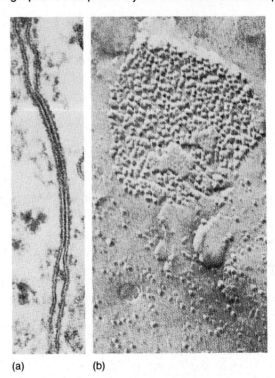

(a)   (b)

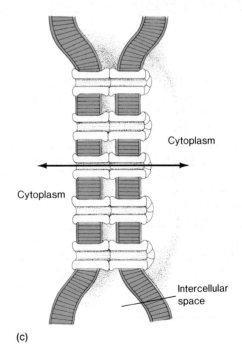

Cytoplasm

Cytoplasm

Intercellular space

(c)

Figuare 7.20 Electron micrograph of a chemical synapse, showing synaptic vesicles at the end of an axon.

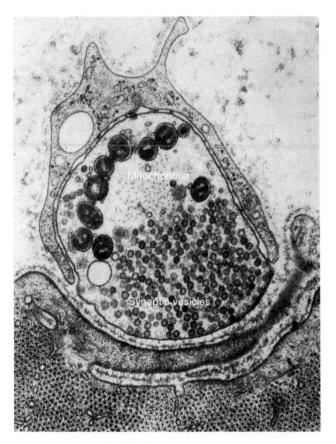

Chemical Synapses

Transmission across the majority of synapses in the nervous system is one-way, and occurs through the release of chemical neurotransmitters from the presynaptic axon endings. These presynaptic endings are known as *varicosities* or *terminal boutons* because of their swollen appearance, and are separated from the postsynaptic cell by a *synaptic cleft* so narrow that it can only be seen with an electron microscope (see figure 7.20).

Neurotransmitter molecules within the presynaptic endings are contained within many small, membrane-enclosed *synaptic vesicles*. In order for the neurotransmitter within these vesicles to be released into the synaptic cleft, the vesicle membrane must fuse with the axon membrane in the process of *exocytosis* (as discussed in chapter 3).

The neurotransmitter is released in *quanta*—that is, in multiples of the amount contained in one vesicle. The number of vesicles that undergo exocytosis is directly related to the frequency of action potentials produced at the presynaptic axon ending.

Release of a neurotransmitter following electrical excitation of the presynaptic endings accounts for most of the 0.5 msec time delay in synaptic transmission. During this time interval there is a sudden, transient inflow of Ca^{++} into the presynaptic endings. This inflow of Ca^{++} is apparently due to opening of Ca^{++} gates in response to electrical excitation and is required for the release of neurotransmitter. The Ca^{++} is believed to activate previously inactive enzymes within the axon terminals, which produces a change in the ratio of different phospholipids within the axon membrane. This may change the fluidity of the membrane and thus allow exocytosis to occur.

Effects of Acetylcholine at the Neuromuscular Junction

The neuromuscular junction—that is, the synapse between a somatic motor neuron and a skeletal muscle cell—is more accessible for study (and hence is better understood) than synapses within the brain. This synapse uses acetylcholine (ACh) as a neurotransmitter that excites the postsynaptic (skeletal muscle) cell. It should be noted, however, that synapses in other parts of the nervous system use ACh as well as other chemicals as neurotransmitters and that some synapses are inhibitory rather than excitatory. A given synapse, however, uses only one neurotransmitter chemical that always has one effect (excitation or inhibition).

At the neuromuscular junction, each synaptic vesicle contains about ten thousand molecules of ACh. Once these molecules are released by exocytosis they quickly diffuse across the narrow synaptic cleft to the membrane of the postsynaptic cell. Here they chemically bond to **receptor proteins** that are built into the postsynaptic membrane. These receptor proteins combine with ACh in a specific manner, analogous to the specific interaction between transport proteins and their substrates.

Acetylcholine is not, however, transported into the postsynaptic cell. Instead, interaction between ACh and its receptor protein causes changes in membrane structure that result in the opening of Na^+ and K^+ gates. These **chemically regulated gates**—which open in response to interaction between a neurotransmitter and its receptor

Figure 7.21 Binding of acetylcholine (ACh) to receptor proteins *(a)* and *(b)* cause opening of chemically regulated gates in the postsynaptic membrane. This results in increased diffusion of Na⁺ and K⁺ through the membrane *(c)*.

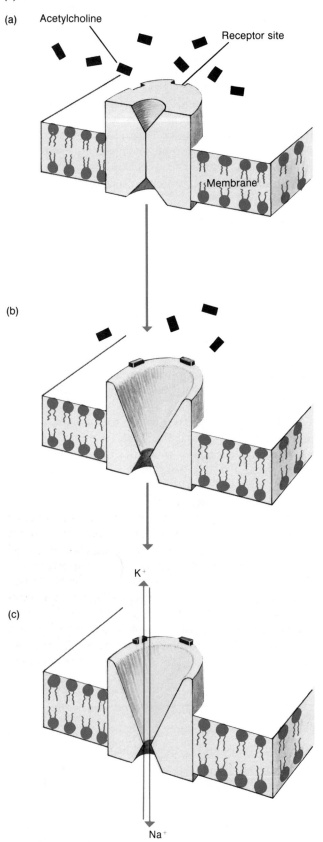

proteins—contrast with the voltage-regulated gates previously discussed in the axon, which open in response to depolarization.

In contrast to voltage-regulated gates, where the outward flow of K^+ occurs after the inward flow of Na^+, chemically regulated gates allow the simultaneous diffusion of Na^+ and K^+ (see figure 7.21). The depolarization effect of Na^+ diffusion predominates because the electrochemical gradient for Na^+ is greater than that for K^+. The outflow of K^+, however, does prevent the "overshoot" characteristic of action potentials—the membrane potential can reach 0 mV, but cannot reverse polarity (as occurs in action potentials). Because of the characteristics of chemically regulated gates, neurotransmitters do not directly produce action potentials. They can only produce depolarization (to a maximum of 0 mV), which may stimulate opening of voltage-regulated gates and action potentials at another site.

Acetylcholinesterase   The bond between ACh and its receptor protein exists for only a brief instant. The ACh-receptor complex quickly dissociates, but can be quickly reformed as long as free ACh is in the vicinity. In order for activity in the postsynaptic cell to be controlled (in this case, for control of skeletal muscle contraction), free ACh must be inactivated very soon after it is released. Inactivation of ACh is accomplished by means of an enzyme called *acetylcholinesterase (AChE),* which is built into the cell membrane near the ACh receptor proteins (see figure 7.22).

The importance of neurotransmitter inactivation in proper neural control is most obvious when this inactivation is prevented by drugs. *Nerve gas,* for example, exerts its odious effects by inhibiting AChE in skeletal muscles. Since ACh is not degraded it can continue to combine with its receptor proteins and can continue to stimulate the postsynaptic cell. This leads to spastic paralysis. Clinically, cholinesterase inhibitors (such as *neostigmine*) are used to enhance the effects of ACh on muscle contraction when neuromuscular transmission is weak, as in myasthenia gravis and in patients treated with curare during anesthesia.

Figure 7.22 Mechanisms of release of acetylcholine (ACh) from presynaptic nerve endings and binding of ACh to receptor proteins (R) in the postsynaptic membrane. Acetylcholine that combines with acetylcholinesterase (AChE) in the postsynaptic membrane is hydrolyzed and thus inactivated.

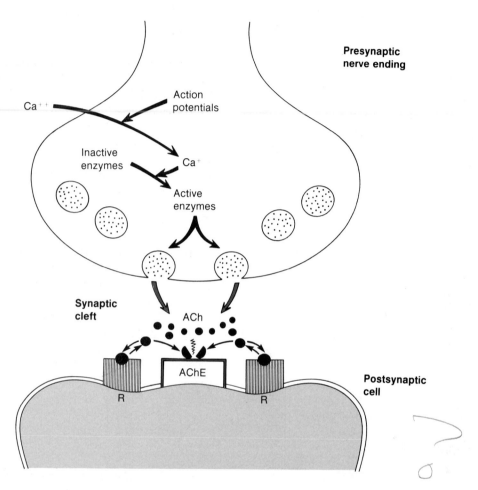

The End-Plate Potential   Interaction of ACh with its receptor proteins, as previously discussed, causes opening of chemically regulated ion gates in the postsynaptic membrane of the skeletal muscle cell. The area of muscle cell membrane affected in this way by ACh released from the axon terminal is known as the *motor end plate*. Depolarizations produced by inward Na^+ diffusion in this region are known as **end-plate potentials.** As previously discussed, these depolarizations cannot exceed 0 mV (they do not "overshoot" to become positive on the inside, unlike action potentials) because K^+ diffuses outwards at the same time that Na^+ diffuses inwards.

Unlike action potentials, end-plate potentials have *no threshold;* a single quanta of ACh (released from a single synaptic vesicle) produces a tiny depolarization of the end plate. These are called "miniature end-plate potentials." When more quanta of ACh are released, the depolarization of the motor end plate is correspondingly greater. End-plate potentials are therefore *graded* in magnitude, unlike action potentials. Since end-plate potentials can be graded, they are capable of *summation*. This is quite different from action potentials, which are separate, all-or-none events.

When the end plate becomes sufficiently depolarized, the depolarization serves as a stimulus for opening of voltage-regulated gates in adjacent areas of the skeletal muscle cell membrane. Opening of voltage-regulated gates in a skeletal muscle cell causes action potentials to be produced. These action potentials in turn serve as stimuli for production of other action potentials in the next area of membrane—skeletal muscle cells conduct action potentials in the same manner as do unmyelinated nerve fibers. As will be discussed in chapter 8, the spread of action potentials in skeletal muscle cells stimulates muscle contraction.

If any stage in the process of neuromuscular transmission is blocked, muscle weakness—sometimes leading to paralysis and death—may result. The drug *curare,* for example, competes with ACh for attachment to the receptor protein and thus reduces the size of the end-plate potential. This drug was first used on poison darts by South American Indians because it caused flaccid paralysis of their victims. Clinically, curare is used as a muscle relaxant during anesthesia and to prevent muscle damage during electroconvulsive shock therapy.

Muscle weakness in the disease *myasthenia gravis* is due to the fact that ACh receptors are blocked by antibodies secreted by the immune system of the affected person. Paralysis in people who eat shellfish poisoned by saxitoxin, which is produced by unicellular organisms that cause red tides, results from inhibition of the chemically regulated gates. In both of these instances, acetylcholine combines with its receptor proteins, but the gates don't open. The effects of these and other poisons on neuromuscular transmission are summarized in table 7.6.

Effects of ACh in Neuron-Neuron Synapses

Within the nervous system, the axon terminals of one neuron typically synapse with the dendrites and cell body of another. The dendrites and cell body thus serve as the "receptive" area of the neuron, and it is in these regions that receptor proteins for neurotransmitters and chemically regulated ion gates are located. The first voltage-regulated gates are located at the beginning of the axon, at the axon hillock. It is here that action potentials are first produced (see figure 7.23).

Release of ACh at these synapses results, through the opening of chemically regulated gates, in the production of graded depolarizations in the dendrites and cell body. These graded depolarizations are known as **excitatory postsynaptic potentials (EPSPs).** Like end-plate potentials, EPSPs can vary in amplitude (are graded), and can be summated. A comparison of EPSPs and action potentials is provided in table 7.7.

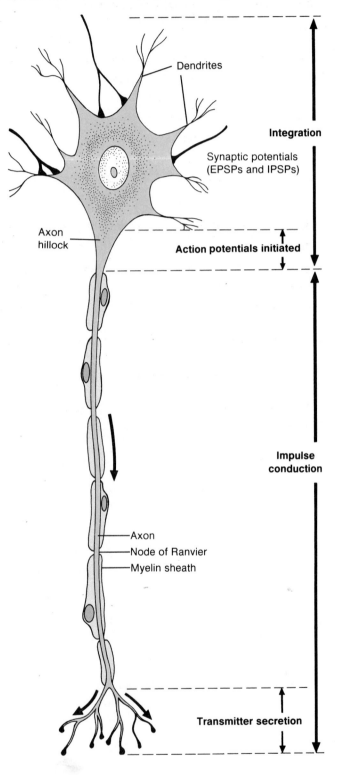

Figure 7.23 Diagram illustrating the functional specialization of different regions in a "typical" neuron.

Table 7.6   Derivation and effect of various drugs that affect neural control of skeletal muscles.

Drug	Origin	Effects
Botulinum toxin	Produced by *Clostridium botulinum* (bacteria)	Inhibits release of acetylcholine (ACh)
Curare	Resin from a South American tree	Prevents interaction of ACh with the postsynaptic receptor protein
α-bungarotoxin	Venom of *Bungarus* snakes	Binds to ACh receptor proteins
Saxitoxin	Red tide *(Gonyaulax)*	Blocks Na^+ channels
Tetrodotoxin	Pufferfish	Blocks Na^+ channels
Nerve gas	Artificial	Inhibits acetylcholinesterase in postsynaptic cell
Prostigmine	Nigerian bean	Inhibits acetylcholinesterase in postsynaptic cell
Strychnine	Seeds of an Asian tree	Prevents IPSPs in spinal cord that inhibit contraction of antagonistic muscles

Table 7.7   Comparison of action potentials with synaptic potentials at the neuromuscular junction.

Characteristic	Action Potential	Excitatory Post-Synaptic Potential
Stimulus for opening of ionic gates	Depolarization	Acetylcholine (ACh)
Initial effect of stimulus	Na^+ gates open	Na^+ and K^+ gates open
Production of repolarization	Opening of K^+ gates	Loss of intracellular positive charges with time and distance
Conduction distance	Not conducted—regenerated over length of axon	1-2mm. A localized potential
Positive feedback between depolarization and opening of Na^+ gates	Present	Absent
Maximum depolarization	$+40$ mV	Close to zero
Summation	No summation—is all-or-none	Summation of EPSPs, producing graded depolarizations
Refractory period	Present	Absent
Effect of drugs	Inhibited by tetrodoxin, not by curare	Inhibited by curare, not by tetrodoxin

Figure 7.24 The graded nature of excitatory postsynaptic potentials is shown, in which stimuli of increasing strength produce increasing amounts of depolarization. When a threshold level of depolarization is produced, action potentials are generated in the axon.

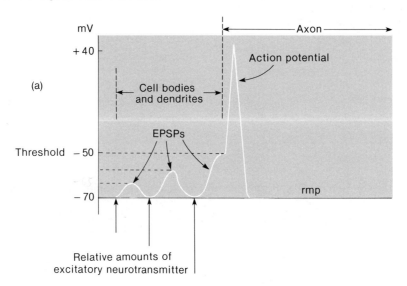

Catecholamine and Other Neurotransmitters

Catecholamines are a group of regulator molecules derived from the amino acid *tyrosine*. These molecules include L-dopa (dihydroxyphenylalanine), dopamine (derived from L-dopa), epinephrine, and norepinephrine. Epinephrine and norepinephrine (which is the same as epinephrine but lacks a CH_3 group) function as both neurotransmitters and as hormones secreted by the adrenal medulla gland.

Like ACh, catecholamine neurotransmitters are released by exocytosis from presynaptic vesicles, and diffuse across the synaptic cleft to interact with specific receptor proteins in the membrane of the postsynaptic cell. The stimulatory effects of catecholamine neurotransmitter action—like those of ACh—are quickly inhibited. Inhibition of ACh stimulation is accomplished by acetylcholinesterase; inhibition of catecholamine action is due to (1) *reuptake* of catecholamines into the presynaptic axon endings; (2) enzymatic degradation of catecholamines in the presynaptic endings by *monoamine oxidase (MAO);* and (3) enzymatic degradation of catecholamines by *catecholamine-O-methyltransferase (COMT)* in the postsynaptic neuron. Drugs that inhibit MAO and COMT activity thus promote the effects of catecholamine action.

Since voltage-regulated gates are absent in the dendrites and cell body, depolarizations produced in these regions as a result of synaptic transmission must be conducted by cable properties to the axon hillock. This causes the amplitude of the EPSP to decrease with distance from its point of origin. If the depolarization is at a threshold level by the time the EPSP reaches the axon hillock it will cause opening of voltage-regulated gates and the production of action potentials.

A skeletal muscle cell receives synaptic input from only one neuron. A neuron, in contrast, may receive synaptic input from as many as one thousand other neurons. So while synaptic input from a single axon must be sufficient to excite a muscle cell, a neuron may require summation of many EPSPs before a threshold depolarization is produced at the axon hillock. Summation, which can only occur between graded synaptic potentials (action potentials can't summate), allows the postsynaptic neuron to "weigh" different combinations of synaptic inputs and respond appropriately.

1. Describe the events that occur between electrical excitation of an axon terminal and the release of ACh.
2. Describe the properties of an EPSP and compare these with the characteristics of an action potential.
3. Describe how EPSPs (or end-plate potentials) produce action potentials in the postsynaptic cell.
4. Describe the function of acetylcholinesterase (AChE) and its physiological significance.

Figure 7.25 Diagram showing the production, release, and reuptake of catecholamine neurotransmitters from presynaptic nerve endings. The transmitters combine with receptor proteins (R) in the postsynaptic membrane. While most of the catecholamines are recaptured by the presynaptic nerve ending, some of the transmitters are inactivated within the postsynaptic cell. COMT (catecholamine = 0-methlytransferase).

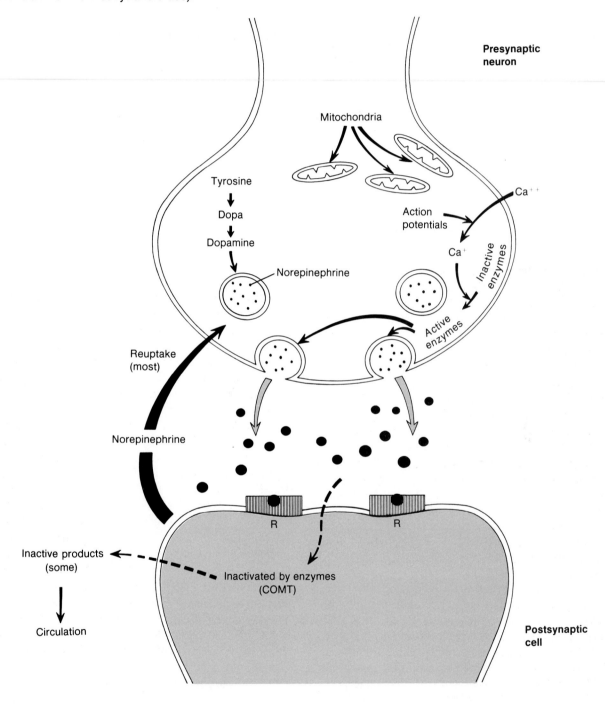

Figure 7.26   Catecholamine neurotransmitters induce the
intracellular production of cyclic AMP (cAMP) in the
postsynaptic cell. The cAMP, in turn, stimulates opening
of ionic gates (short-term effects) and genetic expression
(long-term effects) through the activation of previously
inactive enzymes.

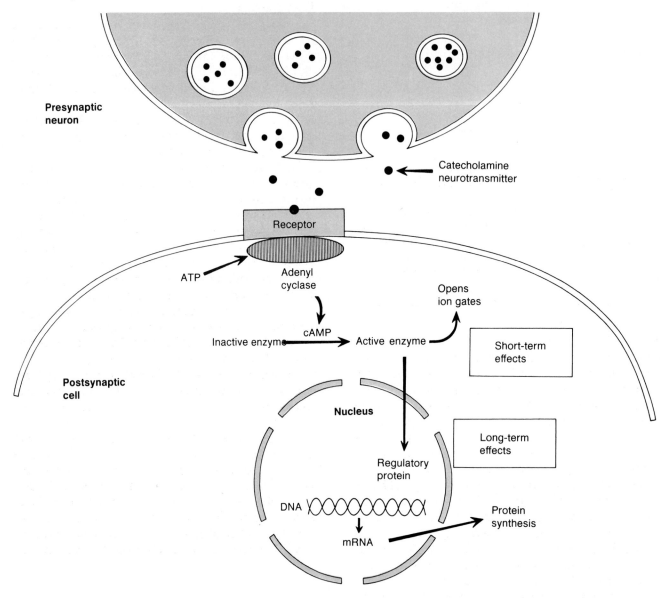

Interaction of catecholamines (and some other) neurotransmitters with their receptors do not directly open ionic gates, in contrast to the more direct effects of ACh on ion permeability. Instead, these neurotransmitters activate an enzyme in the postsynaptic membrane known as *adenyl cyclase*. This enzyme converts ATP into **cyclic AMP (cAMP)** and pyrophosphate (two inorganic phosphates). Cyclic AMP formed within the postsynaptic cell in turn activates other enzymes that open ionic gates in the postsynaptic membrane. The cAMP is thus a **second messenger** in the action of these neurotransmitters, as it is in the action of many hormones (see chapter 19).

Cyclic AMP, in addition to promoting opening of ionic gates (which leads to depolarization of the postsynaptic membrane), may also promote more long-term modifications of the postsynaptic neuron. These long-term modifications may involve cAMP-mediated changes in genetic expression (see figure 7.26).

Table 7.8 Examples of chemicals that are either proven or putative neurotransmitters.

Category	Chemicals
Amines	Acetylcholine GABA (gamma-aminobutyric acid) Histamine Serotonin
Catecholamines	Dopamine Epinephrine Norepinephrine
Amino acids	Aspartic acid Glutamic acid Glycine
Gut-brain polypeptides	Glucagon Insulin Somatostatin Substance P ACTH (adrenocorticotrophic hormone) Angiotensin II Endorphins
Polypeptides (other)	LHRH (luteinizing hormone-releasing hormone) TRH (thyrotophin-releasing hormone) Vasopressin

Putative Neurotransmitters

Many other compounds, in addition to acetylcholine and catecholamines, may function as neurotransmitters. Since complete experimental proof of their transmitter function is not currently available, these compounds are identified as putative neurotransmitters. These putative neurotransmitters include amino acids, such as glycine and glutamic acid, and polypeptide molecules that also serve as hormones of the digestive tract (such as insulin and somatostatin). This latter observation is surprising, and has led to the theory that the brain and gut cells that produce these polypeptides may have the same embryonic origin.

While many of these molecules may function as neurotransmitters in the traditional sense—that is, they may stimulate opening of ionic gates—some may have more subtle effects. Polypeptides, like catecholamines, stimulate production of cAMP and other second messengers, which may modify the responsiveness of the postsynaptic neuron to neurotransmitters. Some of these putative neurotransmitters may thus be *neuromodulators* that do not themselves directly alter ion permeability but, rather, influence the function of neurons in other (and poorly understood) ways.

The Endorphins

The ability of opium and its analogues—that is, the opioids—to relieve pain (promote analgesia) has been known for centuries. Morphine, for example, has long been used for this purpose. The discovery in 1973 of opioid receptor proteins in the brain suggested that the effects of these drugs might be due to stimulation of specific neuron pathways. This implied that opioids—like LSD, mescaline, and other mind-altering drugs—might resemble neurotransmitters produced by the brain.

The analgesic effects of morphine are blocked in a specific manner by a drug called *naloxone*. In the same year that opioid receptor proteins were discovered, it was found that naloxone also blocked the analgesic effect of electrical brain stimulation. Subsequently, evidence suggested that the analgesic effects of hypnosis and acupuncture could also be blocked by naloxone. These experiments indicated that the brain might be producing its own morphinelike analgesic compounds.

These compounds have been identified as a family of chemicals called **endorphins** (for "endogenously produced morphinelike compounds") produced by the brain and pituitary gland. The endorphins include a group of five-amino-acid peptides called **enkephalins,** which may function as neurotransmitters, and a thirty-one-amino-acid polypeptide produced by the pituitary gland called *β-endorphin*.

Endorphins have been shown to block transmission of pain. Current evidence for this includes results obtained both from neurophysiological studies—in which endorphins blocked release of substance P (the chemical transmitter believed to be released by nerve fibers that mediates painful sensations)—and from behavioral studies. The pain threshold of pregnant rats, for example, has been found to decrease when they are treated with naloxone.

Table 7.9 Possible functions of endorphins in the central nervous system.

Proposed Effect	Comments
Analgesia	Analgesia produced by electrical stimulation of brain, hypnosis, and acupuncture may be blocked by naloxone, an opioid antagonist. Naloxone also blocks increased pain threshold in pregnant rats. Endorphins inhibit release of substance P, believed to be the transmitter of pain.
Emotion	Opioid receptors in limbic system (an area involved in emotion) have been demonstrated. Opioids affect the EEG (electroencephalogram) obtained in these areas.
Mental health	Opioids and naloxone (an opioid antagonist) produce behavioral changes in rats that appear similar to human psychotic states. Controversial.
Stress	Endorphins are released from the pituitary together with ACTH (adrenocorticotrophic hormone) during stress reactions.
Obesity	Naloxone abolishes overeating in genetically obese mice, and inhibits stress-induced eating behavior in rats.

Endorphins—like opium and morphine—may also provide pleasant sensations and thus mediate reward or positive reinforcement pathways. Overeating in genetically obese mice, for example, appears to be blocked by naloxone. It has been found that blood levels of β-endorphin are increased in exercise. Some people have suggested that the "joggers high" may thus be due to endorphins. While evidence for this is poor, it does appear that endorphins may promote some type of psychic reward system as well as analgesia.

1. Compare the mechanisms by which catecholamine neurotransmitters are inactivated with the mechanism that inactivates acetylcholine.
2. Describe the role of cyclic AMP in the mechanism of action of catecholamine neurotransmitters.
3. Describe the endorphins and their possible physiological significance.

Synaptic Integration

Since voltage-regulated ion gates are absent in the dendrites and cell body of neurons, the potential changes produced by opening of chemically regulated gates are graded rather than all-or-none. Synaptic potentials can thus add together, or summate. **Spatial summation** occurs because many presynaptic axon terminals converge on a single postsynaptic neuron (see figure 7.27). In spatial summation, synaptic depolarizations (EPSPs) produced at different synapses may summate in the postsynaptic dendrites and cell body. In **temporal summation**, successive activity of presynaptic axon terminals may cause successively produced EPSPs to summate in the postsynaptic neuron.

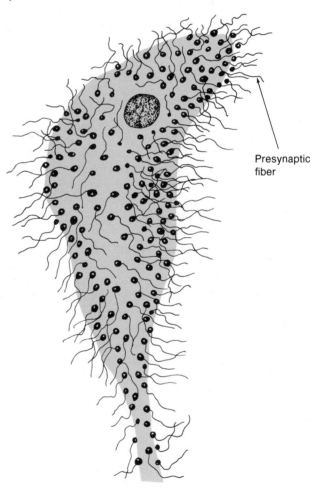

Figure 7.27 Diagram illustrating the convergence of large numbers of presynaptic fibers on the cell body of a spinal motor neuron.

Presynaptic fiber

Figure 7.28 Excitatory postsynaptic potentials (EPSPs) can summate over distance (spatial summation) and time (temporal summation). When summation results in a threshold level of depolarization at the axon hillock, voltage-regulated Na$^+$ gates are opened and an action potential is produced.

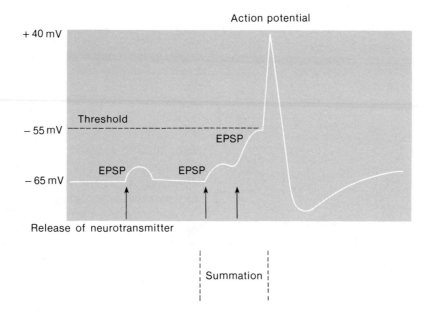

Since there are no voltage-regulated gates in the dendrites and cell body, depolarizations produced in these regions must be conducted by cable properties to the axon hillock. This results in a gradual decay of the depolarization as it spreads through the cell body. Spatial and temporal summation of EPSPs is therefore needed, particularly if the postsynaptic neuron cell body is large, to produce a threshold depolarization (usually about − 55 mV) at the axon hillock. If this threshold depolarization is attained, voltage-regulated gates in the axon hillock will open and action potentials will be produced.

Synaptic Inhibition

While most neurotransmitters depolarize the postsynaptic membrane (produce EPSPs), some transmitters have the opposite effect. These inhibitory neurotransmitters cause *hyperpolarization* of the postsynaptic membrane— they make the inside of the membrane more negative than it is at rest. Since hyperpolarization (from − 65 mV to, for example, − 85 mV) takes the membrane potential farther away from the threshold depolarization required to stimulate action potentials, such hyperpolarizations inhibit activity of the postsynaptic nerve fiber. Hyperpolarizations produced by neurotransmitters are therefore called **inhibitory postsynaptic potentials (IPSPs).**

While acetylcholine produces depolarization of skeletal muscles (thus stimulating production of action potentials and muscle contraction), the same transmitter causes hyperpolarization (IPSPs) in the heart. This is due to the fact that combination of ACh with its receptor protein in the heart causes opening of only K$^+$ gates, and outward diffusion of K$^+$ produces hyperpolarization. The "baseline" membrane potential is thus lowered, so that a longer time is required for the heart cells to reach threshold. The parasympathetic nerves that innervate the heart in this way cause a slowing of the heartbeat.

Acetylcholine is thus excitatory in certain synapses and inhibitory in others. The transmitter chemical called *gamma-aminobutyric acid (GABA),* in contrast, appears to always function as an inhibitor. This neurotransmitter is believed to hyperpolarize central neurons by opening chloride gates. Chloride ion (Cl$^-$), as a result, diffuses into

Figure 7.29 An inhibitory postsynaptic potential (IPSP) makes the inside of the postsynaptic membrane more negative than the resting potential—it hyperpolarizes the membrane. Subsequent or simultaneous excitatory postsynaptic potentials (EPSPs), which are depolarizations, thus must be stronger to reach the threshold required to generate action potentials at the axon hillock.

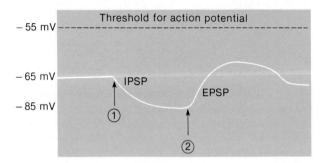

the neuron and makes the membrane potential more negative on the inside. The amino acid *glycine* is also believed to function as an inhibitory neurotransmitter. This transmitter may help in muscular coordination by inhibiting those neurons in the spinal cord that activate antagonistic muscles (as discussed in chapter 8).

Postsynaptic and Presynaptic Inhibition   Inhibition of postsynaptic neuron activity by production of IPSPs is known as **postsynaptic inhibition.** The importance of this type of inhibition is illustrated by the fatal effects of the poison strychnine, which prevents IPSPs within the central nervous system (causing spastic paralysis and death from cessation of breathing).

In **presynaptic inhibition,** the release of neurotransmitter from the axon of one neuron is reduced by a second neuron, which makes an axo-axonic synapse with the first. This synapse partially excites the postsynaptic membrane (the axon membrane), putting it in a partial refractory period. This reduces the frequency of action potentials in this axon, thus reducing the number of synaptic vesicles that undergo exocytosis.

Post-Tetanic Potentiation

When a presynaptic neuron is stimulated continuously, even for as short a time as a few seconds, its ability to excite a postsynaptic neuron is enhanced—or potentiated—when this neuron pathway is subsequently stimulated. This potentiation of synaptic transmission may last for hours or even weeks following the continuous (tetanic) stimulation of the presynaptic neuron.

Post-tetanic potentiation, as this phenomenon is called, is due to increased release of neurotransmitter as a result of previous tetanic stimulation of the presynaptic neuron. Experiments suggest that there is increased Ca^{++} concentration in the axon terminals following tetanic stimulation. Since Ca^{++} is believed to mediate exocytosis of synaptic vesicles in response to action potentials, this observation may explain the improved synaptic efficiency that results. Post-tetanic potentiation may favor transmission along frequently used neural pathways, and thus may represent a mechanism of neural "learning."

There are an estimated 10^{12} neurons in the brain that are interconnected by synapses. Some of these neurons, as previously mentioned, may receive as many as one thousand different synaptic inputs from other neurons. Summation of EPSPs and IPSPs, postsynaptic inhibition, and post-tetanic potentiation are mechanisms that are known to provide integration of synaptic information. Less well-understood mechanisms may produce long-term modifications of neuron responsiveness and contribute to learning and memory. While the complexity of this area of physiology is formidable, current study of less complex neural systems may someday help to explain higher brain functions.

1. Compare the characteristics of synaptic potentials (EPSPs and IPSPs) with those of action potentials. Describe how these characteristics relate to the functions of these two types of activities.
2. Compare EPSPs with IPSPs in terms of mechanism of production and functional significance. Draw a figure illustrating the effects of excitatory and inhibitory neurotransmitters on the membrane potential.
3. Draw a figure of a neuron, showing axodendritic and axosomatic synapses. Indicate in your figure where synaptic potentials and action potentials are produced.
4. Define spatial and temporal summation. Draw a figure that illustrates (1) summation of EPSPs; and (2) summation of EPSPs with IPSPs. Describe how this summation influences the ability of the postsynaptic neuron to produce action potentials.

Summary

Neurons, Neuroglia, and Organization of the Nervous System

I. Each neuron has three functional regions.
  A. The dendrites serve as a "receptive area" for stimulation by other neurons.
  B. The cell body is the nutritive portion of the neuron, which contains the nucleus.
  C. The axon conducts nerve impulses over long distances.

II. There are three types of neurons.
  A. Pseudounipolar neurons, which have a single process that branches to form dendrites and axons, are somatic sensory neurons.
  B. Bipolar neurons have one dendrite and one axon.
  C. Multipolar neurons—with many dendrites and one axon—comprise the majority of neurons in the brain.

III. In addition to neurons, the nervous system contains neuroglia cells.
  A. Schwann cells (in the peripheral nervous system) and oligodendrocytes (in the central nervous system) form a sheath, or neurilemma, around axons.
    1. These neuroglial cells wrap themselves around larger axons to form a myelin sheath.
    2. Myelinated nerve fibers (axons) are wrapped by both a myelin and a neurilemmal sheath.
    3. Gaps between adjacent areas of myelin are called the nodes of Ranvier.
    4. Axons that do not have a myelin sheath, and have only a neurilemmal sheath, are said to be unmyelinated.
  B. Astrocytes contribute to the blood-brain barrier that regulates passage of molecules between blood capillaries and the brain.
  C. Microglial cells are phagocytic.

IV. The nervous system is divided into the central nervous system and the peripheral nervous system.
  A. The central nervous system includes the brain and spinal cord.
  B. The peripheral nervous system includes nerves—collections of axons that leave the central nervous system.
  C. Nerves may consist of sensory and motor axons.
    1. Sensory axons carry nerve impulses into the brain or spinal cord from sensory receptors.
    2. Motor axons carry impulses to effector organs— muscles and glands.
  D. There are two major categories of motor fibers—somatic and autonomic.
    1. Somatic motor fibers innervate skeletal muscles.
    2. Autonomic fibers innervate smooth muscle, cardiac muscle, and glands.
    3. The autonomic system is divided into a sympathetic and a parasympathetic division.

Electrical Activity in Axons

I. At rest, axons have a membrane potential of -65 mV.
  A. The Na^+ concentration outside the axon is higher than it is inside, and the K^+ concentration is higher inside than outside.
  B. At rest, the axon is only slightly permeable to K^+ and is relatively impermeable (but not completely) to Na^+.

II. When a segment of axon is depolarized to a threshold level (about -55 mV), ionic gates open.
  A. These gates are thus said to be voltage regulated.
  B. Voltage-regulated Na^+ gates open first, leading to an inward diffusion of Na^+.
    1. This causes the membrane potential to reverse polarity and change to $+40$ mV.
    2. Inward diffusion of Na^+ causes more depolarization, which increases N^+ permeability even more; this is a position feedback effect.
  C. After a short time delay, voltage-regulated K^+ gates open.
    1. As a result, K^+ diffuses out of the cell membrane.
    2. This helps to repolarize the membrane (restore the resting membrane potential of -65 mV), and is thus an example of negative feedback.
  D. The rapid change in the membrane potential from -65 mV to $+40$ mV and back to -65 mV, due to Na^+ and K^+ diffusion, is called an action potential or nerve impulse.

Action Potentials

I. Action potentials are all-or-none events.
  A. All action potentials have the same amplitude—from -65 mV to about $+40$ mV, regardless of the stimulus strength.
  B. As the stimulus strength is increased above a threshold depolarization, the frequency with which action potentials are produced is increased.
    1. Stimulus strength is thus coded by action potential frequency rather than amplitude.
    2. Action potentials can only be produced at a maximum frequency, because the membrane becomes refractory to stimuli while it is producing an action potential.

3. The absolute refractory period corresponds to the depolarization phase of the action potential; the relative refractory period corresponds to the repolarization phase.

II. Each action potential serves as a stimulus for the production of a new action potential.

   A. Each action potential causes depolarization of the next region of membrane.

   B. Since depolarization produces a new action potential, action potentials are regenerated at each region of the axon.

      1. In an unmyelinated axon, action potentials are produced tenths of a micrometer apart, where voltage-regulated gates are located.

      2. In a myelinated axon, action potentials are only produced at the nodes of Ranvier—this is called saltatory conduction.

   C. Since each action potential is a new all-or-none event, they are conducted without decrement along the axon.

   D. Regeneration of action potentials is needed to conduct impulses over long distances.

      1. The ability of neurons to transmit charges is very poor.

      2. These poor cable properties result in decay of depolarization within one to two millimeters from the site of stimulation.

Synaptic Transmission

I. There are two types of synapses: electrical and chemical.

   A. Electrical synapses, or gap junctions, transmit impulses in two directions from one cell to another.

   B. Chemical synapses transmit in one direction only.

      1. The presynaptic axon endings contain chemical neurotransmitter within synaptic vesicles.

      2. Electrical excitation of these endings cause influx of Ca^{++}, which in turn results in exocytosis and the release of transmitter into the synaptic cleft.

      3. When the neurotransmitter combines with receptor protein within the postsynaptic membrane, chemically regulated ionic gates are opened.

   C. Chemical neurotransmitters are quickly inactivated after they are released.

      1. Acetylcholine (ACh) is inactivated by the enzyme *acetylcholinesterase* (AChE) in the postsynaptic membrane.

      2. Catecholamine neurotransmitters are inactivated mainly by reuptake into the presynaptic endings.

II. Excitatory neurotransmitters produce depolarization of the postsynaptic membrane; this is known as an excitatory postsynaptic potential (EPSP).

   A. Synaptic potentials are graded—the more synaptic vesicles release their content by exocytosis, the greater the depolarization produced in the postsynaptic membrane.

   B. There are no voltage-regulated gates in the dendrites and cell bodies of neurons.

      1. Depolarization in these regions, induced by interactions of transmitters with receptor proteins, decreases in amplitude with distance.

      2. Because of the poor cable properties of the neuron, the depolarization at the axon hillock will be lower in magnitude than that produced at the postsynaptic membrane.

   C. EPSPs can summate over distance and time.

      1. EPSPs produced by convergence of many axon terminals on a single neuron can summate to produce a threshold depolarization (spatial summation).

      2. EPSPs produced at slightly different times can summate (temporal summation).

   D. The first voltage-regulated gates are located at the axon hillock; if depolarization here is sufficient, the first action potential will be produced.

III. Some neurotransmitters produce hyperpolarization of the postsynaptic membrane—this is called an inhibitory postsynaptic potential (IPSP).

   A. IPSPs make the inside of the membrane more negative than it is at rest.

   B. This produces postsynaptic inhibition, since more depolarization is required to reach threshold at the axon hillock.

   C. Hyperpolarization can be produced by opening of K^+ gates only, or by opening of Cl^- gates.

IV. Continuous (tetanic) stimulation of a presynaptic axon can enhance synaptic transmission.

   A. More transmitter, as a result, is released upon subsequent stimulation of the presynaptic axon.

   B. This increased release of transmitter is called post-tetanic potentiation.

Self-Study Quiz

1. A collection of neuron cell bodies located outside the brain and spinal cord is called:
   (a) a tract
   (b) a nerve
   (c) a ganglion
   (d) a nucleus

2. The myelin sheaths of peripheral axons are formed from:
   (a) oligodendrocytes
   (b) Schwann cells
   (c) microglia
   (d) all of these

3. Depolarization of an axon is produced by:
   (a) inward diffusion of Na^+
   (b) inward active transport of Na^+
   (c) outward diffusion of K^+
   (d) active extrusion of K^+

4. Repolarization of an axon during an action potential is produced by:
   (a) inward diffusion of Na^+
   (b) inward active transport of Na^+
   (c) outward diffusion of K^+
   (d) active extrusion of K^+

5. As the strength of a depolarizing stimulus to an axon is increased:
   (a) the amplitude of action potentials is increased
   (b) the duration of action potentials is increased
   (c) the speed with which action potentials are conducted is increased
   (d) the frequency of action potentials is increased

6. Conduction of action potentials in a myelinated nerve fiber is:
   (a) saltatory
   (b) without decrement
   (c) faster than in an unmyelinated fiber
   (d) all of these

7. Which of the following is NOT a characteristic of synaptic potentials?
   (a) they are all-or-none in amplitude
   (b) they decrease in amplitude with distance
   (c) they require a threshold level of depolarization to be produced
   (d) they are graded in amplitude
   (e) *a* and *c*

8. Which of the following is NOT a characteristic of action potentials?
   (a) they are produced by opening of voltage-regulated gates
   (b) they are conducted without decrement along the axon
   (c) Na^+ and K^+ gates open at the same time
   (d) the membrane potential changes from -65 mV to $+40$ mV during the action potential

9. A drug that inactivates acetylcholinesterase would have the following effect:
   (a) cause muscle relaxation
   (b) enhance muscle contraction

10. Postsynaptic inhibition is produced by:
    (a) depolarization of the postsynaptic membrane
    (b) hyperpolarization of the postsynaptic membrane
    (c) axoaxonic synapses
    (d) post-tetanic potentiation

8

Skeletal Muscle: *Mechanisms of Contraction and Neural Control*

Objectives

By studying this chapter, you should be able to:

1. Describe the gross and microscopic structure of skeletal muscles

2. Explain the all-or-none twitch of muscle fibers and relate this to summation and tetanus in a whole muscle

3. Describe how a smooth, graded muscle contraction is produced *in vivo*

4. Distinguish between isometric and isotonic contraction

5. Define a motor unit and describe its functional significance

6. Describe the synaptic pathway in a monosynaptic stretch reflex, in a disynaptic Golgi tendon organ reflex, and in the crossed extensor reflex

7. Describe the physiology of the muscle spindles and the function of alpha and gamma motoneurons

8. Describe the neural pathways involved in the pyramidal and extrapyramidal systems

9. Explain, in terms of the functions of lower and higher motoneurons, how flaccid and spastic paralysis are produced

10. Describe the structure of myofibrils and the structural basis for their banded appearance

11. Explain the sliding filament mechanism of contraction

12. Describe the events that occur during the contraction cycles of cross-bridges

13. Explain the physiological roles of tropomyosin, troponin, and Ca^{++} in muscle contraction

14. Explain the structural and functional differences between fast-, slow-, and intermediate-twitch fibers

Outline

*I*n order to meet the challenges of changing conditions within the internal and external environment, the actions of various **effector organs** are controlled and coordinated by the nervous system. These effectors include cardiac muscle, smooth muscle, and glands—which are discussed in chapter 9—and the skeletal muscle system.

Figure 8.1   Actions of antagonistic muscles in the leg.

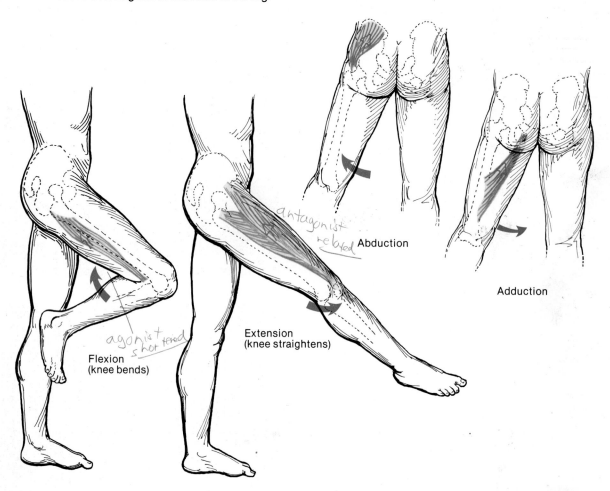

Flexion
(knee bends)

Extension
(knee straightens)

Abduction

Adduction

Table 8.1   Categories of skeletal muscle actions.

Category	Action
Extensor	Increases the angle at a joint
Flexor	Decreases the angle at a joint
Abductor	Moves limb away from the midline of the body
Adductor	Moves limb towards the midline of the body
Levator	Moves insertion upward
Depressor	Moves insertion downward
Rotator	Rotates a bone along its axis
Sphincter	Constricts an opening

Skeletal muscles are usually attached to bone on each end by tough connective-tissue *tendons*. When a muscle shortens during contraction its more movable attachment, known as its *insertion,* is pulled towards its less movable attachment (the *origin*). A variety of different skeletal movements are possible, depending on the type of joint involved and the attachments of the muscles (see table 8.1). When *flexor* muscles contract, for example, they decrease the angle of a joint. Contraction of *extensor* muscles increases the angle of their attached bones at the joint. Flexors and extensors that attach to the same bones are thus *antagonistic muscles.*

The position of the limbs, for example, is determined by the actions of a variety of antagonistic muscles. In addition to the positions of flexion and extension, a limb can be moved away from the midline of the body by *abductor* muscles, and brought inward towards the midline by contraction of *adductor* muscles. In all cases, these skeletal movements are produced by the shortening of the appropriate muscle groups—the *agonists*—while the antagonist muscles remain relaxed.

Figure 8.2 Relationship between muscle fibers and the connective tissues of the tendon, epimysium, perimysium, and endomysium.

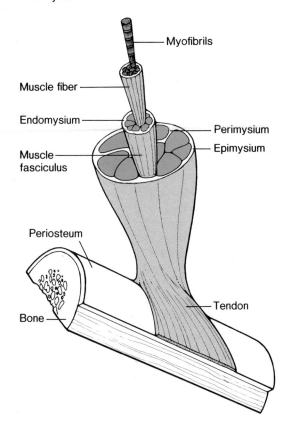

Figure 8.3 Appearance of skeletal muscle fibers in the light microscope (note the striated appearance produced by alternating dark A bands and light I bands).

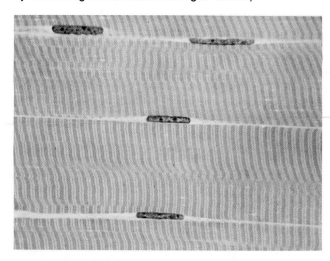

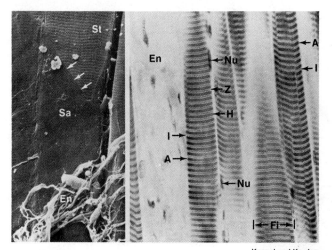

Kessel and Kardon

Structure of Skeletal Muscles

The fibrous connective tissue proteins within the tendons continue in an irregular arrangement around the muscle to form a sheath known as the *epimysium* (epi = above; my = muscle). Connective tissue from this outer sheath extends into the body of the muscle, subdividing it into columns, or *fascicles* (e.g., the "strings" in stringy meat). Each of these fascicles is thus surrounded by its own connective tissue sheath, known as the *perimysium* (peri = around).

Dissection of a muscle fascicle under a microscope reveals that it, in turn, is composed of many **muscle fibers** (or *myofibers*) surrounded by wisps of connective tissue called *endomysium*. These fibers are the muscle cells. Since the connective tissue of the tendons, epimysium, perimysium, and endomysium is continuous, muscle fibers do not normally pull out of the tendons when they contract.

Despite their unusual fiber shape, muscle cells have the same organelles that are present in other cells: mitochondria, intracellular membranes, glycogen granules, and others. Unlike most other cells in the body, skeletal muscle fibers are multinucleate—that is, they contain many nuclei. The most distinctive feature of the microscopic appearance of skeletal muscle fibers, however, is their **striated** appearance. The striations (stripes) are produced by alternating dark and light bands that appear to cross the width of the fiber.

The dark bands are able to polarize visible light (are *anisotropic*), and have therefore been named *A bands*. The light bands do not polarize light (are *isotropic*) and are thus called *I bands*. At high magnification, thin dark lines can be seen in the middle of the I bands. These were once thought to be membranes and were accordingly labeled *Z lines* (for Zwischenscheibe, a German word for membrane).

Figure 8.4 Setup for observing the contractile behavior of isolated gastrocnemius muscle of a frog.

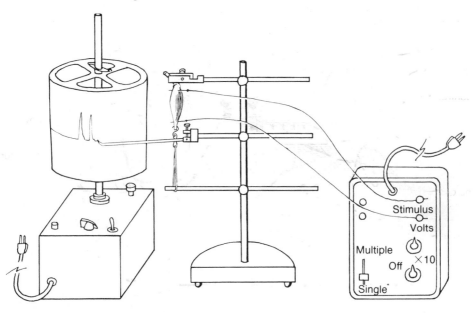

Twitch, Summation, and Tetanus

The contractile behavior of skeletal muscles is more easily studied *in vitro* (outside the body) than *in vivo* (within the body). In these studies a muscle such as the gastrocnemius (calf muscle) of a frog is usually mounted so that one end is fixed and the other is movable. In the classic laboratory studies, the movable end of the muscle directly produces deflections of a pen, which writes on a rotating drum recorder (see figure 8.4). When more modern equipment is used, the mechanical force of muscle contraction is transduced into an electric current, which can be amplified and displayed as pen deflections in a multichannel recorder (a physiograph or polygraph—see figure 8.5).

When the muscle is stimulated with a single electrical pulse of a given voltage it quickly contracts and relaxes. This response is called a **twitch.** Increasing the stimulus voltage increases the strength of the twitch, up to a maximum. The strength of skeletal muscle contraction can thus be *graded,* or varied—an obvious requirement for proper control of skeletal movements. If a second stimulus is delivered immediately after the first, a second twitch will be produced that may partially "ride piggyback" on the first.

This **summation** results from the fact that the second twitch begins before relaxation from the first twitch is complete.

If the stimulus voltage is held constant, and the stimulator is set to deliver automatically an increasing frequency of electric shocks, the relaxation time between successive twitches gets shorter and shorter. Finally, at a particular "fusion frequency" of stimulation, there is no visible relaxation between successive twitches. Contraction is smooth and sustained, very much like normal muscle contraction *in vivo.* This smooth, sustained contraction is called **tetanus** (not to be confused with the disease that has the same name). After the muscle is left in tetanus for a period of time it gradually loses its ability to maintain contraction—the muscle *fatigues.* Muscle fatigue may be due to accumulation of lactic acid and to depletion of ATP.

Similar experiments performed with isolated muscle fibers yield very different results. Isolated muscle fibers do not produce sustained contractions—they can only twitch. The smooth, tetanized contractions of a whole muscle result from rapid summation of twitches from fibers that are stimulated asynchronously. Some fibers contract while others are in the relaxation phase of their twitch.

Figure 8.5 The physiograph Mark III recorder. This is one of many types of electronic recording devices that receive electrical signals produced by transducers and record these signals on moving paper.

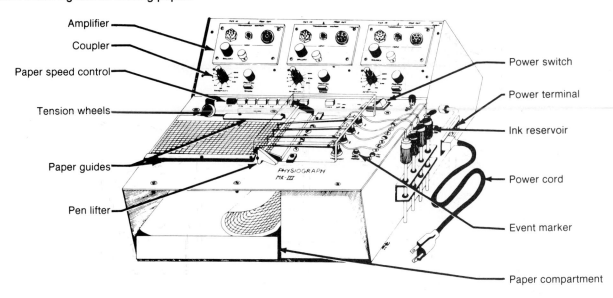

Amplifier
Coupler
Paper speed control
Tension wheels
Paper guides
Pen lifter

Power switch
Power terminal
Ink reservoir
Power cord
Event marker
Paper compartment

Figure 8.6 *(a)* Recording of summation of muscle twitches using physiograph; *(b)* illustration of summation, tetanus, and fatigue in isolated muscle.

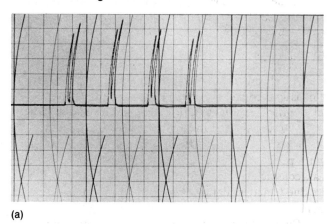

(a)

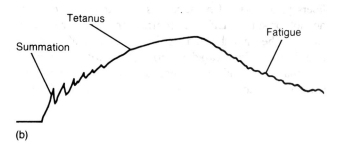

Summation
Tetanus
Fatigue

(b)

When a muscle fiber is stimulated by a motor neuron, its contraction is **all-or-none.** The contraction strength is constant because the amplitude of the action potentials that stimulate it is constant (action potentials are all-or-none—see chapter 7). Graded contractions of whole muscles, such as the gastrocnemius of a frog in the previously described experiment, are produced by variations in the *number* of muscle fibers that are stimulated to contract.

Motor Units

Gradations of muscle twitches *in vitro* result from "recruitment" of increasing numbers of muscle fibers that contribute to the twitch as the stimulating voltage is increased. The electric current stimulates muscle fibers directly. Skeletal muscle fibers *in vivo,* however, are stimulated by somatic motor neurons.

The muscle fibers are stimulated *in vivo* by axons of somatic motor neurons whose cell bodies reside in the ventral grey matter of the spinal cord. A given motor axon has many collateral branches, each of which innervates a single muscle fiber. Each muscle fiber is innervated by only one axon terminal. Each somatic motor neuron, together with all of the muscle fibers that it innervates, is known as a **motor unit.**

Figure 8.7   A motor unit. Note that a single somatic motor neuron makes neuromuscular junctions with a number of muscle fibers.

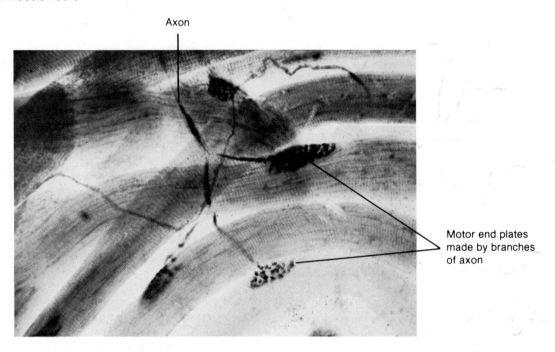

Axon

Motor end plates
made by branches
of axon

Figure 8.8   Diagram illustrating innervation of muscle fibers by different motor units. (Actually, many more muscle fibers would be included in a single motor unit than is shown in this drawing.)

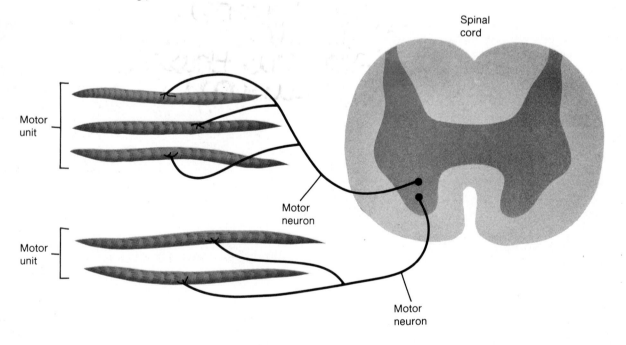

Spinal
cord

Motor
unit

Motor
unit

Motor
neuron

Motor
neuron

Figure 8.9 *(a)* Isometric and *(b)* isotonic contraction.

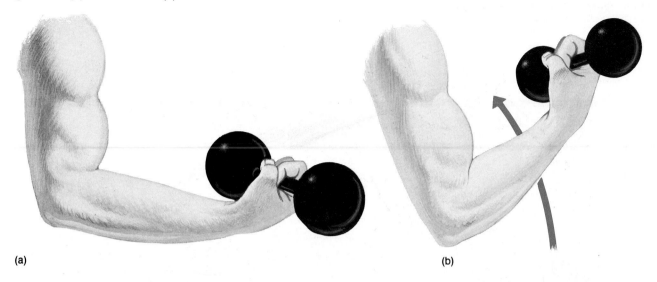

(a)   (b)

Whenever a motor neuron is activated, all of the muscle fibers that it innervates are stimulated to produce all-or-none twitches. Graded contractions of whole muscles *in vivo* are produced by variations in the number of motor neurons that are activated, and smooth, tetanized contractions are produced by rapid, asynchronous stimulation of different motor units.

Fine neural control over the strength of muscle contraction is optimal when there is a large number of small motor units. In the extraocular muscles that position the eyes, for example, the *innervation ratio* of a motor unit is one neuron per twenty-three muscle fibers, on the average. This affords a very fine degree of control. The innervation ratio of the gastrocnemius muscle, in contrast, averages one neuron per one thousand muscle fibers. Stimulation of these motor units results in more powerful contractions at the expense of finer gradations in contraction strength.

All of the motor units controlling the gastrocnemius, however, are not the same size. Innervation ratios vary from 1:100 to 1:2000. Neurons that innervate smaller numbers of muscle fibers have smaller cell bodies, and are stimulated by lower levels of excitatory input (EPSPs—see chapter 7) than the larger neurons that have larger innervation ratios. The smaller motor units, as a result, are the ones that are used most often. The larger motor units are only activated when very forceful contractions are required.

Isotonic and Isometric Contractions

In order for muscle fibers to shorten when they contract, they must generate a force that is greater than the opposing forces that act to prevent movement of the muscle's insertion. Flexion of the forearm, for example, occurs against the force of gravity and the weight of the objects being lifted (see figure 8.9). The tension produced by contraction of each muscle fiber separately is insufficient to overcome these opposing forces, but the combined contractions of large numbers of muscle fibers may be sufficient to overcome the opposing force and flex the forearm. In this case the muscle, and all its fibers, shorten in length.

Contraction that results in muscle shortening is called *isotonic contraction,* so called because the force of contraction remains relatively constant throughout the shortening process (iso = same; tonic = strength). If the opposing forces are too great, or the number of motor units activated is too few to shorten the muscle, however, the contraction is called *isometric* (literally, *same length*).

Isometric contraction can be voluntarily produced, for example by lifting a weight and maintaining the forearm in a partially flexed position. One can then increase the amount of muscle tension produced by recruiting more motor units until the muscle begins to shorten; at this point, isometric contraction is converted to isotonic contraction.

1. Illustrate and verbally describe muscle twitch, summation, and tetanus.
2. Describe how graded muscle contractions result from fiber twitches that are all-or-none, and how a smooth muscle contraction (tetanus) is produced *in vivo.*
3. Illustrate one motor unit with an innervation ratio of 1:5. Describe the functional significance of the innervation ratio.
4. Try to "make a muscle" with your arm and while doing so, feel your *biceps brachii* and your *triceps brachii.* Are these muscles contracting isotonically or isometrically? How is this type of contraction produced (include the antagonism of these two muscle groups in your answer)?

Figure 8.10 Sensory neuron, association neuron (interneuron), and somatic motor neuron at the spinal cord level.

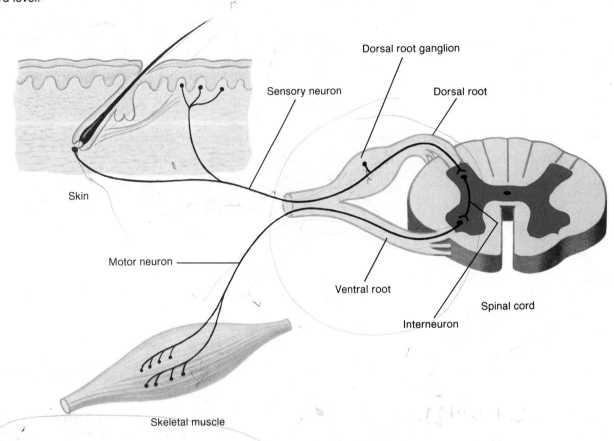

Dorsal root ganglion

Sensory neuron

Dorsal root

Skin

Motor neuron

Ventral root

Spinal cord

Interneuron

Skeletal muscle

Lower Motor Neuron Control of Skeletal Muscle Contraction

The **lower motor neurons** are those previously described in the spinal cord that directly stimulate muscle contraction. The activity of these neurons is influenced by (1) sensory feedback from the muscles and tendons and (2) facilitory and inhibitory effects from **higher motor neurons** in the brain. Lower motor neurons are thus said to be the *final common pathway* by which sensory stimuli and higher brain centers exert control over skeletal movements.

The cell bodies of lower motor neurons are located in the ventral horn of the grey matter of the spinal cord. Axons from these cell bodies leave the ventral side of the spinal cord to form the *ventral roots* of spinal nerves. The *dorsal roots* of spinal nerves contain sensory fibers whose cell bodies are located in the *dorsal root ganglia* (a ganglion, as described in chapter 7, is a collection of cell bodies located outside the CNS). Both sensory *(afferent)* and

motor *(efferent)* fibers join in a common connective tissue sheath to form the spinal nerves at each segment of the spinal cord. In the lumbar region (the small of the back), there are about twelve thousand sensory and six thousand motor fibers per spinal nerve.

About 375,000 cell bodies have been counted in a lumbar segment—a number far larger than can be accounted for by the number of motor neurons. Most of these neurons do not contribute fibers to the spinal nerve, but rather serve as *interneurons* whose fibers conduct impulses up, down, and across the central nervous system. Those fibers that conduct impulses to higher spinal cord segments and the brain form *ascending tracts*. Those fibers that conduct to lower spinal segments contribute to *descending tracts*. And those that cross the midline of the CNS to synapse on the opposite side are part of *commissural tracts*. Interneurons can thus conduct impulses up and down on the same, or *ipsilateral*, side, and can affect neurons on the opposite, or *contralateral*, side of the central nervous system.

Table 8.2 Some terms used to describe neural control of skeletal muscles.

Term	Description
1. Lower motoneurons	Neurons whose axons innervate skeletal muscles—also called the "final common pathway" in the control of skeletal muscles.
2. Higher motoneurons	Neurons in the brain that are involved in the control of skeletal movements; they act by facilitating or inhibiting (usually by way of interneurons) activity of the lower motoneurons.
3. Alpha motoneurons	Lower motoneurons whose fibers innervate ordinary (extrafusal) muscle fibers.
4. Gamma motoneurons	Lower motoneurons whose fibers innervate the muscle spindle fibers (intrafusal fibers).
5. Agonist/antagonist	Pair of muscles or muscle groups that insert on the same bone; the agonist is the muscle of reference.
6. Synergist	A muscle whose action facilitates the action of the agonist.
7. Ipsilateral/contralateral	Ipsilateral refers to the same side, or the side of reference; contralateral is the opposite side.
8. Afferent/efferent	Afferent neurons are sensory; efferent neurons are motor.

Table 8.3 Spindle apparatus content of selected skeletal muscles.

Muscle		Muscle Weight (g)	Average Number of Spindles	Number of Spindles Per Gram Muscle
Gastrocnemius		7.6	35	5
Rectus femoris		8.36	104	12
Tibialis anterior	leg	4.57	71	15
Semitendinosis		6.41	114	18
Soleus		2.49	56	23
Fifth interossei—foot		0.33	29	88
Fifth interossei—hand		0.21	25	119

Muscle Spindle Apparatus and Golgi Tendon Organ

In order for the nervous system to properly control skeletal movements it must receive continuous sensory feedback information concerning the effects of its actions. This sensory information includes (1) tension that the muscle exerts on its tendons, provided by the **Golgi tendon organs** and (2) muscle length, provided by the **muscle spindle apparatus.** The spindle apparatus, so called because it is wider in the center and tapers towards the ends, functions as a length detector. Muscles that require the finest degree of control, such as the muscles of the hand, have the highest density of spindles (see table 8.3).

Each spindle apparatus contains several thin muscle cells, called *intrafusal fibers* (fusus = spindle), packaged within a connective tissue sheath. Like the stronger and more numerous "ordinary" muscle fibers—the *extrafusal fibers*—the spindles insert into tendons on each end of the muscle. Spindles are therefore said to be in parallel with the extrafusal fibers.

The contractile structures (the myofibrils, as discussed in a later section of this chapter) of the intrafusal fibers are located in the tapered ends of the spindle, while the central region contains many nuclei arranged either in a loose aggregate ("bag-type" fibers) or in a row ("chain-type" fibers). These central regions are the most distensible and are therefore the most affected by stretch. Sensory nerve endings located in this central region—called *primary* or *annulospiral endings*—are stimulated by sudden increases in the length of the muscle and spindles. Impulses generated by primary sensory endings are conducted very rapidly (up to 120 meters per second) into the spinal cord. *Secondary,* or *flower spray,* sensory endings are located toward the ends of the spindle. Unlike the primary sensory endings, which respond to sudden stretch of the muscle with a burst of activity, the secondary endings maintain a relatively constant firing rate as long as stretch is maintained.

Figure 8.11   The structure of muscle spindles and their relationship to skeletal muscles.

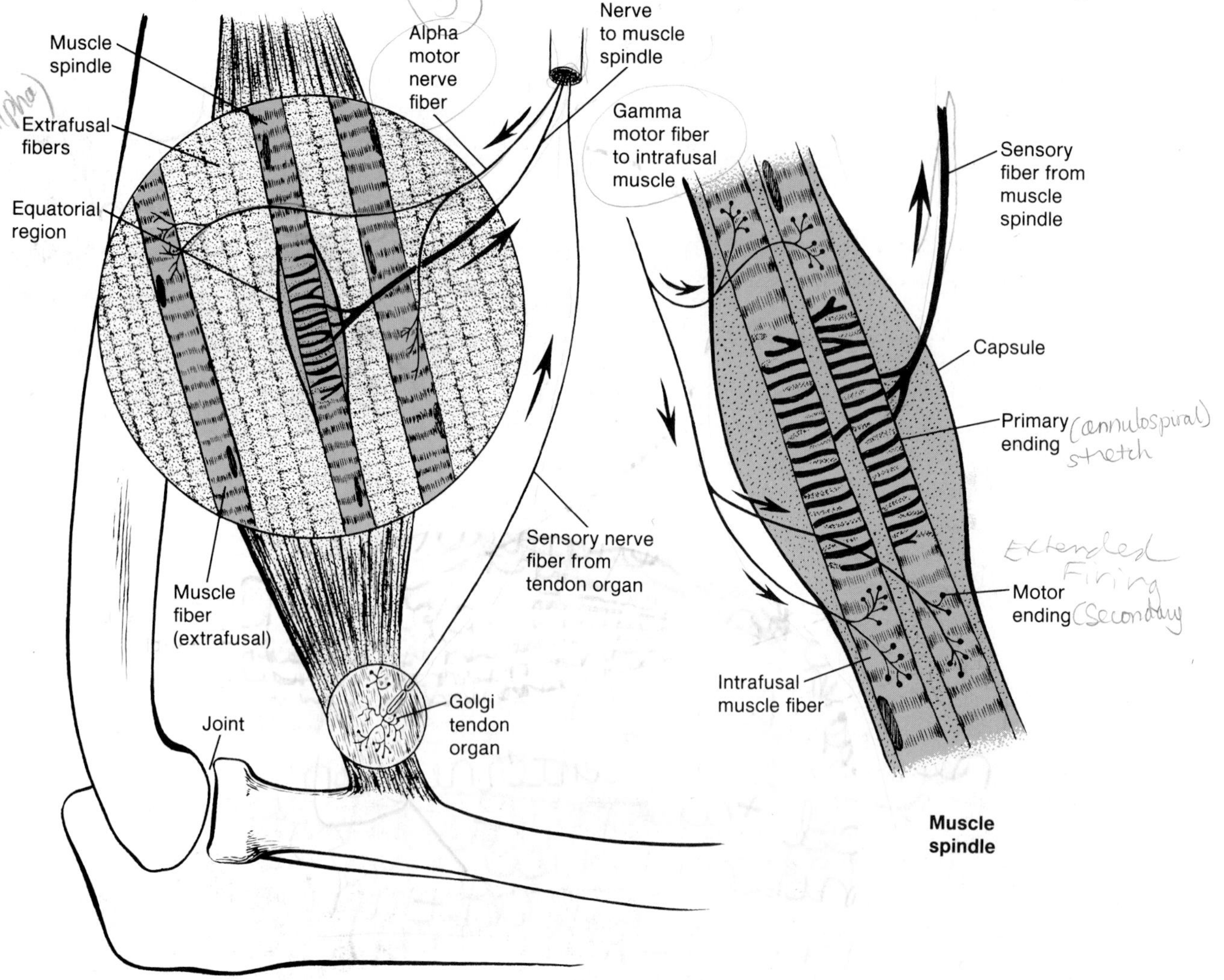

Skeletal Muscle Reflexes

Reflex contraction of skeletal muscles occurs in response to sensory input and is not dependent upon activation of higher motor neurons. The **reflex arc,** which describes the nerve impulse pathway from sensory to motor endings in such reflexes, involves only a few synapses within the CNS. The simplest of all reflexes—the muscle *stretch reflex*—consists of only one synapse within the CNS. The sensory neuron directly synapses with the motor neuron, without involving spinal cord interneurons. The stretch reflex is thus *monosynaptic* in terms of the individual reflex arcs (many sensory neurons, of course, are activated at the same time, leading to activation of many motor neurons).

The Monosynaptic Stretch Reflex

Resting skeletal muscles are maintained at an optimal length (the relationship between resting length and the strength of contraction will be described in a later section) by sensory feedback and nerve stimulation. A muscle that is removed from the body can be easily stretched. Muscles *in vivo,* however, resist passive stretching through activation of stretch reflexes.

The stretch reflex is present in all muscles, but is most dramatic in the extensor muscles of the limbs. The **knee jerk reflex**—the most commonly evoked stretch reflex—is initiated by striking the patellar tendon with a rubber mallet. This stretches the entire body of the muscle and thus passively stretches the spindles within the muscle. Sensory nerves with primary (annulospiral) endings in the spindles are thus activated, which synapse within the ventral grey matter of the spinal cord with *alpha motoneurons.* These relatively large, rapidly conducting (60–90 meters per second) motor nerve fibers stimulate the extrafusal fibers of the extensor muscle, resulting in isotonic contraction and the knee jerk. This is an example of negative feedback—stretching of the muscles (and spindles) stimulates shortening of the muscles (and spindles). These events are summarized in table 8.4 and in figure 8.12.

Figure 8.12 The knee jerk reflex, an example of a monosynaptic stretch reflex.

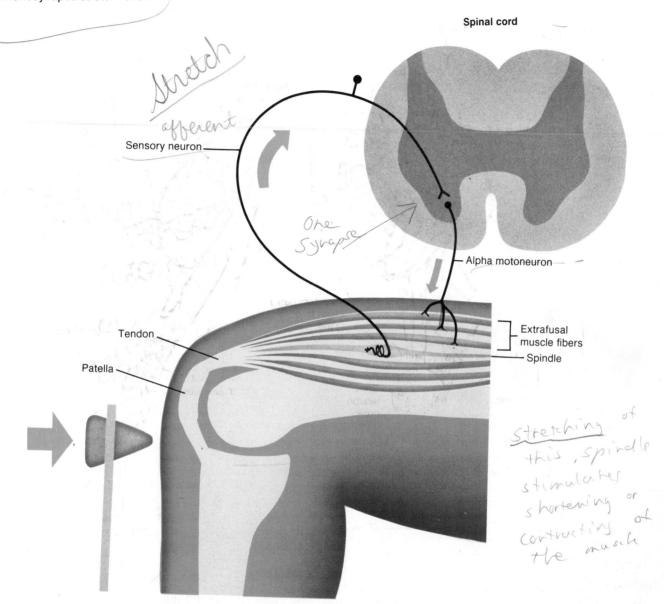

Stretch

afferent

Sensory neuron

Spinal cord

One Synapse

Alpha motoneuron

Tendon

Patella

Extrafusal muscle fibers

Spindle

Stretching of this, spindle stimulates shortening or contracting of the muscle

Table 8.4 Summary of events that occur during a monosynaptic stretch reflex.

1. Passive stretch of a muscle (produced by tapping its tendon) stretches the spindle (intrafusal) fibers.
2. Stretching of a spindle distorts its central (bag or chain) region, which stimulates dendritic endings of sensory nerves.
3. Action potentials are conducted by afferent (sensory) nerve fibers into the spinal cord on the dorsal roots of spinal nerves.
4. Axons of sensory neurons synapse with dendrites and cell bodies of somatic motor neurons located in the ventral horn grey matter of the spinal cord.
5. Efferent nerve impulses in the axons of somatic motor neurons (which form the ventral roots of spinal nerves) are conducted to the ordinary (extrafusal) muscle fibers. These neurons are alpha motoneurons.
6. Release of acetylcholine from the endings of alpha motoneurons stimulates contraction of the extrafusal fibers, and thus of the whole muscle.
7. Contraction of the muscle relieves the stretch of its spindles, thus decreasing electrical activity in the spindle afferent nerve fibers.

Figure 8.13   An increase in muscle tension stimulates activity of sensory nerve endings in the Golgi tendon organ. This sensory input stimulates ⊕ an interneuron, which in turn inhibits ⊖ activity of a motor neuron innervating that muscle. This is therefore a disynaptic reflex.

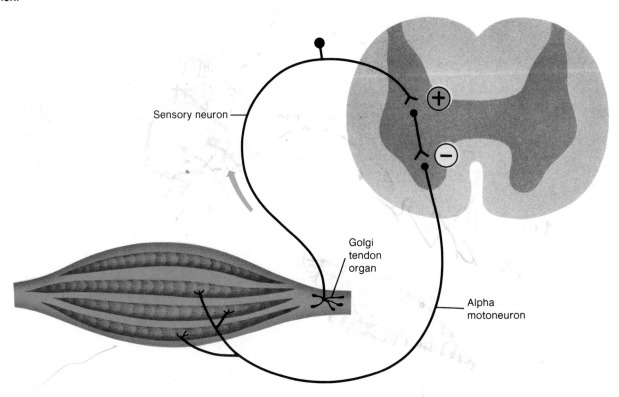

Rapid stretching of skeletal muscles thus stimulates reflex contraction. This can result in painful muscle spasms, as may occur, for example, when muscles are forcefully pulled in the process of setting broken bones. Painful muscle spasms may be avoided in physical exercise by stretching slowly and thereby stimulating mainly the secondary sensory nerve endings. A slower rate of stretch may also allow more time for the inhibitory reflex of Golgi tendon organs to occur.

Golgi Tendon Organs   The Golgi tendon organs continuously monitor tension in the tendons produced by muscle contraction. Sensory neurons from these receptors synapse with spinal cord interneurons, which in turn have *inhibitory synapses* (through production of IPSPs) with motor neurons that innervate the muscle. This inhibitory *disynaptic reflex* (because two synapses are crossed in the CNS) helps prevent excessive muscle contractions and provides a finer control over skeletal movements.

Gamma Motoneurons

The alpha motoneurons, as previously discussed, are rapidly conducting fibers that stimulate the extrafusal muscle fibers. One-third of all efferent fibers in spinal nerves, however, are thin, slowly conducting (10–40 meters per second) **gamma motoneurons** that innervate the intrafusal fibers of the muscle spindles. Activation of gamma motoneurons cannot directly cause muscle (and spindle) shortening because the intrafusal fibers are too few in number and too weak to produce isotonic contraction.

Since the ends of the spindle are embedded within the same tendons as the extrafusal fibers are, the length of the spindle (measured end to end) cannot be shortened if the muscle is not shortened. Gamma motoneuron stimulation of intrafusal fibers instead causes the more distensible middle region of these fibers to be pulled towards the ends; the spindle, in other words, is tightened. This effect of gamma motoneurons is sometimes termed *active stretch* of the spindles, and acts to increase the sensitivity of the spindles to passive stretch.

Figure 8.14 Diagram of reciprocal innervation. Afferent impulses from muscle spindles stimulate alpha motoneurons to the agonist muscle (the extensor) directly, but (via an inhibitory interneuron) inhibit activity in the alpha motoneuron to the antagonist muscle.

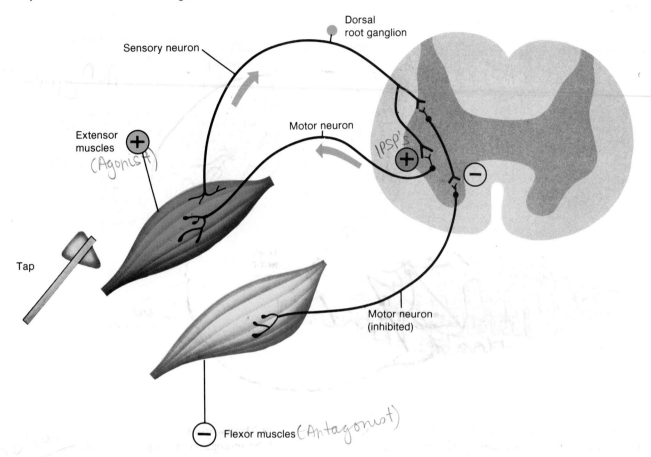

When a muscle is shortened by alpha-motoneuron stimulation of muscle contraction, the spindles are passively shortened along with the entire muscle. If there were no compensations, this shortening would place "slack" in the spindles and make them less sensitive—their function as length detectors would be compromised in the process of muscle shortening. This is prevented by the effects of gamma motoneuron activity, which continuously "take out the slack" in the spindles as the muscle shortens. The sensitivity of the spindles, and their ability to function as length detectors, are thus maintained even during the act of muscle shortening.

Reciprocal Innervation and the Crossed-Extensor Reflex
When a limb is flexed, the antagonistic extensor muscles are passively stretched. Extension of a limb similarly stretches the antagonistic flexor muscles. If the monosynaptic stretch reflexes were not inhibited, reflex contraction of the antagonistic muscles would always interfere with the intended movement. Fortunately, whenever the "intended," or agonist muscles, are stimulated to contract, the alpha motoneurons that stimulate the antagonist muscles are inhibited.

Inhibition of the alpha motoneurons that stimulate antagonistic muscles is accomplished by interneurons, which inhibit motor neurons via inhibitory postsynaptic potentials (IPSPs—that is, by postsynaptic inhibition). In the knee jerk and other stretch reflexes, the sensory neuron that stimulates the motor neuron of the agonist muscle also stimulates, via collateral branches, the interneurons that inhibit the motor neurons of antagonist muscles. This dual stimulatory and inhibitory activity is called **reciprocal innervation** (see figure 8.14).

The stretch reflex, with its reciprocal innervations, involves the muscles of one limb only and is controlled by only one segment of the spinal cord. More complex reflexes involve muscles controlled by a number of spinal cord segments, and affect muscles on the contralateral side of the cord. Such reflexes involve **double reciprocal innervation** of muscles.

Figure 8.15 The crossed extensor reflex, demonstrating double reciprocal innervation.

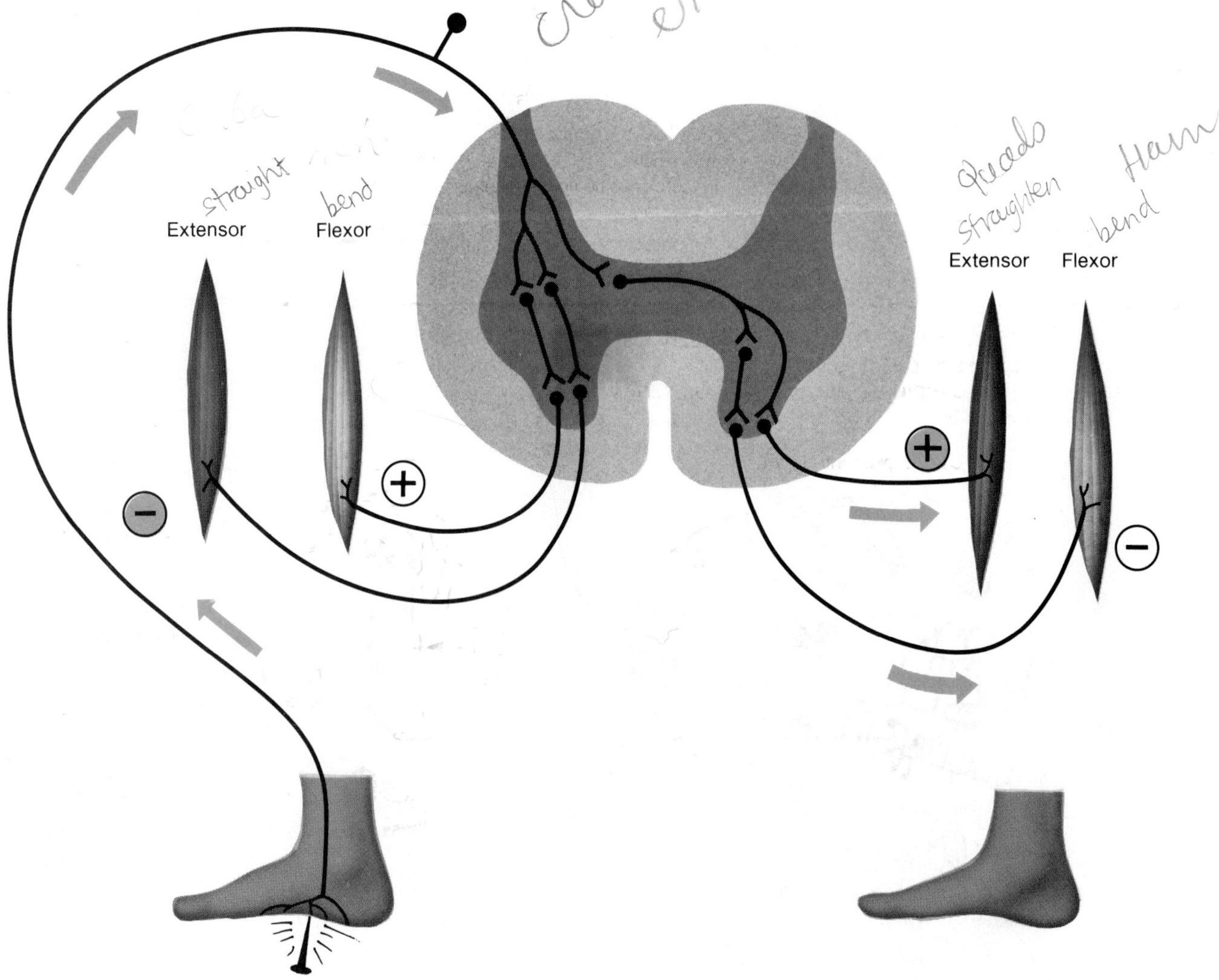

Double reciprocal innervation is illustrated by the **crossed-extensor reflex.** If one steps on a tack with the right foot, for example, this "ipsilateral" foot is withdrawn by contraction of its flexors and relaxation of its extensors. The contralateral left leg, in contrast, extends to help support the body during this withdrawal reflex. The extensors of the left leg contract while its flexors relax. These events are shown in figure 8.15.

---

1. Draw a muscle spindle surrounded by a few extrafusal fibers. Show a cross section of the spinal cord and a sensory and alpha motoneuron in your illustration. Use arrows to show the direction with which impulses are conducted in the reflex arc.
2. Describe all of the events that occur between the time that a knee is struck with a hammer and the time that the leg kicks. Use a flow chart or outline.
3. Describe the role of gamma motoneurons in the control of skeletal movements.

Higher Motor Neuron Control of Skeletal Muscle Contraction

The nature of higher motor neuron control of muscle contraction is illustrated by the different effects of "lesions" (damage) to lower and higher motor neurons. Damage to spinal nerves, or to the cell bodies of lower motor neurons (by poliovirus, for example) produces a **flaccid paralysis** characterized by reduced muscle tone, depressed stretch reflexes, and atrophy. Damage to higher motor neurons at first produces *spinal shock* (in which there is a flaccid paralysis), followed within a few weeks by **spastic paralysis.** Spastic paralysis is characterized by increased muscle tone, exaggerated stretch reflexes, and other signs of hyperactive lower motor neurons.

The immediately produced flaccid paralysis of spinal shock is apparently due to the removal of normal stimulation of lower motor neurons by higher motor neurons.

Table 8.5 Effects of lesions (damage) at different levels in the neural control of skeletal muscles.

Dorsal roots of spinal nerves (sensory)	Coordinated movements difficult; difficulty walking (ataxia).
Transsection of spinal cord	First—*spinal shock*, characterized by lack of stretch reflexes and low muscle tone. After a few weeks—*decerebrate rigidity*, characterized by hyperactive stretch reflexes, flexion of the arms, and extension of the legs (spasticity). Also, forced flexion of hand or foot produces oscillating extension and flexion (clonus).
Cutting of pyramids	Voluntary movements require great effort, and movements lack precision. Low muscle tone on side opposite lesion.
Transsection between midbrain and cerebral cortex	Animal placed on side can right itself, stand, and walk. Movements are clumsy, and there is some rigidity of extensor muscles.
Removal of motor cortex	No paralysis, but movements are uncoordinated.

Figure 8.16 *(a)* The motor area (precentral gyrus) of the cerebral cortex. *(b)* The map shows the disproportionately large representation of face and hands over the rest of the body.

Motor cortex (precentral gyrus)

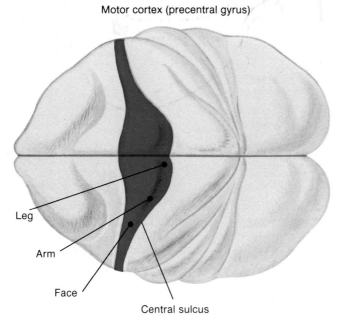

(a)

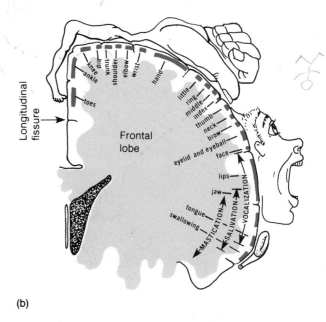

(b)

The spastic paralysis that follows suggests that the higher motor neurons must normally exert an inhibitory effect, as well as an excitatory effect, on the lower motor neurons. When this inhibitory influence is removed by damage to the spinal cord or brain the lower motor neurons become hyperexcitable.

Corticospinal (Pyramidal) Tracts

There are two major groups of descending tracts from the brain: the **pyramidal** (or direct) **tracts** and the **extrapyramidal tracts.** The pyramidal tracts descend directly, without synaptic interruption, from the cerebral cortex to the lower motor neurons. The cell bodies of the neurons that contribute fibers to this tract are located primarily in the *precentral gyrus* (also called the *motor cortex*)—see figure 8.16. Other areas of the cerebral cortex, however, also contribute fibers to this tract.

The neurons that control different body parts are arranged in order (somatotopically) in the motor cortex, as revealed by experiments in which particular areas of this region are stimulated by electrodes. The contractions of muscle groups that result from this procedure allow the motor cortex to be "mapped." This map has a bizarre appearance because the area of cortex devoted to the hands and face are disproportionately larger than the areas devoted to other parts of the body.

Most (about 80 to 90 percent) of the corticospinal fibers decussate (cross) in the brain stem, forming raised structures known as the pyramids; these crossed fiber tracts are thus called pyramidal tracts. The fibers that cross over to the contralateral side form the *lateral corticospinal tracts.* The remaining uncrossed fibers form the *ventral corticospinal tract.* Because of the crossing of fibers from

Figure 8.17   Areas of the brain containing neurons involved in the control of skeletal muscles (higher motor neurons). The thalamus is a relay center between the motor cortex and other brain areas.

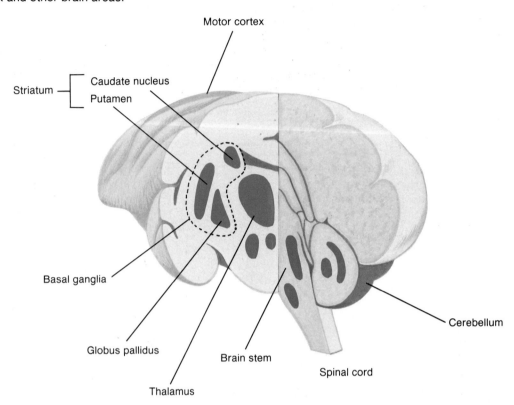

higher motor neurons in the pyramids, the right hemisphere primarily controls the musculature on the left side of the body, while the left hemisphere controls the right musculature. Most of these descending motor tracts synapse with interneurons, which in turn synapse with the lower motor neurons; only about 10 percent of the descending fibers synapse directly with the lower motor neurons.

The corticospinal tracts appear to be particularly important in voluntary movements that require complex interactions between the motor cortex and sensory input. Speech, for example, is impaired when corticospinal tracts are damaged in the thoracic region of the spinal cord, while involuntary breathing continues. Damage to the pyramidal motor system can be detected clinically by a positive *Babinski reflex,* in which stimulation of the sole of the foot causes extension (upward movement) of the toes. This positive Babinski reflex is normal in infants because neural control is not yet fully developed.

Extrapyramidal Tracts
If the pyramidal tracts of an experimental animal are cut, electrical stimulation of the motor cortex and of other areas of the cerebrum can still produce movement. The descending nerve fibers that mediate these movements

must, by definition, be extrapyramidal motor tracts. Unlike the pyramidal tracts, which originate in the cerebral cortex, the extrapyramidal tracts originate in the midbrain and brain stem regions. Electrical stimulation of the *cerebral cortex,* the *cerebellum,* and the *basal ganglia* (also called "cerebral nuclei") indirectly evokes movements because of their synaptic connections with these midbrain and brain stem areas.

The *reticulospinal tract* is the major descending pathway of the extrapyramidal system. Fibers of this tract originate in a loose aggregation of cells and fibers known as the **reticular formation** (reticul = network) in the medulla oblongata and pons areas of the brain stem. Electrical stimulation of the reticular formation either facilitates or inhibits (depending on the area) activity of lower motor neurons. The spastic paralysis that may follow spinal cord injury is believed to be caused by release of gamma motoneurons from the normal inhibitory influence of reticulospinal tract synapses. The brain stem reticular formation, in turn, receives both facilitory and inhibitory input from the cerebrum and the cerebellum (see figure 8.18).

Figure 8.18   Pathways involved in the higher motor neuron control of skeletal muscles.

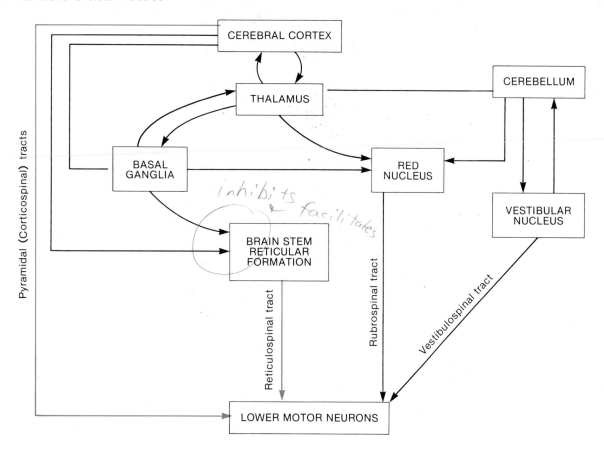

Table 8.6   Descending motor tracts to spinal interneurons and motor neurons.

Tract	Category	Origin	Crossed/Uncrossed
Lateral corticospinal	Pyramidal	Cerebral cortex	Crossed
Anterior corticospinal	Pyramidal	Cerebral cortex	Uncrossed
Rubrospinal	Extrapyramidal	Red nucleus (midbrain)	Crossed
Tectospinal	Extrapyramidal	Superior colliculus (midbrain)	Crossed
Vestibulospinal	Extrapyramidal	Vestibular nuclei (medulla oblongata)	Uncrossed
Reticulospinal	Extrapyramidal	Brain stem reticular formation (medulla and pons)	Crossed

Cerebellum   The cerebellum, like the cerebrum, receives sensory input from muscle spindles and Golgi tendon organs. It also receives fibers from areas of the cerebral cortex devoted to vision, hearing, and equilibrium (the latter areas are called the vestibular nuclei because they process sensory input from the vestibular apparatus in the inner ear).

There are no descending tracts from the cerebellum. The cerebellum can only influence motor activity indirectly, through its output to the vestibular nuclei, red nucleus, and basal ganglia. These structures, in turn, affect lower motor neurons via the vestibulospinal tract, rubrospinal tract, and reticulospinal tract. Damage to the cerebellum interferes with the coordination of movements with spatial judgment. Under or overreaching for an object may occur, followed by *intention tremor,* in which the limb moves back and forth in a pendulum-like motion.

Table 8.7   Some symptoms of higher motor neuron damage.

Babinski reflex—extension of the big toe when the sole of the foot is rubbed along the lateral border

Spastic paralysis—high muscle tone and hyperactive stretch reflexes; flexion of arms and extension of legs

Hemiplegia—paralysis of upper and lower limbs on one side—commonly produced by damage to motor tracts as they pass through internal capsule (such as by cerebrovascular accident—stroke)

Paraplegia—paralysis of lower limbs on both sides, by lower spinal cord damage

Quadriplegia—paralysis of both upper and lower limbs on both sides, by damage to the upper region of the spinal cord or brain

Chorea—random uncontrolled contractions of different muscle groups (such as Saint Vitus Dance), produced by damage to basal ganglia

Resting tremor—shaking of limbs at rest; disappears during voluntary movements; produced by damage to basal ganglia

Intention tremor—oscillations of arm following voluntary reaching movements; produced by damage to cerebellum

Basal Ganglia   The basal ganglia, or cerebral nuclei, include the *caudate nucleus, putamen,* and *globus pallidus.* Often included in this group are other nuclei of the *thalamus, subthalamus, substantia nigra,* and *red nucleus.* Acting directly via the rubrospinal tract and indirectly via synapses in the reticular formation and thalamus, the basal ganglia have profound effects on the activity of lower motor neurons.

The basal ganglia, acting through synapses in the reticular formation particularly, appear normally to exert an inhibitory influence on the activity of lower motor neurons. Damage to the basal nuclei thus results in increased muscle tone. People with such damage display *akinesia* (lack of desire to use the affected limb) and *chorea*—sudden and uncontrolled random movements.

Parkinson's disease (or *paralysis agitans*) is a disorder of the basal ganglia involving degeneration of fibers from the substantia nigra. These fibers, which use dopamine as a neurotransmitter, are required to antagonize the effects of other fibers that use acetylcholine (ACh) as a transmitter. The relative deficiency of dopamine compared to ACh is believed to produce the symptoms of Parkinson's disease, including *resting tremor.* This "shaking" of the limbs tends to disappear during voluntary movements and then reappear when the limb is again at rest.

Parkinson's disease is treated with drugs that block the effects of ACh, and by administration of L-Dopa, which can be converted to dopamine in the brain (dopamine cannot be given directly because it does not cross the blood-brain barrier). Interestingly, excessive administration of L-Dopa may produce an unwanted side effect: the symptoms of schizophrenia! Schizophrenic patients who are treated with drugs that block dopamine action (such as chlorpromazine), conversely, may develop the resting tremor characteristic of Parkinson's disease. From these and other observations, many authorities now believe that schizophrenia may be caused, at least in part, by abnormally high levels of dopamine in the brain.

1. List the structures and function of the pyramidal tracts. Make a similar list for the extrapyramidal tracts.
2. Explain why damage to the right side of the brain primarily affects motor activities on the left side of the body.
3. Describe flaccid and spastic paralysis, and describe how each is produced by damage to the lower and/or higher motor neurons.

Mechanisms of Contraction

When muscle cells are viewed at high magnification, in the electron microscope, each cell is seen to be composed of many subunits known as **myofibrils** (little fibers). These myofibrils are approximately one micrometer (1 μm) in diameter and extend in parallel from one end of the muscle fiber to the other. The myofibrils are so densely packed that other organelles—such as mitochondria and intracellular membranes—are restricted to the narrow cytoplasmic spaces that remain between adjacent myofibrils.

It is the myofibrils that are striated with dark (A) and light (I) bands. The striated appearance of the entire muscle fiber, when seen with a light microscope, is an illusion created by the fact that the dark and light bands of the myofibrils are fairly in line with each other from one side of the fiber to the other. Since the separate myofibrils are not clearly seen at low magnification, the dark and light bands appear to be continuous across the width of the fiber.

Figure 8.19   A skeletal muscle fiber, as seen in longitudinal section in an electron microscope. In this view, the myofibrils are shown in side view, extending from one side of the picture to the other (three myofibrils are labeled for illustration).

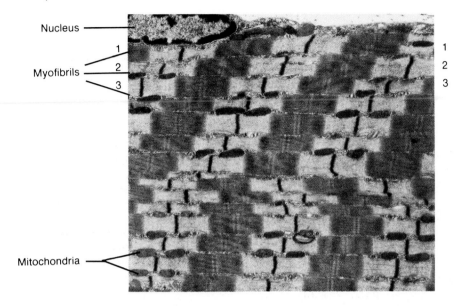

Nucleus

Myofibrils — 1
              2
              3

1
2
3

Mitochondria

Figure 8.20   Electron micrograph of longitudinal section of myofibrils, showing A, H and I bands. Notice how the dark and light bands of each myofibril are stacked in register.

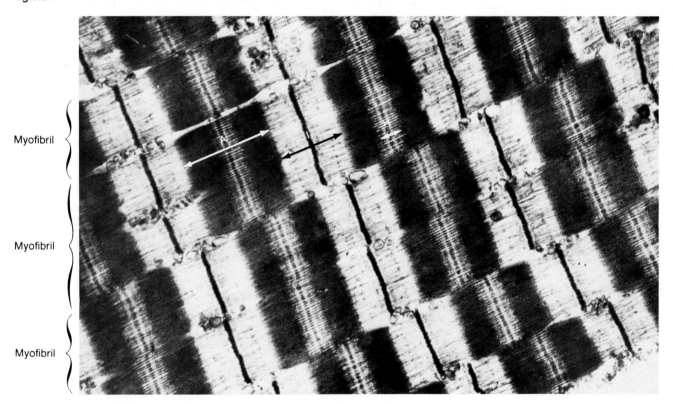

Myofibril

Myofibril

Myofibril

Figure 8.21   *(a)* is low-power (1600X) electron micrograph of a single skeletal muscle fiber showing numerous myofibrils. *(b)* is a higher-power (53,000X) electron micrograph of myofibrils in longitudinal section. Notice the overlapping of thick and thin myofilaments.

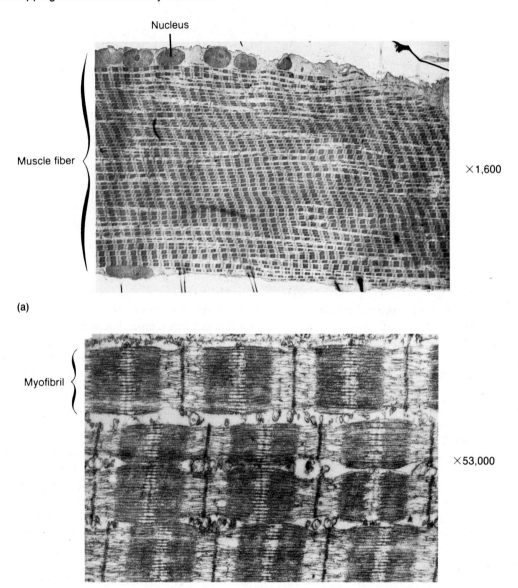

Nucleus

Muscle fiber

×1,600

(a)

Myofibril

×53,000

(b)

(c) shows hexagonal arrangement of thick and thin filaments in cross section (arrows point to cross-bridges) (SR = sarcoplasmic reticulum, M = mitochondria).

Myofibril

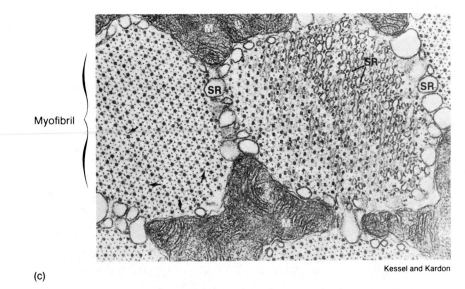

Kessel and Kardon

(c)

When a myofibril is observed at high magnification in longitudinal section (side view), the A bands are seen to contain **thick filaments** (about 110 Å—angstroms—thick; $1Å = 10^{-10}$m) that are stacked in register. It is these thick filaments that give the A band its dark appearance. The lighter I band, in contrast, contains **thin filaments** (about 50–60 Å thick). The thick filaments are composed of the protein **myosin,** while the thin filaments are composed primarily of the protein **actin.**

The I bands within a myofibril extend from the edge of one stack of thick myosin filaments to the edge of the next stack of thick filaments. They are light bands because they contain only thin filaments. Each thin filament, however, extends partway into the A bands on each side (between the thick filaments of each stack). Since thick and thin filaments overlap at the edges of each A band, the edges are darker in appearance than the central region of the A band. These central lighter regions of the A bands are called the *H bands* (for *helle,* a German word for bright). The central H bands thus contain only myosin that is not overlapped with thin filaments.

In the center of each I band is a thin dark Z line. The arrangement of thick and thin filaments between a pair of Z lines forms a repeating pattern that serves as the basic subunit of striated muscle contraction. These subunits, from Z to Z, are known as **sarcomeres.** A longitudinal section of a myofibril thus presents a side view of successive sarcomeres.

This side view is, in a sense, misleading; there are numerous sarcomeres within each myofibril that are out of the plane of the section (and out of the picture). A better appreciation of the three-dimensional structure of a myofibril can be obtained by viewing the myofibril in cross section. In this view, it can be seen that the Z lines are in reality disc shaped and that the thin filaments that penetrate these Z discs surround the thick filaments in a hexagonal arrangement (see figure 8.21c). If one concentrates on a single row of dark thick filaments in this cross section, the alternating pattern of thick and thin filaments seen in longitudinal section becomes apparent.

Figure 8.22 The sliding filament model of contraction. As the filaments slide, the Z lines are brought closer together. The A bands remain the same length during contraction, but the I and H bands get progressively narrower and may eventually become obliterated.

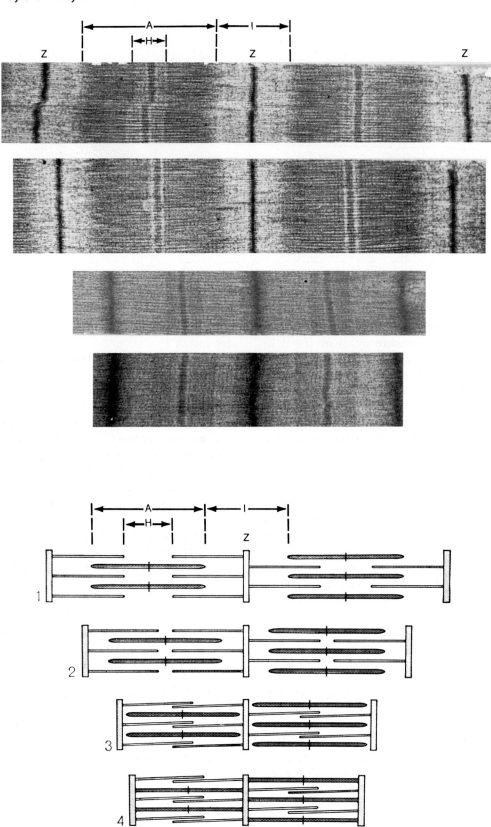

Figure 8.23 Myosin cross-bridges are oriented in opposite directions on either side of a sarcomere.

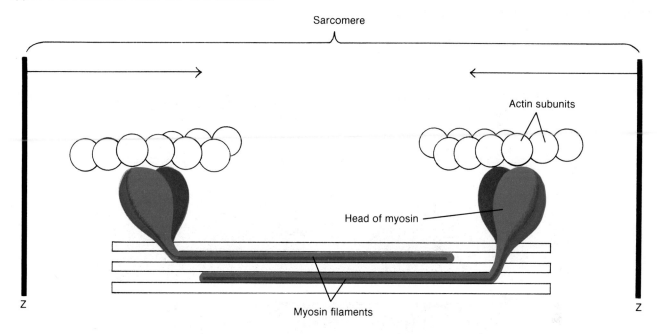

Sliding Filament Theory of Contraction

When a muscle contracts isotonically, it decreases in length as a result of the shortening of its individual fibers. Shortening of the muscle fibers, in turn, is produced by shortening of their myofibrils, which occurs as a result of the shortening of the distance from Z line to Z line. As the sarcomeres shorten in length, however, the A bands do *not* shorten but instead move closer together. The I bands—which represent the distance between A bands of successive sarcomeres—decrease in length.

Close examination reveals that the thick and thin filaments remain the same length during muscle contraction. Shortening of the sarcomeres is produced, not by shortening of the filaments, but rather by the *sliding* of thin filaments over thick filaments. In the process of contraction, the thin filaments on either side of each A band extend deeper and deeper towards the center, producing increasing amounts of overlap with the thick filaments. The central H bands thus get shorter and shorter during contraction (see figure 8.22).

Cross-Bridges Sliding of the filaments is produced by the action of numerous *cross-bridges* that extend out from the myosin towards the actin. These cross-bridges are part of the myosin proteins that extend from the axis of the thick filaments to form "arms" that terminate in globular "heads" (see figure 8.23). The orientation of cross-bridges on one side of a sarcomere is opposite to that on the other side, so that when they attach to actin on each side of the

sarcomere they can pull the actin from each side towards the center.

Relaxed muscles are easily stretched (although this is opposed *in vivo* by reflex contraction), demonstrating that the myosin cross-bridges are not attached to actin when the muscle is at rest. Biochemical evidence has shown that the globular heads of the cross-bridges contain an **ATPase** that binds ATP when the muscle is at rest, but that is incapable of splitting ATP under these conditions. Activation of this ATPase occurs when the myosin cross-bridge attaches to actin.

Attachment of the cross-bridge to actin activates the myosin ATPase; the resultant splitting of ATP provides the energy needed for the *power stroke,* which pulls the thin filaments towards the center of the A bands. The power stroke may be produced by movement of the cross-bridge arm (with the angle of the head constant), or by movement of the head with the arm position fixed. These possible mechanisms are illustrated in figure 8.24.

A single contraction cycle and power stroke of all the cross-bridges in a muscle would shorten the muscle by only about 1 percent of its resting length. Since muscles can shorten up to 60 percent of their resting length, it it obvious that the contraction cycles must be repeated many times. In order for this to occur, the cross-bridges must detach from the actin at the end of a power stroke, reassume their resting orientation, and then reattach to the actin to repeat the cycle.

Figure 8.24 Two possible mechanisms by which cross-bridges could produce sliding of the filaments. In *(a)* the cross-bridge arms are presumed to move on a "hinge" while the heads are fixed in position; in *(b)* the cross-bridge heads are presumed to swivel on "hinges" from a 90° angle to a 45° angle while the cross-bridge arms are stationary.

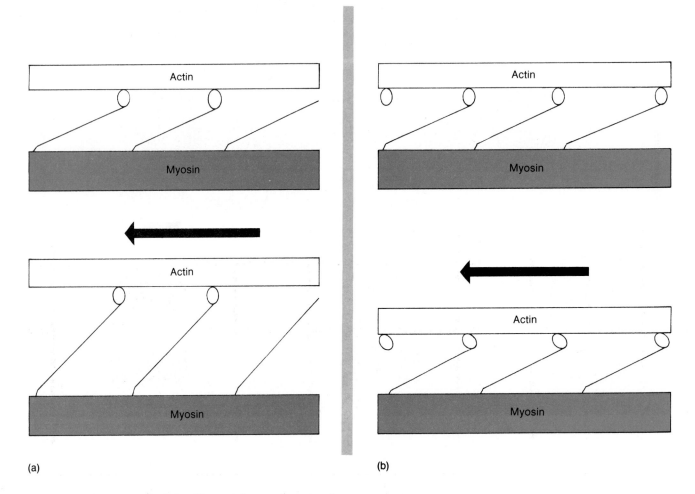

(a)

(b)

Table 8.8 Summary of sliding filament theory of contraction.

1. A myofiber, together with all its myofibrils, shortens by movement of the insertion towards the origin of the muscle.
2. Shortening of the myofibrils is caused by shortening of the sarcomeres—the distance between Z lines (or discs) is reduced.
3. Shortening of the sarcomeres is accomplished by sliding of the myofilaments—each filament remains the same length during contraction.
4. Sliding of the filaments is produced by asynchronous power strokes of myosin cross-bridges, which pull the thin filaments (actin) over the thick filaments (myosin).
5. The A bands remain the same length during contraction, but are pulled towards the origin of the muscle.
6. Adjacent A bands are pulled closer together as the I bands between them shorten.
7. The H bands shorten during contraction as the thin filaments from each end of the sarcomeres are pulled towards the middle.

Detachment of a cross-bridge from actin at the end of a power stroke requires the attachment of a "fresh" ATP to the myosin ATPase. The importance of this process is illustrated by the muscular contracture called *rigor mortis* that occurs due to lack of ATP when the muscle dies. This results in the formation of "rigor complexes" between myosin and actin that cannot detach.

In rigor mortis, all of the cross-bridges are attached to actin at the same time. During normal contraction, however, only about 50 percent of the cross-bridges are attached at any given time. The power strokes are thus not in synchrony, as the strokes of a competitive rowing team are. Rather, they are like the actions of a team engaged in a tug-of-war in which the pulling action of the members is asynchronous. Some cross-bridges are engaged in power strokes at all times during the contraction.

Figure 8.25 Length-tension relationship in skeletal muscles. Maximum relative tension (1.0) is achieved when the muscle is 100–120 percent of its resting length (sarcomere lengths from 2.0 to 2.25 μm). Increases or decreases in muscle (and sarcomere) lengths result in rapid decreases in tension.

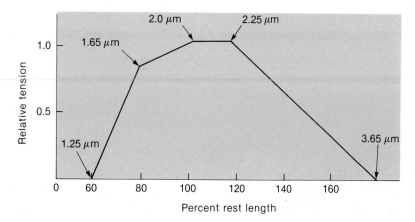

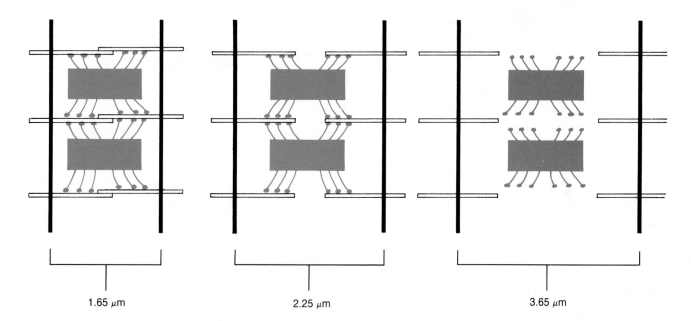

1.65 μm   2.25 μm   3.65 μm

Length-Tension Relationship

The strength of a muscle's contraction is affected by a number of factors. These include the number of muscle fibers within the muscle that are stimulated to contract, the thickness of each muscle fiber (thicker fibers have more myofibrils and thus can exert more power), and the initial length of the muscle fibers when they are at rest.

There is an "ideal" resting length of muscle fibers. When the resting length is more than this ideal, the overlap between actin and myosin is so little that few crossbridges can attach. When the muscle is stretched to the point that there is no overlap of actin with myosin, no crossbridges can attach to the thin filaments and the muscle cannot contract. When the muscle is shortened to about 60 percent of its resting length, the Z lines abut against the thick filaments so that further contraction cannot occur.

The strength of a muscle's contraction can be measured by the force required to prevent it from shortening. Under these isometric conditions, the strength of contraction, or *tension*, can be measured when the muscle length at rest is varied. Maximum tension is produced when the muscle is at its normal resting length *in vivo* (see figure 8.25). This resting length is maintained by reflex contraction in response to passive stretching, as described in an earlier section of this chapter.

Figure 8.26　Relationship of troponin and tropomyosin to actin in the thin filaments. The tropomyosin is attached to actin, while the troponin complex of three subunits is attached to tropomyosin (not directly to actin).

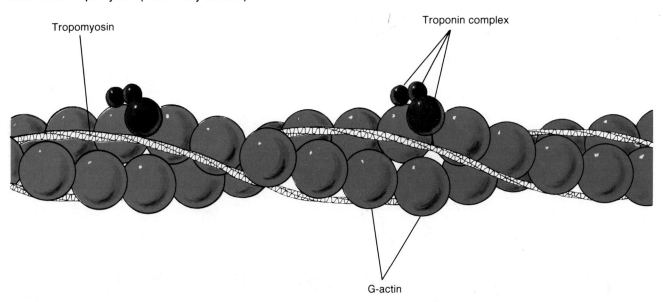

Tropomyosin

Troponin complex

G-actin

Regulation of Contraction

When the cross-bridges attach to actin the myosin ATPase enzymes are activated, resulting in power strokes and muscle contraction. In order for muscle relaxation to occur, the attachment of myosin cross-bridges to actin must be prevented. The regulation of cross-bridge attachment to actin is a function of two proteins found associated with actin in the thin filaments.

The actin filament—or *F-actin*—is a polymer formed of three hundred to four hundred globular subunits *(G-actin)* arranged in a double row and twisted to form a helix (see figure 8.26). A different type of protein, known as *tropomyosin,* lies within the groove between the double row of G-actin. There are forty to sixty tropomyosin molecules per thin filament, with each tropomyosin spanning a distance of approximately seven actin subunits.

Attached to the tropomyosin, rather than directly to the actin, is a third type of protein, called *troponin,* within the thin filaments. Troponin and tropomyosin work together to regulate the attachment of cross-bridges to actin, and thus serve as a switch for muscle contraction and relaxation. In a relaxed muscle the tropomyosin is in a position that physically blocks attachment of cross-bridges to actin. In order for the myosin cross-bridges to attach to actin the tropomyosin must be moved.

Role of Ca^{++} in Muscle Contraction

In a relaxed muscle, when tropomyosin blocks attachment of cross-bridges to actin, the concentration of Ca^{++} in the sarcoplasm (cytoplasm of muscle cells) is very low. When the muscle cell is stimulated to contract, the concentration of Ca^{++} in the sarcoplasm quickly rises above 10^{-6} molar. Some of this Ca^{++} attaches to a subunit of troponin, causing a conformational change that moves the troponin and its attached tropomyosin out of the way so that the cross-bridges can attach to actin (see figure 8.27).

The position of the troponin-tropomyosin complexes in the thin filaments is thus adjustable. When Ca^{++} is not attached to troponin, the tropomyosin is in a position that inhibits attachment of cross-bridges to actin; muscle contraction is prevented. When Ca^{++} attaches to troponin, the troponin-tropomyosin complexes shift position—cross-bridges attach to actin, the myosin ATPase is activated, ATP is split, and muscle contraction occurs. Cross-bridges can continue to attach to actin, produce a power stroke, detach from actin, and continue these contraction cycles as long as Ca^{++} is attached to troponin.

Figure 8.27 Effect of Ca^{++} on myosin cross-bridge attachment to actin. Colored G-actin subunits represent attachment sites for cross-bridges. At rest (1) these sites are blocked by tropomyosin. Attachment of Ca^{++} to troponin *(dark structures)* moves the troponin-tropomyosin so that cross-bridges can attach to G-actin (2). This is followed by hydrolysis of ATP and a power stroke (3). Detachment of the cross-bridges from actin (4) allows the contraction cycle to be repeated.

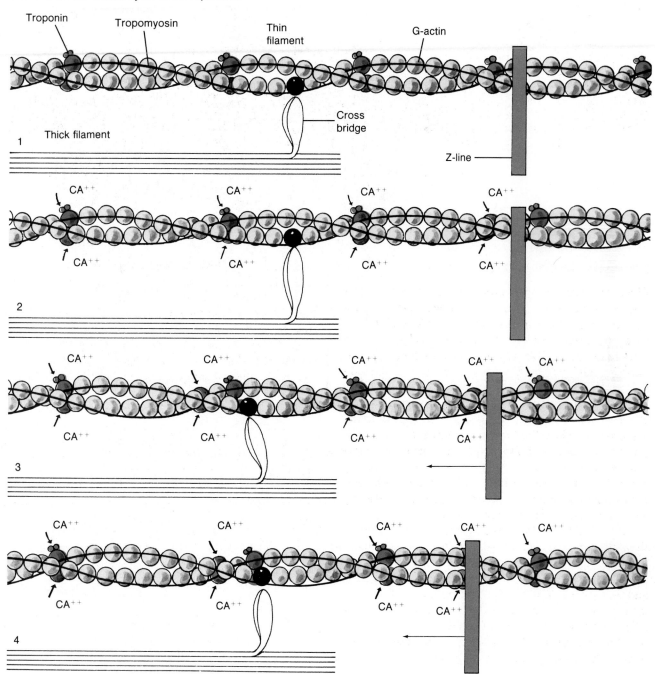

Figure 8.28   Artist's drawing of myofibrils, showing relationships to sarcoplasmic reticulum, transverse tubules, and sarcolemma.

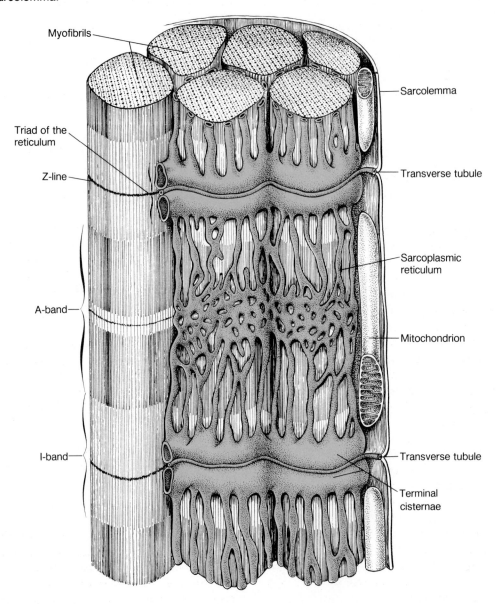

Excitation-Contraction Coupling

Muscle contraction is "turned on" when sufficient amounts of Ca^{++} bind to troponin. This occurs when the Ca^{++} concentration of the sarcoplasm rises above 10^{-6} molar. In order for muscle relaxation to occur, therefore, the Ca^{++} concentration of the sarcoplasm must be lowered below this level. Muscle relaxation is produced by the active transport of Ca^{++} out of the sarcoplasm into the **sarcoplasmic reticulum.** The sarcoplasmic reticulum is a modified endoplasmic reticulum consisting of interconnected sacs and tubes that surround each myofibril within the muscle cell.

Most of the Ca^{++} is stored within expanded portions of the sarcoplasmic reticulum known as *terminal cisternae*. These terminal cisternae are separated by only a very narrow gap from **transverse tubules,** which are narrow membranous "tunnels" that are continuous with the cell membrane (the *sarcolemma*), and which thus open to the extracellular environment through pores in the cell surface. The transverse tubules serve to transmit action potentials from the cell surface into the muscle fiber.

Table 8.9   Summary of events that occur during excitation-contraction coupling.

1. Action potentials in a somatic motor nerve cause release of acetylcholine neurotransmitter at the myoneural junction (one myoneural junction per myofiber).
2. Acetylcholine, through its interaction with receptors in the muscle cell membrane (sarcolemma), produces action potentials that are regenerated across the sarcolemma.
3. The membranes of the transverse tubules (T tubules) are continuous with the sarcolemma, and conduct action potentials deep into the muscle fiber.
4. Action potentials in the T tubules, by a mechanism that is poorly understood, stimulate the release of Ca^{++} from the terminal cisternae of the sarcoplasmic reticulum.
5. Ca^{++} released into the sarcoplasm attaches to troponin, causing a change in its structure.
6. The shape change in troponin causes its attached tropomyosin to shift position in the actin filament, thus exposing bonding sites for the myosin cross-bridges.
7. Attachment of myosin cross-bridges to the exposed sites on actin activates the myosin ATPase, thus hydrolyzing ATP.
8. Hydrolysis of ATP provides energy for the power stroke by which cross-bridges pull the thin filaments over the thick filaments.
9. Attachment of fresh ATP allows the cross-bridges to detach from actin and repeat the contraction cycle as long as Ca^{++} remains attached to troponin.
10. When action potentials stop being produced, the sarcoplasmic reticulum actively accumulates Ca^{++} and tropomyosin moves again to its inhibitory position.

Release of acetylcholine from axon terminals of somatic motor neurons at the neuromuscular junctions, as previously described, causes electrical activation of skeletal muscle fibers. End-plate potentials are produced that generate action potentials. Muscle action potentials, like the action potentials of axons, are all-or-none events that are regenerated along the cell membrane. It must be remembered that action potentials involve the flow of ions between the extracellular and intracellular environments across a cell membrane that separates these two compartments. In muscle cells, therefore, action potentials can be conducted into the interior of the fiber across the membrane of the transverse tubules.

Action potentials in the transverse tubules cause the release of Ca^{++} from the sarcoplasmic reticulum. This, in turn, stimulates muscle contraction. As long as action potentials continue to be produced—which is as long as the neural innervation of the muscle continues to be active—Ca^{++} will remain attached to troponin and cross-bridges will be able to undergo contraction cycles. When action potentials cease, the sarcoplasmic reticulum actively accumulates Ca^{++} and muscle relaxation occurs.

Neural regulation of skeletal muscle contraction is thus mediated by calcium ions. The mechanisms of electrical excitation and the mechanisms of muscle contraction (sliding of the filaments) are "coupled" through adjustments of the sarcoplasmic Ca^{++} concentration. The events that occur in this *excitation-contraction coupling* are summarized in table 8.9.

1. Draw three successive sarcomeres in a myofibril of a resting muscle fiber. Label the myofibril, sarcomeres, A bands, I bands, H bands, and Z lines.
2. Draw three successive sarcomeres in a myofibril of a contracted fiber. Indicate which bands get shorter during contraction and explain how this occurs.
3. Draw a figure of the molecular structure of the thick and thin filaments, showing the cross-bridges, G-actin, troponin, and tropomyosin. Describe the roles of Ca^{++}, troponin, and tropomyosin in muscle contraction.
4. List, in order of occurrence, the events that occur during the contraction cycles of the cross-bridges.
5. Draw a flow chart (using arrows) of the events that occur between release of ACh at the neuromuscular junction and the attachment of Ca^{++} to troponin.

Energy Usage by Skeletal Muscles

Skeletal muscles at rest obtain most of their energy from the aerobic respiration of fatty acids. During exercise, as described in chapter 5, muscle glycogen and blood glucose are also used as energy sources. Energy obtained by cell respiration is used to make ATP, which serves as the immediate source of energy for movement of the cross-bridges and muscle contraction.

During sustained muscle contraction, the utilization of ATP may occur at a faster rate than the rate of ATP production through cell respiration. At these times the

Figure 8.29 Production and utilization of phosphocreatine in muscles.

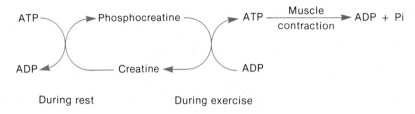

During rest   During exercise

rapid renewal of ATP is accomplished by combination of ADP with phosphate derived from another "high-energy phosphate" compound called **phosphocreatine** (or **creatine phosphate**).

The phosphocreatine concentration within muscle cells is greater than three times the concentration of ATP, and represents a ready reserve of high-energy phosphate that can be donated directly to ADP. During times of rest, the depleted reserve of phosphocreatine can be restored by the reverse reaction—phosphorylation of creatine with phosphate derived from ATP.

The enzyme that transfers phosphate between creatine and ATP is called **creatine kinase.** Skeletal muscle and heart muscle have two different forms of this enzyme (they have different isoenzymes, as described in chapter 4). The skeletal muscle isoenzyme is found to be elevated in the blood of people with *muscular dystrophy* (degenerative disease of skeletal muscles). The plasma concentration of the isoenzyme characteristic of heart muscle is elevated as a result of *myocardial infarction* (see chapter 11), and measurements of this enzyme are thus used as a means of diagnosing this condition.

Fast-, Slow-, and Intermediate-Twitch Fibers
When the cells of a muscle such as the gastrocnemius (calf muscle) are treated with chemicals that stain the fibers in accordance with their content of aerobic respiratory enzymes, three types of fibers are observed (see figure 8.31). The majority of the fibers are large and pale, indicating relatively low amounts of enzyme activity. These types of fibers are surrounded by relatively few capillaries. Scattered within the muscle are fewer small, dark-staining fibers, which have higher concentrations of respiratory enzymes and which are surrounded by more capillaries. Fibers that are intermediate between these two extremes are also observed when the correct staining technique is used.

The thinner fibers, which are richly endowed with respiratory enzymes and capillaries, also contain higher concentrations of the pigment *myoglobin* than do the

Figure 8.30 Creatine kinase in plasma is separated from other proteins and identified by the rate of its movement in an electric field. The height of each peak indicates concentration, and the position of the peaks indicates the isoenzyme form. As can be seen in the *lower figure,* plasma concentration of creatine kinase is elevated in both heart disease (myocardial infarction) and skeletal muscle disease, but the two types of disease may be distinguished in this test by the appearance of a different isoenzyme form in myocardial infarction.

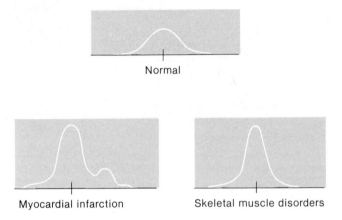

thicker fibers. Myoglobin is a red pigment—hence the term **red fibers** for these cells—which is related to the red hemoglobin pigment in blood. Like hemoglobin, myoglobin can combine with oxygen. The high concentration of myoglobin, high content of aerobic respiratory enzymes, and rich capillary blood supply provide the thinner red fibers with a high capacity for aerobic respiration and with a high resistance to fatigue.

The thicker pale or **white fibers,** in contrast, have a limited capacity for aerobic respiration, but are rich in stored glycogen and glycolytic enzymes that enable them to respire anaerobically. These fibers are thus less resistant to fatigue than the thin fibers. The thicker white fibers, however, contain more myofibrils than the thinner fibers, and have higher concentrations of myosin ATPase (the enzyme found in the cross-bridges). These fibers can thus contract more quickly and develop more power than the thinner fibers.

Figure 8.31 Cross section of muscle showing white, intermediate, and red (the darker) fibers.

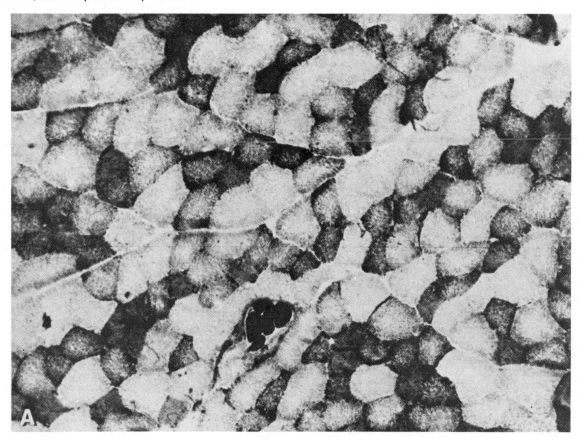

Table 8.10 Characteristics of red, intermediate, and white muscle fibers.

	Red	Intermediate	White
Diameter	Small	Intermediate	Large
Z-line thickness	Wide	Intermediate	Narrow
Glycogen content	Low	Intermediate	High
Resistance to fatigue	High	Intermediate	Low
Capillaries	Many	Many	Few
Myoglobin content	High	High	Low
Respiration type	Aerobic	Aerobic	Anaerobic
Twitch rate	Slow	Fast	Fast
Myosin ATPase content	Low	High	High

Figure 8.32 Comparison of the rates with which maximum tension is developed in three muscles. These include the relatively fast-twitch extraocular *(a)* and gastrocnemius *(b)* muscles, and the slow-twitch soleus muscle *(c)*.

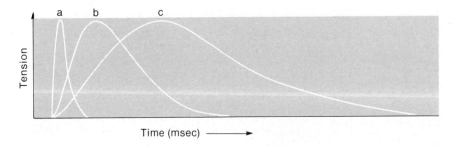

Table 8.11 Summary of the effects of endurance training (long-distance running, swimming, bicycling, etc.) on skeletal muscles.

1. Improved ability to obtain ATP from oxidative phosphorylation
2. Increased size and number of mitochondria
3. Less lactic acid produced per given amount of exercise
4. Increased myoglobin content
5. Increased intramuscular triglyceride content
6. Increased lipoprotein lipase (enzyme needed to utilize lipids from blood)
7. Increased proportion of energy derived from fat, less from carbohydrates
8. Lower rate of glycogen depletion during exercise
9. Improved efficiency in extracting oxygen from blood

Because of their different contents of myosin ATPase, the red, white, and intermediate fibers contract at different speeds. Red fibers have a **slow twitch** while white fibers have a **fast twitch.** The extraocular muscles that position the eyes, for example, contain a high proportion of white fibers and reach maximum force in about 7.3 msec (milliseconds—thousandths of a second). The soleus muscle of the leg, in contrast, which has a high proportion of red slow-twitch fibers, requires about 100 msec to reach maximum tension (see figure 8.32). Muscles like the soleus that are *postural muscles* are well suited for the sustained contractions needed to maintain an upright position; muscles like the gastrocnemius, which contain a high amount of white fast-twitch fibers, are well adapted to produce a great amount of power quickly.

Interestingly, the conduction rate of motor neurons that innervate fast-twitch white muscles is faster (80–90 meters per second) than the conduction rate of nerve fibers to slow-twitch red muscles (60–70 meters per second). The myofiber type (fast or slow twitch) may indeed be determined by the motor neuron that innervates the muscle. When the motor nerves to red and white fibers are switched in experimental animals, the previously fast-twitch fibers become slow and the slow-twitch fibers become fast. As expected from these observations, all of the muscle fibers innervated by the same motor neuron (that are part of the same motor unit) are of the same type.

Adaptations to Exercise

It is currently believed that the fiber types are determined by their innervation; one fiber type cannot, therefore, change into another as a result of exercise training. All three fiber types, however, do adapt to training. Endurance training, for example, causes an increase in the content of myoglobin and aerobic respiratory enzymes of all three types of fibers. Red and intermediate fibers, however, show the greatest changes under these conditions.

Endurance training does not increase the size of muscles. Muscle enlargement is produced only by frequent bouts of muscle contraction against a high resistance—as in weight lifting. As a result of this latter type of training, muscle fibers become thicker and the muscle therefore grows by hypertrophy (increase in cell size). It has, in fact, long been believed that adult skeletal muscles could only grow by hypertrophy and not by hyperplasia (increased cell number). Recent experimental evidence, however, suggests that muscle growth during weight lifting may also be partly due to muscle fiber "splitting"; that is, to longitudinal division of muscle fibers and consequent hyperplasia. Since other recent experiments support the more traditional view that cell numbers remain constant, the proposed ability of skeletal muscles to grow by hyperplasia as a result of weight training is currently very controversial.

1. Draw a figure illustrating the relationship between ATP and creatine phosphate. Describe this relationship verbally.
2. Describe the three different types of muscle fibers in terms of their size, enzyme content, and rates of twitch. Illustrate the differences in rates of twitch with a figure. Describe the functional significance of the different ratios of these fibers in different muscles.
3. Describe the effects of endurance training and strength training on the fiber characteristics of the muscles.

Figure 8.33 Biopsies of gastrocnemius muscles treated in such a way that muscle fibers with a high content of aerobic respiratory enzymes stain most darkly. Sample *(a)* came from an endurance-trained athlete, while the sample *(b)* was taken from a sprinter.

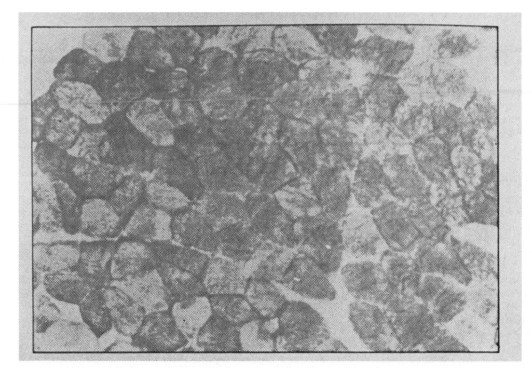

(a)

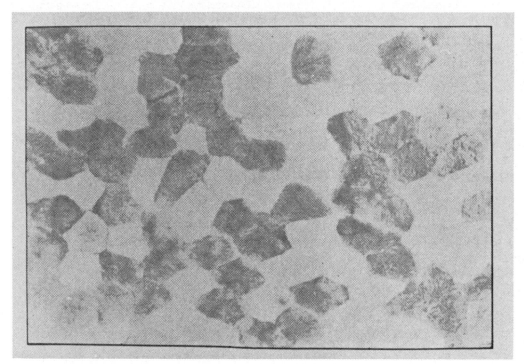

(b)

Summary

Structure and Function of Skeletal Muscles

I. Skeletal muscles are attached by tendons to bones.
   A. Skeletal muscles are composed of separate cells or fibers that are attached in parallel to the tendons.
   B. Skeletal muscle fibers are striated.
      1. The dark striations are called A bands and the light regions are called I bands.
      2. Z lines are located in the middle of each I band.
II. Contraction of separate muscle fibers is all-or-none.
   A. The contraction of separate muscle fibers is called a twitch.
   B. A whole muscle also produces a twitch in response to a single electrical pulse *in vitro*.
      1. The stronger the electric shock, the stronger the muscle twitch—whole muscle can give a graded, rather than an all-or-none, contraction.
      2. The graded contraction of whole muscles is due to different numbers of fibers participating in the contraction.
   C. Summation of fiber twitches can occur so rapidly that the muscle produces a smooth, sustained contraction known as tetanus.
III. Contraction of muscle fibers *in vivo* is stimulated by somatic motor neurons.
   A. Each somatic motor nerve fiber branches to innervate a number of muscle fibers.
   B. The motor neuron and the muscle fibers it innervates is called a motor unit.
      1. When a muscle is composed of many motor units (such as in the hand) there is fine control of muscle contraction.
      2. The large muscles of the leg have relatively few motor units, which are correspondingly large.
      3. Graded contractions are produced by the asynchronous stimulation of different motor units.

IV. When a muscle exerts tension without shortening, the contraction is termed isometric; when shortening does occur, the contraction is isotonic.

Neural Control of Skeletal Muscle

I. The somatic motor neurons that innervate the muscles are called the lower motor neurons.
   A. Alpha motoneurons innervate the ordinary, or extrafusal, muscle fibers—these are the fibers that produce muscle shortening during contraction.
   B. Gamma motoneurons innervate the intrafusal fibers of the muscle spindles.
II. Muscle spindles are length detectors in the muscle.
   A. Spindles consist of several intrafusal fibers wrapped together and in parallel with the extrafusal fibers.
   B. Stretching of the muscle stretches the spindles; this excites sensory endings in the spindle apparatus.
      1. Impulses in the sensory neurons travel into the spinal cord in the dorsal roots of spinal nerves.
      2. The sensory neuron makes a synapse directly with an alpha motoneuron within the spinal cord—this produces a monosynaptic reflex.
      3. The alpha motoneuron stimulates the extrafusal muscle fibers to contract, thus relieving the stretch—this is called the stretch reflex.
   C. Activity of gamma motoneurons tighten the spindles, thus making them more sensitive to stretch and better able to monitor the length of the muscle, even during muscle shortening.
III. The Golgi tendon organs monitor the tension that the muscle exerts on its tendons.
   A. As the tension increases, sensory neurons from Golgi tendon organs inhibit activity of alpha motoneurons.
   B. This is a disynaptic reflex, because the sensory neurons synapse with interneurons, which in turn make inhibitory synapses with motoneurons.

IV. A crossed-extensor reflex occurs when a foot steps on a tack.
   A. Sensory input from the injured foot causes stimulation of flexor muscles and inhibition of the antagonistic extensor muscles.
   B. The sensory input also crosses the spinal cord, to cause stimulation of extensor and inhibition of flexor muscles in the contralateral leg.
V. Neurons in the brain that affect the lower motor neurons are called higher motor neurons.
   A. The fibers of neurons in the precentral gyrus, or motor cortex, descend to the lower motor neurons as the lateral and ventral corticospinal tracts.
      1. Most of these fibers cross to the contralateral side in the brain stem, forming structures called the pyramids; this system is therefore called the pyramidal system.
      2. The left side of the brain thus controls the musculature on the right side, and vice versa.
      3. The pyramidal system is responsible for fine control of voluntary movements.
   B. Other descending motor tracts are part of the extrapyramidal system.
      1. The neurons of the extrapyramidal system make numerous synapses in different areas of the brain, including the midbrain, brain stem, basal ganglia, and cerebellum.
      2. Damage to the cerebellum produces intention tremor.
      3. Degeneration of fibers in the basal ganglia that use dopamine as a transmitter produces Parkinson's disease.

Mechanisms of Contraction

I. Skeletal muscle cells, or fibers, contain structures called myofibrils.
   A. Each myofibril is striated with dark (A) and light (I) bands; there are Z lines in the middle of each I band.
   B. The A bands contain thick filaments composed of myosin.
      1. The edges of each A band also contain thin filaments overlapped with the thick filaments.
      2. The central regions of the A bands contain only thick filaments—these regions are the H bands.
   C. The I bands contain only thin filaments, which are composed primarily of the protein called actin.
   D. Thin filaments are formed of globular actin subunits known as G-actin.
      1. A protein known as tropomyosin is also located periodically in the thin filaments; another protein—troponin—is attached to the tropomyosin.
II. Myosin cross-bridges extend out from the thick filaments to the thin filaments.
   A. The cross-bridges contain ATPase, which binds ATP but which is inactive when the cross-bridges are not attached to actin (when the muscle is at rest).
      1. At rest, tropomyosin prevents attachment of cross-bridges to actin.
      2. During contraction, Ca^{++} attaches to troponin, causing the tropomyosin to shift position.
   B. When the cross-bridge ATPase is activated by attachment to actin, ATP is split and the cross-bridge undergoes a power stroke.
   C. At the end of the power stroke, the cross-bridges bind a new ATP and release from the actin to repeat the contraction cycle.

III. Activity of the cross-bridges causes sliding of the filaments.
   A. The actin on each side of the A bands is pulled towards the center.
   B. The H bands thus get shorter as more actin overlaps with myosin.
   C. The I bands also get shorter as adjacent A bands are pulled closer together.
   D. The A bands stay the same length because the filaments (both thick and thin) do not shorten during muscle contraction.
IV. Calcium ions serve to couple electrical excitation of the muscle fiber with contraction.
   A. When a muscle is at rest, the Ca^{++} concentration of the sarcoplasm is very low and cross-bridges are prevented from attaching to actin.
      1. The Ca^{++} is actively transported into the sarcoplasmic reticulum.
      2. The sarcoplasmic reticulum is a modified endoplasmic reticulum that surrounds the myofibrils.
   B. Action potentials are conducted by transverse tubules into the muscle fiber.
      1. Transverse tubules are invaginations of the cell membrane that almost touch the sarcoplasmic reticulum.
      2. Action potentials in the transverse tubules stimulate the release of Ca^{++} from the sarcoplasmic reticulum.
   C. When action potentials cease, Ca^{++} is removed from the sarcoplasm and stored in the sarcoplasmic reticulum.

V. Aerobic cell respiration is ultimately required for the production of ATP needed for cross-bridge activity.
   A. New ATP can be quickly produced, however, from the combination of ADP with phosphate derived from phosphocreatine.
      1. The phosphocreatine represents a ready reserve of high energy phosphate during sustained muscle contractions
      2. Phosphocreatine is produced at rest from creatine and phosphate is derived from ATP.
   B. There are three types of muscle fibers.
      1. Slow-twitch, thin fibers are adapted for aerobic respiration and are resistant to fatigue.
      2. Fast-twitch thicker fibers are adapted for anaerobic respiration.
      3. Intermediate-twitch fibers are the third type of muscle fibers.
   C. Physical training affects the enzyme content and characteristics of the muscle fibers.
      1. Strength training increases the enzymes of anaerobic respiration and makes the fast and intermediate fibers larger and stronger.
      2. Endurance training increases the capacity for aerobic respiration of the muscle fibers.

Self-Study Quiz

1. A graded muscle contraction is produced by variations in:
   - (a) the strength of the fiber's contraction
   - (b) the number of fibers that are contracting
   - (c) both of these
   - (d) neither of these
2. Sustained muscle contraction (tetanus) is produced by:
   - (a) sustained contraction of the muscle fibers
   - (b) asynchronous twitches of the muscle fibers
   - (c) both of these
   - (d) neither of these
3. Which of the following muscles have motor units with the lowest innervation ratio?
   - (a) leg muscles
   - (b) arm muscles
   - (c) muscles that move the fingers
   - (d) muscles of the torso
4. Stimulation of gamma motoneurons produces:
   - (a) isotonic contraction of intrafusal fibers
   - (b) isometric contraction of intrafusal fibers
   - (c) either isotonic or isometric contraction of intrafusal fibers
   - (d) contraction of extrafusal fibers

5. In a single reflex arc involved in the knee jerk reflex, how many synapses are activated within the spinal cord?
   - (a) thousands
   - (b) hundreds
   - (c) dozens
   - (d) one
6. Spastic paralysis may occur when there is damage to:
   - (a) the lower motor neurons
   - (b) the higher motor neurons
   - (c) either the lower or the higher motor neurons
7. Which of the following statements about the pyramidal motor system is FALSE?
   - (a) nerve fibers descend all the way from the cerebral cortex to the spinal cord without synaptic interruption
   - (b) most of the fibers cross to the contralateral side in the pyramids
   - (c) most of the fibers synapse with interneurons in the spinal cord
   - (d) the pyramidal tracts only activate alpha motoneurons

8. Which of the following statements about Parkinson's disease is TRUE?
   - (a) it is a disease of the pyramidal system
   - (b) one symptom is intention tremor
   - (c) there is a dopamine deficiency in the basal ganglia
   - (d) it is caused by damage to the motor cortex
9. When a skeletal muscle shortens during contraction, which of the following statements is FALSE?
   - (a) the A bands shorten
   - (b) the H bands shorten
   - (c) the I bands shorten
   - (d) the sarcomeres shorten
10. Electrical excitation of a muscle fiber *most directly* causes:
    - (a) movement of tropomyosin
    - (b) attachment of the cross-bridges to actin
    - (c) release of Ca^{++} from the sarcoplasmic reticulum
    - (d) splitting of ATP
11. The energy for muscle contraction is *most directly* obtained from:
    - (a) phosphocreatine
    - (b) ATP
    - (c) anaerobic respiration
    - (d) aerobic respiration

Neural Control of Visceral Function: *The Autonomic Nervous System*

Objectives

By studying this chapter, you should be able to:

1. Describe the structure of cardiac muscle, and explain the functional significance of this structure

2. Compare and contrast the structure and function of skeletal and cardiac muscle

3. Describe the structure of smooth muscle, and the functional characteristics of single and multi-unit smooth muscle

4. Explain how the automatic activity of smooth and cardiac muscle is modified by autonomic nerves

5. Describe the functional characteristics of exocrine glands

6. Describe the structure of the sympathetic and parasympathetic nervous systems

7. List the neurotransmitters of the sympathetic and parasympathetic fibers (pre- and postganglionic)

8. Describe the relationship between the sympathetic system and the adrenal medulla, in terms of structure and function

9. Describe the effects of sympathoadrenal stimulation, and relate these effects to alpha and beta adrenergic receptors in the visceral effector organs

10. Describe the muscarinic effects of acetylcholine released by postganglionic parasympathetic fibers

11. Explain the antagonistic, complementary, and cooperative effects of sympathetic and parasympathetic innervation

12. Describe the higher neural control of the autonomic nervous system

*T*he constancy of the internal environment, or *milieu interieur,* as described by the French physiologist Claude Bernard in 1878, is maintained by a dynamic balance between the effects of various regulatory mechanisms. These include secretions of the endocrine glands (detailed in later chapters) and the actions of the visceral motor, or **autonomic system.** The term *homeostasis* was introduced by Walter B. Cannon in 1929 to describe the internal constancy produced by the effects of these regulatory systems.

Figure 9.1 Cardiac muscle. Note that the cells are short, branched, and striated, and that they are interconnected by intercalated discs.

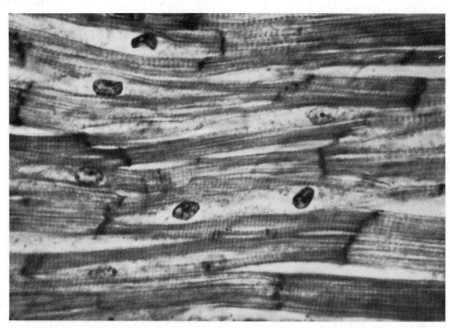

Visceral Effector Organs

Autonomic motor nerves, in contrast to somatic motor nerves, innervate effector organs that are not usually under voluntary control. These effector organs include **cardiac muscle** (the heart), **smooth muscle,** and **glands.** Unlike skeletal muscles, which enter a state of flaccid paralysis when their motor nerves are damaged, the involuntary visceral effectors are somewhat independent of their innervation. Smooth muscles maintain a resting tone in the absence of nerve stimulation, for example. Damage to an autonomic nerve, in fact, makes its target muscle more sensitive than normal to stimulating agents. This phenomenon is called *denervation hypersensitivity.*

In addition to their intrinsic ("built-in") muscle tone, many smooth muscles and the cardiac muscle take their autonomy a step further. These muscles can contract rhythmically, even in the absence of nerve stimulation, in response to electrical waves of depolarization initiated by the muscles themselves. These myogenic (myo = muscle; gen = produce) waves are called *pacemaker potentials.*

Autonomic nerves also maintain a resting "tone," in the sense that they maintain a baseline firing rate that can be either increased or decreased. Such changes in tonic neural activity produce changes in the intrinsic activity of the effector organs. A decrease in excitatory input to the heart, for example, will slow its rate of beat.

Unlike somatic motor neurons—whose neurotransmitter (ACh) always stimulates its effectors (skeletal muscles)—some autonomic nerves release transmitters that inhibit activity of their effectors. An increase in the activity of such an inhibitory nerve to the heart, for example, will slow the heart rate.

Cardiac Muscle

Heart muscle, or *myocardial* cells, like skeletal muscle cells, are striated because they contain actin and myosin filaments arranged in the form of sarcomeres. The long, fibrous skeletal muscle cells, however, are structurally and functionally separated from each other, while myocardial cells are short, branched, and interconnected. Adjacent myocardial cells are joined by electrical synapses, or *gap junctions*. These have an affinity for stain that gives the appearance of **intercalated discs** in the light microscope (see figure 9.1).

Figure 9.2 A hair follicle and associated structures. The piloerector muscle inserts into the follicle and the surrounding connective tissue of the dermis. When this smooth muscle contracts the hair shaft is erected, and the connective tissue is pulled into a "goose bump."

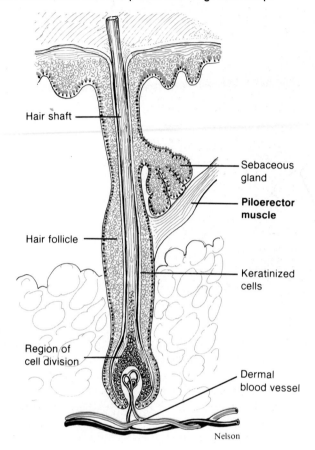

Hair shaft

Sebaceous gland

Piloerector muscle

Hair follicle

Keratinized cells

Region of cell division

Dermal blood vessel

Nelson

Electrical impulses that originate at any point in the mass of myocardial cells (the *myocardium*) can spread to all cells in the mass that are joined by gap junctions. The entire mass therefore is stimulated to produce a simultaneous contraction. Because all cells in the myocardium are electrically joined, the myocardium behaves as a single functional unit, or *functional syncitium*. Unlike skeletal muscles, which produce graded contractions by variations in the numbers of muscle cells stimulated, the heart contracts in an **all-or-none** fashion. It cannot give a graded, half-hearted contraction.

The heart of an experimental animal (such as a frog or turtle) will continue to beat even after it has been removed from the animal. This demonstrates that the beat is an intrinsic property of heart muscle, and not the result of nerve stimulation. Indeed, when the cells of an embryonic heart are separated and kept alive in tissue culture, the separate cells continue to exhibit their own intrinsic pattern of depolarization-contraction, repolarization-relaxation.

These separated myocardial cells can eventually form intercellular connections, which wed them together into a continuous fabric. Although the separated cells have different intrinsic rates of beat, the myocardial mass again contracts and relaxes as a single unit. This is because those cells that have the fastest rhythm initiate electrical waves that stimulate the slower cells before they can depolarize themselves.

Since all of the interconnected cells depolarize and repolarize at about the same time, all experience a refractory period at about the same time. This uniform refractory period prevents the myocardium from conducting any impulses other than those initiated by the fastest cells. For these reasons, the fastest cells in the heart serve as the **pacemaker.** The pacemaker cells are normally grouped together in a specific region of the heart, as described in chapter 11.

Smooth Muscles

Smooth muscles are arranged in circular layers around the walls of blood vessels, bronchioles (small air passages in the lungs), and in the sphincter muscles of the digestive tract. Contraction of these muscles constricts the wall of these tubes and makes their lumen narrower. Both circular and longitudinal smooth muscle layers are found in the tubular digestive tract, the ureters (which transport urine), the vas deferens (which transports semen), and the fallopian tubes (which transport ova). Alternating contraction of circular and longitudinal smooth muscle layers produce **peristaltic waves** that propel the contents of these tubes in one direction.

Contraction of smooth muscles in the spleen, rectum, and urinary bladder help these organs to empty their contents. One of the more interesting smooth muscles, the arrector pili muscle, attaches to hair on one end and to the dermis of the skin on the other. When these piloerector muscles contract, the dermis is raised to produce "goose bumps," and—at least in some locations—the hair stands on end.

Figure 9.3 Electron micrographs of the thick and thin filaments of smooth muscle. *(a)* is a longitudinal section showing a complete long myosin filament (between arrows) (32,000x). *(b)* is a cross-section of a portion of a smooth muscle cell showing the rosette arrangement of thin actin filaments around the thick myosin (large arrow) filaments. The amorphous grey areas are "dense bodies" and are surrounded by 10 mm filaments (small arrows) (19,000x).

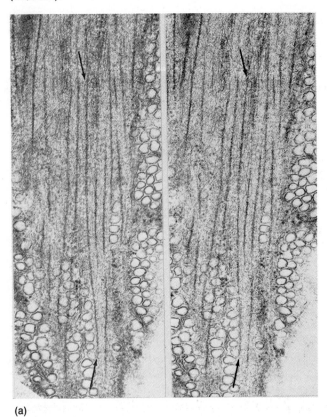

(a)

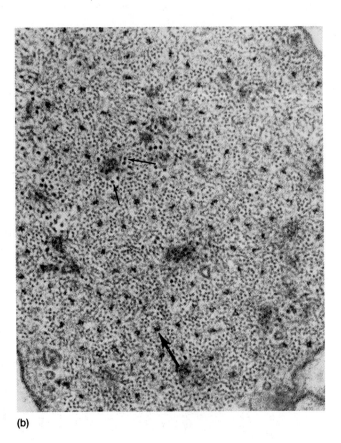

(b)

Smooth muscle cells are widest in the center, which contains a single nucleus, and tapered towards the poles to produce a spindle shape. These cells do not contain sarcomeres and are not striated. They do, however, contain a great amount of actin and some myosin, producing a ratio of thin-to-thick filaments estimated at 16:1 (the ratio in skeletal muscles is 2:1). The myosin filaments in smooth muscles, however, are quite long; this may account for the ability of smooth muscles to develop a great deal of tension.

Mechanism of Contraction   The myosin of smooth muscle cells appear to have cross-bridges, and tropomyosin has been isolated from the thin filaments. Smooth muscles appear to have a Ca^{++} binding protein that is similar to, but not identical with, troponin. Unlike striated muscle in which troponin is associated with the thin filaments, the Ca^{++} binding protein in smooth muscles may be located in the thick myosin filaments.

There are no Z lines in smooth muscle. The thin actin filaments may instead insert into amorphous *dense bodies* in the cytoplasm (see figure 9.3) and into the cell membrane.

As with striated muscles, contraction is triggered by a sharp rise in the Ca^{++} concentration within the cytoplasm. The source of this Ca^{++} is not well established (smooth muscles have little sarcoplasmic reticulum). Since smooth muscle cells are thin compared to most skeletal muscle fibers, the inflow of Ca^{++} across the cell membrane during depolarization may be sufficient to activate even the deepest cross-bridges.

The long length of myosin filaments, and the fact that they are not organized into sarcomeres, may be advantageous for the function of smooth muscles. Smooth muscle must be able to exert tension even when greatly stretched—in the urinary bladder, for example, the smooth muscle cells may be stretched up to 2½ times their resting length. Skeletal muscles, in contrast, lose the ability to contract when the sarcomeres are stretched to the point where actin and myosin no longer overlap.

Table 9.1   Some comparisons of skeletal, cardiac, and smooth muscles.

Skeletal Muscle	Cardiac Muscle	Smooth muscle
Striated; actin and myosin arranged in sarcomeres	Striated; actin and myosin, arranged in sarcomeres	Not striated; more actin than myosin; actin insert into dense bodies and cell membrane
Well-developed sarcoplasmic reticulum and transverse tubules	Moderately developed sarcoplasmic reticulum and transverse tubules	Sarcoplasmic reticulum poorly developed; no transverse tubules
Contains troponin in the thin filaments	Contains troponin in the thin filaments	Contains a Ca^{++} binding protein; may be located in thick filaments
Ca^{++} released into cytoplasm from sarcoplasmic reticulum	Ca^{++} enters cytoplasm from sarcoplasmic reticulum and extracellular fluid	Ca^{++} enters cytoplasm from extracellular fluid, sarcoplasmic reticulum, and perhaps mitochondria
Cannot contract without nerve stimulation; denervation results in muscle atrophy	Can contract without nerve stimulation; action potentials originate in pacemaker cells of heart	Maintains tone in absence of nerve stimulation; visceral smooth muscle produces pacemaker potentials; denervation results in hypersensitivity to stimulation
Muscle fibers stimulated independently; no gap junctions	Gap junctions present as intercalated discs	Gap junctions present in most smooth muscles

Figure 9.4   Autonomic nerve endings in smooth muscle (N = nerve).

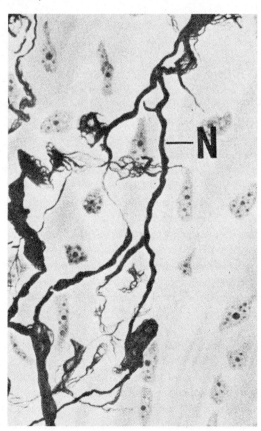

Despite their structural differences, it is currently believed that smooth muscles, like striated muscles, contract by means of a *sliding filament* mechanism. The details of this mechanism in smooth muscles, however, are not well understood.

Regulation of Contraction   There are significant differences between the neural control of skeletal muscles and that of smooth muscles. A skeletal muscle fiber has only one junction with a somatic nerve fiber, and the receptors for the neurotransmitter are localized in the muscle fiber membrane of the neuromuscular junction. The entire surface of smooth muscle cells, in contrast, contains transmitter receptor proteins. The autonomic nerve endings release transmitter from numerous varicosities (bulges) that form *synapses en passant* with the smooth muscle cells (as illustrated in figure 9.4). Transmitter released from these synapses is believed to diffuse to a number of smooth muscle cells (the synaptic cleft is larger than in the somatic neuromuscular junction) and by this means can activate more than one effector cell.

Table 9.2   Some comparisons between single-unit and multi-unit smooth muscles.

	Single-Unit Muscle	Multi-Unit Muscle
Location	Gastrointestinal tract; uterus; ureter; small arteries (arterioles)	Arrector pili muscles of hair follicles; ciliary muscle (attached to lens); iris; vas deferens; large arteries
Origin of Electrical Activity	Spontaneous activity by pacemakers (myogenic)	Not spontaneously active; potentials are neurogenic
Type of Potentials	Action potentials	Graded depolarizations
Response to Stretch	Responds to stretch by contraction; not dependent on nerve stimulation	No inherent response to stretch
Presence of Gap Junctions	Numerous gap junctions join all cells together electrically	Few (if any) gap junctions
Type of Contraction	Slow, sustained contractions	Slow, sustained contractions

Single-Unit and Multi-Unit Smooth Muscles   Although smooth muscles in different regions of the body display a continuous variation in properties, they are often grouped into two convenient functional categories: **single-unit** and **multi-unit.** Single-unit smooth muscles have numerous gap junctions between adjacent cells that wed them together electrically; they thus behave as a single unit. Multi-unit smooth muscles have few, if any, gap junctions; the individual cells must thus be separately stimulated by nerve endings. This is similar to the numerous motor units required for control of skeletal muscles.

Single-unit smooth muscles are found in the gastrointestinal tract, uterus, ureter, and in the walls of small blood vessels. These muscles display *pacemaker* activity in which certain cells stimulate others in the mass. Unlike the heart, in which the pacemaker region is anatomically fixed, different smooth muscle cells can assume the pacemaker function. These single-unit smooth muscles also display intrinsic or *myogenic electrical activity* and contraction in response to passive stretch. Stretch induced by an increase in luminal contents can thus produce myogenic contraction without the necessary involvement of the autonomic nervous system.

The pacemaker cells of single-unit smooth muscles produce *slow waves* of depolarization that generate action potentials. These slow waves usually spread from the outer to the inner layers of smooth muscles, producing action potentials at their peaks of depolarization. The rate at which these slow waves generate action potentials is increased or decreased by the influence of autonomic nerve innervation.

Multi-unit smooth muscles are found in the arrector pili muscles of hair follicles, the ciliary muscle (which adjusts the thickness of the lens in the eye), the iris (which adjusts the opening, or pupil, of the eye), the vas deferens, and in the walls of large blood vessels. Contraction of these muscles depends on depolarizations produced by nerve stimulation.

Exocrine Glands

Exocrine glands are derived from cells of epithelial membranes that cover and line the body surfaces. The secretions of these cells are expressed to the outside of the epithelial membranes (and hence to the outside of the body) through *ducts*. This is in contrast to endocrine glands, which lack ducts and which therefore secrete into capillaries within the body.

The secretory units of exocrine glands may be simple tubes, or may be modified to form clusters of units around branched ducts (see figure 9.5). These clusters or **acini,** are often surrounded by tentacle-like extensions of *myoepithelial cells* that contract and squeeze the secretions through the ducts. The rate of secretion and the action of myoepithelial cells are influenced by hormones and by the autonomic nervous system.

Exocrine Glands of the Skin   Examples of exocrine glands in the skin include the lacrimal (tear) glands, sebaceous glands (which secrete oily sebum into hair follicles), and sweat glands. There are two types of sweat glands. The most numerous *eccrine sweat glands* secrete a dilute salt solution that serves in thermoregulation (evaporation cools the skin). The *apocrine sweat glands,* located in the axilla (underarms) and pubic regions, secrete a protein-rich fluid. This provides nourishment for bacteria that produce the characteristic odor of this type of sweat.

Exocrine Glands of the Digestive Tract   All of the glands that secrete into the digestive tract are exocrine. This is because the lumen of the digestive tract is a part of the external environment, and secretions of these glands go to the outside of the membrane that lines this tract. Mucus glands are located throughout the length of the digestive tract. Other relatively simple glands of the tract include salivary glands, gastric glands, and simple tubular glands in the intestine.

Figure 9.5 Exocrine glands.

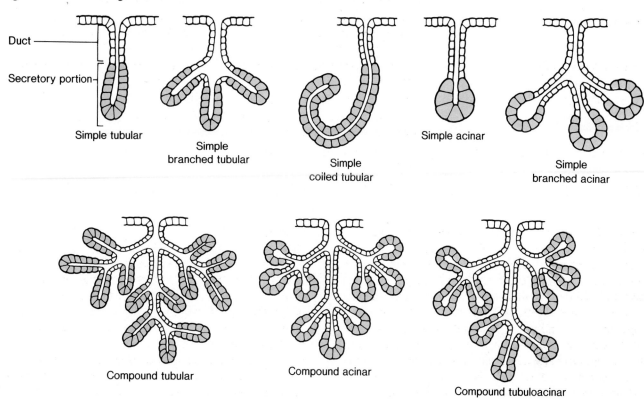

Duct

Secretory portion

Simple tubular

Simple
branched tubular

Simple
coiled tubular

Simple acinar

Simple
branched acinar

Compound tubular

Compound acinar

Compound tubuloacinar

The *liver* and *pancreas* are exocrine (as well as endocrine) glands derived embryologically from the digestive tract. The exocrine secretion of the pancreas is pancreatic juice, containing digestive enzymes and bicarbonate, which is secreted into the small intestine via the pancreatic duct. The liver produces and secretes bile (an emulsifier of fat) into the small intestine via the gallbladder and bile duct.

Exocrine Glands of the Reproductive Tract   The female reproductive tract contains numerous mucus-secreting exocrine glands. The male accessory sexual organs—the *prostate* and *seminal vesicles*—are exocrine glands that contribute to the semen. The testes and ovaries (the gonads) are both endocrine and exocrine glands. They are endocrine because they secrete sex steroid hormones into the blood; they are exocrine because they release gametes (ova and sperm) into the reproductive tracts.

1. Compare and contrast the structure and function of cardiac muscle with that of smooth muscle. Describe the role of the autonomic system in the regulation of the heartbeat.
2. Describe the structure and mechanisms of contraction of smooth muscle in comparison to striated muscle.
3. Describe how regulation by autonomic nerves differs from regulation by somatic motor nerves.
4. Compare the structure and regulation of single-unit smooth muscle with multi-unit smooth muscle.
5. Describe what would happen to the exocrine and endocrine secretions of the pancreas if the pancreatic duct were blocked. Explain your answer in terms of the distinguishing characteristics of endocrine and exocrine secretions.

Table 9.3 Comparisons of the somatic motor system with the autonomic motor system.

Feature	Somatic Motor	Autonomic Motor
Effector Organs	Skeletal Muscles	Cardiac Muscle, Smooth Muscle, and Glands
Presence of Ganglia	No ganglia outside CNS	Cell bodies of postganglionic autonomic fibers located in paravertebral, prevertebral (collateral), and terminal ganglia
Number of Neurons from CNS to Effector	One	Two
Type of Neuromuscular Junction	Specialized motor end plate	No specialization of postsynaptic membrane; all areas of smooth muscle cells contain receptor proteins for neurotransmitters
Effect of Nerve on Muscle	Excitatory only	Either excitatory or inhibitory
Type of Nerve Fibers	Fast-conducting, thick (9–13 μm), and myelinated	Slow conducting; preganglionic fibers, lightly myelinated but thin (3 μm); postganglionic fibers, unmyelinated and very thin (about 1.0 μm)
Effect of Denervation	Flaccid paralysis and atrophy	Much muscle tone and function remains; target cells show denervation hypersensitivity

Figure 9.6 Comparison of somatic motor reflex with autonomic motor reflex. Although each is shown on different sides of the spinal cord for the sake of clarity, both visceral and somatic sensory neurons are found bilaterally in the dorsal roots of spinal nerves, and somatic and autonomic motor fibers are found bilaterally in the ventral roots.

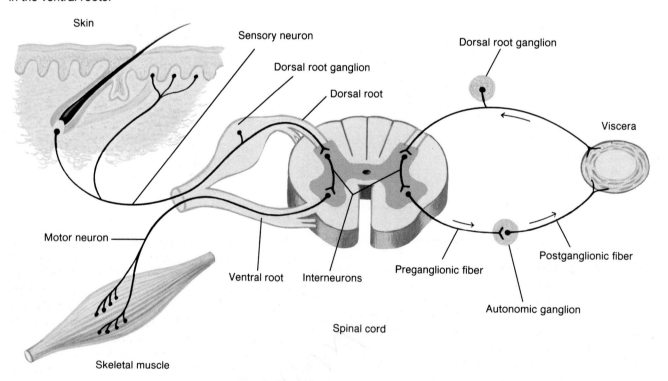

Skin

Sensory neuron

Dorsal root ganglion

Dorsal root ganglion

Dorsal root

Viscera

Motor neuron

Preganglionic fiber

Postganglionic fiber

Ventral root   Interneurons

Autonomic ganglion

Skeletal muscle

Spinal cord

Figure 9.7   The vertebral column and corresponding spinal cord segments.

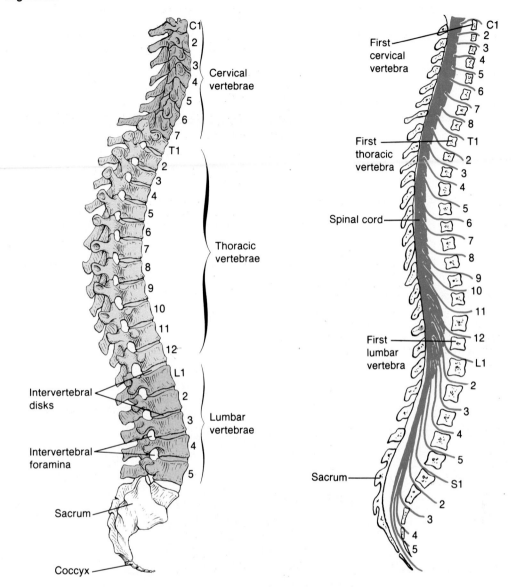

Structure of the Autonomic Nervous System

Unlike somatic motor neurons, which conduct impulses along a single axon from the spinal cord to the neuromuscular junction, the autonomic motor pathway involves two neurons in the efferent flow of impulses. The first of these neurons has its cell body in the grey matter of the brain or spinal cord. The axon of this neuron does not innervate the effector organ, but instead synapses within an *autonomic ganglion* (collection of cell bodies outside the CNS) with a second neuron. The first neuron is thus called a **preganglionic** or **presynaptic** neuron. The axon of the neuron located in the autonomic ganglion innervates the effector organ, and is called a **postganglionic** or **postsynaptic** fiber.

Preganglionic fibers originate in the midbrain and hindbrain, and from the upper thoracic to the fourth sacral levels of the spinal cord (see figure 9.7). Individual autonomic ganglia are located in the head, neck, and abdomen, and within many organs; chains of ganglia also parallel the right and left sides of the spinal cord. The origin of the preganglionic fibers, and the locations of the ganglia, are different in the sympathetic division than they are in the parasympathetic division of the autonomic system.

Figure 9.8   The sympathetic chain of paravertebral ganglia, showing its relationship to the vertebral column and the spinal cord.

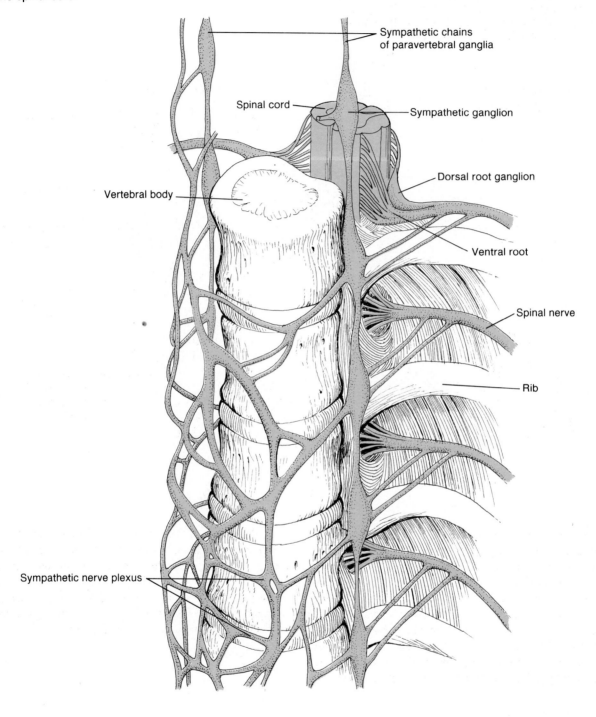

Figure 9.9 Connections between the preganglionic sympathetic fibers, postganglionic sympathetic fibers, and the spinal nerves. Postganglionic fibers that leave the paravertebral ganglia rejoin spinal nerves through the grey rami. Postganglionic fibers that leave the collateral sympathetic ganglia form separate visceral nerves.

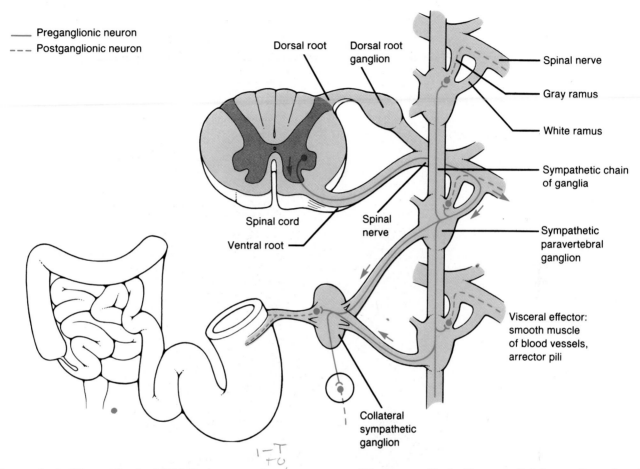

____ Preganglionic neuron
- - - Postganglionic neuron

Dorsal root

Dorsal root ganglion

Spinal nerve

Gray ramus

White ramus

Sympathetic chain of ganglia

Spinal cord

Spinal nerve

Ventral root

Sympathetic paravertebral ganglion

Visceral effector: smooth muscle of blood vessels, arrector pili

Collateral sympathetic ganglion

Sympathetic (Thoracolumbar) Division

The sympathetic system is also called the thoracolumbar division of the autonomic system because its preganglionic fibers exit the spinal cord from the first thoracic to the third lumbar levels. These fibers, together with somatic motor fibers, form the ventral roots of spinal nerves in these regions. Most sympathetic fibers, however, separate from the somatic motor fibers and synapse with postganglionic neurons within the double chain of **paravertebral ganglia** located on either side of the spinal cord.

The preganglionic fibers are lightly myelinated, and therefore are white in appearance. These form *white rami* (branches) from the spinal nerves to the paravertebral ganglia. The postganglionic fibers are thin and unmyelinated, and are therefore grey in appearance. These postganglionic fibers form the *grey rami* as they go from the ganglia to the spinal nerves (see figure 9.9).

Figure 9.10 The cervical sympathetic ganglia.

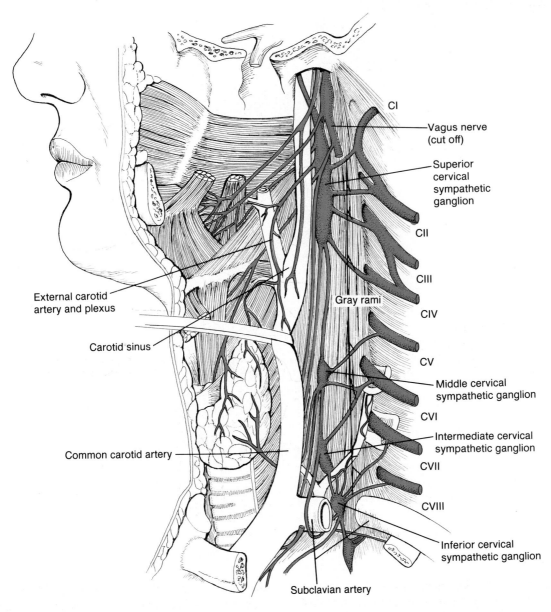

CI

Vagus nerve
(cut off)

Superior
cervical
sympathetic
ganglion

CII

CIII

Gray rami

CIV

External carotid
artery and plexus

Carotid sinus

CV

Middle cervical
sympathetic ganglion

CVI

Intermediate cervical
sympathetic ganglion

CVII

Common carotid artery

CVIII

Inferior cervical
sympathetic ganglion

Subclavian artery

Many preganglionic fibers in the upper thoracic level do not synapse in the paravertebral ganglia, but instead travel upward to synapse in **collateral ganglia** in the neck. The postganglionic fibers from these (the cervical and stellate ganglia—see figure 9.10) innervate smooth muscles and glands in the head and neck. The fibers of other collateral ganglia in the abdomen (the celiac and mesenteric ganglia) innervate the digestive tract (see figure 9.11).

Within the chain of paravertebral ganglia, preganglionic fibers branch to synapse with ganglia located several levels above and below the level at which the preganglionic fiber emerges from the cord. This is an example of *divergence*—one preganglionic fiber may branch to synapse with as many as twenty different postganglionic neurons. *Convergence* is also seen, because a given postganglionic neuron may receive synaptic input from a large number of preganglionic fibers.

The divergence of impulses from the spinal cord to the ganglia, and the convergence of impulses within the ganglia, usually results in the *mass activation* of almost all of the postganglionic fibers. This explains why the sympathetic system is usually activated as a unit, affecting all of its effector organs simultaneously.

Figure 9.11 The collateral sympathetic ganglia: the celiac, superior, and inferior mesenteric ganglia.

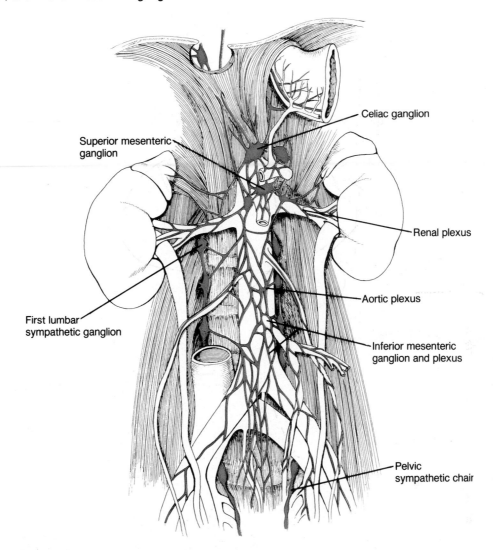

Superior mesenteric
ganglion

First lumbar
sympathetic ganglion

Celiac ganglion

Renal plexus

Aortic plexus

Inferior mesenteric
ganglion and plexus

Pelvic
sympathetic chair

Adrenal Glands   The paired adrenal glands are located above each kidney. Each adrenal is composed of two parts: an outer *cortex* and an inner *medulla*. These two parts are really two functionally different glands with different embryological origins, different hormones, and different regulatory mechanisms. The adrenal cortex secretes steroid hormones; its function will be described in later chapters. The adrenal medulla secretes the hormone **epinephrine** (adrenalin) when it is stimulated by the sympathetic system.

The adrenal medulla is a modified sympathetic ganglion whose cells are derived from postganglionic sympathetic neurons. These cells are innervated by preganglionic fibers originating in the thoracic level of the spinal cord, and the cells of the medulla secrete epinephrine into the blood in response to this neural stimulation. The effects of epinephrine are complementary to those of the transmitter (norepinephrine) released from postganglionic sympathetic nerve endings. For this reason, and because the adrenal medulla is stimulated as part of the mass activation of the sympathetic system, the two are often grouped together as the **sympathoadrenal system.**

Parasympathetic (Craniosacral) Division

The parasympathetic system is also known as the *craniosacral division* of the autonomic system. This is because its preganglionic fibers originate in the brain (specifically, the midbrain and medulla oblongata) and in the second through fourth sacral levels of the spinal cord.

Parasympathetic ganglia, unlike the sympathetic ganglia, are located next to—or actually within—the organs innervated. They are thus called simply **terminal ganglia.** The origins of the preganglionic fibers, and the organs innervated by postganglionic fibers, are summarized in table 9.5.

Table 9.4   The sympathetic (thoracolumbar) system.

Parts of Body Innervated	Spinal Origin of Preganglionic Fibers	Origin of Postganglionic Fibers
Eye	C-8 and T-1	Cervical ganglia
Head and Neck	T-1 to T-4	Cervical ganglia
Heart and Lungs	T-1 to T-5	Upper thoracic (paravertebral) ganglia
Upper Extremity	T-2 to T-9	Lower cervical and upper thoracic (paravertebral) ganglia
Upper Abdominal Viscera	T-4 to T-9	Celiac and superior mesenteric (collateral) ganglia
Adrenal	T-10 and T-11	Adrenal medulla
Urogenital System	T-12 to L-2	Celiac and inferior mesenteric (collateral) ganglia
Lower Extremities	T-9 to L-2	Lumbar and upper sacral (paravertebral) ganglia

Table 9.5   The parasympathetic (craniosacral) system.

Effector Organs	Origin of Preganglionic Fibers	Peripheral Nerve	Location of Terminal Ganglia
Eye (Ciliary and Iris Muscles)	Midbrain (cranial)	Oculomotor (third cranial) nerve	Ciliary ganglion
Lacrimal, Mucous, and Salivary Glands in Head	Medulla oblongata (cranial)	Facial (seventh cranial) nerve	Sphenopalatine and submandibular ganglia
Parotid (Salivary) Gland	Medulla oblongata (cranial)	Glossopharyngeal (ninth cranial) nerve	Otic ganglion
Heart, Lungs, Gastrointestinal Tract, Liver, Pancreas	Medulla oblongata (cranial)	Vagus (tenth cranial) nerve	Terminal ganglia in or near organ
Lower Half of Large Intestine, Rectum, Bladder, Urogenital Organs	S-2 to S-4 (sacral)	Through pelvic spinal nerves	Terminal ganglia near organs

Four of the twelve cranial nerves contain preganglionic parasympathetic fibers. The *oculomotor (third cranial) nerve* contains somatic motor fibers and preganglionic parasympathetic fibers that originate in the oculomotor nuclei of the midbrain. These parasympathetic fibers synapse in the *ciliary ganglion,* whose postganglionic fibers innervate the ciliary muscle and constrictor fibers in the iris of the eye.

Preganglionic parasympathetic fibers that arise in the medulla oblongata travel in the *facial (seventh) cranial nerve* and in the *glossopharyngeal (ninth) cranial nerve* to ganglia located in the head. The ganglia—the *sphenopalatine, submandibular,* and *otic ganglia*—send postganglionic fibers to lacrimal, mucus, and salivary glands.

Other nuclei in the medulla oblongata contribute preganglionic fibers to the very long *tenth cranial,* or *vagus nerve* (the "vagrant" or "wandering" nerve). The preganglionic fibers in the vagus synapse with postganglionic neurons that are actually located *within* the innervated organs. The preganglionic vagus fibers are thus quite long, and provide parasympathetic innervation to the heart, lungs, esophagus, stomach, pancreas, liver, small intestine, and the upper half of the large intestine. Postganglionic parasympathetic fibers arise from terminal ganglia within these organs and synapse with effector cells (smooth muscles and glands).

Preganglionic fibers from the sacral levels of the spinal cord provide parasympathetic innervation to the upper half of the large intestine, rectum, and to the urogenital system. These fibers, like those of the vagus, synapse with terminal ganglia located within the effector organs. Parasympathetic nerves to these organs, and to the organs innervated by the vagus, are thus composed of preganglionic fibers. The sympathetic nerves that innervate these organs, in contrast, consist of postganglionic fibers (see figure 9.12).

Figure 9.12 The autonomic nervous system, including sympathetic (black) and parasympathetic (colored) divisions. Preganglionic sympathetic fibers exit the spinal cord at levels *T-1* through *L-2* (hence the alternate name of *thoracolumbar division*). The parasympathetic system is also called the *craniosacral division* due to the level of origin of its preganglionic fibers.

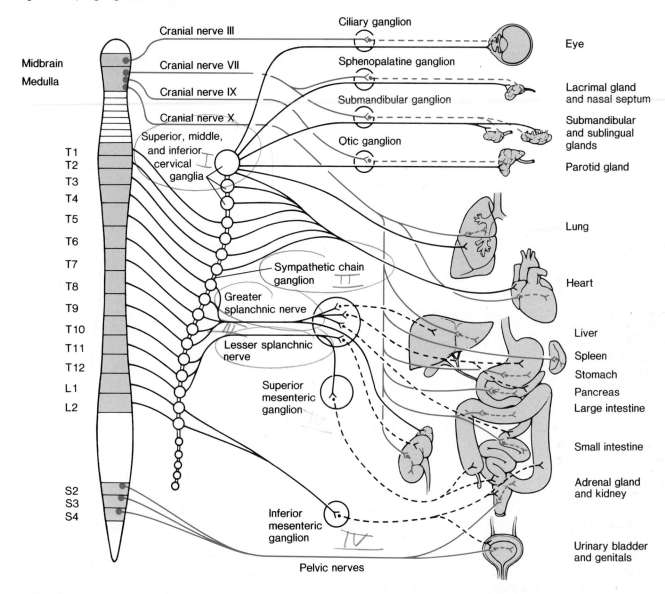

Table 9.6 Some comparisons between the structure of the sympathetic and parasympathetic systems.

	Sympathetic	Parasympathetic
Origin of Preganglionic Outflow	Thoracolumbar levels of spinal cord	Midbrain, hindbrain, and sacral levels of spinal cord
Location of Ganglia	Chain of paravertebral ganglia and prevertebral (collateral) ganglia	Terminal ganglia in or near effector organs
Distribution of Postganglionic Fibers	Throughout the body	Mainly limited to the head and the viscera of the chest, abdomen, and pelvis
Divergence of Impulses from Pre- to Postganglionic Fibers	Great divergence (one preganglionic may activate twenty postganglionic fibers)	Little divergence (one preganglionic only activates a few postganglionic fibers)
Mass Discharge of System as a Whole	Yes	Not normally

Table 9.7 Effects of autonomic nerve stimulation on various visceral effector organs.

Effector Organ	Sympathetic Effect	Parasympathetic Effect
Eye		
Iris (radial muscle)	dilates pupil	——
Iris (sphincter muscle)	——	Constricts pupil
Ciliary muscle	Relaxes (for far vision)	Contracts (for near vision)
Glands		
Lacrimal (tear)	——	Stimulates secretion
Sweat	Stimulates secretion	——
Salivary	Decreases secretion; saliva becomes thick	Increases secretion; Saliva becomes thin
Stomach	——	Stimulates secretion
Intestine	——	Stimulates secretion
Adrenal medulla	Stimulates secretion of hormones	——
Heart		
Rate	Increases	Decreases
Conduction	Increases rate	Decreases rate
Strength	Increases	——
Blood Vessels	Mostly constricts; effects all organs	Dilates in a few organs (eg. penis)
Lungs		
Bronchioles (tubes)	Dilates	Constricts
Mucous glands	Inhibits secretion	Stimulates secretion
Gastrointestinal Tract		
Motility	Inhibits movement	Stimulates movement
Sphincters	Stimulates closing	Inhibits closing
Liver	Stimulates hydrolysis of glycogen	——
Adipose (fat) cells	Stimulates hydrolysis of fat	——
Pancreas	Inhibits exocrine secretions	Stimulates exocrine secretions
Spleen	Stimulates contraction	——
Urinary Bladder	Helps set muscle tone	Stimulates contraction
Piloerector Muscles	Stimulates erection of hair and "goose-bumps"	——
Uterus	If pregnant: contraction If not pregnant: relaxation	——
Penis	Erection; ejaculation	Erection (due to vasodilation)

1. Using a simple line diagram, illustrate the sympathetic pathway (a) from the spinal cord to the heart, and (b) from the spinal cord to the adrenal gland. Label the preganglionic and postganglionic fibers and the ganglion.
2. Describe the mass activation of the sympathetic system, and explain the significance of the term *sympathoadrenal system.*
3. Using a simple line diagram, illustrate the parasympathetic pathway from the brain to the heart. Compare the location of the pre- and postganglionic fibers and the location of their ganglia with those of the sympathetic division.

Functions of the Autonomic Nervous System

The sympathetic and parasympathetic divisions of the autonomic system affect the visceral organs in different ways. Mass activation of the *sympathetic system* prepares the body for intense physical activity in emergencies: the heart rate increases, blood glucose rises, and blood is diverted to the skeletal muscles (away from the visceral organs and skin). These and other effects are listed in table 9.7. The "theme" of the sympathetic system has been aptly summarized in a phrase: **fight or flight.**

Figure 9.13 Neurotransmitters of the autonomic motor system. ACh = acetylcholine; NE = norepinephrine; E = epinephrine. Those nerves that release ACh are called cholinergic; those nerves that release NE are called adrenergic. The adrenal medulla secretes both epinephrine (85%) and norepinephrine (15%) as hormones into the blood.

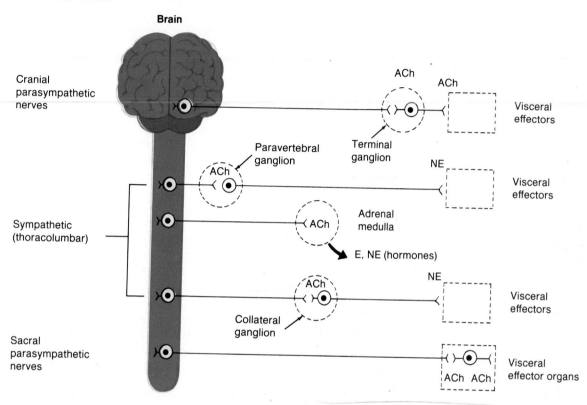

The effects of *parasympathetic nerve* stimulation are in many ways opposite to the effects of sympathetic nerves. The parasympathetic system, however, is not normally activated as a whole. Stimulation of separate parasympathetic nerves can result in slowing of the heart, dilation of visceral blood vessels, and increased activity of the digestive tract (see table 9.7). The different responses of visceral organs to sympathetic and parasympathetic nerve activity is due to the fact that the postganglionic fibers of these two divisions release different neurotransmitters.

Neurotransmitters of the Autonomic System

Acetylcholine (ACh) is the neurotransmitter of all preganglionic fibers (both sympathetic and parasympathetic). Acetylcholine is also the transmitter released by all parasympathetic postganglionic fibers at their synapses with effector cells. Transmission at the autonomic ganglia, and at the synapses of postganglionic parasympathetic nerve fibers, is thus said to be **cholinergic.**

The neurotransmitter released by most postganglionic sympathetic nerve fibers is *norepinephrine* (noradrenalin). Transmission at these synapses is thus said to be **adrenergic.** There are a few exceptions to this rule: some sympathetic fibers that innervate blood vessels in skeletal muscles, and sympathetic fibers to sweat glands, release ACh (are cholinergic).

Figure 9.14 Structure of the catecholamines norepinephrine and epinephrine.

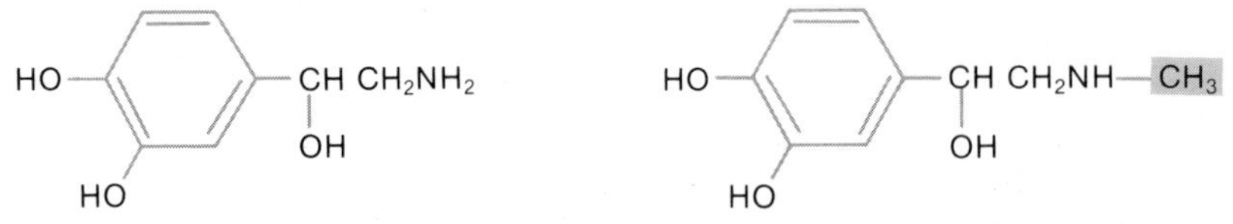

Norepinephrine

Epinephrine

In view of the fact that the cells of the adrenal medulla are derived from postganglionic sympathetic neurons, it is not surprising that the hormones they secrete (normally about 85 percent epinephrine and 15 percent norepinephrine) are similar to the transmitter of postganglionic sympathetic neurons. Epinephrine differs from norepinephrine only in the presence of an additional methyl (CH_3) group—see figure 9.14. Epinephrine, norepinephrine, and dopamine (a transmitter within the CNS) are collectively termed **catecholamines.** The catecholamines, together with serotonin (another transmitter within the CNS), are often referred to as *monoamines* or *biogenic amines.*

Responses to Adrenergic Stimulation

Adrenergic stimulation—by epinephrine in the blood and by norepinephrine released from sympathetic nerve endings—has both excitatory and inhibitory effects. The heart, dilatory muscles of the iris, and the smooth muscles of many blood vessels are stimulated to contract. The smooth muscles of the bronchioles, and of some blood vessels, however, are inhibited from contracting. Adrenergic chemicals therefore cause these structures to dilate.

Since both excitatory and inhibitory effects can be produced in different tissues by the same chemical, the responses clearly depend on the biochemistry of the tissue cells rather than on the intrinsic properties of the chemical. Included in the biochemical differences among the target tissues for catecholamines are differences in the *membrane receptor proteins* for these chemical agents. There are two major classes of these receptor proteins, designated **alpha** (α) and **beta** (β) **receptors.**

Alpha-adrenergic effects are usually excitatory and beta-adrenergic effects are usually inhibitory in the effector organs. There are exceptions to this rule, however. As shown in table 9.8, stimulation of both alpha and beta receptors in the intestine inhibits intestinal activity, and adrenergic stimulation of the heart is mediated by beta receptors.

Drugs have been produced by pharmaceutical companies that either mimic or block the effects of the sympathoadrenal system. Due to slight variations in chemical structure, these drugs may stimulate or inhibit alpha or beta receptors in a relatively specific manner. Propanolol, for example, is a "beta-blocker" often used to slow the heart rate. Other drugs that stimulate beta receptors are used to stimulate the heart and to cause bronchodilation (in the treatment of asthma). Alpha receptor stimulators, which cause vasoconstriction, are used as decongestants, and alpha receptor blockers are sometimes used to help lower high blood pressure.

Responses to Cholinergic Stimulation

Somatic motor neurons, all preganglionic autonomic neurons, and all postganglionic parasympathetic neurons are cholinergic—they use acetylcholine as a neurotransmitter. The cholinergic effects of somatic motor neurons and preganglionic autonomic neurons are always excitatory. The cholinergic effects of postganglionic parasympathetic fibers are usually excitatory, with some notable exceptions—the parasympathetic fibers innervating the heart, for example, cause slowing of the heart rate.

The drug *muscarine,* derived from some poisonous mushrooms, blocks the cholinergic effects of parasympathetic nerves in the heart, smooth muscles, and glands. This drug, however, does not block the effects of somatic motor nerves nor the cholinergic transmission in autonomic ganglia. The acetylcholine receptors in visceral organs are therefore said to promote the *muscarinic effects* of ACh. These effects are identical to those produced by parasympathetic nerve stimulation, and are believed to be mediated by different receptors than those that mediate other cholinergic effects.

Table 9.8 Adrenergic receptors (alpha or beta) found in various organ systems. The effects of epinephrine (E) and norepinephrine (NE) on these receptors is rated on a three-point scale.

Organ (or Organ System)	Receptor	Response	Relative Order of Catecholamine Potency	
			NE	E
Heart	Beta	Increased heart rate; increased contractility; increased conduction velocity	3	2
Blood vessels	Alpha	Constriction	2	1
	Beta†	Dilatation	3	2
Bronchi	Beta	Relaxation	3	2
Intestine				
Muscle	Alpha, beta	Inhibition	2	1
Sphincters	Alpha	Contraction	2	1
Bladder				
Detrusor muscle	Beta	Relaxation	3	2
Trigone muscle of sphincter	Alpha	Contraction	2	1
Uterus	Alpha, beta	Variable§	2	1
Eye				
Radial muscle	Alpha	Contraction	2	1
Ciliary muscle	Beta	Relaxation	3	2

*NE = norepinephrine; E = epinephrine
†Only demonstrated in skeletal muscle and visceral muscle vessels.
†Probably not active in intestine.
§Response is dependent upon stage of menstrual cycle and pregnancy.
Source: Reproduced with permission from Carrier, O., Jr.: Pharmacology of the Peripheral Autonomic Nervous System. Copyright © 1972 by Year Book Medical Publishers, Inc., Chicago.

The muscarinic effects of ACh are specifically inhibited by the drug **atropine,** derived from the deadly nightshade plant *(Atropa belladonna).* Indeed, extracts of this plant were used by women during the middle ages to dilate their pupils (atropine inhibits the parasympathetic stimulation of the iris). This was done to enhance their beauty (belladonna = beautiful woman). Atropine is used clinically today to dilate pupils during eye examinations, to dry mucous membranes of the respiratory system prior to general anesthesia, and to inhibit spasmotic contractions of the lower digestive tract.

The drug nicotine, derived from tobacco plants *(Nicotiana tabacum),* specifically blocks cholinergic transmission of preganglionic fibers in the autonomic ganglia. Drugs that affect transmission in the autonomic ganglia are thus said to have *nicotinic effects,* to distinguish them from drugs that have muscarinic effects in the visceral organs.

Organs with Dual Innervation

Many organs receive a dual innervation—they are innervated by both sympathetic and parasympathetic fibers. When this occurs, the effects of these two divisions on the organ may be antagonistic, complementary, or cooperative.

Antagonistic Effects The effects of sympathetic and parasympathetic stimulation on the pacemaker region of the heart is the best example of antagonism of these two systems. In this case both sympathetic and parasympathetic fibers innervate the same cells. Adrenergic stimulation from sympathetic fibers increase the heart rate, while cholinergic stimulation from parasympathetic fibers inhibit the pacemaker cells and thus decrease the heart rate. Antagonism is also seen in the digestive tract, where sympathetic nerves inhibit and parasympathetic nerves stimulate intestinal movements and secretion.

Figure 9.15 Reciprocal innervation of the iris muscles by the sympathetic and parasympathetic systems. Stimulation of sympathetic nerves produces contraction of the dilator (radial) muscles, which enlarges the pupil. Stimulation of the parasympathetic nerves produces contraction of the constrictor (circular) muscle layer, which makes the pupil smaller.

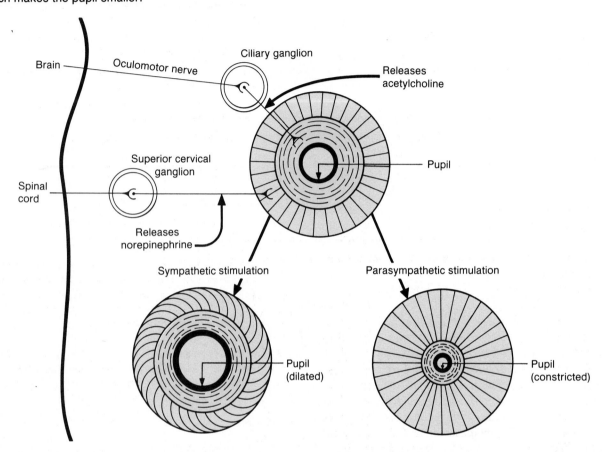

The effects of sympathetic and parasympathetic stimulation on the diameter of the pupil of the eye is analogous to the reciprocal innervation of flexors and extensors by somatic motor neurons. This is because the iris contains antagonstic muscle fibers. Contraction of the radial muscles, which is stimulated by sympathetic nerves, causes dilation; contraction of the circular muscles, which are innervated by parasympathetic nerve endings, causes constriction of the pupils.

Complementary Effects The effects of sympathetic and parasympathetic stimulation on salivary gland secretion are complementary. Secretion of watery saliva is stimulated by parasympathetic nerves, which also stimulate the secretion of other exocrine glands in the digestive tract. Sympathetic nerves stimulate constriction of blood vessels throughout the digestive tract. The decreased blood flow to the salivary glands that results causes a thicker, more viscous saliva to be produced.

Cooperative Effects The effects of sympathetic and parasympathetic stimulation on the urogenital system is cooperative. Erection of the penis, for example, is due to vasodilation that results from parasympathetic nerve stimulation; ejaculation is stimulated by sympathetic nerves. Contraction of the urinary bladder is myogenic (independent of nerve stimulation), but is promoted in part by the action of parasympathetic nerves. This *micturition* (urination) urge and reflex is also enhanced by sympathetic nerve activity, which increases the tone of bladder muscles. Emotional states that are accompanied by high sympathetic nerve activity may thus result in reflex urination at bladder volumes that are normally too low to trigger this reflex.

Organs without Dual Innervation

While most visceral organs are innervated by both sympathetic and parasympathetic nerves, some—including the adrenal medulla, arrector pili muscles, sweat glands, and

Table 9.9 Some reflexes stimulated by sensory input from afferent fibers in the vagus. All of the responses are coordinated by centers in the medulla oblongata.

Organs	Type of Receptors	Reflex Effects
Lungs	Stretch receptors	Inhibits further inhalation; stimulates an increase in cardiac rate and vasodilation
	Type J receptors	Stimulated by pulmonary congestion—produces feelings of breathlessness, and causes a reflex fall in cardiac rate and blood pressure
Aorta	Chemoreceptors	Stimulated by rise in CO_2 and fall in O_2—produces increased rate of breathing, fall in heart rate, and vasoconstriction
	Baroreceptors	Stimulated by increased blood pressure—produces a reflex decrease in heart rate
Heart	Atrial stretch receptors	Inhibits antidiuretic hormone secretion, thus increasing the volume of urine excreted
	Stretch receptors in ventricles	Produces a reflex decrease in heart rate and vasodilation
Gastrointestinal tract	Stretch receptors	Feelings of satiety, discomfort, and pain

most blood vessels—receive only sympathetic innervation. Regulation in these cases is achieved by increases or decreases in the "tone" (firing rate) of the sympathetic fibers. Constriction of blood vessels, for example, is produced by increased sympathetic activity that stimulates alpha receptors; vasodilation results from decreased sympathetic nerve stimulation.

The sympathoadrenal system is required for "nonshivering thermogenesis"—animals deprived of their sympathetic system and adrenals cannot tolerate cold stress. The sympathetic system is also required for proper thermoregulatory responses to heat. During passive heating (by being in a hot room, for example), decreased sympathetic stimulation produces dilation of surface blood vessels in the limbs. This increases surface blood flow and provides better heat radiation. During active heating (exercise), in contrast, there is increased sympathetic activity. This causes constriction of surface blood vessels in the limbs (helping to maintain blood pressure) and stimulation of sweat glands in the trunk.

The eccrine sweat glands in the trunk secrete a watery fluid in response to sympathetic stimulation (the "smelly" apocrine sweat glands only respond to the hormone *epinephrine*). Evaporation of this dilute sweat helps to cool the body. In addition, the eccrine sweat glands secrete, in response to sympathetic nerve stimulation, a chemical called *bradykinin*. Bradykinin stimulates dilation of surface blood vessels near the sweat glands, helping to radiate heat. At the conclusion of exercise, sympathetic stimulation is reduced and blood flow to the surface of the limbs is increased, as in passive heating. Notice that all of these thermoregulatory responses are achieved without the direct involvement of the parasympathetic system.

1. List the neurotransmitters of the preganglionic and postganglionic fibers of the sympathetic and parasympathetic systems.
2. List the effects of sympathoadrenal stimulation on the different visceral organs, and indicate which effects are due to alpha or beta receptor stimulation.
3. Describe the antagonistic effects of sympathetic and parasympathetic activity on the heart, pupils, and digestive tract.
4. Describe the cooperative effects of the sympathetic and parasympathetic systems on the urogenital system.
5. Explain the effects of the drug *atropine* in terms of the actions of the parasympathetic system.

Control of the Autonomic Nervous System by Higher Brain Centers

Some autonomic reflexes occur at the spinal level, where synapses with a number of interneurons are involved. In most autonomic reflexes, however, sensory input is directed to higher brain centers, which regulate activity of descending pathways to preganglionic neurons in the spinal cord.

The **medulla oblongata** of the brain stem is the area that most directly controls activity of the autonomic system. Almost all autonomic responses can be elicited by experimental stimulation of the medulla, which contains centers for the control of the cardiovascular, pulmonary, urogenital, and digestive systems. Much of the sensory input to these centers travels in afferent fibers of the vagus nerve (the vagus is a mixed nerve containing both sensory and motor fibers).

Figure 9.16 *(a)* is sagittal section of the brain, showing area of the hypothalamus. This area is illustrated in *(b)*.

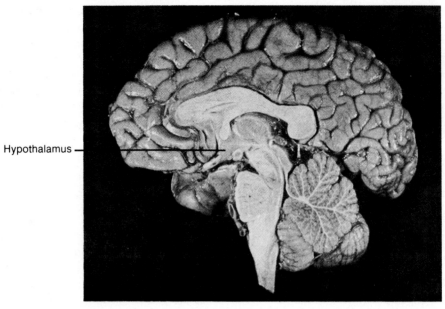

Hypothalamus

(a)

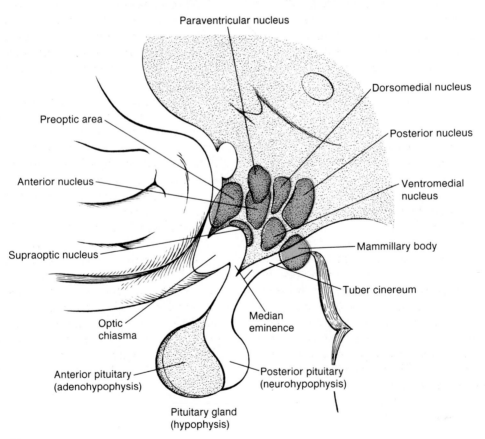

Paraventricular nucleus

Preoptic area

Anterior nucleus

Supraoptic nucleus

Optic chiasma

Anterior pituitary (adenohypophysis)

Pituitary gland (hypophysis)

Median eminence

Posterior pituitary (neurohypophysis)

Dorsomedial nucleus

Posterior nucleus

Ventromedial nucleus

Mammillary body

Tuber cinereum

(b)

The importance of the medulla oblongata, and other higher brain centers, in the control of autonomic reflexes is illustrated by the condition known as *autonomic dysreflexia*. This serious condition, which results in rapid elevations of blood pressure that can lead to stroke (cerebrovascular accident), occurs in 85 percent of people with quadriplegia and in others with spinal cord lesions above the sixth thoracic level.

Lesions to the spinal cord first produces the symptoms of spinal shock, characterized by loss of both skeletal muscle and autonomic reflexes. After a few months both types of reflexes return in an exaggerated state: the skeletal muscles may become spastic, due to absence of higher inhibitory influences, and the visceral organs experience denervation hypersensitivity. Patients in this state have difficulty emptying their bladders and must often be catheterized.

Noxious stimuli, such as overdistension of the urinary bladder, can result in reflex activation of sympathetic nerves below the spinal cord lesion. This produces goose bumps, cold skin, and vasoconstriction at the spinal cord level below the lesion. The rise in blood pressure that results from this vasoconstriction activates pressure receptors that transmit impulses along sensory fibers (in the ninth and tenth cranial nerves) to the medulla. The medulla directs reflex slowing of the heart and vasodilation. Since descending impulses are blocked by the spinal lesion, the skin is warm and moist (due to vasodilation) above the lesion, and cold (due to vasoconstriction) below the lesion.

The Hypothalamus

The hypothalamus is an extremely important region in the brain located just above (superior to) the pituitary gland, which in turn is located above the posterior portion of the roof of the mouth (see figure 9.16). The hypothalamus, by means of efferent fibers to the brain stem and pituitary, and by means of hormones that regulate the anterior pituitary, orchestrates somatic, autonomic, and endocrine responses during various behavioral states.

Experimental stimulation of different areas of the hypothalamus can evoke the autonomic responses characteristic of aggression, sexual behavior, eating, or satiety. Chronic stimulation of the lateral hypothalamus, for example, can make an animal eat and become excessively obese, while stimulation of the medial hypothalamus inhibits eating. Other areas contain osmoreceptors that stimulate thirst and the secretion of antidiuretic hormone (ADH) from the posterior pituitary.

The hypothalamus is also where the body's "thermostat" is located. Experimental cooling of the preoptic-anterior hypothalamus causes shivering (a somatic response), vasoconstriction, and non-shivering thermogenesis (sympathetic responses). Experimental heating of this hypothalamic area results in hyperventilation (due to skeletal muscle contraction), vasodilation, salivation, and sweat gland secretion (autonomic responses).

The coordination of sympathetic and parasympathetic reflexes by the medulla oblongata is thus integrated with the control of somatic and endocrine responses by the hypothalamus. Activities of the hypothalamus are in turn influenced by input from higher brain areas.

Limbic System, Cerebellum, and Cerebrum

The limbic system is a group of fiber tracts and nuclei that form a ring (a limbus) around the brain stem. It includes the cingulate gyrus of the cerebral cortex, the hypothalamus, the fornix (a fiber tract), the hippocampus, and the amygdaloid nucleus (see figure 9.17). These structures were derived early in the course of vertebrate evolution, and were once called the "rhinencephalon" ("smell-brain") because of their importance in the central processing of olfactory information.

These structures are now recognized to be, in higher vertebrates, centers involved in basic emotional drives—such as anger, fear, sex, and hunger—and in short-term memory. Complex circuits between the hypothalamus and other parts of the limbic system contribute visceral responses to emotions.

The autonomic correlates of motion sickness—nausea, sweating, and cardiovascular changes—are abolished by cutting the efferent tracts of the cerebellum. This demonstrates that impulses from the cerebellum to the medulla influence activity of the autonomic system. Experimental and clinical observations have also demonstrated that the frontal and temporal areas of the cerebral cortex influence lower brain areas as part of their involvement in emotion and personality.

One of the most dramatic examples of the role of higher brain areas in personality and emotion is the famous crowbar accident, which occured in 1848. A twenty-five-year-old railroad foreman, Phineas P. Gage, was tamping dynamite down into a hole in a rock with a metal rod when the dynamite exploded. The rod—three feet, seven inches long and one and one-fourth inches thick—was driven through his left eye, passed through his brain, and emerged in the back of his skull.

Figure 9.17   *(a)* is the limbic system. *(b)* shows some of the pathways that interconnect the structures of the limbic system (note: the temporal lobe of the cerebral cortex has been removed).

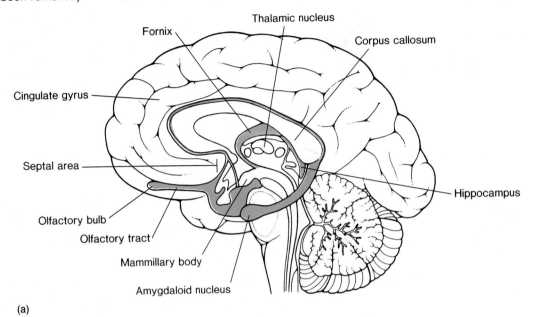

(a)

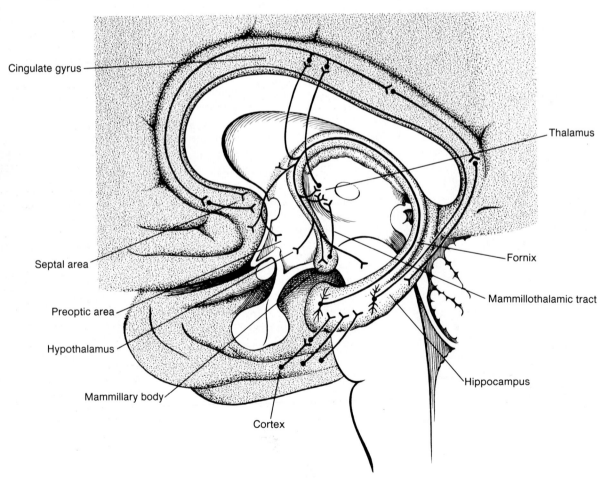

(b)

After a few minutes of convulsions, Gage got up, rode a horse three-quarters of a mile into town, and walked up a long flight of stairs to see a doctor. He recovered well, with no noticeable sensory or motor deficits. His associates, however, noted striking personality changes. Before the accident Gage was a very responsible, capable, and financially prudent man. Afterwards, he appeared to lose social inhibitions, engaged in gross profanity (which he did not do before the accident), and seemed to be tossed about by chance whims. He was eventually fired from his job. His previous friends remarked that he was "no longer Gage."

1. Using a flow diagram (with arrows), describe the reflex pathway by which blood pressure receptors, acting via centers in the medulla oblongata, can cause a reflex slowing of the heart rate.
2. Describe autonomic dysreflexia, and explain the causes of the symptoms associated with this condition.
3. Explain, in terms of the brain regions involved, how autonomic responses—such as flushing, increased heart rate, and so on—are stimulated by various emotional states.

Summary

Visceral Effector Organs

I. Cardiac muscle, like skeletal muscle, is striated.
   A. Cardiac muscle contains sarcomeres, and contracts by means of a sliding filament mechanism.
   B. Unlike skeletal muscle, depolarization of cardiac muscle cells (myocardial cells) is automatic.
      1. Myocardial cells undergo spontaneous depolarization, which produces intrinisic stimulation of contraction.
      2. Adjacent myocardial cells are connected by electrical synapses, or gap junctions—these are part of intercalated discs in the heart muscle.
   C. The cells that have the fastest cycle of depolarization serve as the pacemaker region of the heart.
II. Smooth muscle cells have a spindle shape and lack sarcomeres.
   A. They do, however, contain actin and myosin.
      1. There is a higher ratio of actin to myosin than in skeletal muscle, and the myosin filaments are longer.
      2. The structure of the actin and myosin filaments allows smooth muscle to contract even when greatly stretched.
   B. There are two major categories of smooth muscles.
      1. Single-unit smooth muscles show pacemaker activity, and are interconnected by gap junctions.
      2. Multi-unit smooth muscles have few gap junctions, and are more dependent on nerve stimulation.
III. Exocrine glands secrete their products into ducts leading to the outside of epithelial membranes.
   A. Since epithelial membranes cover the body surface, the secretions of exocrine glands go to the outside of the body.
   B. Glands that secrete to the outer surface of the skin and those that secrete into the digestive tract are exocrine glands.
      1. Glands of the skin include lacrimal, sweat, and sebaceous glands.
      2. Exocrine glands are found throughout the digestive tract; the pancreas and liver secrete pancreatic juice and bile, respectively, into the tract.

Structure of the Autonomic Nervous System

I. There are two neuron types involved in the efferent flow of impulses along an autonomic pathway.
   A. Preganglionic neurons, with their cell bodies in the CNS, conduct impulses out of the brain and spinal cord to autonomic ganglia.
   B. The autonomic ganglia contain cell bodies of postganglionic neurons, whose fibers conduct impulses to synapses with the effector organs (muscles and glands).
II. In the sympathetic (thoracolumbar) division, the preganglionic fibers exit the spinal cord from the first thoracic to the third lumbar levels.
   A. Most of these fibers synapse in a chain of paravertebral ganglia that parallels the spinal cord from the cervical to the coccygeal regions.
      1. Preganglionic fibers from the spinal nerve to these ganglia form the white rami (branches).
      2. Postganglionic fibers from these ganglia to the spinal nerves form the grey rami.
   B. Some preganglionic sympathetic fibers synapse in collateral ganglia in the neck and abdomen.
      1. The collateral ganglia in the neck are the cervical and stellate ganglia.
      2. The celiac, inferior, and superior mesenteric ganglia are collateral ganglia in the abdomen.
   C. There is a great deal of convergence and divergence in the synapses between preganglionic and postganglionic neurons in the sympathetic system, resulting in "mass activation" of this system.

D. The adrenal medulla is derived embryologically from postganglionic sympathetic neurons, and is innervated by preganglionic fibers.
   1. The adrenal medulla secretes its hormone—epinephrine—when the sympathetic system is activated.
   2. Epinephrine is structurally and functionally similar to norepinephrine, the transmitter of postganglionic sympathetic fibers.

III. In the parasympathetic (craniosacral) division, the preganglionic fibers originate in the midbrain, medulla oblongata of the brain stem, and in the sacral levels of the spinal cord.
   A. The fibers that originate in the midbrain travel in the oculomotor (third cranial) nerve to the ciliary ganglion of the eye.
   B. Fibers that originate in the medulla oblongata travel in the facial (seventh), glossopharyngeal (ninth), and vagus (tenth) cranial nerves.
      1. Preganglionic fibers of the vagus innervate the heart, lungs, and digestive tract.
      2. These fibers synapse with postganglionic neurons of terminal ganglia located within the visceral organs.
   C. Preganglionic parasympathetic fibers from the sacral levels of the spinal cord innervate the lower half of the colon, rectum, and the urogenital system.

Functions of the Autonomic System

I. The sympathetic division mediates the visceral responses to "fight or flight."
   A. Sympathetic stimulation causes increased pumping activity of the heart, vasoconstriction of blood vessels in the viscera, increased blood glucose, and increased blood flow to skeletal muscles.

   1. Sympathetic nerve stimulation also causes dilation of the pupils and dilation of the bronchioles.
   2. These responses usually occur together due to mass activation of the sympathetic system.
   B. Different responses to catecholamines (norepinephrine from sympathetic nerve endings and epinephrine from the adrenal medulla) are mediated by different membrane receptor proteins.
      1. Beta receptors mediate the effects of sympathetic nerves on the heart, pupils, and bronchioles.
      2. Alpha receptors mediate adrenergic effects—primarily vasoconstriction in the viscera and skin.

II. Parasympathetic nerve activity stimulates movement and secretions of the digestive tract, vasodilation in the digestive tract and genital regions, constriction of the pupils, and slowing of the heart rate.

III. Some organs receive a "dual innervation" from both sympathetic and parasympathetic fibers.
   A. The effects of these two divisions are antagonistic in the iris, heart, and digestive system.
   B. Sympathetic and parasympathetic effects are complementary and cooperative in some organs.
      1. Erection of the penis is stimulated by parasympathetic nerves, while ejaculation is produced by sympathetic nerve activity.
      2. The micturition (urinating) reflex is aided by both sympathetic and parasympathetic effects.

C. The different actions of sympathetic and parasympathetic nerves are due to different neurotransmitters.
   1. All preganglionic fibers use acetylcholine as a neurotransmitter—they are cholinergic.
   2. All postganglionic parasympathetic fibers are cholinergic.
   3. Most postganglionic sympathetic fibers are adrenergic—they release norepinephrine at their synapses with effector cells.

IV. Arrector pili muscles, eccrine sweat glands, and most blood vessels receive only sympathetic fibers.
   A. An increased activity of these fibers causes vasoconstriction.
   B. A decreased activity of these fibers causes vasodilation.

V. The medulla oblongata coordinates autonomic nerve activity.
   A. Reflex control through the medulla thus coordinates the responses regulating the cardiovascular, genitourinary, digestive, and pulmonary systems.
   B. Higher brain centers—including the hypothalamus, limbic system, and cerebral cortex—influence the activity of the medulla.
      1. The hypothalamus also controls—via the pituitary—activity of much of the endocrine system.
      2. The autonomic and endocrine system are thus affected by, and contribute to, emotional states.

Self-Study Quiz

1. Which of the following statements about cardiac muscle is TRUE?
   (a) it is striated
   (b) it contracts automatically
   (c) its cells are joined by electrical synapses
   (d) all of these
2. Which of the following types of smooth muscle is most like cardiac muscle in its electrical behavior?
   (a) single-unit smooth muscle
   (b) multi-unit smooth muscle
3. When a visceral organ is denervated:
   (a) it ceases to function
   (b) it becomes less sensitive to subsequent stimulation by neurotransmitters
   (c) it becomes hypersensitive to subsequent stimulation
4. The pancreas is both an exocrine and endocrine gland. If its duct is ligated (tied):
   (a) its endocrine secretion would be blocked
   (b) its exocrine secretion would be blocked
   (c) both its endocrine and exocrine secretions would be blocked

5. Parasympathetic ganglia are located
   (a) in a chain parallel to the spinal cord
   (b) in the dorsal roots of spinal nerves
   (c) next to or in the organs innervated
   (d) in the brain
6. The neurotransmitter of preganglionic sympathetic fibers is:
   (a) norepinephrine
   (b) epinephrine
   (c) acetylcholine
   (d) dopamine
7. Which of the following results from stimulation of alpha-adrenergic receptors?
   (a) increased heart rate
   (b) dilation of bronchioles
   (c) dilation of blood vessels
   (d) constriction of blood vessels
8. Which of the following is NOT a result of parasympathetic nerve stimulation?
   (a) increased movement of the digestive tract
   (b) constriction of the pupils
   (c) increased mucus secretion
   (d) constriction of visceral blood vessels

9. The actions of sympathetic and parasympathetic nerves are cooperative in:
   (a) the heart
   (b) the urogenital system
   (c) the digestive system
   (d) the iris
10. Propanolol is a "beta-blocker." It would therefore cause:
    (a) vasodilation
    (b) slowing of the heart rate
    (c) increased blood pressure
11. Atropine blocks the muscarinic effects of ACh. It would therefore cause:
    (a) decreased mucus secretion
    (b) increased heart rate
    (c) decreased movements of the digestive tract
    (d) all of these
12. The area of the brain that is most directly involved in the reflex control of the autonomic system is:
    (a) the hypothalamus
    (b) the cerebral cortex
    (c) the medulla oblongata
    (d) the limbic system

10

Sensory Physiology

Objectives

By studying this chapter, you should be able to:

1. Describe the different categories of sense receptors and how they transduce sensory stimuli into action potentials
2. Explain the *law of specific nerve energies,* and describe the regions of the cerebral cortex where different sensations are perceived
3. Describe the receptors and neural pathways for cutaneous sensation
4. Define sensory acuity, and describe how lateral inhibition enhances acuity
5. Describe the receptors and neural pathways involved in taste and olfaction
6. Describe the structure of the inner ear and the membranous labyrinth
7. Explain how the utricle and saccule provide information about linear acceleration, and how the semicircular canals provide information about angular acceleration

8. Explain how the middle ear ossicles function to transduce sound waves into movements of endolymph in the cochlear duct, and how this in turn is transduced into action potentials by the organ of Corti
9. Describe the structure of the eye and how light is focused on the retina
10. Explain how the lens functions in visual accommodation, and how acuity is affected in presbyopia, myopia, hyperopia, and astigmatism
11. Describe the structure of rhodopsin and how it is changed by the photochemical reaction involved in vision
12. Compare cones and rods in terms of structure, location in the retina, functions, and relative acuity versus sensitivity
13. Describe the function of the tectal system in control of extrinsic and intrinsic eye muscles
14. Describe the geniculostriate system and its projections to the occipital lobe
15. Describe the receptive fields of ganglion cells, lateral geniculate neurons, and neurons in the visual association areas of the cortex

*O*ur perceptions of the world—its textures, colors, and sounds; its warmth, smells, and tastes—are created by the brain from electrochemical nerve impulses delivered to it from sensory receptors. These receptors *transduce* (change) different forms of energy in the "real world" into the energy of nerve impulses, which are conducted into the central nervous system by *sensory neurons.* Different sensory modalities—or qualities of sensation such as sound, light, pressure, and so on—result from differences in neural pathways and synaptic connections. The brain thus interprets impulses in the auditory nerve as sound and in the optic nerve as sight, even though the impulses themselves are identical in the two nerves.

Figure 10.1   Mid-sagittal section of the brain, showing the hindbrain and midbrain structures involved in sensory pathways.

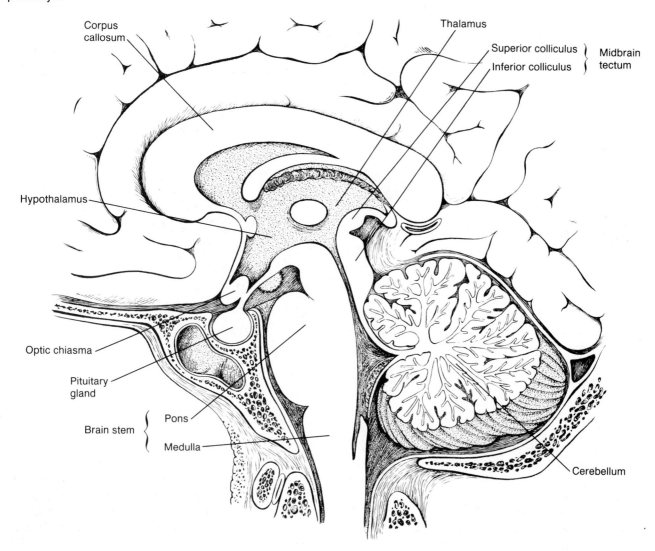

We know, through the use of scientific instruments, that our senses act as energy filters that allow us to perceive only a narrow range of energy. Vision, for example, is limited to light in the visible spectrum (as described in a later section); ultraviolet and infrared light, Xrays, and radio waves, which are the same type of energy as visible light, cannot normally excite the photoreceptors in the eyes. The perception of cold is entirely a product of the nervous system—there is no such thing as cold in the physical world, only varying degrees of heat. The perception of cold, however, has obvious survival value. Though filtered and distorted by the limitations of sensory function, our perceptions of the world allow us to interact effectively with the environment.

The Central Nervous System and Perception

After sensory organs transduce the energy of their sensory stimuli into nerve impulses, this information is conducted into the brain by sensory or ascending *tracts* (a tract is a collection of nerve fibers within the CNS). These sensory fibers may synapse in *nuclei*—collections of neuron cell bodies within the CNS—located in the spinal cord, medulla oblongata, midbrain, and/or thalamus (see figure 10.1) before ultimately projecting to the **cerebral cortex.**

Figure 10.2 Some areas of the cerebral cortex that receive sensory input.

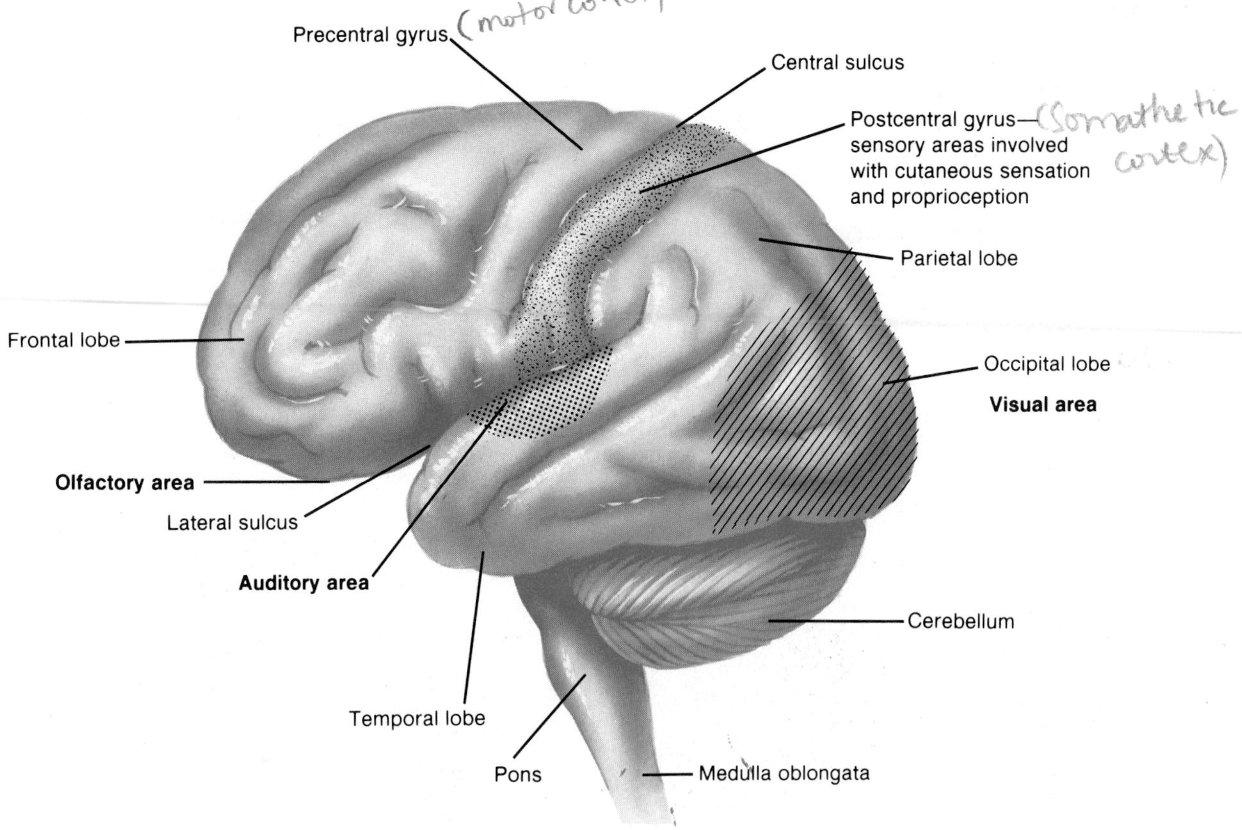

The cerebral cortex is divided by a deep longitudinal fissure into two *cerebral hemispheres* that communicate via a large tract called the **corpus callosum.** Most ascending sensory information crosses over to project to the contralateral hemisphere. Each cerebral hemisphere is highly convoluted—the raised part of each convolution is called a *gyrus,* while the slit between adjacent gyri is called a *sulcus.* Large sulci divide the cortex into four surface **lobes:** the *frontal, parietal, occipital,* and *temporal.* The *central sulcus,* for example, separates the frontal from the parietal lobe; the deep *fissure of Sylvius* divides the temporal lobe from the parietal and frontal lobes.

Anterior to the central sulcus is the precentral gyrus, or motor cortex, as described in chapter 8. The postcentral gyrus, located posterior to the central sulcus, is sometimes called the **somasthetic cortex** because it receives sensory input from the cutaneous receptors and from muscle spindles. The primary visual-processing centers are located in the occipital lobe and the auditory areas are located in the temporal lobe (see figure 10.2).

Categories of Sensory Receptors

Sensory organs can be categorized by their structure and on the basis of different functional criteria. Structurally, the sense organ can be the dendrites of sensory neurons, which are either free (such as those in the skin mediating pain and temperature) or are encapsulated within nonneural structures, such as pressure receptors in the skin (see figure 10.3). Other receptors, including the taste buds, photoreceptors in the eyes, and hair cells in the inner ears, are derived from epithelial cells that synapse with sensory dendrites.

Figure 10.3 Different types of sensory receptors. Free nerve endings *(a)* mediate many cutaneous sensations. Nerve endings that are encapsulated within associated structures are shown in *(b)*. A Pacinian corpuscle *(left)* and a Meissner's corpuscle *(right)* are illustrated. Some receptors, such as the taste bud shown in *(c),* are modified epithelial cells that are innervated by sensory neurons.

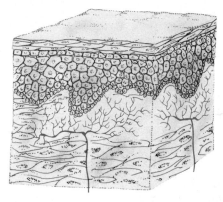

(a)

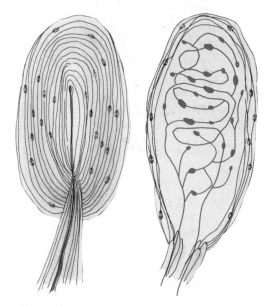

(b)

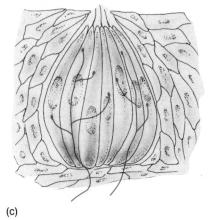

(c)

Functional Categories   Sensory receptors can be grouped according to the type of stimulus energy they transduce. These categories include (1) *chemoreceptors,* such as the taste buds, olfactory epithelium, and the aortic and carotid bodies, which sense chemical changes in the blood; (2) *photoreceptors*—the rods and cones in the retina of the eye; (3) *thermoreceptors,* which respond to heat and cold; and (4) *mechanoreceptors,* which are stimulated by mechanical deformation of the receptor cell membrane—these include touch and pressure receptors in the skin and hair cells within the inner ear. *Nocioreceptors*—or pain receptors—are stimulated by chemicals released from damaged tissue cells, and thus are a type of chemoreceptor.

Receptors can also be grouped according to the type of sensory information they deliver to the brain. *Proprioceptors* include the muscle spindles, Golgi tendon organs, and joint receptors. These provide a sense of body position and allow fine control of skeletal movements. *Cutaneous receptors* include (1) touch and pressure receptors; (2) warmth and cold receptors; and (3) pain receptors. The receptors that mediate sight, hearing, and equilibrium are grouped together as the *special senses.*

Tonic and Phasic Receptors: Sensory Adaptation   Some receptors respond with a burst of activity when a stimulus is first applied, but then quickly decrease their firing rate—adapt to the stimulus—when the stimulus is maintained. Receptors with this response pattern are called *phasic receptors.* Receptors that produce a relatively constant rate of firing as long as the stimulus is maintained are known as *tonic receptors* (see figure 10.4).

Phasic receptors alert us to changes in sensory stimuli, and are in part responsible for the fact that we can cease paying attention to constant stimuli. This ability is called **sensory adaptation.** Odor and touch, for example, adapt rapidly; bathwater feels hotter when we first enter it. Sensations of pain, in contrast, adapt little if at all.

Table 10.1 Classification of receptors based on their normal (or "adequate") stimulus.

Receptor	Normal Stimulus	Mechanisms	Examples
Mechanoreceptors	Mechanical force	Deforms cell membrane of sensory dendrites; or deforms hair cells that activate sensory nerve endings	Cutaneous touch and pressure receptors; vestibular apparatus and cochlea
Pain receptors	Tissue damage	Damaged tissues release chemicals that excite sensory endings	Cutaneous pain receptors
Chemoreceptors	Dissolved chemicals	Chemical interaction affects ionic permeability of sensory cells	Smell and taste (exteroreceptors); osmoreceptors and carotid body chemoreceptors (interoreceptors)
Photoreceptors	Light	Photochemical reaction affects ionic permeability of receptor cell	Rods and cones in retina of eyes

Figure 10.4 · Tonic receptors *(a)* continue to fire at a relatively constant rate as long as the stimulus is maintained. These produce slowly adapting sensations. Phasic receptors *(b)* respond with a burst of action potentials when the stimulus is first applied, but then quickly reduce their rate of firing while the stimulus is maintained. This produces rapidly adapting sensations.

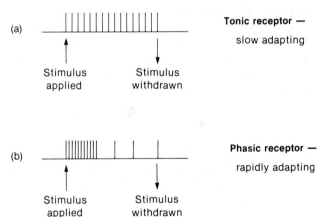

Law of Specific Nerve Energies

Stimulation of a sensory nerve fiber produces only one sensation—touch, cold, pain, and so on. According to the **law of specific nerve energies,** the sensation characteristic of each sensory neuron is that produced by its normal, or *adequate stimulus.* The adequate stimulus for the photoreceptors of the eye, for example, is light. If these receptors are stimulated by some other means—such as by pressure produced by a punch to the eye—a flash of light (the adequate stimulus) may be perceived.

Paradoxical cold provides another example of the law of specific nerve energies. First, a receptor for cold is located by touching the tip of a cold metal rod to the skin. Sensation then gradually disappears as the rod warms to body temperature. Applying the tip of a rod heated to 45°C to the same spot, however, causes the sensation of cold to reappear. This paradoxical cold is produced because the heat slightly damages receptor endings, and by this means produces an "injury current" that stimulates the receptor.

Regardless of how a sensory neuron is stimulated, therefore, only one sensory modality will be perceived. This specificity is due to the synaptic pathways within the brain that are activated by the sensory neuron. The ability of receptors to function as sensory filters, and be stimulated normally by only one type of stimulus (the adequate stimulus), allows the brain to usually perceive the stimulus accurately.

Figure 10.5 Conduction of impulses by a sensory neuron into the spinal cord. The sensory neuron is pseudounipolar (see text for description).

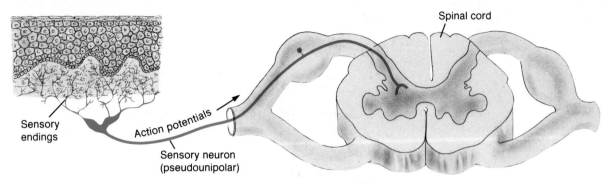

Sensory Neurons and Receptor Potentials

The structure of sensory neurons is unique. They are *pseudounipolar*—only one process extends from the cell body. The term *pseudo* (false) is used because this process splits to form a second one at right angles to the first. This second process is quite long, forming a continuous pathway from the endings in the sensory organ, past the dorsal root ganglia (which contain the cell bodies of sensory neurons), to the intermediate or ventral grey matter of the spinal cord. The structure of a sensory neuron is shown in figure 10.5.

Anatomically, the part of the sensory neuron process that conducts impulses to the cell body is a dendrite, while the part that conducts impulses from the region of the dorsal root ganglia into the spinal cord is an axon. The impulses that originate near the sensory nerve endings, however, are conducted continuously along the sensory process, so that the entire process functions as a single unit. Since the entire length of this process (with the exception of the sensory endings) contains voltage-regulated gates, and can thus produce action potentials, the entire process behaves functionally as an axon.

The electrical behavior of the sensory nerve endings is similar to that of the dendrites of other neurons. Upon stimulation, the dendrites exhibit a local, graded depolarization that increases in proportion to the stimulus intensity. This depolarization is analogous to an excitatory postsynaptic potential (EPSP, as described in chapter 7). In the sensory endings, however, it is known as a **receptor potential** or a **generator potential** because it stimulates opening of voltage-regulated gates and thus serves to generate action potentials (see figure 10.6) in response to sensory stimuli.

1. Our perceptions are products of our brains; they are incompletely and inconstantly related to physical reality. Explain this statement, using examples of vision and perceptions of cold.
2. Define the law of specific nerve energies and the adequate stimulus, and relate these definitions to your answer for question no. 1.
3. Describe sensory adaptation in olfactory and pain receptors. Using a line drawing, relate sensory adaptation to the responses of phasic and tonic receptors.
4. Describe how the magnitude of a sensory stimulus is transduced into a receptor potential and how the magnitude of the receptor potential is coded in the sensory nerve process.

Figure 10.6 Sensory stimuli result in the production of local, graded potential changes known as the receptor or generator potential (number 1 through 4). If the receptor potential reaches a threshold value of depolarization it generates action potentials (number 5) in the sensory neuron.

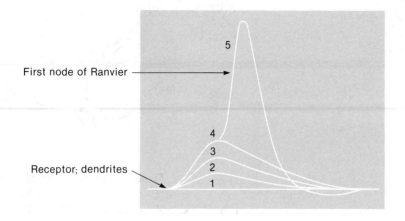

First node of Ranvier

Receptor; dendrites

Table 10.2 Types of cutaneous receptors.

Encapsulated or Free Nerve Endings	Type	Sensation
Encapsulated (dendrites within associated structures)	Pacinian corpuscles; Meissner's corpuscles; Krause's end bulbs; Ruffini's end organs	All serve touch and pressure
Free nerve endings	——	Touch, pressure, heat, cold, pain

Cutaneous Sensations

The cutaneous sensations of touch, pressure, hot and cold, and of pain are mediated by the dendritic nerve endings of different sensory neurons. The receptors for hot, cold, and pain are the naked endings of sensory neurons. Sensations of touch and pressure are mediated by both naked dendritic endings and dendrites that are encapsulated within various structures (see table 10.2). In *Pacinian corpuscles,* for example, the dendritic endings are encased within thirty to fifty onionlike layers of connective tissue (see figure 10.7). These layers absorb some of the pressure when a stimulus is maintained, and thus help to accentuate the phasic response of this receptor.

Figure 10.7 A Pacinian corpuscle.

Figure 10.8   Pathways that lead from the cutaneous receptors and proprioceptors to the postcentral gyrus in the cerebral cortex.

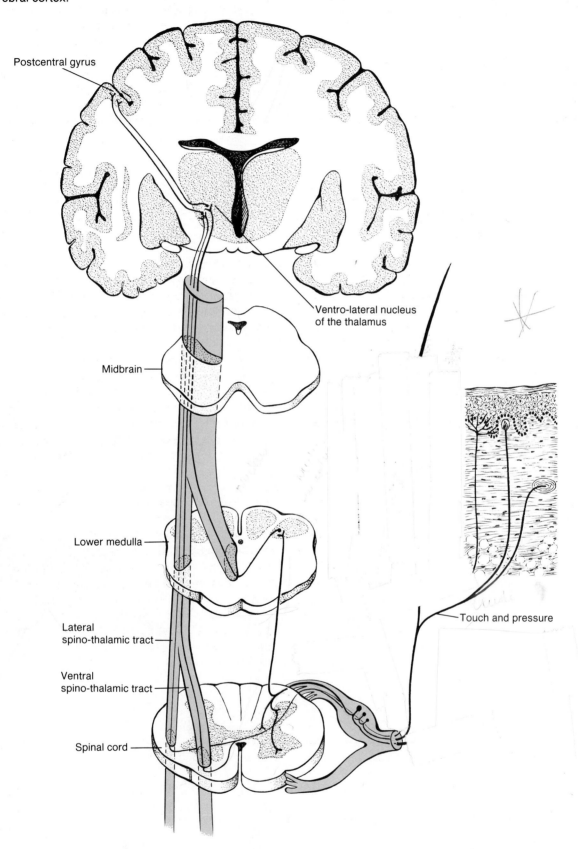

Postcentral gyrus

Ventro-lateral nucleus of the thalamus

Midbrain

Lower medulla

Lateral spino-thalamic tract

Ventral spino-thalamic tract

Spinal cord

Touch and pressure

Figure 10.9 Areas of the somasthetic cortex (postcentral gyrus) devoted to sensation in different parts of the body. Notice that a disproportionately large area of cortex is devoted to the fingers and lips.

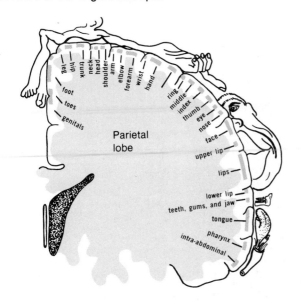

Neural Pathways for Somasthetic Sensations

The conduction pathways for the *somasthetic senses*—a term that includes sensations from cutaneous and proprioceptors—are shown in figure 10.8. *Proprioception* and *pressure* are carried by large, myelinated nerve fibers that ascend in the *dorsal columns* of the spinal cord on the same (ipsilateral) side. These fibers do not synapse until they reach the *medulla oblongata* of the brain stem; fibers that carry these sensations from the feet are thus incredibly long. After synapsing in the medulla with other, second-order sensory neurons, information in the latter neurons crosses over to the contralateral side as it ascends via a fiber tract, called the medial lemniscus, to the *thalamus*. Third-order sensory neurons in the thalamus that receive this input in turn project to the postcentral gyrus.

Sensations of *hot, cold,* and *pain* are carried by thin, unmyelinated sensory neurons into the spinal cord. These synapse with second-order interneurons within the spinal cord, which cross over to the contralateral side and ascend to the brain in the *lateral spinothalamic tract.* Fibers that mediate *touch* and *pressure* ascend in the *ventral spinothalamic tract.* Fibers of both spinothalamic tracts synapse with third-order neurons in the thalamus (bypassing the medulla), which in turn project to the postcentral gyrus. Notice that, in all cases, somasthetic information is carried to the postcentral gyrus in third-order neurons. Also, because of crossing over, somasthetic information from each side of the body is projected to the postcentral gyrus of the contralateral cerebral hemisphere.

All somasthetic information from the same area of the body projects to the same area of the postcentral gyrus, so that a "map" of the body can be drawn on the postcentral gyrus to represent sensory projection points. This map is very distorted, however, because it shows larger areas of cortex devoted to sensation in the face and hands than in other areas in the body (see figure 10.9). This disproportionately larger area of the cortex devoted to the face and hands reflects the fact that there is a higher density of sensory receptors in these regions.

Receptive Fields and Sensory Acuity

The *receptive field* of a neuron serving cutaneous sensation is the area of skin whose stimulation results in changes in the firing rate of the neuron. Changes in the firing rate of primary sensory neurons affect the firing of second- and third-order neurons, which in turn affects the firing of those neurons in the postcentral gyrus that receive input from the third-order neurons. Indirectly, therefore, neurons in the postcentral gyrus can be said to have receptive fields in the skin (as indicated by the sensory map shown in figure 10.9).

The area of each receptive field in the skin varies inversely with the density of receptors in the region. In the back and legs, where a large area of skin is served by relatively few sensory endings, the receptive field of each neuron is correspondingly large. In the fingertips—where a large number of cutaneous receptors serve a small area of skin—the receptive field of each sensory neuron is correspondingly small.

Figure 10.10   The two-point touch threshold test. If each point touches the receptive fields of two different sensory neurons, two separate points of touch will be felt. If both caliper points touch the receptive field of one sensory neuron, only one point of touch will be felt.

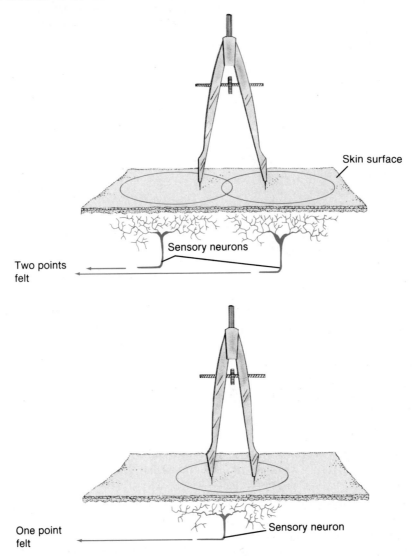

Two-Point Touch Threshold   The approximate size of the receptive fields serving light touch can be measured by the *two-point touch threshold* test. In this procedure, two points of a pair of calipers are lightly touched to the skin at the same time. If the calipers are set sufficiently wide apart, each point will stimulate a different receptive field and a different sensory neuron—two separate points of touch will thus be felt. If the calipers are sufficiently closed, both points will touch the receptive field of only one sensory neuron, and only one point of touch will be felt (see figure 10.10).

The two-point touch threshold, which is the minimum distance that can be distinguished between two points of touch, is a measure of the distance between receptive fields. If the two points of the calipers are closer than this distance, only one "blurred" point of touch can be felt. The two-point touch threshold is thus an indication of tactile *acuity,* or the sharpness (acu = needle) of touch perception.

Table 10.3 The two-point touch threshold for different regions of the body. Source: S. Weinstein, *The Skin Senses,* ed. D. R. Kenshalo (Springfield, Ill.: Charles C Thomas, 1968).

Body Region	Two-Point Touch Threshold (mm)
Big toe	10
Sole of foot	22
Calf	48
Thigh	46
Back	42
Belly	36
Upper arm	47
Forehead	18
Palm of hand	13
Thumb	3
First finger	2

Source: From Weinstein, S., and D. R. Kenshalo (editor), The Skin Senses, 1968. Courtesy of Charles C Thomas, Publisher, Springfield, Illinois.

The high tactile acuity of the fingertips is exploited in the reading of *Braille.* Braille symbols consist of dots that are raised 1 mm up from the page and separated from each other by 2.5 mm, which is slightly above the two-point touch threshold in the fingertips (see table 10.3). Experienced Braille readers can scan words at about the same speed that a sighted person can read aloud—a rate of about 100 words per minute.

Lateral Inhibition

When a blunt object touches the skin a number of receptive fields may be stimulated. Those receptive fields in the center areas where the touch is strongest will be stimulated more than in neighboring fields where the touch is lighter. We do not usually feel a subtle sensory experience of light touch when it surrounds a center of stronger touch, however. Instead, only a single touch is felt, which is somewhat sharper than the actual shape of the blunt object. This sharpening of sensation is due to a process called *lateral inhibition.*

Lateral inhibition, and the sharpening of sensation that results, occurs within the central nervous system. Those sensory neurons whose receptive fields are stimulated most strongly inhibit—via interneurons that pass "laterally" within the CNS—sensory neurons that serve neighboring receptive fields. Lateral inhibition similarly plays a prominent role in pitch discrimination, as described in a later section.

Figure 10.11 When an object touches the skin *(a)*, receptors in the center of the touched skin are stimulated more than neighboring receptors *(b)*. As a result of lateral inhibition within the central nervous system *(c)*, input from these neighboring sensory neurons is reduced. Sensation, as a result, is more sharply localized to the area of skin that was stimulated the most, *(d)*.

(a)

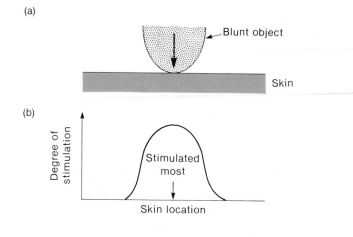

(b)

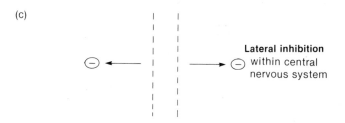

(c)

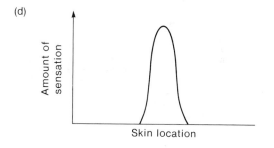

(d)

Taste and Olfaction

Chemoreceptors that respond to chemical changes in the internal environment are called *interoceptors;* those that respond to chemical changes in the external environment are *exteroceptors.* Included in the latter category are *taste receptors,* which respond to chemicals dissolved in food or drink, and *olfactory receptors,* which respond to gaseous molecules in the air. This distinction is somewhat arbitrary, however, because odorant molecules in air must first dissolve in fluid within the olfactory mucosa before the sense of smell can be stimulated.

Taste

Taste receptors are specialized epithelial cells that are grouped together into barrel-shaped arrangements called *taste buds,* located in the epithelium of the tongue. The cells of the taste buds have microvilli at their apical (top) surface, which is exposed to the external environment through a pore in the surface of the taste bud.

Molecules dissolved in saliva at the surface of the tongue interact with receptor molecules in the microvilli of the taste buds. This interaction stimulates the release of a neurotransmitter chemical from the receptor cells, which in turn stimulates sensory nerve endings that innervate the taste buds. Taste buds in the posterior third of the tongue are innervated by the *glossopharyngeal (ninth cranial) nerve;* those in the anterior two-thirds of the tongue are innervated by the *facial (seventh cranial) nerve.*

There are only four basic modalities of taste, which are sensed most acutely in particular regions of the tongue. These are *sweet* (tip of the tongue), *sour* (sides of the tongue), *bitter* (back of the tongue), and *salty* (over most of the tongue). This distribution is illustrated in figure 10.13.

Sour taste is produced by hydrogen ions (H^+); all acids therefore taste sour. Most organic molecules, particularly sugars, taste sweet to varying degrees. Only pure table salt (NaCl) has a pure salty taste—other salts, such as KCl (commonly used in place of NaCl by people with hypertension) taste salty but have bitter overtones. Bitter taste is evoked by quinine and seemingly unrelated molecules.

Figure 10.12   A taste bud.

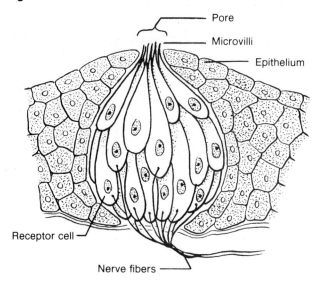

Olfaction

The olfactory receptors are the dendritic endings of the *olfactory (first cranial) nerve,* in association with epithelial supporting cells. Unlike other sensory modalities, which are relayed to the cerebrum from the thalamus, the sense of olfaction is transmitted directly to the olfactory bulb of the cerebral cortex (see figure 10.14).

Unlike taste, which is divisible into only four modalities, many thousands of different odors can be distinguished by people who are trained in this capacity (as in the perfume and wine industries). The molecular basis of olfaction is not understood; although various theories have attempted to explain families of odors on the basis of similarities in molecular shape and/or charges, such attempts have been only partially successful. The extreme sensitivity of olfaction is possibly as amazing as its diversity—at maximum sensitivity, only one odorant molecule is needed to excite an olfactory receptor.

Figure 10.13   Areas of the tongue that are most sensitive to each of the four modalities of taste.

Sweet   Sour   Salty   Bitter

Figure 10.14   The olfactory epithelium contains receptor neurons that synapse with neurons in the olfactory bulb of the brain.

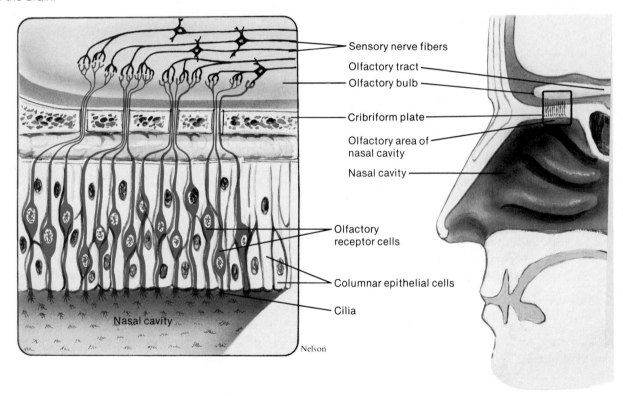

Figure 10.15 Structures within the inner ear include the cochlea and vestibular apparatus. The vestibular apparatus consists of the utricle and saccule (together called the otolith organs) and the three semicircular canals. Each semicircular canal contains a widened area (the ampulla, which is labeled for only one canal) with sensory hair cells.

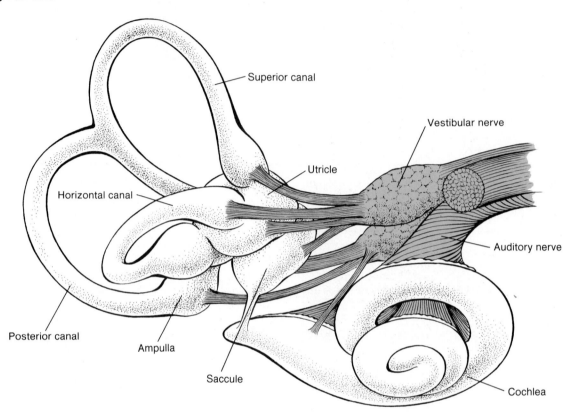

1. Using a flow diagram, describe the neural pathways leading from cutaneous pain and pressure receptors to the postcentral gyrus. Indicate where crossing over occurs.
2. Describe the meaning of the term *sensory acuity,* and explain how acuity is affected by the density of receptive fields and by the process of lateral inhibition.
3. Describe the distribution of taste receptors in the tongue, and explain what effect damage to the facial nerve might have on taste sensation.

The Inner Ear: Vestibular Apparatus

The sense of equilibrium, which provides orientation with respect to gravity, is due to the function of an organ called the **vestibular apparatus.** The vestibular apparatus, together with a snail-like structure called the *cochlea,* which is involved in hearing, form the *inner ear* within the temporal bones of the skull. The vestibular apparatus consists of two parts: (1) the *otolith organs,* which includes the *utricle* and *saccule;* and (2) the *semicircular canals* (see figure 10.15).

Figure 10.16 The endolymph-filled membranous labyrinth *(shaded)* includes the vestibular apparatus and the scala media in the cochlea. The membranous labyrinth is encased within a bony labyrinth *(unshaded)* that is filled with perilymph.

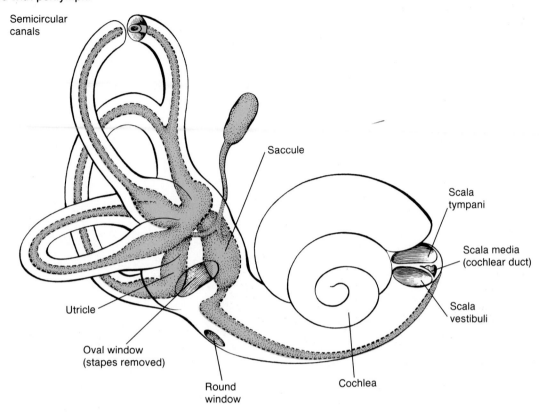

Semicircular canals

Saccule

Scala tympani

Scala media (cochlear duct)

Scala vestibuli

Utricle

Oval window (stapes removed)

Round window

Cochlea

The sensory structures of the vestibular apparatus and cochlea are located within a structure called the **membranous labyrinth,** which is filled with a fluid that is similar in composition to intracellular fluid. This fluid is called *endolymph.* The bony structures surrounding the membranous labyrinth in the inner ear contain a fluid called *perilymph,* which is similar in composition to cerebrospinal fluid. The membranous labyrinth (see figure 10.16), in other words, is filled with endolymph and is surrounded by perilymph.

Sensory Hair Cells of the Vestibular Apparatus

The utricle and saccule provide information about *linear acceleration*—changes in velocity when traveling horizontally or vertically. We therefore have a sense of acceleration and deceleration when riding in a car or when skipping rope. A sense of *rotational* or *angular acceleration* is provided by the semicircular canals, which are oriented in three planes like the faces of a cube. This helps us maintain balance when turning the head, spinning, or tumbling.

Figure 10.17 *(a)* Sensory hair cells in the vestibular apparatus contain hairs (microvilli) and one kinocilium. *(b)* When hair cells are bent in the direction of the kinocilium, the cell membrane is depressed (see *arrow*) and the sensory neuron innervating the hair cell is stimulated. *(c)* When the hairs are bent in a direction opposite to the kinocilium the sensory neuron is inhibited.

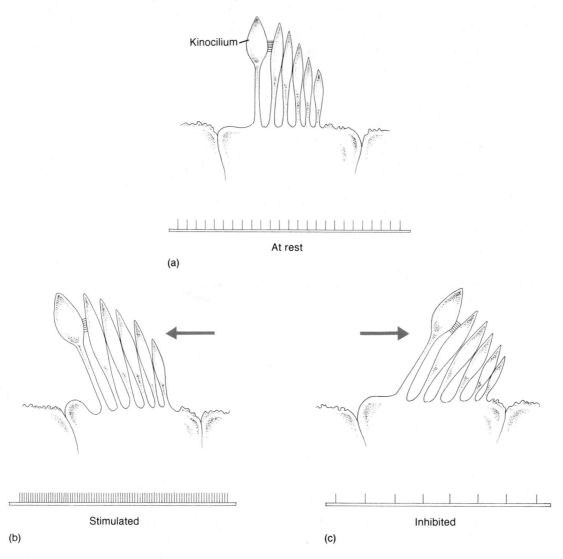

The receptors for equilibrium are modified epithelial cells called *hair cells,* because they contain twenty to fifty "hairs" (actually microvilli) and one cilium, which is called a *kinocilium.* When the hair cells are bent in the direction of the kinocilium, the cell membrane is depressed and becomes depolarized. When the hairs are bent in the opposite direction, the membrane becomes hyperpolarized (as shown in figure 10.17).

Utricle and Saccule

The hair cells of the utricle and saccule protrude into the endolymph-filled membranous labyrinth, with their hairs embedded within a gelatinous **otolith membrane.** The otolith membrane contains many crystals of calcium carbonate from which it derives its name (oto = ear; lith = stone). These stones increase the mass of the membrane, which results in a higher *inertia* (resistance to change in movement).

Figure 10.18 Scanning electron micrograph of hairs and kinocilium within the vestibular apparatus.

When a person is upright, the hairs of the utricle are oriented vertically, while those of the saccule are oriented horizontally into the otolith membrane. During forward acceleration, the otolith membrane lags behind the hair cells, so the hairs are pushed backwards. This is similar to the backwards thrust of the body when a car accelerates rapidly forwards. The inertia of the otolith membrane similarly causes the hairs of the utricle to be pushed upwards when a person jumps from a raised platform. These effects, and the opposite ones that occur when a person accelerates backwards or upwards, allow us to maintain our equilibrium with respect to gravity during linear acceleration.

Figure 10.19 The utricle and saccule contain patches of sensory hair cells (maculae), with the hairs embedded within an otolith membrane. The otolith membrane consists of a gelatinous layer covered with stones of calcium carbonate.

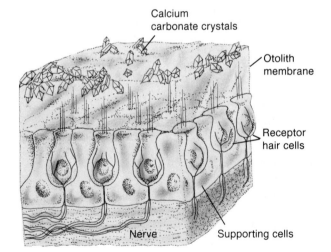

Figure 10.20   Hair cells within the ampulla of the semicircular canals are embedded within a gelatinous membrane called the cupula *(a)*, which protrudes into the endolymph-filled canal. Movement of endolymph during rotation causes the cupula to bend *(b)*, thus stimulating the hair cells.

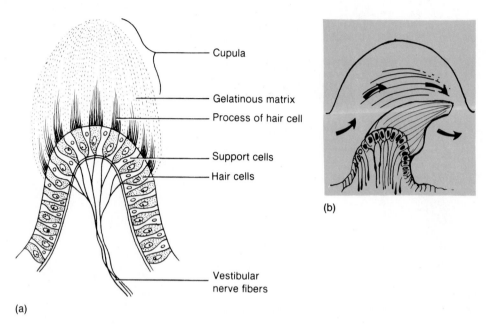

Cupula

Gelatinous matrix
Process of hair cell

Support cells
Hair cells

Vestibular
nerve fibers

(a)

(b)

The Semicircular Canals

The three semicircular canals are oriented at right angles to each other. Each canal contains a bulge, called the *ampulla,* where the sensory hair cells are located. The processes, or "hairs," of these sensory cells are embedded within a gelatinous membrane called the *cupula* (shown in figure 10.20), which projects into the endolymph of the membranous canals. Like a sail in the wind, the cupula can be pushed in one direction or the other by movements of the endolymph.

The endolymph of the semicircular canals serves a function analogous to that of the otolith membrane—it provides inertia so that sensory hairs will be bent in a direction opposite to that of the angular acceleration. As the head rotates to the right, for example, the endolymph causes the cupula to be bent towards the left, thereby stimulating the sensory hair cells.

Neural Pathways   Stimulation of hair cells in the vestibular apparatus activates sensory neurons of the *vestibulocochlear (eighth cranial) nerve.* These fibers transmit impulses to the cerebellum and to the vestibular nuclei of the medulla oblongata. The vestibular nuclei, in turn, send fibers to the occulomotor center of the brain stem and to the spinal cord. Neurons in the occulomotor center control eye movements, and neurons in the spinal cord stimulate movements of the head, neck, and limbs. Movements of the eyes and body produced by these pathways serve to maintain balance and "track" the visual field during rotation.

Figure 10.21 Neural processing involved in maintenance of equilibrium and balance.

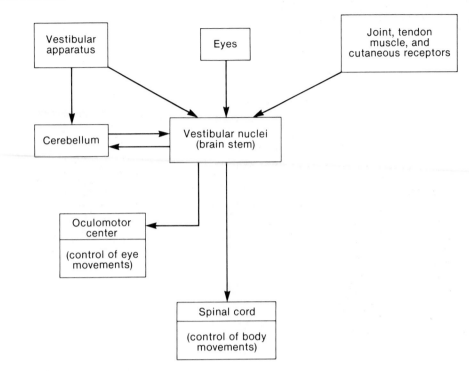

Nystagmus and Vertigo   When a person first begins to spin, the inertia of the endolymph within the semicircular canals causes the cupula to bend in the opposite direction. As the spin continues, however, the inertia of the endolymph is overcome and the cupula straightens. At this time the endolymph and the cupula are moving in the same direction and at the same speed. If movement is suddenly stopped, the greater inertia of the endolymph causes it to continue moving in the previous direction of spin and to bend the cupula in that direction.

Bending of the cupula after movement has stopped affects muscular control of the eyes and body through the neural pathways previously discussed. The eyes slowly drift in the direction of the previous spin, and then are rapidly jerked back to the midline position, producing involuntary oscillations. These movements are called **vestibular nystagmus.** People at this time may feel that they, or the room, are spinning. The loss of equilibrium that results is called *vertigo.* If the vertigo is sufficiently severe, or the person particularly susceptible, the autonomic system may become involved. This can produce dizziness, pallor, sweating, and nausea.

Vestibular nystagmus is one of the symptoms of an inner-ear disease called **Ménière's disease.** The early symptom of this disease is often "ringing in the ears," or *tinnitus.* Since the endolymph of the cochlea and the endolymph of the vestibular apparatus are continuous, through a tiny canal called the duct of Hensen, vestibular symptoms of vertigo and nystagmus often accompany hearing problems in this disease.

Hearing

Sound waves travel in all directions from their source, like ripples in a pond after a stone is dropped. These waves are characterized by their frequency and their intensity. The **frequency,** or distances between crests of the sound waves, is measured in *hertz (Hz),* which is the modern designation for *cycles per second (cps).* The *pitch* of a sound is directly related to its frequency—the higher the frequency of a sound, the higher its pitch.

Figure 10.22 The organs of hearing, including the external ear (pinna and external auditory meatus); middle ear ossicles; and inner ear, which contains the cochlea.

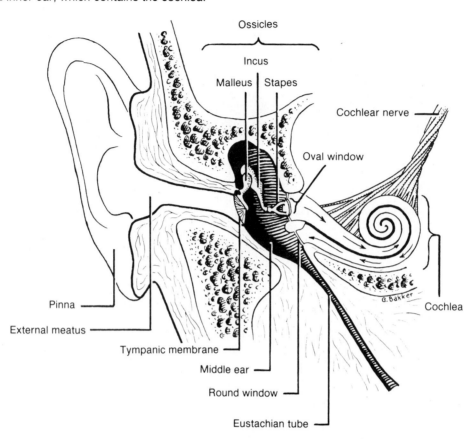

The **intensity,** or loudness of a sound, is directly related to the amplitude of the sound waves. This is measured in units known as *decibels (db)*. A sound that is barely audible—at the threshold of hearing—has an intensity of zero decibels. Every ten decibels indicates a tenfold increase in sound intensity; a sound is ten times higher than threshold at 10 db, 100 times higher at 20 db, a million times higher at 60 db and ten billion times higher at 100 db.

The ear of a trained, young individual can hear sound over a frequency range of 20,000–30,000 Hz, yet can distinguish between two pitches that have only a 0.3 percent difference in frequency. The human ear can detect differences in sound intensities of only 0.1 to 0.5 db, while the range of audible intensities covers twelve order of magnitude (10^{12}), from the barely audible to the limits of painful loudness.

The Outer Ear

Sound waves are funneled by the *pinna,* or *auricle* (flap) into the *external auditory meatus*. These two structures comprise the *outer ear*. The external auditory meatus channels the sound waves (while increasing their intensity) to the eardrum, or *tympanic membrane*. Sound waves in the external auditory meatus produce extremely small vibrations of the tympanic membrane; movements of the eardrum during speech (with an average sound intensity of 60 db) are estimated to be about equal to the diameter of a molecule of hydrogen!

Figure 10.23   Development of otosclerosis in the middle ear immobilizes the stapes, leading to conduction deafness.

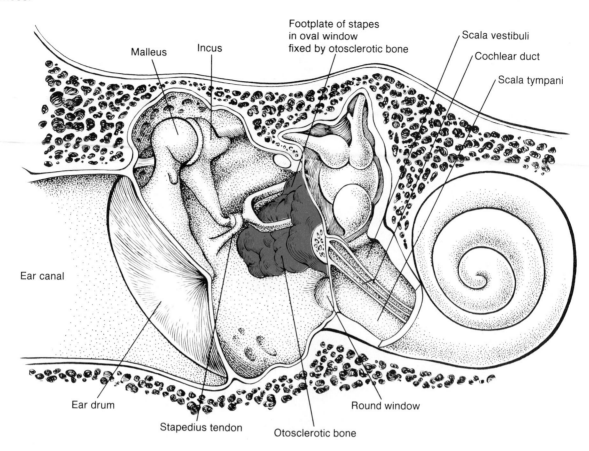

Malleus   Incus   Footplate of stapes in oval window fixed by otosclerotic bone   Scala vestibuli   Cochlear duct   Scala tympani

Ear canal

Ear drum   Round window

Stapedius tendon   Otosclerotic bone

The Middle Ear

The middle ear is the cavity between the tympanic membrane on the outer side and the cochlea on the inner side. Within this cavity are three **middle ear ossicles**—the *malleus* (hammer), *incus* (anvil), and *stapes* (stirrup). The malleus is attached to the tympanic membrane, so that vibrations of this membrane are transmitted, via the malleus and incus, to the stapes. The stapes, in turn, is attached to a membrane in the cochlea called the *oval window,* which thus vibrates in response to vibrations of the tympanic membrane.

The fact that vibrations of the tympanic membrane are transferred through three bones instead of just one affords protection. If the sound is too intense, the ossicles can buckle. This protection is increased by the action of the *stapedius muscle,* which attaches to the neck of the stapes. When sound becomes too loud, this muscle contracts and moves the footplate of the stapes away from the oval window. This action helps to protect against nerve damage within the cochlea.

Damage to tympanic membrane or middle ear ossicles produces **conduction deafness.** This can result from a variety of causes, including *otitus media* and *otosclerosis.* In otitus media, inflammation produces excessive fluid accumulation within the middle ear, which can in turn result in excessive growth of epithelial tissue and damage to the eardrum. This can occur following allergic reactions or respiratory disease. In otosclerosis, bone is resorbed and replaced by "sclerotic bone" that grows over the oval window and immobilizes the footplate of the stapes (see figure 10.23). In conduction deafness these pathological changes hinder transmission of sound waves from air to the cochlea of the inner ear.

Figure 10.24 Cross section of the cochlea showing its three turns and its three compartments—scala vestibuli, cochlear duct (scala media), and scala tympani.

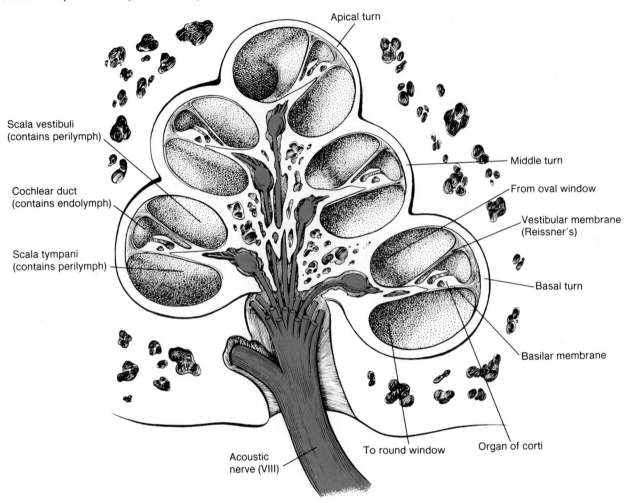

The Cochlea

Vibrations of the stapes and oval window displace endolymph within a part of the membranous labyrinth known as the **cochlear duct** or **scala media.** The latter name indicates that the cochlear duct is the middle part of the snail-shaped cochlea within the inner ear. Like the cochlea as a whole, the cochlear duct coils to form three levels (see figure 10.24), similar to the basal, middle, and apical portions of a snail shell. The part of the cochlea above the cochlear duct is called the *scala vestibuli,* and the part below is called the *scala tympani.* Unlike the central cochlear duct, which contains endolymph, the scala vestibuli and scala tympani are filled with perilymph.

The perilymph of the scala vestibuli and scala tympani are continuous at the apex of the cochlea because the cochlear duct ends blindly, leaving a small space called the *helicotrema* between the end of the cochlear duct and the wall of the cochlea. Vibrations of the round window produced by movements of the stapes cause pressure waves within the scala vestibuli, which pass around the helicotrema to the scala tympani. Movements of perilymph within the scala tympani, in turn, travel to the base of the cochlea where they cause displacement of a membrane called the *round window* into the middle ear cavity (see figure 10.25).

Figure 10.25   The cochlea is illustrated as a straightened structure to show that perilymph in the scala vestibuli and scala tympani is continuous at the helicotrema *(a)*. These two chambers are separated by the blind-ending cochlear duct (scala media). The basilar membrane of the cochlear duct *(b)* is narrower at the base and wider at the apex. Sounds of different frequencies (pitches) produce maximum displacement of the basilar membrane in the regions shown.

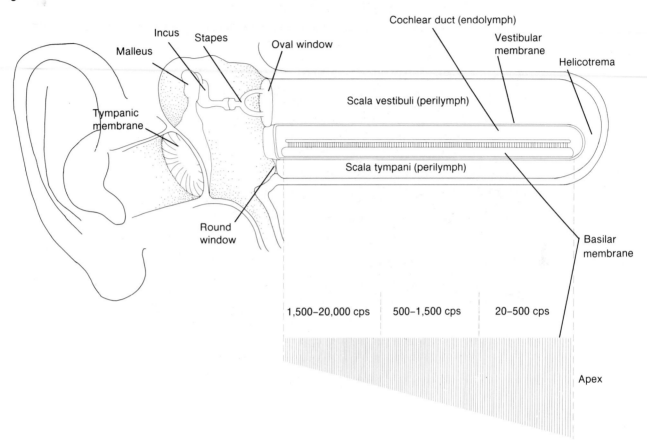

When the sound frequency (pitch) is sufficiently low, there is adequate time for the pressure waves of perilymph within the upper scala vestibuli to travel through the helicotrema to the scala tympani. As the sound frequency increases, however, pressure waves of perilymph within the scala vestibuli do not have time to travel all the way to the apex of the cochlea. Instead, they are transmitted through the *vestibular* (or *Reissner's*) *membrane,* which separates the scala vestibuli from the cochlear duct, and through the *basilar membrane,* which separates the cochlear duct from the scala tympani, to the perilymph of the scala tympani (see figure 10.26). The distance that these pressure waves travel, therefore, decreases as the sound frequency increases.

Figure 10.26 Pressure waves of perilymph in the scala vestibuli cause deflections of the vestibular and basilar membranes. In this way the pressure is transmitted to the scala tympani.

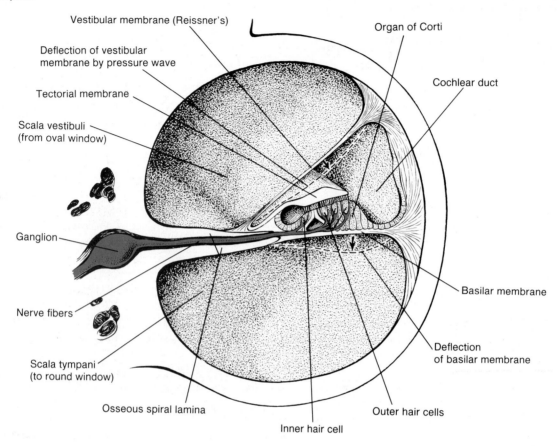

Vestibular membrane (Reissner's)

Deflection of vestibular membrane by pressure wave

Tectorial membrane

Scala vestibuli (from oval window)

Ganglion

Nerve fibers

Scala tympani (to round window)

Osseous spiral lamina

Inner hair cell

Organ of Corti

Cochlear duct

Basilar membrane

Deflection of basilar membrane

Outer hair cells

The Organ of Corti

Movements of perilymph from the scala vestibuli to the scala tympani thus produce displacement of the Reissner's membrane and the basilar membrane. While movement of Reissner's membrane does not directly contribute to hearing, displacement of the basilar membrane is central to pitch discrimination. The basilar membrane is fixed on the inner side of the cochlear wall to a bony ridge and is supported at its free end by a ligament.

Sounds of low pitch (with frequencies below about 50 Hz) cause movements of the entire length of the basilar membrane—from the base to the apex. Higher sound frequencies result in maximum displacement of the basilar membrane closer to its base, as illustrated in figure 10.27.

The sensory cells of hearing are located on the basilar membrane, with their processes or hairs projecting into the endolymph of the cochlear duct. These hair cells are arranged to form one row of inner cells that extend the length of the basilar membrane, and rows of outer hair cells: three rows in the basal turn, four in the middle turn, and five in the apical turn of the cochlea.

Figure 10.27 Sounds of low frequency cause pressure waves of perilymph to pass through the helicotrema. Sounds of higher frequency cause pressure waves to "short cut" through the cochlear duct. This sets up traveling waves in the basilar membrane.

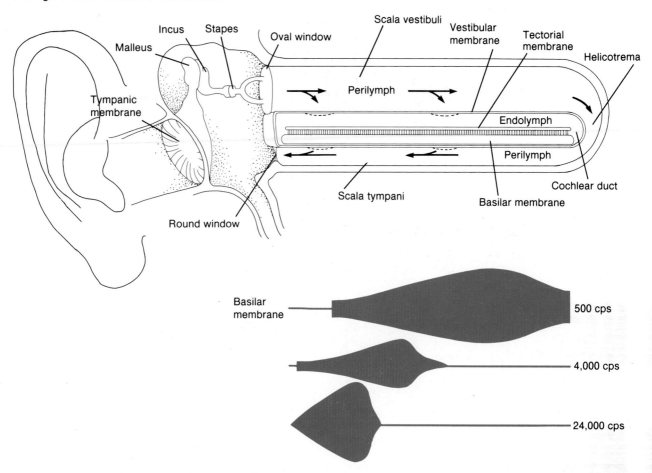

Figure 10.28 Scanning electron micrograph of hair cells in the cochlea.

Kessel and Kardon

The hairs, or microvilli, of the sensory cells are embedded within a gelatinous *tectorial membrane* that overhangs the hair cells within the cochlear duct (see figure 10.29). Displacement of the basilar membrane and hair cells by movements of perilymph causes the microvilli that are embedded in the tectorial membrane to bend. This stimulation excites the sensory cells, which causes the release of an unknown neurotransmitter that excites sensory endings of the cochlear (eighth cranial) nerve.

Figure 10.29   The organ of Corti *(a)* is located within the scala media of the cochlea *(b).*

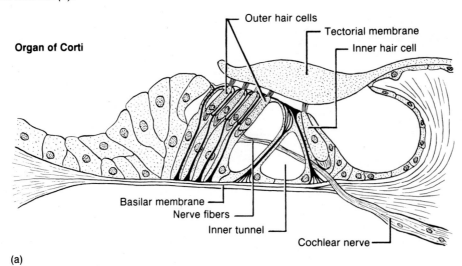

Organ of Corti

Outer hair cells

Tectorial membrane

Inner hair cell

Basilar membrane

Nerve fibers

Inner tunnel

Cochlear nerve

(a)

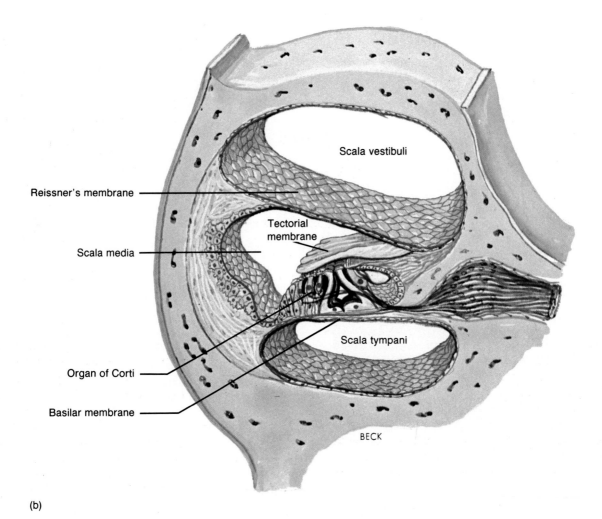

Scala vestibuli

Reissner's membrane

Tectorial membrane

Scala media

Scala tympani

Organ of Corti

Basilar membrane

BECK

(b)

Figure 10.30 Neural pathways from the spiral ganglia of the cochlea to the auditory cortex.

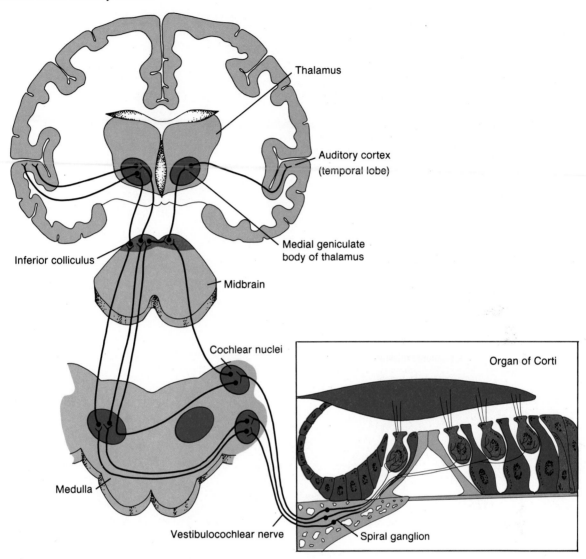

Neural Pathways for Hearing   Sensory neurons in the eighth cranial nerve synapse with neurons in the medulla (see figure 10.30), which project to the inferior colliculus of the midbrain. Neurons in this area in turn project to the thalamus, which sends axons to the auditory cortex of the temporal lobe. By means of this pathway, neurons in different regions of the basilar membrane stimulate neurons in corresponding areas of the auditory cortex; each area of this cortex thus represents a different part of the basilar membrane and a different pitch (see figure 10.31).

Hearing Impairments

There are two major causes of hearing loss: (1) **conductive deafness,** in which transmission of sound waves from air through the middle ear to the oval window is impaired; and (2) **nerve** or **sensory deafness,** in which transmission of nerve impulses anywhere from the cochlea to the auditory cortex is impaired. Conductive deafness can be caused by middle ear damage from otitis media or otosclerosis, as previously discussed. Nerve deafness may result from a wide variety of pathological processes.

Figure 10.31 Sound of different frequencies (pitches) excite different sensory neurons in the cochlea; these in turn send their input to different regions of the auditory cortex.

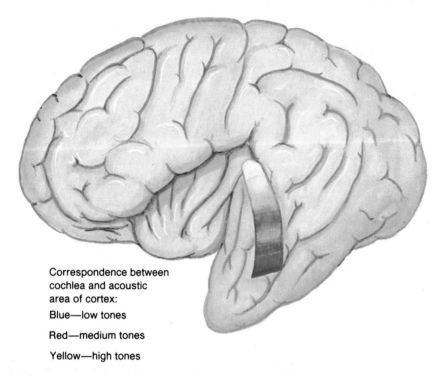

Correspondence between cochlea and acoustic area of cortex:

Blue—low tones

Red—medium tones

Yellow—high tones

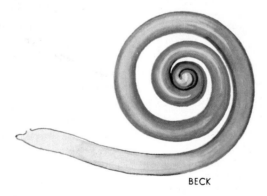

BECK

Hearing loss occurs as a "normal" aspect of aging; while a child can hear high-pitched sounds at 20,000 Hz, the average adult cannot hear sounds well that have frequencies above about 12,000 Hz. Advanced age is usually accompanied by further hearing loss that begins at the higher frequencies and gradually involves lower frequencies. This hearing impairment with age is called **presbycusis.**

The presence of conductive or nerve deafness can often be detected by two simple tests involving the use of tuning forks. In the *Weber test,* the handle of a tuning fork is placed in the midsaggital plane (midline) of the skull, either at the top of the head or the forehead. If a person

has conductive deafness that is worse in one ear than the other, the sound will appear loudest in the poorer ear. This is because the sound waves produced by the tuning fork are conducted through bone while distracting sounds in the room can't be heard. If the person has nerve deafness, the sound will appear loudest in the better ear.

In the *Rinne test,* the handle of a tuning fork is placed against the mastoid process of the temporal bone, behind the outer ear. The tines of the tuning fork are then moved in front of the pinna. If conduction deafness is present, the sound will appear louder when the tuning fork is against the bone (because of bone conduction) than it will when placed in front of the ear.

Figure 10.32 The electromagnetic spectrum *(top)* is shown in Angström units (1Å = 10⁻¹⁰ meter). The visible spectrum comprises only a small range of this spectrum *(bottom)*, shown in nanometer units (1 nm = 10⁻⁹ meter).

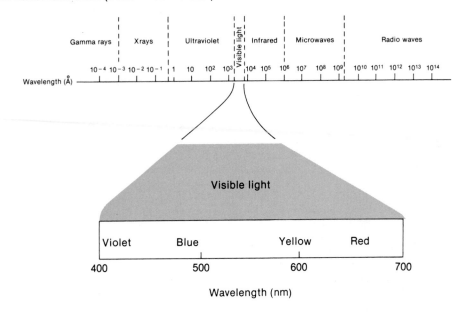

Conduction deafness impairs hearing at all sound frequencies. Sensory deafness, in contrast, often impairs the ability to hear some pitches more than others. These impairments can be detected by a technique called *audiometry,* in which the threshold intensity of different pitches are determined. The ability to hear speech is particularly affected by hearing loss in the higher frequencies. People with these types of impairments can be helped by *hearing aids,* which amplify sounds and conduct the sound waves through bone to the inner ear.

1. Explain how the otolith organs maintain a sense of balance during forward and downward acceleration. Include in your answer the action of the otolith membrane and the bending of the hair cell processes.
2. Explain the events that occur in the semicircular canals when a gymnast flips or a ballerina twirls. Describe the effect on eye movements that results, and how nystagmus can occur when the flip or twirl suddenly stops.
3. Use a flow chart to describe how sound waves in air within the external auditory meatus are transduced into movements of the basilar membrane.
4. Explain how movements of the basilar membrane of the organ of Corti can code for different sound frequencies (pitches).

The Eyes and Vision

The eyes transduce energy in the *electromagnetic spectrum* (see figure 10.32) into nerve impulses. Only a limited part of this spectrum can excite the photoreceptors—electromagnetic energy with wavelengths between 400 and 700 nanometers (nm) comprise *visible light.* Light of longer wavelengths, which are in the infrared regions of the spectrum, do not have sufficient energy to excite the receptors but are felt as heat. Ultraviolet light, which has shorter wavelengths and more energy than visible light, is filtered out by the yellow color of the eye's lens. Honeybees—and people who have had their lens removed—can see light in the ultraviolet range.

The outermost layer of the eye is a tough coat of connective tissue called the *sclera.* This can be seen externally as the white of the eyes. The tissue of the sclera is continuous with the transparent *cornea.* Light passes through the cornea to enter the *anterior chamber* of the eye. Light then passes through an opening called the *pupil,* within a pigmented (colored) muscle called the *iris.* Light that passes through the pupil enters the *lens.*

Figure 10.33 Structure of the eye.

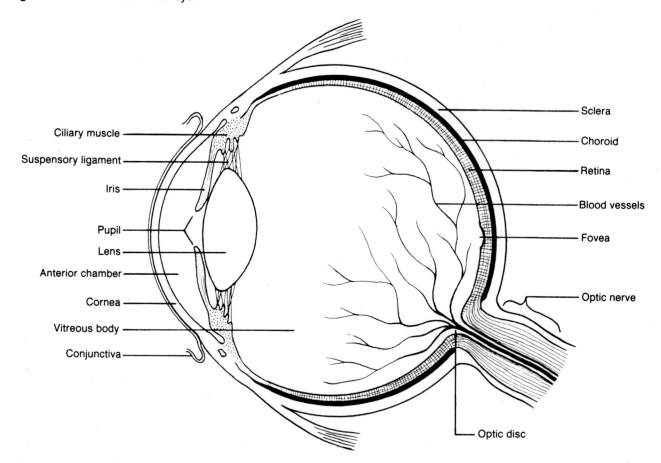

Ciliary muscle

Suspensory ligament

Iris

Pupil

Lens

Anterior chamber

Cornea

Vitreous body

Conjunctiva

Sclera

Choroid

Retina

Blood vessels

Fovea

Optic nerve

Optic disc

The iris is like the diaphragm of a camera, which can increase or decrease the diameter of its aperture (the pupil) to admit more or less light. Pupillar constriction is produced by contraction of circular muscles within the iris; dilation is produced by contraction of radial muscles. Constriction of the pupils results from parasympathetic stimulation, while dilation results from sympathetic stimulation (as described in chapter 9). Variations in the diameter of the pupil are similar in effect to variations in the "f-stop" of a camera. Analogies between the parts of a camera and the eye are summarized in table 10.4.

The posterior part of the iris contains a pigmented epithelium that gives the eye its color. The color of the eye is determined by the amount of pigment—blue eyes have the least pigment, brown eyes have more, and black eyes have the greatest amount of pigment. Albinos, who have a congenital defect in the ability to produce melanin pigment, have eyes that appear pink because the absence of pigment allows blood vessels to be seen.

Table 10.4 Comparisons between eye structures and analogous structures in a camera.

Eye Structure	Analogous Camera Structure
Cornea and lens	Lens system
Iris and pupil	Variable aperture system
Eyelid	Lens cap
Sclera	Camera frame
Pigment epithelium and choroid	Black interior of camera
Retina*	Film

*Since neural processing begins in the retina, this eye structure may also be considered analogous (in part) to the photographer!

Figure 10.34 Illustration of the anterior portion of the eye. The iris divides the anterior part into an anterior and posterior chamber, which is separated by the zonular fibers and lens from the vitreous body posteriorly. Zonular fibers attach to processes in the ciliary body.

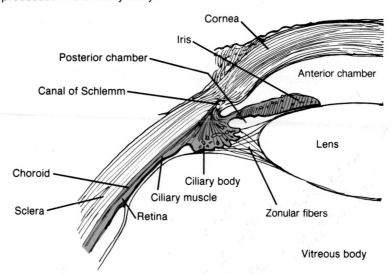

The lens is suspended from a muscular process called the **ciliary body,** which is connected to the sclera and encircles the lens. *Zonular fibers* (zon = girdle) suspend the lens from the ciliary body, forming a *suspensory ligament* that supports the lens. The space bounded by the iris anteriorly, and the ciliary body and lens posteriorly, is called the *posterior chamber* (as distinct from the anterior chamber between the iris and the cornea—see figure 10.34).

The anterior and posterior chambers are filled with a fluid called the **aqueous humor.** This fluid is produced by the ciliary body, secreted into the posterior chamber, and enters the anterior chamber through the pupil. Aqueous humor is drained from the anterior chamber into the *canal of Schlemm,* which returns this fluid to the venous blood. Inadequate drainage of aqueous humor can lead to excessive accumulation of fluid, which in turn results in increased intraocular pressure (a condition called *glaucoma*). This may cause serious damage to the retina.

The portion of the eye located behind the lens is filled with a thick, viscous substance known as the **vitreous body.** Light from the lens that passes through the vitreous body enters the neural layer, which contains photoreceptors, at the back of the eye. This neural layer is called the **retina.** Light that passes through the retina is absorbed by a darkly pigmented *choroid* layer underneath. While passing through the retina, some of this light stimulates photoreceptors, which in turn activate neurons. Neurons in the retina contribute fibers that are gathered together at a region called the *optic disc* (see figure 10.36) to exit the retina as the optic nerve. The optic disc is also the site of entry and exit of blood vessels.

Refraction and Accommodation

Light that passes from a medium of one density into a medium of a different density is *refracted,* or bent. The degree of refraction depends on the comparative densities of the two media, as indicated by their *refractive index.* The refractive index of air is set at 1.00; the refractive index of the cornea, in comparison, is 1.38, and the refractive index of the lens is 1.45. Since the greatest difference in refractive index occurs at the air-cornea interface, the light is refracted most at the cornea.

Figure 10.35 The fluid that fills the anterior and posterior chambers is called aqueous humor. This fluid is formed by the ciliary body and enters the posterior chamber. Aqueous humor is normally drained from the anterior chamber *(a)* through the canal of Schlemm. If this drainage is blocked *(b),* accumulation of aqueous humor results in increased intraocular pressure and glaucoma.

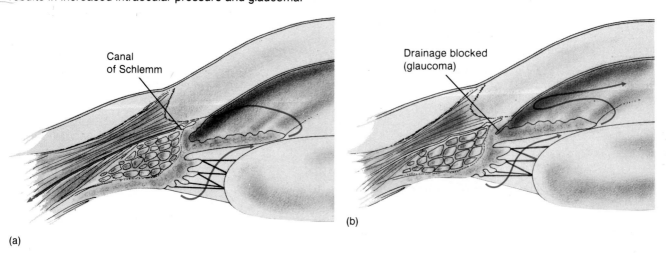

Canal
of Schlemm

Drainage blocked
(glaucoma)

(b)

(a)

Figure 10.36 The eye as seen through an ophthalmoscope.

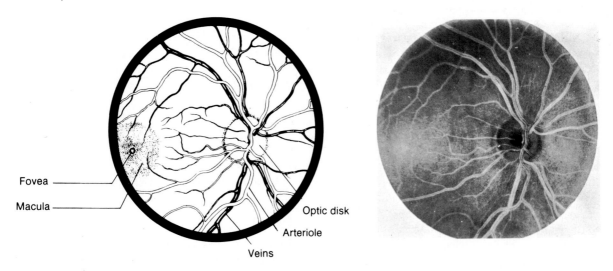

Fovea

Macula

Optic disk

Arteriole

Veins

The degree of refraction also depends on the curvature of the interface between two media. The curvature of the cornea is constant, however, while the curvature of the lens can be varied. The refractive properties of the lens can thus provide fine control for focusing light on the retina. As a result of light refraction, the image formed on the retina is upside down and right to left (see figure 10.37).

The *visual field*—which is the part of the external world projected onto the retina—is thus reversed in each eye. The cornea and lens focus the right part of the visual field on the left half of the retina of each eye, while the left half of the visual field is focused on the right half of each retina (see figure 10.38). The medial (or nasal) half-retina of the left eye therefore receives the same image as the lateral (or temporal) half of the right eye. The nasal half-retina of the right eye receives the same image as the temporal half-retina of the left eye.

Figure 10.37   Refraction of light in the cornea and lens
produces an upside-down image in the retina.

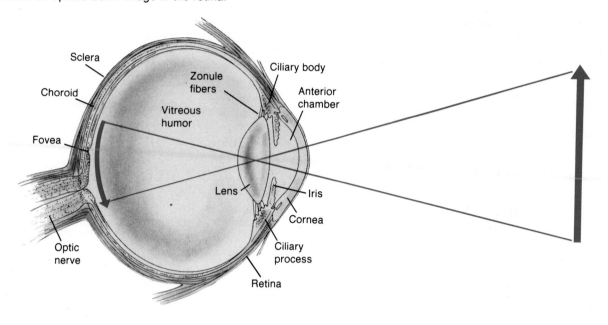

Figure 10.38   Refraction of light in the cornea and lens
produces a right-to-left image on the retina. The left side
of the visual field is projected to the right side of each
retina, while the right side of each visual field is projected
to the left half of each retina.

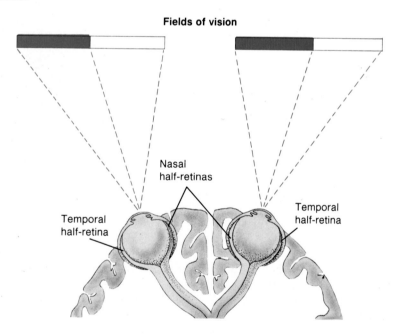

Figure 10.39 Accommodation. When an object is twenty feet or more from the eye the ciliary muscle is relaxed *(a)*, and the lens is stretched thin by tension of the suspensory ligament. This allows the image of the object to be focused on the retina. When the object is closer than twenty feet, the ciliary body contracts. This places slack in the suspensory ligament, allowing the lens to become rounder and focus the image on the retina.

Figure 10.40 In a normal eye *(a)*, parallel rays of light are brought to a focus on the retina by refraction in the cornea and lens. If the eye is too long, as in myopia *(b)*, the focus is in front of the retina. This can be corrected by a concave lens. If the eye is too short *(c)*, as in hyperopia, the focus is behind the retina. This is corrected by a convex lens.

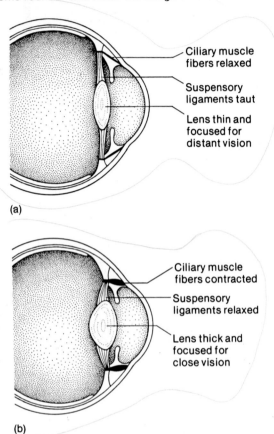

(a)

(b)

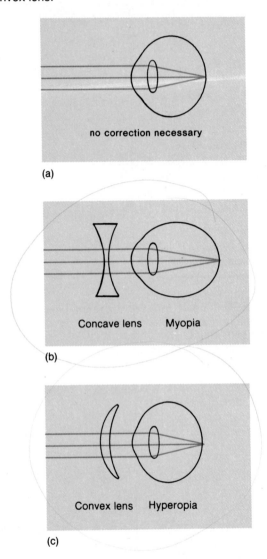

(a)

(b)

(c)

Accommodation

When a normal eye views an object, parallel rays of light are refracted to a point, or *focus,* on the retina (see figure 10.40). If the degree of refraction were to remain constant, movement of the object closer to or farther from the eye would cause corresponding movement of the focal point, so that the focus would either be behind or in front of the retina.

The ability of the eyes to keep the image focused on the retina as the distance between the eye and object is changed, is called **accommodation.** When the object is twenty feet or more from a normal eye, the image is focused on the retina and the lens is in its most flat, least convex form. In this state the muscles of the ciliary body, from which the lens is suspended by zonular fibers, are relaxed. As the object moves closer to the eyes the muscles of the ciliary body contract. This muscular contraction brings the ciliary body closer to the lens, which reduces

tension in the zonular fibers suspending the lens. When the tension is reduced, the lens becomes more round and convex as a result of its inherent elasticity (see figure 10.39).

Presbyopia As an object is brought ever closer to the eyes, therefore, the convexity and refractive power of the lens increases and the image remains in focus on the retina. There is, however, a limit to the ability of the eyes to accommodate as an object gets close to the eyes. This *near point of vision* increases with age as a result of the loss of lens elasticity; a printed page must be held farther from the eyes as a result. This loss of accommodating ability due to loss of lens elasticity with age is called *presbyopia* (presby = old).

Visual Acuity

Visual acuity refers to the sharpness of vision. The sharpness of an image depends on the *resolving power* of the visual system—that is, on the ability of the visual system to distinguish (resolve) two closely spaced dots. The better the resolving power of the system is, the closer together these dots can be and still be seen as separate; when the resolving power of the system is exceeded, the dots are blurred together as a single image.

Myopia and Hyperopia   When a person with normal visual acuity stands twenty feet from a *Snellen eye chart* (so that accommodation is not a factor influencing acuity), the line of letters marked "20/20" can be read. If a person has **myopia** (nearsightedness) this line will appear blurred because the focus of this image will be in front of the retina. This is usually caused by the fact that the eyeball is too long. Myopia is corrected by glasses with concave lenses that cause the light rays to diverge; the focus is thus pushed back to the retina (see figure 10.40).

If the eyeballs are too short, the line marked "20/20" will appear blurred because the focus of the image will be behind the retina; the object must thus be placed farther from the eyes to be seen clearly. This condition is called **hyperopia** (farsightedness). Hyperopia is corrected by glasses with convex lenses that increase the convergence of light so that the focus is brought closer to the lens and falls on the retina.

Astigmatism   The curvature of the cornea and lens is not perfectly symmetrical, so that light passing through some parts of these structures may be refracted to a different degree than light passing through other parts. When the asymmetry of the cornea and/or lens is significant, the person is said to have **astigmatism.** If a person with astigmatism views a circle, the image of the circle will not appear clear in all 360 degrees; the parts of the circle that appear blurred can thus be used to map the astigmatism. This condition is corrected by cylindrical lenses that compensate for the asymmetry in the cornea or lens of the eye.

1. Using a line diagram, explain why an inverse image is produced on the retina. Also explain how the image in one eye corresponds to the image in the other eye.
2. Using a line diagram, show how parallel rays of light are brought to a focus on the retina. Explain how this focus is maintained as the distance from the object to the eye is increased or decreased (that is, explain accommodation).
3. Use a line diagram to show how a blurred image is produced in presbyopia when an object is brought too close to the eyes. Relate this condition to myopia and hyperopia.

Figure 10.41   The layers of the retina. The retina is inverted, so that light must pass through various layers of nerve cells before reaching the photoreceptors (rods and cones).

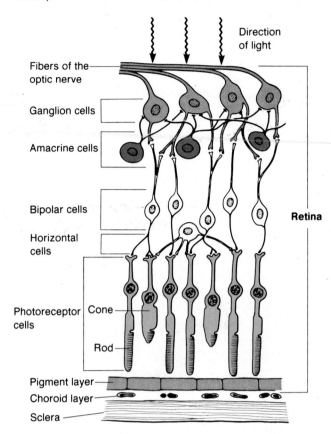

The Retina

The retina consists of a *pigment epithelium,* photoreceptors called *rods* and *cones,* and of layers of *neurons.* The neural layers of the retina are actually a forward extension of the brain. In this sense the optic nerve can be considered a tract, and indeed the myelin sheaths of its fibers are derived from oligodendrocytes (like other CNS nerve fibers) rather than from Schwann cells.

Since the retina is an extension of the brain, the neural cell layers face outwards, toward the incoming light. Light, in other words, must pass through several neural layers before striking the photoreceptors (see figure 10.41). The photoreceptors then excite the neurons so that nerve impulses are conducted outward in the retina.

Figure 10.42   The blind spot. Hold the drawing about twenty inches from your face, with the left eye closed and your right eye focused on the circle. Slowly move the drawing closer to your face until the cross disappears. This occurs because the image of the cross is focused on the optic disc, where photoreceptors are absent.

The outer layers of neurons that contribute axons to the optic nerve are called *ganglion cells*. This layer receives synaptic input from *bipolar cells* underneath, which are in turn activated by the rods and cones in response to light. In addition to the flow of information from photoreceptors to bipolar cells to ganglion cells, there are neurons called *horizontal cells* that synapse with several photoreceptors (see figure 10.41), and neurons called *amacrine cells* that synapse with several ganglion cells.

The Blind Spot   The axons of the ganglion cell layer gather together and exit the retina at the optic nerve in the region of the optic disc. Since there are no photoreceptors in the optic disc, an image that falls on this part of the retina cannot be seen. This results in a *blind spot* in the visual field of each eye. The blind spot can easily be demonstrated as described in figure 10.42.

Lateral Inhibition   There are approximately 130 million rods and cones, but only about one million ganglion cells in the retina. This convergence of neural information, together with the effects of lateral associations made by horizontal cells and amacrine cells, allows the retina to begin processing visual information before relaying it to other areas of the brain. *Lateral inhibition* within the retina, for example, can exaggerate brightness contrast between two adjacent parts of an image and thus enhance the sharpness of boundaries. The significance of this effect is demonstrated by observing that the shape of a white paper cutout is clearer if it is placed against a black background.

Visual Pigments

The photoreceptors—rods and cones—are activated when light produces a chemical change in molecules of pigment contained within the receptor cells. Rods, for example, contain a purple pigment known as **rhodopsin.** This pigment appears purple (a combination of red and blue) because it transmits light in the red and blue regions of the spectrum, while absorbing light energy in the green region. The wavelength of the light that is absorbed best—the *absorption maximum*—is about 500 nm (corresponding to yellowing-green).

Cars and other objects that are yellowish green in color are seen more easily at night (when rods are used for vision) than are red objects. This is because red light is not well absorbed by rhodopsin, and only absorbed light can produce the photochemical reactions that result in vision. In response to absorbed light, rhodopsin dissociates into its two component parts: a pigment called **retinaldehyde** (or retinene) and a protein called **opsin.**

Retinaldehyde can exist in two possible configurations (shapes)—one known as the *all-trans* form and one called the *11-cis* form (see figure 10.43). The all-trans form is the most stable, but only the 11-cis form is found attached to opsin. In response to absorbed light energy, the 11-cis retinaldehyde is converted to its all-trans form, causing it to dissociate from the opsin. This dissociation reaction in response to light initiates changes in ionic permeability of the rod cell membrane and ultimately results in the production of nerve impulses in the ganglion cells.

Dark Adaptation   The photochemical dissociation of rhodopsin is called the *bleaching reaction.* Rods are thus bleached by light; in the dark, new rhodopsin is made, using either the all-trans retinaldehyde left from the bleaching reaction or using *vitamin A,* which is a precursor of retinaldehyde. This is why eating carrots (which are rich in vitamin A) can improve night vision in people who are vitamin A deficient. In the dark-adapted eye, there is a higher content of rhodopsin in the rods, and sight is possible under very dim illumination (as occurs after a few minutes in a dark theatre). Under these conditions of low illumination, rods provide black-and-white vision.

Figure 10.43 The photopigment rhodopsin consists of the protein opsin combined with 11-cis retinaldehyde. In response to light the retinaldehyde is converted to a different form, called "all-trans," and dissociates from the opsin. This photochemical reaction induces changes in ionic permeability that ultimately results in stimulation of ganglion cells in the retina.

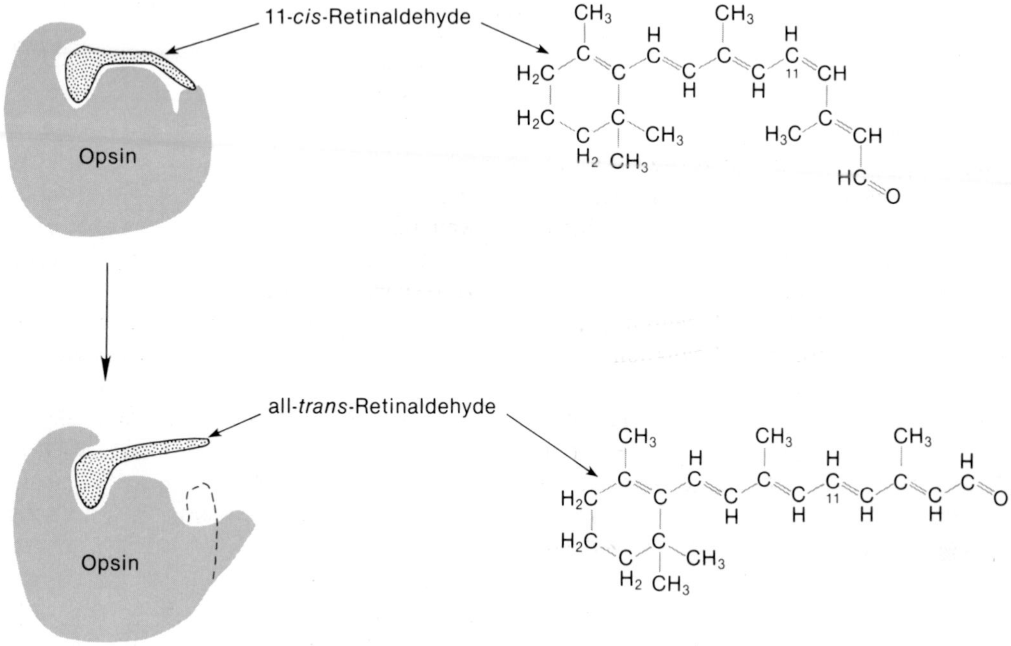

Color Vision   Cones are less sensitive than rods to light, but provide color vision and greater visual acuity (as described in the next section). During the day, therefore, the high light intensity bleaches out the rods, and color vision with high acuity is provided by the cones. There appear to be three types of cones: blue cones, green cones, and red cones. Each type contains a different pigment protein than the other types of cones, all of which differ from the opsin protein contained in rods. These differences in pigment proteins produce differences in the absorption maximum of each pigment (see figure 10.44). Green cones, for example, have an absorption maximum in the green region of the spectrum (at 530 nm), while red cones have an absorption maximum at 625 nm in the red region.

Since rods do not absorb red light but red cones do, a red light in a photographic darkroom allows vision (because of the red cones), but does not cause bleaching of the rods. When the light is turned off, therefore, the eyes will still be dark adapted and the person will be able to see (because of the rods).

Figure 10.44 There are three types of cones. Each type of cone contains retinaldehyde combined with a different type of protein, producing a pigment that absorbs light maximally at a different wavelength. Color vision, according to the trichromatic theory, is produced by activity of these blue cones, green cones, and red cones.

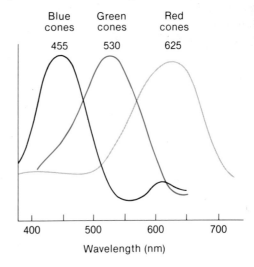

Figure 10.45   When the eyes "track" an object, the image is cast upon the fovea centralis of the retina. The fovea is literally a "pit" formed by parting of the neural layers, so that light falls directly on the photoreceptors (cones) in this region.

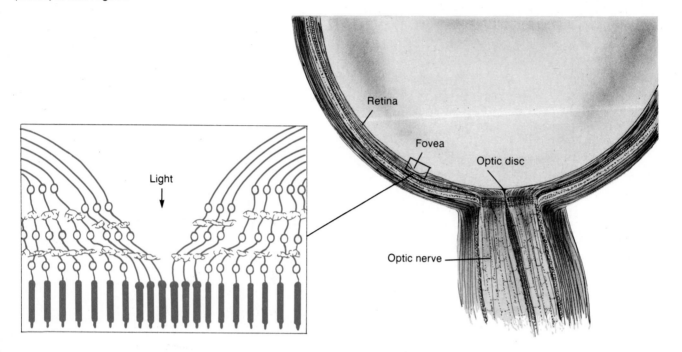

Our ability to see a multitude of colors with only three types of cones is called the **trichromatic theory** of color vision. (*Color blindness* is due to congenital lack of one or more types of cones.) In red-green color blindness, for example, red and green appear the same because one of these two types of cones is missing. People with this condition must rely on other cues—such as relative intensity of light and position—to tell the differences between red and green traffic signals.

Visual Acuity and Sensitivity

While reading, or similarly focusing visual attention on objects in daylight, each eye is oriented so that the image falls within a tiny area of the retina called the **fovea centralis.** The fovea is a pinhead-sized pit (fovea = pit) within a yellow area of the retina called the *macula lutea.* The pit is formed as a result of displacement of neural layers around the fovea, so that light falls directly on photoreceptors in this region (see figure 10.45), while light falling on other areas must pass through layers of neurons, as previously described.

The photoreceptors are distributed in such a way that the fovea contains only cones, while more peripheral regions of the retina contain a mixture of rods and cones. Approximately four thousand cones in the fovea provide input to approximately four thousand ganglion cells; each ganglion cell in this region therefore has a private line to the visual field. Each ganglion cell thus receives input from an area of retina corresponding to the diameter of one cone (about 2 μm). Peripheral to the fovea, however, many rods synapse with a single bipolar cell, and many bipolar cells synapse with a single ganglion cell. A single ganglion cell outside the fovea thus may receive input from large numbers of rods, corresponding to an area of about 1 mm on the retina.

Figure 10.46 Since bipolar cells receive input from convergence of many rods *(a),* and since a number of such bipolar cells converge on a single ganglion cell, rods provide high sensitivity to low levels of light at the expense of visual acuity. The 1:1:1 ratio of cones to bipolar cells to ganglion cells, *(b),* in contrast, provides high visual acuity, but low sensitivity.

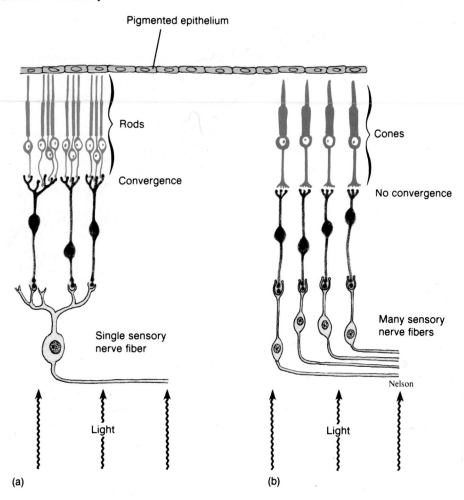

Pigmented epithelium

Rods

Convergence

Single sensory nerve fiber

Light

(a)

Cones

No convergence

Many sensory nerve fibers

Nelson

Light

(b)

Since each cone in the fovea has a private line to a ganglion cell, and since each ganglion cell receives input from only a tiny region of the retina, visual acuity is greatest and sensitivity to low light is poorest when light falls on the fovea. In dim light, only the rods are activated, and vision is best out of the corners of the eye when the image falls away from the fovea. Under these conditions, convergence of many rods on a single bipolar cell and convergence of many bipolar cells on a single ganglion cell increase sensitivity to dim light at the expense of visual acuity. Night vision is therefore less distinct than day vision.

Neural Pathways from the Retina

As a result of light refraction by the cornea and lens, the right half of the visual field is projected to the left half of the retina of both eyes (the temporal half of the left retina and the nasal half of the right retina); the left half of the visual field is projected to the right half of the retina of both eyes. The temporal half of the left retina and the nasal half of the right retina therefore see the same image.

Figure 10.47 Neural pathway leading from the retina to the lateral geniculate body to the visual cortex. As a result of the crossing of optic fibers, the visual cortex of each cerebral hemisphere receives input from the opposite (contralateral) visual field.

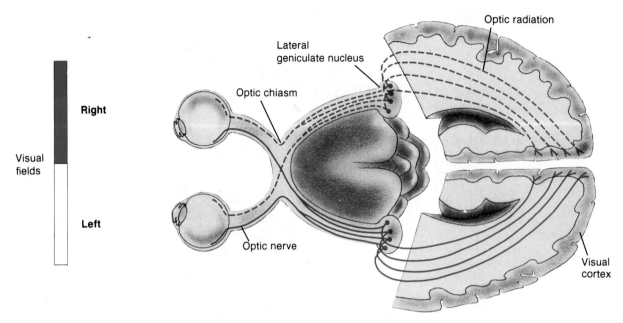

Axons from ganglion cells in the left retina pass to the left **lateral geniculate body** of the thalamus. Axons from ganglion cells in the nasal half of the right retina cross (decussate) in the *optic chiasm* to synapse also in the left lateral geniculate body. The left lateral geniculate therefore receives input from both eyes that relates to the right half of the visual field (see figure 10.47).

The right lateral geniculate body, similarly, receives input from both eyes relating to the left half of the visual field. Neurons in both lateral geniculate bodies of the thalamus in turn project to the **striate cortex** of the occipital lobe in the cerebral cortex. This area is also called area 17, in reference to a numbering system developed by Brodmann. Neurons in area 17 synapse with neurons in areas 18 and 19 of the occipital lobe (see figure 10.48).

Approximately 70 to 80 percent of the axons from the retina pass to the lateral geniculate bodies and to the striate cortex. This **geniculostriate system** is involved in perception of the visual field. Put another way, the geniculostriate system is needed for answering the question, What is it? Approximately 20 to 30 percent of the fibers from the retina, however, follow a different path to the *superior colliculus* of the midbrain (also called the *optic tectum*). Axons from the superior colliculus activate motor pathways leading to eye and body movements. The **tectal system,** in other words, is needed for answering the question, Where is it?

The Superior Colliculus and Eye Movements

Neural pathways from the superior colliculus to motor neurons in the spinal cord help mediate the "startle" response to the sight of an unexpected intruder. Other nerve fibers from the superior colliculus stimulate the **extrinsic eye muscles** (see table 10.5), which are the striated muscles that move the eyes.

There are two types of eye movements coordinated by the superior colliculus. *Smooth pursuit movements* track moving objects and keep the image focused on the fovea centralis. *Saccadic eye movements* are short (lasting 20 to 50 msec), jerky movements that occur while the eyes appear to be still. These saccadic movements are believed to be important in maintaining visual acuity.

The tectal system is also involved in the control of the intrinsic eye muscles—the iris and the muscles of the ciliary body. Shining a light into one eye stimulates the *pupillary reflex* in which both pupils constrict. This is caused by activation of parasympathetic neurons by fibers from the superior colliculus. Postganglionic neurons in the ciliary ganglia behind the eyes, in turn, stimulate constrictor fibers in the iris. Contraction of the ciliary body during *accommodation* also involves stimulation of the superior colliculus.

Figure 10.48 The striate cortex (area 17) and the visual association (areas 18 and 19).

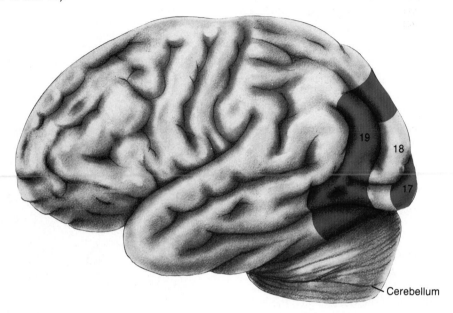

Cerebellum

Table 10.5 Muscles of the eye.

Extrinsic Muscles (Striated)

Superior rectus	Oculomotor nerve (III)	Rotates eye upward and toward midline
Inferior rectus	Oculomotor nerve (III)	Rotates eye downward and toward midline
Medial rectus	Oculomotor nerve (III)	Rotates eye toward midline
Lateral rectus	Abducens nerve (VI)	Rotates eye away from midline
Superior oblique	Trochlear nerve (IV)	Rotates eye downward and away from midline
Inferior oblique	Oculomotor nerve (III)	Rotates eye upward and away from midline

Intrinsic Muscles (Smooth)

Ciliary muscles	Oculomotor nerve (III) parasympathetic fibers	Causes suspensory ligaments to relax
Iris, circular muscles	Oculomotor nerve (III) parasympathetic fibers	Causes size of pupil to decrease
Iris, radial muscles	Sympathetic fibers	Causes size of pupil to increase

Neural Processing of Visual Information

Light that is cast on the retina directly affects the activity of photoreceptors, and indirectly affects neural activity in bipolar and ganglion cells. The part of the visual field that affects activity of a particular ganglion cell can be considered to be its *receptive field*. Since each cone in the fovea has a private line to a ganglion cell, the receptive fields of these ganglion cells are equal to the width of one cone (about 2 μm). Ganglion cells in more peripheral parts of the retina receive input from hundreds of photoreceptors, and thus are influenced by a larger area of the retina (about 1 mm in diameter).

Ganglion Cell Receptive Fields

Studies of electrical activity of ganglion cells have yielded some surprising results. In the dark, each ganglion cell discharges spontaneously at a slow rate. When the room lights are turned on, the firing rate of many (but not all) ganglion cells increases slightly. A small spot of light that is directed at the center of some ganglion cell's receptive fields, however, stimulates a large increase in firing rate. A small spot of light can thus be a more effective stimulus than larger areas of light.

When the spot of light is moved only a short distance away from the center of the receptive field the ganglion cell responds in an opposite manner. The ganglion cell that was stimulated with light at the center of its receptive field is inhibited by light in the periphery of its field. The response produced by light in the center and by light in the surround of the visual field is *antagonistic*. Those ganglion cells that are stimulated by light at the center of their visual fields are said to have **on-center fields.** Ganglion cells that are inhibited by light in the center and stimulated by light in the surround have **off-center fields.**

The reason wide illumination of the retina has less effect than pinpoint illumination is now clear; diffuse illumination gives the ganglion cell conflicting orders—on and off. Because of the antagonism between the center and surround of ganglion cell receptive fields, the activity of each ganglion is a result of the *difference in light intensity* between the center and surround of its visual field. This helps to accentuate the contours of images at the expense of information about absolute brightness.

Lateral Geniculate Bodies

Each of the two lateral geniculate bodies receives input from ganglion cells in both eyes. The right lateral geniculate receives input from the right half of each retina (corresponding to the left half of the visual field); the left lateral geniculate receives input from the left half of each retina (corresponding to the right half of the visual field).

Figure 10.49 The lateral geniculate nucleus. Each lateral geniculate consists of six layers (numbered 1 through 6 in this figure). Each of these layers receives input from only one eye, with right and left eyes alternating. An arrow through these six layers (see figure) of the left lateral geniculate, for example, encounters corresponding projections from a part of the right visual field in right and left eyes, alternatively, as it passes from the outer to the inner layers.

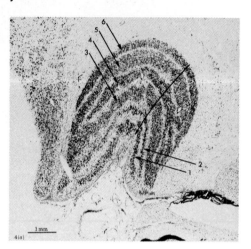

Figure 10.50 Cortical neurons described as simple cells have rectangular receptive fields that are best stimulated by slits of light of particular orientations. This may be due to the fact that these simple cells receive input from ganglion cells that have circular receptive fields along a particular line.

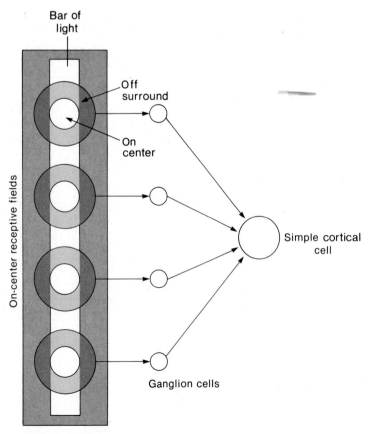

Figure 10.51 Simple cells are best stimulated by a slit or bar of light along a particular orientation within a particular region of the receptive field. The behavior of two different cortical cells *(a)* and *(b)* is illustrated*.

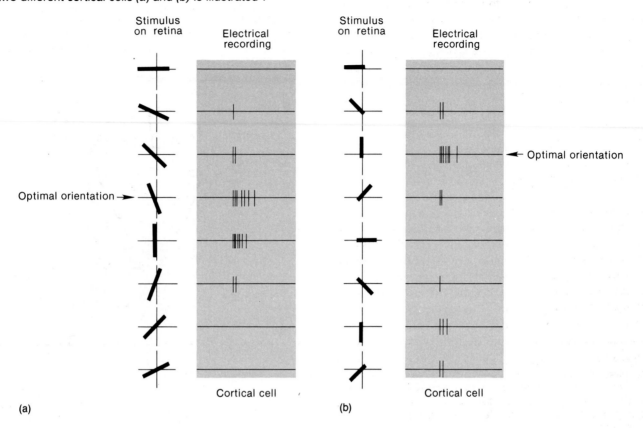

(a)

(b)

Each neuron in the lateral geniculate, however, is activated only by input from one eye. Neurons that are activated by ganglion cells from the left eye are in separate layers within the lateral geniculate from those that are activated by the right eye (see figure 10.49).

The receptive field of each ganglion cell, as previously described, is the part of the retina it "sees" through its photoreceptor input. The receptive field of lateral geniculate neurons, similarly, is the part of the retina it "sees" through its ganglion cell input. Experiments in which the lateral geniculate receptive fields are mapped with a spot of light reveal that they are circular, with an antagonistic center and surround, much like the ganglion cell receptive fields.

The Cerebral Cortex

Projections of nerve fibers from the lateral geniculate bodies to area 17 of the occipital lobe form the *optic radiation*. These fiber projections give area 17 a striped or striated appearance; this area is consequently also known as the striate cortex (see figure 10.48). Neurons in area 17, in turn, project to areas 18 and 19 of the occipital lobe.

Cortical neurons in areas 17, 18, and 19 are thus stimulated indirectly by light on the retina. On the basis of their stimulus requirements, these cortical neurons are classified as simple, complex, and hypercomplex.

Simple Neurons   The receptive fields of simple cortical neurons are rectangular rather than circular. This results from the fact that they receive input from lateral geniculate neurons whose receptive fields are aligned in a particular way (as illustrated in figure 10.50). Simple cortical neurons are best stimulated by a slit or bar of light that is located in a precise part of the visual field (of either eye) at a precise orientation (see figure 10.51).

The striate cortex (area 17) contains simple, complex, and hypercomplex neurons. The other visual association areas, designated areas 18 and 19, contain only complex and hypercomplex cells. Complex neurons receive input from simple cells, and hypercomplex neurons receive input from complex cells.

Complex and Hypercomplex Neurons Complex cells respond best to straight lines with a specific orientation that move in a particular direction through the receptive field. Unlike simple cells, complex cells do not require that the stimulus have a particular position within the receptive field. Hypercomplex cells require that the stimulus be of a particular length, or have a particular bend or corner.

The dotlike information from ganglion and lateral geniculate cells is thus transformed in the occipital lobe into information about edges—their position, length, orientation, and movement. Although this represents a high degree of abstraction, the visual association areas of the occipital lobe probably serve as only an early stage in the integration of visual information.

Higher Processing of Visual Information

In order for visual information to have meaning it must be associated with past experience and integrated with information from other senses. Some of this higher processing occurs in the **inferior temporal lobes** of the cortex. Experimental removal of these areas from monkeys impairs their ability to remember visual tasks that they previously learned, and hinders their ability to associate visual images with the significance of the object. Monkeys with their inferior temporal lobes removed, for example, will handle a snake without fear. The symptoms produced by loss of the inferior temporal lobes are known as the *Kluver-Bucy syndrome*.

In an attempt to reduce the symptoms of severe epilepsy, surgeons at one time cut the corpus callosum in some patients. This fiber tract, as previously described, transmits impulses between the right and left cerebral hemispheres. The right cerebral hemisphere of patients with such *split brains*, therefore, receives sensory information from only the left half of the external world. The left hemisphere, similarly cut off from communication with the right hemisphere, receives sensory information from only the right half of the external world.

Experiments with split-brain patients have revealed that the functions of the two hemispheres are different. This is true even though each hemisphere would normally receive input from both halves of the external world through the corpus callosum. If the sensory image of an object, such as a key, is delivered to only the left hemisphere (by showing it to only the right visual field), the object can be named. If the object is presented to the right cerebral cortex, the person knows what the object is but cannot name it. Experiments such as this suggest that (in right-handed people) the left hemisphere is needed for language, while the right hemisphere is responsible for pattern recognition (see figure 10.52).

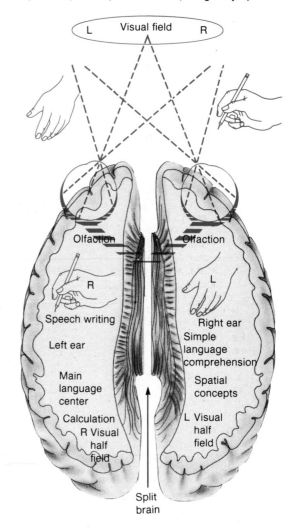

Figure 10.52 Different functions of the right and left cerebral hemispheres, as revealed by experiments with people who have had the tract connecting the two hemispheres (the corpus callosum) surgically split.

1. List the different layers of the retina, and describe the path of light and of nerve activity through these layers.
2. Describe the photochemical reaction that leads to activation of rods, and the role of vitamin A in vision.
3. Compare the architecture of the fovea centralis with more peripheral regions of the retina, and relate your answer to visual acuity in daytime vision and sensitivity in nighttime vision.
4. Describe the response characteristics of ganglion cells to light on the retina, and explain why a small spot of light is a more effective stimulus than general illumination of the retina.

Summary

Cutaneous Sensations: Taste and Olfaction

I. The cutaneous sensations include touch, pressure, pain, hot, and cold.
   A. Touch and pressure are mediated by mechanoreceptors.
      1. Mechanoreceptors are stimulated by mechanical deformation of the receptor cell membrane.
      2. Touch and pressure receptors may be either free sensory nerve endings or nerve endings that are encapsulated within associated structures.
   B. Stimulation of a particular sensory neuron produces a specific modality of sensation.
      1. This is called the law of specific nerve energies.
      2. The modality of sensation evoked is dependent on the central connections of the sensory neuron.
      3. Each receptor is most sensitive to its normal or adequate stimulus that produces the modality of sensation characteristic for that stimulus.
II. The acuity of sensation depends on receptor density and lateral inhibition.
   A. Skin in the fingers and lips have a high density of touch receptors.
   B. When an object touches the skin, sensation is sharpened by lateral inhibition.
      1. Sensory input from receptors that are stimulated the most inhibit sensory neurons from neighboring areas of skin.
      2. This lateral inhibition helps to localize the sensation by attenuating sensory "noise."

III. Taste and olfaction are produced by interaction of dissolved chemicals with chemoreceptors.
   A. Taste receptors are modified epithelial cells organized into taste buds.
   B. There are four modalities of taste sensation—sweet, salty, sour, and bitter.
   C. Olfactory receptors are sensory nerve endings.
      1. These nerve endings are associated with supporting epithelial cells in the olfactory epithelium.
      2. There are thousands of different identifiable odors.
IV. Cutaneous senses, taste, and olfaction adapt rapidly.
   A. These receptors are phasic—they respond maximally to the onset of a stimulus, but decrease their firing rate when the stimulus is maintained.
   B. These receptors therefore sense changes in stimuli, not the absolute levels of stimulation.

Hearing and Equilibrium

I. The senses of hearing and equilibrium are mediated by receptors in the inner ear.
   A. The receptors for hearing and equilibrium are located in the membranous labyrinth.
      1. The membranous labyrinth includes the vestibular apparatus and the scala media of the cochlea.
      2. These structures are filled with a fluid called endolymph.
   B. The membranous labyrinth is encased within a bony labyrinth.
      1. The space between the membranous labyrinth and bone is filled with perilymph.
      2. The bony labyrinth includes the scala vestibuli and scala tympani of the cochlea.

II. The vestibular apparatus includes the otolith organs and the semicircular canals.
   A. The otolith organs are the utriculus and sacculus.
      1. These structures contain patches, or maculae, of sensory hair cells.
      2. The hairs are embedded in a gelatinous membrane containing colified structures called otoliths.
   B. The semicircular canals contain hair cells in the ampullae, or widened areas.
      1. The hair cells are embedded in a gelatinous membrane called a cupula.
      2. The cupula extends into the endolymph fluid within the semicircular canals.
   C. Hair cells are stimulated by bending of the hairs in the direction of the kinocilium.
      1. This depresses the cell membrane and stimulates depolarization.
      2. Bending of the hairs in the opposite direction causes hyperpolarization and inhibits sensory nerve activity.
   D. The otolith organs provide a sense of linear acceleration, while the semicircular canals provide a sense of angular acceleration.
III. Sound waves are funneled by the pinna and external auditory meatus to the tympanic membrane.
   A. Movements of the tympanic membrane are transformed into movements of the oval window of the cochlea via the three middle ear ossicles.
      1. Movements of the tympanic membrane cause movements of the malleus, incus, and stapes.
      2. Middle ear damage, as in otitis media and otosclerosis, can lead to conduction deafness.

B. Movements of the stapes against the oval window sets up pressure waves of perilymph within the scala vestibuli of the cochlea.

C. The pressure induced by the inward bulging of the oval window is compensated by the outward bulging of the round window (in the scala tympani) a split second later.

D. The scala vestibuli and scala tympani are separated by the scala media or cochlear duct.
1. This is a blind-ending sac—after it ends (at the helicotrema), the scala vestibuli and tympani become continuous with each other.
2. The cochlear duct is filled with endolymph, which is continuous with the endolymph of the vestibular apparatus.

V. The cochlear duct contains the organ of Corti, located on the basilar membrane.
A. The basilar membrane separates the cochlear duct from the scala tympani.
1. Sensory hair cells are organized in rows on the basilar membrane, with their hairs projecting into an overhanging tectorial membrane.
2. The association of basilar membrane, hair cells, and tectorial membrane is called the organ of Corti.

B. Pressure waves of perilymph in the scala vestibuli cause the basilar membrane to vibrate.
1. Sounds of high pitch (frequency) cause maximum vibration near the base of the membrane.
2. Sounds of low pitch cause maximum displacement of the basilar membrane near its apex—at very low pitch, the entire membrane vibrates.

C. Movement of the basilar membrane causes sensory hair cells embedded in the tectorial membrane to bend.
1. Bending of the hair cells stimulates sensory fibers of the cochlear nerve.
2. Fibers from particular regions of the cochlea, representing specific pitches, are projected to corresponding parts of the auditory cortex.

Vision

I. Refraction of light in the cornea and lens brings the image to a focus on the retina.
A. The degree of refraction is adjusted by the curvature of the lens.
1. When the lens is in its most flattened, least convex form the focus is farthest from the lens.
2. When the lens is in its most convex form the focus is closest to the lens.

B. The curvature of the lens is regulated by tension of the zonular fibers of the suspensory ligament, which is regulated by the ciliary body.
1. When an object is twenty feet or more from the eye the ciliary body is relaxed, and tension is greatest in the zonular fibers.
2. As the object moves closer to the eye the ciliary muscle contracts—this slackens the zonular fibers and allows the lens to become more convex.
3. The ability of the eye to maintain a focus on the retina as the distance from the eye to the object changes is called accommodation.

C. Refraction of light causes the image to be upside down and right to left on the retina.
1. The right side of the visual field is projected to the left side of the retina of each eye.
2. The right side of the retina of each eye receives the image of the left half of the visual field.

II. The retina contains a pigment epithelium, photoreceptors, and nerve cells.
A. When the eyes are fixed on an object, its image falls on a small pit in the retina called the fovea centralis.
1. In the fovea, the layers of neurons are pushed aside so that light can fall directly on the photoreceptors.
2. The only photoreceptors in the fovea are cones.
3. The cones in the fovea have a ''direct line'' to ganglion cells, which contribute axons to the optic nerve.

B. Photoreceptors in more peripheral parts of the retina are rods and cones.
1. Many rods converge on a single bipolar nerve cell; many bipolar cells converge on a single ganglion cell.
2. Rods provide black-and-white vision under conditions of low illumination (at night); cones provide color vision and high visual acuity during the day.

III. Rods contain a pigment called rhodopsin.
A. Rhodopsin consists of a vitamin A derivative known as retinaldehyde and a protein called opsin.
B. Retinaldehyde is in its 11-cis form when it is attached to opsin.
1. In response to absorption of light energy, the retinaldehyde is converted to a different form known as the all-trans form.
2. This photochemical reaction results in the dissociation of retinaldehyde from opsin, in a reaction called bleaching.
3. The bleaching reaction ultimately gives rise to production of nerve impulses in the ganglion cells.

C. Cones also contain retinaldehyde, but have different pigment proteins.
1. As a result of their different ospins, cones contain pigments that absorb light in different regions of the visible spectrum.
2. There are three types of cones that have three types of pigments (based on their absorption of light): red, green, and blue.

IV. Neural pathways from the retina to the superior colliculus in the midbrain help regulate eye and body movements.

V. Most fibers from the retina project to the lateral geniculate body, and from there to the striate cortex.
   A. Fibers from the left half of the right eye, and from the right half of the left eye cross at the optic chiasm before synapsing in the lateral geniculate.
      1. The right lateral geniculate therefore receives fibers from the right side of the right eye and the right side of the left eye.
      2. This corresponds to the left half of the visual field; the converse holds for the left lateral geniculate (which "sees" the right visual field).
   B. Neurons of the lateral geniculate bodies project to area 17 of the occipital lobe, also called the striate cortex.
   C. Neurons in area 17, in turn, have connections with adjacent cortical areas 18 and 19 (the visual association areas).

VI. Each neuron in the visual pathway is best activated by a characteristic type of stimulus within its receptive field on the retina.
   A. Many ganglion cells are best activated by a small spot of light in the center of their receptive field.
      1. These cells are said to have "on-center" receptive fields.
      2. If light is directed to the area immediately surrounding these on-centers, activity of the ganglion cell is inhibited.
      3. The ganglion cells therefore have a receptive field with an "on-center" and an "off-surround."
   B. Other ganglion cells have receptive fields with an off-center and an on-surround.

   C. Receptive fields of lateral geniculate cells are like those of ganglion cells—circular with antagonistic centers and surrounds.
   D. The lateral geniculate projects to the striate cortex (area 17) and synapses with simple cells.
      1. Simple cells have rectangular receptive fields that can be activated by a line or bar of a particular orientation in a particular position in the receptive field.
      2. A cortical cell receives input from corresponding visual fields of both eyes.
      3. Simple cells converge on complex cells, which in turn synapse with hypercomplex cells.
      4. Cortical neurons are best activated by moving edges with a defined length or angle.
   E. Visual integration involves other areas of the cerebral cortex, including the inferior temporal lobes.

Self-Study Quiz

1. Receptors that are fast adapting are:
   (a) tonic
   (b) phasic
2. Cutaneous receptive fields are smallest in:
   (a) the fingertips
   (b) the back
   (c) the thighs
   (d) the arms
3. The process of lateral inhibition:
   (a) increases sensitivity of receptors
   (b) promotes sensory adaptation
   (c) increases sensory acuity
   (d) prevents adjacent receptors from being stimulated
4. The receptors for taste are:
   (a) naked sensory nerve endings
   (b) encapsulated sensory nerve endings
   (c) modified epithelial cells
5. The utricle and saccule:
   (a) are otolith organs
   (b) are located in the middle ear
   (c) provide a sense of linear acceleration
   (d) both a and c
   (e) both b and c

6. Stimulation of hair cells in the semicircular canals results from movement of:
   (a) endolymph
   (b) perilymph
   (c) the otolith membranes
7. Since fibers of the optic nerve that originate in the nasal halves of each retina cross at the optic chiasma, each lateral geniculate receives input from:
   (a) both the right and left sides of the visual field of both eyes
   (b) the ipsilateral visual field of both eyes
   (c) the contralateral visual field of both eyes
   (d) the ipsilateral field of one eye and the contralateral field of the other eye
8. When a person with normal vision views an object from a distance of at least 20 feet:
   (a) the ciliary muscles are relaxed
   (b) the suspensory ligament is tight
   (c) the lens is in its most flat, least convex shape
   (d) all of these

9. The lens is in its most convex form:
   (a) when viewing objects at a great distance
   (b) at the near point of vision
10. Glasses with concave lenses help correct:
    (a) presbyopia
    (b) myopia
    (c) hyperopia
    (d) astigmatism
11. Parasympathetic nerves that stimulate constriction of the iris (in the pupillary reflex) are activated by neurons in:
    (a) the lateral geniculate
    (b) the superior colliculus
    (c) inferior colliculus
    (d) striate cortex
12. A bar of light in a specific part of the retina, with a particular length and orientation, is the most effective stimulus for:
    (a) ganglion cells
    (b) lateral geniculate cells
    (c) simple cortical cells
    (d) complex cortical cells

The Heart and Circulation

Objectives

By studying this chapter, you should be able to:

1. Describe the structure of arteries and veins

2. Explain the forces that produce blood flow in arteries and veins

3. Describe the structure of capillaries and their functional significance

4. Describe the path of blood flow through the heart and the action of the heart valves

5. Describe the pulmonary and systemic circulations

6. Explain the origin of the heartbeat, and describe the conduction pathway for electrical activity in the heart

7. Describe the electrocardiograph, and relate these waves to events occurring in the heart

8. Describe the relationship between myocardial action potentials and the heart's contraction, and explain how this relationship insures that the heart cannot sustain a contraction or normally continue to conduct waves in circus rhythms

9. Describe the origin of the heart sounds, and relate these to the changes in electrical activity and pressure within the heart

10. Describe the causes of some heart murmurs

11. Explain the causes of tachycardia, bradycardia, flutter, fibrillation, and heart block

12. Describe atherosclerosis, and explain how this condition may be produced and what effects it may have on heart function

13. Describe the effects of myocardial ischemia and myocardial infarction

In humans, as in all vertebrate animals, blood circulates within a closed system of blood vessels. This circulation requires the action of a muscular pump—the **heart**—that creates the pressure "head" required to move blood through the **arteries.** Blood travels away from the heart in arteries, and returns to the heart in **veins.** The system is said to be "closed" because arteries and veins are continuous with each other through smaller vessels.

Arteries branch extensively to form a "tree" of ever-smaller vessels. Those that are microscopic in diameter are called *arterioles*. Conversely, microscopic-sized *venules* merge together to form ever-larger vessels that empty into the large veins. Blood usually passes from the arterial to the venous systems through *capillaries,* which are the thinnest and most numerous blood vessels. All exchanges of fluid, nutrients, and wastes between blood and tissues occurs across the walls of capillaries.

Structure of the Heart

The heart is divided into four chambers. The right and left **atria** (singular, *atrium*) receive blood from the venous system; the right and left **ventricles** pump blood into the arterial system. The right atrium and ventricle (sometimes called the *right pump*) are separated from the left atrium and ventricle (the *left pump*) by a muscular wall, or *septum*. This septum normally prevents mixture of blood from the two sides of the heart.

Pulmonary and Systemic Circulations

Blood that has become partially depleted of its oxygen content and increased in carbon dioxide content, as a result of gas exchange across tissue capillaries, returns to the right atrium. This blood then enters the right ventricle, which pumps it into the *pulmonary trunk* and *pulmonary arteries*. The pulmonary artery branches to transport blood to the lungs, where gas exchange occurs between lung capillaries and the air sacs (alveoli) of the lungs. Oxygen diffuses from the air to the capillary blood, while carbon dioxide diffuses in the opposite direction.

The blood that returns to the left atrium by way of the *pulmonary veins* is therefore enriched in oxygen and partially depleted in carbon dioxide. The path of blood from the heart (right ventricle), through the lungs, and back to the heart (left atrium) completes one circuit: the **pulmonary circulation.**

Oxygen-rich blood in the left atrium enters the left ventricle and is pumped into a very large, elastic artery—the *aorta*. The aorta ascends for a short distance, makes a U turn, and then descends through the thoracic (chest) and abdominal cavities. Arterial branches from the aorta supply oxygen-rich blood to all of the organ systems and are thus part of the **systemic circulation.**

Figure 11.1   Arteries carry blood away from the heart; veins return blood to the heart. The arteries in the systemic circulation contain oxygen-rich blood that has passed through the lungs. Veins in the systemic circulation carry oxygen-poor blood that has passed through capillaries in the body (the body cells extract oxygen from capillary blood). This oxygen-poor blood is pumped to the lungs in the pulmonary artery and oxygen-rich blood returns to the heart in the pulmonary vein.

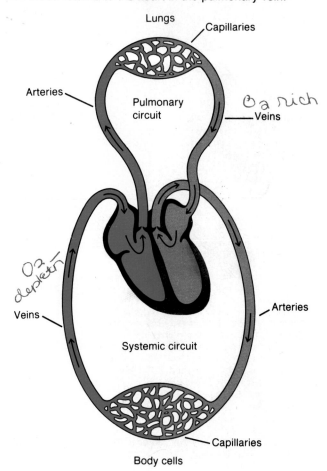

Figure 11.2 Pulmonary and systemic circulations. Notice that the liver receives oxygen-rich blood from the hepatic artery and also venous blood from the gastrointestinal tract. This venous blood, which contains the products of digestion, is modified as it passes through the liver on its way to the hepatic vein. This unusual pattern of circulation—from intestinal capillaries to veins to liver capillaries and then again to veins—is called a *portal system* (hence the name *hepatic portal vein*).

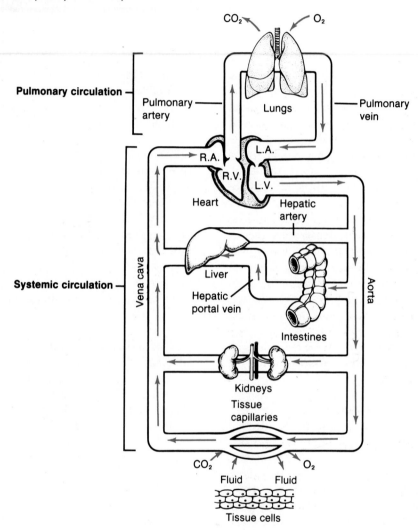

Table 11.1   Summary of the pulmonary and systemic circulations.

	Source	Arteries	O₂ Content of Arteries	Veins	Termination
Pulmonary Circulation	Right ventricle	Pulmonary arteries	Low	Pulmonary veins	Left atrium
Systemic Circulation	Left ventricle	Aorta and its branches	High	Superior and inferior vena cavae and their branches*	Right atrium

*Blood from the coronary circulation does not enter the vena cava, but instead returns directly to the right atrium via the coronary sinus.

As a result of cellular respiration, the oxygen concentration is lower and the carbon dioxide concentration is higher in the tissues than in the capillary blood. Blood that drains into the systemic veins is thus partially depleted in oxygen and increased in carbon dioxide content. These veins ultimately empty into two large veins—the *superior* and *inferior vena cavae*—that return the oxygen-poor blood to the right atrium. This completes the systemic circulation: from the heart (left ventricle), through the organ systems, and back to the heart (right atrium).

The numerous small muscular arteries and arterioles of the systemic circulation present greater resistance to blood flow than that in the pulmonary circulation. The amount of work performed by the left ventricle is greater (by a factor of 5 to 7) than that performed by the right ventricle, because the rate of blood flow through the systemic circulation must be matched to the flow rate of the pulmonary circulation despite the differences in resistance. It is not surprising, therefore, that the muscular wall of the left ventricle is thicker (8–10 mm) than that of the right ventricle (2–3 mm).

Atrioventricular and Semilunar Valves

Although adjacent myocardial cells are joined together mechanically and electrically by intercalated discs, the atria and ventricles are separated into two functional units by a sheet of connective tissue located between them. Embedded within this sheet of tissue are one-way **atrioventricular (A-V) valves.** The A-V valve located between the right atrium and right ventricle has three flaps, and is therefore called the *tricuspid valve.* The A-V valve between the left atrium and left ventricle has two flaps and is thus called the *bicuspid valve;* this is also known as the *mitral valve.*

The A-V valves allow blood to flow from atria to ventricles, but normally prevent the backflow of blood into the atria. Opening and closing of these valves occur as a result of pressure differences between the atria and ventricles. When the ventricles are relaxed, the venous return of blood to the atria causes the pressure in the atria to exceed that in the ventricles. The A-V valves therefore open, allowing blood to enter the ventricles. As the ventricles contract, the intraventricular pressure rises above the pressure in the atria and pushes the A-V valves closed.

There is a danger, however, that the high pressure produced by contraction of the ventricles could push the valve flaps too much and evert them. This is normally prevented by contraction of *papillary muscles* within the ventricles, which are connected to the A-V valve flaps by the *chordae tendinae.* Contraction of papillary muscles occurs at the same time as contraction of the muscular walls of the ventricles and serves to keep the valve flaps tightly closed.

One-way **semilunar valves** located at the origin of the pulmonary artery and aorta open during ventricular contraction, allowing blood to enter the pulmonary and systemic circulations. These valves close during ventricular relaxation, when the pressure in the arteries is greater than the pressure in the ventricles, and thus prevent the backflow of blood into the ventricles.

---

1. Using a flow diagram (arrows), describe the pathway of the pulmonary circulation. Indicate the relative amounts of oxygen and carbon dioxide in the vessels involved.
2. Use a flow diagram to describe the systemic circulation, and indicate the relative amounts of oxygen and carbon dioxide in the blood vessels. Explain the significance of the statement that this is a "higher-resistance" pathway than the pulmonary circulation.
3. Name the heart valves and the valves of the pulmonary artery and aorta. Describe how these valves insure a one-way flow of blood.

Figure 11.3 Diagram of the structure of the heart *(a)* and photograph of the aortic and pulmonary semilunar valves *(b)*.

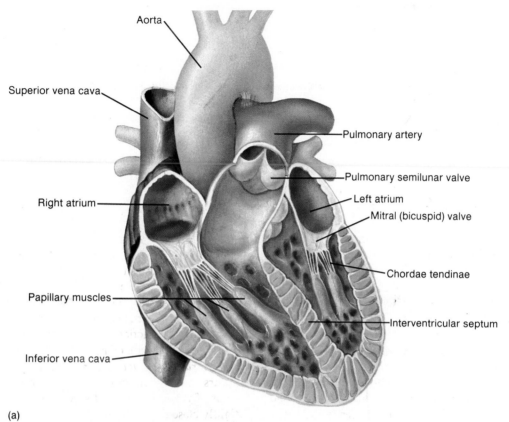

(a)

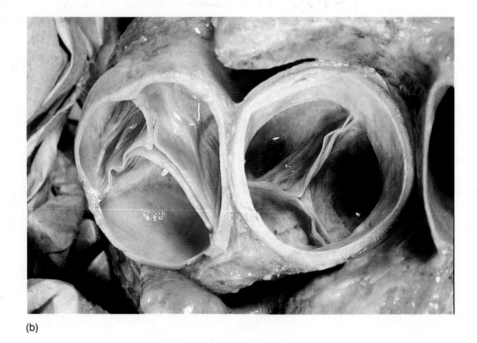

(b)

Figure 11.4   The three layers, or "tunics," of an artery.
The tunica intima consists of an inner epithelial
membrane (the endothelium), subendothelial connective
tissue, and a layer of elastic tissue (the elastica interna).
The elastic tissue appears *black* in the special stain
shown in *(b)*. The tunica media consists of smooth
muscle and the tunica adventitia is composed of
connective tissue.

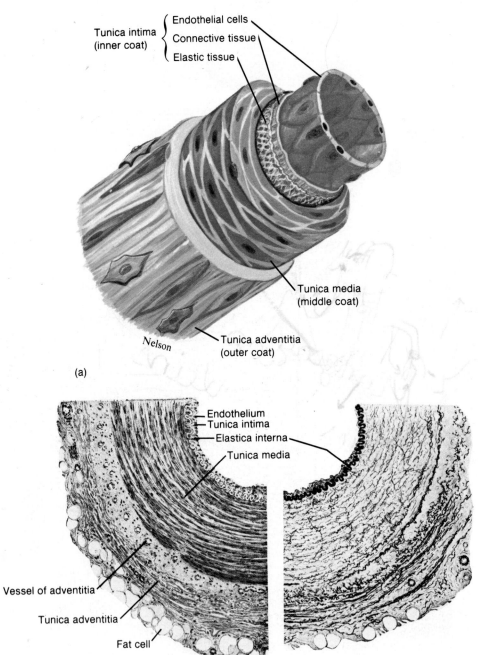

Figure 11.5 *(a)* is a diagram of a cross section of a small artery and vein. *(b)* is a scanning electron micrograph of a medium-sized artery *(MA)* and medium-sized vein *(MV)*.

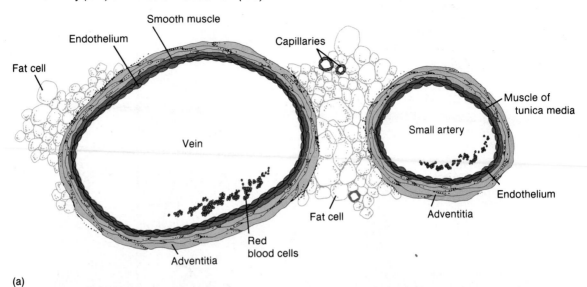

(a)

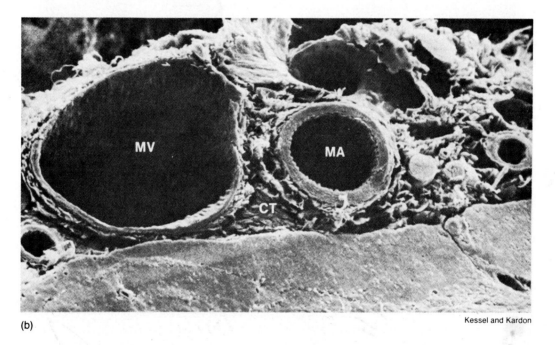

(b)

Kessel and Kardon

Blood Vessels

The walls of arteries and veins are composed of three coats, or "tunics." The outermost layer is the *tunica adventitia,* the middle layer is the *tunica media,* and the inner layer is the *tunica intima.* The tunica adventitia is composed of connective tissue. The tunica media is composed primarily of smooth muscle. The tunica intima consists of three parts: (1) an innermost simple squamous epithelium—the *endothelium*—that lines the lumen of all blood vessels; (2) a connective tissue layer; and (3) a layer of elastic fibers, called elastin, that forms the internal elastic lamina (see figure 11.4).

While both arteries and veins have the same basic structure, the two types of vessels can be distinguished when seen in cross section under the microscope. Arteries have relatively more muscle for their size than do veins and appear more round, while veins are usually partially collapsed. This is due to the fact that veins are not usually filled to their capacity; they can be stretched when they receive more blood and thus function as reservoirs, or "capacitance" vessels.

Figure 11.6 Action of the one-way venous valves. Contraction of skeletal muscles helps to pump blood towards the heart, but is prevented from pushing blood away from the heart by closure of the venous valves.

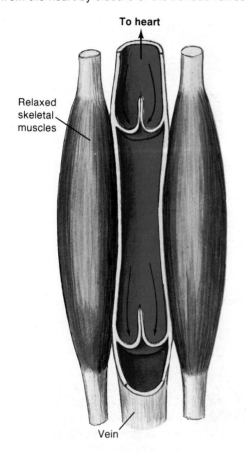

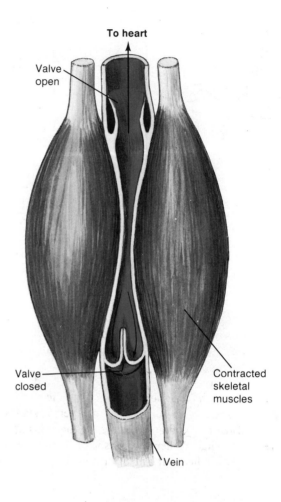

Veins

The average pressure in the veins, or *venous pressure,* is only 2 mm Hg (millimeters of mercury), compared to a much higher average arterial pressure of 100 mm Hg. These pressures represent the hydrostatic pressure that the blood exerts on the walls of the vessels, and the numbers indicate the differences from atmospheric pressure. A blood pressure of 2 mm Hg thus indicates that the pressure that the blood exerts on the wall of the vein is 2 mm Hg higher than the atmospheric pressure.

The low venous pressure is insufficient to return blood to the heart, particularly from the lower limbs. The veins, however, pass between skeletal muscle groups that produce a massaging action as they contract (see figure 11.6). As the veins are squeezed by contracting skeletal muscles, one-way flow of blood to the heart is insured by the presence of **venous valves.** The ability of these valves to prevent the flow of blood away from the heart was demonstrated in the seventeenth century by William Harvey (as shown in figure 11.7). After applying a tourniquet to a subject's arm, Harvey found that he could push the blood in a bulging vein towards the heart but not in the reverse direction.

Figure 11.7 Classical demonstration by William Harvey of the existence of venous valves that prevent the flow of blood away from the heart.

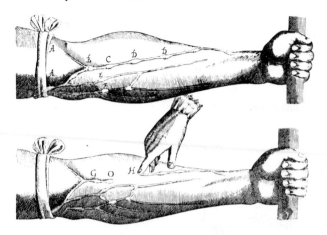

The effect of the massaging action of skeletal muscles on venous blood flow is often described as the **skeletal muscle pump.** During walking, for example, the movements of the foot—described as dorsiflexion and plantar flexion—activate the soleus muscle pump. This effect can be produced in bedridden people by upwards and downwards manipulations of the feet.

The rate of venous return to the heart is thus dependent, in large part, on the action of skeletal muscle pumps. When these pumps are less active, as when a person stands still or is bedridden, blood accumulates in the veins and causes them to bulge. When the person is more active, blood returns to the heart at a faster rate and less is left in the venous system.

Arteries

The aorta and other large arteries contain numerous layers of elastin fibers between smooth muscle cells in the tunica media. These large arteries expand when the hydrostatic pressure of the blood rises as a result of the heart's contraction; they recoil, like a stretched rubber band, when the blood pressure falls during relaxation of the heart. This elastic recoil helps to produce a smoother, less pulsatile flow of blood through the smaller arteries and arterioles.

Figure 11.8 Wedge of a large elastic artery, the aorta, stained so that elastic fibers appear black. Note the presence of numerous layers of elastin fibers in the tunica media.

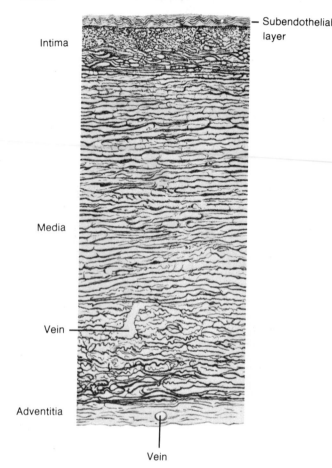

The small arteries and arterioles are less elastic than the larger arteries, and have a thicker layer of smooth muscle for their radius. Unlike the large *elastic arteries,* therefore, the smaller *muscular arteries* retain almost the same diameter as the hydrostatic pressure of the blood rises and falls due to the heart's pumping activity. Since arterioles and small muscular arteries have a narrow lumen, they provide the greatest resistance to blood flow through the arterial system. These small muscular arteries and arterioles are therefore called the *resistance vessels* of the circulatory system.

Figure 11.9   The microcirculation. Metarterioles provide
a path of least resistance between arterioles and venules.
Precapillary sphincter muscles regulate the flow of blood
through the capillaries.

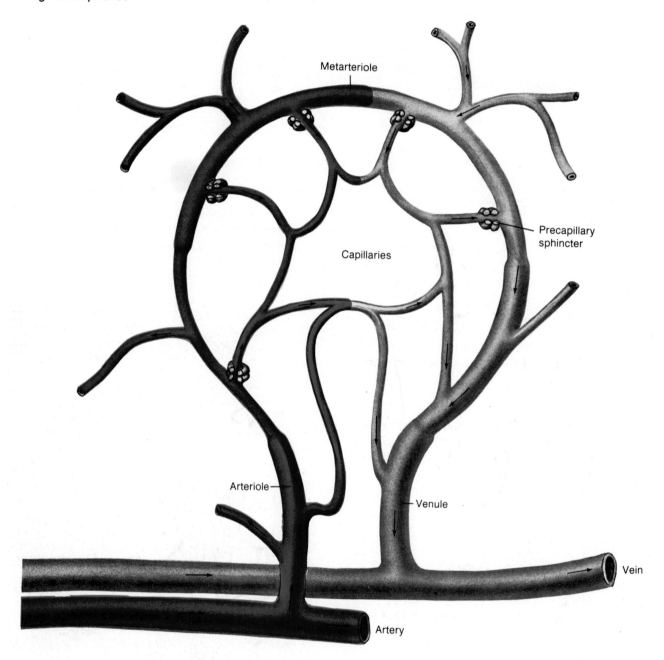

Table 11.2 Characteristics of the vascular supply to the mesenteries in a dog.

Kind of Vessel	Diameter (mm)	Number	Total Cross Sectional Area (cm²)	Length (cm)	Total Volume (cm³)
Aorta	10	1	0.8	40	30
Large arteries	3	40	3.0	20	60
Main artery branches	1	600	5.0	10	50
Terminal branches	0.6	1,800	5.0	1	25
Arterioles	0.02	40,000,000	125	0.2	25
Capillaries	0.008	1,200,000,000	600	0.1	60
Venules	0.03	80,000,000	570	0.2	110
Terminal veins	1.5	1,800	30	1	30
Main venous branches	2.4	600	27	10	270
Large veins	6.0	40	11	20	220
Vena cava	12.5	1	1.2	40	50

Capillaries through Vena cava braced together: 740. Total: 930.

Source: Reprinted with permission of Macmillan Publishing Company from *Animal Physiology: Principles and Adaptations* by Malcom S. Gordon. Copyright © 1977 by Malcom S. Gordon.

Small muscular arteries that are 100 μm or less in diameter branch to form smaller arterioles (20–30 μm in diameter). In some tissues, blood from arterioles can enter venules directly through *arteriovenous anastomoses* (from a Greek word meaning "coming together"). In most cases, however, blood from arterioles passes into capillaries. Capillaries are the narrowest of blood vessels (7–10 μm in diameter), and serve as the "business end" of the circulatory system in which exchange of gases and nutrients with the tissues occurs.

Capillaries

The arterial system branches extensively to deliver blood to over forty billion capillaries in the body. The extensiveness of these branches is indicated by the fact that all tissue cells are located within a distance of only 60–80 μm of a capillary, and by the fact that capillaries provide a total surface area of one thousand square miles for diffusion between blood and tissue fluid.

Despite their large number, capillaries contain only about 250 ml of blood at any time, out of a total blood volume of about 5,000 ml (most is contained within the venous system). The amount of blood flowing through a particular capillary bed is determined in part by the action of *precapillary sphincter muscles*. These muscles allow only 5 to 10 percent of the capillary beds in skeletal muscles, for example, to be open at rest. Blood flow to an organ is regulated by the action of these precapillary sphincters and by the degree of resistance to blood flow (due to constriction or dilation) provided by the small arteries and arterioles in the organ.

Figure 11.10 Electron micrograph of a capillary in a coronary vessel. Notice the thin intercellular channel and the fact that the capillary wall is only one cell thick. Arrows show some of the many pinocytotic vesicles.

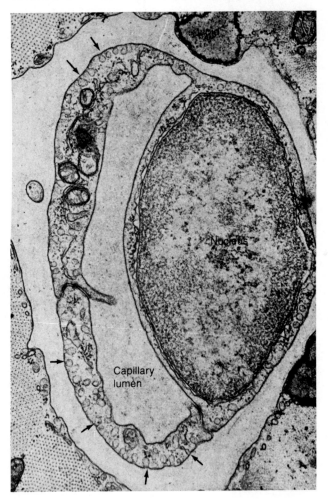

Nucleus

Capillary lumen

Figure 11.11   Diagrams of continuous, fenestrated, and discontinuous capillaries as they appear in the electron microscope. This classification is derived from the continuity of the endothelial layer.

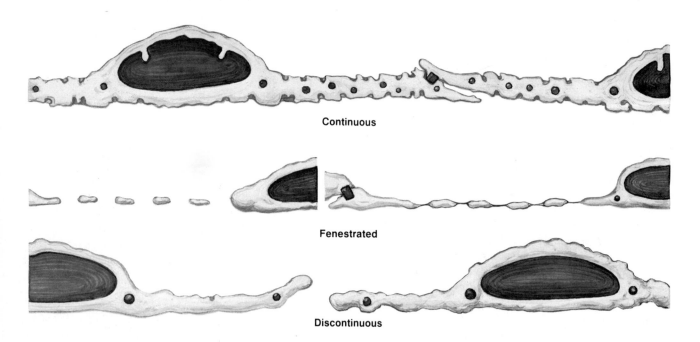

Continuous

Fenestrated

Discontinuous

Unlike the vessels of the arterial and venous sytems, the walls of capillaries are composed of only one cell layer—a simple squamous epithelium, or endothelium. The absence of smooth muscle and connective tissue layers permits a more rapid transport of materials between the blood and tissues.

Types of Capillaries   Different organs contain different types of capillaries, which are distinguished by significant differences in structure. These capillary types include those that are *continuous,* in terms of their endothelial lining, and those that are *discontinuous.*

Continuous capillaries are those in which adjacent epithelial cells are closely joined together. These are found in the muscles, lungs, adipose (fat) tissue, and in the central nervous system. Continuous capillaries in the CNS lack intercellular channels (producing a blood-brain barrier—see chapter 7); continuous capillaries in other organs have narrow intercellular channels (about 40–45 Å in width) that allow passage of molecules other than protein between capillary blood and tissue fluid.

Examination of endothelial cells with the electron microscope has revealed the presence of pinocytotic vesicles, which suggests that intracellular transport of materials may occur across the capillary walls. This type of transport appears to be the only mechanism of capillary exchange available within the central nervous system, and may account in part for the selective nature of the "blood-brain barrier."

The kidneys, endocrine glands, and intestine have discontinuous, or *fenestrated, capillaries,* characterized by wide intercellular pores (800–1,000 Å) that are covered by a layer of mucoprotein that may serve as a diaphragm. In the bone marrow, liver, and spleen, the distance between endothelial cells is so great that the capillaries appear as little cavities *(sinusoids)* in the organs.

1. Describe the basic structural pattern of arteries and veins. Describe how arteries and veins differ in structure and how these differences contribute to the resistance function of arteries and the capacitance function of veins.
2. Describe the functional significance of the "skeletal muscle pump" and illustrate the action of venous valves.
3. Explain the functions of capillaries, and describe the structural differences between capillaries in different organs.

The Cardiac Cycle

The cardiac cycle refers to the repeating pattern of contraction and relaxation in the heart. The phase of contraction is called **systole,** and the phase of relaxation is called **diastole.** When these terms are used alone they refer to contraction and relaxation of the ventricles. It should be noted, however, that the atria also contract and relax—there is an atrial systole and diastole. Atrial contraction occurs during diastole (when the ventricles are relaxed); when the ventricles contract during systole, the atria are relaxed.

The heart thus has a two-step pumping action. The right and left atria contract almost simultaneously, followed about 0.1 second later by contraction of the right and left ventricles. During the time when both the atria and ventricles are relaxed, the venous return of blood fills the atria and—because the A-V valves are open—partially fills the ventricles. It has been estimated that the ventricles are about 80 percent filled with blood even before the atria contract. Contraction of the atria adds the final one-fifth to the *end-diastolic volume* of blood in the ventricles.

Contraction of the ventricles in systole ejects about two-thirds of the blood in the ventricles, leaving one-third of the initial amount as the *end-systolic volume.* The ventricles then fill with blood during the next diastole. At an average **cardiac rate** of 75 beats per minute, each cycle lasts 0.8 second; 0.5 second is spent in diastole, and systole takes 0.3 second.

Figure 11.12 The cardiac cycle of ventricular systole and diastole. Contraction of the atria occurs in the last 0.1 second of diastole. Relaxation of the atria occurs during ventricular systole. The durations of systole and diastole given are accurate for a cardiac rate of seventy-five beats per minute.

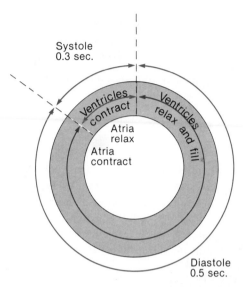

Diastole begins as the ventricles relax, and ends when the ventricles contract immediately after the atria contract. This period of relaxation and filling is shortened as the cardiac rate increases. The time interval between two systoles, in other words, is shortened at high cardiac rates. If the cardiac rate increases from 75 beats per minute to 180 beats per minute, for example, the time of each cycle is shortened from 0.8 second (60 sec/min ÷ 75 beats per minute) to 0.33 second (60 sec/min ÷ 180 beats per minute). The length of diastole is shortened from 0.5 second to 0.13 second, and the time spent in systole is also slightly shortened from 0.3 second to 0.2 second. These changes are illustrated in figure 11.13.

Figure 11.13 Increased heart rate is achieved primarily through a decreased time in diastole, which can therefore be likened to "spacer segments" between successive heartbeats. The length of time spent in systole also decreases slightly at high cardiac rates.

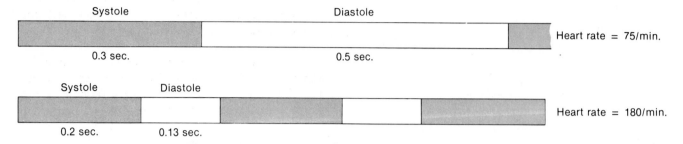

Pressure Changes during the Cardiac Cycle

When the heart is in diastole, the blood pressure in the systemic arteries averages about 80 mm Hg. Contraction of the left ventricle raises the intraventricular pressure to about 120 mm Hg. This sharp rise in pressure first causes the A-V valves to close (during the phase of *isovolumic contraction*, before blood is ejected), and then opens the semilunar valves as blood enters the arteries. Blood thus flows through the arteries as a result of the pressure gradient created by contraction of the ventricles. Arterial blood pressure, as a result of ventricular systole, rises from 80 to 120 mm Hg in the systemic circulation (maximum pressure in the pulmonary circulation is much less—25 mm Hg).

When the ventricles relax in diastole, their pressure falls to zero millimeters of mercury (remember that these numbers are in reference to atmospheric pressure—0 mm Hg is equal to atmospheric pressure). Since the pressure in systemic arteries at this time is again 80 mm Hg, blood would flow back into the ventricles were it not for the one-way action of the semilunar valves that close under the back-pressure of the blood.

Heart Sounds

Closing of the A-V and semilunar valves produces sounds that can be heard at the surface of the chest with a stethoscope. These sounds are often described phonetically as *lub-dub*. The "lub," or **first sound,** is produced by closing of the A-V valves; the "dub," or **second sound,** is produced by closing of the semilunar valves. The first sound is thus heard when the ventricles contract at systole, and the second sound is heard when the ventricles relax at diastole.

Figure 11.14 Relationship between the heart sounds and the intraventricular pressure and volume. Closing of the A-V valves occurs during the early part of contraction, when the intraventricular pressure rises prior to ejection of blood. Closing of the semilunar valves occurs at the beginning of ventricular relaxation, just prior to filling. The first and second heart sounds thus appear during the stages of isovolumic contraction and isovolumic relaxation (iso = same).

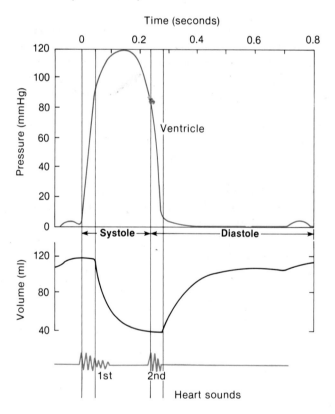

Figure 11.15 Routine stethoscope positions for listening to the heart sounds.

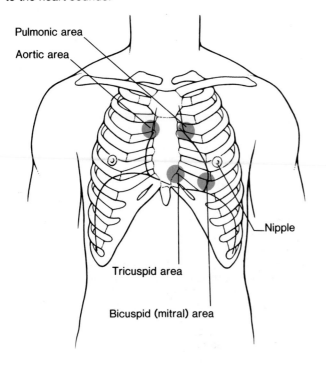

Pulmonic area

Aortic area

Nipple

Tricuspid area

Bicuspid (mitral) area

The first sound may, particularly during inhalation, be heard to split into tricuspid and mitral components. Closing of the tricuspid is best heard at the fourth right intercostal space (between the ribs), while closing of the valve is best heard at the apex of the heart (see figure 11.15). The second sound may also be "split" under certain conditions. Closing of the pulmonary and aortic semilunar valves is best heard at the second left and right intercostal spaces, respectively.

Heart Murmurs   Abnormal heart sounds, or *heart murmurs,* can be produced by abnormal patterns of blood flow in the heart. One of the major causes of murmurs are defective heart valves. Defective heart valves can be congenital (inherited) or, most commonly, they can occur as a result of *rheumatic endocarditis* associated with rheumatic fever. In this disease, the valves become damaged by antibodies made in response to infection by streptococcus bacteria (the same bacteria that produce strep throat).

In *mitral stenosis,* for example, the mitral valve becomes thickened and calcified (stenosis = narrowing). This can impair the blood flow from the left atrium to the left ventricle. Accumulation of blood in the left atrium may cause a rise in left atrial and pulmonary vein pressure, resulting in pulmonary hypertension (high blood pressure). As a compensation for the increased pulmonary pressure, the right ventricle grows thicker and stronger.

Stenosis of the aortic semilunar valve may interfere with ejection of blood from the left ventricle into the aorta. This results in an increase in end-systolic volume, and eventually in hypertrophy (growth) of the left ventricle as a compensation in order to eject more blood through the constricted valve.

Valves are said to be *incompetent* when they do not close properly, and murmurs may be produced as blood regurgitates through the valve flaps. One important cause of incompetent A-V valves is damage to the papillary muscles. When this occurs, the tension in the chordae tendinae may not be sufficient to prevent the valve from everting as pressure in the ventricle rises during systole.

Murmurs can also be produced by the flow of blood through *septal defects*—holes in the septum between the right and left sides of the heart. These are usually congenital, and may occur either in the interventricular septum or in the septum between the atria (see figure 11.16).

Foramen Ovale and Ductus Arteriosus   A fetus obtains its oxygen from its mother's blood by diffusion through the placenta to the fetal blood. The fetal lungs are not active in gas exchange and are partially collapsed (although fetuses do perform respiratory movements). The resistance to blood flow is thus high in the pulmonary circulation, which favors the *shunting* (diversion) of blood along lower resistance pathways. There are two such pathways available in the fetus: (1) the **foramen ovale** (oval hole), which is an opening between the right and left atria and (2) the **ductus arteriosus.**

Figure 11.16 Abnormal patterns of blood flow due to septal defects. Left-to-right shunting of blood is shown because the left pump is normally at a higher pressure than the right pump. Under abnormal conditions, however, the pressure in the right atrium may exceed that of the left, causing right-to-left shunting of blood through a septal defect in the atria.

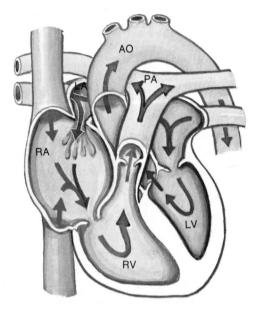

Septal defect
in atria

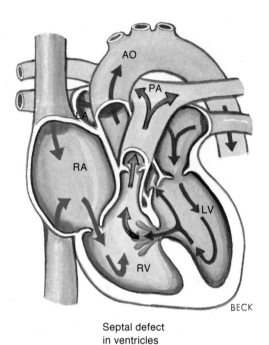

Septal defect
in ventricles

Some of the blood in the right atrium can pass through the foramen ovale to the left atrium, and in this way bypass the right ventricle and pulmonary circuit. Much of the blood that is pumped by the right ventricle can still avoid the pulmonary circulation by passing through the ductus arteriosus, which connects the pulmonary trunk with the aorta (see figure 11.17). In this way, blood is diverted away from the lungs towards the systemic circulation and the placenta.

A tissue flap normally closes the foramen ovale after birth. When the newborn baby takes its first breath, the rising concentrations of blood oxygen stimulate muscular contraction of the ductus arteriosus, sealing it shut (the closed ductus is eventually transformed into a ligament). If these pathways remain open (are *patent*) after birth, however, they can cause right-to-left shunting of blood where oxygen-poor blood on the right side is mixed with oxygen-rich blood in the left side. Since blood low in oxygen has a bluish appearance when seen through skin and mucous membranes, this can produce *cyanosis* (cyano = blue).

1. Using a figure or an outline, describe the sequence of events that occurs during the cardiac cycle. Indicate when atrial and ventricular filling occur, and when atrial and ventricular contraction occur.
2. Describe, in words, how the pressure in the left ventricle and in the systemic arteries varies during the cardiac cycle.
3. Draw a figure to illustrate the pressure variations described in no. 2, and indicate in your figure when the A-V and semilunar valves close. Discuss the origin of the heart sounds.

Electrical Activity of the Heart

Each myocardial cell in the heart has the potential for spontaneous electrical activity; each has an intrinsic rhythm. In the normal heart, however, spontaneous electrical activity is limited to a small region (about $2 \times 5 \times 15$ mm) in the right atrium near the opening of the vena cava. This region is called the **sinoatrial** (or **S-A) node,** and serves as the normal *pacemaker* of the heart.

Figure 11.17 Flow of blood through a patent (open) ductus arteriosus.

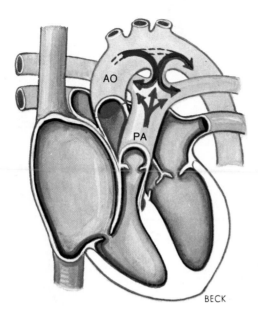

BECK

Spontaneous electrical activity in the pacemaker region results from slow decay of the resting membrane potential, due to inward diffusion of Na⁺ and Ca⁺⁺, during diastole. This gradual *diastolic depolarization* results in the decay of the membrane potential from close to −90 mV (millivolt) to the threshold level required to open voltage-regulated Na⁺ gates. The rapid inward diffusion of Na⁺ that results changes the membrane potential to +30 mV. This part of the cardiac action potential is identical to that which occurs in skeletal muscles and nerves.

Following the rapid reversal of the membrane potential polarity, the voltage quickly declines to about −10 to −20 mV. Unlike the action potential of skeletal muscle fibers and axons, however, this level of depolarization is maintained in myocardial cells for 200–300 msec (milliseconds) before repolarization. This "plateau phase" contrasts with the short action potential of nerve fibers (which last only about 3 msec), and is produced by a slow inward diffusion of Ca⁺⁺ that balances a slow outward diffusion of cations. Rapid repolarization at the end of the plateau phase is achieved, as in nerve fibers, by opening of K⁺ gates and the rapid outward diffusion of K⁺ that results.

Figure 11.18 Action potential in a myocardial cell from the ventricles. The plateau phase of the action potential is maintained by a slow inward diffusion of Ca⁺⁺. The cardiac action potential, as a result, has a duration that is about one hundred times longer than the "spike potential" of an axon.

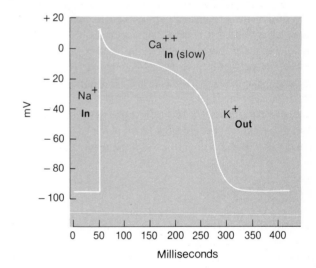

Figure 11.19 The conduction tissue of the heart. The appearance of action potentials in the S-A node, atria, and ventricles is also shown.

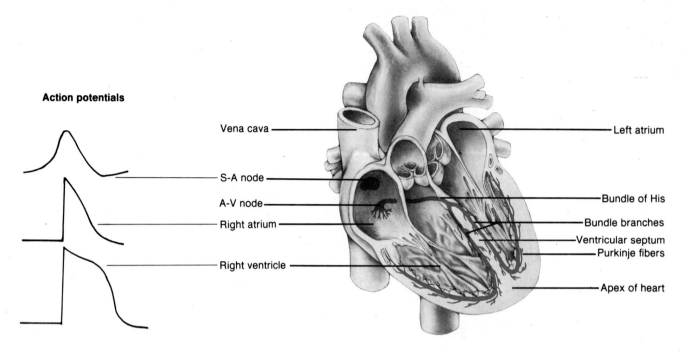

Action potentials

Vena cava

S-A node

A-V node

Right atrium

Right ventricle

Left atrium

Bundle of His

Bundle branches

Ventricular septum

Purkinje fibers

Apex of heart

The myocardial cells of the S-A node function as the pacemaker because their spontaneous rate of diastolic depolarization reaches threshold before that of other myocardial cells. While the resting membrane potential of other myocardial cells also gradually decays during diastole, the rate of this decay is slower than that of the S-A node. They are therefore stimulated to contract by action potentials originating in the S-A node before they can generate their own action potentials. If action potentials from the S-A node are prevented from reaching a region of the heart, the cells of this region will generate their own action potentials and become an *ectopic* ("out-of-place") *pacemaker* (also called an *ectopic focus*).

Conductive Tissue in the Heart

Action potentials in the S-A node spread very quickly— at a rate of 0.8 to 1.0 m per second—across the myocardial cells of both atria. These impulses, however, cannot spread directly across myocardial cells into the ventricles. They must instead be conducted by specialized tissue, derived from heart muscle, that traverses the connective tissue sheet between the atria and ventricles. This specialized conductive tissue includes the **A-V node, bundle of His,** and **Purkinje fibers.**

Slow conduction of impulses (0.03 to 0.05 m per second) through the A-V node accounts for over half of the time delay between excitation of the ventricles and atria. After the impulses spread through the A-V node, the conduction rate increases greatly in the bundle of His and reaches very high velocities (5 m per second) in the Purkinje fibers. The bundle of His, and its right and left

Figure 11.20 The time course for the myocardial action potential is compared with the duration of contraction. Notice that the long action potential results in correspondingly long absolute refractory period (ARP) and relative refractory period (RRP). These refractory periods last almost as long as the twitch, so that the myocardial cells cannot be stimulated a second time until they have finished their contraction from the first stimulus.

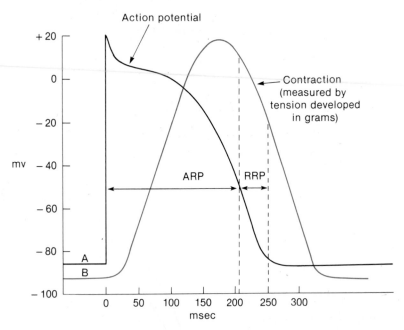

branches, conduct impulses along the interventricular septum to the apex of the heart; the Purkinje fibers conduct the impulses into the muscular walls of the ventricles. As a result of this rapid conduction of impulses, ventricular contraction begins about 0.1–0.2 second after contraction of the atria.

Unlike skeletal muscles, the heart cannot sustain a contraction—it can only twitch. The twitch of the heart, however, lasts almost 300 msec, compared to a skeletal muscle twitch of 20–100 msec. Since the heart cannot sustain a contraction, its rhythmic pumping action of contraction and relaxation is insured. This results from the long duration of myocardial action potentials, which are accompanied by correspondingly *long refractory periods* (about 250 msec). Since the duration of the refractory periods is approximately as long as the duration of contraction, the heart cannot be stimulated to contract again until it has relaxed from its previous electrical stimulus.

The Electrocardiogram (ECG)

A pair of surface electrodes placed directly on the heart will record a repeating pattern of potential changes. As action potentials spread from the atria to the ventricles, the voltage measured between these two electrodes will vary in a way that provides a "picture" of the electrical activity of the heart. The nature of this picture can be varied by changing the position of the recording electrodes; different positions provide different perspectives, enabling an observer to gain a more complete picture of the electrical events.

The body is a good conductor of electricity because tissue fluids contain high concentrations of ions that can move (creating a current) in response to potential differences. Potential differences generated by the heart are thus conducted to the body surface, where they can be recorded as an **electrocardiogram** (ECG or EKG).

Figure 11.21 The electrocardiogram indicates the conduction of electrical impulses through the heart *(a)* and measures and records both the intensity of this electrical activity (in millivolts) and the time intervals involved *(b)*.

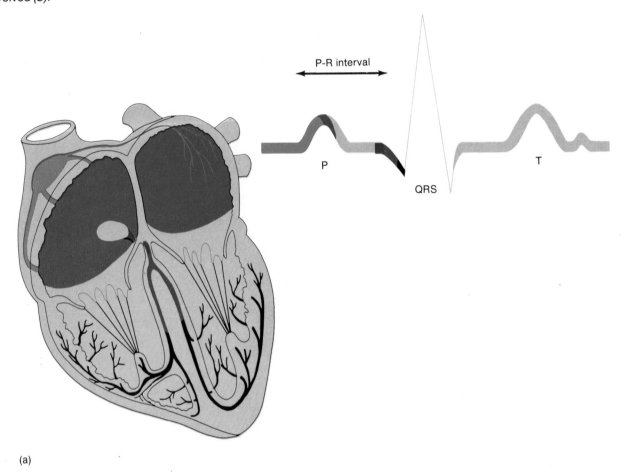

(a)

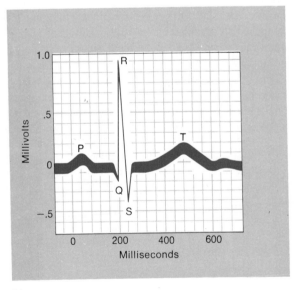

(b)

Figure 11.22   Placement of the bipolar limb leads and the exploratory electrode for the unipolar chest leads in an electrocardiogram (ECG).

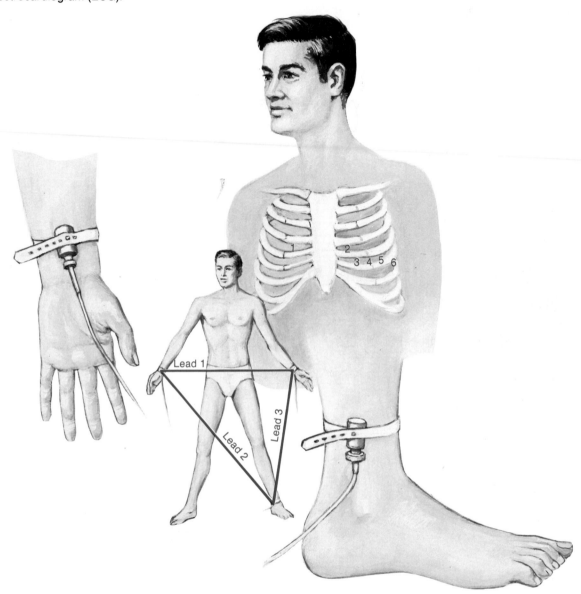

There are two major types of ECG recording electrodes or "leads." The *bipolar limb leads* record the voltage between electrodes placed on the wrists and the legs. These bipolar leads include lead I (right arm to left arm), lead II (right arm to left leg), and lead III (left arm to left leg). In the *unipolar leads,* voltage is recorded between a single "exploratory electrode" placed on the body and an electrode that is built into the electrocardiograph and maintained at zero potential (ground).

The unipolar limb leads are placed on the right arm, left arm, and left leg, and are labeled AVR, AVL, and AVF respectively (the *A* stands for *augmented*). The unipolar chest leads are labeled one through six, starting from the midline position (see figure 11.22). There are thus a total of twelve standard ECG leads that "view" the

Figure 11.23 The conduction of electrical impulses in the heart, as indicated by the electrocardiogram (ECG). The direction of the arrows in *(e)* indicates that depolarization of the ventricles occurs from the inside (endocardium) out (to the epicardium), while the arrows in *(g)* indicate that repolarization of the ventricles occurs in the opposite direction.

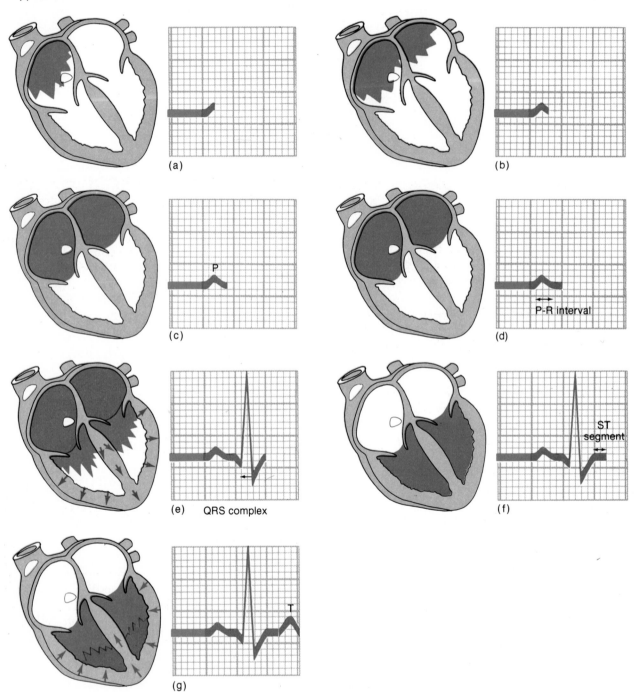

(a)

(b)

(c)

(d)

(e)   QRS complex

(f)   ST segment

(g)

P

P-R interval

T

Table 11.3 The electrocardiograph (ECG) leads.

Name of Lead	Placement of Electrodes
Bipolar limb leads	
I	Right arm and left arm
II	Right arm and left leg
III	Left arm and left leg
Unipolar limb leads	
AVR	Right arm
AVL	Left arm
AVF	Left leg
Unipolar chest leads	
V₁	4th intercostal space right of sternum
V₂	4th intercostal space left of sternum
V₃	5th intercostal space to the left of the sternum
V₄	5th intercostal space in line with the middle of the clavicle (collarbone)
V₅	5th intercostal space to the left of V₄
V₆	5th intercostal space in line with the middle of the axilla (underarm)

changing pattern of the heart's electrical activity from different perspectives. This is important because certain abnormalities are best seen with particular leads, and may not be visible at all with other leads.

Each cardiac cycle produces three distinct ECG waves. The first, or **P wave**, occurs as a result of atrial depolarization. The second, or **QRS wave**, represents depolarization of the ventricles. Atrial repolarization occurs during this time but is "hidden" by the larger electrical changes occurring in the ventricles. The last ECG wave—the **T wave**—is produced by repolarization of the ventricles.

Correlation of ECG with Heart Sounds   Depolarization of the ventricles, as indicated by the QRS wave, stimulates contraction by promoting the uptake of Ca⁺⁺ into the region of the sarcomeres. The QRS wave is thus seen to occur at the beginning of systole. The rise in intraventricular pressure that results causes the A-V valves to close, so that the first heart sound (lub) is produced at about the same time that the QRS wave occurs (see figure 11.24).

Repolarization of the ventricles, as indicated by the T wave, occurs at the same time that the ventricles relax at the beginning of diastole. The resulting fall in intraventricular pressure causes the aortic and pulmonary semilunar valves to close, so that the second heart sound (dub) is produced shortly after the T wave is seen in an electrocardiogram.

Figure 11.24 Relationship between changes in intraventricular pressure and the electrocardiogram during the cardiac cycle. The *QRS* wave (representing depolarization of the ventricles) occurs at the beginning of systole, while the *T* wave (representing repolarization of the ventricles) occurs at the beginning of diastole.

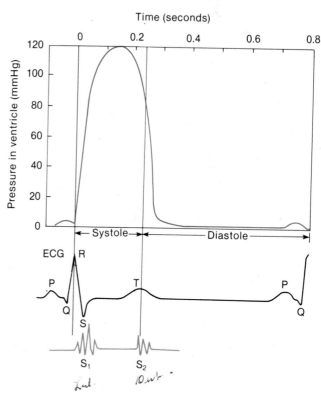

Figure 11.25 In *(a)* the heartbeat is paced by the normal pacemaker—the S-A node (hence the name *sinus rhythm*). This can be abnormally slow (bradycardia—46 beats per minute in this example) or fast (tachycardia—136 beats per minute in this example). Compare the pattern of tachycardia in *(a)* with that in *(b)*. The tachycardia in *(b)* is produced by an ectopic pacemaker in the ventricles. This dangerous condition can quickly lead to ventricular fibrillation (also shown in *b*).

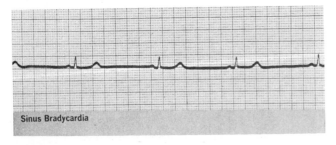

Sinus Bradycardia

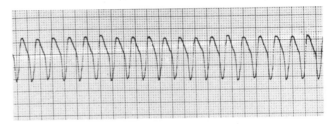

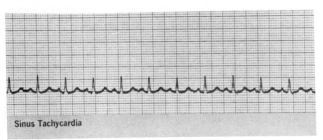

Sinus Tachycardia

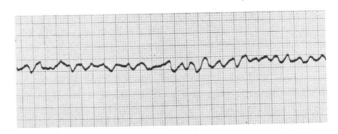

(a)　　　　　　　　　　　　　　　　(b)

Arrhythmias Detected by the Electrocardiograph Arrhythmias, or abnormal heart rhythms, can be detected and described by the abnormal ECG tracings they produce. While proper clinical interpretation of electrocardiograms requires more information than is presented in this chapter, some knowledge of abnormal rhythms is interesting in itself, and is useful in gaining an understanding of normal physiology.

Since a heartbeat occurs whenever a normal QRS complex is seen, and since the ECG chartpaper moves at a known speed so that its "X-axis" indicates time, the cardiac rate (beats per minute) can be easily obtained from examination of the ECG recording. A cardiac rate slower than 60 beats per minute indicates **bradycardia;** a rate faster than 100 beats per minute is described as **tachycardia.**

Both bradycardia and tachycardia can occur normally. Endurance-trained athletes, for example, commonly have a slower heart rate than the general population. This *athlete's bradycardia* occurs as a result of higher levels of parasympathetic inhibition of the S-A node, and is a beneficial adaptation. Activation of the sympathetic system, during exercise or emergencies ("fight or flight"), causes a normal tachycardia to occur.

Abnormal tachycardia occurs when a person is at rest. This may result from abnormally fast pacing by the atria, due to drugs or to the development of abnormally fast *ectopic pacemakers*—cells located outside the S-A node that assume a pacemaker function. This abnormal atrial tachycardia thus differs from normal "sinus" (S-A node) tachycardia. *Ventricular tachycardia* results when abnormally fast ectopic pacemakers in the ventricles cause them to beat rapidly and independently of the atria. This is very dangerous because it can quickly degenerate into a lethal condition known as ventricular fibrillation.

Flutter and Fibrillation   Extremely rapid rates of electrical excitation and contraction of either the atria or the ventricles may occur. In *flutter,* the contractions are very rapid (200–300 per minute) but are coordinated. In *fibrillation,* contractions of different groups of myocardial fibers occur at different times so that a coordinated pumping action of the chambers is impossible.

Atrial flutter usually degenerates quickly into *atrial fibrillation.* This causes the pumping action of the atria to stop. Since the ventricles fill to about 80 percent of their end-diastolic volume before atrial contraction normally occurs, however, the heart is still able to eject a sufficient quantity of blood into the circulation. People who have atrial fibrillation can thus live for many years. People who have *ventricular fibrillation,* in contrast, can live for only a few minutes before the brain and heart—which are very dependent upon oxygen for their metabolism—cease to function.

Fibrillation is caused by a continuous recycling of electrical waves, known as **circus rhythms,** through the myocardium. This recycling is normally prevented by the fact that the entire myocardium enters a simultaneous refractory period (due to the long duration of action potentials, as previously discussed). If some cells emerge from their refractory period before others, however, electrical waves can be continuously regenerated and conducted. Recycling of electrical waves along continuously changing pathways produces uncoordinated contraction and an impotent pumping action.

Circus rhythms are thus produced whenever impulses can be conducted without interruption by non-refractory tissue. This may occur when the conduction pathway is longer than normal, as in a dilated heart. It can also be produced by an electric shock delivered during the middle of the T wave, when different myocardial cells are in different stages of recovery from their refractory period. Finally, circus rhythms and fibrillation may be produced by damage to the myocardium, which slows the normal rate of impulse conduction.

Fibrillation can sometimes be stopped by a strong electric shock delivered to the chest. This procedure is called **electrical defibrillation.** The electric shock depolarizes all the myocardial cells at the same time, causing them all to enter a refractory state. Conduction of circus rhythms thus stops, and—within a couple of minutes—the S-A node can begin to stimulate contraction in a normal fashion. This does not correct the initial problem that caused circus rhythms and fibrillation, of course, but it does keep the person alive long enough to take other corrective measures.

Heart Block   The time interval between the beginning of atrial depolarization—indicated by the P wave—and the beginning of ventricular depolarization (as shown by the Q part of the QRS complex) is called the *P-R interval.* In the normal heart this time interval is 0.12 to 0.20 second. Damage to the A-V node causes slowing of impulse conduction and is reflected by changes in the P-R interval. This condition is known as A-V *heart block.*

First-degree heart block occurs when the rate of impulse conduction through the A-V node (as reflected by the P-R interval) is greater than 0.20 second. **Second-degree heart block** occurs when the A-V node is damaged so severely that only one out of every two, three, or four atrial electrical waves can pass through to the ventricles. This is indicated in an ECG by the presence of P waves without associated QRS waves.

In **third-degree,** or **complete, heart block,** none of the atrial waves can pass through the A-V node to the ventricles. The atria are paced by the S-A node (follow a normal "sinus rhythm"), but the ventricles are paced by a different, ectopic pacemaker (usually located in the bundle of His or Purkinje fibers). Since the S-A node is the normal pacemaker by virtue of the fact that it has the fastest cycle of electrical activity, the ectopic pacemaker in the ventricles causes the ventricles to beat at an abnormally slow rate. The bradycardia that results is usually corrected by insertion of an artificial pacemaker.

1. Using a line diagram, illustrate a myocardial action potential and the time course for myocardial contraction. Describe how the relationship between these two events prevents the heart from sustaining a contraction and how it normally prevents circus rhythms of electrical activity.
2. Draw an ECG and label the waves. Indicate the electrical events in the heart that produce these waves.
3. Draw a figure illustrating the relationship between ECG waves and the heart sounds. Explain this relationship.
4. Using a flow chart (arrows), describe the pathway of electrical conduction of the heart, starting with the S-A node. Explain how damage to the A-V node affects this conduction pathway and the ECG.
5. Describe normal and pathological examples of bradycardia and tachycardia. Also describe how flutter and fibrillation are produced.

Figure 11.26   Atrioventricular (A-V) heart block. In first degree block the P-R interval is greater than 0.20 second (in the example above, the P-R interval is 0.26–0.28 second). In second degree block P waves are seen that are not accompanied by QRS waves; in this example, the atria are beating 90 times per minute (as represented by the P waves), while the ventricles are beating at 50 times per minute (as represented by the QRS waves). In third degree block the ventricles are paced independantly of the atria by an ectopic pacemaker. Ventricular depolarization (QRS) and repolarization (T) therefore have a variable position in the electrocardiogram in relationship to the P waves (atrial depolarization).

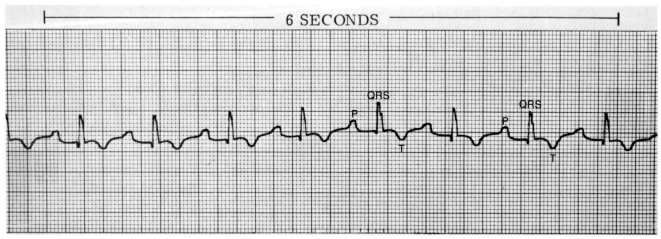

First degree A-V block

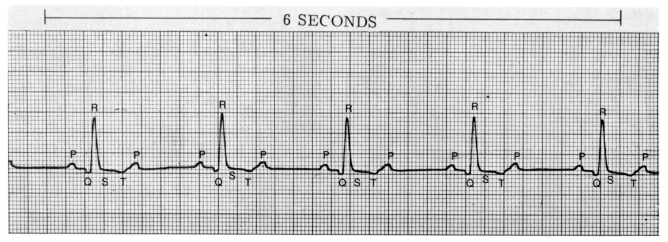

Second degree A-V block

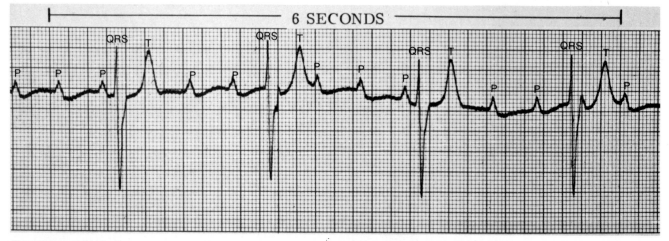

Third degree A-V block

Figure 11.27 Photograph of atherosclerotic plaque protruding into the lumen of an artery.

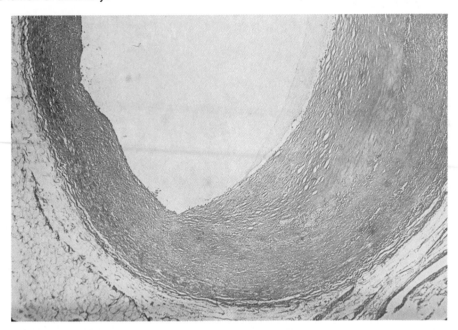

Atherosclerosis and Ischemic Heart Disease

Atherosclerosis is the most common form of arteriosclerosis (hardening of the arteries). In atherosclerosis, localized *plaques* or *atheromas* (athera = gruel) protrude into the lumen of the artery (see figure 11.27) and thus reduce blood flow. The plaques additionally serve as sites for *thrombus* (blood clot) formation, which can further occlude the blood supply to an organ (see figure 11.28).

It is currently believed that atheromas begin as benign tumors of smooth muscle cells that migrate from the tunica media to the intima and proliferate by cell division. In later stages, these cells accumulate cholesterol and other lipids, giving them a "foamy cell" appearance. In fully developed plaque, cholesterol and other lipids accumulate outside the smooth muscle cells, and these accumulations can become calcified. Damage to the endothelium that covers the plaque results in exposure of subendothelial tissue to the blood. This may contribute to growth of the plaque and to the formation of blood clots.

Factors that have been suggested to cause the initial migration and proliferation of smooth muscle cells in the early formation of plaque include (1) molecules released from platelets (components of blood described in chapter 13); (2) derivatives of cholesterol; and (3) plasma lipoproteins, which serve as "carriers" of cholesterol. Risk factors in the development of atherosclerosis include advanced age, smoking, and hypertension. Hypertension may contribute to the development of atherosclerosis by damaging the endothelium. Estrogen (female sex hormone) appears to decrease the risk of atherosclerosis, so that the incidence of this disease in women before the age of menopause is lower than in men of comparable age.

Cholesterol and Plasma Lipoproteins

It has long been known that high blood cholesterol is associated with increased risk of atherosclerosis. Blood cholesterol levels, however, do not necessarily reflect the amount of cholesterol ingested in the diet as was once believed. This is partly due to the fact that many organs (including blood vessels) can manufacture cholesterol. Also, it is now known that the total cholesterol concentration in the blood is not as good a predictor of the risk factor for atherosclerosis as is the amount of cholesterol attached to particular protein carriers.

Figure 11.28   The lumen (opening) of a human coronary artery is partially occluded by an atherosclerotic plaque *(a)* and almost completely occluded by a thrombus *(b)*.

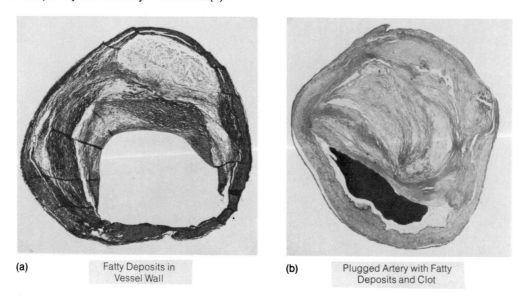

(a)   Fatty Deposits in Vessel Wall

(b)   Plugged Artery with Fatty Deposits and Clot

Figure 11.29   Diagram of a fully developed atherosclerotic plaque. Notice the ulcerated endothelium that exposes the underlying connective tissue.

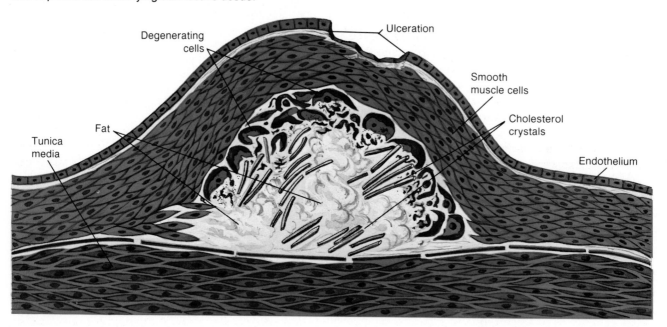

Table 11.4 Relationship between the plasma concentration of high-density lipoproteins (HDL) and the risk of coronary heart disease (CHD).

CHD Risk and High Density Lipoprotein Cholesterol

HDL Cholesterol	Relative Risk of CHD	
mg/dl	Men	Women
25	2.00	—
30	1.82	—
35	1.49	—
40	1.22	1.94
45	1.00	1.55
50	0.82	1.25
55	0.67	1.00
60	0.55	0.80
65	0.45	0.64
70	—	0.52
75	—	—

Source: W. B. Kannel, W. P. Castell, and T. Gordon, *Annals of Internal Medicine* 90 (1979):85

Since cholesterol is a lipid, very little can be dissolved in blood plasma; most travel in the blood attached to protein carriers known as lipoproteins. There are three types of these carriers, based upon their relative densities as indicated by the rate at which they travel towards the bottom of a test tube during centrifugation. These are the *very-low-density lipoproteins (VLDL),* the *low-density lipoproteins (LDL),* and the *high-density lipoproteins (HDL).*

The tendency towards atherosclerosis appears to increase with increasing amounts of cholesterol attached to VLDL and LDL carriers. These carriers are believed to transport cholesterol to the artery walls. The risk of atherosclerosis, however, is more accurately predicted by measuring HDL-cholesterol. The risk of atherosclerosis is *inversely* related to the HDL-cholesterol concentration; high levels of HDL-cholesterol appear to *protect* the artery from atherosclerosis. The reasons for this are not well understood, but it has been suggested that the HDL carriers may transport cholesterol away from the artery.

The concentration of HDL-cholesterol appears to be higher, and the risk of atherosclerosis lower, in people who exercise regularly. The HDL-cholesterol concentration, for example, is higher in marathon runners than in joggers and is higher in joggers than in inactive men. Women in general have higher HDL-cholesterol concentrations and a lower risk of atherosclerosis than men. This is believed to be due to a protective effect of estrogen.

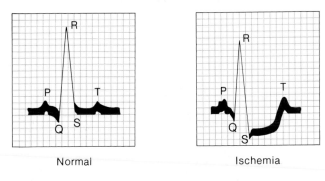

Figure 11.30 Depression of the S-T segment of the electrocardiogram as a result of myocardial ischemia.

Ischemic Heart Disease

A tissue is said to be *ischemic* when it receives an inadequate supply of oxygen because of inadequate blood flow. The most common cause of myocardial ischemia is atherosclerosis of the coronary vessels that supply blood to the heart muscles (the coronary circulation will be described in chapter 13). Because of its contribution to heart disease, and to cerebrovascular accident (CVA, or stroke) when the cerebral vessels are involved, atherosclerosis is the leading cause of death in the United States.

The adequacy of blood flow to a tissue is relative—it depends on the metabolic requirements of the tissue for oxygen. An obstruction in a coronary artery, for example, may allow sufficient blood flow at rest but may produce ischemia when the heart is stressed by exercise or emotional conditions.

Myocardial ischemia is associated with increased concentrations of lactic acid in the blood, produced by anaerobic respiration of the ischemic tissue. This condition often causes substernal pain that may also be referred to the left shoulder and arm areas—a referred pain called **angina pectoris.** People with angina frequently take nitroglycerin or related drugs that help to relieve the ischemia and the pain. These drugs are effective because they stimulate vasodilation, which improves circulation to the heart and decreases the work that the heart must perform to eject blood into the arteries.

Myocardial cells are adapted to respire aerobically, as described in chapter 13, and are unable to respire anaerobically for more than a few minutes. If ischemia and anaerobic respiration continue for more than a few minutes, *necrosis* (cellular death) occurs in the cells furthest removed from the source of blood and oxygen. A sudden, irreversible ischemic injury of this kind is termed a **myocardial infarction,** or MI.

Myocardial ischemia may be detected by changes in the S-T segment of the electrocardiogram (see figure 11.30). The diagnosis of myocardial infarction is aided by

Table 11.5 Changes in enzyme activity in plasma following a myocardial infarction.

Serum Enzyme	Earliest Increase (hr)	Maximum Concentration (hr)	Return to Normal (days)	Amplitude of Increase (× normal)
Creatine phosphokinase	3–6	24–36	3	7
Malate dehydrogenase	4–6	24–48	5	4
Glutamate oxaloacetate transaminase	6–8	24–48	4–6	5
Lactate dehydrogenase	10–12	48–72	11	3
α-Hydroxybutyric dehydrogenase	10–12	48–72	13	3–4
Aldolase	6–8	24–48	4	4

Source: From Montgomery, Rex, Dryer, Robert L., Conway, Thomas W., and Spector, Arthur A.: Biochemistry: a case-oriented approach, ed. 4, St. Louis, 1983, The C. V. Mosby Co.

measurement of the concentration of enzymes in the blood that are released by the infarcted tissue. Plasma concentrations of *creatine phosphokinase (CPK),* for example, increase within three to six hours after the onset of symptoms and return to normal after three days. Plasma levels of *lactate dehydrogenase (LDH)* reach a peak within forty-eight to seventy-two hours after the onset of symptoms and remain elevated for about eleven days.

1. Describe the structure of an atheroma, and explain how it can cause heart disease and cerebrovascular accident (stroke).
2. Explain how cholesterol is carried in the plasma and how the concentrations of cholesterol carriers are related to the risk of developing atherosclerosis.
3. Explain how angina pectoris is produced and the relationship of this symptom to conditions in the heart.

Summary

Blood Vessel and Heart Structure

I. Arteries carry blood away from the heart; veins return blood to the heart.
   A. Both arteries and veins contain three layers of tissues.
      1. The tunica adventitia is the outermost layer of connective tissue.
      2. The tunica media consists of smooth muscle.
      3. The tunica intima consists of a small connective tissue layer and an innermost epithelial lining, called the endothelium.
   B. Large arteries are elastic; medium-sized and small arteries are muscular.
      1. Elastic arteries expand as the blood pressure rises and then recoil as the pressure drops.
      2. Muscular arteries present a resistance to blood flow.
   C. Veins have less muscle than do arteries, and are able to expand as they are filled with blood.
      1. Veins thus have a reservoir function and are called capacitance vessels.
      2. The amount of blood returned to the heart depends, in part, on the squeezing of veins by contracting skeletal muscles—this is the "skeletal muscle pump."
      3. The presence of venous valves helps to insure a one-way flow of blood to the heart.
   D. The smallest branches of arteries are called arterioles; the smallest veins are called venules.

II. Capillaries are the smallest and most numerous type of blood vessels.
   A. Capillaries usually receive blood from arterioles and empty blood into venules.
   B. Each capillary is only one endothelial cell layer thick.
      1. Exchanges of fluid, nutrients, and wastes between blood and tissue fluid occur across the walls of capillaries.
      2. Such exchange may occur between adjacent endothelial cells or through the cells.

III. The heart consists of four chambers.
   A. The right and left ventricles pump blood into arteries.
      1. The right ventricle pumps blood into the pulmonary artery; this begins the pulmonary circulation.

2. The left ventricle pumps into the aorta; this begins the systemic circulation.

B. Blood from the lungs, which is rich in oxygen, returns to the left atrium; this completes the pulmonary circulation.

C. Blood from the body tissues returns via the vena cavae to the right atrium; this completes the systemic circulation.

D. A septum (wall) separates the right and left sides of the heart, so that oxygen-rich blood on the left does not normally mix with oxygen-poor blood on the right side.

The Cardiac Cycle

I. The relaxation phase of the ventricles is called diastole; the phase of contraction is called systole.

A. Contraction of the atria (atrial systole) occurs at the end of ventricular diastole.

1. Venous blood fills the atria while they are at rest.

2. Blood from the atria enters the ventricles during the entire diastolic phase, so that the ventricles are almost 80 percent filled by the time the atria contract.

B. Contraction of the ventricles raises the intraventricular pressure.

1. This rise in pressure causes the atrioventricular (A-V) valves—tricuspid on the right and bicuspid or mitral on the left—to close.

2. Closing of the A-V valves prevents blood from entering the atria.

3. The rise in intraventricular pressure also causes the semilunar valves that guard the entrance to the aorta and to the pulmonary trunk to open.

4. The ventricles normally eject about two-thirds of their end-diastolic volume at systole.

C. As the ventricles relax at the beginning of diastole, the intraventricular pressure falls below the pressure in the arteries.

1. This causes the semilunar valves to close.

2. In this way, blood from the arteries does not normally enter the ventricles.

II. Closing of the A-V and semilunar valves produces sounds that can be heard in the chest.

A. The first sound, or ''lub,'' is produced by closing of the A-V valves.

B. The second sound, or ''dub,'' is produced by closing of the semilunar valves.

C. The first sound thus occurs at the beginning of systole, while the second sound occurs at the beginning of diastole.

D. The period of diastole—as indicated by the time between dub and lub—shortens as the heart beats faster.

1. At a resting cardiac rate of 75 beats per minute, the length of diastole is 0.5 second.

2. At a cardiac rate of 180 beats per minute, diastole is shortened to 0.13 second.

III. Abnormal heart sounds are called murmurs.

A. Murmurs may be produced by defective heart valves and by septal defects.

B. One cause of a septal defect is a patent (open) foramen ovale; a patent ductus arteriosus can also cause murmurs.

1. The foramen ovale and ductus arteriosus are shunts normally present in fetuses that divert blood from the pulmonary to the systemic circulations.

2. These shunts normally close by birth.

Electrical Activity of the Heart

I. The sinoatrial (S-A) node, found in the right atrium, is the normal pacemaker of the heart.

A. This area of tissue undergoes spontaneous depolarization during diastole at a faster rate than do other areas of the heart.

1. If other areas assume the role of pacemakers, they are called ectopic pacemakers or ectopic foci.

2. Impulses originating in the S-A node spread through the right and left atria by conduction across the intercalated discs.

B. Impulses are conducted into the ventricles by specialized conducting tissue.

A. The atrioventricular (A-V) node is located near the junction of the interventricular septum with the atria.

1. This is the slowest-conducting tissue.

2. Impulses from the A-V node are transmitted to the bundle of His.

3. Branches of the bundle of His conduct the impulses along the interventricular septum to the apex of the heart.

4. Impulses are transmitted from the bundle of His into the ventricular muscle by Purkinje fibers.

C. Because the myocardial cells have long action potentials, and correspondingly long refractory periods, the conduction of impulses normally stops after the chambers have been stimulated.

1. If some cells finish their refractory period before others, the impulse may reenter the tissue and be continuously conducted.

2. This abnormal pattern of conduction is called circus rhythms.

3. Circus rhythms can cause flutter (heart rate from 200–300 per minute) or fibrillation (uncoordinated contraction with no pumping action).

II. Electrodes placed on the surface of the body can record the electrical activity of the heart.

A. This recording is called an electrocardiogram.

B. The ECG recording can be used to measure the heart rate and the P-R interval.

1. A slow heart rate is called bradycardia; a fast rate is tachycardia.

2. An abnormally long P-R interval can indicate damage to the A-V node; this is called heart block.

C. The ECG recording can also detect flutter, fibrillation, and other problems associated with abnormal electrical activity.

D. Since the QRS wave is produced by ventricular depolarization, it appears at about the same time the ventricles contract and produce the first heart sound.

E. Ventricular repolarization, represented by the T wave, occurs while the ventricles are relaxing and the second heart sound is produced.

Atherosclerosis and Ischemic Heart Disease

I. Atherosclerosis is a type of arteriosclerosis (hardening of the arteries) in which growths, or atheromas, protrude into the lumen of arteries.

A. Atheromas are composed of smooth muscle cells, cholesterol, and other lipids; they can become calcified.

B. Atheromas occlude the lumen of arteries and serve as sites for the formation of thrombi (clots), which can further occlude the lumen and reduce blood flow.

C. Reduced blood flow thus produced in the heart and in the cerebral circulation is the leading cause of death in the United States.

II. The risk of developing atherosclerosis is related, in part, to the relative concentrations of plasma protein carriers of cholesterol.

A. A high concentration of HDL (high-density lipoproteins) cholesterol compared to LDL cholesterol is beneficial.

1. The HDL carriers appear to confer some protection against atherosclerosis.

2. Women and people who exercise regularly have higher ratios of HDL/LDL carriers and a lower risk of atherosclerosis.

III. Ischemia refers to inadequate blood supply to an organ.

A. Myocardial ischemia may be produced by occlusion of coronary blood flow by atherosclerotic plaques.

B. The ischemic heart tissue must respire anaerobically.

1. This results in liberation of lactic acid and certain characteristic enzymes into the blood, as well as in characteristic changes in the ECG pattern.

2. Myocardial ischemia is associated with a referred pain called angina pectoris.

C. If the ischemia continues, areas of cells may die and produce an infarct.

D. Myocardial infarction can damage different parts of the heart; circus rhythms and fibrillation may result.

Self-Study Quiz

1. Which of the following statements is FALSE?
   (a) Most of the total blood volume is contained in veins.
   (b) Capillaries have a greater total surface area than any other type of vessel.
   (c) Exchanges between blood and tissue fluid occur across the walls of venules.
   (d) Small arteries and arterioles present great resistance to blood flow.

2. All arteries in the body contain oxygen-rich blood with the exception of:
   (a) the aorta
   (b) the pulmonary artery
   (c) the renal artery
   (d) the coronary arteries

3. The ''lub,'' or first heart sound, is produced by closing of:
   (a) the aortic semilunar valve
   (b) the pulmonary semilunar valve
   (c) the tricuspid valve
   (d) the bicuspid valve
   (e) both A-V valves

4. The first heart sound is produced at:
   (a) the beginning of systole
   (b) the end of systole
   (c) the beginning of diastole
   (d) the end of diastole

5. Changes in the cardiac rate primarily reflect changes in the duration of:
   (a) systole
   (b) diastole

6. The QRS wave of an ECG is produced by:
   (a) depolarization of the atria
   (b) repolarization of the atria
   (c) depolarization of the ventricles
   (d) repolarization of the ventricles

7. The second heart sound immediately follows the occurrence of:
   (a) the P wave
   (b) the QRS wave
   (c) the T wave

8. The cells that normally have the fastest rate of spontaneous diastolic depolarization are located in:
   (a) the S-A node
   (b) the A-V node
   (c) the bundle of His
   (d) the Purkinje fibers

9. Which of the following statements is TRUE?
   (a) The heart can produce a graded contraction.
   (b) The heart can produce a sustained contraction.
   (c) The action potentials produced at each cardiac cycle normally travel around the heart in circus rhythms.
   (d) All of the myocardial cells in the ventricles are normally in a refractory period at the same time.

10. An ischemic injury to the heart that results in death of some myocardial cells is:
    (a) angina pectoris
    (b) a myocardial infarction
    (c) fibrillation
    (d) heart block

12

Cardiac Output, Blood Flow, and Blood Pressure

Objectives

By studying this chapter, you should be able to:

1. Define cardiac output and describe how it is affected by cardiac rate and stroke volume
2. Explain the action of autonomic nerves on cardiac rate and strength of ventricular contraction
3. Explain the relationship between end-diastolic volume, myocardial stretch, and the strength of ventricular contraction (Frank-Starling law)
4. Describe the factors that affect the venous return of blood to the heart
5. Explain how tissue fluid is formed and how it is returned to the capillary blood
6. Describe the relationship between the lymphatic vessels and the vascular system
7. Explain how antidiuretic hormone helps to regulate the blood volume, blood osmolality, and the blood pressure

8. Explain the role of aldosterone in the regulation of blood volume and pressure
9. Describe the renin-angiotensin system and its significance in cardiovascular regulation
10. Explain the relationship of blood flow, arterial pressure, and vascular resistance
11. Define total peripheral resistance, and explain how vascular resistance is regulated
12. Describe how resistance and blood flow changes as blood goes from large arteries to capillaries
13. Describe the baroreceptor reflex and explain its significance in blood pressure regulation
14. Explain how the sounds of Korotkoff are produced and how these sounds are used to measure blood pressure
15. Describe how blood pressure is affected by changes in total peripheral resistance, blood volume, and cardiac rate

*G*as exchange across capillaries in the lungs and tissues, delivery of nutrients and removal of wastes, and maintenance of tissue fluid and electrolyte balance in the body are absolutely dependent upon the circulation of the blood. The rate of **blood flow** through an organ, in turn, depends in part on the arterial **blood pressure,** the **total blood volume,** and on the frictional **vascular resistance** to flow. These factors also affect the pumping action of the heart, as measured by the **cardiac output.** The pumping of the heart, in turn, creates the pressure head that acts as the major driving force of the circulation. Mechanisms that regulate cardiac output, total blood volume, and vascular resistance are thus interrelated through their effects on blood flow. These mechanisms and their relationships will be described in this chapter.

Figure 12.1 Effects of sympathetic and parasympathetic activity on the pacemaker function of the S-A node. Sympathetic stimulation causes action potentials to be generated more rapidly, and parasympathetic inhibition causes action potentials to be generated more slowly, than would occur in the absence of autonomic influences *(shaded areas).*

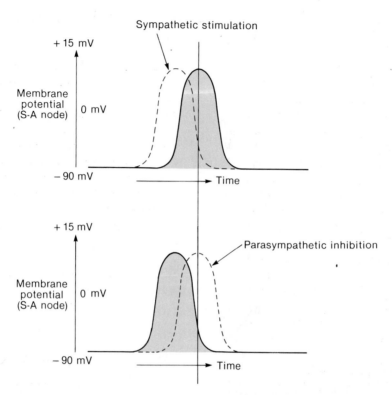

Cardiac Output

The cardiac output is equal to the milliliters of blood pumped per minute by each ventricle. The average resting *cardiac rate* is 70 beats per minute; the average *stroke volume* (milliliters of blood pumped per beat by each ventricle) is 70–80 milliliters per beat. The product of these two variables gives an average cardiac output of 5.5 L per minute:

$$\text{Cardiac output} = \text{Stroke volume} \times \text{Cardiac rate}$$
$$\frac{ml}{min} = \frac{ml}{beat} \times \frac{beats}{min}$$

The *total blood volume* is also equal to about 5–6 liters. This implies that the heart pumps the equivalent of the total blood volume each minute under resting conditions. An increase in cardiac output, as occurs during exercise, must thus be accompanied by an increased rate of blood flow through the circulation. This is accomplished, in part, by changes in factors that regulate the cardiac rate and stroke volume.

Regulation of Cardiac Rate

In the complete absence of neural influences, the heart will continue to beat according to the automatic rhythm set by the S-A node pacemaker. As described in chapter 11, this automatic rhythm is produced by the spontaneous decay of the resting membrane potential in the pacemaker cells to a threshold value. When this automatic depolarization reaches a threshold level, voltage-regulated ionic gates are opened; action potentials are produced and contraction is stimulated.

Normally, however, sympathetic and vagus (parasympathetic) nerve fibers to the heart are continuously active. Norepinephrine released from sympathetic nerve endings, as well as other catecholamines (such as epinephrine secreted by the adrenal medulla), stimulates an increase in the spontaneous rate of firing of the S-A node. Acetylcholine released from parasympathetic endings hyperpolarizes the S-A node and thus decreases the rate of its spontaneous firing. The actual pace set by the S-A node at any time depends on the net effect of these antagonistic influences.

Table 12.1 Effects of autonomic nerve activity on the heart.

Region Affected	Sympathetic Nerve Effects	Parasympathetic Nerve Effects
S-A node	Increased rate of diastolic depolarization; increased cardiac rate	Decreased rate of diastolic depolarization; decreased cardiac rate
A-V node	Increased conduction rate	Decreased conduction rate
Atrial muscle	Increased strength of contraction	Decreased strength of contraction
Ventricular muscle	Increased strength of contraction	No significant effect

Autonomic innervation of the S-A node represents the major means by which cardiac rate is regulated. Autonomic activity does, however, affect cardiac rate to some degree by other mechanisms as well. Sympathetic endings in the musculature of the atria and ventricles, for example, increase the strength of contraction and decrease slightly the time spent in systole at high cardiac rates (from 0.3 second at 75 beats per minute to 0.2 second at 180 beats per minute). Autonomic fibers also innervate the A-V node (the slowest conducting tissue in the heart): sympathetic fibers increase—and parasympathetic fibers decrease—its conduction rate.

Changes in cardiac rate result from changes in autonomic nerve activity. In exercise, for example, the cardiac rate increases as a result of decreased vagus nerve inhibition and increased sympathetic stimulation. The resting bradycardia of endurance-trained athletes (described in chapter 11) is due to increased vagus and decreased sympathetic nerve activity. These changes are coordinated by the **cardiac control centers** in the *medulla oblongata* of the brain stem. There may be separate cardioaccelerator and cardioinhibitory centers (this is currently controversial) in the medulla that are influenced by higher brain centers and by sensory feedback from pressure receptors *(baroreceptors)*. The reflex control of cardiac rate (illustrated in figure 12.19) will be discussed in conjunction with blood pressure regulation in a later section.

Regulation of Stroke Volume

The stroke volume, or amount of blood ejected by each ventricle per beat, is regulated by three variables. These are (1) the *end-diastolic volume (EDV),* which is the volume of blood in the ventricles at the end of diastole just prior to contraction; (2) the average or mean aortic blood pressure, which is essentially the same as the *mean arterial pressure;* and (3) the *contractility,* or strength of ventricular contraction.

The amount of blood in the ventricles available to be pumped—the EDV—is the work load imposed on the ventricular muscle prior to contraction. The EDV is thus often called a **preload.** In order to eject this blood, the pressure generated by ventricular contraction must exceed the aortic pressure, as measured by the mean arterial pressure. The arterial pressure thus represents an impedance to the ejection of blood from the ventricles, or an **afterload** of work imposed on the ventricles after contraction has begun.

Stroke volume is thus inversely proportional to the afterload; an increase in mean arterial pressure produces a decrease in stroke volume. This effect can, however, be at least partially compensated by an increased strength of ventricular contraction. Stroke volume is also increased when there is an increase in EDV (preload):

$$\text{Stroke volume } \alpha \begin{cases} \text{End diastolic volume (preload)} \\ \text{Mean arterial pressure (afterload)} \\ \text{Contraction strength (contractility)} \end{cases}$$

α means "is proportional to"

The proportion of the EDV that is ejected against a given afterload depends upon the strength of ventricular contraction. Normally, contraction strength is sufficient to eject 70–80 ml of blood out of a total end-diastolic volume of 110–130 ml. The *ejection fraction* is thus about two-thirds, or 0.67. Amazingly, this fraction remains relatively constant as the EDV increases. This means that the stroke volume must increase proportionately with the increase in EDV, which implies that the strength of ventricular contraction must correspondingly increase.

Intrinsic Control of Contractility Two physiologists, Frank and Starling, demonstrated that the strength of ventricular contraction (contractility) varies directly with the end-diastolic volume. Even in experiments where the heart is removed from the body (and thus is not subject to neural or hormonal regulation), and where the heart is filled with blood flowing from a reservoir, an increase in EDV results in increased contraction strength, and therefore in increased stroke volume. This relationship between EDV, contraction strength, and stroke volume is thus a built-in or *intrinsic* property of heart muscle.

Figure 12.2 The Frank-Starling mechanism (law of the heart). When the heart muscle is subject to increased amounts of stretch it contracts more strongly. As a result of the increased contraction strength (shown as tension), the time required to reach maximum contraction is the same regardless of the degree of stretch.

Resting sarcomere lengths

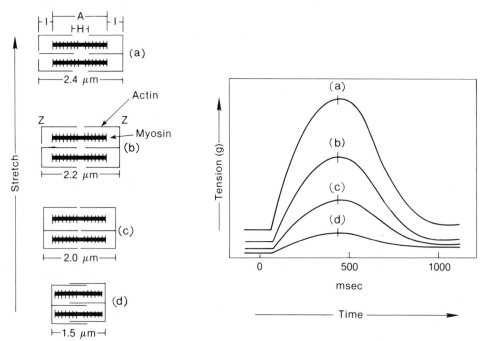

Frank-Starling Law of the Heart   The intrinsic control of contractility and stroke volume is due to variations in the degree to which the myocardium is stretched by the EDV. As the EDV rises, the myocardium is increasingly stretched and as a result, contracts more forcefully. This mechanism is known as the **Frank-Starling law of the heart.**

Stretch can also improve the contractility of skeletal muscle. The resting length of skeletal muscles, however, is close to ideal, so that significant stretching decreases contraction strength (see chapter 8). This is not true of the heart. Prior to filling with blood during diastole, the sarcomere lengths of myocardial cells are only about 1.5 μm. At this length the actin filaments from each side overlap in the middle, and the cells can contract only weakly.

As the ventricles fill with blood the myocardium stretches, so that the actin filaments overlap with myosin only at the edges of the A bands (see figure 12.2). This allows more force to be developed during contraction. Since this more advantageous overlapping of actin and myosin is produced by stretching of the ventricles, and since the degree of stretching is controlled by the degree of filling (the end-diastolic volume), the strength of contraction is intrinsically adjusted so that more blood will be ejected as the EDV rises.

Figure 12.3 Relationship between stroke volume and pressure in the aorta (afterload). As the aortic pressure increases, stroke volume decreases (shown by the negative slope of the graph lines). At any particular aortic pressure, however, stroke volume increases as the end-diastolic volume (EDV)—the "preload," or amount of stretch—increases. The strength of contraction, and thus the stroke volume, can be further increased by sympathetic nerve stimulation (see text for details).

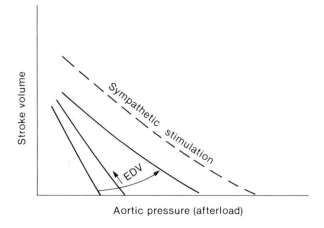

Figure 12.4 Regulation of cardiac output. Factors that stimulate cardiac output are shown as solid arrows; factors that inhibit cardiac output are shown as dotted arrows.

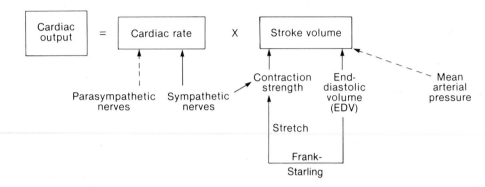

Extrinsic Control of Contractility   At any given degree of stretch, the strength of ventricular contraction depends on the activity of the sympathoadrenal system. Norepinephrine from sympathetic endings and epinephrine from the adrenal medulla produce a catecholamine-induced increase in contraction strength. This is often called a *positive inotropic* effect of catecholamines. The stroke volume is thus directly proportional to the end-diastolic volume and degree of sympathoadrenal stimulation; it is inversely proportional to the afterload (mean arterial pressure)—see figures 12.3 and 12.4.

The cardiac output is thus affected in two ways by activity of the sympathoadrenal system: the positive inotropic effect on contraction strength that increases stroke volume, and the *positive chronotropic effect* (chrono = time) that increases cardiac rate. The parasympathetic system has a *negative chronotropic effect* (decreases cardiac rate), but does not affect the contraction strength of the ventricles. These extrinsic controls are superimposed on the intrinsic control of cardiac rate by the S-A node and the intrinsic control of the stroke volume by the Frank-Starling mechanism.

Venous Return

The end-diastolic volume, and thus the stroke volume and cardiac output, is controlled by factors that affect the **venous return** of blood to the heart. The rate at which the atria are filled with venous blood depends on the total blood volume (discussed in the next section) and the venous pressure, which serves as the driving force for the return of blood to the heart.

Veins have relatively thinner, less muscular walls than do arteries and thus have a higher *compliance*. This means that a given amount of pressure will cause more distension (expansion) in veins than in arteries, so that the veins can hold more blood. Approximately two-thirds of the total blood volume is located in the veins. Veins are therefore called *capacitance vessels,* by analogy with capacitors in electronics that can accumulate electric charges. Muscular arteries and arterioles expand less under pressure (are less compliant) and thus are called *resistance vessels.*

Although veins contain almost 70 percent of the total blood volume, the mean venous pressure is only 2 mm Hg, compared to a much higher mean arterial pressure of 90–100 mm Hg. The lower venous pressure is due in part to a pressure drop between arteries and capillaries (discussed in a later section) and in part to the high venous compliance.

The venous pressure is highest in the venules (10 mm Hg) and lowest at the junction of the vena cavae with the right atrium (0 mm Hg). In addition to this pressure difference, the venous return to the heart is increased by (1) sympathetic nerve activity; (2) the skeletal muscle pump, which constricts veins during skeletal muscle contraction (described in chapter 11); and (3) the pressure difference between the thoracic (chest) and abdominal cavities.

Sympathetic stimulation of the veins causes some contraction of the smooth muscle walls in veins. This does not cause much constriction of the lumen of these vessels (in contrast to the response of muscular arteries and arterioles to sympathetic stimulation); it does, however, make the veins less compliant. This reduced venous compliance helps to raise venous pressure and aid the return of venous blood to the heart.

Figure 12.5 Variables that affect venous return and thus end-diastolic volume. Direct relationships are indicated by solid arrows; inverse relationships are shown with dashed arrows.

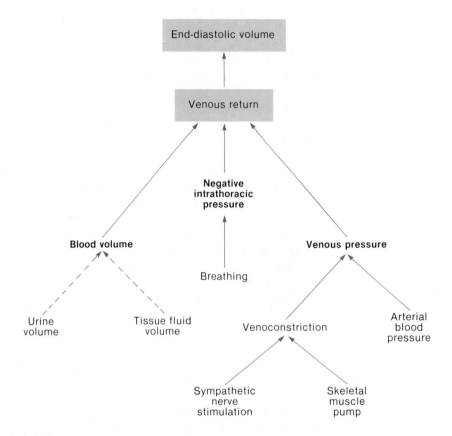

Contraction of skeletal muscles functions as a "pump" by virtue of its squeezing action on veins, as described in chapter 11. Contraction of the diaphragm (a muscular sheet separating the thoracic and abdominal cavities) during inhalation also improves venous return. This results from the fact that, as the diaphragm contracts, it lowers to increase the thoracic volume and decrease the abdominal volume. This creates a partial vacuum in the thoracic cavity and a higher pressure in the abdominal cavity; the pressure difference thus produced favors blood flow from abdominal to thoracic veins. A summary of the factors affecting venous return, and therefore affecting end-diastolic volume, is shown in figure 12.5.

1. Draw a flow chart showing how end-diastolic volume, myocardial stretch, contractility, and stroke volume are related. Explain this relationship.
2. Explain how cardiac output is influenced by autonomic nerve innervation of the heart. Include chronotropic and inotropic effects in your answer. Also, explain the effects of sympathetic stimulation of veins on cardiac output.
3. During physical exercise the venous return of blood to the heart is increased. Describe three mechanisms that contribute to this increased venous return.
4. Define the terms *preload* and *afterload.* Explain how these concepts might relate to the function of the right ventricle in a person who has high blood pressure in the pulmonary circulation.
5. Draw a flow chart showing how an increase in venous return can result in an increase in cardiac output.

Figure 12.6 The daily intake and excretion of body water and its distribution between different intracellular and extracellular compartments.

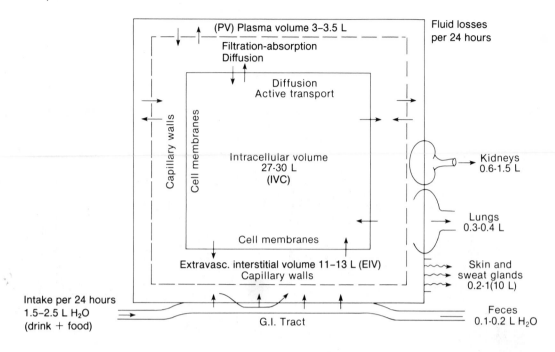

Blood Volume

The end-diastolic volume, and thus the stroke volume and cardiac output, are influenced in part by the total blood volume. This volume represents one part, or "compartment," of the total body water. Approximately two-thirds of the total body water is located within cells—in the *intracellular compartment*. The remaining one-third is in the *extracellular compartment*. This extracellular fluid is normally distributed so that about 80 percent is in the tissues—as *tissue*, or *interstitial*, *fluid*—and 20 percent is in the blood *plasma*.

The distribution of water—and thus the relative volumes—of these compartments is determined by a balance of opposing forces. Blood pressure, for example, promotes formation of tissue fluid at the expense of plasma, while osmotic forces draw water from tissues into the vascular system. The total volume of intracellular and extracellular fluid is normally maintained constant by a balance between water loss (mainly as urine) and water gain. Mechanisms that affect drinking, urine volume, and the distribution of water between plasma and tissue fluid thus help regulate blood volume, and by this means help regulate cardiac output and blood flow.

Exchange of Fluid between Capillaries and Tissues

The distribution of extracellular fluid between the plasma and interstitial compartments is in a state of dynamic equilibrium. Tissue fluid is not normally a "stagnant pond," but is rather a continuously circulating medium, formed from and returning to the vascular system. In this way, the tissue cells receive a continuously fresh supply of glucose and other plasma solutes that are filtered through tiny endothelial channels in the capillary wall.

Filtration results from the blood pressure within the capillaries. This hydrostatic pressure, which is exerted against the inner capillary wall, is equal to about 30 mm Hg at the arteriolar end of systemic capillaries. As a result of fluid loss through filtration, the pressure decreases to 10–15 mm Hg at the venular ends of the capillaries. This results in an average capillary blood pressure (near the middle) of 17 mm Hg. It should be noted that the values of these pressures are experimentally derived and not to be considered absolute.

The **net filtration pressure** is equal to the pressure difference between the inside and the outside of the capillary wall. If the hydrostatic pressure of tissue fluid were equal to the capillary blood pressure, as an extreme example,

Figure 12.7   Tissue or interstitial fluid is formed by filtration as a result of blood pressure at the arteriolar ends of capillaries and returned to the venular ends of capillaries by the colloid osmotic pressure of plasma proteins (see text for complete description).

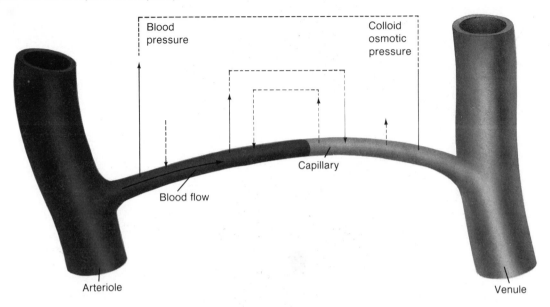

Blood pressure

Colloid osmotic pressure

Capillary

Blood flow

Arteriole

Venule

there would be no filtration. The normal value of the tissue hydrostatic pressure is controversial currently. Most investigators believe that this value is actually a *negative pressure,* meaning that it is lower than the pressure of the atmosphere. If the tissue fluid hydrostatic pressure is −6.5 mm Hg, as believed by some authorities, the net filtration pressure will be equal to 17 mm Hg minus (−6.5 mm Hg), or 23.5 mm Hg.

Glucose, comparably sized organic molecules, inorganic salts, and ions are filtered along with water through the capillary channels. The concentration of these substances in tissue fluid is thus the same as in plasma. The protein concentration of tissue fluid, however, is only about 2 g per 100 ml, which is significantly less than the protein concentration of plasma (6–8 g per 100 ml). This difference is due to the fact that filtration of proteins is restricted by the capillary pores. The osmotic pressure exerted by plasma proteins—called the **colloid osmotic pressure** or **oncotic pressure** of the plasma—is therefore greater than the colloid osmotic pressure of tissue fluid.

The colloid osmotic pressure of plasma averages 28 mm Hg, while that of tissue fluid averages 5 mm Ng. The difference between these two values (23 mm Hg) represents the net pressure favoring osmosis of water into the capillaries. If these values are correct, this pressure is lower than the filtration pressure at the arteriolar ends of capillaries, and greater than the filtration pressure at the venular ends of capillaries. Fluid and solutes would thus be filtered out of the capillaries at the arteriolar ends, while fluid would be returned by osmosis to the venular ends of the capillaries.

This "classic" view of capillary dynamics has recently been challenged by some investigators who believe that the net filtration force exceeds the osmotic return of water, even at the venular ends of capillaries that are open. It should be recalled that capillary beds can be opened and closed by the action of precapillary sphincter muscles. The osmotic return of water, according to this view, exceeds filtration when capillaries are closed. Net filtration, in summary, would occur in open capillaries, while net absorption of water would occur in closed capillaries.

By either proposed mechanism, plasma and tissue fluid would be continuously interchanged through the action of filtration and colloid osmotic pressure forces. The return of fluid to the vascular system by osmosis, however, does not exactly equal the amount filtered. According to some estimates, 20 L of filtrate are formed per day, while only about 18 L are returned to the capillaries by osmosis. The difference—about 2 L per day—is ultimately returned to the vascular system by way of *lymphatic vessels.*

Figure 12.8 The lymphatic system returns excess tissue fluid to the venous system.

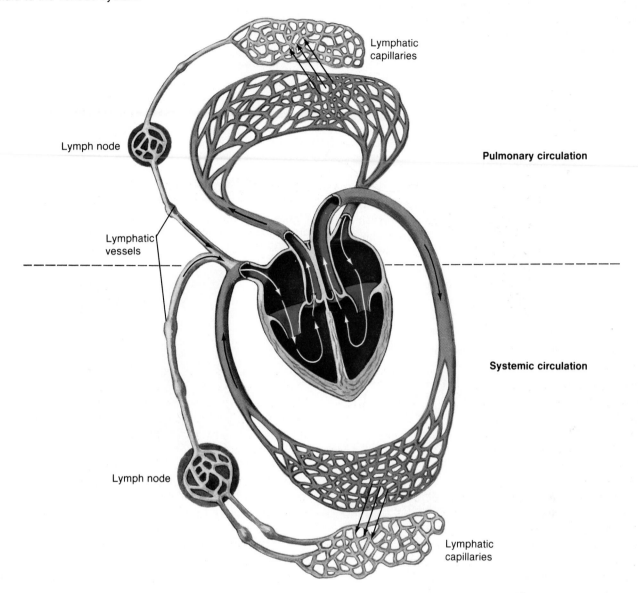

Lymphatic capillaries

Lymph node

Pulmonary circulation

Lymphatic vessels

Systemic circulation

Lymph node

Lymphatic capillaries

Lymphatic System and Edema

Excessive accumulation of tissue fluid and filtered proteins is normally prevented by drainage into highly permeable, blind-ending lymphatic capillaries located in the connective tissues. Tissue fluid that enters these lymphatic capillaries is called *lymph*, and it is drained into *lymphatic vessels*. These vessels, like veins, have one-way valves, and skeletal muscle contraction serves as a pump by squeezing lymphatic vessels as well as veins. In addition to the pressure generated by the skeletal muscle pump, lymphatic vessels display automatic, rhythmic contraction that helps to move the lymph.

Lymph is eventually transported to the *right lymphatic duct* and the *thoracic duct*, which drain into the large veins (see figure 12.8). It is estimated that this system returns approximately 2 L of lymph per day to the vascular system. As lymph passes through this system it percolates through *lymph nodes*, which contain antibody-producing lymphocytes (a type of white blood cell). The functions of lymph nodes and white blood cells will be described in detail in chapter 17.

Table 12.2   The causes of edema.

Variables That Cause Edema	Effects
Increased blood pressure	Increases filtration pressure so that more tissue fluid is formed at the arteriolar ends of capillaries.
Increased tissue protein concentration	Decreases osmosis of water into the venular ends of capillaries. Usually a localized tissue edema due to leakage of plasma proteins through capillaries during inflammation and allergic reactions.
Decreased plasma protein concentration	Decreases osmosis of water into the venular ends of capillaries. May be caused by liver disease (which can be associated with insufficient plasma protein production), kidney disease (due to leakage of plasma protein into urine), or protein malnutrition.
Obstruction of lymphatic vessels	Infections by filaria roundworms (nematodes) transmitted by a certain species of mosquito block lymphatic drainage, causing edema and tremendous swelling of the affected areas (a condition called elephantiasis).

Causes of Edema   Excessive accumulation of tissue fluid is known as **edema.** This condition is normally prevented by a proper balance between capillary filtration and osmotic uptake of water and by proper lymphatic drainage. Edema may thus result from (1) *high blood pressure,* which increases capillary pressure and causes excessive filtration; (2) *leakage of plasma proteins into tissue fluid,* which causes reduced osmotic flow of water into the capillaries (this occurs during inflammation and allergic reactions as a result of increased capillary permeability); (3) *decreased plasma protein concentration* as a result of liver disease (the liver makes most of the plasma proteins), or as a result of kidney disease in which proteins are excreted in the urine; and (4) *obstruction of the lymphatic drainage.*

In the tropical disease *filariasis,* mosquitoes transmit a nematode worm parasite to humans. The larvae of these worms invade lymphatic vessels and block lymphatic drainage. The edema that results can be so severe (particularly in the legs), that the tissues swell to produce an elephantlike appearance—a condition aptly termed *elephantiasis.*

Regulation of Blood Volume by the Kidneys

The formation of urine begins in the same manner as formation of tissue fluid—by filtration of plasma through capillary channels. The kidneys produce about 180 L per day of blood filtrate, but since there is only 5.5 L of blood in the body, it is clear that most of this filtrate must be returned to the vascular system and recycled. Only about 1–2 L per day of urine is excreted; 98–99 percent of the amount filtered is *reabsorbed* back into the vascular system.

Figure 12.9   Parasitic larvae that block lymphatic drainage produce tissue edema, resulting in elephantiasis.

The volume of urine excreted can be varied by changes in the reabsorption of filtrate. If 99 percent of the filtrate is reabsorbed for example, 1 percent must be excreted; decreasing the reabsorption by only one percent doubles the urine volume. Such changes can have a profound effect on total blood volume. The percent of the kidney filtrate reabsorbed—and thus the urine and blood volumes—is regulated by hormones. While kidney physiology will be discussed in detail in chapter 15, some information about renal regulation of blood volume is required for a proper understanding of cardiovascular function.

Figure 12.10 Role of antidiuretic hormone (ADH) in the regulation of blood osmolality and volume. Both dehydration and salt ingestion stimulate drinking and increased water retention. In dehydration, this provides a negative feedback mechanism (shown by dashed arrows) that helps to counter the loss of blood volume. As a result of salt ingestion, this same mechanism produces a rise in blood volume.

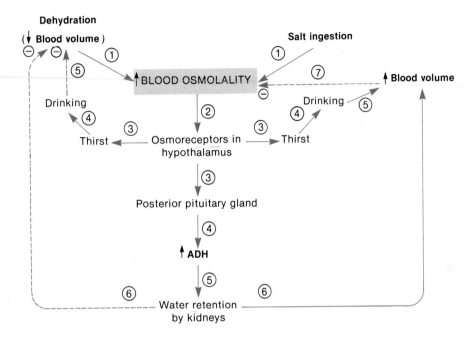

Regulation by Antidiuretic Hormone (ADH)   One of the major hormones involved in regulation of blood volume is **antidiuretic hormone (ADH),** also known as **vasopressin.** This hormone is produced by neurons in the hypothalamus, transported by axons into the posterior pituitary gland and released from this storage gland in response to hypothalamic stimulation. Secretion of ADH from the posterior pituitary occurs when neurons called *osmoreceptors* in the hypothalamus detect an increase in plasma osmolality or osmotic pressure (indices of plasma concentration, as described in chapter 6).

An increase in plasma osmolality occurs when the plasma becomes more concentrated. This can be produced either by *dehydration* or by *excessive salt intake.* Stimulation of osmoreceptors produces sensations of thirst (leading to increased water intake) and increased secretions of ADH from the posterior pituitary. Through mechanisms that will be discussed in chapter 15, ADH causes increased reabsorption of filtrate. A smaller volume of urine is thus excreted.

A person who is dehydrated, or who eats excessive amounts of salt, thus drinks more and urinates less. This raises the blood volume and in the process, dilutes the plasma to lower its previously elevated osmolality. The raised blood volume is, in a sense, a by-product of the negative feedback loop that acts to maintain a relatively constant plasma osmolality. The rise in blood volume that results from this mechanism is extremely important in a dehydrated person with low blood volume and pressure. In a person who eats excessive amounts of salt, however, the increased blood volume could produce an abnormally high blood pressure (as will be described in a later section of this chapter).

Drinking excessive amounts of water, without eating excessive amounts of salt, does not result in high blood pressure. The water does enter the blood from the intestine, and momentarily raises the blood volume; at the same time, however, it dilutes the blood. This decreases the plasma osmolality and thus inhibits ADH secretion. With less ADH there is less reabsorption of filtrate in the kidneys—a larger volume of urine is excreted. Water is therefore a *diuretic*—a substance that promotes urine formation—because it inhibits secretion of antidiuretic hormone.

Figure 12.11 The Renin-angiotensin-aldosterone system. Negative feedback effects are shown with dashed arrows. Numbers indicate the sequence of cause-and-effect steps.

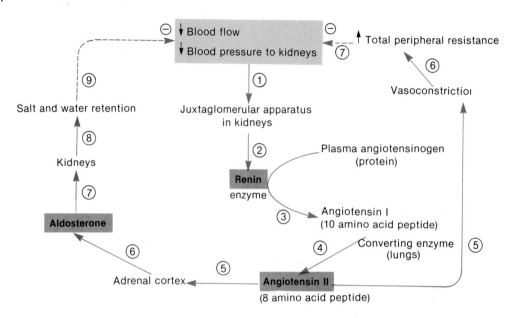

Regulation by Aldosterone   Since Na^+ and Cl^- are easily filtered in the kidneys, a mechanism must exist to promote reabsorption and retention of salt when the dietary salt intake is too low. Throughout most of human history, salt was in short supply and thus was highly valued. Moorish merchants in the sixth century traded an ounce of salt for an ounce of gold; salt cakes were used as money in Abyssinia; part of the Roman soldier's pay was given in salt—it is from this practice that the word *salary* (sal = salt) is derived. The phrase "worth his salt" is derived from the fact that Greeks and Romans sometimes purchased slaves with salt.

Aldosterone, a steroid hormone secreted by the adrenal cortex (the outer part of the adrenal glands), stimulates retention of salt by the kidneys. Aldosterone is thus a "salt-retaining hormone." Retention of salt, however, indirectly promotes retention of water (in part, by the action of ADH as previously discussed). The action of aldosterone thus causes an increase in blood volume, but—unlike the effects of ADH—does not cause a change in plasma osmolality. Secretion of aldosterone is stimulated during salt deprivation, when the blood volume and pressure are reduced. The adrenal cortex is not directly stimulated by these conditions, however; an intermediate mechanism is required.

The Renin-Angiotensin System   When the blood flow and pressure are reduced in the renal artery (as they would be in the low blood volume state of salt deprivation), a group of cells in the kidneys called the *juxtaglomerular apparatus* secretes the enzyme *renin* into the blood. This enzyme cleaves a ten-amino-acid polypeptide called *angiotensin I* from a plasma protein known as angiotensinogen. Angiotensin I is biologically inactive, but as it passes through different organs (primarily the lungs) a *converting enzyme* removes two amino acids. This leaves an eight-amino-acid polypeptide known as *angiotensin II*. Conditions of salt deprivation, low blood volume, and low blood pressure, in summary, cause increased production of angiotensin II in the blood.

Angiotensin II has numerous effects that cause a rise in blood pressure. This rise in pressure is due to both constriction of muscular arteries and arterioles (vasoconstriction), and to increases in blood volume. Vasoconstriction is produced directly by the effects of angiotensin II on blood vessels, and is produced indirectly by augmentation of sympathetic nerve activity by angiotensin II (causing increased release of norepinephrine in the blood vessels).

Table 12.3 Estimated distribution of the cardiac output at rest.

Organs	Blood flow	
	Milliliters per minute	Percent Total
Gastrointestinal tract and liver	1,400	24
Kidneys	1,100	19
Brain	750	13
Heart	250	4
Skeletal muscles	1,200	21
Skin	500	9
Other organs	600	10
Total organs	5,800	100

Source: From Wade, O. L., and J. M. Bishop, *Cardiac Output and Regional Blood Flow.* Copyright © 1962 Blackwell Scientific Publications, Ltd., England. Used with permission.

Angiotensin II promotes a rise in blood volume by means of two mechanisms: (1) thirst centers in the hypothalamus are stimulated by angiotensin II, so that more water is ingested; and (2) secretion of aldosterone by the adrenal cortex is stimulated by angiotensin II, so that more salt and water are retained by the kidneys. The relationship between angiotensin II and aldosterone is sometimes described as the *renin-angiotensin-aldosterone system.*

This mechanism can also work in the opposite direction: high salt intake, leading to high blood volume and pressure, normally causes inhibition of angiotensin II production and aldosterone secretion. With less aldosterone, less salt is retained by the kidneys and more is excreted in the urine. Unfortunately, many people with chronically high blood pressure (hypertension) have normal or even elevated levels of angiotensin II and aldosterone. In these cases, the intake of salt must be lowered to match the impaired ability to excrete salt in the urine.

Summary The total blood volume is influenced by factors that affect the distribution of water between the plasma and tissue fluid, and by factors that regulate excretion of plasma water in the urine. The latter include mechanisms that control the secretion of ADH and mechanisms that control aldosterone secretion. Through these mechanisms, blood volume may be maintained during dehydration and can be elevated as a result of excessive salt ingestion. This in turn influences blood pressure and cardiac output. Cardiac output, blood pressure, and blood flow are also affected by vascular resistance, as described in the next section.

1. Describe the composition of tissue fluid. Using a diagram, illustrate how tissue fluid is formed and how it is returned to the vascular system.
2. Define edema and describe four different mechanisms that can produce this condition.
3. Draw a flow chart to illustrate the effects of dehydration on blood volume and urine volume. How are blood volume, venous return, cardiac output, and blood pressure affected by severe dehydration?
4. Draw a flow chart to illustrate how salt deprivation can lead to increased thirst and increased salt and water retention. Describe how a high-salt diet can contribute to hypertension.

Vascular Resistance and Blood Flow

The amount of blood that the heart pumps per minute is equal to the rate of venous return, and thus is equal to the rate of blood flow through the entire circulation. The cardiac output of 5–6 L per minute, however, is not distributed equally to the different organs. At rest, the blood flow rate is about 2,500 ml per minute through the liver and kidneys, 1,000 ml per minute through the skeletal muscles, 750 ml per minute through the brain, and 250 ml per minute through the heart. These organs thus receive a total of 4.5 L per minute; the balance of the cardiac output (500–1,000 ml per minute) is distributed to the other organs.

Figure 12.12 The flow of blood in the systemic circulation is ultimately dependent on the pressure difference ($\triangle P$) between the origin of the flow (mean pressure of about 100 mm Hg in the aorta) and the end of the circuit—zero mm Hg in the vena cava where it joins the right atrium *(RA)*; *(LA* = left atrium; *RV* = right ventricle; *LV* = left ventricle).

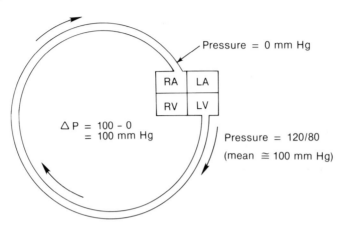

Pressure = 0 mm Hg

$\triangle P = 100 - 0$
$= 100$ mm Hg

Pressure = 120/80
(mean $\cong$ 100 mm Hg)

Physical Laws Describing Blood Flow

The flow of blood through the vascular system, like the flow of any fluid through a tube, depends in part on the difference in pressure at the two ends of the tube. If the pressure at both ends is the same, there will be no flow. If the pressure at one end is greater than at the other, blood (or other fluid) will flow from the region of higher to the region of lower pressure. The rate of flow is proportional to the pressure difference ($P_1 - P_2$) between the two ends of the tube. The term *pressure difference* is abbreviated ΔP, in which the Greek letter Δ *(delta)* means "change in."

If the systemic circulation is pictured as a single tube leading from and back to the heart, blood flow through this system would occur as a result of the pressure difference between the beginning of the tube (the aorta) and the end of the tube (the junction of the vena cavae with the right atrium). The average or mean arterial pressure is 90–100 mm Hg; the pressure at the right atrium is 0 mm Hg. The "pressure head" or "driving pressure" is therefore about 100–0 = 100 mm Hg. Thus:

$$\Delta P = \text{mean arterial pressure} \cong 100 \text{ mm Hg}$$

$$\cong \text{ means "approximately equal to"}$$

Blood flow rate is directly proportional to the pressure difference between the two ends of the tube (ΔP), but is *inversely proportional* to the frictional resistance of blood flow through the vessels. Inverse proportionality is expressed by showing one of the factors in the denominator

of a fraction, since a fraction *decreases* when the denominator *increases:*

$$\text{Blood flow} \quad \alpha \quad \frac{\Delta P}{\text{Resistance}}$$

The *resistance* to blood flow through a vessel is directly proportional to the length of the vessel and to the viscosity of the blood (the "thickness," or ability of the molecules to "slip over" each other). Vascular resistance is inversely proportional to the fourth power of the radius of the vessel:

$$\text{Resistance} \quad \alpha \quad \frac{L\eta}{r^4}$$

$$\text{where: } L = \text{length of vessel}$$
$$\eta = \text{viscosity of blood}$$
$$r = \text{radius of vessel}$$

If one vessel has twice the radius of another, and if all other factors are the same, flow in the larger vessel would thus be sixteen times greater (2^4) than in the smaller vessel. When physical constants are added to the relationship, the exact rate of flow can be calculated according to **Poiseuille's law:**

$$\text{Blood flow} = \frac{\Delta P \, r^4 \, (\pi)}{\eta \, L \, (8)}$$

It should be noted that Poiseuille's law is valid only for the non-pulsatile flow of homogenous fluids through a tube; it is therefore not a completely accurate description of the flow of blood through arteries. The relationships,

Figure 12.13 Blood pressure in different vessels of the systemic circulation.

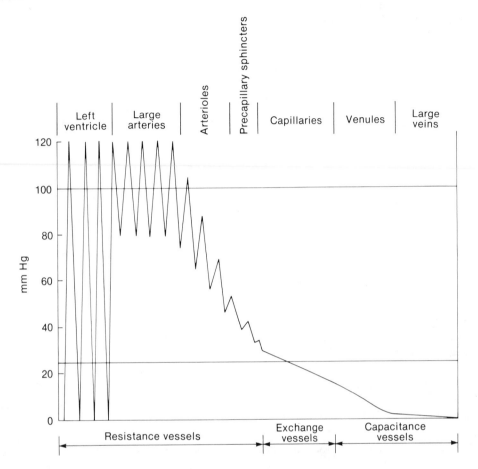

however, are valid: blood flow will increase with arterial pressure and decrease with vascular resistance. Since arterial blood pressure, vessel length, and blood viscosity do not vary significantly in normal physiology, the primary factor regulating blood flow must be the radius of the vessels.

If all other factors are constant, constriction of a vessel to half its previous radius will increase its resistance and decrease its flow rate by a factor of 16 (as previously discussed). Since the effect of changes in radius on changes in flow rate are magnified by a power of four, blood can be diverted from one organ system to another by varying degrees of vasoconstriction and vasodilation.

Sources of Vascular Resistance   Small muscular arteries and arterioles provide most of the vascular resistance to blood flow. The large elastic arteries, in contrast, contribute relatively little to the vascular resistance because of their shorter lengths (when compared to the combined lengths of smaller arteries) and their ability to expand under pressure. The diameter of the aorta, for example, increases by 6 percent and the diameter of the pulmonary

artery increases by 10 percent during systole. Elastic recoil of these arteries during diastole provides an additional pressure rise or "push." This helps to dampen the pressure fluctuations between systole and diastole in the smaller arteries and arterioles, and produces a more continuous flow of blood in the capillaries.

While individual capillaries are the thinnest of vessels, their large number combines to produce a huge cross-sectional area. This causes a large drop in blood pressure and flow rate, which allows more time for the exchange of molecules by diffusion between capillary blood and tissue fluid. The large cross-sectional area of capillary beds results in a much lower vascular resistance than that provided by the arterioles that "feed" these capillary beds.

Total Peripheral Resistance   The sum of all the vascular resistances—due primarily to muscular arteries and arterioles—within the systemic (or pulmonary) circulations is called the **total peripheral resistance.** It is important to realize that the separate resistances that contribute to this total, represented by the arterial supply to each organ, are not "in series" with each other. With very few exceptions

Figure 12.14 Diagram of the systemic and pulmonary circulations. Notice that, with few exceptions (such as the blood flow in the renal circulation) the flow of arterial blood is in series rather than in parallel (arterial blood does not usually flow from one organ to another).

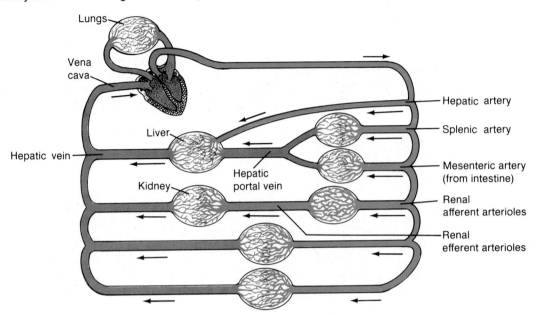

(such as in the kidney), arterial blood only passes through one set of resistance vessels (arterioles) before returning to the heart. Since one region is not "downstream" from another in terms of its arterial supply, changes in resistance within one organ directly affects blood flow in only that organ.

The arterial supply to different organs, and the vascular resistances within these organs, are thus said to be "in parallel." If all other factors remained constant, constriction of arterioles within one organ would increase the resistance and decrease the blood flow only to that organ; vasodilation would, conversely, increase blood flow to that organ.

Vasodilation in a large organ could, however, significantly decrease the total peripheral resistance. Since arterial blood pressure is partially dependent on total peripheral resistance (as described in a later section), the "driving pressure" (ΔP) for blood flow through all organs could be reduced. In order for vasodilation to increase blood flow in one organ system, therefore, there must be simultaneous vasoconstriction in other areas, so that the total peripheral resistance and arterial blood pressure are not reduced. During exercise, for example, blood flow to the exercising muscles is increased by dilation of arterioles in the muscles, together with constriction of arterioles in the viscera and skin. Arterial blood is thus diverted from the viscera and skin to the exercising muscles (see figure 12.15).

Extrinsic Regulation of Blood Flow

The term *extrinsic regulation* refers to control by the autonomic nervous system and endocrine system. *Angiotensin II*, for example, directly stimulates vascular smooth muscle to produce a generalized vasoconstriction. Angiotensin II therefore increases total peripheral resistance and blood pressure, in addition to its indirect effects on blood volume (via stimulation of aldosterone secretion). Antidiuretic hormone (ADH) also has a vasoconstrictor effect in anesthetized animals—hence its alternate name of *vasopressin*. The significance of this vasoconstrictor effect in normal human physiology, however, is presently controversial.

Regulation by Sympathetic Nerves Stimulation of the sympathoadrenal system produces an increase in cardiac output (as previously discussed) and an increase in total peripheral resistance. The increased total peripheral resistance is due to stimulation of alpha-adrenergic receptors (see chapter 9) in vascular smooth muscle by norepinephrine and, to a lesser degree, by epinephrine. This results in vasoconstriction in the viscera and skin. Catecholamine (norepinephrine and epinephrine)-induced vasoconstriction raises peripheral resistance and blood pressure.

Figure 12.15 Relationships between blood flow, vessel radius, and resistance. In *(a)* the resistance and blood flow is equally divided between two branches of a vessel. In *(b)* a doubling of the radius of one branch and a halving of the radius of the other produces a sixteenfold increase in blood flow in the former and a sixteenfold decrease of blood flow in the latter. The total vascular resistance in *(a)* is equal to that of *(b)*.

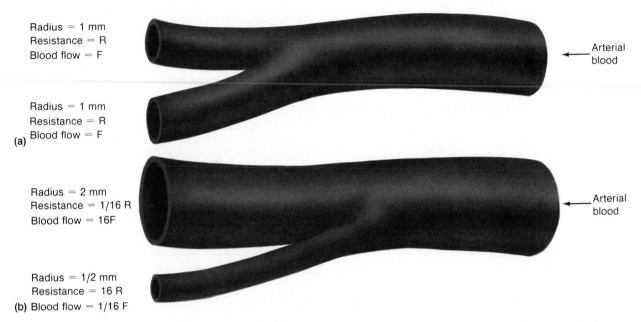

Radius = 1 mm
Resistance = R
Blood flow = F

← Arterial blood

Radius = 1 mm
Resistance = R
(a) Blood flow = F

Radius = 2 mm
Resistance = 1/16 R
Blood flow = 16F

← Arterial blood

Radius = 1/2 mm
Resistance = 16 R
(b) Blood flow = 1/16 F

Even when a person is calm, the sympathoadrenal system is active to a certain degree and helps set the "tone" of vascular smooth muscle. In this state, adrenergic (or norepinephrine-releasing) nerve fibers cause a basal level of vasoconstriction in skeletal muscles and throughout the body. During the *defense reaction* ("fight or flight"), and just prior to exercise, these *adrenergic fibers* increase in activity within the gastrointestinal tract, kidneys, and skin. This raises resistance and decreases blood flow in these organs. Adrenergic fibers in skeletal muscles, however, decrease their activity during the defense reaction.

Skeletal muscles also receive post-ganglionic sympathetic fibers that release acetylcholine (ACh) as a neurotransmitter. During the defense reaction, these *cholinergic sympathetic fibers* increase in activity as the activity of adrenergic fibers decreases. These cholinergic sympathetic fibers cause dilation of blood vessels in skeletal muscles. Thus, while the total peripheral resistance and blood pressure is increased as a result of adrenergic sympathetic fiber activity, the blood flow to skeletal muscles is increased as a result of vasodilation induced by cholinergic sympathetic fibers. Blood flow is thus diverted to skeletal muscles under emergency conditions, which may produce an "extra edge" for the skeletal muscle's response to the emergency.

Table 12.4 Extrinsic control of vascular resistance and blood flow.

Extrinsic Agent	Effect	Comments
Sympathetic nerves		
Alpha-adrenergic	Vasoconstriction	Occurs throughout the body. This is the dominant effect of sympathetic nerve stimulation on the vascular system.
Beta-adrenergic	Vasodilation	Some activity in arterioles in skeletal muscles and coronary vessels, but effects are masked by dominant alpha-receptor-mediated constriction.
Cholinergic	Vasodilation	Localized to arterioles in skeletal muscles; are only activated during defense (fight or flight) reaction.
Parasympathetic nerves	Vasodilation	Effects primarily restricted to gastrointestinal tract, external genitals, and salivary glands. Has little effect on total peripheral resistance.
Angiotensin II	Vasoconstriction	A powerful vasoconstrictor produced as a result of secretion of renin from the kidneys. May function to help maintain adequate filtration pressure in kidneys when systemic blood flow and pressure is reduced.
Vasopressin (ADH)	Vasoconstriction	Although the effects of this hormone on vascular resistance and blood pressure in anesthetized animals is well documented, the importance of these effects in conscious humans is controversial.
Histamine	Vasodilation	Promotes localized vasodilation during inflammation and allergic reactions.
Bradykinens	Vasodilation	Bradykinens are polypeptides secreted by sweat glands that promote local vasodilation.
Prostaglandins	Vasodilation or vasoconstriction	Prostaglandins are cyclic fatty acids that can be produced by most tissues, including blood vessel walls. Prostaglandin I_2 is a vasodilator, while thromboxane A_2 is a vasoconstrictor. The physiological significance of these effects is presently controversial.

Parasympathetic Control of Blood Flow Parasympathetic endings in arterioles are always cholinergic and always promote vasodilation. This innervation, however, is restricted to the blood vessels in the digestive tract, external genitals, and salivary glands. Because of this limited distribution, the parasympathetic system is less important than the sympathetic system in the control of total peripheral resistance. Increased blood flow to vessels in organs innervated by parasympathetic nerves is produced by both an increase in parasympathetic and a decrease in sympathetic nerve activity. The extrinsic control of blood flow is summarized in table 12.4.

Intrinsic Control of Blood Flow

Extrinsic control mechanisms, including the effects of autonomic nerves and hormones, affect resistance and flow in many regions of the body. In contrast to these more generalized effects, intrinsic ("built-in") mechanisms within individual tissues provide a more localized regulation of vascular resistance and blood flow. The major mechanisms that mediate this intrinsic control of blood flow are of two types: (1) *myogenic* and (2) *metabolic*.

Myogenic Control Mechanisms If the arterial blood pressure and flow through an organ is inadequate—if the organ is inadequately *perfused* with blood—the metabolism of the organ cannot be maintained beyond a limited period of time. Excessively high blood pressure can also be dangerous, particularly in the brain, because this may cause fine blood vessels to burst (leading to cerebrovascular accident—CVA or stroke).

Changes in systemic pressure are compensated for in the brain and in some other organs by appropriate responses of vascular smooth muscle. A decrease in systemic arterial pressure causes cerebral blood vessels to dilate, so that adequate rates of blood flow may be maintained despite the decreased pressure. High blood pressure, in contrast, causes cerebral vessels to constrict so that finer vessels "downstream" are protected from the elevated pressure. These responses are "myogenic"; changes in arterial pressure (and thus changes in stretch of the vessels) affect the vascular smooth muscle directly. The autonomic nervous system is not involved in these responses.

Metabolic Control Mechanisms Local vasodilation within the systemic circulation can occur as a result of the chemical environment created by the organ's metabolism. The localized chemical conditions that promote vasodilation include (1) *decreased oxygen* concentrations that result from increases in cell metabolism; (2) *increased carbon dioxide concentration;* (3) *decreased tissue pH* as a result of accumulation of carbonic acid (formed from CO_2), lactic acid, and other metabolic products; and (4) release of other tissue substances, including K^+ and *adenosine*. The importance of each of these factors varies in different organs.

The vasodilator effects that occur in response to tissue metabolism can be demonstrated by constricting the blood supply to an area for a short time and then removing the

Table 12.5   The intrinsic control of vascular resistance and blood flow.

Category	Agent (↑ = increase; ↓ = decrease)	Comments
Myogenic	↑ Blood pressure	Stretching of the arterial wall as the blood pressure rises directly stimulates increased smooth muscle tone (vasoconstriction).
Metabolic	↓ Oxygen ↑ Carbon dioxide ↓ pH ↑ Adenosine ↑ K^+	Local changes in gas and metabolite concentrations act directly on vascular smooth muscle walls to produce vasodilation in the systemic circulation. Importance of different agents varies in different organs.
Myogenic	———	Helps to maintain relatively constant rates of blood flow and pressure within an organ despite changes in systemic arterial pressure (autoregulation).
Metabolic	———	Aids in autoregulation of blood flow. Also helps to shunt increased amounts of blood to organs with higher metabolic rates (active hyperemia).

constriction. The constriction allows metabolic products to accumulate in the regions normally served by the constricted vessels. When the constriction is removed and blood flow resumes, the metabolic products that have accumulated cause vasodilation; the tissue thus appears red. This response is called *reactive hyperemia.* A similar increase in blood flow occurs in skeletal muscles and other organs as a result of increased metabolism. These responses are called *active hyperemia.*

Intrinsic metabolic and myogenic regulation allows organs a degree of independence in the control of their own blood flow rates. Some organs—the brain and kidneys in particular—have the ability to maintain remarkably constant flow rates despite wide fluctuations in systemic blood pressure. This ability is termed **autoregulation.** Intrinsic control mechanisms are summarized in table 12.5, and the control of blood flow in specific organs is discussed more completely in chapter 13.

1. Write Poiseuille's law and explain, in words, the relationship between blood flow rate, arterial blood pressure, and vascular resistance.
2. Explain why muscular arteries and arterioles provide more vascular resistance to flow than the large elastic arteries and the capillary beds.
3. Draw a line diagram to illustrate how changes in arterial diameter affects blood flow, assuming arterial pressure remains constant. Explain how total peripheral resistance and arterial pressure can remain relatively constant when arterioles in skeletal muscles dilate.
4. Describe the effects of *(a)* sympathetic adrenergic fibers, *(b)* sympathetic cholinergic fibers, and *(c)* parasympathetic fibers on total peripheral resistance and on blood flow through specific regions of the body.
5. Define autoregulation, and explain how this is accomplished through myogenic and metabolic regulation of blood flow.

Figure 12.16   A constriction increases blood pressure upstream (analogous to the arterial pressure) and decreases pressure downstream (analogous to capillary and venous pressure).

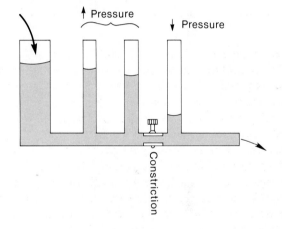

Blood Pressure

Resistance to flow in the arterial system is greatest in the arterioles because these vessels have the smallest diameter. Although the total blood flow through a system of arterioles must be equal to the flow in the larger vessels that gave rise to the arterioles, the narrow diameter of arterioles reduces the flow rate within each according to Poiseuille's law. Blood flow rate and pressure is thus reduced in the capillaries, which are located "downstream" of the high resistance imposed by the arterioles. The blood pressure "upstream" of the arterioles—in the medium and large arteries—is correspondingly increased (see figure 12.16).

Figure 12.17 As blood passes from the aorta to smaller arteries, arterioles, capillaries, and to the venous system, the cross-sectional area increases as the pressure decreases.

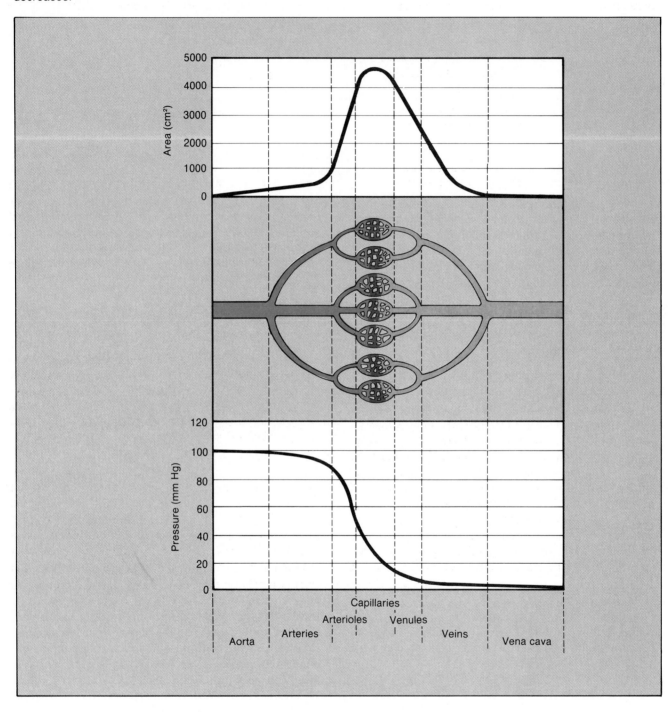

Figure 12.18 Action potential frequency in sensory nerve fibers from baroreceptors in the carotid sinus and aortic arch. As the blood pressure increases, the baroreceptors become increasingly stretched. This results in an increase in the frequency of action potentials that are transmitted to the cardiac and vasomotor control centers in the medulla oblongata.

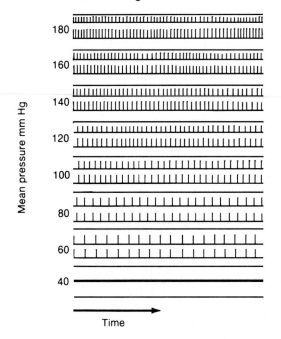

cardiac output. This may be due to elevations in cardiac rate, or to increased stroke volume as a result of increased blood volume or other causes. The three most important variables affecting arterial blood pressure are *cardiac rate, blood volume,* and *total peripheral resistance.* An increase in any of these three, if not compensated by a decrease in the other variables, will result in an increased blood pressure.

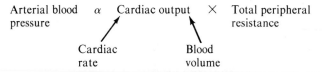

Blood pressure can thus be regulated to a large degree by the kidneys, which control blood volume, and by the sympathoadrenal system. Increased activity of the sympathoadrenal system can raise blood pressure by stimulating vasoconstriction of arterioles (thus raising total peripheral resistance) and by promoting increased cardiac output (largely by stimulating an increase in cardiac rate). Renal and autonomic regulation of blood pressure is controlled by the brain.

Regulation of Blood Pressure

In order for blood pressure to be maintained within normal limits, specialized receptors for pressure are needed. These **baroreceptors** (baro = pressure) are stretch receptors located in the *aortic arch* and in the *carotid sinuses.* An increase in pressure causes the arterial wall of these regions to stretch and stimulate activity of sensory nerve endings. A fall in pressure below the normal range, in contrast, causes a decrease in the frequency of action potentials produced in these sensory neurons.

Sensory nerve activity from the baroreceptors ascends via the vagus nerve to the medulla oblongata of the brain and *vasomotor control centers* in the medulla, which direct the autonomic system to respond appropriately. Changes in sympathetic and parasympathetic activity, directed by the cardiac center (separate cardiostimulatory and inhibitory centers may exist), affect the cardiac rate and contractility. Changes in sympathetic nerve activity

The blood pressure and flow rate within capillaries is further reduced by the fact that the total cross-sectional area of arterioles and capillaries is much greater than the cross-sectional area of the medium and large arteries (see figure 12.17). Variations in the diameter of arterioles, due to vasoconstriction and vasodilation, thus simultaneously affect both blood flow through capillaries and the "upstream" or **systemic arterial pressure.**

An increase in total peripheral resistance, due to vasoconstriction of arterioles, can raise arterial blood pressure. Blood pressure can also be raised by an increase in

Figure 12.19   The baroreceptor reflex. Sensory stimuli
from baroreceptors in the carotid sinus and aortic arch,
acting via control centers in the medulla oblongata, affect
activity of sympathetic and parasympathetic nerve fibers
in the heart.

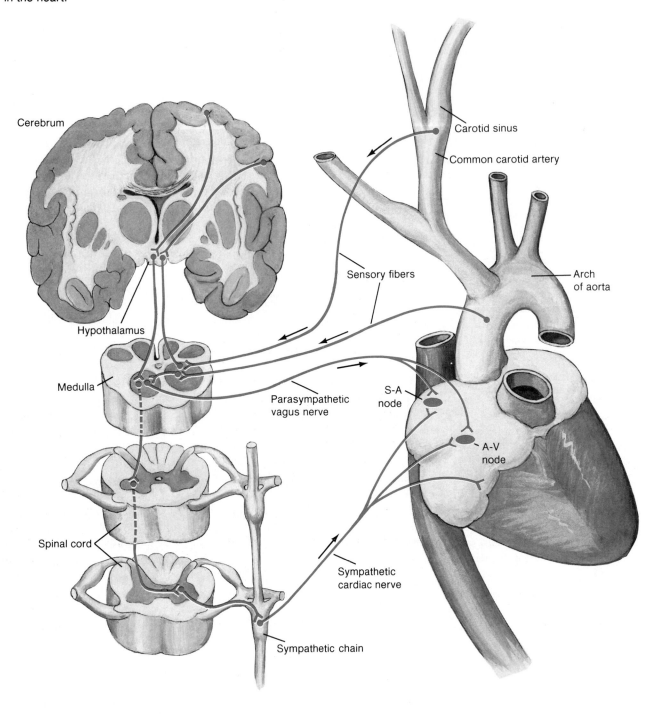

Figure 12.20 Compensations for the upright posture induced by the baroreceptor reflex.

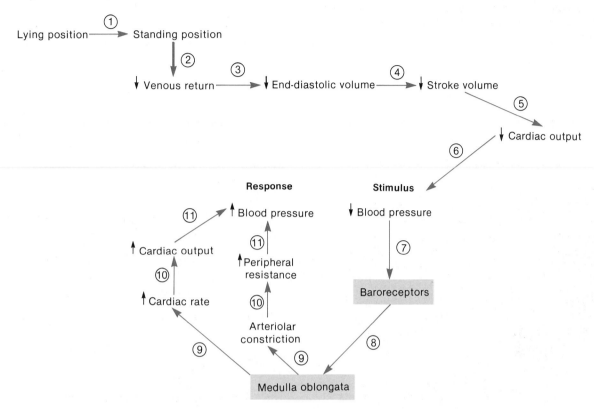

directed by the vasomotor center promotes vasoconstriction or vasodilation, and thus affects total peripheral resistance.

Baroreceptor Reflex When a person goes from a lying to a standing position, there is a shift of 500–700 ml of blood from the veins of the thoracic cavity to veins in the lower extremities, which expand to contain the extra volume of blood. This "pooling" of blood reduces the effective blood volume by decreasing the venous return and cardiac output. The resulting fall in blood pressure is almost immediately compensated by the baroreceptor reflex. Baroreceptor information to the medulla oblongata stimulates sympathetic and inhibits parasympathetic nerve activity, resulting in increased cardiac rate and vasoconstriction.

An effective baroreceptor reflex helps to maintain adequate blood flow to the brain upon standing. If the baroreceptor sensitivity is reduced, as can occur in arteriosclerosis and hypertension, an uncompensated fall in pressure may occur when a person stands up too rapidly. This condition—called *postural* or *orthostatic hypotension* (hypotension = low blood pressure)—can make a person feel dizzy or even faint because of inadequate perfusion of the brain.

The baroreceptor reflex can also mediate the opposite response. When the blood pressure rises above an individual's normal range, the baroreceptor reflex causes a slowing of the cardiac rate and vasodilation (due to decreased sympathetic and increased parasympathetic nerve activity). Manual massage of the carotid sinus, a procedure sometimes performed by physicians to reduce tachycardia and lower blood pressure, also evokes this reflex. Interestingly, this procedure was used by the ancient Greeks to induce anesthesia (the word *carotid* is derived from the Greek word for sleep). Carotid massage should be used cautiously, however, because it can cause the heart to stop as a result of excessive parasympathetic inhibition.

Activity of the cardiac and vasomotor centers in the medulla is influenced by higher brain centers (such as the hypothalamus) as well as by sensory feedback from the baroreceptors. Emotionally induced fainting, for example, results from a fall in blood pressure (due to decreased cardiac rate and peripheral resistance) that occurs as a result of higher brain influences on the medulla. Such higher brain influences may also be responsible for the fact that cardiac rate increases during exercise, *despite* the fact that blood pressure also increases.

Other Reflexes Controlling Blood Pressure   Reflex control of cardiac rate, blood volume, and peripheral resistance is evoked by receptors other than the arterial baroreceptors. Stretch receptors, for example, are also located in the left atrium of the heart. These receptors are activated by increased venous return to the heart, and produce (1) reflex tachycardia, as a result of increased sympathetic nerve activity, and (2) inhibition of antidiuretic hormone (ADH) secretion from the posterior pituitary, resulting in a larger volume of urine excretion and a lowering of blood volume. The reduced blood volume and the tachycardia that result help to lower the end-diastolic volume and, in the process, act to lower the blood pressure.

Measurement of Blood Pressure

Stephen Hales (1677–1761) accomplished the first documented measurement of blood pressure by inserting a cannula (small tube) into the artery of a horse and measuring the height to which blood would rise in the vertical tube. Modern clinical blood pressure measurements, fortunately, are less direct. The indirect, or *auscultatory method* (auscul = listen to) of blood pressure measurement is based on the correlation between blood pressure and arterial sounds.

In this indirect method, an inflatable rubber bladder within a cloth cuff is wrapped around the upper arm and a stethoscope is applied over the brachial artery (see figure 12.22). The artery is normally silent before inflation of the cuff because blood normally travels in a smooth, **laminar flow** through the vessels. The term *laminar* means *layered*—blood in the central axial stream moves the fastest, while blood flowing closer to the artery wall moves more slowly; there is little transverse movement between these layers that would produce mixing.

The laminar flow that normally occurs in arteries produces little vibration, and is thus silent. When the artery is pinched, however, blood flow through the constriction can become turbulent. This causes the artery to vibrate and produce sounds. In order for the pressure cuff to constrict the brachial artery, the cuff pressure must be greater than the diastolic pressure, which is the lowest pressure within the artery that opposes the cuff pressure. If the cuff pressure was greater than the systolic pressure (the highest arterial blood pressure), the artery would be pinched shut and thus would be silent.

Figure 12.21   Use of a pressure cuff and sphygomanometer to measure blood pressure.

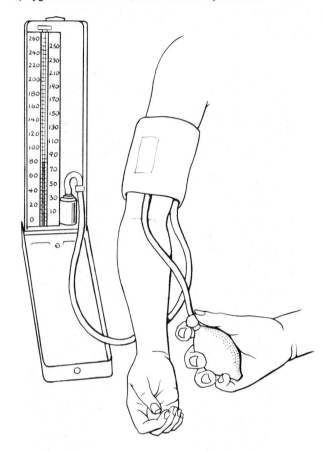

Suppose that a person has a systolic pressure of 120 mm Hg and a diastolic pressure of 80 mm Hg (the average normal values). When the cuff pressure is between 80 and 120 mm Hg the artery will be closed during diastole—because the cuff pressure exceeds the diastolic pressure—but will open during systole because the cuff pressure is less than the systolic pressure. As the artery begins to open, **turbulent flow** of blood through the constriction will create vibrations that are heard as the **sounds of Korotkoff**. These sounds will be heard at every systole, usually as a "tapping" sound because laminar flow and silence resumes with every diastole. These sounds, it should be noted, are *not* produced by closing of the heart valves—if they were, the artery would not be silent (which it usually is) before the pressure cuff is inflated.

Figure 12.22 Korotkoff sounds are produced by the turbulent flow of blood through the partially constricted brachial artery. This occurs when the cuff pressure is greater than the diastolic pressure but less than the systolic pressure.

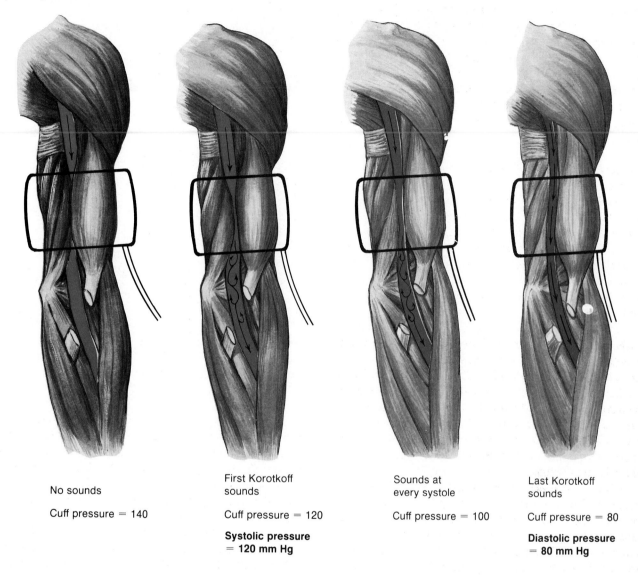

No sounds

Cuff pressure = 140

First Korotkoff sounds

Cuff pressure = 120

Systolic pressure = 120 mm Hg

Sounds at every systole

Cuff pressure = 100

Last Korotkoff sounds

Cuff pressure = 80

Diastolic pressure = 80 mm Hg

Blood pressure = 120/80

Figure 12.23   The indirect, or auscultory method of blood-pressure measurement. Korotkoff sounds, produced by turbulent blood flow through a constricted artery, occur whenever the cuff pressure is less than the systolic blood pressure and greater than the diastolic blood pressure. As a result, the first Korotkoff sound is heard when the cuff pressure is equal to the systolic blood pressure, and the last sound is heard when the cuff pressure and diastolic blood pressures are equal.

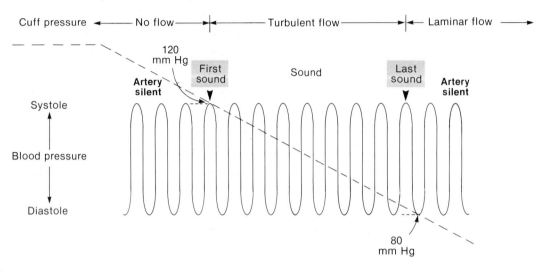

Figure 12.24   The five phases of blood-pressure measurement.

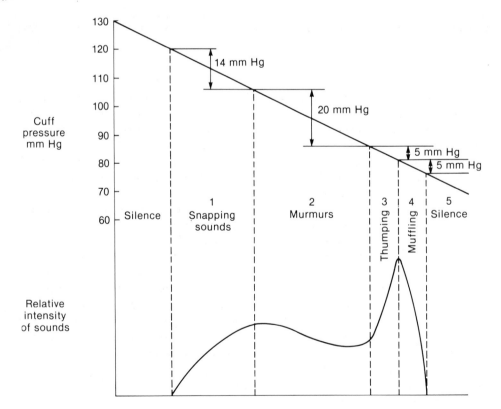

Table 12.6 Normal arterial blood pressure at different ages.

Age	Systolic Men	Systolic Women	Diastolic Men	Diastolic Women	Age	Systolic Men	Systolic Women	Diastolic Men	Diastolic Women
1 day	70				16 years	118	116	73	72
3 days	72				17 years	121	116	74	72
9 days	73				18 years	120	116	74	72
3 weeks	77				19 years	122	115	75	71
3 months	86				20–24 years	123	116	76	72
6–12 months	89	93	60	62	25–29 years	125	117	78	74
1 year	96	95	66	65	30–34 years	126	120	79	75
2 years	99	92	64	60	35–39 years	127	124	80	78
3 years	100	100	67	64	40–44 years	129	127	81	80
4 years	99	99	65	66	45–49 years	130	131	82	82
5 years	92	92	62	62	50–54 years	135	137	83	84
6 years	94	94	64	64	55–59 years	138	139	84	84
7 years	97	97	65	66	60–64 years	142	144	85	85
8 years	100	100	67	68	65–69 years	143	154	83	85
9 years	101	101	68	69	70–74 years	145	159	82	85
10 years	103	103	69	70	75–79 years	146	158	81	84
11 years	104	104	70	71	80–84 years	145	157	82	83
12 years	106	106	71	72	85–89 years	145	154	79	82
13 years	108	108	72	73	90–94 years	145	150	78	79
14 years	110	110	73	74	95–106 years	145	149	78	81
15 years	112	112	75	76					

Data compiled and published in K. Diem and C. Lentner, eds., *Scientific Tables*, Geigy Pharmaceuticals, Division of Ciba-Geigy Corporation, Ardsley, New York, 1975.

The cuff is usually inflated to produce a pressure greater than the systolic pressure so that the artery is pinched off and silent. The pressure in the cuff is read from an attached meter called **sphygmomanometer.** A valve is then turned to allow the release of air from the cuff, causing a gradual decrease in cuff pressure. When the cuff pressure is equal to the systolic pressure, the *first sound* of Korotkoff will be heard as blood passes in a turbulent flow through the constricted opening in the artery.

As long as the cuff pressure is still greater than the diastolic pressure, a Korotkoff sound will be heard at every systole, as previously described. When the cuff pressure, which acts to constrict the artery, becomes equal to the diastolic blood pressure (which pushes against the artery wall from the inside), the artery remains open. Laminar flow thus resumes and the artery is again silent. The *last sound* of Korotkoff thus occurs when the cuff pressure is equal to the diastolic pressure.

Different phases in the measurement of blood pressure are identified on the basis of the quality of the Korotkoff sounds (see figure 12.24). In some people, the Korotkoff sounds don't disappear even when the cuff pressure is reduced to zero (zero pressure, remember, means that it is equal to atmospheric pressure). In these cases— and often routinely—the onset of muffling of the sounds (phase 4 in figure 12.24) is used rather than the onset of silence (phase 5) as an indication of diastolic pressure. Normal blood pressure values are shown in table 12.6.

Pulse Pressure and Mean Arterial Pressure

The **pulse pressure** is equal to the difference between the systolic and diastolic pressures. If a person has a blood pressure of 120/80 (systolic/diastolic), therefore, the pulse pressure would equal 40 mm Hg. This value is significant because it represents the pressure difference between the "heart side" of an artery and the "capillary side" of the artery when the heart contracts. The pulse pressure is thus the driving pressure or ΔP that moves the arterial blood towards the capillaries.

Pulse pressure = Systolic pressure − Diastolic pressure

Table 12.7 Effects of different variables on blood pressure measurements.

Variable	Diastolic Pressure	Systolic Pressure	Pulse Pressure
↑Cardiac rate	Increases (more)	Increases (less)	Decreases
↑Peripheral resistance	Increases (more)	Increases (less)	Decreases
↑Blood volume*	Increases (less)	Increases (more)	Increases
↑Stroke volume	Increases (less)	Increases (more)	Increases

*Conversely, a fall in blood volume (as in dehydration) produces a decrease in pulse pressure.

The **mean arterial pressure** represents an average of the systolic and diastolic blood pressures. This is usually calculated by adding one-third of the pulse pressure to the diastolic pressure. If a person has a blood pressure of 120/80, for example, the mean arterial pressure will be 80 + ⅓(40) = 93 mm Hg.

Mean arterial pressure = Diastolic pressure + ⅓ × Pulse pressure

A rise in total peripheral resistance and cardiac rate increases the diastolic pressure more than it increases the systolic pressure. When the baroreceptor reflex is activated by going from a lying to a standing position, for example, the diastolic pressure usually increases by 5–10 mm Hg, while the systolic pressure is either unchanged or slightly reduced (as a result of decreased venous return). People with hypertension (high blood pressure) who usually have elevated total peripheral resistance and cardiac rates, likewise have a greater increase in diastolic than in systolic pressure. Dehydration or blood loss results in decreased cardiac output and thus in decreased pulse pressure.

An increase in cardiac output, in contrast, raises the systolic pressure more than it raises the diastolic pressure (though both pressures do rise). This occurs during exercise, for example, when the blood pressure may rise to values as high as 200/100 (yielding a pulse pressure of 100 mm Hg). The effects of different variables on blood pressure are summarized in table 12.7.

Hypertension

Approximately 20 percent of all adults in the United States have *hypertension*—blood pressure in excess of the normal range for the person's age and sex, or in excess of the somewhat arbitrary "borderline" pressure of 140/90 that is often used. Hypertension that is a result of ("secondary to") known disease processes is logically called *secondary hypertension*. These comprise only about 10 percent of the hypertensive population. Hypertension that is the result of complex and poorly understood disease processes is not-so-logically called *primary* or *essential hypertension*.

Diseases of the kidneys (such as glomerulonephritis and pyelonephritis) and disease of the renal arteries (such as atherosclerosis), can cause secondary hypertension as a result of increased blood volume. This occurs because kidney damage severely restricts urine production. More commonly, reduction of renal blood flow can raise blood pressure via activation of the renin-angiotensin system. Experiments in which the renal artery is clipped, for example, produces hypertension that is associated (at least initially) with elevated renin secretion. These and other causes of secondary hypertension are summarized in table 12.8.

Essential Hypertension

The vast majority of people with hypertension have essential hypertension—high blood pressure that cannot be attributed to an obvious cause. An increased total peripheral resistance is a universal characteristic of people with hypertension. Cardiac rate and cardiac output are elevated in many, but not all, of these cases.

Secretion of renin—which is correlated with angiotensin II production, aldosterone secretion, and blood volume—is likewise variable. While some people with essential hypertension have low renin secretion, most have either normal or elevated secretion of renin from the kidneys. Renin secretion in the normal range is inappropriate in people with hypertension, since high blood pressure should inhibit renin secretion. Since renin secretion leads to angiotensin II production and aldosterone secretion, people with high renin secretion would have inappropriately high retention of salt and high blood volumes.

Since the peripheral resistance, cardiac rate, and blood volume are not all affected to the same degree in different people with essential hypertension, it appears that this condition may have a variety of causes. Our understanding of essential hypertension is further clouded by the observation that some people are more susceptible than others. Essential hypertension "runs in families," is more common in males than in females, blacks than whites, and in obese than in thin people.

Table 12.8 Some possible causes of secondary hypertension.

System Involved	Examples	Mechanisms
Kidneys	Kidney disease	Decreased urine formation
	Renal artery disease	Secretion of vasoactive chemicals
Endocrine	Excess catecholamines (tumor of adrenal medulla)	Increased cardiac output and total peripheral resistance
	Excess aldosterone (Conn's disease)	Excess salt and water retention by the kidneys
Nervous	Increased intracranial pressure	Activation of sympathoadrenal system
	Damage to vasomotor center	Activation of sympathoadrenal system
Cardiovascular	Complete heart block; patent ductus arteriosus	Increased stroke volume
	Arteriosclerosis of aorta; coarctation of aorta	Decreased distensibility of aorta

Figure 12.25 Sequence of events that have been proposed as possible causes of essential hypertension.

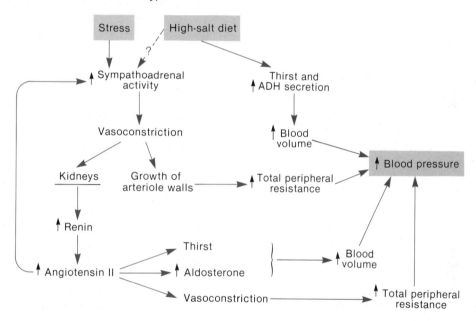

Stress and High-Salt Diet   The primary cause of essential hypertension is unknown and is probably variable. Two likely candidates for this role in genetically susceptible people, however, are stress and a high-salt diet. Frequent activation of the sympathoadrenal system, according to some authorities, would raise the total peripheral resistance by stimulating vasoconstriction (an alpha-adrenergic effect), and would increase cardiac output (a beta-adrenergic effect). This might stimulate hyperplasia (growth by cell division) of the vessel's smooth muscle wall, decreasing the diameter of the lumen. The total peripheral resistance would thus be high even in the absence of sympathetic stimulation. Decreased diameter of the renal artery could secondarily reduce blood flow to the kidneys and stimulate renin secretion. The essential hypertension would then be complicated by excessive angiotensin II production, leading (through stimulation of aldosterone secretion) to excessive salt and water retention. The person would then have elevated peripheral resistance, blood volume, and cardiac output, as commonly seen in people with essential hypertension.

Other lines of evidence suggest that a high-salt diet may be the primary cause of essential hypertension in individuals who—perhaps for genetic reasons—have difficulty excreting salt in the urine. Rats that are genetically salt sensitive have an increased blood volume, as expected, but also show an increased peripheral resistance for unknown reasons. Some of the possible mechanisms that may produce essential hypertension are illustrated in figure 12.25.

Table 12.9 Examples and mechanisms of action of some anti-hypertensive drugs.

Category of Drugs	Examples	Mechanisms
Extracellular fluid volume depletors	Thiazide diuretics	Increase volume of urine excreted, thus lowering blood volume
Sympathoadrenal system inhibitors	Clonidine; alpha-methyldopa	Acts on brain to decrease sympathoadrenal stimulation
	Guanethidine; reserpine	Depletes norepinephrine from sympathetic nerve endings
	Propanolol	Blocks beta-adrenergic receptors, decreasing cardiac output and/or renin secretion
	Phentolamine	Blocks alpha-adrenergic receptors, decreasing sympathetic vasoconstriction
Vasodilators	Hydralazine; sodium nitroprusside	Causes vasodilation by acting directly on vascular smooth muscle

Dangers of Hypertension If other factors remain constant, blood flow increases as arterial blood pressure increases. People with hypertension thus can have adequate perfusion of their organs with blood until the hypertension causes vascular damage. Hypertension, as a result, is usually symptomless until a dangerous amount of vascular damage is produced.

Hypertension is dangerous because (1) high arterial pressure increases the afterload, making it more difficult for the ventricles to eject blood; this increases the amount of work that the heart must perform and may result in pathological changes in heart structure (as described in chapter 13); (2) high pressure may damage cerebral blood vessels, leading to cerebrovascular accident (stroke); (3) hypertension can contribute to the development of atherosclerosis, which can itself lead to heart disease and stroke as previously described.

Treatment of Hypertension Hypertension is usually treated by restricting salt intake and by taking drugs that have a variety of actions. Most commonly, these drugs are *diuretics* that increase urine production, thus decreasing blood volume and pressure. Sympathetic-blocking drugs are also often used; drugs that block beta-adrenergic effects ("beta-blockers", such as propanolol) decrease cardiac rate. Various vasodilators (see table 12.9) may also be used to decrease total peripheral resistance.

Overview

Adequate blood flow through an organ depends upon maintenance of an adequate blood volume and an adequate arterial blood pressure. Since sodium is the major extracellular solute and contributes most to the osmotic pressure of the plasma, ingestion, retention, and excretion of salt (NaCl) greatly affect blood volume and pressure. These factors are controlled largely through the actions of two hormones: ADH and aldosterone.

Blood exerts a hydrostatic pressure against the walls of blood vessels, which in capillaries serves as the filtration force in the formation of tissue fluid and urine. In order to move blood through the arterial system, the pumping of the heart must produce a pressure gradient, which is equal to the difference between systolic and diastolic pressures. The rate of blood flow through any system of vessels is directly related to this pressure difference, and is inversely related to the frictional resistance to flow in the vessels. Since this resistance is raised by vasoconstriction and decreased by vasodilation, blood flow through any organ can be adjusted by variations in the caliber of the resistance vessels—mainly, of the muscular arteries and arterioles.

The pressure gradient in the arterial system is dependent in part on the vascular resistance—which affects the diastolic pressure—and on the cardiac output, which determines the pressure rise during systole. The cardiac output, in turn, is regulated by variations in cardiac rate and stroke volume. In this chapter the mechanisms that regulate cardiac rate, stroke volume, and other determinants of blood flow have been described.

1. Explain why arterioles provide most of the total peripheral resistance. Describe the effect of arteriole constriction on blood flow through capillaries and on the systemic arterial pressure.
2. Describe how the sounds of Korotkoff are produced and how these sounds are used to measure blood pressure.
3. Using a flow chart, describe how the baroreceptor reflex helps to compensate for a fall in arterial pressure. Explain why a person who is severely dehydrated will have a rapid pulse.
4. Define pulse pressure, and describe how it is changed during exercise and when a person suffers from severe blood loss.
5. Explain, using flow charts, how stress and a high-salt diet can contribute to hypertension.

Summary

Cardiac Output and Venous Return

I. Cardiac output is equal to the product of cardiac rate and stroke volume.
   A. Cardiac rate is the number of beats per minute; stroke volume is the amount of blood ejected by each ventricle per beat.
   B. The cardiac rate is set intrinsically by the S-A node, but is modified by autonomic influences.
      1. Sympathoadrenal stimulation increases cardiac rate by increasing the diastolic depolarization of the S-A node.
      2. Acetylcholine released from parasympathetic nerves decreases the cardiac rate.
II. The stroke volume is regulated both intrinsically and extrinsically.
   A. Intrinsic regulation occurs by means of the Frank-Starling mechanism.
      1. Variations in end-diastolic volume produce variations in the stretch of the myocardium.
      2. As the myocardium is stretched, actin and myosin are brought to a more advantageous position with respect to each other.
      3. An increase in end-diastolic volume thus produces an increase in the strength of ventricular contraction.
      4. In this way, the ventricles are able to adjust the strength of their contraction to eject about two-thirds of the end-diastolic volume.
   B. Extrinsic control of stroke volume is due to the effects of sympathoadrenal stimulation.
      1. Catecholamines (norepinephrine and epinephrine) increase contraction strength.
      2. This is called a positive inotropic effect, in contrast to the positive chronotropic effect of catecholamines on cardiac rate.
   C. Other factors also affect stroke volume.
      1. The mean arterial pressure provides an impedance to the ejection of blood; this is called the afterload (in contrast to end-diastolic volume, which is called the preload).
      2. If other factors are constant, stroke volume is directly proportional to venous return (because the venous return determines the end-diastolic volume).
III. The venous return of blood to the atria is dependent on the venous blood pressure and on the blood volume.
   A. The venous pressure is much lower than the arterial pressure for two reasons.
      1. There is a large pressure drop as blood goes through arterioles and capillaries.
      2. Veins are very compliant—they can expand to accept large quantities of blood.
         a. As a result of their high compliance, veins usually contain almost 70 percent of the total blood volume.
   B. Venous pressure is increased by sympathetic stimulation and by activity of the skeletal muscle pump.
      1. Inhalation serves to lower the pressure in the large thoracic veins.
      2. As a result of the fact that venous pressure is lower in the thoracic cavity than in other regions, the pressure difference helps to return blood to the heart.

Blood Volume

I. The volume of plasma and tissue fluid comprise the extracellular fluid volume of the body.
   A. Tissue fluid is formed by filtration from plasma.
      1. Since the hydrostatic pressure of the blood in capillaries is greater than the hydrostatic pressure of tissue fluid, there is a net filtration pressure.
      2. In addition to plasma water, such solutes as glucose, salts, and inorganic ions are filtered into the tissue fluid.
      3. Proteins, however, are restricted from passing through the channels between adjacent endothelial cells of the capillaries.
   B. Since the plasma protein concentration is greater than the tissue fluid protein concentration, the greater osmotic pressure of the plasma draws water back into the capillaries.
      1. Osmotic pressure exerted by proteins is called colloid osmotic pressure.
      2. The amount of fluid returned to the capillaries is less than the amount filtered; the balance is returned to the vascular system by lymphatic vessels.
   C. Edema refers to abnormal accumulation of tissue fluid.
      1. This can be due to excessive filtration pressure or inadequate plasma protein concentration.
      2. Blockage of lymphatic vessels can also produce edema.

II. Blood volume is controlled, to a large degree, by factors that regulate urine volume.
  A. Urine, like tissue fluid, is produced as a plasma filtrate.
    1. Most of this filtrate is returned to the vascular system—a process called reabsorption.
    2. The proportion of the kidney filtrate excreted as urine and the proportion reabsorbed is adjusted by two hormones—antidiuretic hormone (ADH) and aldosterone.
  B. ADH is produced by the hypothalamus and secreted by the posterior pituitary when the osmotic pressure of the blood rises.
    1. A rise in blood osmolality is produced by dehydration and by excessive salt intake.
    2. ADH promotes water reabsorption in the kidneys, thus diluting the blood and raising the blood volume.
  C. Aldosterone is a steroid hormone secreted by the adrenal cortex.
    1. Aldosterone stimulates the renal reabsorption of salt and water, thus raising the blood volume and pressure.
    2. Secretion of aldosterone is stimulated by angiotensin II.
  D. A fall in blood volume and pressure stimulates the juxtaglomerular apparatus of the kidneys to secrete the enzyme renin.
    1. Renin catalyzes the formation of angiotensin I from angiotensinogen.
    2. Angiotensin II is formed from angiotensin I.
    3. Angiotensin II stimulates vasoconstriction and aldosterone secretion; both actions help to raise blood pressure.

Vascular Resistance, Blood Flow, and Blood Pressure

I. The major factor regulating vascular resistance is the diameter of blood vessels.
  A. Resistance is inversely proportional to the fourth power of the radius of the vessel.
    1. Vasoconstriction thus raises resistance, while vasodilation lowers resistance.
    2. The small arteries and arterioles provide the most resistance to blood flow.
  B. Blood flow rate is inversely proportional to resistance and directly proportional to arterial pressure.
    1. Vasoconstriction, which increases resistance, thus reduces the blood flow into capillaries.
    2. Vasodilation increases blood flow, provided that the arterial pressure remains unchanged.
    3. Arterial pressure is partially dependent on total peripheral resistance, so when arterioles in some organs dilate to increase blood flow, arterioles in other organs must constrict.
  C. Arterial pressure provides the driving force for blood flow into capillaries.
    1. Arterial pressure is directly proportional to total peripheral resistance and to cardiac output.
    2. By dilating arterioles in one organ and constricting arterioles in other organs, blood can be diverted to the organs that need it most.

II. Arterial blood pressure is commonly measured by the indirect, or auscultatory, method.
  A. The arteries are normally silent because blood travels in a laminar flow.
  B. Applying a constriction to the artery causes blood to pass through the constriction in a turbulent flow.
    1. This causes sounds known as the sounds of Korotkoff.
    2. Korotkoff sounds are produced when the cuff pressure is equal to or less than the systolic pressure, but greater than the diastolic pressure.
  C. The first Korotkoff sound is produced when the cuff pressure equals the systolic pressure; the last sound is heard when the cuff pressure equals the diastolic pressure.
  D. The pulse pressure is the difference between systolic and diastolic pressure, and represents the pressure difference that causes blood to flow towards the capillary beds.

III. The baroreceptor reflex helps to maintain adequate blood pressure and thus adequate perfusion of organs with blood.
  A. A fall in arterial pressure is sensed by baroreceptors located in the aortic arch and carotid sinus.
  B. Sensory feedback from baroreceptors affects the cardiac and vasomotor centers in the medulla oblongata.
  C. These brain centers stimulate increased sympathetic and decreased parasympathetic activity.
  D. This reflex is activated when a person stands up rapidly, and results in an increased cardiac rate and vasoconstriction.

IV. Hypertension may be caused by excessive blood volume, excessive cardiac rate, or excessive total peripheral resistance.
  A. Secondary hypertension is produced by known disease processes.
  B. Primary or essential hypertension is the most common form of chronic high blood pressure.
    1. The cause of essential hypertension is unknown.
    2. Hypertension is often treated with diuretics and drugs that block sympathetic stimulation.

Self-Study Quiz

1. According to the Frank-Starling law, the strength of ventricular contraction is:
   (a) directly proportional to the end-diastolic volume
   (b) inversely proportional to the end-diastolic volume
   (c) independent of the end-diastolic volume

2. In the absence of compensations, stroke volume will DECREASE when:
   (a) blood volume increases
   (b) venous return increases
   (c) contractility increases
   (d) arterial blood pressure increases

3. The venous return to the heart is increased by:
   (a) increased blood volume
   (b) increased skeletal muscle contractions
   (c) increased breathing
   (d) increased venous pressure
   (e) all of these

4. Which of the following statements about tissue fluid is FALSE?
   (a) it contains the same glucose and salt concentration as plasma
   (b) it contains a lower protein concentration than plasma
   (c) its colloid osmotic pressure is greater than that of plasma
   (d) its hydrostatic pressure is less than that of plasma

5. Edema may be caused by:
   (a) high blood pressure
   (b) decreased plasma protein concentration
   (c) leakage of plasma proteins into tissue fluid
   (d) blockage of lymphatic vessels
   (e) all of these

6. Secretion of ADH is stimulated by:
   (a) increased blood pressure
   (b) increased blood osmotic pressure
   (c) increased water intake
   (d) all of these

7. Both ADH and aldosterone act to:
   (a) increase urine volume
   (b) increase blood volume
   (c) increase total peripheral resistance
   (d) all of these

8. Angiotensin II stimulates:
   (a) vasoconstriction
   (b) aldosterone secretion
   (c) thirst
   (d) sympathetic nerve activity
   (e) all of these

9. The greatest resistance to flow occurs in:
   (a) the large arteries
   (b) the medium-sized arteries
   (c) the arterioles
   (d) the capillaries

10. If a vessel were to dilate to twice its previous radius, and if pressure remained constant, blood flow through this vessel would:
    (a) increase by a factor of 16
    (b) increase by a factor of 4
    (c) increase by a factor of 2
    (d) decrease by a factor of 2

11. The sounds of Korotkoff are produced by:
    (a) closing of the semilunar valves
    (b) closing of the A-V valves
    (c) turbulent flow of blood
    (d) elastic recoil of the aorta

12. As a person goes from a lying to a standing position:
    (a) baroreceptors are stimulated
    (b) sympathetic nerves are activated
    (c) pulse rate is increased
    (d) all of these

13

Cardiovascular Adaptations to Stress: *Distribution of Blood Flow and Blood-Clotting Mechanisms*

Objectives

By studying this chapter, you should be able to:

1. Describe coronary blood flow during systole and diastole, and explain how the coronary circulation is regulated at rest and during exercise

2. Describe the changes that occur in the distribution of blood flow in the body during exercise

3. Explain the mechanisms that contribute to increased cardiac output during exercise

4. Describe the autoregulation of cerebral blood flow

5. Describe the cutaneous circulation, and explain how circulation in the skin is regulated

6. Describe the compensatory responses of the cardiovascular system to circulatory shock and congestive heart failure

7. Describe the roles of platelets in the formation of blood clots

8. Explain the intrinsic and extrinsic pathways in the formation of fibrin

9. Explain how clotting is retarded by anticoagulants and by various genetic disorders

10. Describe the formation of plasmin and its function in dissolution of blood clots

*S*urvival requires that the heart and brain receive adequate rates of blood flow at all times. The ability of skeletal muscles to respond quickly in emergencies, and to maintain continued high levels of activity, may also be critically important for survival. During such times, high rates of blood flow to the muscles must be maintained without compromising blood flow to the heart and brain. These adjustments can be made in part by reduced blood flow to visceral organs (digestive tract, liver, and kidneys), and in part, by reduced blood flow to the skin. These mechanisms divert blood to the heart, brain, and skeletal muscles, while at the same time other mechanisms cause an increase in cardiac output and total blood flow.

In addition to exercise and other physiological stresses, various pathological processes—such as heart disease and shock—can challenge the cardiovascular system. Survival under these conditions depends upon adequate compensatory responses that maintain proper rates of blood flow to the different organs.

Each organ is dependent upon capillary blood flow for delivery of oxygen and removal of wastes. The capillaries of the body only contain 4 to 5 percent of the total blood volume, but the rate of capillary flow within each organ can be significantly increased by opening of precapillary sphincters and by vasodilation of the arterioles that supply the organ. A review of the distribution of blood in the different types of vessels is shown in figure 13.1.

Figure 13.1 Distribution of blood within the circulatory system at rest.

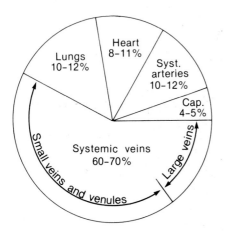

In the absence of physical exercise, the liver, kidneys, and skeletal muscles receive about two-thirds of the total blood flow. When blood flow is expressed per unit weight of tissue, it can be seen that the heart and kidneys each receive an extraordinarily high rate of blood flow. In contrast, skeletal muscles—which may comprise 45 percent of the body mass—have a relatively low rate of flow per unit weight. Low rates of blood flow through muscles and some other organs is due to a high resting vascular resistance. At maximal vasodilation, blood flow to the heart can increase five times, flow to skeletal muscles can increase twenty times, and flow to the skin can increase to over 100 times the resting values (see figure 13.2).

The Coronary Circulation

The muscular wall of the heart (myocardium) is richly supplied with blood by branches of the right and left **coronary arteries**, which diverge from the aorta just behind the mitral valve flaps. The heart is therefore the first organ to receive oxygen-rich blood in the systemic circulation. The coronary arteries branch extensively (see figure 13.3). Some of these branches interconnect (have anastomoses), so that a given area of the heart may receive blood from more than one pathway. These alternate vascular pathways are termed *collateral circulation*.

Figure 13.2 Blood flow in milliliters per minute per 100 grams of tissue and in total (bottom figures in liters per minute) at rest and during maximal vasodilation.

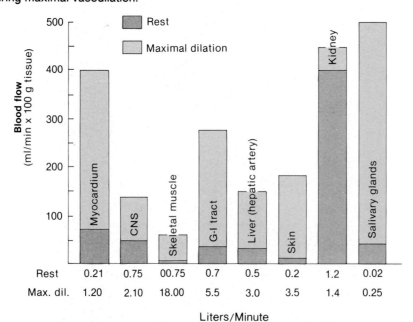

Figure 13.3   The coronary arteries.

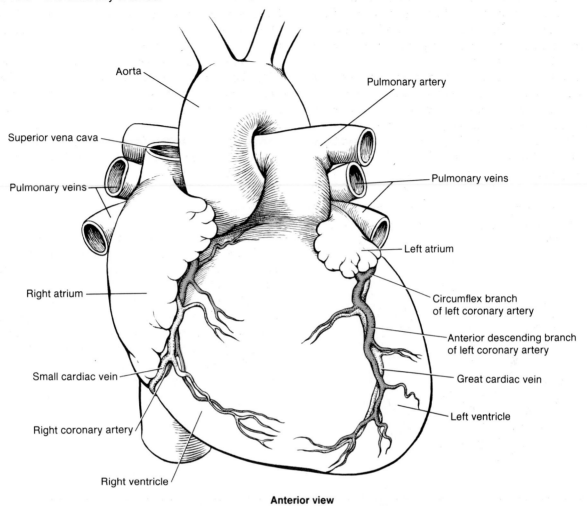

Aorta

Pulmonary artery

Superior vena cava

Pulmonary veins

Pulmonary veins

Left atrium

Right atrium

Circumflex branch
of left coronary artery

Anterior descending branch
of left coronary artery

Small cardiac vein

Great cardiac vein

Right coronary artery

Left ventricle

Right ventricle

Anterior view

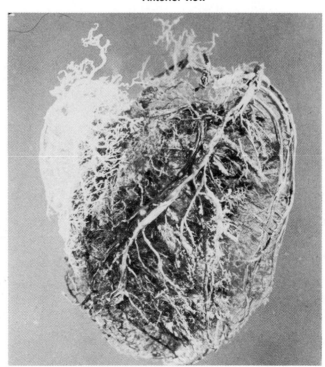

Table 13.1   Some vascular and metabolic comparisons between heart and skeletal muscle.

	Cardiac Muscle	Skeletal Muscle
Number of capillaries	4X	1X
Mean blood flow	10–20X	1X
Myolemma (sarcolemma)	Thin, low resistance	Thicker, higher resistance
Sarcomere length	1X	1.7X
Glycogen concentration	Maintained	Depleted by fasting, diabetes
Glycolytic enzyme systems	1X	2X, strongly developed
Creatine phosphate concentration	1X	6X
Anaerobic energy production	Beating: 2 min Arrested: 30–90 min	Up to 40% total energy
Lactic acid production	Terminal mechanism	Frequent when incurring O_2 debt
Ability to incure O_2 debt	1X	4X
Increased O_2 requirement primarily by:	Increased flow	Increased extraction
Oxygen consumption	3X	1X
Oxygen extraction at rest	Near maximal	Significant reserve
Increased O_2 consumption with increased work	2X	30X
Myoglobin	Present	Present in red skeletal muscle
Mitochondria	Abundant, giant	Fewer, smaller
Krebs cycle enzymes	2–3X	1X
Cytochrome-C	6X	1X
Myosin ATPase activity	1X	3X

Source: Brachfeld, N. The Physiology of Muscular Exercise, *Primary Cardiology*, June, 1979, p. 112. Reproduced with permission.

Aerobic Requirements of the Heart

The coronary arteries supply an enormous number of capillaries, which are packed within the myocardium at a density of about 2,500–4,000 per cubic millimeter of tissue. Fast-twitch skeletal muscle, in contrast, has a capillary density of 300–400 per cubic millimeter. Each myocardial cell, as a consequence, is within 10 μm of a capillary (compared to an average distance in other organs of 60–80 μm). Exchange of gases by diffusion between myocardial cells and capillary blood thus occurs very quickly.

Contraction of the myocardium squeezes the coronary arteries. Unlike blood flow in all other vessels, flow in the coronary arteries thus decreases in systole and increases in diastole. The myocardium, however, contains large amounts of *myoglobin* that normally release sufficient oxygen during systole to maintain aerobic respiration. Since coronary blood flow, and the loading of myoglobin with oxygen, occurs during diastole, the shortening of diastole that occurs at high cardiac rates can reduce oxygen delivery to the heart.

In addition to containing large amounts of myoglobin, heart muscle contains numerous mitochondria and aerobic respiratory enzymes. This indicates that—even more than slow-twitch skeletal muscles—the heart is extremely specialized for aerobic respiration. The normal heart can respire aerobically even during heavy exercise, when the heart's metabolic need for oxygen may increase to five times the oxygen consumption at rest. This increased oxygen requirement is met by a corresponding increase in coronary blood flow, from 60–80 ml per minute per 100 g tissue at rest, up to 400 ml per minute per 100 g tissue during exercise.

Figure 13.4 Arteriogram of the left coronary artery in a patient when the ECG was normal *(a)* and when the ECG showed evidence of myocardial ischemia *(b)*. Notice that a coronary artery spasm (see arrow in *b*) appears to accompany the ischemia.

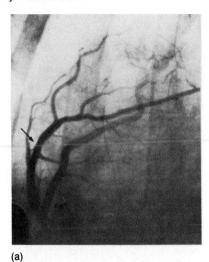

(a)

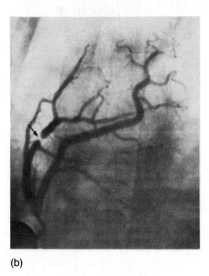

(b)

Figure 13.5 Diagram of coronary artery bypass surgery.

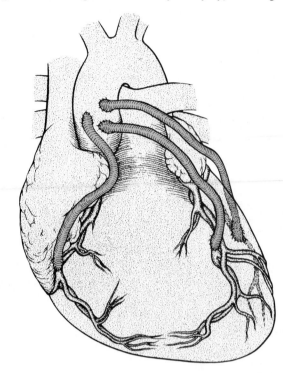

Regulation of Coronary Blood Flow

Sympathetic nerve fibers, through stimulation of alpha-adrenergic receptors in the coronary arteries, contribute to the high vascular resistance of the coronary arteries at rest. Vasodilation of coronary vessels may be due in part to sympathetic nerve stimulation of beta receptors. Most of the vasodilation of the coronary arteries that occurs during exercise, however, is due to *local metabolic control mechanisms*. As the metabolism of the myocardium increases, local accumulation of carbon dioxide acts directly on the vascular smooth muscles to cause vasodilation. There is also evidence that K^+ and adenosine released by myocardial cells may contribute to vasodilation of the coronary arteries.

Occlusion of Coronary Flow   Under abnormal conditions, the heart's metabolic need for oxygen may be greater than the amount of oxygen supplied by the coronary blood flow. This may be due to (1) muscular spasm of the coronary arteries; (2) atherosclerotic plaques that reduce blood flow; and (3) a thrombus (blood clot), often produced by atherosclerotic damage, which may occlude a small coronary artery. The effects of myocardial ischemia (inadequate blood flow) are discussed in chapter 11.

Occlusion of the coronary arteries can be visualized by a technique known as *selective coronary arteriography*. In this procedure, a catheter (plastic tube) is inserted into a brachial or femoral artery all the way to the openings of the coronary arteries in the aorta, and radiographic contrast material is injected. The picture thus obtained (see figure 13.4)—called an *angiogram*—shows the exact site of the obstruction.

Coronary Bypass Surgery   If the occlusion is sufficiently great, a coronary bypass operation may be performed. In this procedure a length of blood vessel, usually obtained from the saphenous vein in the legs, is sutured to the aorta and to the coronary artery at a location beyond the site of the occlusion (see figure 13.5).

Table 13.2 Changes in skeletal muscle blood flow under conditions of rest and exercise.

Condition	Blood Flow (ml/min)	Mechanism
Rest	1,000	High adrenergic sympathetic stimulation of vascular alpha-receptors causing vasoconstriction
Beginning exercise	Increased	Dilation of arterioles in skeletal muscles due to cholinergic sympathetic nerve activity
Heavy exercise	20,000	Fall in alpha-adrenergic activity Increased sympathetic cholinergic activity Increased metabolic rate of exercising muscles

Cardiovascular Adaptations to Exercise

The arterioles in skeletal muscles, like those of the coronary circulation, have a high vascular resistance at rest as a result of adrenergic sympathetic stimulation. This produces a relatively low rate of blood flow (4–6 ml per minute per 100 g tissue), but because muscles have such a large mass this still accounts for 20 to 25 percent of the total blood flow in the body.

In addition to adrenergic fibers (those that release norepinephrine), there are also sympathetic cholinergic fibers in skeletal muscles. Release of acetylcholine (ACh) from these fibers stimulates vasodilation just prior to exercise when the fight or flight reaction is produced. The increased blood flow that results from this vasodilation may provide an "extra edge" that improves skeletal muscle performance once exercise begins.

Changes in Blood Flow during Exercise

As exercise progresses, the vasodilation and increased skeletal muscle blood flow that occurs is due almost entirely to intrinsic metabolic control. The high metabolic rate of skeletal muscles during exercise causes local changes such as increased carbon dioxide concentrations, decreased pH (due to carbonic acid and lactic acid), decreased oxygen, increased extracellular K^+, and secretion of adenosine. Like the intrinsic control of the coronary circulation, these changes cause vasodilation of arterioles in skeletal muscle. Vasodilation of skeletal muscle arterioles decreases the vascular resistance and thus increases blood flow. This effect is combined with "recruitment" of capillaries by the opening of precapillary sphincter muscles (only 5 to 10 percent of skeletal muscle capillaries are open at rest). As a result of these changes, skeletal muscles can receive as much as 90 percent of the total blood flow in the body during maximal exercise.

While the vascular resistance in skeletal muscles decreases during exercise as a result of metabolic vasodilation, the resistance to flow through visceral organs increases. This occurs as a result of vasoconstriction stimulated by adrenergic sympathetic fibers. As a result of these changes, blood flow to the muscles increases fifteen to twenty times, while flow to the viscera decreases by 20 to 30 percent. Blood flow to the heart, as previously discussed, can increase up to five times the resting level. Blood flow to the brain does not appear to change significantly during exercise (see figure 13.6).

Changes in Cardiac Output during Exercise

There is a decrease in total peripheral resistance during exercise, because vasodilation in the skeletal muscles, skin, and heart more than compensates for the vasoconstriction in the visceral organs. This decreased total peripheral resistance makes it easier for the ventricles to eject blood (by decreasing the after load). The ability of the ventricles to eject blood is further augmented by an increased contraction strength (contractility), due to sympathoadrenal stimulation. As a result of these changes, the ejection fraction (percent of the end-diastolic volume ejected) can increase from 67 percent at rest to as much as 90 percent during heavy exercise.

Figure 13.6 Distribution of blood flow (cardiac output) during rest and heavy work. At rest, the cardiac output is 5 L per minute *(bottom of figure)*; during heavy work the cardiac output increases to 25 L per minute *(top of figure)*. During rest, for example, the brain receives 15 percent of 5 L per minute (= 750 ml/min.), while during exercise it receives 3–4 percent of 25 L per minute. (0.03 × 25 = 750 ml/min.). Flow to the skeletal muscles increases more than twentyfold, because the total cardiac output increases (from 5L/min. to 25L/min.) and because the percent of the total received by the muscles increases from 15 percent to 80 percent.

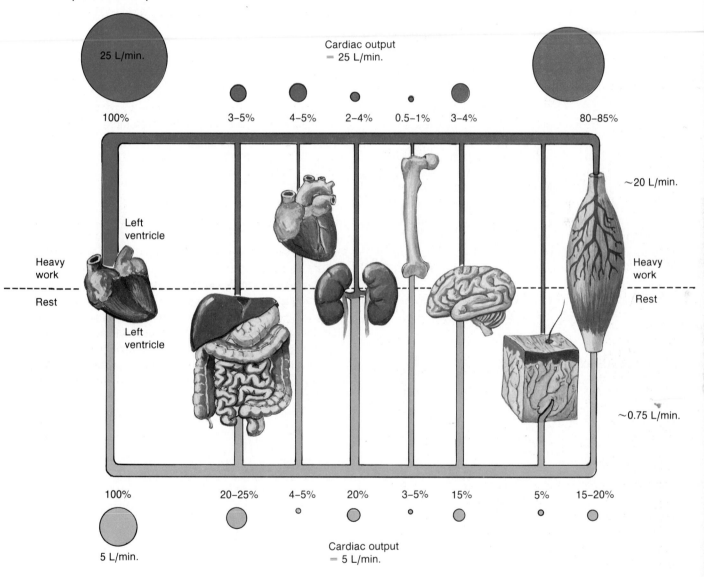

Figure 13.7   Cardiovascular adaptations to exercise.

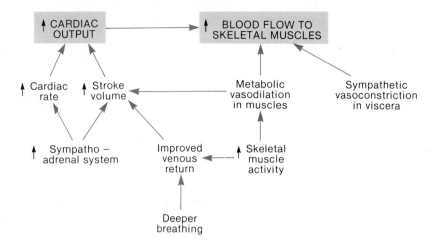

Since the cardiac rate is increased in exercise, the "filling time" between ventricular contractions in diastole is correspondingly reduced. Despite this, the end-diastolic volume is not significantly changed. This is because the rate of venous return during exercise is increased as a result of improved action of the skeletal muscle pump and increased respiratory movements. Stroke volume as well as cardiac rate thus increases during exercise. The cardiac output may therefore rise from 5 L per minute at rest to as high as 25 L per minute (or 35–40 L per minute in exceptional athletes) during strenuous activity. There is thus an increase in total blood flow as well as in the proportion of the total blood flow diverted to muscles during exercise.

Extraction of Blood Oxygen by Skeletal Muscles

High skeletal muscle metabolism during exercise may result in an eightfold to twentyfold increase in oxygen consumption. Despite this, the oxygen concentration in systemic arteries is not normally decreased, because of simultaneous changes in the rate and depth of breathing. The oxygen concentration in venous blood, however, can be reduced to one-half or one-third resting levels by exercising muscles. This indicates that the muscles are extracting a much higher proportion of the oxygen transported to them in the arterial blood.

Table 13.3   Effect of aging on the maximum cardiac rate.

Age	Maximum Cardiac Rate
20–29	190 beats/min
30–39	160 beats/min
40–49	150 beats/min
50–59	140 beats/min
60 and above	130 beats/min

Cardiovascular Effects of Physical Training

Cardiovascular responses vary in different types of exercises. The degree to which intrinsic vasodilation lowers peripheral resistance and the degree to which the skeletal muscle pumps improve venous return, for example, are directly related to the mass of skeletal muscles that is exercised. Cardiovascular adaptations are also influenced by the degree of effort involved, by the length of time this effort is maintained, and by the frequency of exercise training.

In *isometric* exercise, where muscle shortening during contraction is prevented, there is less enhancement of venous return and stroke volume than there is in isotonic exercise, and less of a decrease in peripheral resistance. This results in a greater increase in cardiac rate and in diastolic pressure, and thus in greater cardiac work during isometric exercise compared to isotonic exercise.

Table 13.4 Cardiovascular changes that occur during exercise.

Variable	Change	Mechanisms
Cardiac output	Increased	Cardiac rate and stroke volume increased.
Cardiac rate	Increased	Increased sympathetic nerve activity; decreased activity of the vagus.
Stroke volume	Increased	Increased myocardial contractility due to stimulation by sympathoadrenal system; decreased total peripheral resistance.
Total peripheral resistance	Decreased	Vasodilation of arterioles in skeletal muscles (and in skin when thermoregulatory adjustments are needed).
Arterial blood pressure	Increased	Increased systolic and pulse pressure due primarily to increased cardiac output; diastolic pressure rises less due to decreased total peripheral resistance.
End-diastolic volume	Unchanged	Decreased filling time at high cardiac rates is compensated by increased venous pressure, increased activity of the skeletal muscle pump, and decreased intrathoracic pressure aiding the venous return.
Blood flow to heart and muscles	Increased	Increased muscle metabolism produces intrinsic vasodilation; aided by increased cardiac output and increased vascular resistance in visceral organs.
Blood flow to visceral organs	—	Vasoconstriction in digestive tract, liver, and kidneys due to sympathetic nerve stimulation.
Bloc		Metabolic heat produced by exercising muscles produces reflex (involving hypothalamus) that reduces sympathetic constriction of arteriovenous shunts and arterioles.
Blood		Autoregulation of cerebral vessels maintains constant cerebral blood flow despite increased arterial blood pressure.

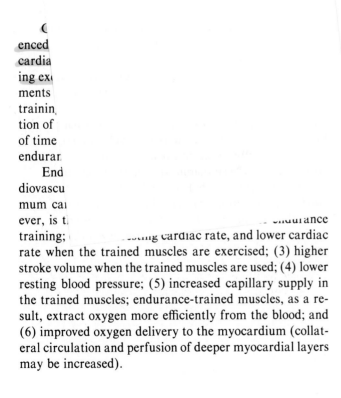

enced
cardia
ing ex
ments
trainin
tion of
of time
endurar
End
diovascu
mum ca
ever, is t
training;   cardiac rate, and lower cardiac
rate when the trained muscles are exercised; (3) higher
stroke volume when the trained muscles are used; (4) lower
resting blood pressure; (5) increased capillary supply in
the trained muscles; endurance-trained muscles, as a result, extract oxygen more efficiently from the blood; and
(6) improved oxygen delivery to the myocardium (collateral circulation and perfusion of deeper myocardial layers
may be increased).

1. Describe the effects of sympathetic nerves and of intrinsic metabolic control mechanisms on vascular resistance and blood flow in skeletal muscles and in the coronary circulation during rest and exercise.
2. Using a flow chart, describe the effects of exercise on blood flow to muscles and visceral organs, total peripheral resistance, venous return, and cardiac output.
3. Describe coronary blood flow during systole and diastole, and explain why oxygen delivery to the myocardium may be reduced at high cardiac rates.
4. Compare the effects of isometric exercise of the biceps muscles of the arms with isotonic exercise of many muscles (during swimming, for example) on venous return, total peripheral resistance, and cardiac rate. Also explain the long-term effects of these two forms of training on cardiac output, resting cardiac rate, and blood pressure.

Figure 13.8 *(a)* illustrates some of the vascular supply of the brain. *(b)* is an angiogram taken by injecting radiopaque contrast medium into the internal carotid artery. The arrow points to the middle cerebral artery.

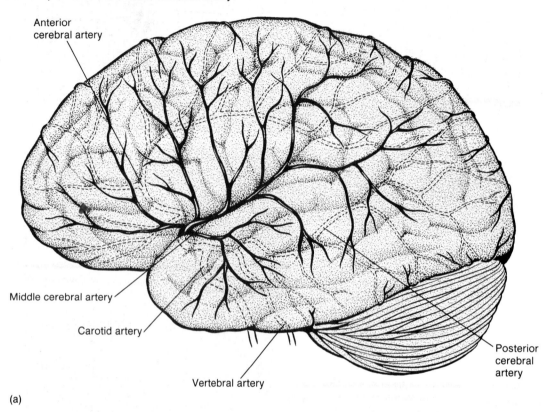

Anterior cerebral artery

Middle cerebral artery

Carotid artery

Vertebral artery

Posterior cerebral artery

(a)

Cerebral Circulation

Of all the organs in the body, the brain is the least able to tolerate a reduction in blood flow. When the brain is deprived of oxygen for a few seconds the person will lose consciousness; irreversible brain injury will occur after a few minutes. For these reasons, the cerebral blood flow is normally maintained remarkably constant at about 750 milliliters per minute. This amounts to about 15 percent of the total cardiac output at rest.

Unlike the coronary and skeletal muscle blood flow, cerebral blood flow is not normally influenced by sympathetic nerve activity. Only when the mean arterial pressure rises to about 200 mm Hg do sympathetic nerves cause a significant degree of vasoconstriction in the cerebral circulation. This vasoconstriction helps to protect small, thin-walled arterioles from bursting under the pressure, and thus helps to prevent cerebrovascular accident (stroke).

In the normal range of arterial pressures, cerebral blood flow is regulated almost exclusively by intrinsic mechanisms. These mechanisms help to insure a constant rate of blood flow despite changes in systemic arterial pressure—a process called **autoregulation.** Autoregulation of blood flow in the cerebral circulation is achieved by both myogenic and metabolic control mechanisms.

Myogenic regulation occurs when there is variation in systemic arterial pressure. The cerebral arteries exhibit a myogenic dilation when the blood pressure falls and a myogenic constriction when the pressure rises. Constriction and dilation, in other words, occur as a direct vascular smooth muscle response (a "myogenic" response) to changes in pressure. This helps to maintain a constant flow rate during the normal pressure variations that occur during rest, exercise, and emotional states.

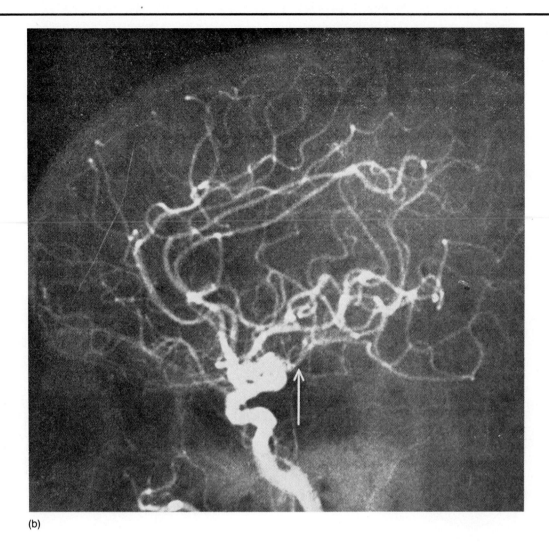

(b)

The cerebral vessels are also sensitive to the carbon dioxide concentration of arterial blood. When the carbon dioxide concentration rises, as a result of inadequate ventilation (hypoventilation), the cerebral arterioles dilate. This is believed to be due to associated decreases in the pH of cerebrospinal fluid, rather than to a direct effect of CO_2 on the cerebral vessels. Conversely, when the arterial CO_2 concentrations fall below normal (during hyperventilation), the cerebral vessels constrict. The resulting decrease in cerebral blood flow is responsible for the dizziness that occurs during hyperventilation.

The cerebral arterioles are exquisitely sensitive to local changes in metabolic activity, so that those brain regions with the highest metabolic activity get the most blood. Indeed, areas of the brain that control specific processes have been mapped by the changing patterns of blood flow that result when these areas are activated. Vision and hearing, for example, increase blood flow to the appropriate sensory areas of the cerebral cortex, while motor activities such as movements of the eyes, arms, and the organs of speech result in different patterns of blood flow (see figure 13.9).

Figure 13.9   Computerized picture of blood flow distribution in the brain after injecting the carotid artery with a radioactive isotope. Images show the distribution of the percent changes from the resting flow map. In *(a) (left)* the subject followed a moving object with his eyes. High activity is seen over the precentral gyrus (the somatomotor cortex) and over the occipital lobe of the brain. In *(a) (right)* the subject listened to spoken words. Notice that high activity is seen over the temporal lobe (the auditory cortex). In *(b) (left)* the subject moved his hand in the form of rhythmical fist clenching on the side of the body opposite to the cerebral hemisphere being studied. It should be noted that with skilled movements the supplementary motor area above and anterior to the primary sensory-motor hand area also show uncreased activity. In *(b) (right)* the average effect in nine subjects of counting to twenty was high activity seen over the mouth area of the motor cortex, the supplementary motor area, and over the auditory cortex.

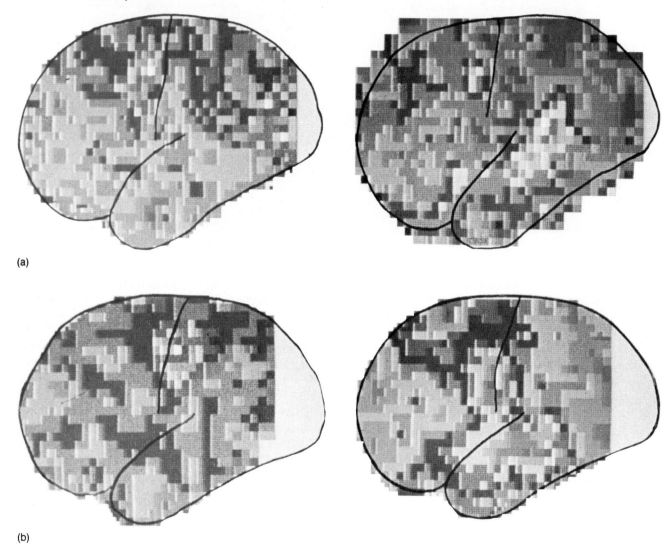

(a)

(b)

Figure 13.10 Circulation in the skin showing arteriovenous anastomoses or shunts that allow blood to be diverted directly from arteriole to venule, thus bypassing superficial capillary loops.

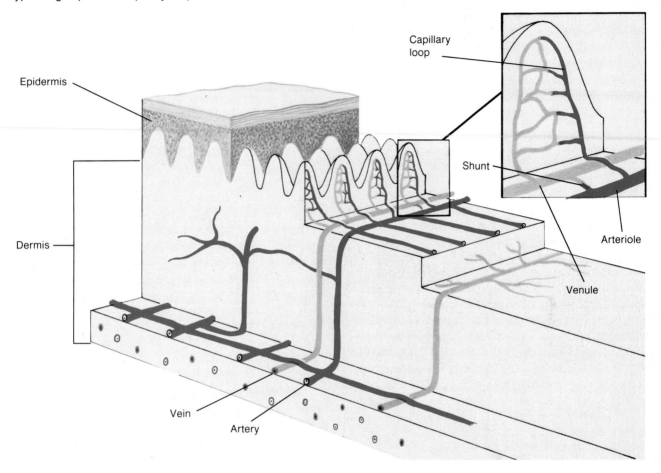

Blood Flow to the Skin

The skin is the outer covering of the body, and as such serves a variety of functions. These functions include defense against disease-causing organisms (pathogens) that would otherwise invade the body and protection against changes in deep body temperature. The small thickness and large size of the skin (1.0–1.5 mm thick; 1.7–1.8 square meters in surface area) make it an effective radiator of heat when the body temperature is greater than the ambient (outside) temperature. Transfer of heat from the body to the external environment is greatly aided by the flow of warm blood through capillary loops near the surface of the skin.

Blood flow through the skin is adjusted to maintain deep body temperature at about 37°C (98.6°F). When the body needs to lose heat, blood is diverted towards superficial capillary loops so that the heat can be radiated to the external environment. When heat must be conserved, unique *arteriovenous anastomoses* in the cutaneous vessels shunt blood from arterioles directly to deep venules, thus bypassing the superficial capillary loops. Opening of these shunts, and constriction of cutaneous arterioles that results from sympathetic nerve stimulation, reduces cutaneous blood flow when the ambient temperature is cold.

Skin can tolerate an extremely low blood flow rate in cold weather because its metabolic rate decreases when the ambient temperature decreases. In cold weather, therefore, the skin requires less blood. In very cold weather, however, blood flow to the skin can be so severely restricted that the tissue does die. This causes *frostbite*. Blood flow to the skin can vary from less than 20 ml per minute at maximal vasoconstriction to as much as 3–4 L per minute at maximal vasodilation.

As the temperature warms, cutaneous arterioles in the hands and feet dilate as a result of decreased sympathetic nerve activity. Continued warming causes dilation of arterioles in other areas of the skin. If the resulting increase in cutaneous blood flow is not sufficient to cool the body, secretion of sweat glands may be stimulated. Sweat helps to cool the body as it evaporates from the surface of the skin. Also, sweat glands secrete *bradykinin*, a polypeptide that stimulates vasodilation. This increases blood flow to the skin and to the sweat glands, so that larger volumes of more dilute sweat are produced.

Under usual conditions of ambient temperature, the cutaneous vascular resistance is high and the blood flow is low when a person is not exercising. In the pre-exercise state of "fight or flight" sympathetic nerve activity reduces blood flow still further. During exercise, however, the need to maintain a constant deep body temperature apparently takes precedence over the need to maintain an adequate blood pressure. As the body temperature rises during exercise, vasodilation in cutaneous vessels occurs together with vasodilation in skeletal muscles, producing an even greater lowering of total peripheral resistance. If exercise is performed in hot and humid weather, and if restrictive clothing is worn that increases skin temperature and cutaneous vasodilation, a dangerously low blood pressure may be produced after exercise has ceased and cardiac output has declined. People have lost consciousness—and even have died—as a result.

Changes in blood flow to the skin occur as a result of changes in sympathetic nerve activity. Since activity of the sympathetic system is controlled by the brain (see chapter 9), emotional states, acting through the control centers in the medulla oblongata, can thus affect sympathetic activity and cutaneous blood flow. During fear reactions, for example, vasoconstriction in the skin together with activation of sweat glands can produce a "cold sweat." The physiology of blushing is presently not well understood.

1. Define autoregulation, and describe how this is accomplished in the cerebral circulation.
2. Describe, using a flow chart, how hyperventilation leads to dizziness.
3. Describe how cutaneous blood flow is adjusted to maintain a constant deep body temperature.

Circulatory Shock and Congestive Heart Failure

Autonomic reflexes and intrinsic control mechanisms function to maintain adequate tissue perfusion when the cardiovascular system is challenged with physiological stresses. The cardiovascular system responds effectively to the stress of exercise, as described in this chapter, and to such other challenges as the redistribution of blood volume that occurs when a person goes from a lying to a standing position. When cardiovascular responses are insufficient to maintain adequate tissue perfusion, *circulatory shock* may occur.

Circulatory Shock
Circulatory shock occurs when there is inadequate blood flow and/or oxygen utilization by the tissues. Some of the signs of shock (see table 13.5) are a result of inadequate tissue perfusion; other signs of shock are produced by cardiovascular responses that help to compensate for the poor tissue perfusion (see table 13.6). When these compensatory responses are effective, they—together with emergency medical care—are able to reestablish adequate tissue perfusion. In some cases, however, neither physiological control mechanisms nor medical treatment can halt the progression to irreversible shock and death.

Hypovolemic Shock　Shock due to low blood volume may be caused by hemorrhage (bleeding), dehydration, or burns. This can result in decreased blood pressure, decreased venous return, and decreased cardiac output. In response to these changes, baroreceptors in the carotid sinus and aortic arch, and stretch receptors in the heart, mediate a reflex response through the autonomic system. Activation of the sympathoadrenal system stimulates cardiac rate and contractility. Sympathetic nerves also stimulate vasoconstriction in the skin, digestive tract, kidneys, and muscles.

Table 13.5 Signs of shock.

	Early Sign	Late Sign
Blood pressure	Decreased pulse pressure Increased diastolic pressure	Decreased systolic pressure
Urine	Decreased Na$^+$ concentration Increased osmolality	Decreased volume
Blood pH	Increased pH (alkalosis) due to hyperventilation	Decreased pH (acidosis) due to "metabolic" acids
Effects of poor tissue perfusion	Slight restlessness; occasionally warm, dry skin	Cold, clammy skin "Cloudy" senses

Source: R. F. Wilson, ed., *Principles and Techniques of Critical Care*, vol. 1, Upjohn Company, 1977.

Table 13.6 Cardiovascular reflexes that help to compensate for circulatory shock.

Organ(s)	Compensatory Mechanisms
Heart	Sympathoadrenal stimulation produces increased cardiac rate and increased stroke volume, due to "positive inotropic effect" on myocardial contractility (strength of contraction).
Digestive tract and skin	Decreased blood flow due to vasoconstriction as a result of sympathetic nerve stimulation (alpha-adrenergic effect).
Kidneys	Decreased urine production as a result of sympathetic-nerve-induced constriction of renal arterioles. Increased salt and water retention due to increased aldosterone and antidiuretic hormone (ADH) secretion.

Since the resistance in the coronary and cerebral circulations is not increased, blood is diverted to the heart and brain at the expense of other organs. Interestingly, a similar response occurs in diving mammals, and to a lesser degree in Japanese pearl divers, during prolonged submersion. These responses help to deliver blood to the two organs that have the highest requirements for aerobic metabolism.

Vasoconstriction in organs other than the brain and heart raises total peripheral resistance, which helps (along with the reflex increase in cardiac rate) to compensate for the drop in blood pressure due to low blood volume. Constriction of arterioles also decreases capillary blood flow and capillary filtration pressure. Less filtrate is formed, as a result, while the osmotic return of fluid to the capillary is increased because of increased plasma colloid osmotic pressure. The blood volume is thus raised at the expense of tissue fluid volume. Blood volume is also conserved by decreased urine production, which occurs as a result of vasoconstriction in the kidneys and the water-conserving effects of ADH and aldosterone, which are secreted in increased amounts during shock.

Other Causes of Circulatory Shock   A rapid fall in blood pressure occurs in *anaphylactic shock* as a result of severe allergic reactions (usually to bee stings or penicillin). This occurs because of widespread release of histamine, which causes vasodilation and decreased total peripheral resistance. A rapid fall in blood pressure similarly occurs in *neurogenic shock*, in which sympathetic "tone" is decreased, usually because of upper spinal cord damage or by spinal anesthesia. *Cardiogenic shock* results from cardiac failure, as defined by a cardiac output that is inadequate to maintain tissue perfusion.

Congestive Heart Failure

Cardiac failure occurs when the cardiac output is insufficient to supply the blood flow required by the body. This may be due to heart disease—resulting from myocardial infarction or congenital defects—or to hypertension, which increases the workload of the heart. The most common causes of "left pump" failure are myocardial infarction, arotic valve stenosis, and incompetence of the aortic and mitral valves. Atrial fibrillation in the presence of ventricular disease may precipitate heart failure, but atrial fibrillation alone is not immediately life threatening (as described in chapter 11). Failure of the "right pump" is usually caused by prior failure of the left pump.

Table 13.7 Cardiac output and distribution of blood flow in cardiac failure during rest and exercise. Absolute values are expressed per square meter (m²) of body surface area.

	Normal		Cardiac Failure	
	Rest	Exercise	Rest	Exercise
Cardiac output				
(L/min./m.²)	3.0	6.0	1.5	2.3
Blood flows (% of cardiac output)				
Muscle	20	50	36	60
Splanchnic	25	12	25	10
Renal	20	9	10	4
Cerebral	12	6	12	12
Coronary	4	4	9	10
Skin	9	15	3	1
Blood flows (absolute values in ml./min./m.²)				
Muscle	600	3,000	540	1,380
Renal	600	540	150	92
Coronary	120	240	135	230

Source: D. T. Mason, *Modern Concepts in Cardiovascular Disease* 36 (1967): 25. By permission of the American Heart Association, Inc.

Heart failure can also result from disturbance in the electrolyte concentrations of the blood. Excessive plasma K^+ concentration decreases the resting membrane potential of myocardial cells; low blood Ca^{++} reduces excitation-contraction coupling. High blood K^+ and low blood Ca^{++} can thus cause the heart to stop in diastole. Conversely, low blood K^+ and high blood Ca^{++} can arrest the heart in systole.

The term *congestive* is often used in describing heart failure because of the increased venous volume and pressure that results. Failure of the left pump, for example, raises the left atrial pressure and produces pulmonary congestion and edema. This causes shortness of breath, while the reduced cardiac output causes fatigue. Failure of the right pump results in increased right atrial pressure, which produces congestion and edema in the systemic circulation. The effects of cardiac failure on blood distribution are summarized in table 13.7.

The compensatory responses that occur during heart failure are similar to those that occur during hypovolemic shock. Activation of the sympathoadrenal system stimulates cardiac rate, contractility of the ventricles (see figure 13.11), and constriction of peripheral arterioles. As in hypovolemic shock, these responses are accompanied by decreased urine production due to renal vasoconstriction and increased secretion of ADH and aldosterone.

Figure 13.11 Relationship between end-diastolic volume and ventricular pressure (a measure of contraction strength). At a given end-diastolic volume the ventricular pressure is decreased during heart failure and increased by sympathetic stimulation. Heart failure may be compensated by sympathetic stimulation.

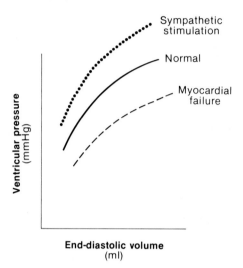

As a result of the compensatory responses, chronically low cardiac output is associated with elevated blood volume and dilation and hypertrophy of the ventricles. These changes can themselves be dangerous. Elevated blood volume places a work overload on the heart, and the dilated ventricles have a higher metabolic requirement for oxygen. These problems are often treated with drugs that increase myocardial contractility (such as digitalis), drugs that are vasodilators (such as nitroglycerin), and diuretic drugs that lower blood volume by increasing the volume of urine produced.

1. Using a flow chart to show cause-and-effect, explain why a person in hypovolemic shock may have a fast pulse and cold, clammy skin.
2. Describe the compensatory mechanisms that act to raise blood volume during cardiovascular shock.
3. Describe congestive heart failure, and explain the compensatory responses that occur during this condition.

Figure 13.12 Blood cells become packed at the bottom of the test tube when whole blood is centrifuged, leaving the fluid plasma at the top of the tube. Red blood cells are the most abundant of the blood cells—white blood cells and platelets form only a thin, light-colored "buffy coat" at the interface between the packed red blood cells and the plasma.

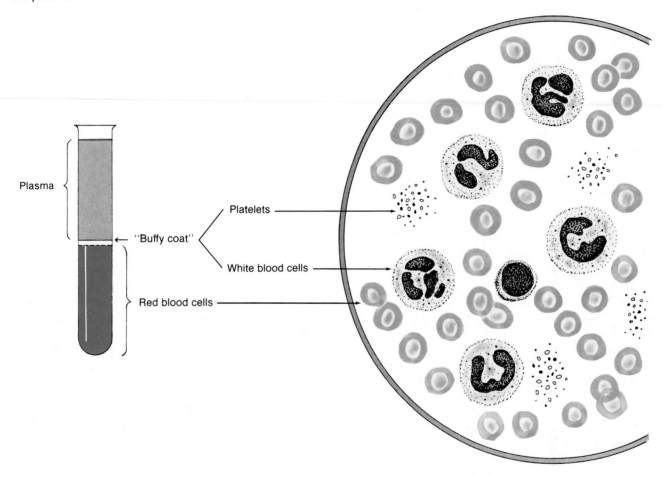

Blood Composition and Clotting Mechanisms

If a sample of blood is centrifuged, so that the cells are packed at the bottom side of the tube and the extracellular fluid—**plasma**—is on top, it will be seen that the packed **red blood cells**, or **erythrocytes**, comprise about 37 to 52 percent of the sample volume. This ratio is known as the *hematocrit*. The value of this measurement is significant because red blood cells function to transport oxygen (this function of blood is discussed in chapter 15).

Above the packed red blood cells, there is a thin "buffy coat" of packed **white blood cells (leukocytes)** and **platelets.** White blood cells help provide immunity to infection, as described in chapter 17. Platelets function in blood clotting.

Plasma contains salts and ions, as well as organic molecules such as metabolites, hormones, enzymes, and antibodies. The composition, volume, and pH of plasma is maintained within normal limits by a variety of physiological control mechanisms. Soluble clotting factors in plasma act together with platelets to form blood clots, and thus help prevent loss of blood volume when vessels are damaged.

Table 13.8 Representative normal blood values.

Measurement	Normal Range
Blood volume	80–85 ml/kg body weight
Blood osmolality	285–295 mOsm
Blood pH	7.35–7.45
Enzymes	
Creatine phosphokinase	Female: 5–35 mU/ml
(CPK)	Male: 5–55 mU/ml
Lactic dehydrogenase (LDH)	60–120 U/ml
Phosphatase (acid)	Female: 0.01–0.56 Sigma U/ml
	Male: 0.13–0.63 Sigma U/ml
Hematology values	
Hematocrit	Female: 37–48%
	Male: 45–52%
Hemoglobin	Female: 12–16 g/100ml
	Male: 13–18 g/100ml
Red blood cell count	4.2–5.9 million/mm³
White blood cell count	4,300–10,880/mm³
Hormones	
Testosterone	Male: 300–1,100 ng/100ml
	Female: 25–90 ng/100ml
Adrenocorticotrophic	15–70 pg/ml
Hormone (ACTH)	
Growth hormone	Children: over 10 ng/ml
	Adult male: below 5 ng/ml
Insulin	6–26 µU/ml (fasting)
Ions	
Bicarbonate	24–30 mmol/l
Calcium	2.1–2.6 mmol/l
Chloride	100–106 mmol/l
Potassium	3.5–5.0 mmol/l
Sodium	135–145 mmol/l
Organic molecules (other)	
Cholesterol	120–220 mg/100ml
Glucose	70–110 mg/100ml (fasting)
Lactic acid	0.6–1.8 mmol/l
Protein (total)	6.0–8.4 g/100ml
Triglyceride	40–150 mg/100ml
Urea nitrogen	8–25 mg/100ml
Uric acid	3–7 mg/100ml

Source: *New England Journal of Medicine*, 1980.

Functions of Platelets

Platelets are not whole cells; rather, they are cytoplasmic fragments of large, multinucleate cells in the bone marrow known as *megakaryocytes*. In the absence of vessel damage, platelets are repelled from each other and from the endothelial lining of the vessels. Repulsion of platelets from an intact endothelium is believed to be due to *prostacyclin*, a derivative of prostaglandins produced within the endothelium. Mechanisms that prevent platelets from sticking to the blood vessels and to each other are obviously needed to prevent inappropriate blood clotting.

Damage to the endothelium exposes subendothelial connective tissue to the blood. Platelets are able to stick to exposed collagen proteins, which form the major supporting fibers of connective tissues. Platelets that stick to collagen *degranulate*, as their content of secretory granules release numerous chemicals, including ADP (adenosine diphosphate), serotonin, and prostaglandins. This event is also known as the **platelet release reaction.**

Effect of ADP Release of ADP makes other platelets "sticky," so that they adhere to those that are stuck to the collagen. This second layer of platelets, in turn, undergoes a release reaction, so that the ADP that is secreted causes additional platelets to aggregate at the site of the injury. This produces a *platelet plug* in the damaged vessel, which is strengthened by activation of plasma clotting factors (described in the next section).

Effect of Aspirin In order to undergo a release reaction, production of a prostaglandin known as *thromboxane A_2* within the platelets is required. *Aspirin* inhibits conversion of arachidonic acid (a cyclic fatty acid) into prostaglandins, and thus inhibits the release reaction and consequent formation of a platelet plug. Ingestion of excessive amounts of aspirin can thus significantly prolong bleeding time. This is why blood donors and women in the last trimester of pregnancy are advised to avoid aspirin.

Clotting Factors: Formation of Fibrin

The platelet plug is strengthened by a meshwork of insoluble protein fibers known as **fibrin** (see figure 13.13). Blood clots therefore contain platelets, fibrin, and often trapped red blood cells that give the clot a red color (clots formed in arteries generally lack red blood cells and are grey in appearance). Finally, contraction of the platelets in the process of *clot retraction* forms a more compact and effective plug. Fluid squeezed from the clot as it retracts is called **serum**, which is defined as plasma without fibrinogen (the soluble precursor of fibrin).

Figure 13.13 Scanning electron micrograph of a red blood cell caught in a web of fibrin.

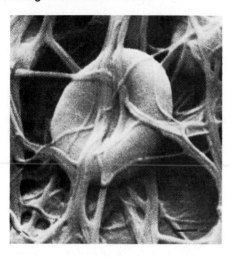

Table 13.9 The plasma clotting factors.

Factor I	Fibrinogen
Factor II	Prothrombin
Factor III	Tissue thromboplastin
Factor IV	Calcium
Factor V	Proaccelerin
Factor VII	Proconvertin; SPCA
Factor VIII	Antihemophilic factor (AHF)
Factor IX	Plasma thromboplastin component (PTC), Christmas factor
Factor X	Stuart-Prower factor
Factor XI	Plasma thromboplastin antecedent (PTA)
Factor XII	Hageman factor
Factor XIII	Fibrin stabilizing factor

Figure 13.14 Sequence of events leading to platelet aggregation and the formation of a blood clot.

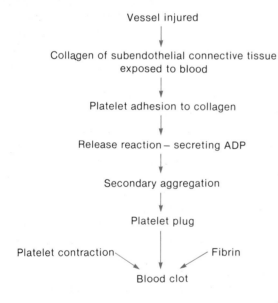

There are two pathways that result in conversion of fibrinogen into fibrin. Blood left in a test tube will clot without the addition of any external chemicals; the pathway that produces this clot is thus called the *intrinsic pathway*. Damaged tissue, however, releases a chemical that initiates a "shortcut" in the formation of fibrin. This pathway of fibrin formation is thus called the *extrinsic pathway*.

Intrinsic Pathway   The intrinsic clotting pathway begins when collagen proteins in the subendothelial tissue of a vessel (or the glass walls of a tube) activate a soluble plasma protein called *Hageman factor*, or *factor XII*. Active factor XII (XII_a) is a protein-digesting enzyme, or protease, that activates another plasma protein—*factor XI*—by cleaving part of its inactive precursor. Activated factor XI, in turn, activates *factor IX*.

Combination of active factor IX with *factor VIII* and with platelet phospholipids activates *factor X*. A complex is then formed between active *factor X, factor V*, and platelet phospholipids that convert *prothrombin* (inactive factor II) to *thrombin* (factor II_a).

Thrombin is a protease that converts the soluble protein fibrinogen (factor I) into fibrin monomers. These monomers, finally, are joined together to form the insoluble protein fibrinogen by the action of *factor XIII* (which is also activated by thrombin). Calcium ions (Ca^{++}) are involved in a number of these reactions (see figure 13.15).

Figure 13.15   The intrinsic and extrinsic pathways leading to the formation of fibrin (see text for details).

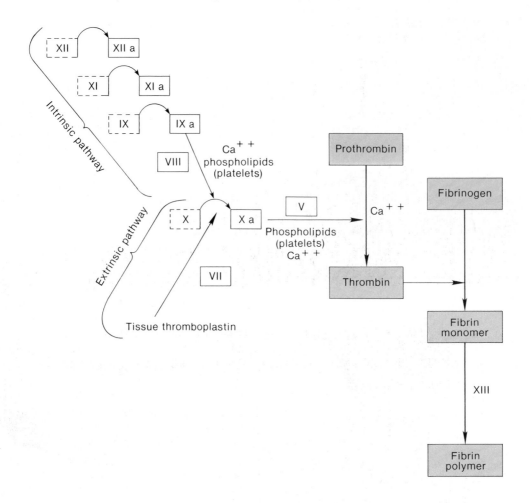

Extrinsic Pathway   Formation of fibrin can occur more rapidly as a result of release of *tissue thromboplastin* from damaged cells. Thromboplastin combines with factor VII and Ca^{++} to form a complex that activates factor X. From this point, the extrinsic pathway is identical to the intrinsic pathway; both lead to conversion of prothrombin to thrombin and conversion of fibrinogen to fibrin.

Anticoagulants and Hereditary Clotting Disorders

Blood placed in a test tube will clot due to activation of Hageman factor and initiation of the intrinsic pathway. If this blood is centrifuged, the clot and trapped cells will become packed at the bottom of the tube as a gelatinous mass, leaving serum at the top of the tube. In order to prevent clotting, and obtain plasma instead of serum, anticoagulants must be added to the tube.

Addition of *citrate* to a test tube of blood prevents clotting by combining with Ca^{++}. A mucoprotein called *heparin* can also be added to the tube to prevent clotting. Heparin activates a plasma protein called *antithrombin III*, which combines with and inactivates thrombin. Heparin is often given to patients to prevent clotting during various medical procedures. The *coumarin anticoagulants* prevent clotting by competing with *vitamin K*, which is needed for formation of factors II, VII, IX, and X in the liver. In contrast to the immediate effects of heparin, coumarin must be given to a patient for several days to be effective.

Anticoagulants and dietary deficiency in vitamin K result in *acquired disorders* of the clotting system. There are also several genetic diseases that result in impaired clotting ability. Examples of such *hereditary disorders* include two different genetic defects in factor VIII. A defect in one subunit of factor VIII prevents this factor from participating in the intrinsic clotting pathway. This genetic disease is known as *hemophilia A*, and is prevalent in the royal families of Europe.

Figure 13.16 Role of activated factor XII in initiating pathways leading to dissolution of blood clots and vasodilation.

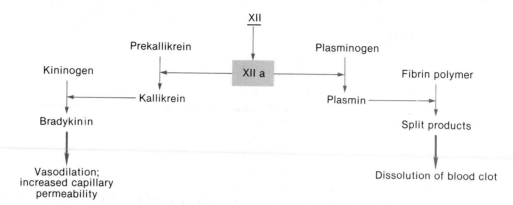

Table 13.10 Some acquired and inherited defects in the clotting mechanism.

Category	Cause of Disorder	Comments
Acquired clotting disorders	Vitamin K deficiency	Inadequate formation of prothrombin and other clotting factors in the liver
	Aspirin	Inhibits prostaglandin production, resulting in defective platelet release reaction
Anticoagulants	Coumarin	Competes with the action of vitamin K
	Heparin	Inhibits activity of thrombin
	Citrate	Combines with Ca^{++} and thus inhibits the activity of many clotting factors
Inherited clotting disorders	Hemophilia A (defective factor VIII_{AHF})	Recessive trait carried on X chromosome; results in delayed formation of fibrin
	Von Willebrand's disease (defective factor VIII_{VWF})	Dominant trait carried on autosomal chromosome; impaired ability of platelets to adhere to collagen in subendothelial connective tissue
	Hemophilia B (defective factor IX), also called Christmas disease	Recessive trait carried on X chromosome; results in delayed formation of fibrin

A defect in another subunit of factor VIII results in *Von Willebrand's disease*. In this condition, rapidly circulating platelets are unable to stick to collagen. The mechanism by which this subunit of factor VIII normally promotes platelet adhesion to collagen is not presently known.

Dissolution of Clots

As the damaged blood vessel wall is repaired, activated Hageman factor initiates another sequence of reactions that ultimately lead to the conversion of *plasminogen* to *plasmin* (also called *fibrinolysin*). Plasmin is an enzyme that digests fibrin into "split products," thus promoting dissolution of the clot.

Activated Hageman factor also initiates another pathway that results in the formation of a molecule called *bradykinin*. Bradykinin stimulates vasodilation, which helps to counter the vasoconstrictor effects of thromboxane A_2 that is secreted by aggregating platelets during the release reaction.

1. Describe the platelet release reaction and its significance. Explain why ingestion of aspirin can prolong clotting time.
2. Explain why defective factor VIII results in hemophilia; describe the effect of this genetic defect on the intrinsic and extrinsic clotting pathways.
3. Explain why people with vitamin K deficiency, or people treated with coumarins, have difficulty clotting. Also explain how heparin and citrate act on the clotting system.

Summary

Coronary Circulation; Adaptations to Exercise

I. Coronary arteries deliver systemic arterial blood to the myocardium.
   A. Contraction of the heart in systole squeezes the coronary arteries so that blood flow is greatest during diastole.
   B. During systole, the myocardium receives oxygen from myoglobin, which in turn loads with oxygen during diastole.
   C. The coronary arteries normally deliver sufficient oxygen, even during exercise, to maintain aerobic respiration in the myocardium.
      1. At very fast cardiac rates, however, blood flow is reduced because of the correspondingly shorter diastolic phases.
      2. If coronary vessels are occluded, bypass surgery may have to be performed.
II. The metabolic requirements of the heart and skeletal muscles increase during exercise.
   A. During rest, blood flow through the coronary and skeletal muscle arterioles is relatively low because of a high resting vascular resistance due to sympathetic nerve stimulation.
   B. As exercise progresses, metabolic products in the heart and skeletal muscles (such as CO_2) accumulate within the tissues.
   C. The metabolic products act intrinsically (within the tissue) to stimulate vasodilation and decrease vascular resistance.
   D. Blood flow to the heart and exercising muscles, as a result, can increase up to five and twenty times, respectively, the flow rates at rest.
      1. Vascular resistance in the viscera increases during exercise as a result of adrenergic sympathetic stimulation.
      2. Despite vasoconstriction in the viscera, during exercise of large muscle groups the total peripheral resistance decreases.
III. In addition to diversion of blood from the viscera to the heart and muscles, total blood flow increases during exercise.

A. Cardiac output can increase from a resting value of about 5 L per minute to about 25 L per minute in heavy exercise.
B. The increased cardiac output is due to both increased cardiac rate and increased stroke volume.
   1. Despite the increased cardiac rate, end-diastolic volume is not reduced because of improved venous return to the heart during exercise.
   2. Increased myocardial contractility and decreased after load contribute to the rise in stroke volume during exercise.
C. Exercising muscles extract more oxygen from the blood.
   1. Arterial oxygen concentrations do not decrease, however, because breathing also increases.
   2. Trained muscles are able to extract more oxygen from the blood than untrained muscles during exercise.

Cerebral and Cutaneous Circulation

I. Within normal limits, cerebral blood flow is maintained constant in the face of fluctuating systemic arterial pressures.
   A. This autoregulation of blood flow is due to intrinsic metabolic and myogenic mechanisms.
   B. Cerebral vascular smooth muscle dilates as the systemic pressure falls and constricts as the pressure rises—this is a myogenic control of blood flow.
   C. Cerebral vessels are also sensitive to metabolic products.
      1. A rise in blood CO_2 causes cerebral vasodilation; a fall in blood CO_2 (due to hyperventilation) causes cerebral vasoconstriction.
      2. Changing patterns of metabolism within the brain intrinsically directs blood flow to those areas that are most active.
II. Cutaneous blood flow is regulated in large part by adrenergic sympathetic nerve activity.

A. During cold weather, and during the fight or flight reaction, sympathetic nerves cause cutaneous vasoconstriction.
   1. This raises total peripheral resistance and blood pressure while also preventing body heat from being lost from the blood through the skin.
   2. The skin is able to tolerate very low rates of blood flow because its metabolism decreases as the ambient temperature gets colder.
B. Body temperature is also regulated by arteriovenous shunts in the cutaneous circulation.
   1. When the ambient temperature is warm, these shunts are closed and blood in arterioles passes into superficial capillary loops to radiate heat.
   2. When the ambient temperature is cold, opening of these shunts diverts blood from arterioles directly to deep venules, thus bypassing the superficial capillary loops and conserving heat.

Circulatory Shock, Congestive Heart Failure, and Blood Clotting

I. Circulatory shock occurs when there is inadequate tissue perfusion with blood and/or inadequate oxygen utilization by the tissues to meet their metabolic needs.
   A. Hypovolemic shock occurs when the blood volume is reduced by hemorrhage, dehydration, or burns.
   B. Hypovolemic shock causes a reduction in venous return and cardiac output, leading to a fall in blood pressure.
      1. A fall in blood volume and pressure stimulates the baroreceptor reflex and atrial stretch reflex, causing activation of the sympathoadrenal system.
      2. Increased sympathetic stimulation produces increased cardiac rate, increased myocardial contractility, and vasoconstriction in the viscera and skin.

3. Blood, as a result, is preferentially shunted to the heart and brain.
4. Vasoconstriction of arterioles in the kidneys, combined with increased secretion of ADH and aldosterone, helps to reduce urine volume and thus to conserve blood volume.
5. There is also a net return of tissue fluid to the vascular system that helps to expand the blood volume.

C. Cardiogenic shock is produced by failure to maintain an adequate cardiac output.
1. Heart failure may result from weakness of the myocardium, cardiac arrhythmias, electrolyte imbalance, or excessive arterial blood pressure.
2. The term *congestive heart failure* is used to indicate the venous congestion that may result when the ventricles fail to eject a sufficient output.

II. Blood clotting is usually initiated by damage to the endothelium of a blood vessel.
A. Circulating blood platelets stick to exposed collagen, but not to an intact endothelium.
B. The platelets that stick to collagen undergo a release reaction that releases ADP and other molecules.
C. ADP makes other platelets "sticky," so they aggregate at the site of the damaged vessel.
D. Exposure of collagen to plasma activates Hageman factor, a circulating plasma protein.
1. Activated Hageman factor leads to a cascade of reactions, in which one factor helps to activate another.
2. This is called the intrinsic clotting pathway.
E. The extrinsic clotting pathway is initiated by secretion of tissue thromboplastin from damaged tissue cells.

III. The extrinsic and intrinsic pathways both lead to conversion of prothrombin to thrombin.

A. The extrinsic pathway is shorter than the intrinsic pathway because fewer steps are involved.
B. Thrombin converts fibrinogen to fibrin, which is a fibrous protein that forms a supporting meshwork around the platelet plug.
C. Clot retraction makes a more compact plug, as serum (plasma minus fibrinogen) is squeezed from the contracting plug.
D. The clotting process can be interrupted by a number of mechanisms.
1. Aspirin inhibits the platelet release reaction.
2. Vitamin K is needed by the liver to form various clotting factors; coumarin anticoagulants compete with vitamin K.
3. Citric acid is an anticoagulant because it binds Ca^{++}, which is a necessary cofactor in some of the activation steps.
4. Heparin is a mucoprotein that indirectly inactivates thrombin, and thus prevents clotting.

Self-Study Quiz

1. Vasodilation in heart and skeletal muscles during exercise is primarily due to the effects of:
   (a) alpha-adrenergic stimulation
   (b) beta-adrenergic stimulation
   (c) cholinergic stimulation
   (d) products released during metabolism of tissue cells
2. Blood flow in the coronary circulation is:
   (a) increased during systole
   (b) increased during diastole
   (c) constant throughout the cardiac cycle
3. During exercise, there is an increase in all of the following EXCEPT:
   (a) total peripheral resistance
   (b) cardiac rate
   (c) stroke volume
   (d) ejection fraction
   (e) venous return
4. Blood flow in the cerebral circulation:
   (a) varies with the systemic arterial pressure

   (b) is regulated primarily by the sympathetic system
   (c) is maintained constant within physiological limits
   (d) is increased during exercise
5. Which of the following organs is able to tolerate the greatest restriction in blood flow?
   (a) brain
   (b) heart
   (c) skeletal muscles
   (d) skin
6. Arteriovenous shunts in the skin:
   (a) divert blood to superficial capillary loops
   (b) are closed when the ambient temperature is very cold
   (c) are closed when the deep body temperature rises above 37°C
   (d) all of these
7. A person in hypovolemic shock will have:
   (a) low blood pressure
   (b) rapid pulse
   (c) low urine production
   (d) vasoconstriction in the skin
   (e) all of these

8. The ability of platelets to stick together is promoted by:
   (a) ATP
   (b) ADP
   (c) cyclic AMP
   (d) adenosine
9. A defect in factor VIII could prevent:
   (a) the intrinsic pathway of fibrin formation
   (b) the extrinsic pathway of fibrin formation
   (c) the ability of platelets to stick to collagen
   (d) both (a) and (b)
   (e) both (a) and (c)
10. Vitamin K is needed for blood clotting because it:
   (a) is required by the liver for production of several clotting factors
   (b) is needed for activation of clotting factors in the plasma
   (c) is required for the platelet release reaction
   (d) converts fibrin monomers into fibrin polymer

14 Respiration: *Ventilation and Gas Exchange in the Lungs*

Objectives

By studying this chapter, you should be able to:

1. Describe the structure and function of the conducting and respiratory zones of the lungs and how gas exchange occurs

2. Describe, in terms of Boyle's law, how changes in intrapulmonary pressure result in inspiration and expiration

3. Describe how changes in thoracic volume are produced during normal and forced ventilation and how this affects intrapulmonary and intrapleural pressures

4. Explain how the transpulmonary pressure normally prevents the lungs from collapsing and why lung collapse occurs in a pneumothorax

5. Define lung compliance, and explain how it is affected by pulmonary fibrosis and emphysema

6. Describe how the surface tension of alveoli changes during breathing, and describe the role of lung surfactant

7. Differentiate between obstructive and restrictive pulmonary disease, and describe how these conditions affect pulmonary function testing.

8. Explain how the partial pressures of a gas mixture are calculated and how they are affected by changes in water vapor pressure and altitude

9. Describe how the P_{O_2} and P_{CO_2} of blood is measured, and explain the significance of these measurements in terms of oxygen content of blood and the delivery of oxygen to the tissues

10. Describe the significance of the ventilation/perfusion ratios in the lungs

*T*he term *respiration* includes three separate but related functions: (1) **ventilation** (breathing); (2) **gas exchange,** between the air and blood in the lungs and between the blood and tissues; and (3) **oxygen utilization** by the tissues in the energy-liberating reactions of cell respiration (as described in chapter 5). Ventilation and the exchange of gases (oxygen and carbon dioxide) between the air and blood are together called *external respiration*. Gas exchange between the blood and tissues and oxygen utilization by the tissues are known together as *internal respiration*.

Figure 14.1   Diagram showing relationship between lung alveoli and pulmonary capillaries.

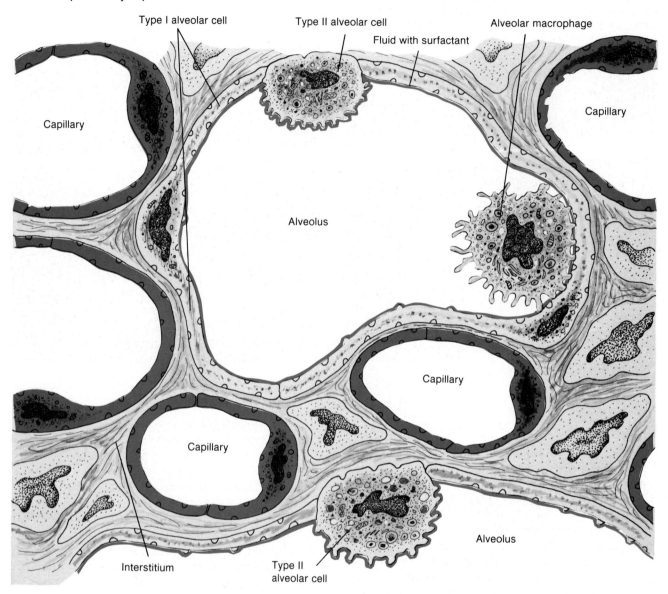

Type I alveolar cell

Type II alveolar cell

Alveolar macrophage

Fluid with surfactant

Capillary

Capillary

Alveolus

Capillary

Capillary

Capillary

Interstitium

Type II alveolar cell

Alveolus

Ventilation is the mechanical process that moves air into and out of the lungs. Since air in the lungs has a higher oxygen concentration than the blood, oxygen diffuses from air to blood. Carbon dioxide, similarly, moves from the blood to the air within the lungs by diffusion. As a result of this gas exchange, the inspired air contains more oxygen and less carbon dioxide than the expired air. More importantly, blood leaving the lungs (in the pulmonary veins) contains a higher oxygen and a lower carbon dioxide concentration than the blood delivered to the lungs in the pulmonary artery. This results from the fact that the lungs function to bring the blood into gaseous equilibrium with the air.

Gas exchange between the air and blood occurs entirely by diffusion through lung tissue. This diffusion occurs very rapidly because there is a high surface area within the lungs and a very short diffusion distance between blood and air. The fact that blood in the pulmonary veins and systemic arteries is almost in complete equilibrium with the inspired air is testimony to the high efficiency of normal lung function.

Figure 14.2 *(a)* is a scanning electron micrograph showing lung alveoli and a small bronchiole. *(b)* shows the alveoli under higher power (the arrow points to an alveolar pore through which air can pass from one alveolus to another).

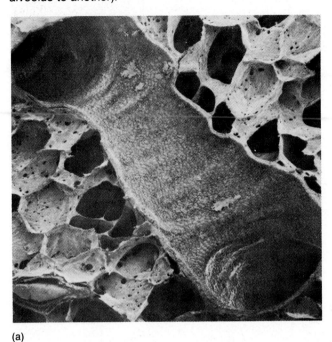

(a)

(b)

Structure of the Respiratory System

Gas exchange in the lungs occurs across about 300 million tiny (0.25–0.50 mm in diameter) "air sacs" known as **alveoli.** The enormous number of these structures provides a high surface area—60–80 square meters, or about 760 square feet—for diffusion of gases. The diffusion rate is further increased by the fact that each alveolus is only one cell-layer thick, so that the total "air-blood barrier" is only two cells across (an alveolar cell and a capillary endothelial cell), or about 2 μm. This is an average distance because the type II alveolar cells are thicker than the type I cells (see figure 14.1).

Alveoli are polyhedral in shape and are usually clustered together, like the units of a honeycomb. Air within one member of a cluster can enter other members through tiny pores. These clusters of alveoli usually occur at the ends of *respiratory bronchioles,* which are the very thin air tubes that end blindly in alveolar sacs. Individual alveoli also occur as separate outpouchings along the length of respiratory bronchioles. Although the distance between each respiratory bronchiole and its terminal alveoli is only about 0.5 mm, these units together comprise most of the mass of the lungs.

The air passages of the respiratory system are divided into two functional zones. The **respiratory zone** is the region where gas exchange occurs, and it therefore includes the respiratory bronchioles (because they contain separate outpouchings of alveoli) and the terminal clusters of alveolar sacs. The **conducting zone** includes all of the anatomical structures through which air passes before reaching the respiratory zone.

Air enters the respiratory bronchioles from *terminal bronchioles,* which are narrow airways formed from many successive divisions of the right and left *primary bronchi.* These two large air passages, in turn, are continuous with the *trachea,* or windpipe, which is located in the neck in front of the esophagus (a muscular tube carrying food to the stomach). The trachea is a sturdy, non-muscular tube supported by rings of cartilage. (An opening called a tracheotomy is sometimes created surgically to aid ventilation).

Air enters the trachea from the *pharynx,* which is the cavity located behind the palate that receives the contents of both the oral and nasal passages. In order for air to enter or leave the trachea and lungs, however, it must pass through a valvelike opening called the *glottis* between the vocal cords. The vocal cords are part of the *larynx,* or voice box, which guards the entrance to the trachea. The Adam's apple in the neck is produced by a protruding part of the larynx.

The conducting zone of the respiratory system, in summary, consists of the mouth, nose, pharynx, larynx, trachea, primary bronchi, and all successive branchings of the bronchioles up to and including the terminal bronchioles. In addition to conducting air into the respiratory zone, these structures serve two additional functions: *humidification* of the inspired air and *filtration* and *cleaning.*

Regardless of the temperature and humidity of the atmosphere, when the inspired air reaches the respiratory zone it is at a temperature of 37°C (body temperature), and it is saturated with water vapor. The first function is needed to maintain a constant internal body temperature, and the latter function is needed to protect delicate lung tissue from desiccation.

Figure 14.3 The conducting and respiratory zones of the respiratory system.

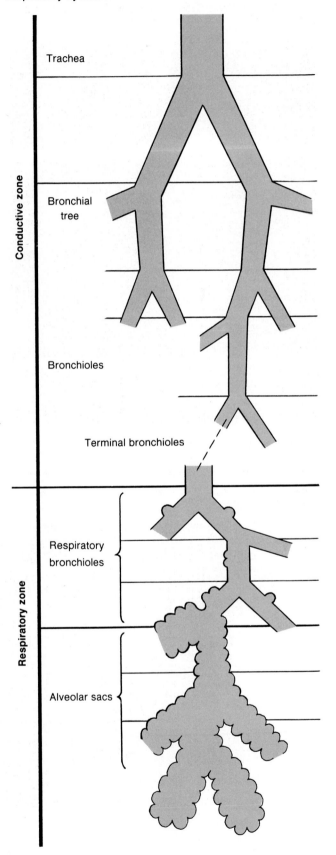

Figure 14.4 *(a)* is a plastic cast of the conducting airways from the trachea to the terminal bronchioles. The relationship between these airways and the structure of the lungs is illustrated in *(b)*.

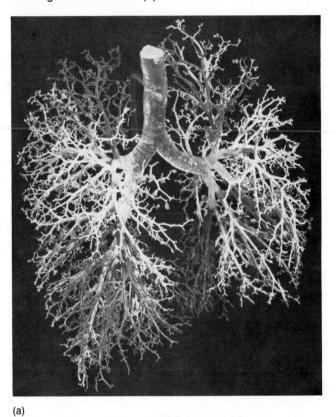

(a)

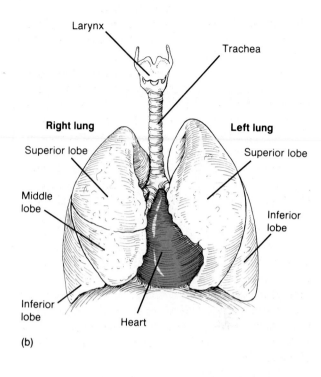

(b)

Figure 14.5 Photograph of the larynx showing the true and false vocal cords and the glottis.

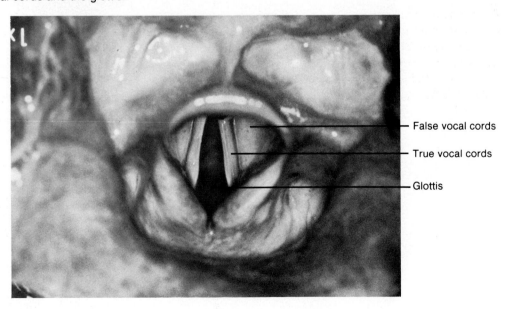

Figure 14.6 Scanning electron micrograph of bronchial wall showing cilia, which help to cleanse the lung by moving trapped particles.

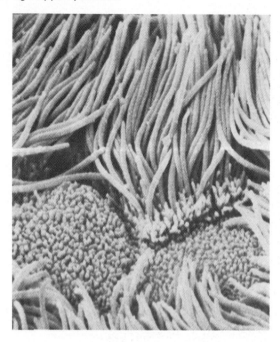

Mucus secreted by cells of the conducting zone serves to trap small particles in the inspired air, and so performs a filtration function. This mucus is moved along at a rate of 1–2 centimeters per minute by cilia projecting from the tops of epithelial cells that line the conducting zone. There are about three hundred cilia per cell that beat in a coordinated fashion to move mucus towards the pharynx, where it can either be swallowed or expectorated.

As a result of this filtration function, particles larger than about 6 μm do not normally enter the respiratory zone of the lungs. The importance of this function is evidenced by the disease called *black lung* occurring in miners who inhale too much carbon dust, and who therefore develop pulmonary fibrosis (as described in a later section). The alveoli themselves are normally kept clean by the action of *macrophages* (literally, "big eaters") that reside in the alveoli (see figure 14.7). The cleansing action of cilia and macrophages in the lungs have been shown to be damaged by cigarette smoke.

Figure 14.7 Horse lung. A pulmonary alveolar macrophage occupies the center of the micrograph and a type II alveolar epithelial cell is on the left. The macrophage has blunt pseudopodia and fine filamentous projections (filopodia). Depressions, presumably related to phagocytosis, are seen on the surface of the cell. (Field width twenty-four microns.)

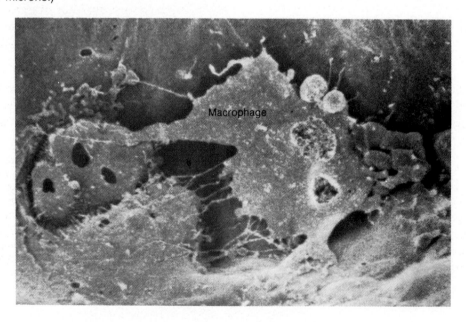

Macrophage

Figure 14.8 Cross section of the thoracic cavity showing the mediastinum and pleural membranes.

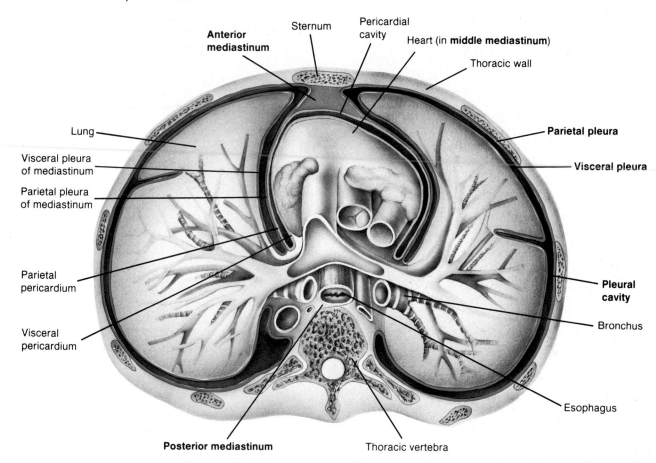

Anterior mediastinum · Sternum · Pericardial cavity · Heart (in **middle mediastinum**) · Thoracic wall · Lung · **Parietal pleura** · Visceral pleura of mediastinum · **Visceral pleura** · Parietal pleura of mediastinum · Parietal pericardium · **Pleural cavity** · Bronchus · Visceral pericardium · Esophagus · **Posterior mediastinum** · Thoracic vertebra

Thoracic Cavity

The *diaphragm,* a dome-shaped sheet of striated muscle, divides the body cavity into two parts. The area below the diaphragm is the *abdominal cavity,* and contains the liver, pancreas, gastrointestinal tract, spleen, genitourinary tract, and other organs. The chest, or *thoracic cavity,* above the diaphragm contains the heart, large blood vessels, trachea, esophagus, and thymus gland in the central region, and is filled elsewhere by the right and left lungs.

The structures in the central region—or *mediastinum*—are enveloped by a double layer of wet epithelial membranes called the *pleural membranes.* One membrane of this double layer is continuous with the *parietal pleural membrane,* a wet epithelial membrane that lines the inside of the thoracic wall. The other layer is continuous with the *visceral pleural membranes* that cover the surface of the lungs (see figure 14.8).

The lungs normally fill the thoracic cavity so that the visceral pleural membranes covering the lungs are pushed against the parietal pleural membrane lining the thoracic wall. There is normally little or no air between the visceral and parietal pleural membranes. There is, however, a "potential space"—called the *intrapleural space*—that can become a real space if a lung collapses. The normal position of the lungs in the thoracic cavity is shown in the radiograph in figure 14.9.

---

1. Describe the structures involved in gas exchange in the lungs, and describe how this process occurs.
2. Describe the structures involved in the conducting zone of the respiratory system, and list its functions.
3. Describe how each lung is packaged separately in pleural membranes, and describe the relationship between the visceral and parietal pleural membranes.

Figure 14.9   Radiographic (X ray) views of the chest of a
normal female *(a)* and a normal male *(b)*.

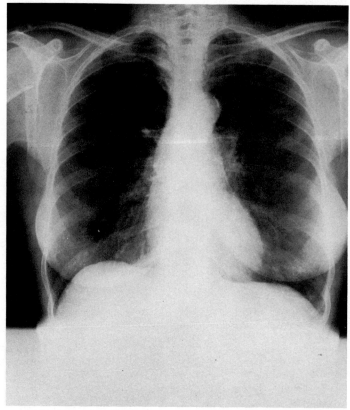

(a)

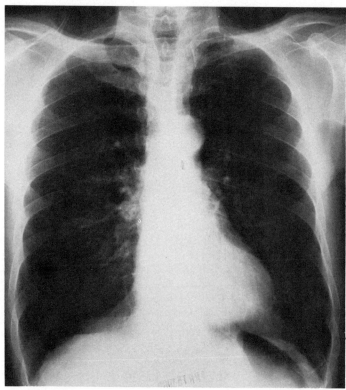

(b)

Figure 14.10 Examples of atmospheric pressure (shown by arrows) causing glass slides to "stick" to each other. The slides stick together because the atmospheric pressure is greater than the air pressure between the two slides.

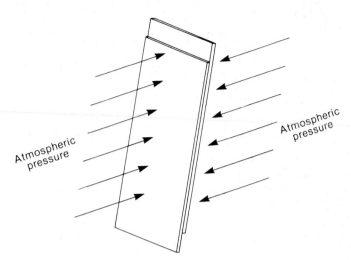

Two wet glass slides

Ventilation

Movement of air through the conducting zone to the terminal bronchioles occurs as a result of the pressure difference between the two ends of the airways. Air flow through bronchioles, like blood flow through blood vessels (see chapter 12), is directly proportional to the pressure head and inversely proportional to the frictional resistance to flow. Air flow through the respiratory bronchioles, like blood flow through arterioles and capillaries, is very slow because of the large cross-sectional areas of these structures. There is, as a result, an abrupt slowdown of air velocity between the conducting and respiratory zones of the lungs. Small dust particles therefore tend to settle at the ends of terminal bronchioles.

Intrapulmonary and Intrapleural Pressures

Air enters the lungs during inspiration because the atmospheric pressure is greater than the **intrapulmonary,** or **intra-alveolar pressure.** Since the atmospheric pressure does not normally change, the intrapulmonary pressure must fall below atmospheric pressure during inspiration. A pressure below that of the atmosphere is called *subatmospheric pressure* or *negative pressure.* During quiet breathing, for example, the intrapulmonary pressure may become 3 mm Hg less than the pressure of the atmosphere. This subatmospheric pressure is commonly shown as −3 mm Hg (hence the name *negative pressure*). Expiration, conversely, occurs when the intrapulmonary pressure is greater than the atmospheric pressure. During quiet breathing, the intrapulmonary pressure may rise to at least +3 mm Hg.

Boyle's Law   Changes in intrapulmonary pressure occur as a result of changes in lung volume. This follows from *Boyle's law,* which describes the fact that *the pressure of a gas is inversely proportional to its volume.* Inspiration thus occurs when the lung volume increases and the intrapulmonary pressure decreases to subatmospheric levels. Expiration occurs when the lung volume decreases and the intrapulmonary pressure increases above the atmospheric pressure. Changes in lung volume occur as a result of changes in thoracic volume produced by muscle contraction (as described in a later section).

Intrapleural Pressure   The lungs "stick" to the chest, and follow the respiratory movements of the chest wall during breathing. This is because there is normally no gas in the intrapleural space. The intrapleural pressure is therefore lower than the intrapulmonary pressure, so that the lungs are pushed against the chest wall by the pressure difference. Like two wet pieces of glass that are pressed together by the atmosphere (with a pressure of 14.7 pounds per square inch at sea level), the lungs can slide along the chest wall but cannot normally be pulled away from it (see figure 14.10).

The difference between the intrapulmonary pressure and the intrapleural pressure—the pressure difference across the lung wall—is called the **transpulmonary pressure.** This pressure difference keeps the lungs against the chest wall and acts to expand the lungs as the thoracic

Table 14.1 Intrapulmonary and intrapleural pressures in normal, quiet breathing, and the transpulmonary pressure (intrapulmonary minus intrapleural pressure) acting to expand the lungs.

	Inspiration	Expiration
Intrapulmonary pressure (mm Hg)	−3 (to zero)	+3 (to zero)
Intrapleural pressure (mm Hg)	−6	−3
Transpulmonary pressure (mm Hg)	+3	+6

Note: Pressures indicate mm Hg below or above atmospheric pressure. Intrapleural pressure is normally always negative (subatmospheric).

volume increases. The elastic properties of the lungs, however (as described in the next section), provide a resistance to distension, so that lung volume always increases less than thoracic volume during inspiration. In accordance with Boyle's law, the intrathoracic and intrapleural pressures thus fall to lower values than the intrapulmonary pressure during inhalation. The intrapleural pressure therefore remains less than the intrapulmonary pressure—even during forced inhalation, when the intrapulmonary pressure may fall to as low as −20 mm Hg—and the lungs remain against the chest wall regardless of the inspiratory effort.

Lung Compliance

The lungs are very distensible—they are, in fact, about 100 times more distensible than a toy balloon. Another term for distensibility is **compliance,** which is defined as the change in lung volume per unit change in transpulmonary pressure. The transpulmonary pressure, as previously described, is the pressure difference across the lung wall (intrapulmonary minus intrapleural pressure) that acts to expand the lungs as the thoracic volume is increased during inspiration. The ability of the lungs to expand, and the volume of air inspired at a given transpulmonary pressure (that is, the compliance of the lungs), is altered under certain conditions.

The compliance of the lungs is reduced whenever there is increased resistance to distension. If the lungs were filled with concrete (as an extreme example), a given transpulmonary pressure would produce no increase in lung volume and no air would enter; the compliance of the lungs would be zero. Infiltration of lung tissue with connective tissue proteins—a condition called *pulmonary fibrosis*—similarly decreases lung compliance (see figure 14.11). In *emphysema,* in which alveolar tissue is destroyed, the lungs are less resistant to distension and have a greater than normal compliance.

Figure 14.11 The volume of air inspired as a function of the transpulmonary pressure (the pressure acting to distend the lungs, which is equal to the difference between the intrapulmonary pressure and the intrapleural pressure). The slope of each line (change in volume per unit change in pressure) in the linear regions is called the *compliance* and is a measure of the distensibility of the lungs. Distensibility is increased in emphysema because of destruction of lung tissue, and decreased in fibrosis.

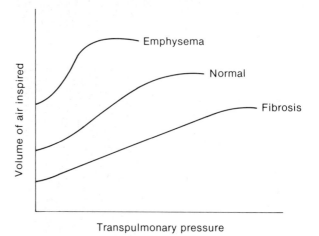

Elasticity of the Lungs   The term *elasticity* refers to the tendency of a structure to return to its initial size after being distended. Since the lungs are very elastic, due to a high content of elastin proteins, they resist distension. This elastic resistance is responsible for the fact that the lungs expand slightly less than the thorax during inhalation, so that the intrapulmonary pressure is always greater than the intrapleural pressure (as previously described).

The elastic nature of lung tissue is revealed when air enters the intrapleural space (as a result of an open chest wound, for example). This condition is called a **pneumothorax.** As air enters the intrapleural space, the intrapleural pressure rises until it is equal to the atmospheric pressure. When the intrapleural pressure is the same as the intrapulmonary pressure, the lungs can no longer expand because the transpulmonary pressure becomes zero. Not only do the lungs fail to expand during inspiration, they actually collapse away from the chest wall as a result of their elastic recoil (see figure 14.12).

Surface Tension   The forces that act to resist lung distension include elastic resistance and the *surface tension* of fluid in the alveoli. Although alveoli are relatively dry, they do contain a very thin film of fluid, much like soap bubbles. Surface tension is created by the fact that water molecules at the surface are attracted more to other water molecules than to air. As a result, the surface water molecules are pulled tightly together by attractive forces from underneath (see figure 14.13).

Figure 14.12 Pneumothorax of the right lung. The right side of the thorax appears uniformly dark because it is filled with air; the spaces between ribs is also greater than on the left due to release from the elastic tension of the lungs. The left lung appears denser (less dark) because of shunting of blood from the right to the left lung.

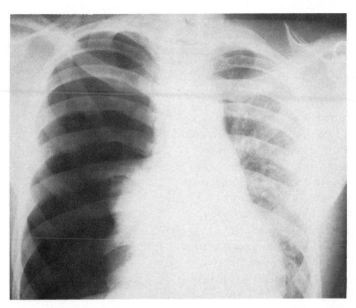

Figure 14.13 Water molecules at the surface have a greater attraction for other water molecules than for air. The surface molecules are thus attracted to each other and pulled tightly together by the attractive forces of water underneath. This produces surface tension.

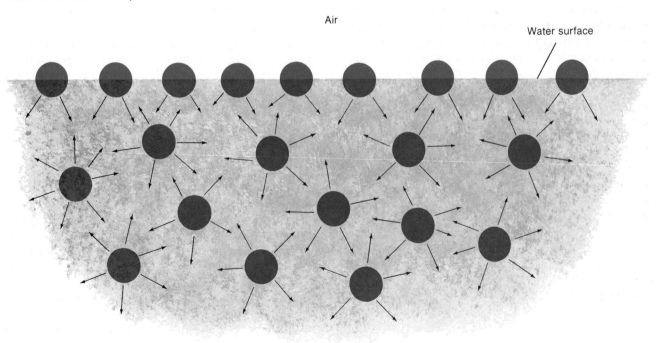

When water molecules are arranged in a bubble, the pressure generated by their surface tension is directed inwards to the center (similar to the effects of tightening a slipknot). According to the **law of LaPlace,** the magnitude of this pressure is inversely proportional to the radius of the bubble

$$\text{Pressure} = \frac{4 \times \text{surface tension}}{\text{radius}}$$

As in tightening a slipknot, the pressure generated by surface tension should make soap bubbles and alveoli collapse to a smaller radius. According to the law of LaPlace, as the alveoli become smaller their surface tension pressure should increase, so that a positive feedback, or "avalanche-like," effect is created. Since some alveoli are normally smaller than others, and since alveoli decrease in size during expiration, one might expect that smaller alveoli would collapse and empty their air into larger alveoli. The reasons why this does not normally occur are discussed in the next section.

Surfactant and the Respiratory Distress Syndrome

Alveolar fluid contains a phospholipid known as dipalmitoyl lecithin, probably attached to a protein, that functions to lower surface tension. This compound is called **lung surfactant** (a contraction of the term *surface active agent*). Because of the presence of surfactant, the surface tension in the alveoli is lower at any given lung volume than would be expected if surfactant were absent. Further, the ability of surfactant to lower surface tension improves as the alveoli get smaller during expiration. Surfactant thus helps to prevent the alveoli from collapsing as a result of surface tension forces. Even after a forceful expiration, the alveoli remain open and a *residual volume* of air remains in the lungs. Since the alveoli don't collapse, it is easier to inflate them at the next inspiration (studies reveal that lungs inflate by increases in the size of alveoli that are already open, rather than by recruitment of collapsed alveoli).

Figure 14.14 According to the law of LaPlace, the surface tension *(T)* force tending to collapse the alveoli should be greater in the smaller alveolus *(on the right)* than in the larger alveolus *(left).* This implies that (without surfactant) smaller alveoli would collapse and empty their air into larger alveoli.

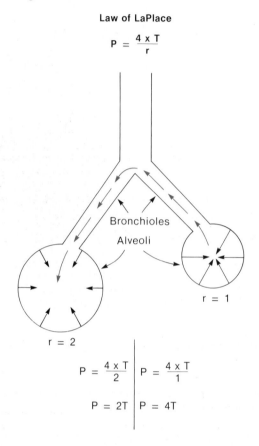

Law of LaPlace

$$P = \frac{4 \times T}{r}$$

Bronchioles

Alveoli

r = 1

r = 2

$$P = \frac{4 \times T}{2} \qquad P = \frac{4 \times T}{1}$$

$$P = 2T \qquad P = 4T$$

Surfactant is produced by type II alveolar cells (see figure 14.15) in late fetal life. Since surfactant does not start to be produced until about the eighth month, premature babies are sometimes born with lungs that lack sufficient surfactant and that are collapsed as a result of the high surface tension. This condition is called **respiratory distress syndrome (RDS).** It is also called **hyaline membrane disease** because the high surface tension causes plasma fluid to leak into the alveoli, producing a glistening "membrane" appearance (as well as pulmonary edema). This condition does not occur in all premature babies—the rate of lung development depends on hormonal conditions (thyroxine and hydrocortisone, primarily) and on genetic factors.

Figure 14.15 Production of pulmonary surfactant by
type II alveolar cells. Surfactant appears to be composed
of a derivative of lecithin combined with protein.

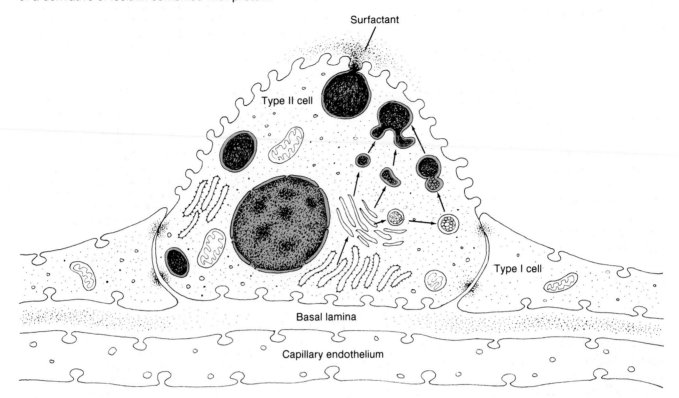

Even under normal conditions, the first breath of life
is a difficult one because the newborn must overcome large
surface tension forces to inflate its partially collapsed alveoli. The transpulmonary pressure required for the first
breath is fifteen to twenty times that required for subsequent breaths, and an infant with RDS must duplicate
this effort with every breath. Fortunately, many babies
with RDS can be saved by mechanical ventilators that
keep them alive long enough for their lungs to mature and
manufacture sufficient surfactant.

1. Explain the reason that the lungs remain against the
   chest wall, and describe what happens in a
   pneumothorax.
2. Describe inspiration and expiration in terms of Boyle's
   law.
3. Define lung compliance, and use a graph to illustrate
   normal compliance. Explain how compliance is affected
   by pulmonary fibrosis and emphysema.
4. Define elasticity, and describe how the elastic
   resistance of the lungs affects the intrapleural pressure.
5. Describe the law of LaPlace, and explain how lung
   surfactant normally prevents the collapse of alveoli.

Table 14.2   Summary of the mechanisms involved in normal, quiet ventilation and forced ventilation.

	Inspiration	Expiration
Normal quiet breathing	Contraction of diaphragm and internal intercostal muscles increases thoracic and lung volume, decreasing intrapulmonary pressure to about −3 mm Hg.	Relaxation of diaphragm and external intercostals, plus elastic recoil of lungs, decreases lung volume and increases intrapulmonary pressure to about +3 mm Hg.
Forced ventilation	Inspiration aided by contraction of accessory muscles such as the scalenes and sternocleidomastoid. Intrapulmonary pressure decreased to −20 mm Hg or less.	Expiration aided by contraction of abdominal muscles and internal intercostal muscles. Intrapulmonary pressure rises to +30 mm Hg or more.

Mechanics of Breathing and Pulmonary Diseases

The mechanisms involved in normal quiet breathing and in forced ventilation are summarized in table 14.2. In quiet breathing, the respiratory rate is about twelve breaths per minute, and the volume per breath (the tidal volume—described in a later section) is about 500 ml. The *total minute volume* is, therefore, 12 × 500 = 6,000 ml per minute. About 150 ml per breath, however, remains in the conducting zone, which is also called the anatomical dead space because air in this region is unavailable for gas exchange. The *total alveolar ventilation* is therefore equal to 12 × (500 − 150) = 4,200 ml per minute in quiet breathing.

Normal and Forced Ventilation

In normal, quiet breathing, inspiration occurs as a result of contraction of two groups of muscles—the *diaphragm* and the *external intercostal muscles*. The diaphragm is a sheet of striated muscle whose convex surface protrudes into the thoracic cavity when the muscle is relaxed. Contraction makes the diaphragm assume a more flattened shape so that the thoracic volume is increased at the expense of the abdomen. The abdomen may thus bulge outwards during inspiration as lowering of the diaphragm pushes the viscera against the abdominal muscles. The external intercostal muscles run downwards and forwards between the ribs. When these muscles contract, the rib cage is rotated upwards and outwards, expanding the thoracic volume.

As a result of increased lung volume, intrapulmonary pressure decreases in quiet inspiration to −3 mm Hg. As air enters the lungs, the intrapulmonary pressure gradually rises to 0 mm Hg (equal to the atmospheric pressure). In quiet expiration, elastic recoil of the thorax and lungs causes a rapid decrease in lung volume and increase in intrapulmonary pressure to about +3 mm Hg. Air, as a result, is forced out of the lungs until—at the end of expiration—the intrapulmonary pressure again equals the atmospheric pressure.

Figure 14.16   Change in lung volume, as shown by radiographs, during expiration *(A)* and inspiration *(B)*. The increase in lung volume during full inspiration is shown by comparison with the lung volume in full expiration (shown by dashed lines).

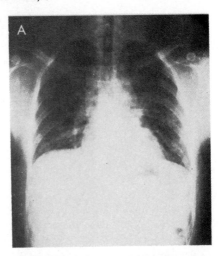

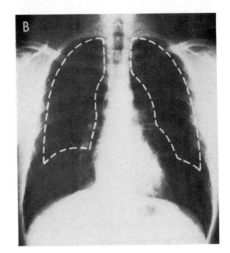

Figure 14.17   After the end of expiration, before the next inspiration, the intrapulmonary pressure equals the atmospheric pressure. The intrapleural pressure is subatmospheric (negative), because of elastic recoil of the lungs and chest wall. During inspiration, the intrathoracic volume increases as a result of lowering of the diaphragm and contraction of the external intercostal muscles. This creates a subatmospheric pressure within the lung airspaces, causing air to enter the lungs.

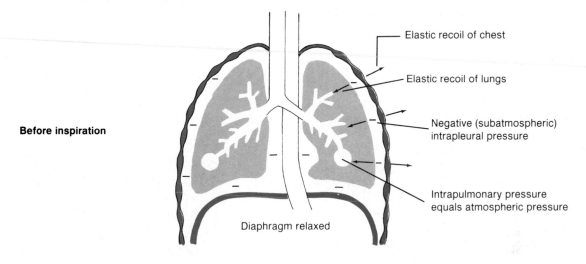

Before inspiration

Elastic recoil of chest

Elastic recoil of lungs

Negative (subatmospheric) intrapleural pressure

Intrapulmonary pressure equals atmospheric pressure

Diaphragm relaxed

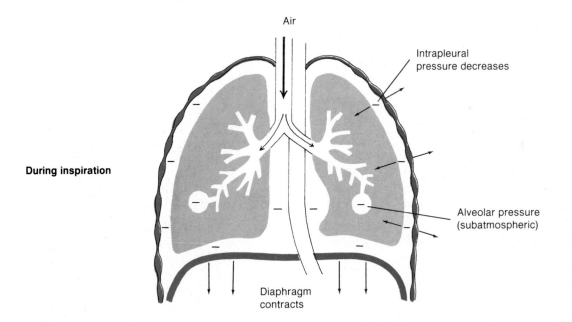

Air

During inspiration

Intrapleural pressure decreases

Alveolar pressure (subatmospheric)

Diaphragm contracts

Figure 14.18   During forced expiration, contraction of the abdominal muscles and internal intercostals combine with the passive elastic recoil of the lungs and chest wall to expel air from the lungs. During forced expiration, the intrapleural pressure can rise to above the pressure of the atmosphere (+20 mm Hg in this example). The alveolar pressure is greater than this (+25 mm Hg) because of the additional effect of elastic recoil of the lungs. The pressure within the bronchioles "downstream" from the alveoli is less, however, due to frictional resistance. At the point where the alveolar pressure is greater than the pressure within the bronchioles the bronchioles become compressed.

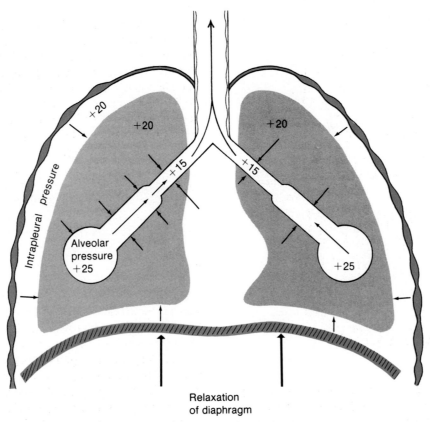

Relaxation
of diaphragm

When breathing is performed more forcefully, the intrapulmonary pressure can vary between inspiration and expiration from −20 mm Hg to +30 mm Hg. These large pressure changes follow from large changes in thoracic volume, which are produced by contractions of *accessory respiratory muscles.* The accessory muscles that aid inspiration include those that insert in the upper ribs and sternum (breastbone), such as the *scalenes* and *sternocleidomastoid.* In order for the rate and depth of breathing to be increased, however, accessory muscles that aid expiration must also be activated. These include the *internal intercostal muscles,* which are antagonistic to the external intercostal muscles, and various abdominal muscles. Contraction of abdominal muscles pushes the viscera against the underside of the diaphragm, forcing it up into the thoracic cavity.

Maximum Expiratory Flow Rate   As the expiratory effort increases, the expiratory flow rate also increases—but only up to a maximum value. This observation results from the fact that the frictional resistance to air flow—or *airway resistance*—causes a decrease in pressure "downstream" from the alveoli. The pressure in the bronchioles therefore becomes less than the alveolar pressure outside the bronchioles. The bronchioles thus become compressed (see figure 14.18). A further increase in expiratory effort, which increases intra-alveolar pressure, does not cause a further increase in expiratory flow rate because compression of the bronchioles is increased correspondingly.

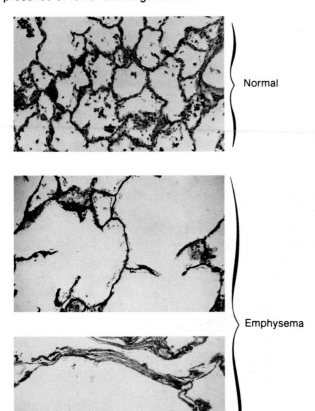

Figure 14.19 Photomicrographs of tissue from a normal lung and from the lung of a person with emphysema. In emphysema, lung tissue is destroyed, resulting in the presence of fewer and larger alveoli.

Normal

Emphysema

Emphysema   In the disease *emphysema,* in which lung tissue is destroyed (see figure 14.19), alveoli become fewer in number and larger in size than normal. This results in a decreased surface area for gas exchange, and in a decreased ability of the weakened bronchioles to remain open during expiration. Collapse of bronchioles as a result of the compression force during expiration produces *air trapping,* which further decreases the efficiency of gas exchange in the alveoli.

Pulmonary Diseases

Pulmonary disorders are classified as **obstructive** when there is increased resistance to air flow in the bronchioles, and are classified as **restrictive** when the alveolar tissue is damaged. Asthma and bronchitis are usually just obstructive; chronic bronchitis and emphysema are often both obstructive and restrictive. Pulmonary fibrosis, in contrast, is a purely restrictive disorder.

Obstructive Disorders   Obstruction of air flow through the bronchioles may occur as a result of excessive mucous secretion, inflammation, and/or contraction of smooth muscles in the bronchioles. *Asthma* results from bronchiolar constriction, which increases airway resistance and makes breathing difficult. Constriction of bronchiolar smooth muscle may occur as a result of ACh released from parasympathetic nerve endings, or from local release of chemicals such as histamine. Notice that the effects of ACh and histamine on bronchiolar smooth muscles (constriction) is opposite to the dilation effect that these chemicals have on systemic blood vessels.

Catecholamines, such as epinephrine, stimulate bronchiolar dilation. At one time, asthmatics were given epinephrine (as an inhaled spray) to reduce airway resistance, but this treatment produced undesirable side effects such as tachycardia. More specific drugs that produce the bronchodilator action of epinephrine without the undesirable side effects of adrenergic stimulation are now available.

Chronic Obstructive Pulmonary Disease   Increased airway resistance can also be produced by excessive mucus, as occurs in bronchitis. Chronic bronchitis (of long duration, as opposed to that which is shorter term, or "acute") may cause progressive pathological changes that are similar to those that occur in emphysema. Chronic bronchitis and emphysema are the most common causes of respiratory failure, and together are called *chronic obstructive pulmonary disease (COPD)*.

In emphysema, a term meaning "filled with air," lung tissue is destroyed. This results in decreased surface area for gas exchange and air trapping, as previously described. Other pathological changes include edema, inflammation, and hyperplasia (increased cell number), zones of pulmonary fibrosis, pneumonia (infection by *Streptococcus pneumoniae* bacteria or various viruses), pulmonary emboli (traveling blood clots) and heart failure. Patients with severe emphysema may eventually develop *cor pulmonale*—pulmonary hypertension with hypertrophy (and eventual failure) of the right ventricle.

Table 14.3 Pulmonary disorders caused by inhalation of noxious agents containing particles less than 6 μm in size that can lodge in the alveoli.

Condition	Agent	Pulmonary Disorders
Silicosis	Silica	Fibrosis, obstructive emphysema, tuberculosis, cor pulmonale
Anthracosis (Black lung)	Coal dust	Similar to silicosis (most coal dust contains silica)
Asbestosis	Asbestos fibers	Diffuse fibrosis, cor pulmonale, pulmonary malignancy

Figure 14.20 A spirometer (Collins 9L respirometer) used to measure lung volumes and capacities.

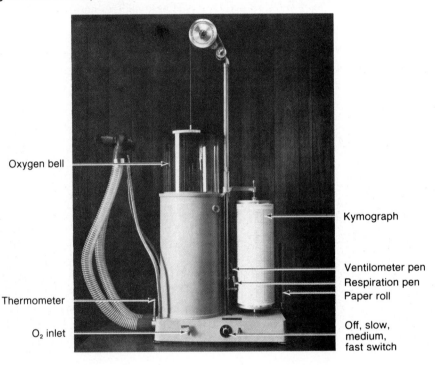

There are different types of emphysema. The most common type occurs almost exclusively in people who have smoked cigarettes heavily over a period of years. A component of cigarette smoke apparently stimulates secretion of proteolytic (protein-digesting) enzymes from macrophages and leukocytes that destroy lung tissues. A less common type of emphysema results from the genetic inability to produce a plasma protein called alpha-l-antitrypsin. This protein normally inhibits proteolytic enzymes such as trypsin, and thus normally protects the lungs against the effects of such enzymes that are released from alveolar macrophages.

Pulmonary Fibrosis   Under certain conditions, for reasons that are poorly understood, lung damage leads to pulmonary fibrosis instead of emphysema. Fibrosis can result, for example, from inhalation of particles less than 6 μm in size that are able to accumulate in the respiratory zone of the lungs. Included in this category is *anthracosis,* or black lung, which is produced from inhalation of carbon particles from coal dust (see table 14.3).

Table 14.4 Definitions of terms used to describe lung volumes and capacities.

Term	Definition
Lung volumes	The four non-overlapping components of the total lung capacity
Tidal volume	The volume of gas inspired or expired in an unforced respiratory cycle
Inspiratory reserve volume	The maximum volume of gas that can be inspired from the end of a tidal inspiration
Expiratory reserve volume	The maximum volume of gas that can be expired from the end of a tidal expiration
Residual volume	The volume of gas remaining in the lungs after a maximum expiration
Lung capacities	Measurements that are the sum of two or more lung volumes
Total lung capacity	The total amount of gas in the lungs at the end of a maximum inspiration
Vital capacity	The maximum amount of gas that can be expired after a maximum inspiration
Inspiratory capacity	The maximum amount of gas that can be inspired at the end of a tidal expiration
Functional residual capacity	The amount of gas remaining in the lungs at the end of a tidal expiration

Figure 14.21 Spirogram showing lung volumes and capacities at rest.

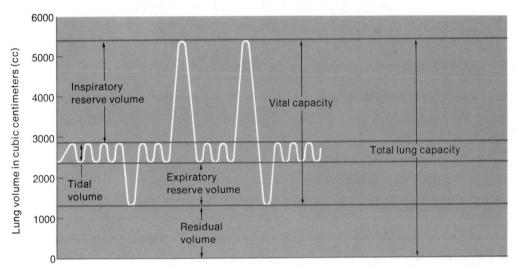

Pulmonary Function Tests

Pulmonary function may be assessed clinically by means of a technique known as *spirometry*. In this procedure a subject breathes in a closed system in which air is trapped within a light plastic bell floating in water. The bell moves up when the subject exhales, and down when the subject inhales. Movements of the bell cause corresponding movements of a pen, which traces a record of the breathing on a rotating drum recorder (see figure 14.20).

The appearance of a spirogram taken during quiet, resting breathing is shown in figure 14.21, and the various lung volumes and capacities are defined in table 14.4. During quiet breathing, the amount of air exhaled in each breath (the *tidal volume*) is equal to about 500 ml, which is approximately 12 percent of the *vital capacity* (the maximum amount of air that can be exhaled after a maximum inhalation). Multiplying the tidal volume at rest times the number of breaths per minute yields a *total minute volume* of about 6 L per minute. During exercise, the tidal volume can increase to as much as 50 percent of the vital capacity, and the number of breaths per minute likewise increases to produce a total minute volume as high as 100–200 L per minute.

Figure 14.22   Illustration of the one-second forced expiratory volume (FEV$_{1.0}$) spirometry test for detecting obstructive pulmonary disorders.

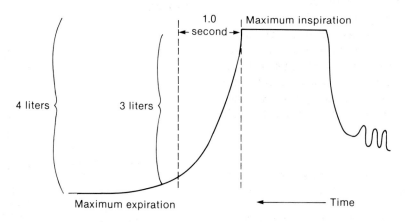

$$FEV_{1.0} = \frac{3L}{4L} \times 100\% = 75\%$$

Table 14.5   Definitions of some terms used to describe ventilation.

Term	Definition
Air spaces	Alveolar ducts, alveolar sacs, and alveoli.
Airways	Structures that conduct air from the mouth and nose to the respiratory bronchioles.
Alveolar ventilation	Removal and replacement of gas in alveoli. Equal to the tidal volume minus the volume of dead space.
Anatomical dead space	Volume of the conducting airways to the zone where gas exchange occurs.
Apnea	Cessation of breathing.
Dyspnea	Unpleasant subjective feeling of difficult or labored breathing.
Eupnea	Normal, comfortable breathing at rest.
Hyperventilation	An alveolar ventilation that is excessive in relation to metabolic rate. Results in abnormally low alveolar CO_2.
Hypoventilation	An alveolar ventilation that is low in relation to metabolic rate. Results in abnormally high alveolar CO_2.
Physiological dead space	Combination of anatomical dead space and underventilated alveoli that do not contribute normally to blood-gas exchange.
Pneumothorax	Presence of gas in the pleural space (the space between the visceral and parietal pleural membranes) causing lung collapse.
Torr	Synonymous with millimeters of mercury (760 mm Hg = 760 Torr).

Spirometry is useful in the diagnosis of lung diseases. In purely restrictive disorders, such as pulmonary fibrosis for example, the vital capacity is reduced below the normal value. In emphysema—because of air trapping—the *residual volume* (amount of air left in the lungs after a maximum exhalation) is increased at the expense of the vital capacity.

In disorders that are only obstructive, such as asthma, the vital capacity is normal because lung tissue is not damaged. Bronchiolar constriction, and the increased airway resistance that results, however, makes expiration more difficult and take a longer time. Obstructive disorders are thus diagnosed by tests that measure the rate of expiration. One such test is the *forced expiratory volume (FEV)*, in which the percent of the vital capacity that can be exhaled in the first second ($FEV_{1.0}$), second ($FEV_{2.0}$) and third ($FEV_{3.0}$) seconds are measured. An $FEV_{1.0}$ that is significantly less than 75 to 85 percent suggests the presence of obstructive pulmonary disease.

People with a pulmonary disorder frequently complain of *dyspnea,* which is a subjective feeling of "shortness of breath." Dyspnea may occur even when alveolar ventilation is normal, and is not necessarily correlated with increased ventilation, because people can breathe very hard during exercise without experiencing dyspnea. Some of the terms used to describe ventilation are defined in table 14.5.

1. Explain why the alveolar ventilation is less than the total minute volume and how both values change during exercise.
2. Compare the mechanisms involved in quiet breathing with those required for forced inspiration and expiration. Explain why the expiratory flow rate cannot be increased above a maximum value.
3. Describe the causes and characteristics of emphysema and how this condition affects the results of pulmonary function tests.
4. Describe the causes and characteristics of asthma and how this condition affects the results of pulmonary function tests.

Gas Exchange in the Lungs

The atmosphere is an ocean of gas that exerts pressure on all objects within it. The amount of this pressure can be measured with a glass U-tube filled with fluid. One end of the U-tube is exposed to the atmosphere, while the other side is continuous with a sealed vacuum tube. Since the atmosphere presses on the exposed side, but not on the side connected to the vacuum tube, atmospheric pressure pushes fluid in the U-tube up on the vacuum side to a height determined by the atmospheric pressure and the density of the fluid. Water, for example, would be pushed up to a height of 33.9 feet (10,332 mm) at sea level, while mercury (Hg)—which is more dense—is raised to a height of 760 mm. As a matter of convenience, therefore, devices used to measure atmospheric pressure (barometers) use mercury rather than water. The atmospheric pressure at sea level is thus said to be equal to *760 mm Hg (or 760 Torr),* which is also described as a pressure of *one atmosphere.*

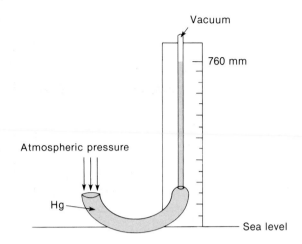

Figure 14.23 Atmospheric pressure at sea level can push a column of mercury to a height of 760 millimeters. This is also described as 760 torr, or one atmosphere pressure.

According to **Dalton's law,** the total pressure of a gas mixture (such as air) is equal to the sum of the pressures that each gas in the mixture would exert independently. The pressure that each gas would exert independently is equal to the product of the total pressure times the fractional composition of the gas in the mixture. The total pressure of the gas mixture is thus equal to the sum of the partial pressures of the constituent gases. Since oxygen comprises about 21 percent of the atmosphere, for example, its partial pressure (abbreviated P_{O_2}) is 21 percent of 760, or about 159 mm Hg. Since nitrogen comprises about 79 percent of the atmosphere, its partial pressure is equal to $0.79 \times 760 = 601$ mm Hg. These two gases thus contribute about 99 percent of the total pressure of 760 mm Hg:

$$P_{\text{dry atmosphere}} = P_{N_2} + P_{O_2} + P_{CO_2} = 760 \text{ mm Hg}$$

Effect of Altitude   As one goes to a higher altitude, the total atmospheric pressure and the partial pressure of the constituent gases decreases. At Denver, for example (5,000 feet above sea level), the atmospheric pressure is decreased to 619 mm Hg, and the P_{O_2} is therefore reduced to $619 \times 0.21 = 130$ mm Hg. At the peak of Mount Everest (at 29,000 feet), the P_{O_2} is only 42 mm Hg. As one descends below sea level, as in ocean diving, the total pressure increases by one atmosphere for every 33 feet. At 33 feet therefore, the pressure equals $2 \times 760 = 1,520$ mm Hg. At 66 feet, the pressure equals three atmospheres.

Effect of Water Vapor Pressure Inspired air contains variable amounts of moisture. By the time the air has passed into the respiratory zone of the lungs, however, it is normally saturated with water vapor (has a relative humidity of 100 percent). The capacity of air to contain water vapor depends on its temperature; since the temperature of the respiratory zone is constant at 37°C, its water vapor pressure is also constant (47 mm Hg).

Water vapor, like the other constituent gases, contributes a partial pressure to the total atmospheric pressure. Since the total atmospheric pressure is constant (depending only on the height of the air mass), the water vapor "dilutes" the contribution of other gases to the total pressure:

$$P_{wet\ atmosphere} = P_{N_2} + P_{O_2} + P_{CO_2} + P_{H_2O}$$

Figure 14.24 Partial pressures of gases in the inspired air and in the alveolar air.

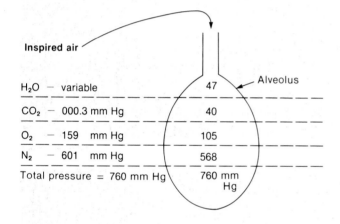

Inspired air		Alveolus
H_2O — variable	47	
CO_2 — 000.3 mm Hg	40	
O_2 — 159 mm Hg	105	
N_2 — 601 mm Hg	568	
Total pressure = 760 mm Hg	760 mm Hg	

When the effect of water vapor pressure is considered, the partial pressure of oxygen in the inspired air is decreased at sea level to:

$$P_{O_2}\ (sea\ level) = 0.21 \times (760 - 47) = 150\ mm\ Hg$$

As a result of gas exchange in the alveoli, the P_{O_2} of *alveolar air* is further reduced to about 105 mm Hg. A comparison of the partial pressures of the inspired air with the partial pressures of alveolar air is shown in figure 14.24.

Partial Pressures of Gases in Blood

The enormous surface area of alveoli and the short diffusion distance between the alveolar air and the capillary blood help to bring the blood into gaseous equilibrium with the alveolar air. This function is further aided by the fact that each alveolus is surrounded by so many capillaries that they form an almost continuous sheet of blood around the alveoli (see figure 14.25).

When a fluid and a gas, such as blood and alveolar air, are at equilibrium, the amount of gas dissolved in the fluid reaches a maximum value. The amount of gas dissolved in the fluid under these conditions depends, according to **Henry's law,** on (1) The solubility of the gas in the fluid, which is a physical constant; (2) the temperature of the fluid—more oxygen can be dissolved in cold water than in warm water; and (3) the partial pressure of the gas.

Blood Gas Measurements Since the temperature of the blood does not vary significantly, the amount of dissolved gas depends directly on its partial pressure. When water—or plasma—is brought into equilibrium with air at a P_{O_2} of 100 mm Hg, for example, the fluid will contain 0.3 ml of O_2 per 100 ml of fluid (at 37°C). If the P_{O_2} of the air were reduced by half, the amount of dissolved oxygen would also be reduced by half.

Table 14.6 Some of the physical laws that are important in ventilation and gas exchange.

Physical Law	Description
Boyle's law	The volume of a gas is inversely proportional to its pressure when the temperature and mass are constant (PV = constant).
Dalton's law	The total pressure of a gas mixture is equal to the sum of the partial pressures of its constituent gases. The partial pressure of each gas is the pressure it would exert if it alone occupied the total volume of the mixture.
Graham's law	The rate of diffusion of a gas through a liquid is directly proportional to its solubility and inversely proportional to its density (or gram molecular weight).
Henry's law	The weight of a gas dissolved in a liquid at a given temperature is proportional to the partial pressure of the gas.
LaPlace's law	The inward pressure tending to collapse a bubble is equal to four times its surface tension divided by the radius of the bubble $\left(P = \dfrac{4 \times surface\ tension}{radius}\right)$

Figure 14.25 Scanning electron micrographs of plastic casts of alveolar capillaries. In the upper photograph only the capillaries are visible. In the lower photograph, plastic also entered the alveoli so that the relationship between alveoli and capillaries can be seen.

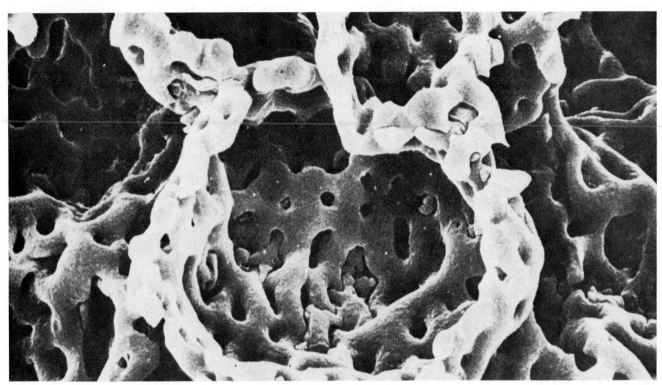

Kessel and Kardon

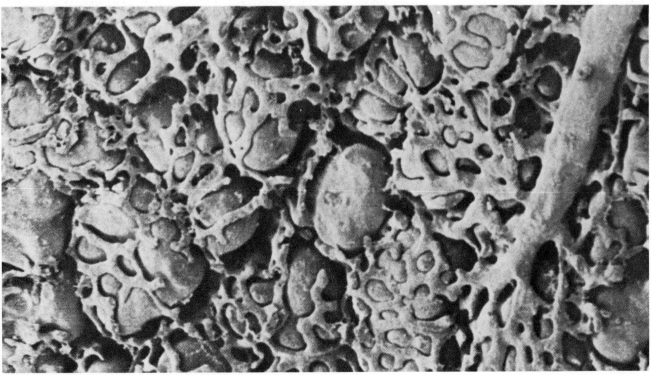

Kessel and Kardon

Figure 14.26 Blood-gas measurements using a P_{O_2} electrode. In *(a)* the electrical current generated by the oxygen electrode is calibrated so that the needle of the blood-gas machine points to the P_{O_2} of the gas with which the fluid is in equilibrium. Once standardized in this way, the electrode can be inserted in a fluid such as blood, and the P_{O_2} of this solution can be measured.

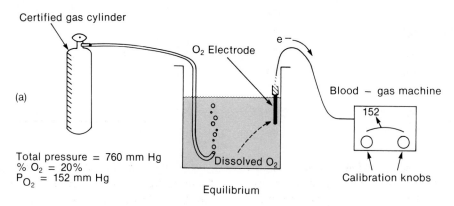

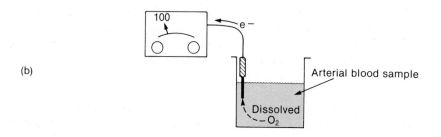

Measurement of the oxygen content of blood (in ml O_2 per 100 ml blood) is a laborious procedure. Fortunately, an **oxygen electrode** has been developed that produces an electric current *in proportion to the amount of dissolved oxygen*. If oxygen is bubbled into a fluid, the current produced by an oxygen electrode in that fluid will increase until it reaches a maximum value. At this point the fluid is saturated with oxygen. The actual amount of dissolved oxygen under these circumstances is not known, although it can easily be determined from solubility tables.

When the oxygen electrode produces a maximum current (as may be seen by deflection of a needle on a scale), it is known that the fluid is in equilibrium with the gas and no further amounts of oxygen can be dissolved at the P_{O_2} of the gas that is used. As a matter of convenience, it can now be said that *the fluid has the same P_{O_2} as the gas*. If it is known that the gas mixture has a P_{O_2} of 150 mm Hg, for example, the deflection of the needle produced by the oxygen electrode can also be calibrated at 150 mm Hg.

If the oxygen electrode is next inserted into an unknown sample of blood, the P_{O_2} of that sample can be read directly from the previously calibrated scale. Suppose, as illustrated in figure 14.26, the blood sample has a P_{O_2} of 100 mm Hg. Since alveolar air has a P_{O_2} of about 105 mm Hg (as shown in figure 14.24), this reading indicates that the blood is almost in complete equilibrium with the alveolar air.

The oxygen electrode is sensitive only to the oxygen that is physically dissolved in water or plasma—it cannot respond to oxygen that is "hidden" inside red blood cells. Most of the oxygen in blood, however, is located in the red blood cells attached to hemoglobin (as described in chapter 15). Thus, while the oxygen content of water or plasma can be determined from solubility tables if the P_{O_2} is known, the oxygen content of whole blood depends on both its P_{O_2} and its red blood cell and hemoglobin content. At a P_{O_2} of about 100 mm Hg, whole blood normally contains 20 ml O_2 per 100 ml blood; of this amount, only 0.3 ml O_3 is dissolved in the plasma—the rest (19.7 ml O_2) is within the red blood cells. This does not affect P_{O_2}—if the cells were removed by centrifugation, the remaining plasma would still have a P_{O_2} of 100 mm Hg.

Table 14.7   The effect of altitude on P_{O_2}.

Changes in P_{O_2} at Various Altitudes

Altitude (Feet above Sea Level)	Atmospheric Pressure (mm Hg)	P_{O_2} in Air (mm Hg)	P_{O_2} in Alveoli (mm Hg)	P_{O_2} in Arterial Blood (mm Hg)
0	760	159	105	100
2,000	707	148	97	92
4,000	656	137	90	85
6,000	609	127	84	79
8,000	564	118	79	74
10,000	523	109	74	69
20,000	349	73	40	35
30,000	226	47	21	19

Significance of Blood P_{O_2} and P_{CO_2} Measurements   Since blood P_{O_2} measurements are not directly affected by oxygen in red blood cells, the P_{O_2} does not provide a measurement of the oxygen content of blood. It does, however, provide a good index of *lung function.* If the inspired air has a normal P_{O_2} but the arterial P_{O_2} is below normal, for example, gas exchange in the lungs must be impaired. Measurements of arterial P_{O_2} thus provide valuable information in the treatment of people with pulmonary disease, during surgery when breathing may be depressed by anesthesia, and in the care of premature babies with respiratory distress syndrome.

When the lungs are functioning properly, the P_{O_2} of systemic arterial blood is only about 5 mm Hg less than the P_{O_2} of the alveolar air. This value is affected by both the functioning of the lungs and by the P_{O_2} of the inspired air. As one goes to a higher altitude, the total atmospheric pressure—and thus the P_{O_2} of the inspired air—decreases (see table 14.7). Normal arterial P_{O_2} is thus lower for residents of Albuquerque or Salt Lake City than for people who live closer to sea level.

At the normal arterial P_{O_2} of about 100 mm Hg, hemoglobin is almost completely loaded with oxygen. Since this accounts for almost all the oxygen that blood can carry (the amount dissolved in plasma is negligible compared to the amount attached to hemoglobin), an increase in blood P_{O_2} has no significant effect on the total oxygen content of blood. Although negligible in terms of total oxygen content, an increase in P_{O_2} can significantly increase the amount of oxygen dissolved in plasma. If the P_{O_2} doubles, for example, the amount of dissolved oxygen also doubles. Since the oxygen carried by red blood cells must first dissolve in plasma before it can reach the tissue cells, the *rate of oxygen delivery* to the tissues would double under these circumstances. For these reasons, breathing from a tank of 100 percent oxygen (with a P_{O_2} of 760 mm Hg) would significantly increase oxygen delivery to the tissues, though it would have little effect on the total oxygen content of the blood.

An electrode that develops a current in response to dissolved carbon dioxide has also been developed, so that the P_{CO_2} of blood can be measured together with its P_{O_2}. Blood in the systemic veins, which is delivered to the lungs in the pulmonary artery, has a P_{O_2} of 40 mm Hg and a P_{CO_2} of 46 mm Hg. After gas exchange in the alveoli of the lungs, blood in the pulmonary vein and systemic arteries has a P_{O_2} of about 100 mm Hg and a P_{CO_2} of 40 mm Hg. As will be described in chapter 15, the P_{CO_2} of arterial blood is an important factor in the regulation of breathing.

Pulmonary Circulation and Ventilation/Perfusion Ratios
In the fetus, the pulmonary circulation has a high vascular resistance as a result of the fact that the lungs are partially collapsed. This high vascular resistance helps to shunt blood from the right to the left atrium through the foramen ovale, and from the pulmonary artery to the aorta through the ductus arteriosus (described in chapter 11). After birth, the foramen ovale and ductus arteriosus close, and the vascular resistance of the pulmonary circulation falls sharply. This fall in vascular resistance at birth is due to (1) opening of vessels as a result of the subatmospheric intrapulmonary pressure and physical stretching of the lungs during inspiration and (2) to dilation of pulmonary arterioles in response to increased alveolar P_{O_2} (this will be described in a later section).

Pulmonary Circulation   In the adult, the right ventricle (like the left) has an output of about 5.5 L per minute. The rate of blood flow through the pulmonary circulation is thus equal to the flow rate through the systemic circulation. Blood flow rate, as described in chapter 12, is directly proportional to the pressure difference between the two ends of the vessel and inversely proportional to the vascular resistance. In the systemic circulation, the mean arterial pressure is 90–100 mm Hg, and the pressure of the right atrium is 0 mm Hg; the pressure difference is thus about 100 mm Hg. The mean pressure of the

Figure 14.27 The P_{O_2} and P_{CO_2} of blood as a result of gas exchange in the lung alveoli and gas exchange between systemic capillaries and body cells.

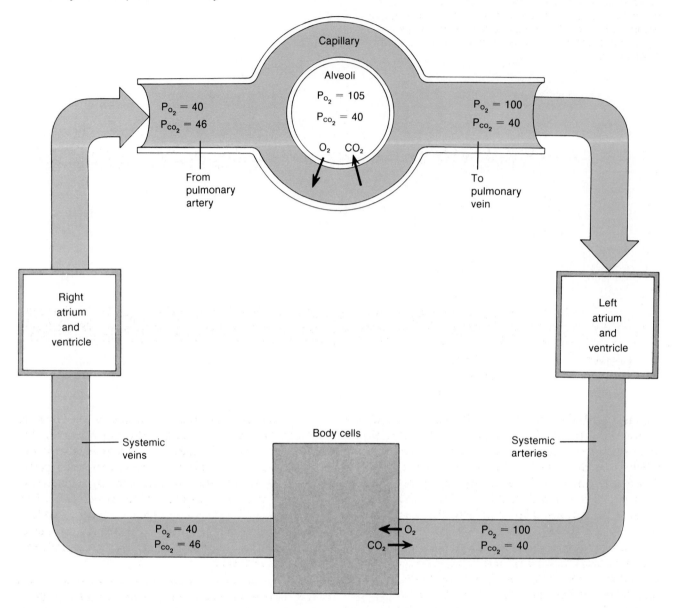

pulmonary artery, in contrast, is only 15 mm Hg and the pressure of the left atrium is 5 mm Hg. The driving pressure in the pulmonary circulation is thus 15 − 5, or 10, mm Hg.

Since the driving pressure in the pulmonary circulation is only one-tenth that of the systemic circulation, and since the flow rates are equal, it follows that the pulmonary vascular resistance must be one-tenth that of the systemic vascular resistance. The pulmonary circulation, in other words, is a low resistance, low pressure pathway. The

low pulmonary blood pressure helps to protect against *pulmonary edema* (there is less filtration pressure in pulmonary capillaries than in systemic capillaries), which is a dangerous condition that can impede ventilation and gas exchange. Pulmonary edema can occur when there is pulmonary hypertension produced, for example, by left heart failure.

Ventilation/Perfusion Ratios   Pulmonary arterioles constrict when the alveolar P_{O_2} is low, and dilate as the alveolar P_{O_2} is raised. This response is opposite to that of systemic arterioles, which dilate in response to low tissue

Figure 14.28   Ventilation, blood flow, and ventilation/perfusion ratios at the apex and base of the lungs. The ratios indicate that the apex is relatively overventilated and the base underventilated in relation to their blood flows. Such uneven matching of ventilation to perfusion results in the fact that the blood leaving the lungs has a P_{O_2} that is slightly less (by 4–5 mm Hg) than the P_{O_2} of alveolar air.

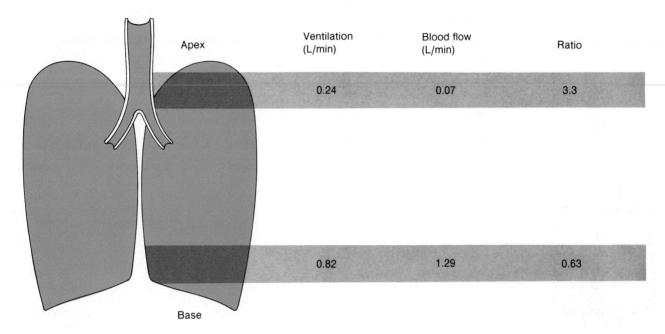

P_{O_2} (as described in chapter 13). Dilation of systemic arterioles when the P_{O_2} is low helps to supply more blood and oxygen to the tissues; constriction of pulmonary arterioles when the alveolar P_{O_2} is low helps to decrease blood flow to alveoli that are inadequately ventilated.

Constriction of pulmonary arterioles where the alveolar P_{O_2} is low, and dilation of arterioles where the alveolar P_{O_2} is high, helps to *match ventilation to perfusion* (the term *perfusion* refers to blood flow). If this autoregulation of blood flow did not occur, blood from less-well-ventilated alveoli (with a low P_{O_2}) would mix with blood from well-ventilated alveoli, and the blood leaving the lungs would have a lowered P_{O_2} as a result of this dilution effect.

Dilution of the P_{O_2} of pulmonary vein blood actually does occur to some degree despite autoregulation of pulmonary blood flow. When a person stands upright, gravity causes a greater blood flow to the base of the lungs than to the apex (top). Ventilation likewise decreases from base to apex, but this decrease is not proportionate to the decrease in blood flow. The *ventilation/perfusion ratio* at the apex is thus high (0.24 L air per minute divided by 0.07 L blood per minute equals a ratio of 3.4 to 1.0), while at the base of the lungs it is low (0.82 L air per minute divided by 1.29 L blood per minute equals a ratio of 0.6 to 1.0). This is illustrated in figure 14.28.

Functionally, the alveoli at the apex of the lungs are thus overventilated and are actually larger, compared to alveoli at the base. The apices of the lungs, in other words, are under-perfused. This mismatch of ventilation/perfusion ratios is responsible for the 5 mm Hg difference in P_{O_2} between alveolar air and arterial blood. This difference is not significant because, as previously described (and to be described in more detail in chapter 15), the hemoglobin of normal arterial blood is almost completely saturated with oxygen.

1. Explain why air at sea level (and dry) has a P_{O_2} of 159 mm Hg, and why the normal alveolar air at sea level has a P_{O_2} of 105 mm Hg.
2. Explain the relationship between arterial P_{O_2} and *(a)* lung function; *(b)* altitude; *(c)* oxygen capacity of whole blood; and *(d)* the rate of oxygen delivery to the tissues.
3. Suppose water and blood were bubbled with a gas mixture with a P_{O_2} of 100 mm Hg until no more oxygen could be dissolved. Describe the P_{O_2} and oxygen contents of these two samples. If a sample of anemic blood with a hemoglobin concentration that is half normal was treated in the same way, what would be its P_{O_2} and oxygen content?
4. Describe the ventilation/perfusion ratios of the lungs, and explain why systemic arterial blood has a slightly lower P_{O_2} than alveolar air.

Summary

Structure and Function of the Respiratory System

I. The respiratory zone of the lungs is the region where gas exchange occurs.
  A. Gas exchange between inspired air and blood occurs across the walls of alveoli.
   1. Alveoli are blind-ended sacs that are only one cell layer thick.
   2. Alveoli occur separately as outpouchings of respiratory bronchioles and as terminal structures at the ends of these bronchioles.
   3. In the process of gas exchange, blood is enriched in oxygen and depleted in carbon dioxide.
  B. Exchange of oxygen and carbon dioxide between alveolar air and blood occurs by diffusion.
II. The conducting zone includes all of the structures that deliver air to the respiratory zone.
  A. The conducting zone (or anatomical dead space) includes the mouth, nose, pharynx, larynx, trachea, bronchi, and all bronchioles that do not have alveoli.
  B. The conducting zone humidifies the air so that it has a relative humidity of 100 percent, and filters the air before entering the respiratory zone.
III. Inspiration and expiration occur as a result of changes in lung volume.
  A. As lung volume increases during inspiration, the intrapulmonary (or intra-alveolar) pressure falls to subatmospheric levels.
   1. Increases in lung volume occur as a result of increases in thoracic volume.
   2. As the intrapulmonary pressure becomes subatmospheric, air enters the lungs.
  B. Expiration occurs when the lung volume decreases and the intrapulmonary pressure rises above the atmospheric pressure.
  C. Changes in thoracic volume cause changes in lung volume because the lungs are normally stuck to the chest wall.
   1. The lungs stick to the chest wall because the intrapulmonary pressure is greater than the intrapleural pressure.
   2. Since the visceral and parietal pleural membranes are normally against each other, and since the lung volume is normally less than the thoracic volume, the intrapulmonary pressure is always greater than the intrapleural pressure.
  D. Changes in thoracic volume occur as a result of muscle contraction.
   1. Quiet inspiration occurs as a result of contraction of the diaphragm and the external intercostal muscles.
   2. Quiet expiration occurs as a result of passive elastic recoil of the thorax and lungs as the inspiratory muscles relax.
   3. Forced inspiration and expiration occur as a result of contraction of accessory respiratory muscles.

Lung Compliance, Elasticity, and Surface Tension

I. Compliance is a measure of lung distensibility.
  A. The force that distends the lungs is the transpulmonary pressure (intrapulmonary minus intrapleural pressure).
  B. The compliance of the lungs is the amount of volume change produced at a given amount of transpulmonary pressure.
   1. With a given distending force (transpulmonary pressure), lungs with normal compliance inhale more air than lungs with reduced compliance.
   2. Compliance is reduced in pulmonary fibrosis and edema, and increased in emphysema.
II. Lung elasticity and surface tension act to resist distension.
  A. The resistance of the lungs to distension results in a higher intrapulmonary pressure than intrapleural pressure during breathing.
  B. When the transpulmonary pressure becomes zero, as in a pneumothorax, the elastic recoil of the lungs causes lung collapse.
  C. Alveolar surface tension generates a pressure that tends to make the alveoli collapse.
   1. According to the law of LaPlace, the surface tension is greater in smaller alveoli.
   2. Lung collapse is normally prevented by a phospholipid surfactant.
   3. Premature newborns who lack pulmonary surfactant develop respiratory distress syndrome (hyaline membrane disease).
III. Pulmonary diseases are classified as obstructive and restrictive.
  A. Obstructive diseases (asthma and bronchitis) are those in which airway resistance is abnormally increased.
  B. Restrictive disorders (such as fibrosis, edema, and emphysema) are those in which the alveoli are damaged.
  C. Pulmonary function tests help to detect these disorders.
   1. In restrictive disorders, the vital capacity is reduced.
   2. In obstructive disorders, the forced expiratory volume is reduced.
  D. Chronic bronchitis and emphysema are classified as a chronic obstructive pulmonary disease (COPD), which can also become restrictive.

Gas Exchange in the Lungs

I. The lungs bring blood into gaseous equilibrium with alveolar air.
  A. The partial oxygen pressure (P_{O_2}) of the atmosphere is equal to the percent of oxygen composition times the total atmospheric pressure (if the air is dry).

1. In the alveoli the air has a relative humidity of 100 percent, so the water vapor pressure (47 mm Hg) must be subtracted from the total pressure in calculations of partial pressures.
2. The atmospheric pressure decreases with altitude and increases by one atmosphere with every thirty-three feet below sea level.

B. The alveolar P_{O_2} at sea level is about 105 mm Hg.
C. By diffusion of oxygen from alveoli to blood, pulmonary vein and systemic arterial blood has a P_{O_2} of about 100 mm Hg.
   1. The P_{O_2} of a fluid is directly related to the amount of dissolved oxygen.
   2. Since whole blood contains red blood cells with hemoglobin that binds oxygen, whole blood contains much more oxygen than just the amount dissolved in plasma.

D. The P_{O_2} of arterial blood is used as an index of lung function.
E. Gas exchange in the lungs also maintains the systemic arterial P_{CO_2} at about 40 mm Hg.
II. Adequate gas exchange depends on the ventilation/perfusion ratio.
A. Ventilation is disproportionately greater than perfusion in the apex of the lungs, and disproportionately less than perfusion in the base of the lungs.
B. Although the ventilation/perfusion ratio throughout the lungs is not perfect, arterial P_{O_2} is only about 5 mm Hg less than alveolar P_{O_2}.

Self-Study Quiz

1. During normal inspiration, which of the following statements is TRUE?
   (a) the lungs inflate because the chest expands
   (b) the chest expands because the lungs inflate
2. Which of the following statements about intrapulmonary and intrapleural pressure is TRUE?
   (a) the intrapulmonary pressure is always subatmospheric
   (b) the intrapleural pressure is always greater than the intrapulmonary pressure
   (c) the intrapulmonary pressure is always higher than the intrapleural pressure
   (d) the intrapleural pressure equals the atmospheric pressure
3. If the transpulmonary pressure equals zero:
   (a) a pneumothorax has probably occurred
   (b) the lungs cannot inflate
   (c) elastic recoil causes the lungs to collapse
   (d) all of these

4. If the lungs have an abnormally low compliance:
   (a) the person may have pulmonary fibrosis or edema
   (b) the lungs are less distensible than normal
   (c) a larger than normal transpulmonary pressure is needed to inhale a normal amount of air
   (d) all of these
5. According to the law of LaPlace, the tendency for alveoli to collapse due to pressure generated by surface tension is greater in:
   (a) larger alveoli
   (b) smaller alveoli
6. The maximum amount of air that can be expired after a maximum inspiration is:
   (a) the tidal volume
   (b) the forced expiratory volume
   (c) the vital capacity
   (d) the maximum expiratory flow rate
7. Which of the following statements about emphysema is FALSE?
   (a) the lung compliance is reduced
   (b) the alveolar surface area is reduced
   (c) there is air trapping
   (d) the vital capacity is reduced

8. If a person who breathes normal air at sea level has an arterial P_{O_2} significantly below normal, one can conclude that:
   (a) the person has anemia
   (b) the person has impaired lung function
   (c) both (a) and (b)
   (d) neither (a) nor (b)
9. If blood were water (without red blood cells and hemoglobin), and if lung function were normal:
   (a) the arterial P_{O_2} would be normal
   (b) the amount of oxygen contained in arterial blood would be normal
   (c) both (a) and (b)
   (d) neither (a) nor (b)
10. If a person were to dive with scuba equipment to a depth of sixty-six feet, which of the following statements would be FALSE?
    (a) arterial P_{O_2} would be three times normal
    (b) the oxygen content of plasma would be three times normal
    (c) the oxygen content of whole arterial blood would be three times normal

Respiration: *Blood Gas Transport and Regulation of Breathing*

Objectives

By studying this chapter, you should be able to:

1. Explain the relationship between P_{O_2} and the interconversion of oxyhemoglobin and deoxyhemoglobin

2. Describe how changes in pH and temperature affect the unloading of oxygen by hemoglobin, and describe the functional significance of 2,3-DPG in oxygen transport

3. Describe how hemoglobin and red blood cell production are regulated and how this regulation helps to maintain the oxygen-carrying capacity of the blood

4. Explain the effect of ventilation on the acid-base balance of the blood

5. Describe the chloride shift and how bicarbonate and carbonic acid are formed and function to maintain acid-base balance

6. Define respiratory and metabolic acidosis and alkalosis, and explain how disturbances in one component can be partially compensated by changes in the other component

7. Describe the central control of breathing and how ventilation is regulated by peripheral chemoreceptor input

8. Explain how changes in blood P_{CO_2}, pH, and P_{O_2} affect ventilation in normal individuals at sea level, in normal people at high altitude, and in people with emphysema

9. Describe how the respiratory system adapts during exercise and during acclimatization to high altitude

10. Define oxygen toxity and nitrogen narcosis, and describe how decompression sickness can be produced

*T**he* lungs, through the processes of ventilation and gas exchange with the pulmonary blood, maintain the P_{O_2} and P_{CO_2} of systemic arterial blood within normal limits. A normal P_{O_2} is needed for complete loading of hemoglobin with oxygen, and thus for adequate delivery of oxygen to the tissues. A normal P_{CO_2} is required for maintenance of acid-base balance of the blood; breathing helps to maintain a normal blood pH. This occurs because carbon dioxide can combine with water to form carbonic acid, as will be described later in this chapter. Since breathing regulates the P_{O_2}, P_{CO_2}, and pH of arterial blood, it is not surprising—in view of the principles of negative feedback—that changes in these factors in turn affect the rate and depth of ventilation. The topics of oxygen and carbon dioxide transport are thus considered together with the regulation of breathing in this chapter.

Figure 15.1   Plasma and whole blood that are brought into equilibrium with the same gas mixture have the same P_{O_2} and thus the same amount of dissolved oxygen molecules (shown as black dots). The oxygen content of whole blood, however, is much higher than that of plasma because of the binding of oxygen to hemoglobin.

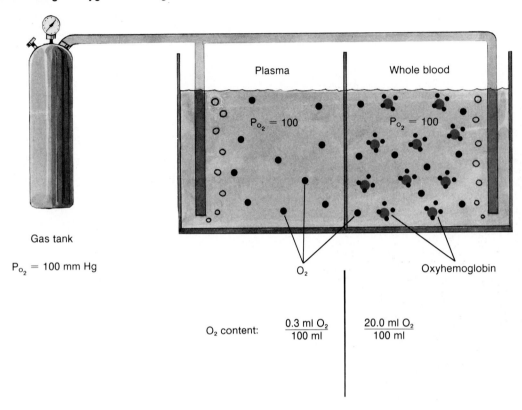

Gas tank

P_{O_2} = 100 mm Hg

O_2 content: $\dfrac{0.3 \text{ ml } O_2}{100 \text{ ml}}$ $\dfrac{20.0 \text{ ml } O_2}{100 \text{ ml}}$

Hemoglobin and Oxygen Transport

If the lungs are functioning properly, blood leaving in the pulmonary vein would have a normal P_{O_2} (100 mm Hg at sea level) regardless of its hemoglobin content. This is because, as described in chapter 14, the P_{O_2} indicates the amount of oxygen dissolved in plasma and is not affected by oxygen attached to hemoglobin. At a P_{O_2} of 100 mm Hg and a temperature of 37°C, plasma contains 0.3 ml O_2 per 100 ml. Blood with a normal hemoglobin concentration, however, contains about 20 ml O_2 per 100 ml. The ability of the blood to transport oxygen thus depends not only on lung function but also on the hemoglobin concentration.

Hemoglobin

Most of the oxygen in blood is contained within the red blood cells (erythrocytes) chemically bonded to **hemoglobin.** Each hemoglobin molecule consists of (1) a protein part—the globin—composed of four polypeptide chains; and (2) four nitrogen-containing, disc-shaped organic pigment molecules called *hemes.* The structure of hemoglobin is illustrated in figure 15.2.

The protein part of hemoglobin is composed of two identical *alpha chains* that are each 141 amino acids long, and two identical *beta chains* that are each 146 amino acids long. Each of the four polypeptide chains is combined with one heme group. In the center of each heme group is a single atom of iron, which can combine with one molecule of oxygen (O_2). One hemoglobin can thus combine with four molecules of oxygen; since there are about 280 million hemoglobin molecules per red blood cell, each red blood cell can carry over a billion molecules of oxygen.

Figure 15.2 The three-dimensional structure of hemoglobin is illustrated in *(a)*, in which the two alpha and two beta polypeptide chains are shown. The four heme groups are shown as flat structures with iron *(the dark spheres)* in the centers. The chemical structure of heme is shown in *(b)*.

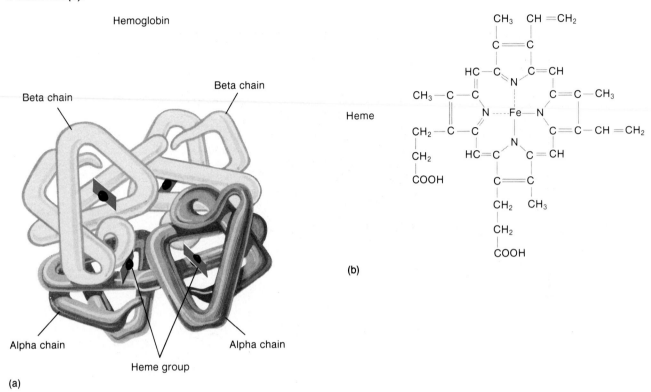

Hemoglobin

Beta chain

Beta chain

Alpha chain

Alpha chain

Heme group

(a)

Heme

(b)

Normal heme contains iron in the reduced form (Fe^{++}, or ferrous iron). In this form, the iron can share electrons and bond to oxygen to form **oxyhemoglobin.** When oxyhemoglobin dissociates to release oxygen to the tissues, the heme iron is still in the Fe^{++} form, and part of **reduced** (or **deoxy**) **hemoglobin.** The term *oxyhemoglobin* is thus not equivalent to *oxidized* hemoglobin or **methemoglobin.** This latter form of hemoglobin contains iron in the oxidized state (Fe^{+++}—the term *oxidized* is used to indicate loss of an electron). Methemoglobin, containing oxidized iron (Fe^{+++}, or ferric iron) lacks the electron it needs to bond to oxygen. Blood normally contains only a small amount of methemoglobin, which cannot participate in oxygen transport.

Carboxyhemoglobin is another abnormal form of hemoglobin in which the reduced heme is combined with carbon monoxide gas (*not* carbon dioxide) instead of oxygen. Since the bond between reduced heme and carbon monoxide is about 210 times stronger than the bond between heme and oxygen, carbon monoxide tends to displace oxygen in hemoglobin and remain attached as carboxyhemoglobin passes through systemic capillaries.

As a result of the competition between carbon monoxide and oxygen for binding sites in the heme groups, carbon monoxide poisoning produces a functional anemia in which the total oxygen content of the blood is reduced. If the percent of carboxyhemoglobin is 30 percent, for example, the amount of oxyhemoglobin would be reduced accordingly (from 97 percent in normal arterial blood to 67 percent in this instance). Such severe carbon monoxide poisoning may be detected by the bright cherry-red color of carboxyhemoglobin, as seen through the skin or mucous membranes.

According to the Federal Clean Air Act of 1971, an active, healthy non-smoker should not have a carboxyhemoglobin saturation (percent carboxyhemoglobin) greater than 1.5 percent. The carboxyhemoglobin levels found in residents of many cities, however, are significantly above this value (see table 15.1).

Table 15.1 Percent carboxyhemoglobin values of non-smokers and cigarette smokers in selected cities (values are averages of the data reported.) These measurements apply only to the blood samples taken during the dates indicated and cannot be used to rate "smogginess" in these cities (other components of smog were not measured). It would be expected that carboxyhemoglobin saturations would be different if blood samples were taken at different times (blood samples during the smoggiest months in Los Angeles, for example—usually July, August, and September—were not obtained). The Clean Air Act of 1971 sets a goal of 1.5 percent carboxyhemoglobin in the blood of active non-smokers.

Cities	Percent Carboxyhemoglobin				
	Average for Nonsmokers		Average for Smokers		Dates Sampled
	Median	*90% Range*	*Median*	*90% Range*	
Chicago	1.99	1.36–3.2	5.81	2.73–9.4	Nov., 1970
Los Angeles	1.88	1.22–3.04	6.13	2.17–9.58	Jan., Feb., May, June, 1972
Milwaukee	1.26	0.5–3.86	5.10	1.03–9.67	April–Dec., 1969 Jan.–Dec., 1970 April, 1971
New York	1.38	0.71–2.40	4.86	1.44–8.18	Dec., 1970 Oct.–Dec., 1971

Source: "Carboxyhemoglobin Concentrations in Blood from Donor in Chicago, Milwaukee, New York, and Los Angeles," Stewart, R. D. et al., *Science* Vol. 182, pp. 1,362–1,364, Table 1, 28 December 1973. Copyright 1973 by the American Association for the Advancement of Science.

Hemoglobin Concentration The *oxygen-carrying capacity* of whole blood is determined by its concentration of normal hemoglobin. Blood with a normal hemoglobin concentration of 15 g per 100 ml can carry as much as 20 ml O_2 per 100 ml, as previously described. If the hemoglobin concentration is below normal—a condition called **anemia**—the oxygen concentration of the blood is reduced accordingly. A hemoglobin concentration that is reduced to two-thirds normal (10 g per 100 ml), for example, can only carry 13 ml O_2 per 100 ml. Conversely, when the hemoglobin concentration is increased above the normal range—as occurs in **polycythemia** (high red blood cell count)—the oxygen-carrying capacity of blood is increased correspondingly. This occurs as an adaptation to life at a high altitude, as will be discussed in a later section.

Production of hemoglobin and red blood cells within bone marrow is controlled by a hormone called **erythropoietin.** This hormone is produced by the kidneys. Secretion of erythropoietin, and thus red blood cell production, is stimulated when delivery of oxygen to the kidneys and other organs is lower than normal. Red blood cell production is also promoted by androgens, which explains why the hemoglobin concentration in men averages 1–2 g per 100 ml higher than that in women.

Reaction between Reduced Hemoglobin and Oxygen
Reduced hemoglobin plus oxygen combine to form oxyhemoglobin. Oxyhemoglobin, in turn, dissociates to yield reduced hemoglobin (also called deoxyhemoglobin) and free oxygen. The former reaction is called the *loading reaction* and occurs in the lungs; the latter reaction is the *unloading reaction* that occurs in the systemic capillaries.

Loading and unloading can thus be shown as a reversible reaction:

$$\text{Deoxyhemoglobin} + O_2 \underset{\text{(tissues)}}{\overset{\text{(lungs)}}{\rightleftarrows}} \text{Oxyhemoglobin}$$

The extent that the reaction will go in each direction is determined by two factors. These are (1) the P_{O_2} of the blood, and (2) the *affinity* or bond strength between hemoglobin and oxygen. High P_{O_2} drives the reaction to the right, and since P_{O_2} is high in the lungs, hemoglobin loads with oxygen to form oxyhemoglobin. Low P_{O_2} in the systemic capillaries drives the reaction to the left, so that some of the oxygen that is bound to hemoglobin is released to the tissues.

The loading of oxygen with hemoglobin in the lungs, and the amount of oxygen unloading that occurs in the tissue capillaries, is also dependent on the bond strength (affinity) of hemoglobin for oxygen. A very strong bond would favor loading, but inhibit unloading; a weak bond, on the other hand, would increase oxygen unloading to the tissues. The bond strength between hemoglobin and oxygen is normally strong enough so that 97 percent of the hemoglobin leaving the lungs is in the form of oxyhemoglobin, yet the bond is sufficiently weak so that adequate amounts of oxygen are unloaded to sustain aerobic metabolism in the tissues.

The Oxyhemoglobin Dissociation Curve
Blood in the systemic arteries, at a P_{O_2} of 100 mm Hg, has a *percent oxyhemoglobin saturation* of 97 percent (which means that 97 percent of the hemoglobin is in the form of oxyhemoglobin). This blood is delivered to the systemic capillaries, where oxygen diffuses into the tissue cells and is consumed in aerobic respiration. Blood leaving

Table 15.2 The relationship between percent oxyhemoglobin saturation and P_{O_2}
(at pH = 7.40 and temperature = 37°C).

P_{O_2}(mm Hg)	100	80	61	45	40	36	30	26	23	21	19
Percent Oxyhemoglobin	97	95	90	80	75	70	60	50	40	35	30
		Arterial blood			Venous blood						

Figure 15.3 The percent oxyhemoglobin saturation and the blood oxygen content are shown at different values of P_{O_2}. Notice that there is about a 25 percent decrease in percent oxyhemoglobin as blood passes through the tissue from arteries to veins, resulting in the unloading of approximately 5 ml O_2 per 100 ml to the tissues.

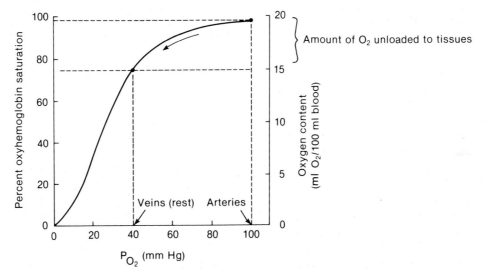

the capillaries in the systemic veins is thus depleted in oxygen; it has a P_{O_2} of about 40 mm Hg, and a percent oxyhemoglobin saturation of about 75 percent. Since blood entering the tissues contains 97 percent oxyhemoglobin, and blood leaving the tissues contains 75 percent oxyhemoglobin, about 22 percent of the oxyhemoglobin that entered the capillaries must have unloaded its oxygen to the tissues.

A graphic illustration of the percent oxyhemoglobin saturation at different values of P_{O_2} is called an **oxyhemoglobin dissociation curve**. The values in this graph are obtained by subjecting samples of blood *in vitro* outside the body to different partial oxygen pressures. The percent oxyhemoglobin saturation obtained by this procedure, however, indicates what the percent unloading would be *in vivo* (within the body) with a given difference in arterial and venous P_{O_2} values.

Figure 15.3 shows the difference between arterial and venous P_{O_2} and percent oxyhemoglobin saturation at rest. Under these conditions, as shown by the graph, only 22 percent of the oxyhemoglobin unloads its oxygen to the tissues, leaving about 75 percent oxyhemoglobin in venous blood. This relatively small amount of oxygen unloading is sufficient to meet the aerobic requirements of the tissues at rest, and accounts for the color difference between arterial and venous blood. Arterial blood is a bright red, while venous blood is a dark scarlet due to its higher content of deoxyhemoglobin.

The relatively large amount of oxyhemoglobin remaining in venous blood at rest functions as an oxygen reservoir. If a person stops breathing, there is a sufficient reserve of oxygen in the blood to keep the brain and heart alive for approximately four to five minutes in the absence of cardiopulmonary resuscitation (CPR) techniques. This reserve supply of oxygen can also be tapped when the tissue's requirements for oxygen are raised.

Figure 15.4 A decrease in blood pH (an increase in H$^+$ concentration) decreases the affinity of hemoglobin for oxygen at each P$_{O_2}$ value, resulting in a "shift to the right" of the oxyhemoglobin dissociation curve. A curve that is shifted to the right has a lower percent oxyhemoglobin saturation at each P$_{O_2}$, but the effect is more marked at lower P$_{O_2}$ values. This effect is called the *Bohr effect.*

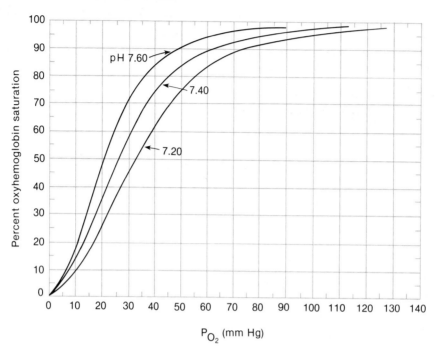

Effect of pH and Temperature on Oxygen Unloading

The oxyhemoglobin dissociation curve is S-shaped, or *sigmoidal.* The fact that it is relatively flat at high P$_{O_2}$ values indicates that changes in P$_{O_2}$ within this range have little effect on the loading reaction. One would have to ascend to as high as 20,000 feet, for example, before the oxyhemoglobin saturation of arterial blood would decrease from 97 percent to 70 percent. At more common elevations the percent oxyhemoglobin saturation would not be significantly different from the 97 percent that occurs at sea level.

At the steep part of the sigmoidal curve, however, small changes in P$_{O_2}$ result in large differences in percent oxyhemoglobin saturation. A slight decrease in venous P$_{O_2}$, therefore—as might be produced by increased tissue oxygen consumption during exercise—produces a large drop in percent oxyhemoglobin saturation. Since the arterial saturation is not changed, this means that the amount of oxygen unloading in the tissues would be greatly increased as a result of the fall in tissue (and venous) P$_{O_2}$.

Changes in Hemoglobin Affinity for Oxygen: Effect of pH

In addition to the effect of tissue and venous P$_{O_2}$ on oxygen unloading, changes in the affinity of hemoglobin for oxygen can affect the unloading reaction. The strength of the bond between hemoglobin and oxygen is weakened by a decrease in pH (increased acidity), resulting in increased unloading of oxygen to the tissues. This is called the **Bohr effect.** The Bohr effect, in which acidotic conditions result in decreased percent oxyhemoglobin saturation, is illustrated by a *shift to the right* of the oxyhemoglobin dissociation curve (see figure 15.4). The graph illustrates that, at a given P$_{O_2}$ for venous blood, there is a lower percent oxyhemoglobin saturation in the curve to the right. Since the percent of saturation of arterial blood is much less affected, a shift to the right of the graph indicates a higher percent unloading of oxygen to the tissues.

Protons (H$^+$) bond to hemoglobin; hemoglobin therefore acts as a buffer in the red blood cells, as will be described in a later section. The ability of deoxyhemoglobin to bond H$^+$ is greater than that of oxyhemoglobin. Conversely, bonding of H$^+$ to deoxyhemoglobin tends to stabilize the deoxyhemoglobin form. Therefore, when there is a higher than normal H$^+$ concentration in the blood—a condition called *acidosis*—the relative proportion of deoxyhemoglobin is increased while the percent oxyhemoglobin is decreased. This explains the fact that more oxyhemoglobin is converted to deoxyhemoglobin in systemic capillaries under acidotic conditions, so that more oxygen is unloaded to the tissues (the Bohr effect).

Table 15.3 Effect of pH on hemoglobin affinity for oxygen and oxyhemoglobin "unloading" (the difference between arterial and venous blood oxygen content). Oxygen content is indicated as ml O_2 per 100 ml blood.

pH	Affinity	Arterial O_2 Content per 100 ml	Venous O_2 Content per 100 ml	O_2 Unloaded to Tissues per 100 ml
7.40	Normal	19.8 ml O_2	14.8 ml O_2	5.0 ml O_2
7.60	Increased	20.0 ml O_2	17.0 ml O_2	3.0 ml O_2
7.20	Decreased	19.2 ml O_2	12.6 ml O_2	6.6 ml O_2

Figure 15.5 The oxyhemoglobin dissociation curve is shifted to the right as the temperature increases, indicating a lowered affinity of hemoglobin for oxygen at each P_{O_2}. This effect, like the Bohr effect (see figure 15.8), is more marked at lower P_{O_2} values.

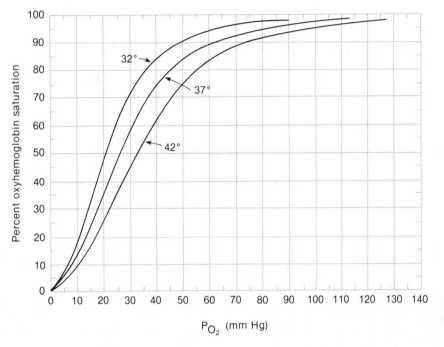

Effect of Temperature on Hemoglobin Affinity for Oxygen   Under normal conditions of blood pH, the affinity of hemoglobin for oxygen is inversely related to the temperature of the blood. At a higher than normal temperature, the bond between hemoglobin and oxygen is weakened. This means that at the P_{O_2} of venous blood there is a lower than normal percent oxyhemoglobin, indicating that more of the oxyhemoglobin that entered the capillaries unloaded their content of oxygen. The effect of elevated blood temperature is thus similar to the effect of decreased blood pH, and is shown as a shift to the right of the oxyhemoglobin dissociation curve (see figure 15.5).

Summary   Under normal resting conditions at sea level, where the arterial P_{O_2} is 100 mm Hg and the venous P_{O_2} is 40 mm Hg, the oxyhemoglobin saturation of arterial and venous blood is 97 percent and 75 percent, respectively. This indicates that about 22 percent of the oxyhemoglobin entering tissue capillaries has unloaded its oxygen. Since this arterial blood contains about 20 ml O_2 per 100 ml, and venous blood contains about 15.5 ml O_2 per 100 ml, approximately 4.5 ml O_2 per 100 ml must be unloaded in the systemic capillaries.

Table 15.4 Factors that affect affinity of hemoglobin for oxygen and the position of the oxyhemoglobin dissociation curve.

Factor	Affinity	Position of Curve	Comments
↓ pH	Decreased	Shift to the right	Called the Bohr effect. Increases oxygen delivery during hypercapnia.
↑ P_{CO_2}			
↑ Temperature	Decreased	Shift to the right	Increases oxygen unloading during exercise and fever.
↑ 2,3-DPG	Decreased	Shift to the right	Increases oxygen unloading when there is a decrease in total hemoglobin or total oxygen content. An adaptation to anemia and high-altitude living.

During exercise or other conditions in which the pH falls and/or the temperature rises in the tissues, the bond strength between hemoglobin and oxygen is weakened. This means that more oxygen is unloaded to the tissues under these conditions, so that the P_{O_2} and oxygen content of venous blood is decreased. A weakened affinity of hemoglobin for oxygen is shown graphically as a shift to the right of the oxyhemoglobin dissociation curve. It should be noted that opposite conditions—a rise in pH and a fall in temperature—produce the opposite effects. Less oxygen is unloaded to the tissues under these conditions, and the oxyhemoglobin dissociation curve is displaced to the left of normal.

Effect of 2,3-DPG on Oxygen Unloading

Mature red blood cells lack both nuclei and mitochondria. They therefore cannot use the oxygen they carry, and must rely on anaerobic respiration to meet their energy needs. Red blood cells are also unique in that they produce large amounts of **2,3-diphosphoglyceric acid (2,3-DPG)** as a "side product" of glycolysis.

Production of 2,3-DPG is inhibited by oxyhemoglobin. When the oxyhemoglobin concentration of red blood cells is decreased, as in life at a high altitude and in anemia, 2,3-DPG is produced in increased amounts. This compound combines with deoxyhemoglobin and makes it more stable. At the P_{O_2} of tissues and venous blood, therefore, a higher proportion of oxyhemoglobin will be converted to deoxyhemoglobin by unloading its oxygen. Increased concentrations of 2,3-DPG in red blood cells thus increase oxygen unloading and shift the oxyhemoglobin dissociation curve to the right.

The importance of 2,3-DPG within red blood cells is now recognized in blood banking. Old, stored red blood cells can lose their ability to produce 2,3-DPG as they lose their ability to respire anaerobically. Modern techniques for blood storage therefore include the addition of energy substrates for respiration and the inclusion of precursors needed in the production of 2,3-DPG.

Anemia When the total blood hemoglobin concentration is reduced below normal in anemia, each red blood cell produces increased amounts of 2,3-DPG. A normal hemoglobin concentration of 15 g per 100 ml unloads about 4.5 ml O_2 per 100 ml at rest, as previously described. If the hemoglobin concentration were reduced by half, one might expect that the tissues would receive only half the normal amount of oxygen (2.25 ml O_2 per 100 ml). It has been shown, however, that as much as 3.3 ml O_2 per 100 ml are unloaded to the tissues under these conditions. This occurs as a result of increased 2,3-DPG production and thus of decreased affinity of hemoglobin for oxygen.

Hemoglobin F The effects of 2,3-DPG are also important in the transfer of oxygen from maternal to fetal blood. Normal adult hemoglobin in the mother—called *hemoglobin A*—is able to bond to 2,3-DPG, and as a result has a reduced affinity for oxygen. Fetal hemoglobin (*hemoglobin F*), in contrast, cannot bond to 2,3-DPG and thus has a higher affinity for oxygen at a given P_{O_2} than does hemoglobin A. Since hemoglobin F can have a higher percent oxyhemoglobin saturation than hemoglobin A at a given P_{O_2}, gas exchange in the placenta between maternal and fetal blood results in the transfer of oxygen to the fetal blood.

Inherited Defects in Hemoglobin Structure and Function

There are a number of hemoglobin diseases that are produced by inherited (congenital) defects in the protein part of hemoglobin. **Sickle-cell anemia**—a disease affecting 8–11 percent of the black population of the United States—for example, is caused by an abnormal form of hemoglobin called *hemoglobin S*. Hemoglobin S differs from normal hemoglobin A in only one amino acid: valine is substituted for glutamic acid in position 6 on the beta chains. This amino acid substitution is caused by a single base change in the region of DNA that codes for the beta chains (see table 15.5).

Under conditions of low blood P_{O_2}, hemoglobin S comes out of solution and cross links to form a "paracrystalline gel" within the red blood cells. This causes the characteristic sickle shape of red blood cells and makes them less flexible. Since red blood cells must be able to bend in the middle to pass through many narrow capillaries, a decrease in their flexibility may cause them to block small blood channels and produce organ ischemia. The decreased solubility of hemoglobin S in solutions of low P_{O_2} is used in the diagnosis of sickle-cell anemia and sickle-cell trait (the carrier state, in which a person has the genes for both hemoglobin A and hemoglobin S).

Table 15.5 DNA base triplets, mRNA codons, and amino acids in position 6 of the beta chains in normal hemoglobin A and in two mutant hemoglobin forms.

Hemoglobin	DNA Base Triplet	mRNA Codon	Amino Acid in Position 6
A	CTT	GAA	Glutamic acid
S	CAT	GUA	Valine
C	TTT	AAA	Lysine

Figure 15.6 *(a)* shows the turbidity test for sickle-cell anemia and sickle-cell trait. At low P_{O_2} the hemoglobin *S* comes out of solution and makes the solution appear cloudy *(right and left tubes),* while normal hemoglobin *A* remains dissolved, producing a clear solution. *(b)* shows a sickled red blood cell as seen in the light microscope. *(c)* are normal and *(d)* are sickled red blood cells seen in the scanning electron microscope.

(a)

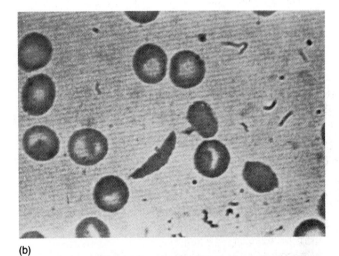

(b)

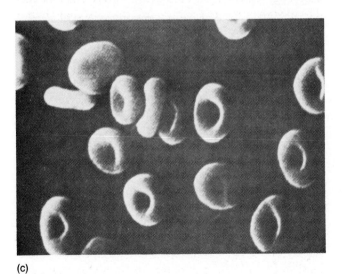

(c)

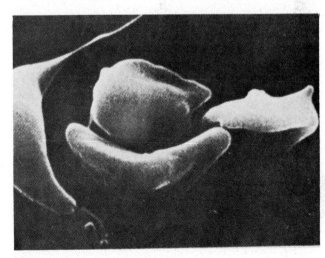

(d)

Figure 15.7   Comparison of the dissociation curves for hemoglobin and for myoglobin. At the P_{O_2} of venous blood, the myoglobin retains almost all of its oxygen, indicating a higher affinity than hemoglobin for oxygen. The myoglobin does, however, release its oxygen at the very low P_{O_2} values found inside the mitochondria.

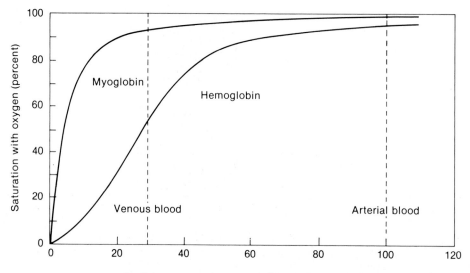

Partial pressure of oxygen (millimeters of mercury)

Thalassemia is a hemoglobin disease found predominately among people of Mediterranean ancestry. In this disease the person has an abnormal form of hemoglobin with a very high affinity for oxygen. In β-thalassemia, for example, an adult retains large amounts of hemoglobin F (fetal hemoglobin). Since hemoglobin F has a higher affinity for oxygen than hemoglobin A, an abnormally low amount of oxygen is unloaded to the tissues.

Some types of abnormal hemoglobins have been shown to be advantageous in the environments in which they evolved. A person who is a carrier for sickle-cell anemia, for example (and who therefore has both hemoglobin A and hemoglobin S), has a high resistance to malaria. This is because the parasite that causes this disease cannot live in red blood cells that contain hemoglobin S.

Muscle Myoglobin
Myoglobin is a red pigment found exclusively in striated muscle fibers (not in blood). In particular, slow-twitch, aerobically respiring skeletal fibers and cardiac fibers are rich in myoglobin. Myoglobin is similar to hemoglobin, but it has one rather than four hemes and therefore can carry only one molecule of oxygen.

As a result of the difference in structure between myoglobin and hemoglobin, myoglobin has a higher affinity for oxygen than hemoglobin. At the P_{O_2} for venous blood (40 mm Hg), for example, only 10 percent of the

oxygen attached to myoglobin is released, compared to a 22 percent oxyhemoglobin dissociation as previously described. The dissociation curve for myoglobin is shown in figure 15.7.

Myoglobin discharges its oxygen at the very low P_{O_2} found within mitochondria (about 1 mm Hg), and so may act as a "middleman" in the exchange of oxygen between blood and mitochondria. Reserves of oxygen attached to myoglobin may be particularly important in cardiac muscle, where blood flow is occluded during systole as the coronary arteries are squeezed shut. Myoglobin may discharge its oxygen during systole and become recharged with oxygen as coronary blood flow is resumed during diastole.

1. Explain how the loading and unloading reactions of hemoglobin with oxygen are influenced by P_{O_2}. Describe how a lowering of tissue (and venous) P_{O_2} affects the amount of oxygen unloaded. Illustrate this with an oxyhemoglobin dissociation curve.

2. Explain how the unloading of oxygen by oxyhemoglobin is affected by the affinity of hemoglobin for oxygen. Describe the effects of pH and temperature on oxygen transport, and illustrate this with an oxyhemoglobin dissociation curve.

3. Describe the effects of 2,3-DPG on oxygen transport. Explain how this mechanism helps during acclimatization to life at a high altitude and for compensation in anemia.

Figure 15.8 Illustration of carbon dioxide transport by the blood and the "chloride shift" (see text for complete description). Carbon dioxide is transported in three forms: as dissolved CO_2 gas, attached to hemoglobin as carbaminohemoglobin, and as carbonic acid and bicarbonate.

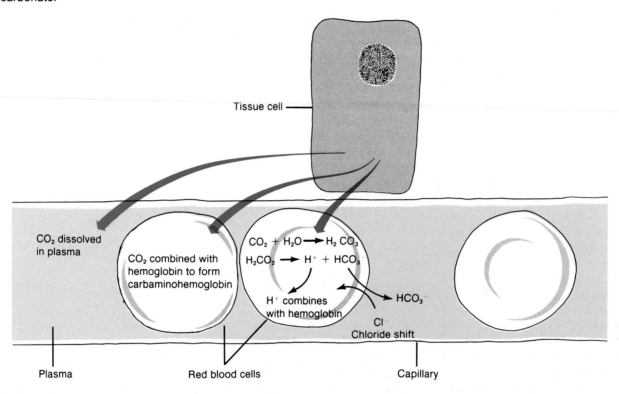

Transport of Carbon Dioxide and Acid-Base Balance

Carbon dioxide is carried by the blood in three forms. These include (1) *dissolved CO_2*; since carbon dioxide is about twenty-one times more soluble than oxygen in water, a substantial proportion (about one-tenth) of the total blood CO_2 is dissolved in plasma; (2) *carbaminohemoglobin;* about one-fifth of the total blood CO_2 is carried attached to an amino acid in hemoglobin (this should not be confused with the combination of carbon monoxide with heme in carboxyhemoglobin); and (3) *carbonic acid* and *bicarbonate,* which account for most of the CO_2 carried by the blood.

Carbon dioxide is able to combine with water to form carbonic acid. This reaction occurs spontaneously in the plasma at a slow rate, but occurs much more rapidly within red blood cells due to the catalytic action of the enzyme **carbonic anhydrase**. Since this enzyme is confined to the red blood cells, most of the carbonic acid is produced there rather than in the plasma. Formation of carbonic acid from CO_2 and water is favored by the high P_{CO_2} found in tissue capillaries.

$$CO_2 + H_2O \xrightarrow[\text{(high } P_{CO_2})]{\text{carbonic anhydrase}} \underset{\text{carbonic acid}}{H_2CO_3}$$

The Chloride Shift

As a result of catalysis by carbonic anhydrase within red blood cells, large amounts of carbonic acid are produced as blood passes through systemic capillaries. Buildup of carbonic acid concentrations within red blood cells favors dissociation of these molecules into H^+ (protons, which contribute to the acidity of the solution) and HCO_3^- (bicarbonate):

$$\underset{\text{carbonic acid}}{H_2CO_3} \longrightarrow \underset{\text{bicarbonate}}{H^+ + HCO_3^-}$$

The H^+ released by dissociation of carbonic acid is largely buffered by combination with deoxyhemoglobin within the red blood cells. While the unbuffered H^+ is free to diffuse out of the red blood cells, more bicarbonate diffuses outward into the plasma than does H^+. As a result of the "trapping" of H^+ within the red blood cell by attachment to hemoglobin, and the outward diffusion of HCO_3^-, the inside of the red blood cells gain a net positive charge. This attracts chloride ions (Cl^-), which move into the red blood cells as HCO_3^- moves out. This exchange of anions as blood moves through the tissue capillaries is called the **chloride shift** (see figure 15.8).

Figure 15.9   Carbon dioxide is released from the blood as it travels through the pulmonary capillaries. During this time a "reverse chloride shift" occurs and carbonic acid is transformed into CO_2 and H_2O.

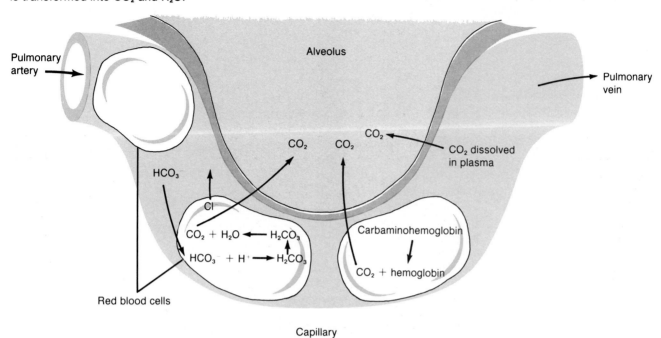

When blood reaches the pulmonary capillaries, the P_{CO_2} of the blood is decreased and deoxyhemoglobin is converted to oxyhemoglobin. Since oxyhemoglobin has a lower affinity for H^+ than does deoxyhemoglobin (as previously described), H^+ is released within the red blood cells. This attracts HCO_3^- from the plasma, which combines with H^+ to form carbonic acid:

$$H^+ + HCO_3^- \longrightarrow H_2CO_3$$

Under conditions of low P_{CO_2}, carbonic anhydrase catalyzes the conversion of carbonic acid to CO_2 and water:

$$H_2CO_3 \xrightarrow[\text{(low } P_{CO_2})]{\text{carbonic anhydrase}} CO_2 + H_2O$$

In summary, carbon dioxide produced by tissue cells is converted within systemic capillaries to carbonic acid and bicarbonate. These events require the action of carbonic anhydrase within red blood cells and are known as the chloride shift. A reverse chloride shift operates in the pulmonary capillaries to convert carbonic acid and bicarbonate to CO_2 gas, which is eliminated in the expired breath. The P_{CO_2}, carbonic acid, and bicarbonate concentrations in the systemic arteries are thus maintained relatively constant by normal ventilation.

Ventilation and Acid-Base Balance

Normal systemic arterial blood has a pH of 7.35 to 7.45 (or 7.40 ± 0.05). Put another way, arterial blood has a H^+ concentration of about $10^{-7.4}$ molar. Some of these H^+ are derived from carbonic aid and some are derived from nonvolatile *metabolic acids* (fatty acids, ketone bodies, lactic acid, and others) that cannot be eliminated in the expired breath. It should be recalled that an acid is a molecule that can dissociate and release H^+ to solution.

Under normal conditions, much of the H^+ released by metabolic acids do not affect blood pH because these H^+ combine with HCO_3^-, and are thereby removed from solution. *Bicarbonate is the major buffer in the plasma,* and acts to maintain a blood pH of 7.4 despite the constant production of metabolic acids by the tissues. In this buffering process, some of the HCO_3^- released from the red blood cells during the chloride shift is converted into H_2CO_3 in the plasma. Normally, however, there is still a *buffer reserve* of free bicarbonate that can help protect against unusually large additions of metabolic acids to the blood.

Figure 15.10 Bicarbonate released into the plasma from red blood cells functions to buffer H^+ produced by the ionization of metabolic acids (lactic acid, fatty acids, ketone bodies, and others).

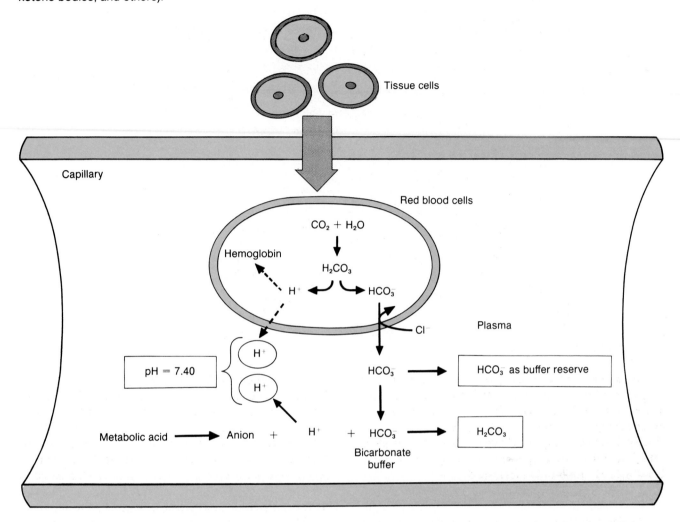

Normal plasma therefore contains free bicarbonate, carbonic acid, and a H^+ concentration indicated by a pH of 7.4. Carbonic acid produced by the buffering reaction would not normally ionize. If the blood pH should rise above normal (indicating a fall in H^+ concentration), however, dissociation of H_2CO_3 in plasma would add additional H^+ to the blood and help to compensate for this condition. Carbonic acid and bicarbonate are thus said to function as a *buffer pair*.

Acidosis and Alkalosis   A fall in blood pH below 7.35 is called *acidosis*, because the pH is to the acid-side of normal. The blood is not acidic; a blood pH of 7.2, for example, represents serious acidosis. Similarly, a rise in blood pH above 7.45 is known as *alkalosis*. There are two components to acid base imbalance: *respiratory* and *metabolic*. Respiratory acidosis and alkalosis are due to abnormal amounts of carbonic acid, while metabolic acidosis and alkalosis result from abnormal amounts of H^+ derived from metabolic acids.

Table 15.6   Definition of terms used to describe acid-base balance.

Terms	Definitions
Acidosis, respiratory	Increased carbon dioxide retention (due to hypoventilation), which can result in the accumulation of carbonic acid and thus a fall in blood pH below normal.
Acidosis, metabolic	Increased production of "nonvolatile" acids such as lactic acid, fatty acids, and ketone bodies, or loss of blood bicarbonate (such as by diarrhea) resulting in a fall in blood pH below normal.
Alkalosis, respiratory	A rise in blood pH due to loss of CO_2 and carbonic acid (through hyperventilation).
Alkalosis, metabolic	A rise in blood pH produced by loss of nonvolatile acids (as in excessive vomiting) or by excessive accumulation of bicarbonate base.
Compensated acidosis or alkalosis	Metabolic acidosis or alkalosis is partially compensated by opposite changes in blood carbonic acid levels (through changes in ventilation). Respiratory acidosis or alkalosis is partially compensated by increased retention or excretion of bicarbonate in the urine.

Table 15.7   The effect of lung function on blood acid-base balance.

Condition	pH	P_{CO_2}	Ventilation	Cause or Compensation
Normal	7.35–7.45	39–41 mm Hg	Normal	Not applicable
Respiratory acidosis	Low	High	Hypoventilation	Cause of the acidosis
Respiratory alkalosis	High	Low	Hyperventilation	Cause of the alkalosis
Metabolic acidosis	Low	Low	Hyperventilation	Compensation for acidosis
Metabolic alkalosis	High	High	Hypoventilation	Compensation for alkalosis

Ventilation is normally adjusted to keep pace with the metabolic rate, by mechanisms that will be described in a later section. *Hypoventilation* occurs when the rate of CO_2 production exceeds the rate at which it is "blown off" in ventilation. Under these conditions, carbonic acid production in the red blood cells is excessively high compared to ventilation, and **respiratory acidosis** occurs. In *hyperventilation,* similarly, depletion of carbonic acid raises the blood pH and **respiratory alkalosis** occurs.

Metabolic acidosis occurs when production of nonvolatile acids is abnormally increased. In uncontrolled diabetes mellitus, for example, ketone bodies (derived from fatty acids) may accumulate and produce ketoacidosis. In order for metabolic acidosis to occur, however, the buffer reserve of bicarbonate must first be depleted. Until that time the blood pH remains normal—ketosis can occur, for example, without ketoacidosis. **Metabolic alkalosis** is less common than metabolic acidosis. This can occur when there is loss of acidic gastric juice by vomiting, or when there is excessive intake of bicarbonate (from stomach antacids, or from an intravenous infusion).

Compensations for Acidosis and Alkalosis   A change in blood pH produced by alterations in either the respiratory or metabolic component of acid-base balance can be partially compensated by a change in the other component. Metabolic acidosis, for example, is partially compensated by hyperventilation, which causes a decrease in carbonic acid. This occurs because the receptors that regulate breathing are sensitive to H^+. People who have metabolic acidosis would thus have a low pH (by definition) and a low P_{CO_2} as a result of hyperventilation. Metabolic alkalosis is partially compensated by retention of carbonic acid due to hypoventilation (see table 15.7).

A person with respiratory acidosis, by definition, has a low blood pH and a high blood P_{CO_2} due to hypoventilation. This condition is partially compensated by production and retention of bicarbonate by the kidneys. The role of the kidneys in acid-base balance will be discussed in chapter 16.

1. Describe the reaction catalyzed by carbonic anhydrase within red blood cells in *(a)* systemic capillaries, and *(b)* pulmonary capillaries. Explain why the reactions go in different directions.
2. Write an equation showing the buffering of H^+ from metabolic acids by bicarbonate. Describe how this reaction helps to maintain a normal blood pH.
3. Describe how respiratory acidosis and alkalosis are produced and what would happen to the HCO_3^- concentration of plasma under these conditions.
4. Describe how metabolic acidosis and alkalosis are produced and what would happen to the HCO_3^- concentration of plasma under these conditions.
5. Explain why a person with ketoacidosis would hyperventilate and how this could partially compensate for the acidosis.

Figure 15.11 Approximate locations of the brain stem respiratory centers.

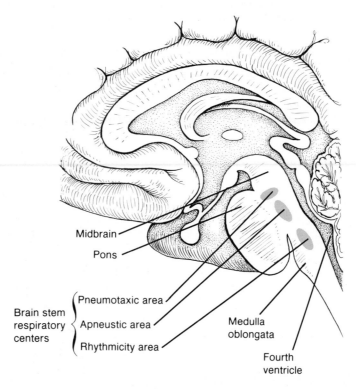

Midbrain

Pons

Brain stem respiratory centers
{
Pneumotaxic area

Apneustic area

Rhythmicity area
}

Medulla oblongata

Fourth ventricle

Regulation of Breathing

Inspiration and expiration are produced by contraction and relaxation of the diaphragm and external intercostal muscles during quiet breathing, and by contraction of accessory respiratory muscles during forced breathing (as described in chapter 14). Contraction and relaxation of these striated muscles are directly controlled by somatic motor neurons in the spinal cord. Activity of these lower motor neurons, in turn, is directly controlled by the *respiratory centers* in the medulla oblongata and pons areas of the brain stem.

Respiratory Centers

A loose aggregation of neurons in the reticular formation of the medulla oblongata form the center that controls rhythmic ventilation. Discharges of some of these neurons produce inspiration, while discharges of other neurons produce expiration. Activity of inspiratory neurons and expiratory neurons vary in a reciprocal way, so that a rhythmic pattern of breathing is produced. The cycle of inspiration and expiration is thus built in, or intrinsic to the neural activity of the medulla.

Neural activity of the medullary respiratory center, however, is influenced by centers in the pons area of the brain stem. As a result of research in which different areas of the pons are selectively destroyed, two respiratory control centers in the pons have been identified. One area—called the *pneumotaxic area* of the pons—apparently promotes activity of the inspiratory neurons of the medulla. The other pontine area—known as the *apneustic center*—seems to inhibit inspiration (see figures 15.11 and 15.13).

Figure 15.12 The peripheral chemoreceptors (aortic and carotid bodies) regulate the brain stem respiratory centers by means of sensory nerve stimulation.

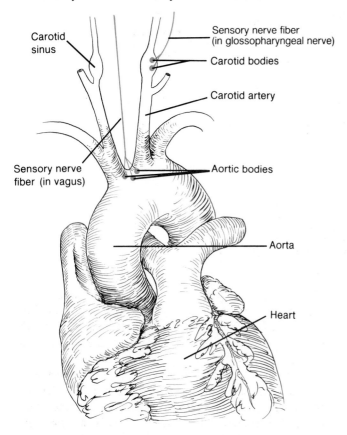

Carotid sinus

Sensory nerve fiber (in glossopharyngeal nerve)

Carotid bodies

Carotid artery

Sensory nerve fiber (in vagus)

Aortic bodies

Aorta

Heart

The brain stem respiratory centers produce rhythmic breathing even in the absence of other neural input. This intrinsic respiratory pattern produced by the brain stem centers, however, is modified by input from higher brain centers and by input from receptors sensitive to the chemical composition of the blood. The influence of higher brain centers on brain stem control of respiration is evidenced by the fact that one can voluntarily hyperventilate and hypoventilate (as in breath holding). The inability to hold one's breath longer than thirty to sixty seconds is due to reflex breathing in response to sensory input from *chemoreceptors*.

Figure 15.13 Schematic illustration of the control of ventilation by the central nervous system. The feedback effects of pulmonary stretch receptors and "irritant" receptors are not shown.

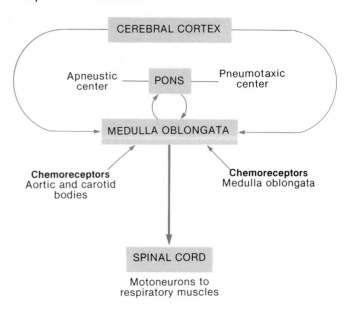

CEREBRAL CORTEX

Apneustic center — PONS — Pneumotaxic center

MEDULLA OBLONGATA

Chemoreceptors
Aortic and carotid bodies

Chemoreceptors
Medulla oblongata

SPINAL CORD

Motoneurons to respiratory muscles

There are two groups of chemoreceptors that respond to changes in blood P_{CO_2}, pH, and P_{O_2}. These are the **central chemoreceptors** in the medulla and the **peripheral chemoreceptors.** The peripheral chemoreceptors include the *aortic bodies*, located in the aortic arch, and the *carotid bodies*, which are located in the common carotid arteries (see figure 15.12). These peripheral chemoreceptors control breathing indirectly via sensory nerve fibers to the medulla. The aortic bodies send sensory information to the medulla in the tenth cranial (vagus) nerve; the carotid bodies stimulate sensory fibers in the ninth cranial (glossopharyngeal) nerve.

Effects of Blood P_{CO_2} and pH on Ventilation

Chemoreceptor input to the brain stem respiratory control centers modifies the rate and depth of breathing so that, under normal conditions, arterial P_{CO_2}, pH, and P_{O_2} remain relatively constant. If hypoventilation (as in breath holding) occurs, P_{CO_2} quickly rises and pH falls due to accumulation of carbonic acid. The oxygen content of the blood decreases much more slowly, because there is a large "reservoir" of oxygen attached to hemoglobin (venous blood normally contains about 75 percent oxyhemoglobin, as previously described). During hyperventilation, conversely, blood P_{CO_2} quickly falls and pH rises due to excessive elimination of carbonic acid. The oxygen content of the blood, on the other hand, is not significantly increased by hyperventilation (hemoglobin in arterial blood is 97 percent saturated during normal ventilation).

Figure 15.14 Negative feedback control of ventilation through changes in blood P_{CO_2} and pH.

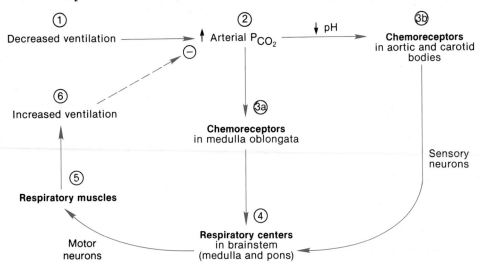

Table 15.8 The effect of ventilation, as measured by total minute volume (breathing rate × tidal volume), on the P_{CO_2} of arterial blood.

Total Minute Volume	Arterial P_{CO_2}	Type of Ventilation
2 L/min	80 mm Hg	Hypoventilation
4–5 L/min	40 mm Hg	Normal ventilation
8 L/min	20–25 mm Hg	Hyperventilation

The blood P_{CO_2} and pH are therefore more immediately affected by changes in ventilation than is the oxygen content. Indeed, changes in P_{CO_2} provide a sensitive index of ventilation, as shown in table 15.8. In view of these facts, it is not surprising that changes in P_{CO_2} provide the most potent stimulus for reflex control of ventilation. The rate and depth of ventilation, in other words, is adjusted to maintain a constant arterial P_{CO_2}; proper oxygenation of the blood occurs naturally as a side product of this reflex control.

The rate and depth of ventilation is normally adjusted to maintain an arterial P_{CO_2} of 40 mm Hg. Hypoventilation causes a rise in P_{CO_2}—a condition called *hypercapnia* (or hypercarbia). Hyperventilation, conversely, results in *hypocapnia*. Chemoreceptor regulation of breathing in response to changes in P_{CO_2} is illustrated in figure 15.14.

Chemoreceptors in the Medulla The chemoreceptors most affected by changes in arterial P_{CO_2} are located in the ventral area of the medulla, near the exit of the ninth and tenth cranial nerves. These chemoreceptor neurons are anatomically separate from, but electrically connected to, the neurons of the respiratory control center.

An increase in arterial P_{CO_2} causes a rise in the H^+ concentration of the blood as a result of increased carbonic acid concentrations. The H^+ in the blood, however, cannot cross the blood-brain barrier (as described in chapter 7), and so cannot influence the medullary chemoreceptors. Carbon dioxide in arterial blood can cross the blood-brain barrier, and through formation of carbonic acid, can lower the pH of cerebrospinal fluid. Since a fall in the pH of cerebrospinal fluid stimulates the medullary chemoreceptors, ventilation is indirectly stimulated by a rise in arterial P_{CO_2}.

Peripheral Chemoreceptors The aortic and carotid bodies are stimulated directly by a rise in the H^+ concentration of the arterial blood. This is normally produced by an increase in P_{CO_2} and carbonic acid. Retention of CO_2 during hypoventilation thus stimulates the medullary chemoreceptors through a lowering of cerebrospinal fluid pH, and stimulates peripheral chemoreceptors through a lowering of blood pH. Since the peripheral chemoreceptors are stimulated by H^+, a fall in blood pH produced by metabolic acidosis can cause hyperventilation and respiratory alkalosis, as previously described.

Table 15.9   Definitions of some terms used to describe blood oxygen and carbon dioxide levels.

Term	Definition
Hypoxemia	The oxygen content or P_{O_2} is lower than normal in arterial blood.
Hypoxia	The oxygen content or P_{O_2} is lower than normal in the lungs, blood, or tissues. This is a more general term than hypoxemia. Tissues can be hypoxic, for example, even though there is no hypoxemia (as when the blood flow is occluded).
Hypercapnia or hypercarbia	The P_{CO_2} of systemic arteries is higher than 40 mm Hg. Usually occurs when the ventilation is inadequate for a given metabolic rate (hypoventilation). Antonyms are *hypocapnia* or *hypocarbia* (usually produced by hyperventilation).

Table 15.10   The sensitivity of chemoreceptors to changes in blood gases and pH.

Stimulus	Chemoreceptor(s)	Comments
↑ P_{CO_2}	Medulla oblongata	Medullary chemoreceptors are sensitive to the pH of cerebrospinal fluid (CSF). Diffusion of CO_2 from the blood into the CSF lowers the pH of CSF by forming carbonic acid.
↓ pH	Aortic and carotid bodies	Peripheral chemoreceptors are stimulated by decreased blood pH independent of the effect of blood CO_2. Chemoreceptors in the medulla are not affected by changes in blood pH because H^+ cannot cross the blood-brain barrier.
↓ P_{O_2}	Carotid bodies	Low blood P_{O_2} (hypoxemia) augments the chemoreceptor response to blood P_{CO_2}, and can stimulate ventilation directly when the P_{O_2} falls below 50 mm Hg.

People who hyperventilate during psychological stress are sometimes told to breathe into a paper bag, so that they re-breathe their expired air that is enriched in CO_2. This advice may at first seem incorrect, since increased CO_2 stimulates breathing. It should be realized, however, that hyperventilation under these conditions produces hypocapnea, which may be corrected by re-breathing the expired air. As described in chapter 13, low blood CO_2 stimulates cerebral vasoconstriction and ischemia. In addition to producing symptoms of dizziness, the ischemia can lead to local metabolic acidosis that, through stimulation of medullary chemoreceptors, can cause further hyperventilation. Breathing into a paper bag can thus relieve the hypocapnea and stop the hyperventilation.

Effect of Blood P_{O_2} on Ventilation

Under normal conditions, blood P_{O_2} affects breathing only indirectly, by influencing the chemoreceptor sensitivity to changes in P_{CO_2}. Chemoreceptor sensitivity to P_{CO_2} is augmented by a low P_{O_2} and is decreased by a high P_{O_2}. If the blood P_{O_2} is raised by breathing 100 percent oxygen, therefore, the breath can be held longer because the response to increased P_{CO_2} is blunted.

When the P_{CO_2} is held constant by experimental techniques, the P_{O_2} of arterial blood must fall from 100 mm Hg to below 50 mm Hg before ventilation is significantly stimulated. This stimulation is apparently due to a direct effect of P_{O_2}—not oxygen content—on the *carotid bodies*. Since this degree of *hypoxemia* (low-blood oxygen) does not normally occur, even in breath holding, P_{O_2} does not normally exert this direct effect on breathing.

When there is chronic retention of carbon dioxide, as in *emphysema* due to air trapping (see chapter 14), the chemoreceptors eventually lose their ability to respond to increases in arterial P_{CO_2}. The abnormally high P_{CO_2}, however, enhances the sensitivity of the carotid bodies to a fall in P_{O_2}. Breathing in people with emphysema may thus be stimulated by a *hypoxic drive* (see table 15.9) rather than by increases in blood P_{CO_2}.

When people with emphysema are given oxygen therapy, the resulting rise in arterial P_{O_2} may cause them to "forget to breathe" because their hypoxic drive is satisfied. This can result in further carbon dioxide retention and respiratory acidosis. Oxygen therapy in these patients must thus be performed carefully, with a gas mixture that is only slightly enriched in oxygen (25 to 30 percent oxygen, compared to the 21 percent oxygen composition of normal air). After therapy, the patients must consciously breathe at regular intervals for a period of time.

Table 15.11 Pulmonary receptors and reflexes served by sensory nerve fibers in the vagus (tenth cranial nerve).

Stimulus	Receptors	Comments
Stretch of lungs at inspiration	Stretch receptors	The Hering-Breuer reflex. This reflex is believed to be important in the control of breathing in the newborn but not in the adult at normal tidal volumes.
Stretch of lungs at expiration	Stretch receptors	Like the Hering-Breuer reflex, this is not believed to be significant in adults at normal tidal volumes.
Pulmonary congestion	Type J (juxta-capillary) receptors	May produce feelings of dyspnea at high altitudes and during severe exercise.
Irritation	"Irritant receptors"	Causes reflex constriction of bronchioles in response to irritation from smoke, smog, and other noxious agents.

Pulmonary Stretch and Irritant Reflexes

The lungs contain various types of receptors that influence the brain stem respiratory control centers via sensory fibers in the vagus. Irritant receptors in the lungs, for example, stimulate reflex constriction of the bronchioles in response to noxious agents in smoke and smog. Similarly, sneezing, sniffing, and coughing may be stimulated by irritant receptors in the nose, larynx, and trachea.

The **Hering-Breuer reflex** is stimulated by pulmonary stretch receptors. Activation of these receptors during inspiration inhibits the respiratory control centers, making further inspiration increasingly difficult. This helps to prevent undue distension of the lungs and may contribute to the smoothness of the ventilation cycles. A similar inhibitory reflex may occur during expiration. The Hering-Breuer reflex appears to be important in the control of normal ventilation in the newborn. Pulmonary stretch receptors in adults, however, are probably not active at normal tidal volumes (500 ml per breath), but may contribute to respiratory control at high tidal volumes.

Breathing during Sleep

Dreaming occurs several times a night during short phases called *rapid-eye-movement (REM) sleep*. During these times the eyes move rapidly, bodily movements often occur, and the person is in a generally stressful state (as evidenced by high rates of gastric juice secretion during these periods in people with gastric ulcers). The electroencephalogram (EEG), or "brain waves," are irregular and desynchronized during REM sleep. During non-REM sleep, in contrast, the EEG shows a regular synchronized wave pattern and the body is quiet. As might be expected, the breathing pattern is quite different during REM sleep than during non-REM sleep.

Slow, regular breathing during non-REM sleep results from chemoreceptor control of ventilation. Ventilation rate is slow because metabolic rate is slow. Breathing during REM sleep, in contrast, is less sensitive to chemoreceptor stimulation (unless hypoxia develops), so that arterial P_{CO_2} can rise from 40 mm Hg to 46 mm Hg. The brain stem centers at this time, however, are responsive to input from the cerebral cortex; breathing patterns are thus affected by the content of dreams.

Sleep Apnea As a result of the differences in respiratory control, *apnea* (cessation of breathing) is more likely to occur during REM than during non-REM sleep. It is not uncommon, in fact, for people to have periods of sleep apnea for up to twenty seconds during these times. Such normal, short periods of apnea are not dangerous unless the person has cardiac or pulmonary disease. There are some people, however, who suffer from much more severe forms of sleep apnea.

Patients with *Pickwickian syndrome* have severe sleep apnea accompanied by obesity (hence the name of this syndrome), which can lead to heart failure. People who have had their ninth and tenth cranial nerves severed can develop *Ondine's curse,* in which peripheral chemoreceptor input to the respiratory centers is abolished. These people must be artificially ventilated during sleep or else they may "forget to breathe" during non-REM sleep when ventilation is largely controlled by chemoreceptor input.

Sudden Infant Death Syndrome (SIDS) This is an especially tragic form of sleep apnea that claims the lives of about ten thousand babies annually in the United States. Victims of this condition are apparently healthy two-to-five-month-old babies who die in their sleep without apparent reason—hence the layman's term of *crib death.* These deaths seem to be caused by failure of the respiratory control mechanisms in the brain stem and/or by failure of the carotid bodies to be stimulated by hypoxia.

Figure 15.15 Cheyne-Stokes breathing, showing progressively increasing and decreasing tidal volumes separated by periods of apnea (indicated by the straight lines).

Tidal volume
(ml)

Time

Cheyne-Stokes Breathing

Abnormal breathing patterns often appear prior to death caused by brain damage or heart disease. The most common of these abnormal patterns is *Cheyne-Stokes breathing,* in which the depth of breathing progressively increases and then progressively decreases. These cycles of increasing and decreasing tidal volumes (see figure 15.15) may be followed by periods of apnea of varying durations.

Cheyne-Stokes breathing may be caused by neurological damage, or by insufficient rates of oxygen delivery to the brain. The latter may result from heart disease, for example, or from a brain tumor that diverts a large part of the vascular supply from the respiratory centers. Ventilation seems to be more sensitive than normal to P_{CO_2} because arterial P_{CO_2} is maintained at less than 40 mm Hg during Cheyne-Stokes breathing.

1. Describe the effects of voluntary hyperventilation and of breath holding on arterial P_{CO_2}, pH, and oxygen content. Indicate the relative degree of change in these values during hyperventilation and breath holding.
2. Using a flow chart to describe a negative feedback loop, explain how the P_{CO_2} of arterial blood is regulated by ventilation.
3. Explain the effect of increased arterial P_{CO_2} on (a) chemoreceptors in the medulla oblongata; (b) chemoreceptors in the aortic and carotid bodies.
4. Explain why a person suffering from ketoacidosis would hyperventilate and what effects an intravenous infusion of bicarbonate would have on ventilation.
5. Describe the effects of arterial P_{O_2} on breathing in a healthy person and in a person suffering from emphysema.

Figure 15.16 Effect of moderate and heavy exercise on arterial blood gases and pH. Notice that there are no changes in these measurements during the first several minutes of moderate and heavy exercise and that only the P_{CO_2} changes (actually decreases) during more prolonged exercise.

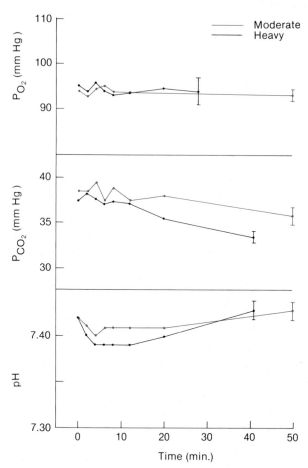

Time (min.)

The Effect of Exercise, Altitude, and Diving on Respiratory Function

Immediately upon exercise, the rate and depth of breathing increase to produce a total minute volume that is many times the resting value. The increased ventilation—particularly in endurance-trained athletes—is exquisitely matched to the simultaneous increase in oxygen consumption and carbon dioxide production by the exercising muscles. Arterial blood P_{O_2}, P_{CO_2}, and pH thus remain surprisingly constant during exercise (see figure 15.16).

It is tempting to suppose that ventilation increases during exercise as a result of increased CO_2 production by the exercising muscles. The evidence (as shown in figure 15.16), however, suggests that this is not so, because these values do not change significantly during exercise. Ventilation does, of course, increase during exercise, but the mechanisms responsible appear to be quite complex.

Table 15.12 Changes in respiratory function during exercise.

Variable	Change	Comments
Ventilation	Increased	This is not hyperventilation because ventilation is matched to increased metabolic rate. Mechanisms responsible for increased ventilation are not well understood.
Blood gases	No change	Blood gas measurements during light, moderate, and heavy exercise show little change because ventilation is increased to match increased muscle oxygen consumption and carbon dioxide production.
Oxygen delivery to muscles	Increased	Although the total oxygen content and P_{O_2} do not increase during exercise, there is an increased rate of blood flow to the exercising muscles.
Oxygen extraction by muscles	Increased	Increased oxygen consumption lowers the tissue P_{O_2} and lowers the affinity of hemoglobin for oxygen (due to the effect of increased temperature). More oxygen, as a result, is unloaded so that venous blood contains a lower oxyhemoglobin saturation than at rest. This effect is enhanced by endurance training.

Two groups of mechanisms—*neurogenic* and *humoral*—have been proposed to explain the increased ventilation that occurs during exercise. Possible neurogenic mechanisms include the following: (1) sensory nerve activity from the exercising limbs may stimulate the respiratory muscles, either directly or via the brain stem respiratory centers; and/or (2) input from the cerebral cortex may stimulate the brain stem centers to modify the rate of ventilation. These neurogenic theories help to explain the immediate increase in ventilation that occurs at the beginning of exercise.

Rapid and deep ventilation continues after exercise has stopped, suggesting that "humoral" (chemical) factors in the blood may also stimulate ventilation during exercise. Since the P_{O_2}, P_{CO_2}, and pH of blood samples from exercising subjects are within the normal range, these humoral theories propose that (1) the P_{CO_2} and pH in the region of the chemoreceptors may be different from these values in the "downstream" blood where samples are taken for measurement, and that (2) there may be cyclic variations in these values that stimulate the chemoreceptors, but these cycles offset each other to produce constant values of the blood samples taken for measurement. Evidence suggests that both of these mechanisms are involved in the *hyperpnea* (increased ventilation) that immediately follows exercise.

Anaerobic Threshold and Endurance Training
The ability of the cardiopulmonary system to deliver adequate amounts of oxygen to the exercising muscles may be insufficient for the first minute or so of exercise, as a result of the time lag required to make proper cardiovascular adjustments. During this time, therefore, the muscles respire anaerobically and a "stitch in the side"—possibly due to hypoxia of the diaphragm—may develop. After the cardiovascular adjustments have been made, a person may experience a "second wind" when the muscles receive sufficient oxygen for their needs.

Continued heavy exercise can cause a person to reach the **anaerobic threshold**, which is the maximum amount of oxygen consumption that can be attained before blood lactic acid levels rise as a result of anaerobic respiration. This rise in lactic acid levels is not due to malfunction of the cardiopulmonary system—indeed, the arterial oxyhemoglobin saturation remains at 97 percent, blood flow is as high as it can be, and venous blood draining the muscles still contains oxygen. The anaerobic threshold, however, is higher in endurance-trained athletes than it is in other people.

The rise in blood lactic acid that occurs when the anaerobic threshold is exceeded is due to the inability of the exercising muscles to increase their oxygen consumption any further once a given metabolic rate is obtained. Endurance training increases the skeletal muscle content of myoglobin, mitochondria, and Krebs cycle enzymes. These muscles, therefore, are able to utilize more of the oxygen delivered to them by the arterial blood. At a given level of exercise, in other words, the venous blood draining the muscles in endurance-trained people contains a lower percent oxyhemoglobin saturation than that in other people. The effect of exercise and endurance training on respiratory function is summarized in table 15.12.

Acclimatization to High Altitude
When a person who is from a region near sea level moves to a significantly higher altitude, several adjustments in the respiratory system must be made to compensate for the decreased atmospheric pressure and P_{O_2} at the higher altitude. These adjustments include changes in ventilation, hemoglobin affinity for oxygen, and in total hemoglobin concentration.

Table 15.13   Changes in respiratory function during acclimatization to high altitude.

Variable	Change	Comments
Partial oxygen pressure	Decreased	Due to decreased total atmospheric pressure.
Percent oxyhemoglobin saturation	Decreased	Due to lower P_{O_2} in pulmonary capillaries.
Ventilation	Increased	Due at first to low arterial P_{O_2}; later to changes in blood P_{CO_2} as chemoreceptor sensitivity increases.
Total hemoglobin	Increased	Due to stimulation by erythropoietin. Increased hemoglobin raises oxygen capacity of blood to partially or completely compensate for the reduced partial pressure.
Oxyhemoglobin affinity	Decreased	Due to increased DPG within the red blood cells. Results in a higher percent unloading of oxygen to the tissues, which may partially or completely compensate for the reduced arterial oxyhemoglobin saturation.

Changes in Ventilation   When the arterial P_{O_2} decreases below 50 mm Hg the carotid bodies stimulate an increase in ventilation. After a few days at high altitude the total minute volume becomes stabilized at about 2–3 L per minute more than the total minute volume at sea level. Since this *hyperventilation* occurs in response to low arterial P_{O_2}, the arterial P_{CO_2} decreases from 40 mm Hg (its value at sea level) to about 29 mm Hg.

After a few days, breathing again becomes regulated by arterial P_{CO_2} rather than by P_{O_2}, despite the fact that hyperventilation continues and P_{CO_2} remains lower than the normal value at sea level. This is because chemoreceptor sensitivity to P_{CO_2} gradually increases at higher altitude. This is partially due to the fact that hypoxemia by itself enhances chemoreceptor sensitivity to P_{CO_2}. Also, the bicarbonate concentration of cerebrospinal fluid becomes reduced, so that lower levels of blood CO_2 are needed to produce a decrease in cerebrospinal fluid pH and stimulate chemoreceptors in the medulla.

Hyperventilation cannot, of course, increase blood P_{O_2} above that of the inspired air. The P_{O_2} of arterial blood is therefore low at high altitudes. In the Peruvian Andes, for example, normal arterial P_{O_2} is reduced to 45 mm Hg (compared to 100 mm Hg at sea level). Loading of hemoglobin with oxygen is therefore incomplete, producing an oxyhemoglobin saturation of only 81 percent compared to the normal sea level value of 97 percent.

Decreased Hemoglobin Affinity   Normal arterial blood at sea level only unloads about 22 percent of its oxygen to the tissues at rest; the percent oxyhemoglobin saturation is reduced from 97 percent in arterial blood to 75 percent in venous blood. As a partial compensation for the decrease in oxygen content at high altitude, the affinity of hemoglobin for oxygen is reduced, so that a higher proportion of the oxyhemoglobin in arterial blood releases its oxygen to the tissues. This occurs because the low oxyhemoglobin content of red blood cells stimulates production of 2,3-DPG, which in turn decreases hemoglobin affinity for oxygen as previously described.

Increased Hemoglobin and Red Blood Cell Production   In response to tissue hypoxia, the kidneys secrete *renal erythropoietic factor,* which reacts with a plasma protein to produce the hormone **erythropoietin**. Erythropoietin stimulates the bone marrow to increase its production of hemoglobin and red blood cells. In the Peruvian Andes, for example, people have a total hemoglobin concentration of 19.8 g per 100 ml (compared to the normal sea level value of 15 g per 100 ml). Although the percent oxyhemoglobin saturation is still lower than at sea level, the total oxygen content of the blood is actually greater—22.4 ml O_2 per 100 ml compared to a sea level value of about 20 ml O_2 per 100 ml.

Acclimatization responses do not, however, compensate for the untoward effects of hypoxia on the cardiovascular system. Low alveolar P_{O_2}, as described in chapter 14, stimulates vasoconstriction of the pulmonary arterioles. While this response is valuable for matching ventilation to perfusion, the chronically low alveolar P_{O_2} and resulting vasoconstriction at high altitudes can produce *pulmonary hypertension.* A high percentage of people who are born to life at a high altitude show hypertrophy of the right heart as a result of their elevated pulmonary artery pressure.

Oxygen Toxicity, Decompression Sickness, and Nitrogen Narcosis

The total atmospheric pressure increases by one atmosphere (760 mm Hg) for every 10 m, or 33 feet, that one descends below sea level. In a 10 m dive, therefore, the partial pressures and amounts of dissolved gases are twice that at sea level. At 66 feet they are three times, and at 100 feet they are four times the values at sea level. The increased amounts of dissolved oxygen and nitrogen under these conditions can have serious effects on the body.

Oxygen Toxicity While breathing 100 percent oxygen at one or two atmospheres pressure (with a P_{O_2} of 760–1520 mm Hg) can be safely tolerated for a few hours, higher partial oxygen pressures can be very dangerous. *Oxygen toxicity* develops when the P_{O_2} rises above about 2.5 atmospheres. This is apparently caused by oxidation of enzymes and by other destructive changes that can damage the nervous system and lead to coma and death. For these reasons, divers commonly use gas mixtures in which oxygen is "diluted" with inert gases such as nitrogen (as in ordinary air) or helium.

Hyperbaric oxygen—oxygen at greater than 1 atmosphere pressure—is often used to treat conditions such as carbon monoxide poisoning, hyaline membrane disease, circulatory shock, and gas gangrene. Before the dangers of oxygen toxicity were realized these hyperbaric oxygen treatments sometimes resulted in tragedy. Particularly tragic were the cases of *retrolental fibroplasia,* in which retinal damage and blindness resulted from hyperbaric oxygen treatment of premature babies with hyaline membrane disease.

Nitrogen Narcosis Although at sea level nitrogen is a physiologically inert gas in which about one liter of nitrogen is dissolved in the body fluids, larger amounts of dissolved nitrogen under hyperbaric conditions have deleterious effects. Since it takes time for the nitrogen to dissolve, these effects usually don't appear until the person has remained submerged for over an hour. *Nitrogen narcosis* resembles alcohol intoxication—depending on the depth of the dive, the diver may experience "rapture of the deep" or may become so drowsy that he or she is totally incapacitated.

Decompression Sickness The amount of nitrogen that can remain dissolved decreases as the partial pressure decreases when a diver ascends to sea level. If the diver surfaces slowly, a large amount of nitrogen can be eliminated by diffusion through the alveoli and expiration. If decompression occurs too rapidly, however, bubbles of nitrogen gas (N_2) can form in the blood and block small blood channels, producing muscle and joint pain as well as more serious nervous system damage. These effects are known as *decompression sickness* or the bends.

The cabins of airplanes that fly long distances at high altitudes (30,000 to 40,000 feet) are pressurized so that passengers and crew are not exposed to the very low atmospheric pressures of these altitudes. If a cabin were to become rapidly depressurized at high altitude, much less nitrogen could remain dissolved at the greatly lowered atmospheric pressure. People in this situation, like divers who ascend too rapidly, would thus experience decompression sickness.

1. Describe the mechanisms that have been proposed to explain the increased ventilation that occurs during exercise. Also, explain why endurance-trained athletes have a higher anaerobic threshold than other people.
2. Explain the mechanisms responsible for hyperventilation during acclimatization to life at a high altitude. Also explain how the oxygen-transport system helps compensate for the lowered P_{O_2} at high altitudes.
3. Define the terms *oxygen toxicity, nitrogen narcosis,* and *decompression sickness.* Explain how decompression sickness is produced.

Summary

Hemoglobin and Oxygen Transport

I. Hemoglobin consists of four polypeptide chains and four hemes.
   A. Each heme in reduced, or deoxyhemoglobin can bond to one molecule of oxygen to form oxyhemoglobin.
      1. This is called the loading reaction, and occurs at the high P_{O_2} in the pulmonary capillaries.
      2. The iron in heme must be in the reduced (Fe^{++}) form to combine with oxygen.
   B. At the lower P_{O_2} of the tissue capillaries, some of the oxyhemoglobin dissociates to release its oxygen.
      1. This is called the unloading reaction.

2. Since arterial blood contains 97 percent oxyhemoglobin and venous blood contains 75 percent oxyhemoglobin under resting conditions, about 22 percent of the oxyhemoglobin unloads its oxygen in the tissues.
   C. During exercise, a higher proportion of the oxygen in arterial blood is extracted by the exercising muscles, so that venous percent oxyhemoglobin saturation and P_{O_2} are lower than at rest.
II. The affinity, or bond strength, between hemoglobin and oxygen is variable.

A. At any given tissue (and venous P_{O_2}), the proportion of oxygen unloading depends on pH, temperature, and the red blood cell content of 2,3-DPG.
B. Hemoglobin affinity for oxygen is lowered by decreases in pH and by increases in temperature.
   1. These conditions can occur in the tissues during exercise.
   2. When hemoglobin affinity is lower, a higher proportion of the oxyhemoglobin entering tissue capillaries unloads its oxygen.

C. A decrease in red blood cell content of oxyhemoglobin stimulates production of 2,3-diphosphoglyceric acid (2,3-DPG).
   1. This occurs in anemia, when total hemoglobin, and thus total oxygen content of the blood, is decreased below normal.
   2. This also occurs during acclimatization to life at a high altitude.
   3. Increasing concentrations of 2,3-DPG within red blood cells decrease hemoglobin affinity for oxygen and promote greater oxygen unloading.
D. Decreased hemoglobin affinity for oxygen is shown as a shift to the right in the oxyhemoglobin dissociation curve.
III. The rate of red blood cell and hemoglobin production by bone marrow is stimulated by the hormone *erythropoietin*.
A. Erythropoietin is secreted by the kidneys in response to tissue hypoxia.
   1. This occurs as an adaptation to life at a high altitude where the P_{O_2} of arterial blood and percent oxyhemoglobin saturation are low.
   2. The kidneys actually secrete renal erythropoietic factor, which reacts with a plasma protein to form erythropoietin.
B. Androgens also stimulate hemoglobin and red blood cell production, which is why these values in males average higher than in females.

Carbon Dioxide and Acid-Base Balance
I. Carbon dioxide is carried by the blood in three forms.
A. About 10 percent of the blood carbon dioxide is carried dissolved in plasma.
B. About 20 percent of the carbon dioxide is carried attached to an amino group in hemoglobin in the form of carbaminohemoglobin.
C. Most of the carbon dioxide is carried in the form of bicarbonate and carbonic acid.

II. The reactions that lead to conversion of CO_2 to bicarbonate and carbonic acid are known as the chloride shift.
A. Carbonic anhydrase within the red blood cells of tissue capillaries catalyze the formation of carbonic acid from CO_2 and H_2O.
B. Carbonic acid formed in the red blood cells ionizes.
   1. Most of the H^+ released combines with and is buffered by deoxyhemoglobin.
   2. Bicarbonate (HCO_3^-) diffuses into the plasma, as chloride diffuses into the red blood cells.
III. Bicarbonate and carbonic acid function as a buffer pair.
A. Bicarbonate in the plasma combines with H^+ released from metabolic acids such as ketone bodies and lactic acid.
   1. Carbonic acid is thus formed in the plasma as the concentration of free bicarbonate is decreased.
   2. This buffering reaction helps to keep the pH of arterial blood within the range of 7.35 to 7.45.
B. If the blood should lack a normal concentration of H^+, the carbonic acid in plasma can ionize to help stabilize the pH.
C. When the H^+ concentration of arterial blood is greater than normal, the person is said to have acidosis; when it is less than normal, the person has alkalosis.
IV. Acidosis and alkalosis can be caused by either a respiratory or a metabolic disturbance.
A. Hypoventilation causes CO_2 retention, increased carbonic acid concentrations, and thus respiratory acidosis.
B. Hyperventilation, conversely, causes respiratory alkalosis.
C. Abnormal production of metabolic acids can produce metabolic acidosis.
   1. In this condition, the buffer reserve of bicarbonate is depleted.
   2. Acidosis stimulates hyperventilation (and respiratory alkalosis) which partially compensates the metabolic acidosis.

Regulation of Breathing
I. The respiratory control centers are located in the brain stem.
A. Neurons in the medulla oblongata produce the cyclic breathing pattern.
B. Two centers in the pons—the pneumotaxic and apneustic centers—modify the respiratory rhythm set by the medulla.
C. Input from higher brain centers in the cerebral cortex allow voluntary control of breathing to a certain degree.
D. Input from chemoreceptors make ventilation responsive to changes in arterial P_{CO_2}, pH, and P_{O_2}.
II. Changes in arterial P_{CO_2} provide the most important cue for regulating breathing.
A. This is a logical control system because changes in arterial P_{CO_2} are proportional to changes in ventilation.
B. Chemoreceptors in the medulla are stimulated by a rise in arterial P_{CO_2}.
   1. Carbon dioxide from the blood, through formation of carbonic acid in cerebrospinal fluid, lowers the pH of cerebrospinal fluid.
   2. Chemoreceptors in the medulla are directly stimulated by H^+ in cerebrospinal fluid, but are not affected by H^+ in plasma.
C. Chemoreceptors in the aortic and carotid bodies are stimulated by H^+ in plasma.
   1. Since increased P_{CO_2} produces increased amounts of carbonic acid in the blood, breathing is stimulated by the aortic and carotid bodies as well as by the chemoreceptors in the medulla.
   2. The aortic and carotid bodies stimulate hyperventilation when there is metabolic acidosis.
III. Arterial P_{O_2} can affect breathing both indirectly and directly.
A. Low P_{O_2} makes the chemoreceptors more sensitive to CO_2; high P_{O_2} has the opposite effect.

B. Arterial P_{O_2} must fall below 50 mm Hg before it can serve as a direct stimulus for breathing.
  1. Since the arterial P_{O_2} at sea level is 100 mm Hg it does not directly affect breathing.
  2. At high altitude, however, and in people with emphysema, P_{O_2} can become an important stimulus for breathing.

IV. Stretch receptors in the lungs may also modify breathing.
  A. In the Hering-Breuer reflex, stretch receptors inhibit inspiration and thus prevent undue distension of the lungs.

B. A similar inhibitory effect may occur during expiration.
C. The Hering-Breuer reflex appears to be important in newborns, and in adults when the tidal volumes are higher than the normal resting values.

V. The mechanisms that promote increased ventilation during exercise are complex.
  A. Measurements of arterial blood gases and pH during exercise do not show significant changes.
    1. Such changes may nevertheless occur because humoral (chemical) factors seem to be involved in the hyperpnea that follows exercise.

  2. Neurogenic factors may also be involved, particularly at the beginning of exercise.
  B. Blood gases in arterial blood remain normal even when the anaerobic threshold is reached.
    1. The anaerobic threshold occurs when the muscles cannot extract enough oxygen from the arterial blood to meet their aerobic requirements.
    2. Muscles of endurance-trained athletes are able to extract more oxygen, so these people have higher anaerobic thresholds than others.

Self-Study Quiz

1. Which of the following would be most affected by a decrease in the affinity of hemoglobin for oxygen?
   (a) arterial P_{O_2}
   (b) arterial percent oxyhemoglobin saturation
   (c) venous oxyhemoglobin saturation
   (d) arterial P_{CO_2}

2. If a person with normal lung function were to hyperventilate for several seconds, there would be a significant:
   (a) increase in arterial P_{O_2}
   (b) decrease in arterial P_{CO_2}
   (c) increase in arterial percent oxyhemoglobin saturation
   (d) decrease in arterial pH

3. Erythropoietin is produced by:
   (a) the kidneys
   (b) the liver
   (c) the lungs
   (d) the bone marrow

4. The affinity of hemoglobin for oxygen is decreased under conditions of:
   (a) acidosis
   (b) fever
   (c) anemia
   (d) acclimatization to high altitude
   (e) all of these

5. Most of the carbon dioxide in the blood is carried in the form of:
   (a) dissolved CO_2
   (b) carbaminohemoglobin
   (c) bicarbonate and carbonic acid
   (d) carboxyhemoglobin

6. The bicarbonate concentration of the blood would be decreased during:
   (a) metabolic acidosis
   (b) respiratory acidosis
   (c) metabolic alkalosis
   (d) respiratory alkalosis

7. The chemoreceptors in the medulla are stimulated by:
   (a) CO_2 directly
   (b) H^+ from the blood
   (c) H^+ in cerebrospinal fluid that are derived from blood CO_2
   (d) decreased arterial P_{O_2}

8. Rhythmic control of breathing is produced by activity of inspiratory and expiratory neurons in:
   (a) the pons
   (b) the medulla oblongata
   (c) the cerebral cortex
   (d) the aortic and carotid bodies

9. Which of the following occurs during hypoxemia?
   (a) increased ventilation
   (b) increased production of 2,3-DPG
   (c) increased secretion of erythropoietin
   (d) all of these

10. During exercise, which of the following statements is TRUE?
    (a) the percent oxyhemoglobin saturation of arterial blood is decreased
    (b) the percent oxyhemoglobin saturation of venous blood is decreased
    (c) arterial P_{CO_2} is measurably increased
    (d) arterial pH is measurably decreased

Physiology of the Kidneys

Objectives

By studying this chapter you should be able to:

1. Describe the structure of the kidneys and the structure of the tubular and vascular components of the nephron

2. Describe how glomerular filtration is regulated, and explain why the ultrafiltrate has a low protein concentration

3. Explain the mechanisms of and significance of salt and water reabsorption in the proximal tubules

4. Explain how the loop of Henle functions in countercurrent multiplication to produce a hypertonic renal medulla

5. Explain the significance of a hypertonic medulla and the possible contribution of urea to this hypertonicity

6. Describe how the final urine volume is regulated by the effects of antidiuretic hormone on the collecting ducts

7. Describe how clearance rates are measured and the effects of reabsorption and secretion on clearance rates

8. Explain how glomerular filtration rate and total renal blood flow are measured by inulin and PAH clearance rates

9. Describe the mechanisms for glucose and amino acid reabsorption

10. Explain how sodium reabsorption and K^+ secretion in the distal tubules are affected by aldosterone and how aldosterone secretion is regulated by blood Na^+ and K^+

11. Describe the structure of the juxtaglomerular apparatus, and describe the mechanisms that regulate renin secretion

12. Explain the relationship between Na^+ reabsorption and K^+ secretion and how diuretics can cause loss of K^+

13. Describe the interaction between plasma K^+ and H^+ concentrations and how this affects tubular secretion of these ions

14. Describe the respiratory and metabolic components of acid-base balance, and explain why the metabolic component is represented by the plasma bicarbonate concentration

15. Explain how the kidneys reabsorb and produce bicarbonate and how H^+ is buffered in the urine

16. Describe the different mechanisms by which substances can act as diuretics

*T*he primary function of the kidneys is regulation of the extracellular fluid (plasma and tissue fluid) environment in the body. This function is accomplished through the formation of urine, which is a modified filtrate of plasma. In the process of urine formation, the kidneys regulate the (1) volume of blood plasma, and thus contribute significantly to regulation of blood pressure; (2) concentration of waste products in the blood; (3) concentration of electrolytes (Na^+, K^+, HCO_3^-, Ca^{++}, and PO_4^{-3}) in the plasma; and (4) the pH of plasma.

Figure 16.1 Anatomical locations of the kidneys, ureters, and urinary bladder.

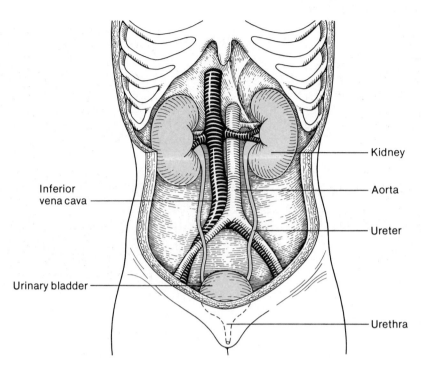

Structure of the Kidneys

The paired right and left kidneys are located in the abdominal cavity below the diaphragm and the liver. Each kidney in an adult weighs approximately 160–175 g, and is about 10–12 cm long and 5–6 cm wide—about the size of a fist. Urine produced in the kidneys is drained into the *renal pelvis* (= basin), and from there it is channeled via two long ducts—the *ureters*—to the single *urinary bladder* (see figure 16.1).

Urine enters the renal pelvis through funnel-shaped structures known as the *calyces*, which collect the urine from *renal pyramids* that form the inner **medulla** of the kidneys. The outer part of the kidneys is called the **cortex,** and is covered by a tough connective tissue capsule.

Blood enters each kidney in the *renal artery*, which divides in the pelvis to form *interlobular arteries* that extend out to the border between the renal medulla and cortex. At this border, the interlobular arteries divide to produce *arcuate arteries* (see figure 16.3). Perpendicular branches from the arcuate arteries extend into the cortex, where they end in tufts of capillaries called **glomeruli**. Each glomerulus contains twenty to forty capillaries and forms part of the functional unit of the kidneys—the *nephrons*. There are over a million nephrons in each kidney.

Figure 16.2 The interior structure of the kidneys. The medulla is in the form of renal pyramids.

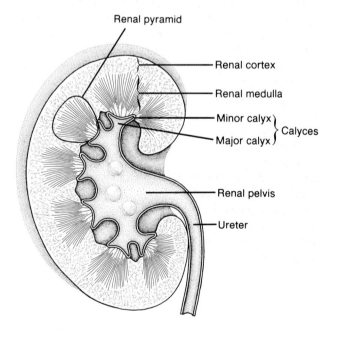

Figure 16.3 Vascular structure of the kidneys. *(a)* is an illustration of the major arterial supply, and *(b)* is a scanning electron micrograph of the glomeruli.

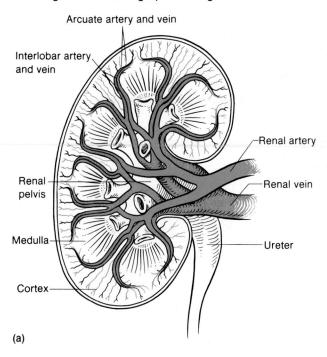

Arcuate artery and vein

Interlobar artery and vein

Renal artery

Renal vein

Renal pelvis

Medulla

Ureter

Cortex

(a)

(b)

Kessel and Kardon

Nephron Tubules

Each nephron consists of vascular and tubular components that function together to produce urine. A system of little tubes, or *tubules,* transport plasma filtrate formed by glomeruli in the renal cortex to the calyces in the renal pelvis. These tubules are not simple conduits, however. By means of various transport processes (described in later sections), the plasma filtrate that enters the tubules in the cortex is highly modified in both volume and composition. The fluid that enters the tubules is plasma filtrate; the fluid that leaves the tubules is urine.

Each glomerulus is surrounded and encapsulated by an expanded portion of the tubule called **Bowman's capsule**. Bowman's capsule is blind-ending (closed); it surrounds the glomerulus much like a soft balloon would surround a fist pushed into it. Although Bowman's capsule is blind-ending, it contains pores (described in a later section) that allow the entry of plasma filtrate formed by the glomerulus.

Fluid flows from Bowman's capsule into the **proximal convoluted tubule,** which is also located in the renal cortex. The next segment of tubule forms a hairpin loop called the **loop of Henle,** which extends into the renal medulla. Fluid within the tubule enters the medulla in the *descending limb* of the loop, and reenters the cortex in the *ascending limb* of the loop. Now back in the cortex, fluid enters the **distal convoluted tubule.** This part is continuous with a short, straight segment known as the *collecting tubule,* which drains into a **collecting duct**. The collecting duct plunges through the medulla to empty its fluid (now called urine) into the calyces.

Figure 16.4   Segments of the renal nephron. Drawings at *left* show microscopic appearance of cross sections of each segment.

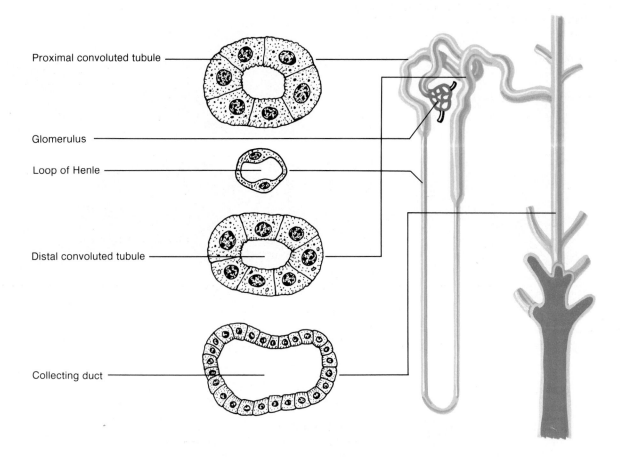

Proximal convoluted tubule

Glomerulus

Loop of Henle

Distal convoluted tubule

Collecting duct

Nephrons that originate in the inner one-third of the cortex—called *juxtamedullary nephrons*—have longer loops of Henle than the *cortical nephrons* that originate in the outer two-thirds of the cortex. In addition, the thin segments of the loops (see figure 16.5) are proportionately longer in the juxtamedullary nephrons. The functional significance of these differences will be described later in this chapter.

Vascular Component of Nephrons

Glomeruli are the only capillary beds in the body that are drained by arterioles rather than by venules. Blood is delivered to the glomeruli by **afferent arterioles** and is carried from the glomeruli in **efferent arterioles**. These vessels are called arterioles because they have thicker, more muscular walls than venules (thus providing more resistance to blood flow) and because they deliver blood to a *second* capillary bed—the **peritubular capillaries** that surround the tubules of the nephron. Blood from the peritubular capillaries drains into venules and leaves the kidney in the renal vein.

Figure 16.5   All nephrons begin in the cortex, but the appearance of nephrons in the outer two-thirds of the cortex *(right)* is different than those in the inner one-third of the cortex (the juxtamedullary nephrons). Notice that the juxtamedullary nephrons have very long loops of Henle that extend into the inner medulla.

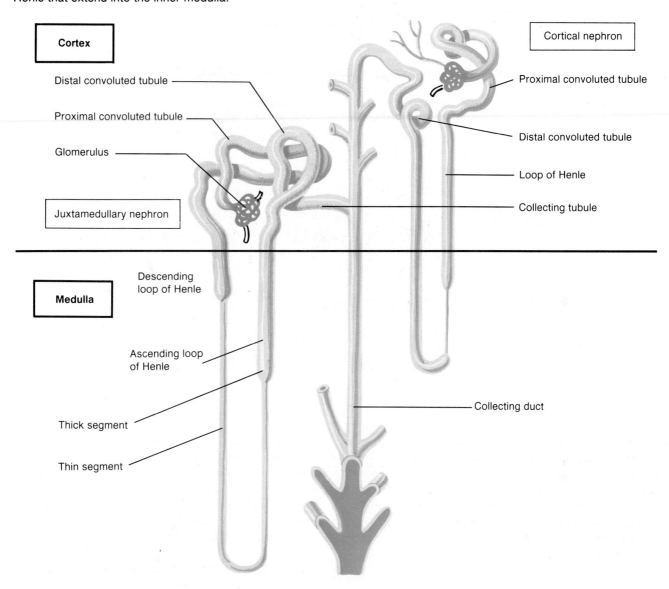

Figure 16.6 A simplified illustration of blood flow from glomerulus to efferent arteriole, to peritubular capillaries, to the venous drainage of the kidneys.

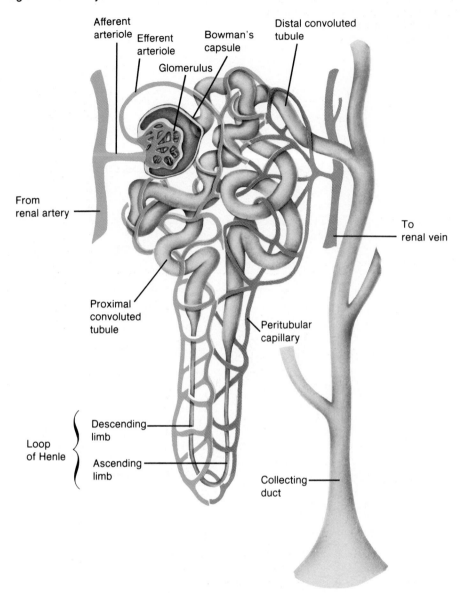

An illustration of the vascular supply of a single nephron (as shown in figure 16.6) can be somewhat misleading because the peritubular capillaries around most tubule segments are derived from the efferent arterioles of several glomeruli (see figure 16.7). Also, the vascular supply of juxtamedullary nephrons differs from that of the outer cortical nephrons. Efferent arterioles from juxtamedullary nephrons produce branches that descend deep into the medulla and form hairpin loops in parallel with the long loops of Henle in these nephrons. The vessels that form these loops—called the **vasa recta**—are very important in the function of juxtamedullary nephrons.

Glomerular Filtration

The glomerular capillaries have extremely large pores (200–500 A in diameter) and are therefore said to be *fenestrated,* in comparison with the continuous capillaries found in organs such as skeletal muscles. As a result of these large pores, glomerular capillaries are one hundred to four hundred times more permeable to plasma water and dissolved solutes than are capillaries of skeletal muscles. The large pores are freely permeable to plasma solutes, but exclude the passage of "formed elements"—red and white blood cells and platelets—because of their large size.

Figure 16.7 The vascular supply of cortical *(right)* and juxtamedullary nephrons *(left)*. Notice that the vascular supply of each nephron is derived from a number of glomeruli, and that long loops of blood vessels—the vasa recta—extend down into the medulla along the loops of Henle of juxtamedullary nephrons.

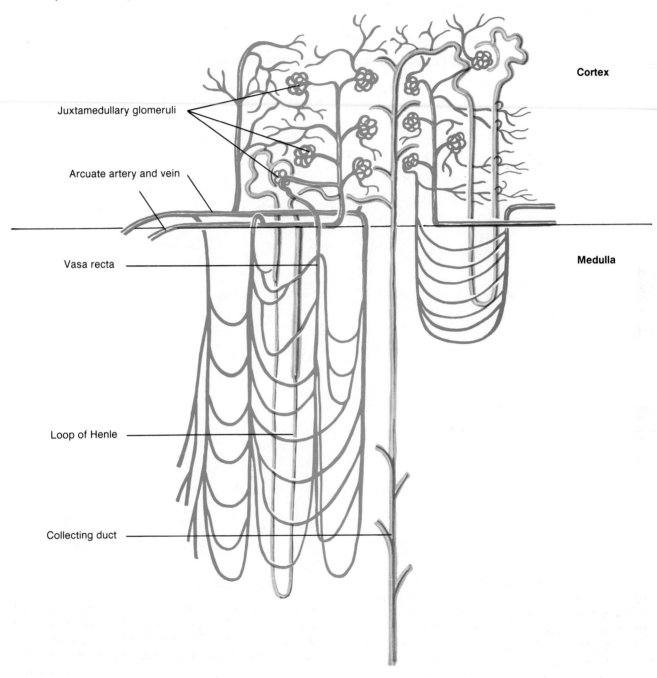

Juxtamedullary glomeruli

Arcuate artery and vein

Vasa recta

Loop of Henle

Collecting duct

Cortex

Medulla

Figure 16.8 The relationship between glomerular capillaries and the inner layer of Bowman's capsule is illustrated in *(a)*. This inner layer of Bowman's capsule is composed of podocytes, as shown in the scanning electron micrograph in *(b)*. Very fine extensions of these podocytes form foot processes, or pedicels, that interdigitate around the glomerular capillaries. Spaces between adjacent pedicels form the "filtration slits."

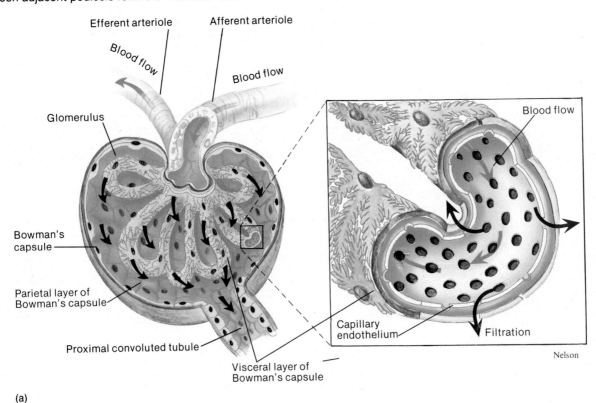

(a)

Before the filtrate can enter Bowman's capsule it must pass through the capillary endothelial cell pores, the basement membrane of these cells (a thin layer of glycoproteins immediately outside the endothelial cells), and the inner epithelial layer of Bowman's capsule. The inner layer of Bowman's capsule is composed of unique cells called *podocytes* ("foot cells") with numerous cytoplasmic extensions known as *pedicels* ("foot processes"). These pedicels—to mix analogies—interdigitate like fingers wrapped around the glomerular capillaries. The narrow slits between adjacent pedicels provide the passageways through which filtered molecules must pass to enter the interior of Bowman's capsule.

While the glomerular capillary pores are apparently large enough to permit passage of proteins, the fluid that enters Bowman's capsule is normally almost completely free of plasma proteins. This exclusion of plasma proteins from the filtrate is partially a result of their negative charge, which hinders their passage through the negatively charged glycoproteins in the basement membrane of the capillaries. The large size and negative charge of plasma proteins may also restrict their movement through the filtration slits between pedicels in the inner epithelial membrane of Bowman's capsule (see figure 16.9).

(b)

Figure 16.9   Illustration *(a)* and electron micrograph *(b)* of the "filtration barrier" between the capillary lumen and the cavity of Bowman's capsule.

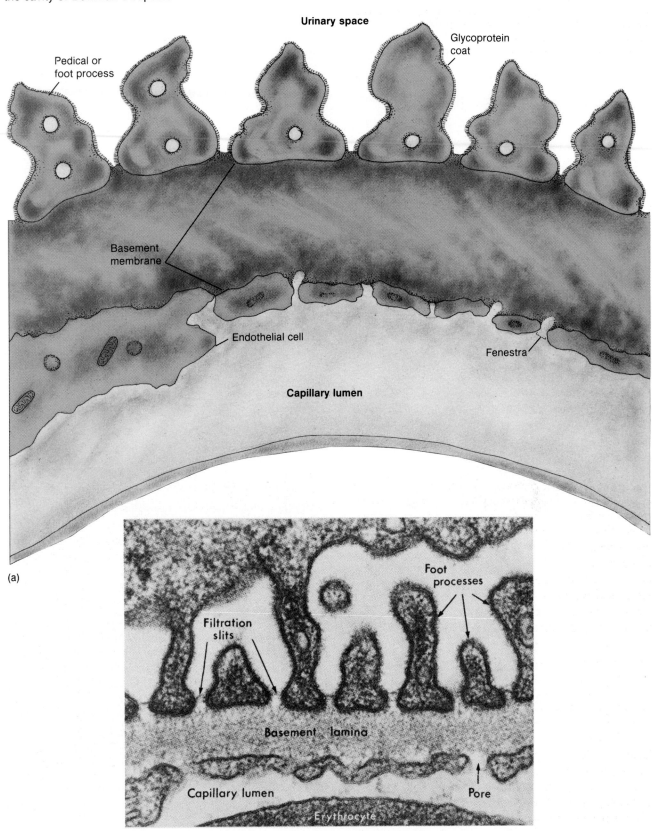

Figure 16.10 Formation of glomerular ultrafiltrate. Proteins *(large circles)* are not filtered, but smaller plasma solutes *(dots)* easily enter the glomerular ultrafiltrate.

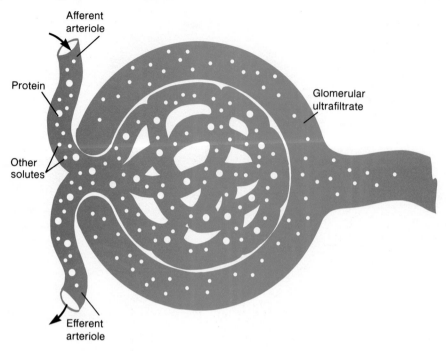

The Glomerular Ultrafiltrate

The fluid that enters Bowman's capsule is called an *ultrafiltrate* because it is formed under pressure (the hydrostatic pressure of the blood). This is similar to the formation of tissue fluid by other capillary beds, as described in chapter 12. The force favoring filtration is opposed by a counter force developed by the hydrostatic pressure of fluid in Bowman's capsule. Also, since the protein concentration of tubular fluid is low (less than 2–5 mg per 100 ml) compared to that of plasma (6–8 g per 100 ml), the colloid osmotic pressure of the plasma promotes osmosis of fluid from Bowman's capsule into the glomerular capillaries. When these two opposing pressures are subtracted from the hydrostatic pressure of the glomerular capillaries, a *net filtration pressure* of approximately 10 mm Hg is obtained.

Because glomerular capillaries are extremely permeable and have a high surface area, this modest net filtration pressure produces an extraordinarily large volume of filtrate. The **glomerular filtration rate (GFR)** averages 115 ml per minute in women and 125 ml per minute in men.

This is equivalent to 7.5 L per hour or 180 L per day (about 45 gallons)! Since the total blood volume averages about 5 L, this means that the total blood volume is filtered into the renal tubules every forty minutes. Most of the filtered water must obviously be immediately returned to the vascular system or people would literally urinate to death within several minutes.

Regulation of Glomerular Filtration Rate

Vasoconstriction, or dilation, of afferent arterioles affects the rate of blood flow to the glomerulus, and thus affects the glomerular filtration rate. Changes in the diameter of the afferent arterioles result from both extrinsic (sympathetic nerve) and intrinsic regulatory mechanisms.

Sympathetic Nerve Effects An increase in sympathetic nerve activity, as occurs during the fight or flight reaction and exercise, stimulates constriction of afferent arterioles. This is an alpha-adrenergic effect (as described in chapter 9), which helps to preserve blood volume and to divert blood to the muscles and heart. A similar effect occurs during cardiovascular shock, in which sympathetic nerve activity stimulates vasoconstriction; the decreased GFR and thus decreased urine formation that result help to compensate for the rapid drop in blood pressure under these circumstances.

Figure 16.11 Effect of increased sympathetic nerve activity on kidney function and other physiological processes.

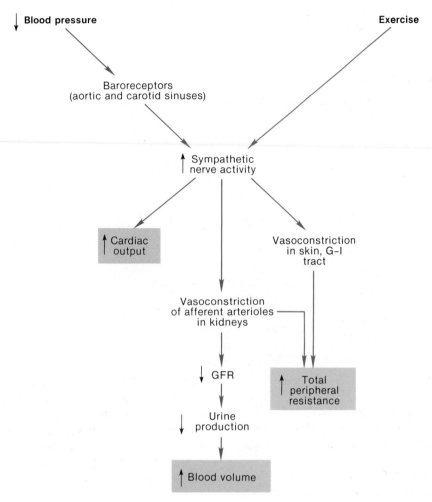

Renal Autoregulation When the direct effect of sympathetic nerves on the afferent arterioles is removed—by cutting the renal sympathetic nerve, for example—the effect of systemic blood pressure on GFR can be observed. Under these conditions, surprisingly, the GFR remains relatively constant despite changes in mean arterial pressure within a range of 70–180 mm Hg (normal mean arterial pressure averages 100 mm Hg). The ability of the kidneys to maintain a relatively constant GFR in the face of fluctuating blood pressures is called *renal autoregulation.*

Renal autoregulation results from the effects of locally produced chemicals on the afferent arterioles (effects on efferent arterioles are believed to be of secondary importance). When systemic arterial pressure falls towards a mean of 70 mm Hg the afferent arterioles dilate, and when the pressure rises the afferent arterioles constrict. Since blood flow is directly proportional to mean arterial pressure and inversely proportional to vascular resistance, changes in resistance due to constriction and dilation of afferent arterioles compensate for changes in blood pressure. Blood flow to the glomeruli and GFR can thus remain constant within the "autoregulatory range" of blood pressure values.

Table 16.1   Regulation of glomerular filtration rate (GFR).

Regulation	Stimulus	Afferent Arteriole	GFR
Sympathetic nerves	Activation by aortic and carotid baroreceptors or by higher brain centers	Constricts	Decreases
Autoregulation	Decreased blood pressure	Dilates	No change
Autoregulation	Increased blood pressure	Constricts	No change

Figure 16.12   Plasma water and its dissolved solutes (except proteins) enter the glomerular ultrafiltrate, but most of these filtered molecules are reabsorbed. The term *reabsorption* refers to the transport of molecules out of the tubular filtrate back into the blood.

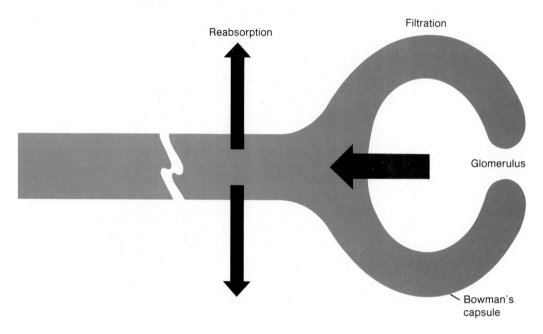

The mechanisms responsible for the autoregulatory ability of the kidneys are not well understood. According to one theory, angiotensin II—a potent vasoconstrictor produced when the enzyme renin is secreted from the kidneys (as described in chapter 12)—may act locally in the renal cortex and contribute to autoregulation. This theory, however, has been seriously challenged. Prostaglandins, a type of cyclic fatty acid produced by the kidneys and other organs, promotes vasodilation; these compounds may also contribute to renal autoregulation.

1. Draw a figure of the renal nephron. Label all the parts, and indicate with a dotted line which parts are located in the cortex and which parts are located in the medulla.
2. Using a line diagram, describe the vascular component of the nephron. Show peritubular capillaries and vasa recta in your illustration.
3. Describe the forces that affect formation of glomerular ultrafiltrate, and explain why this filtrate has a low concentration of proteins.

Reabsorption of Salt and Water

Although about 180 L per day of glomerular ultrafiltrate are produced, the kidneys normally excrete only 1–2 L per day of urine. Approximately 99 percent of the filtrate must thus be returned to the vascular system, while 1 percent is excreted as urine. The urine volume, however, can be varied according to the needs of the body. In severe dehydration, when the body needs to conserve water, only 0.3 ml per minute or 400 ml per day of urine are produced. In contrast, when a well-hydrated person drinks a liter or more of water, urine volume increases to 16 ml per minute (the equivalent of 23 L per day if this was continued for twenty-four hours). A volume of 400 ml per day of urine is needed to excrete the amount of metabolic wastes produced by the body. This is called the *obligatory water loss*. When water in excess of this amount is added to the blood (by drinking and by aerobic respiration), the extra water is excreted to produce a larger volume of more dilute urine.

Figure 16.13 Illustration of the appearance of tubule cells in the electron microscope. Molecules that are reabsorbed pass through the tubule cells from the apical membrane *(facing the filtrate)* to the basolateral membrane *(facing the blood).*

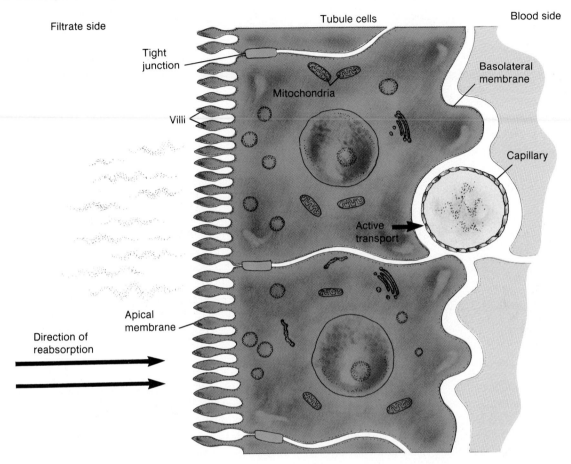

Regardless of the body's state of hydration, it is clear that most of the filtered water must be returned to the vascular system to maintain blood volume and pressure. The return of filtered molecules from the tubules to the blood is called **reabsorption**. It is important to realize that transport of water always occurs passively by *osmosis;* there is no such thing as active transport of water. A concentration gradient must thus be created between tubular fluid and blood that favors the osmotic return of water to the vascular system.

Reabsorption in the Proximal Tubule

Since all plasma solutes, with the exception of proteins, are able to enter the glomerular ultrafiltrate freely, the osmolality of the filtrate is essentially the same as that of the plasma (300 milliosmoles per liter, or 300 mOsm). The filtrate is thus said to be isosmotic with the plasma. Osmosis therefore cannot occur unless the concentration of plasma in the peritubular capillaries and the concentration of fluid in the proximal convoluted tubule are altered by active transport processes. This is achieved by the active transport of Na^+ from the filtrate to the peritubular blood.

Active and Passive Transport   The epithelial cells that compose the wall of the proximal tubule have two exposed sides. The "apical" sides of these cells face the tubular fluid in the lumen. The "basolateral" sides face the outside of the tubule, and thus are near the peritubular capillaries. The term *basolateral* is used because the membranes of adjacent epithelial cells are fused only near their apical surface so that the lateral borders of the cells, like their basal surfaces, are exposed to peritubular capillaries (see figure 16.13).

Figure 16.14   Mechanisms of salt and water reabsorption in the proximal tubule. Sodium is actively transported out of the filtrate and chloride follows passively by electrical attraction. Water follows the salt out of the tubular filtrate into the peritubular capillaries by osmosis.

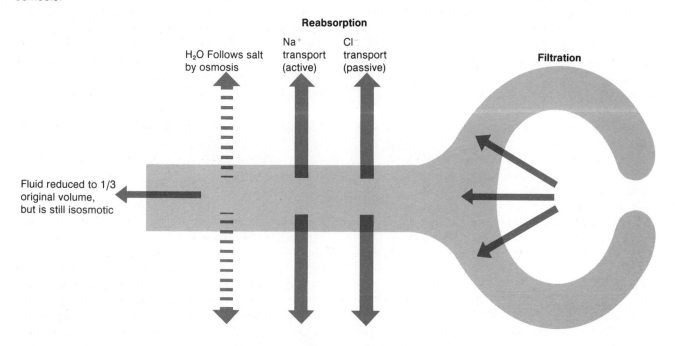

The concentration of Na$^+$ in the glomerular ultrafiltrate, and thus in the fluid entering the proximal tubule, is the same as in plasma and the same as in the tissue fluid surrounding the outside of the tubules. The tubule cells that separate filtrate from tissue fluid, however, have a lower Na$^+$ concentration. As discussed in chapter 6, this lower concentration is partially due to low permeability of cell membranes to Na$^+$, and partially due to action of active transport Na$^+$ "pumps," which extrude Na$^+$ from the cell.

In the cells of the proximal tubule, the active carriers of Na$^+$ are located only in the basolateral membrane. As a result, Na$^+$ that diffuses across the apical membrane from the tubular fluid is extruded across the basolateral membrane into the surrounding tissue fluid. This raises the Na$^+$ concentration of tissue fluid and results in diffusion of Na$^+$ into adjacent peritubular capillaries.

Reabsorption of Na$^+$ in the proximal tubule is thus dependent on *active transport*. As a result of the active movement of Na$^+$ from tubular fluid to tissue fluid, an electrical gradient is created that allows Cl$^-$ to follow the Na$^+$. Movement of Cl$^-$ in this case occurs by *passive transport*. These transport processes create an osmotic gradient because they raise the osmolality and osmotic pressure of the surrounding tissue fluid above that of the tubular fluid. Since the membranes of the proximal tubule cells are permeable to water, movement of water occurs by *osmosis* from the tubular fluid to the surrounding tissue fluid, and from there to the peritubular capillaries.

Significance of Proximal Tubule Reabsorption
Approximately 65 percent of the salt and water in the original glomerular ultrafiltrate is reabsorbed across the proximal tubule and returned to the vascular system. The volume of tubular fluid remaining to flow into the loop of Henle is reduced accordingly, but this fluid is still isosmotic with the blood (has a concentration of 300 mOsm). This results from the fact that the cell membranes in the proximal tubule are freely permeable to water so that water and salt are removed in proportionate amounts.

An additional 15 percent of the filtered water and 25 percent of the filtered salt are returned to the vascular system as a result of processes occurring in the loop of Henle (described in the next section). This reabsorption, like that which occurs across the proximal tubule, occurs constantly regardless of the person's state of hydration. Unlike reabsorption in later regions of the tubules, it is not subject to hormonal regulation. Approximately 85 percent of the filtered salt and water is therefore reabsorbed in a constant, unregulated fashion in the early regions of the nephron (proximal tubule and loop of Henle). This reabsorption is very costly in terms of energy expenditures, accounting for as much as 6 percent of the calories consumed by the body at rest.

Since 85 percent of the original glomerular ultrafiltrate is immediately reabsorbed in the early region of the nephron (mainly across the proximal tubule), only 15 percent of the initial filtrate remains to enter the distal convoluted tubule and collecting duct. This is still a large volume of fluid—15 percent $\times$ GFR (180 L per day) = 27 L per day—that must be reabsorbed to varying degrees in accordance with the body's state of hydration. This "fine tuning" of the percent reabsorption and urine volume is accomplished by the action of hormones on the later regions of the nephron.

The Countercurrent Multiplier System
Water cannot be actively transported across the tubule wall, and cannot be passively transported if the tubular fluid and surrounding tissue fluid remain isotonic to each other. In order for water to be reabsorbed by osmosis the surrounding tissue fluid must be hypertonic. The osmotic pressure of tissue fluid in the renal medulla is, in fact, raised to over four times that of plasma. This results partly from the fact that the tubule bends; the geometry of the loop of Henle allows interaction to occur between the descending and ascending limbs. Since the ascending limb is the active partner in this interaction, its properties will be described here before those of the descending limb.

Ascending Limb of the Loop of Henle   Chloride ion (Cl^-) is actively extruded from the ascending limb of the loop, and Na^+ follows the chloride passively. This is the reverse of salt transport in the proximal tubule, but the effects are the same—salt is pumped out of the tubule and enters the surrounding tissue fluid. Unlike the proximal tubule, however, the walls of the ascending limb are *not permeable to water*. The tubular fluid thus becomes increasingly dilute as it ascends towards the cortex, while the tissue fluid around the loops of Henle in the medulla becomes increasingly more concentrated. By means of these processes, the tubular fluid that enters the distal tubule in the cortex is hypotonic (with a concentration of 100 mOsm), while the tissue fluid in the medulla becomes hypertonic.

Descending Limb of the Loop of Henle   The deeper regions of the medulla, around the tips of the loops of juxtamedullary nephrons, reach a concentration of 1200–1400 mOsm. In order to reach this high a concentration, the salt pumped out of the ascending limbs must accumulate in the tissue fluid. This occurs as a result of the properties of the descending limb and as a result of the fact that blood vessels around the loop do not carry back all of the extruded salt to the general circulation.

The descending limb does not actively transport salt. Instead it is *passively permeable* to water and perhaps also to salt (this is currently controversial). The wall of the descending limb is like a porous plastic membrane—water and perhaps salt are free to diffuse according to their concentration gradients. Since the renal medulla is hypertonic to the fluid entering the descending limb (which initially has a concentration of 300 mOsm, as previously described), water moves by osmosis out of the descending limb and is removed by peritubular capillaries. Similarly, the concentration of salt is higher in the medullary tissue fluid than it is in the descending limb so that salt may diffuse into the tubular fluid.

Figure 16.15 The countercurrent multiplier system. Active extrusion of Cl⁻ followed by Na⁺ from the ascending limb makes the surrounding tissue fluid more concentrated. This concentration is multiplied by the fact that the descending limb is passively permeable so that its fluid increases in concentration as the surrounding tissue fluid becomes more concentrated. The transport properties of the loop and their effect on tubular fluid concentration is shown in (a). The values of these changes in osmolality, together with the effect on surrounding tissue fluid concentration, are shown in (b).

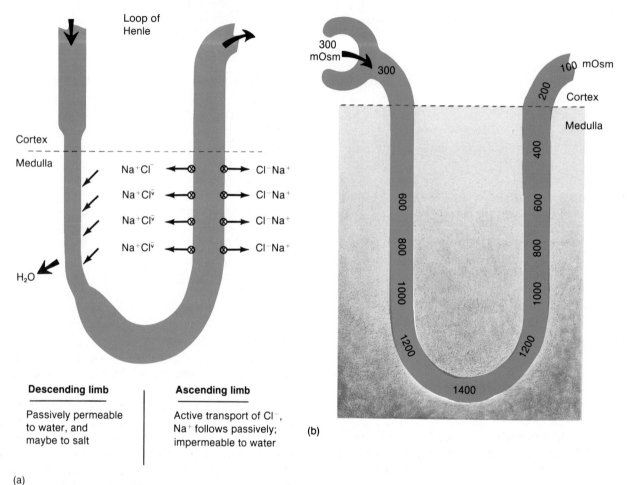

Descending limb

Passively permeable to water, and maybe to salt

Ascending limb

Active transport of Cl⁻, Na⁺ follows passively; impermeable to water

(a)

(b)

As a result of these passive transport processes in the descending limb, the fluid that "rounds the bend" at the tip of the loop has the same osmolality as the surrounding tissue fluid (1200–1400 mOsm). There is therefore a higher salt concentration arriving in the ascending limb than there would be if the descending limb delivered isotonic fluid. Salt transport by the ascending limb is increased accordingly, so that the "saltiness" of the tissue fluid in the medulla is multiplied (see figure 16.15).

Countercurrent Multiplication Countercurrent flow (flow in opposite directions) in the ascending and descending limbs and the close proximity of the two limbs allow interaction to occur. Since the concentration of tubular fluid in the descending limb reflects the concentration of surrounding tissue fluid, and since the concentration of this tissue fluid is raised by active extrusion of salt from the ascending limb, a *positive feedback* mechanism is created. The more salt the ascending limb extrudes, the more concentrated will be the fluid that returns to it from the descending limb. The positive feedback mechanism multiplies the concentration of tissue fluid and descending limb

Figure 16.16 Countercurrent exchange in the vasa recta. Diffusion of salt and water first into and then out of these blood vessels helps to maintain "saltiness" (hypertonicity) of the interstitial fluid in the renal medulla (numbers indicate osmolality).

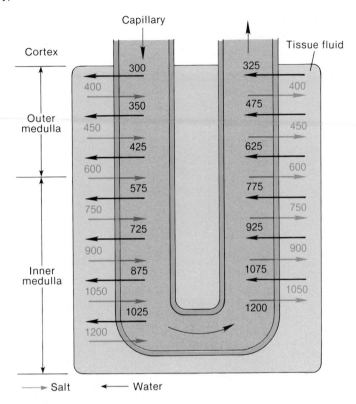

fluid that can be achieved by the active extrusion of salt from the ascending limb. This positive feedback mechanism is thus called the *countercurrent multiplier system.*

The countercurrent multiplier system traps some of the filtered salt that enters the loop of Henle in the tissue fluid of the renal medulla. This results in a gradually increasing concentration of renal tissue fluid from cortex to inner medulla; the osmolality of tissue fluid increases from 300 mOsm (isotonic) in the cortex to 1200–1400 mOsm in the deepest part of the medulla.

Vasa Recta In order for the countercurrent multiplier system to be effective only a small amount of salt that is extruded from the ascending limbs must be carried away by the vascular system. Hypertonicity of the renal medulla is maintained by *countercurrent exchange* of salt and water in the vasa recta. These vessels, as previously described, form long loops that parallel the long loops of Henle of juxtamedullary nephrons. Salt that diffuses into the blood as it descends these loops subsequently diffuses out of the blood in the ascending segments of these vessels. This countercurrent exchange helps to maintain the saltiness of the medullary tissue fluid (see figure 16.16).

Figure 16.17   According to some authorities, urea diffuses out of the collecting duct and contributes significantly to the concentration of interstitial fluid in the renal medulla. Active transport of Cl^- out of the thick segments of the ascending limbs also contributes to the hypertonicity of the medulla so that water is reabsorbed by osmosis from the collecting ducts.

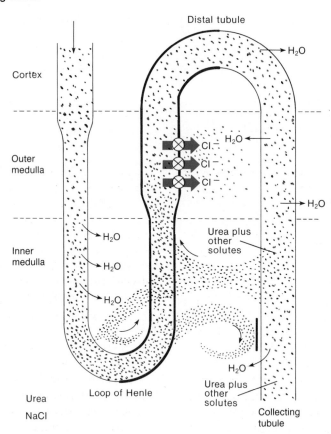

Possible Effects of Urea   Experimental evidence suggests that active transport of Cl^- may occur only in the thick segments of the ascending limbs. The thin segments of the ascending limbs, which are located in the deeper regions of the medulla, may not be able to extrude salt actively. Since salt does leave the thin segments of the ascending limb, this must occur by diffusion. It is known, however, that in the deeper regions of the medulla the tissue fluid around the thin segments of the ascending limb is at least as concentrated as in the tubular fluid. Some investigators therefore conclude that molecules other than salt—specifically, urea—contribute to the hypertonicity of the tissue fluid.

If urea does contribute to the osmolality of medullary tissue fluid, the salt concentration of the tissue fluid around the thin segments may indeed be less than the salt concentration of tubular fluid, and outward diffusion of salt may therefore occur across the wall of the thin segments. According to some authorities, accumulation of urea in the medullary tissue fluid may result from the recycling of urea between the collecting duct and the loop of Henle (see figure 16.17).

Table 16.2 Properties of different nephron segments in regard to the concentrating-diluting mechanisms of the kidney.

Nephron Segment	Active Transport	Passive Transport		
		Salt	Water	Urea
Proximal tubule	Na⁺	Cl⁻	Yes	Yes
Descending limb of Henle's loop	None	Maybe	Yes	No
Thin segment of ascending limb*	Cl⁻ or None	Na⁺ or NaCl	No	Yes
Thick segment of ascending limb	Cl⁻	Na⁺	No	No
Distal tubule	Na⁺	No	No	No
Collecting duct**	Slight Na⁺	No	Yes (ADH) or Slight (no ADH)	Yes

*According to one theory, the thin segment of the ascending limb actively transports Cl⁻, and Na⁺ follows passively; according to another theory, the thin segment does not actively extrude salt. Both theories are presented in this table.
**The permeability of the collecting duct to water depends on the presence of ADH.

The Collecting Duct: Effect of Antidiuretic Hormone

As a result of the recycling of salt between the ascending and descending limbs of the loop of Henle, and possibly of the recycling of urea between the collecting duct and the loop of Henle, the medullary tissue fluid is made very hypertonic. The collecting ducts must transport their fluid through this hypertonic environment in order to empty their contents of urine into the calyces. Since the distal convoluted tubules are impermeable to water, the fluid that enters the collecting ducts in the cortex is hypotonic (with a concentration of 100 mOsm) as a result of active salt transport by the ascending limbs of the loops.

The walls of the collecting ducts are *permeable to water but not to salt*. Since the surrounding tissue fluid in the renal medulla is very hypertonic, as a result of the countercurrent multiplier system, water is drawn out of the collecting ducts by osmosis. This water does not dilute the surrounding fluid because it is transported by capillaries to the general circulation. In this way, most of the water remaining in the tubules after reabsorption in the proximal tubules is returned to the vascular system.

Antidiuretic Hormone (ADH) The osmotic gradient created by the countercurrent multiplier system provides the force for water reabsorption through the collecting ducts. The rate of this reabsorption, however, is determined by the permeability of the collecting duct cell membranes to water. The permeability of the collecting duct to water is determined by the concentration of *antidiuretic hormone (ADH)* in the blood. When the concentration of ADH is increased, the collecting ducts are made more permeable to water, and more water is reabsorbed. A decrease in ADH, conversely, results in less reabsorption of water and thus in excretion of a larger volume of more dilute urine.

Secretion of ADH from the posterior pituitary is stimulated by osmoreceptors in the hypothalamus, which respond to increased plasma osmotic pressure. During dehydration, therefore, when the plasma becomes more concentrated, increased secretion of ADH promotes increased permeability of the collecting ducts to water. In severe dehydration only the minimum amount of water needed to eliminate the body's wastes is excreted. The *obligatory water loss* is due to the fact that fluid in the collecting ducts cannot become more concentrated than the surrounding tissue fluid. At a maximum urinary concentration of 1200–1400 mOsm, a volume of 400 ml of urine is excreted per day. Under these conditions, therefore, 99.8 percent of the initial glomerular ultrafiltrate volume (180 L per day) is reabsorbed.

Figure 16.18   The countercurrent multiplier system in the loop of Henle and countercurrent exchange in the vasa recta help create a hypertonic renal medulla. Under the influence of antidiuretic hormone (ADH), the collecting duct is permeable to water so that water is drawn by osmosis out into the hypertonic renal medulla and into the peritubular capillaries.

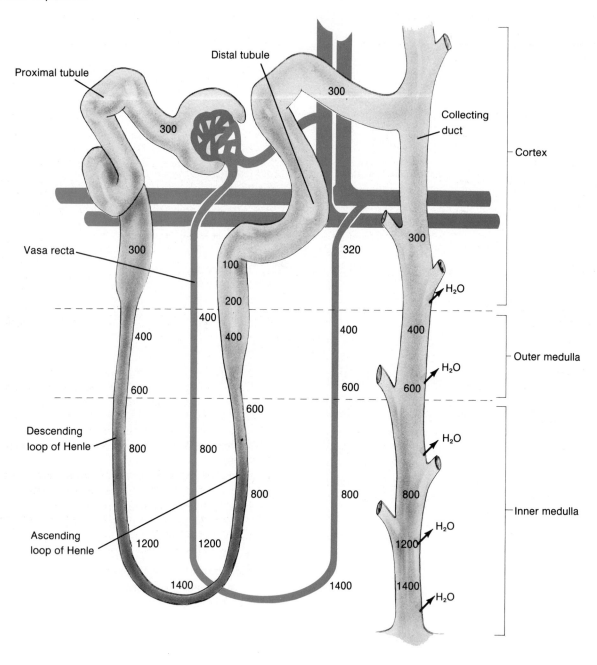

Table 16.3 Antidiuretic hormone secretion and action.

Stimulus	Receptors	Secretion of ADH	Effects On	
			Urine Volume	Blood
↑ Osmolality (dehydration)	Osmoreceptors in hypothalamus	Increased	Decreased	Increased water retention; decreased blood osmolality
↓ Osmolality	Osmoreceptors in hypothalamus	Decreased	Increased	Water loss increases blood osmolality
↑ Blood volume	Stretch receptors in left atrium	Decreased	Increased	Decreased blood volume
↓ Blood volume	Stretch receptors in left atrium	Increased	Decreased	Increased blood volume

A person in a state of normal hydration excretes about 1.5 L per day of urine, indicating that 99.2 percent of the glomerular ultrafiltrate volume is reabsorbed. Notice that small changes in percent reabsorption translate into large changes in urine volume. Increasing water ingestion, and thus decreasing ADH secretion, results in correspondingly larger volumes of urine excretion. Mechanisms controlling ADH secretion were discussed in chapter 12 and are reviewed in table 16.3. It should be noted that even in the complete absence of ADH, some water is still reabsorbed through the collecting ducts.

Diabetes Insipidus Since ADH is produced by the hypothalamus and transported down fiber tracts to the posterior pituitary, damage to the hypothalamus, the fiber tracts, or to the posterior pituitary can result in loss of ADH secretion. This condition is called *diabetes insipidus*. More rarely, a person's pituitary may secrete ADH but the nephron tubules are unable to respond to the hormone. This variant is called nephrogenic diabetes insipidus. In either case, the low permeability of the collecting ducts to water results in excretion of a large volume (usually 5–10 L per day) of dilute (about 100 mOsm) urine. About 96 percent of the filtered water is reabsorbed under these conditions, but such large volumes of water are lost in the urine that serious dehydration can result if the person fails to drink correspondingly large volumes of fluids.

Diuretics Diuretics are substances that increase urine production. The most commonly ingested diuretics are those that act by inhibiting antidiuretic hormone secretion; these include water (which inhibits activity of osmoreceptors by diluting the blood), caffeine and related molecules, and ethyl alcohol. Caffeine and alcohol inhibit ADH secretion by acting directly on the hypothalamus.

The most common clinical diuretics—prescribed most often for the treatment of high blood pressure—are the thiazides and furosemide. These diuretics are believed to act by inhibiting active Cl⁻ transport by the ascending limbs of Henle's loops. This inhibition decreases the effectiveness of the countercurrent multiplier system. The renal medulla becomes less hypertonic as a result, and less water can be reabsorbed through the collecting ducts. The actions and side effects of these and other diuretics will be discussed in a later section.

1. Describe the mechanisms for salt and water reabsorption in the proximal tubule, and illustrate this process with a line diagram.
2. Compare the transport of Na⁺, Cl⁻, and water across the walls of the proximal tubule, ascending and descending limbs of the loop of Henle, and collecting ducts.
3. Explain the interaction of the ascending and descending limbs of the loop and how this interaction results in a hypertonic renal medulla.
4. Describe how variations in ADH secretion affect the volume and concentration of urine.

Figure 16.19 Secretion refers to the active transport of substances from the peritubular capillaries into the tubular fluid. This transport is in a direction opposite to that of reabsorption.

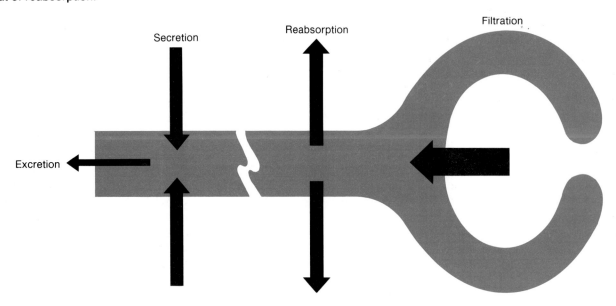

Table 16.4 The effects of filtration, reabsorption, and secretion on renal clearance rates.

Term	Means	Effect on Renal Clearance
Filtered	A substance enters the glomerular ultrafiltrate	Some or all of filtered substance may enter the urine and be "cleared" from the blood.
Reabsorbed	The transport of a substance from the filtrate, through tubular cells, and into the blood	Reabsorption decreases the rate at which a substance is cleared; clearance rate is less than the glomerular filtration rate (GFR).
Secreted	The transport of a substance from peritubular blood, through tubular cells, and into the filtrate	When a substance is secreted by the nephrons its clearance rate is greater than the GFR.

Renal Clearance Rates

One of the major functions of the kidneys is excretion of waste products such as urea, creatinine, and other molecules. These molecules are filtered through the glomerulus into Bowman's capsule along with water, salt, and other plasma solutes (except proteins). In addition, some waste products (such as creatinine) and many foreign molecules (such as antibiotics) can gain access to the urine by a process called **secretion.** Secretion is the opposite of reabsorption by active transport. Molecules that are secreted are transported actively from the peritubular capillaries, across the tubule wall, and into the tubular fluid. In this way, molecules that were not filtered from the blood in the glomerulus can still be excreted in the urine (see figure 16.19).

While most (about 99 percent) of the filtered water is returned to the vascular system by reabsorption, most of the wastes that are filtered, and all those that are secreted, will be eliminated in the urine. The concentration of these substances in the renal vein leaving the kidneys is therefore lower than their concentrations in the blood that entered the kidneys in the renal artery. Some of the blood that passes through the kidneys, in other words, is "cleared" of these waste products. This is known as **renal plasma clearance.**

The quantity of a substance excreted in the urine within a given period of time depends on the (1) quantity *filtered* through the glomeruli into the tubular fluid; (2) quantity *secreted* from the unfiltered blood in peritubular capillaries by active transport into the tubules; and the (3) quantity *reabsorbed* by transport from the tubules into the peritubular blood. Filtration and secretion increase renal plasma clearance; reabsorption decreases the amount excreted and thus decreases the clearance rate (see table 16.4).

Renal Clearance of Inulin: Measurement of GFR

If a substance is neither reabsorbed nor secreted by the tubules, the amount excreted per minute in the urine will obviously be equal to the amount that is filtered out of the glomeruli. There does not seem to be a single substance produced by the body, however, that is neither reabsorbed nor secreted to some degree. Plants such as artichokes, dahlias, onions, and garlic, fortunately, do produce such a compound. This compound—a polymer of the monosaccharide fructose—is *inulin*. Once injected into the blood, inulin is filtered by the glomeruli, and the amount of inulin excreted per minute is exactly equal to the amount that was filtered per minute. The clearance rate of inulin can thus be used clinically to measure the glomerular filtration rate.

If the concentration of inulin in the urine is measured, and the rate of urine formation is determined, the rate of inulin excretion in the urine can easily be calculated:

$$\textit{Quantity excreted per minute} = V \times U$$
$$\left(\frac{mg}{min}\right) \qquad \left(\frac{ml}{min}\right)\left(\frac{mg}{ml}\right)$$

Where: V = rate of urine formation
U = inulin concentration in urine

In order to calculate the rate at which any substance is filtered by the glomeruli (in mg per minute), the **glomerular filtration rate** (**GFR**—given in ml per minute) can be multiplied by the concentration of that substance in the blood (in mg per ml):

$$\textit{Quantity filtered} = GFR \times P$$
$$\left(\frac{mg}{min}\right) \qquad \left(\frac{ml}{min}\right) \quad \left(\frac{mg}{ml}\right)$$

Where P = inulin concentration in plasma

This calculation cannot be performed at this time because the GFR is not yet known. It is known, however, that the rate of inulin excretion is equal to the rate of its filtration (because inulin is neither reabsorbed nor secreted). Therefore:

$$GFR \times P \quad = \quad V \times U$$
(amount filtered) (amount excreted)

If this last equation is solved for the glomerular filtration rate:

$$GFR_{(ml/min)} = \frac{V_{(ml/min)} \times U_{(mg/ml)}}{P_{(mg/ml)}}$$

Suppose, for example, that inulin is infused into a vein and its concentrations in the urine and plasma are found to be 30 mg per ml and 0.5 mg per ml, respectively. If the rate of urine formation is 2 ml per minute:

$$GFR = \frac{2 \text{ ml/min} \times 30 \text{ mg/ml}}{0.5 \text{ mg/ml}} = 120 \text{ ml/min}$$

This equation simply states that, at a plasma inulin concentration of 0.5 mg per ml, 120 ml of plasma must have been filtered in order to excrete the measured amount of 60 mg per minute in the urine. The GFR tends to remain relatively constant, as previously described, with an average of 115 ml per minute for women and 125 ml per minute for men. If a well-hydrated person excreted twice the volume of urine per minute, therefore, the plasma concentration would be decreased accordingly and the GFR would remain—in this example—near 120 ml per minute.

Creatinine Measuring GFR, and thus kidney function, by inulin clearance is a simple and accurate procedure. It is, however, inconvenient and uncomfortable for the patient; inulin must be infused at a constant rate into a vein, and the ureter must be cannulated. Fortunately, simple measurement of creatinine concentrations in the plasma provide an index of kidney function that can determine if an inulin clearance test is needed.

Creatinine is produced at relatively constant rates as a waste product of muscle creatine (described in chapter 8). Creatinine is not reabsorbed by the tubules, and only a small amount is secreted. Although the excretion rate of creatinine is therefore slightly higher than that of inulin, the excretion rate is closely related to the filtration rate. A decrease in the GFR below normal would thus cause the plasma creatinine concentration to rise above the normal range, and the degree of this rise would be proportionate to the fall in GFR.

Figure 16.20 Renal clearance of inulin. Inulin is present in blood entering the glomeruli *(a)*, and some of this blood, together with its dissolved inulin, is filtered *(b)*. All of this filtered inulin enters the urine, whereas most of the filtered water is returned to the vascular system (is reabsorbed). Blood leaving the kidneys in the renal vein therefore contains less inulin than blood that entered the kidneys in the renal artery *(c)*. Since inulin is filtered but neither reabsorbed nor secreted, the inulin clearance rate equals the glomerular filtration rate (GFR).

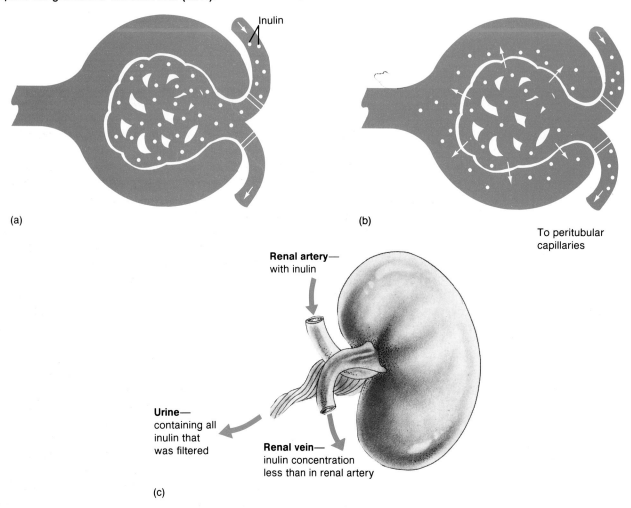

Inulin

(a)

(b)

To peritubular capillaries

Renal artery— with inulin

Urine— containing all inulin that was filtered

Renal vein— inulin concentration less than in renal artery

(c)

Clearance Rates   If one knows the amount of inulin excreted per minute (the excretion rate, in mg per minute) and the plasma inulin concentration, one can calculate the volume of plasma that was filtered per minute (the GFR). Since 120 ml per minute are filtered (in the previous example), the amount of inulin contained in 120 ml of plasma is excreted per minute. Since creatinine is filtered like inulin, but is also secreted to a slight degree, the excretion rate of creatinine is slightly greater than the filtration rate. The amount of creatinine excreted per minute, in other words, is greater than the amount contained in the 120 ml of plasma that was filtered.

If a substance is reabsorbed to some degree, conversely, the excretion rate will be less than the amount contained in the 120 ml of plasma that was filtered. In order to compare the renal "handling" of various substances, in terms of their reabsorption or secretion, the **renal plasma clearance rate** of these substances can be calculated using the same formula used for determination of the GFR:

$$Clearance\ rate = \frac{V \times U}{P}$$

Where: V = urine volume per minute
U = urine concentration of the substance
P = plasma concentration of the substance

Table 16.5 Renal "handling" of different plasma molecules.

If Substance Is:	Example	Concentration in Renal Vein	Renal Clearance Rate
Not filtered	Proteins	Same as in renal artery	Zero
Filtered, not reabsorbed nor secreted	Inulin	Less than in renal artery	Equal to GFR (115–125 ml/min)
Filtered, partially reabsorbed	Urea	Less than in renal artery	Less than GFR
Filtered, completely reabsorbed	Glucose	Same as in renal artery	Zero
Filtered and secreted	PAH	Less than in renal artery; approaches zero	Greater than GFR; equal to total plasma flow rate (~625 ml/min)
Filtered, reabsorbed, and secreted	K^+	Variable	Variable

In the case of inulin, which is filtered but neither reabsorbed nor secreted, the clearance rate is equal to the glomerular filtration rate. If a substance is secreted, its clearance rate will be greater than the GFR; if it is reabsorbed, its clearance rate will be less than the GFR (see table 16.5). The clearance rate is not the same as the excretion rate (mg excreted per minute). Rather, it indicates the amount of plasma that originally contained the quantity of a substance excreted in a minute's time (and is thus shown in ml per minute units).

Clearance of Urea

Urea is a waste product of amino acid metabolism that is secreted by the liver into the blood (as described in chapter 5). Though a waste product, a significant proportion of the filtered urea (40–60 percent) is reabsorbed passively, as a result of the high permeability of the tubule membranes to this compound. Not all of the urea contained in the filtered plasma, therefore, is cleared. Using the formula for renal clearance rate previously described:

If: $V = 2$ ml/min
  $U = 7.5$ mg/ml of urea
  $P = 0.2$ mg/ml of urea
Urea clearance rate $= \dfrac{2 \text{ ml/min} \times 7.5 \text{ mg/ml}}{0.2 \text{ mg/ml}} = 75$ ml/min

The amount of urea excreted per minute, in other words, was originally contained in 75 ml of plasma. Since 120 ml of plasma were filtered (as determined by the inulin clearance rate), approximately 40 percent of the filtered urea must have been reabsorbed. One might suppose that the GFR could be determined by dividing the urea clearance rate by 60 percent. This procedure is not reliable, however, because the percent urea reabsorption is variable.

Clearance of PAH: Measurement of Renal Blood Flow

It is important to realize that not all of the blood delivered to the glomeruli is filtered; most of the glomerular blood passes straight through to the efferent arterioles and peritubular capillaries. The inulin and urea in this unfiltered blood is not excreted, but it returns instead to the general circulation through the renal vein. The process of *secretion,* or active transport from the peritubular capillaries to the tubular fluid, can help eliminate certain solutes from this unfiltered blood. Creatinine is secreted to a slight degree, for example, so that its clearance rate is 5–10 percent greater than the glomerular filtration rate.

The clearance rate of inulin thus measures the volume of plasma filtered by glomeruli in both kidneys–it does not measure the total renal blood flow. In order to measure total renal blood flow, one must measure the clearance rate of a compound that is cleared from all of the blood entering the kidneys (filtered plus unfiltered blood). In this case, the amount of the compound excreted per minute would be equal to the total plasma flow per minute. *Para-aminohippuric acid (PAH)* is almost completely eliminated (by filtration and secretion) from the blood entering the kidneys; its clearance rate is thus a measure of the rate of total renal blood flow (see figure 16.21).

The normal PAH clearance rate has been found to average 625 ml per minute. Since the glomerular filtration rate in an average person is approximately 120 ml per minute, this indicates that about 20 percent of the renal blood flow is filtered. The remaining 80 percent (of the 625 ml per minute) contains solutes that cannot enter the urine unless they are secreted by active transport into the tubules.

Figure 16.21 Some of the para-aminohippuric acid (PAH) in glomerular blood *(a)* is filtered into Bowman's capsules *(b)*. The PAH present in the unfiltered blood is secreted from peritubular capillaries into the nephron *(c)*, so that all of the blood leaving the kidneys is free of PAH *(d)*. The clearance rate of PAH therefore equals the total plasma flow to the glomeruli.

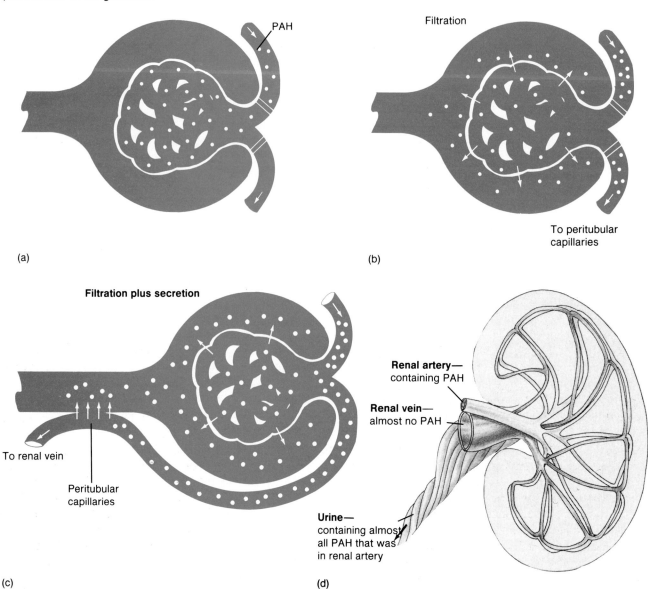

PAH

Filtration

To peritubular capillaries

(a)

(b)

Filtration plus secretion

To renal vein

Peritubular capillaries

(c)

Renal artery— containing PAH

Renal vein— almost no PAH

Urine— containing almost all PAH that was in renal artery

(d)

Effect of Hematocrit on Blood Flow Calculations   The amount of PAH excreted per minute is equal to the amount contained in all of the plasma flowing to the kidneys per minute. In order to determine how much whole blood this represents, one must take into account the volume occupied by blood cells (mainly erythrocytes). The fraction of whole blood volume occupied by blood cells is the *hematocrit*. If the hematocrit is 45, for example, blood cells occupy 45 percent of the whole blood volume, and plasma accounts for the remaining 55 percent of the volume (obtained by subtracting 45 from 100 percent). The total renal blood flow can therefore be calculated by dividing the PAH clearance rate by the relative proportion of plasma:

$$\text{Renal blood flow} = \frac{\text{plasma flow (the PAH clearance rate)}}{1 - \text{hematocrit}}$$

If the hematocrit is 45:

$$\text{Renal blood flow} = \frac{625 \text{ ml/min}}{0.55} = 1.1 \text{ L/min}$$

Figure 16.22 Reabsorption of glucose in the proximal tubule. Glucose is reabsorbed by cotransport with Na^+ from the tubular fluid and then transported by Na^+-independent carriers through the basolateral membrane into the peritubular blood. By this means normally all of the filtered glucose is reabsorbed.

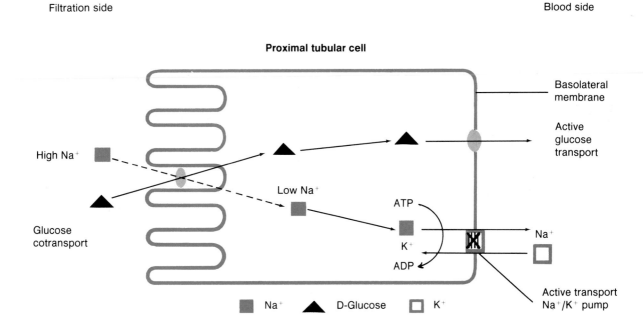

Filtration side

Blood side

Proximal tubular cell

High Na^+

Glucose cotransport

Low Na^+

ATP

K⁺

ADP

Basolateral membrane

Active glucose transport

Na^+

Active transport Na^+/K^+ pump

◼ Na^+ ▲ D-Glucose   ◻ K^+

The total renal blood flow under these conditions is thus about 1.1 L per minute. Since the total blood flow in the body (the cardiac output) at rest is about 5 L per minute, this means that the kidneys receive approximately 20–25 percent of the total cardiac output per minute.

Penicillin and Diodrast   Many compounds other than creatinine and PAH are secreted to variable degrees, and therefore have clearance rates greater than the GFR. Penicillin, for example, is rapidly removed from the blood because it is secreted as well as filtered. When penicillin is in short supply, or when high concentrations are needed in the blood, another drug is given that inhibits tubular secretion. This drug thus increases the concentration of penicillin in the blood and enhances its potency.

The ability of the kidneys to be visualized in radiographs is improved by injection of *Diodrast,* a material that absorbs X rays and that is secreted into the tubules. Many drugs that are injected into the body, as well as endogenously produced compounds such as steroid hormones, are inactivated in the liver by chemical transformations and are rapidly cleared from the blood by active secretion in the nephrons.

Reabsorption of Glucose and Amino Acids

Glucose and amino acids dissolved in plasma are easily filtered through the glomeruli into the tubules. These molecules are not normally lost in the urine, however, and their concentration in the blood leaving the kidneys (in the renal vein) is the same as in the arterial blood entering the kidneys. One can thus conclude that these compounds are normally completely reabsorbed.

Reabsorption of glucose and amino acids from the tubular fluid, across the wall of the proximal tubules, and into the peritubular capillaries is an energy-requiring process. The energy required for movement of these compounds into the tubule cells is provided by co-transport (see chapter 6) with Na^+, which diffuses down its electrochemical gradient when it passes through the apical membranes. The glucose and amino acids appear to share carriers with Na^+ in the apical membrane. Extrusion of glucose and amino acids from the other side of the cell (across the basolateral membranes) occurs by means of a Na^+—independent active transport carrier (see figure 16.22).

Carrier-mediated transport displays the property of *saturation.* This means that when the transported molecule (such as glucose) is present in sufficiently high concentrations, all of the carriers become "busy," and the transport rate reaches a maximal value. The concentration of transported molecules needed to just saturate the carriers, and to just achieve the maximal transport rate, is called the **transport maximum** (abbreviated T_m).

Table 16.6   Inherited diseases associated with the presence of specific amino acids in the urine.

Disease	Cause of Disease	Effect of Defect	Treatment
Cystinuria	Renal carriers for cystine and related amino acids are defective.	Kidney stones	Bicarbonate and diuretic administration
Hartnup disease	Renal carrier for tryptophane are defective.	Decreased NAD and NADP within body cells	Nicotinamide administration
Homocystinuria	Enzyme defect results in excessive blood levels of homocystine.	Speech defects, mental retardation	Diet low in methionine, high in cystine
Phenylketonuria	Enzyme defect results in excessive accumulation of phenylalanine and in urinary excretion of phenylpyruvic acid.	Severe mental retardation	Diet low in phenylalanine

The carriers for glucose and amino acids in the renal tubules are not normally saturated and so are able to remove the filtered molecules completely. The T_m for glucose, for example, averages 375 mg per minute; this is well above the rate at which glucose is normally delivered to the tubules. The rate of glucose delivery can be calculated by multiplying the plasma glucose concentration (about 1 mg per ml) by the GFR (about 125 ml per minute). Approximately 125 mg per minute of glucose are thus delivered to the proximal tubule, while 375 mg per minute are required to reach saturation.

Glycosuria   Glucose appears in the urine—a condition called glycosuria—when more glucose passes through the tubules than can be reabsorbed. This occurs when the plasma glucose concentration reaches 180–200 mg per 100 ml. Since the rate of glucose delivery under these conditions (2 mg per ml × 125 ml per minute = 250 mg per minute) is still below the average T_m for glucose (375 mg per minute), one must conclude that some nephrons have considerably lower T_m values—perhaps due to fewer carriers—than the average.

The **renal plasma threshold** is the minimum plasma concentration of a substance that results in the excretion of that substance in the urine. Excretion of glucose, for example, occurs at a plasma glucose concentration of 180–200 mg per 100 ml. Glucose is normally absent from the urine because plasma glucose concentrations normally remain below this threshold value (fasting levels average 100 mg per 100 ml). It should be noted that a higher than normal plasma glucose concentration—*hyperglycemia*—can produce glycosuria only when the threshold is reached. A fasting plasma glucose concentration of 130 mg per 100 ml, for example, indicates hyperglycemia but does not result in glucose "spilling" into the urine.

Fasting hyperglycemia is caused by inadequate secretion or action of the hormone *insulin,* as will be described in more detail in chapter 20. When this hyperglycemia results in glycosuria, the disease is called **diabetes mellitus.** The term *diabetes* refers to "passing through" or "spilling over" into the urine; mellitus refers to sugar. A person with uncontrolled diabetes mellitus also has *polyuria* (a large volume of urine), because the excreted glucose carries water with it as a result of the osmotic pressure it generates in the tubules. This condition should not be confused with diabetes insipidus, in which a large volume of very dilute urine is excreted as a result of inadequate ADH secretion.

Excretion of Amino Acids   The appearance of amino acids in the urine, unlike the appearance of glucose, is not due to excessively high concentrations in the blood. Rather, the excretion of specific amino acids in the urine indicates that there is a genetic disease in which the carriers are either missing or defective. Since different classes of amino acids are transported by different carriers, the types of amino acids that "spill" into the urine in these genetic diseases are very specific (see table 16.6).

Reabsorption of Calcium and Phosphate

Unlike the reabsorption of glucose and amino acids, the transport maximum of Ca^{++} and PO_4^{-3} is very close to the normal levels of these ions in the blood. A relatively slight increase in plasma Ca^{++} (produced by a calcium-rich meal) is compensated directly by "spilling over" of the excess Ca^{++} into the urine. Also unlike the reabsorption of glucose and amino acids, the reabsorption of Ca^{++} and PO_4^{-3} is regulated by hormones.

Table 16.7 Filtration, reabsorption, and excretion of electrolytes in a twenty-four-hour period. Values are given in milliequivalents (mEq—millimoles multiplied by a valence of one) and in grams.

Electrolyte	Filtered (mEq)	Reabsorbed (mEq)	Excreted (mEq)	Percent Reabsorbed	Percent Excreted
Sodium	25,200 (580 g)	25,110 (578 g)	90 (2 g)	99.64	0.36
Potassium	756 (30 g)	666 (26 g)	90 (4 g)	88.10	11.90
Chloride	18,000 (637 g)	17,880 (633 g)	120 (4 g)	99.33	0.67
Bicarbonate	4,320 (264 g)	4,320 (264 g)	0	100.00	0

Homeostasis of blood Ca^{++} levels is extremely important because of the role of Ca^{++} in muscle contraction and because the Ca^{++}/Mg^{++} ratio in extracellular fluid affects the stability of cell membranes. Low blood calcium, for example, can produce muscle tetany by increasing the excitability of nerve and muscle cell membranes. Reabsorption of Ca^{++} in the distal tubule is stimulated by *parathyroid hormone (PTH)*, secreted from the parathyroid glands. Reabsorption of PO_4^{-3} is inhibited by this hormone, but is stimulated by *1,25-dihydroxy vitamin D_3* (derived from vitamin D). The actions of these two hormones on calcium homeostasis is discussed in detail in chapter 20.

---

1. Define clearance rate and describe how it is measured.
2. Describe how the clearance rate is affected by reabsorption and by secretion. Give examples of molecules that are *(a)* not filtered; *(b)* filtered, but neither reabsorbed nor secreted; *(c)* filtered and completely reabsorbed; *(d)* filtered and partially reabsorbed; *(e)* filtered and secreted.
3. Describe the significance of the inulin clearance rate and PAH clearance rate measurements.
4. Define transport maximum and renal plasma threshold. Explain why people with diabetes mellitus have glycosuria.

Renal Control of Electrolyte Balance

The kidneys help regulate the concentrations of plasma electrolytes—sodium, potassium, calcium, chloride, bicarbonate, and phosphate—by matching their urinary excretion to the amounts ingested (see table 16.7). Control of plasma Na^+ is important in the regulation of blood volume and pressure; control of plasma K^+ is required to maintain proper function of cardiac and skeletal muscles. Regulation of Na^+/K^+ balance is also intimately related to renal control of acid-base balance of the blood.

Regulation of Na^+/K^+ Balance: Role of Aldosterone

Approximately 90 percent of the filtered Na^+ and K^+ is reabsorbed in the early part of the nephron before the filtrate reaches the distal tubule. This reabsorption, as previously described, occurs at a constant rate and is not subject to hormonal regulation. The final concentration of Na^+ and K^+ in the urine is varied according to the needs of the body by processes that occur in the **distal tubule.** These processes are regulated by a steroid hormone secreted from the adrenal cortex—**aldosterone.**

Sodium Reabsorption Although 90 percent of the filtered sodium is reabsorbed in the early region of the nephron, the amount left in the filtrate delivered to the distal tubule is still quite large. In the absence of aldosterone, 80 percent of this amount is automatically reabsorbed through the wall of the distal tubule into the peritubular blood. This is 8 percent of the original amount filtered. The amount of sodium excreted is thus 2 percent of the amount filtered. While the percentage seems small, the actual amount of sodium this represents is an impressive 30 g per day excreted in the urine. When aldosterone is secreted in maximal amounts, in contrast, all of the sodium delivered to the distal tubule is reabsorbed. Under these conditions urine contains no Na^+ at all.

Figure 16.23 Potassium is almost completely reabsorbed in the proximal tubule, but under aldosterone stimulation, is secreted into the distal tubule. All of the K⁺ in urine is derived from secretion rather than from filtration.

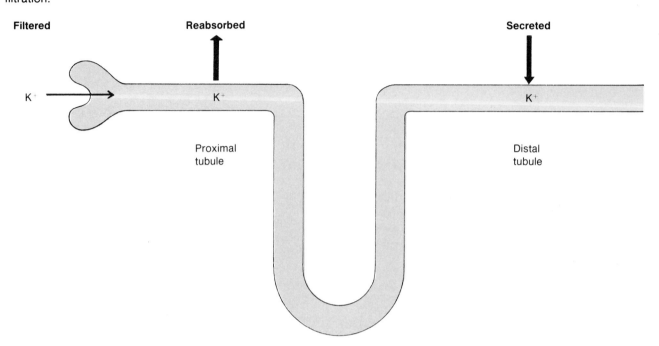

Potassium Secretion About 90 percent of the filtered K⁺, as previously described, is reabsorbed in the early regions of the nephron (primarily across the proximal tubule). In the absence of aldosterone, all of the filtered K⁺ that is not reabsorbed in the proximal tubule is reabsorbed in the distal tubule. In the absence of aldosterone, therefore, no K⁺ is excreted in the urine. The presence of aldosterone stimulates *secretion of K⁺*—active transport of K⁺ from the peritubular blood into the distal tubule. This secretion is the only means by which K⁺ can be eliminated in the urine. When aldosterone secretion is maximal, as much as fifty times more K⁺ is excreted in the urine (because of active transport from blood to tubular fluid) than was originally filtered through the glomeruli.

Summary Aldosterone promotes sodium retention and potassium loss from the blood by stimulating reabsorption of Na⁺ and secretion of K⁺ across the wall of the distal tubules. Since aldosterone promotes retention of Na⁺, it contributes to an increased blood volume and pressure. This occurs because the reabsorbed Na⁺ raises the sodium concentrations of the surrounding tissue fluid and peritubular blood, and thus promotes water retention by osmosis. This water is reabsorbed across the walls of the collecting tubules and collecting ducts, not across the walls of the distal convoluted tubules (which are impermeable to water).

Since urine provides the major route for excretion of K⁺, the body cannot get rid of excess K⁺ in the absence of aldosterone-stimulated secretion of K⁺ into the distal tubules. Indeed, when both adrenal glands are removed from an experimental animal, the *hyperkalemia* (high blood K⁺) that results can produce fatal cardiac arrythmias. Abnormally low plasma K⁺ concentrations, as might result from excess aldosterone secretion, produces muscular weakness and cramps.

Control of Aldosterone Secretion

Since aldosterone promotes Na⁺ retention and K⁺ loss, one might predict (on the basis of negative feedback) that aldosterone secretion would be increased when there is a low Na⁺ or a high K⁺ concentration in the plasma. This is indeed the case. Increases in plasma K⁺ *directly* stimulates secretion of aldosterone from the adrenal cortex. Decreases in plasma Na⁺ concentrations also promote aldosterone secretion, but they do so indirectly.

The Juxtaglomerular Apparatus The juxtaglomerular apparatus is the region in each nephron where the afferent arteriole and distal tubule come into contact (see figure 16.24). The microscopic appearance of the afferent arteriole and distal tubule in this small region differs from

Figure 16.24 The juxtaglomerular apparatus *(a)* includes the region of contact of the afferent arteriole with the distal tubule. The afferent arterioles in this region contain granular cells with renin, and the distal tubule cells in contact with the granular cells form an area called the *macula densa (b)*. The granular cells of the afferent arteriole are innervated by renal sympathetic nerve fibers.

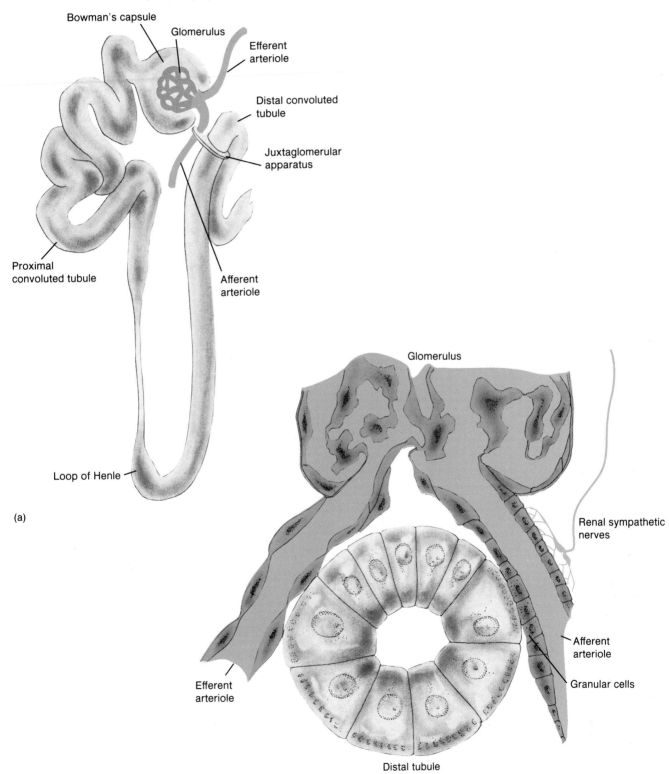

Figure 16.25 The sequence of events by which a low sodium (salt) intake leads to increased sodium reabsorption by the kidneys. The *dotted arrow* and *negative sign* show the completion of the negative feedback loop.

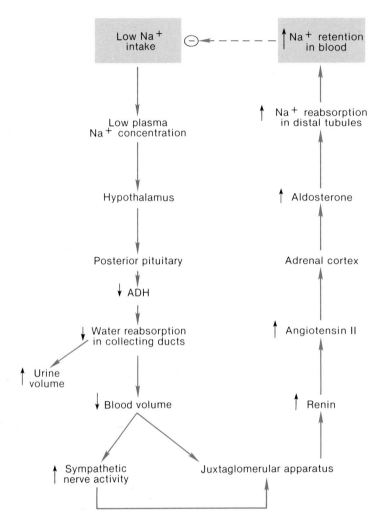

their appearance in other regions. **Granular cells** within the afferent arteriole secrete the enzyme *renin* into the blood; this enzyme catalyzes the conversion of angiotensinogen (a protein) into angiotensin I (a ten-amino acid polypeptide). The region of the distal tubule in contact with the granular cells is called the **macula densa.**

Secretion of renin into the blood results in the formation of angiotensin I, which is converted to *angiotensin II* (an eight-amino acid polypeptide) as blood passes through the lungs and other organs. Angiotensin II, in addition to several other effects (discussed in chapter 12), stimulates the adrenal cortex to secrete aldosterone. Secretion of renin from the granular cells of the juxtaglomerular apparatus is thus said to initiate the **renin-angiotensin-aldosterone system.** Conditions that result in increased renin secretion by this means result in increased aldosterone secretion, and therefore in increased reabsorption of Na^+ in the distal tubules.

Regulation of Renin Secretion   A fall in plasma Na^+ concentration is always accompanied, within a short time, by a fall in blood volume. This is because ADH secretion is inhibited by the decreased plasma concentration (osmolality); with less ADH, more of the filtered water passing through the collecting ducts is excreted in the urine. The fall in blood volume and renal blood flow that results cause increased renin secretion. This is believed to be due in part to a *direct effect* of blood flow on the granular cells, which may function as stretch receptors (and thus as baroreceptors) in the afferent arterioles.

Renin secretion is also stimulated by *sympathetic nerve* activity. There are two proposed mechanisms by which this may occur: (1) beta-adrenergic stimulation may directly stimulate renin secretion from the granular cells;

Table 16.8   Regulation of renin and aldosterone secretion.

Stimulus	Effect on Renin Secretion	Mechanisms	Angiotensin II Production	Aldosterone Secretion
↓ Na⁺	Increased	Low blood volume stimulates renal baroreceptors; granular cells release renin.	Increased	Increased
↑ Na⁺	Decreased	Increased blood volume inhibits baroreceptors; increased Na⁺ in distal tubule acts via macula densa to inhibit release of renin from granular cells.	Decreased	Decreased
↑ K⁺	None	Not applicable	Not changed	Increased
↑ Sympathetic nerve activity	Increased	α-adrenergic effect stimulates constriction of afferent arterioles; β-adrenergic effect stimulates renin secretion directly.	Increased	Increased

and/or (2) alpha-adrenergic activity, which stimulates constriction of afferent arterioles, may indirectly promote renin secretion by decreasing blood flow to the granular cells. It should be recalled (from chapter 12) that the sympathetic system is activated by a fall in total blood volume and pressure, as a result of reflexes such as the aortic and carotid sinus baroreceptor reflex.

If there is inadequate sodium intake, therefore, the fall in plasma volume that results acts, via a direct effect on the juxtaglomerular apparatus and via sympathetic stimulation of granular cells, to increase renin secretion. The increased renin secretion acts, via increased production of angiotensin II, to stimulate aldosterone secretion. In consequence, there is increased Na⁺ reabsorption in the distal tubules and decreased Na⁺ excretion in the urine. This negative feedback system is illustrated in figure 16.25.

Role of the Macula Densa   When Na⁺ is ingested in large amounts, the secretion of aldosterone is reduced. This follows from the effects of blood volume and pressure on renin secretion, as previously discussed. Experiments suggest, however, that high plasma Na⁺ concentrations inhibit renin secretion even when the blood volume is not raised. Many people believe that the macula densa is responsible for this inhibition of renin secretion. The macula densa is the region of distal tubule in contact with the granular cells of the juxtaglomerular apparatus.

According to the proposed mechanism, the cells of the macula densa respond to Na⁺ within the filtrate delivered to the distal tubule. When the plasma Na⁺ concentration is raised, the rate of Na⁺ delivery to the distal tubule is also increased. Through an effect on the macula densa, this increase in filtered Na⁺ may inhibit the granular cells from secreting renin. Aldosterone secretion thus decreases, and since less Na⁺ is reabsorbed in the distal tubule, more Na⁺ is excreted in the urine. The regulation of renin and aldosterone secretion is summarized in table 16.8.

Relationship between Na⁺, K⁺, and H⁺

Hormones, as a general rule, alter the rate of already existing processes. In the absence of aldosterone the distal tubules reabsorb 80 percent of the Na⁺ delivered to them, as previously discussed; aldosterone can increase this reabsorption to 100 percent. When increased amounts of Na⁺ arrive at the distal tubule, there is a proportionate increase in Na⁺ reabsorption (even though the macula densa may promote inhibition of aldosterone secretion).

When a person takes diuretics such as the thiazides and furosemide, which inhibit active transport of Cl⁻ (and thus passive transport of Na⁺) out of the ascending limbs of Henle's loops, the amount of Na⁺ arriving at the distal tubules is increased. The rate of Na⁺ reabsorption in the distal tubules is thus increased as well. Reabsorption of Na⁺ in the distal tubules appears to stimulate secretion of K⁺—these two ions may share the same carrier, but the ratio of Na⁺ reabsorbed to K⁺ secreted does not appear to be constant.

Reabsorption of Na⁺ in the distal tubule promotes some water retention through the walls of the collecting tubules and ducts, but not enough water is retained in this way to counteract the effectiveness of the diuretic drug. The K⁺ secretion that occurs under these circumstances, however, may present a serious side effect of the action of these clinical diuretics. The increased K⁺ secretion may cause *hypokalemia* (low plasma K⁺), which often must be compensated by increased ingestion of potassium. People who take these diuretics for treatment of high blood pressure usually are on a low-sodium diet and supplement their meals with potassium chloride (KCl).

Figure 16.26 In the distal tubule, K⁺ and H⁺ are secreted in exchange for Na⁺. High concentrations of H⁺ may therefore decrease K⁺ secretion, and vice versa.

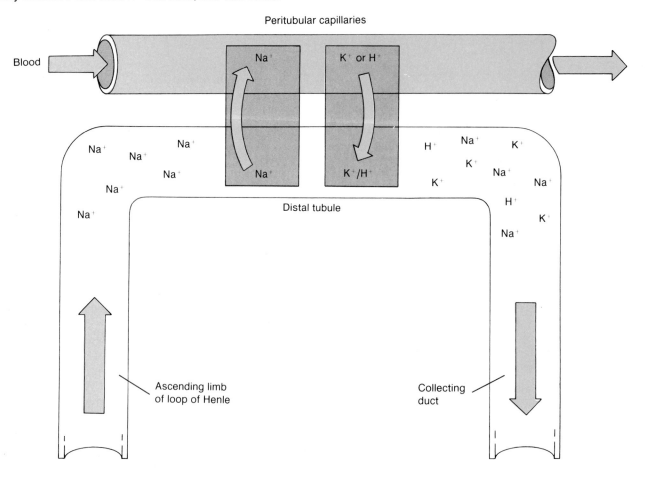

Relationship between K⁺ and H⁺ The plasma K⁺ concentration indirectly affects the plasma H⁺ concentration—that is, the pH. Changes in plasma pH likewise affect the K⁺ concentration of the blood. These effects serve to stabilize the *ratio* of K⁺ to H⁺. When the extracellular H⁺ concentration increases, for example, some of the H⁺ moves into the tissue cells and causes cellular K⁺ to diffuse outward into the extracellular fluid. The plasma concentration of H⁺ is thus decreased while the K⁺ increases, helping to maintain the same ratio of these ions that existed before the increased H⁺ concentrations.

This exchange of H⁺ for K⁺ also occurs in the cells of the distal tubule when the plasma pH is lowered. Under these conditions the tubule cells contain more H⁺ and less K⁺ than before. When these cells reabsorb Na⁺ from the filtrate, they can replace it by secreting either K⁺ or H⁺ (they usually secrete both ions). When the tubule cells contain increased amounts of H⁺ as a result of the fall in plasma pH, they secrete increased amounts of H⁺ at the expense of K⁺. This helps to lower the plasma H⁺ concentration and make the urine more acidic. As a side effect, however, less K⁺ is secreted; the plasma concentration of K⁺ may thus be increased under acidotic conditions.

Hyperkalemia (high blood K⁺) may thus occur in metabolic acidosis. Metabolic alkalosis can have the opposite effect, resulting in lowered plasma K⁺ concentrations. This lowering of plasma K⁺ as a result of alkalosis is sometimes induced intentionally by infusion of bicarbonate in patients with heart or renal failure, who are susceptible to hyperkalemia as a result of decreased glomerular filtration.

Table 16.9 Classification of metabolic and respiratory components of acidosis and alkalosis.

P_{CO_2}	HCO_3^-	Condition	Causes
Normal	Low	Metabolic acidosis	Increased production of "nonvolatile" acids (lactic acid, ketone bodies, and others), or loss of HCO_3^- in diarrhea
Normal	High	Metabolic alkalosis	Vomiting of gastric acid; hypokalemia; excessive steroid administration
Low	Normal	Respiratory alkalosis	Hyperventilation
High	Normal	Respiratory acidosis	Hypoventilation

Aldosterone appears to stimulate secretion of H^+ as well as K^+ into the distal tubules. Abnormally high aldosterone secretion (in *primary aldosteronism,* or *Conn's syndrome*) therefore results in both hypokalemia and metabolic alkalosis. Conversely, abnormally low aldosterone secretion (as in *Addison's disease*) can produce hyperkalemia and metabolic acidosis.

---

1. Describe how aldosterone secretion is affected by the *(a)* plasma K^+ concentration, and *(b)* plasma Na^+ concentration. Describe the effect of increased renin secretion on aldosterone secretion and on Na^+ and K^+ excretion in the urine.
2. Describe the mechanisms by which changes in plasma Na^+ concentrations result in changes in renin secretion. Draw a negative feedback loop to illustrate these mechanisms.
3. Explain why people who take diuretic drugs may suffer from hypokalemia and why they must supplement their diets with potassium.
4. Explain the mechanisms involved in the lowering of plasma K^+ concentrations by intravenous infusion of bicarbonate.

Renal Control of Acid-Base Balance

Normal arterial blood has a pH of 7.40 ± 0.05. It should be recalled that the pH number is *inversely* related to the H^+ concentration. An increase in H^+, derived from carbonic acid or the nonvolatile metabolic acids, can lower blood pH. An increase in H^+ derived from metabolic acids, however, is normally prevented from changing the blood pH by *bicarbonate buffer,* which combines with, and thus removes from solution, the excess H^+. The pH is inversely related to the logarithm of the H^+ concentration—when the H^+ concentration doubles, the pH number decreases by 0.3 (which is the log of 2). A doubling of the normal plasma H^+ would therefore decrease the blood pH to 7.10.

Respiratory and Metabolic Components of Acid-Base Balance Acid-base balance has both respiratory and metabolic components, as described in chapter 15. The respiratory component describes the effect of ventilation on arterial P_{CO_2}, and thus on production of carbonic acid. The metabolic component describes the effect of nonvolatile metabolic acids—such as lactic acid, fatty acids, and ketone bodies—on blood pH. Since these acids are normally buffered by bicarbonate, the metabolic component can be described in terms of the free HCO_3^- concentration. An increase in metabolic acids "uses up" free bicarbonate (as HCO_3^- is converted to H_2CO_3) and is thus associated with a fall in plasma HCO_3^- concentrations. A decrease in metabolic acids, conversely, is associated with a rise in free HCO_3^-.

Since the respiratory component of acid-base balance is represented by the plasma P_{CO_2} and carbonic acid concentrations, and the metabolic component is represented by the free bicarbonate concentrations, the study of acid-base balance can be considerably simplified. A normal plasma pH of 7.40 is obtained when the ratio of bicarbonate to carbonic acid concentrations is twenty to one:

$$pH \quad \alpha \quad \frac{HCO_3^- \text{ concentration}}{H_2CO_3 \text{ concentration}} = \frac{20}{1} = 7.40$$

(α means "is proportional to")

A change in this ratio results in an abnormal blood pH. Pure respiratory acidosis or alkalosis occurs when the HCO_3^- concentration is normal but the P_{CO_2} and H_2CO_3 concentration is altered. Pure metabolic acidosis or alkalosis occurs when the P_{CO_2} (and H_2CO_3 concentration) is normal but the HCO_3^- concentration is abnormal. This classification is summarized in table 16.9.

Table 16.10 When the bicarbonate (HCO_3^-) concentration is constant, the blood pH is determined by the P_{CO_2}. In this case a pH less than 7.35 is called respiratory acidosis, and a pH greater than 7.45 is called respiratory alkalosis.

P_{CO_2} (mm Hg)	H_2CO_3 (mEq/L)	HCO_3^- (mEq/L)	HCO_3^-/H_2CO_3 Ratio	pH	Condition
20	0.6	24	40/1	7.70	Respiratory alkalosis
30	0.9	24	26.7/1	7.53	Respiratory alkalosis
40	1.2	24	20/1	7.40	Normal
50	1.5	24	16/1	7.30	Respiratory acidosis
60	1.8	24	13.3/1	7.22	Respiratory acidosis

Figure 16.27 Mechanism of bicarbonate reabsorption (see text for description).

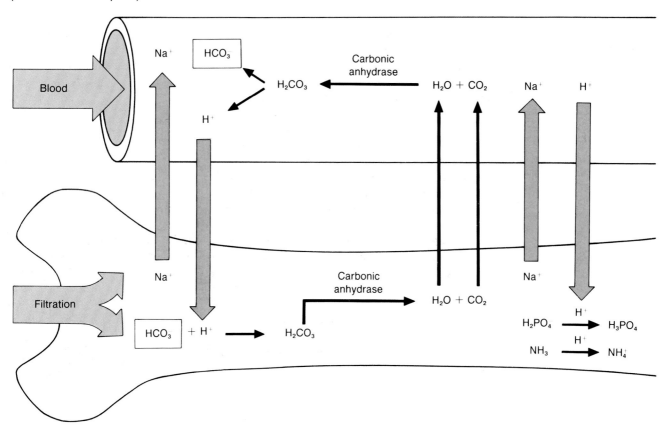

A normal blood pH is produced when the P_{CO_2} is 40 mm Hg and the bicarbonate concentration is 24 milliequivalents (mEq) per liter (one mEq is equal to one millimole times the valence of the ion—since the valence of bicarbonate is 1, in this case mEq = mmole). Table 16.10 shows how changes in P_{CO_2} alter the HCO_3^- to H_2CO_3 ratio and thus affect blood pH when the bicarbonate concentration is normal. These changes are produced by hyperventilation and hypoventilation, and result in respiratory alkalosis and respiratory acidosis, respectively.

A normal plasma P_{CO_2} is maintained by proper lung function, as described in chapter 15. A normal plasma bicarbonate concentration of 24 mEq per L is maintained, despite continuous production of metabolic acids, by proper kidney function. The kidneys normally reabsorb all of the filtered bicarbonate. During metabolic acidosis the kidneys can produce and add new bicarbonate to the blood, and during metabolic alkalosis they can excrete excess bicarbonate. Acid-base balance is thus maintained by both lung and kidney function:

$$\text{pH} \;\alpha\; \frac{HCO_3^-}{H_2CO_3} \quad \text{or} \quad \frac{\text{kidney function}}{\text{lung function}}$$

Table 16.11 Categories of disturbances in acid-base balance, including those that involve mixtures of respiratory and metabolic components. These mixtures can serve as partial compensations for the primary disturbance.

P_{CO_2} (mmHg)	Bicarbonate (mEq/L)		
	Less than 21	21–26	More than 26
More than 45	Combined metabolic and respiratory acidosis	Respiratory acidosis	Metabolic alkalosis and respiratory acidosis
35–45	Metabolic acidosis	Normal	Metabolic alkalosis
Less than 35	Metabolic acidosis and respiratory alkalosis	Respiratory alkalosis	Combined metabolic and respiratory alkalosis

Mechanisms of Renal Acid-Base Regulation

The kidneys help regulate the blood pH by excreting H^+ in the urine and by retention and production of bicarbonate. These two mechanisms are interdependent—reabsorption of bicarbonate occurs as a result of filtration and secretion of H^+. The kidneys normally reabsorb all of the filtered bicarbonate and excrete about 40–80 mEq of H^+ per day. Normal urine, therefore, is free of bicarbonate and is slightly acidic (with an average pH of 6).

The apical membranes of tubule cells are impermeable to filtered bicarbonate. Reabsorption of bicarbonate must therefore occur indirectly. When the urine is acidic, HCO_3^- combines with H^+ (which was secreted into the tubule) to form carbonic acid. Carbonic acid in the filtrate is then converted to CO_2 and H_2O by the action of *carbonic anhydrase*. This enzyme is located within the apical cell membrane, which forms microvilli that project into the filtrate. Notice that the reaction that occurs in the tubule is the same one that occurs within red blood cells of pulmonary capillaries, as described in chapter 15.

Carbon dioxide, unlike bicarbonate, can easily pass from the filtrate, through the tubule cells, and enter the blood. Once inside red blood cells the CO_2 combines with water to form carbonic acid; this reaction is also catalyzed by carbonic anhydrase and results in the addition of HCO_3^- and H^+ to the blood (see figure 16.27).

Production of Bicarbonate in Tubules

The tubule cell cytoplasm also contains carbonic anhydrase. Some of the CO_2 that enters the tubule cells from the filtrate can thus be converted to carbonic acid, which can in turn dissociate to HCO_3^- and H^+ within the tubule cells. Under acidotic conditions the H^+ produced in this way is secreted back into the filtrate, while the bicarbonate diffuses across the basolateral membranes and enters the blood. The kidneys in this way effectively add "new" bicarbonate to the blood, because the bicarbonate that enters the blood through this process is not accompanied by H^+.

During alkalosis there is less H^+ secreted into the filtrate. Since reabsorption of filtered bicarbonate requires combination with H^+ to form carbonic acid, less bicarbonate is reabsorbed. This can result in urinary excretion of bicarbonate, which helps to partially compensate for the alkalosis. In this way disturbances in acid-base balance caused by respiratory problems can be partially compensated by changes in plasma bicarbonate concentrations. Metabolic acidosis or alkalosis—in which changes in bicarbonate concentrations occur as the primary disturbance—can be similarly compensated in part by changes in ventilation. These effects result in deviations from the normal range of both P_{CO_2} and plasma bicarbonate concentrations (see table 16.11).

Urinary Buffers

In order for H^+ to be excreted in the urine without making the urine damagingly (and perhaps painfully) acidic, the acid must be buffered. Bicarbonate cannot serve this function because it is normally completely reabsorbed. Instead, the urine pH is usually prevented from falling below a value of about 4.6 by the buffering action of phosphates (mainly $H_2PO_4^-$) and ammonia (NH_3). Phosphate enters the urine by filtration, as previously described. Ammonia (whose presence is strongly evident in a diaper pail or kitty litter box) is produced in the tubule cells by deamination of amino acids. These molecules buffer H^+ as described by the following equations:

$$NH_3 + H^+ \longrightarrow NH_4^+ \text{ (ammonium ion)}$$
$$H_2PO_4^- + H^+ \longrightarrow H_3PO_4$$

1. Suppose a person has an arterial blood pH of 7.25 and a P_{CO_2} of 40 mm Hg. Describe the bicarbonate concentration (low, high, or normal) and explain the reason this would be true.
2. Explain how the kidneys reabsorb bicarbonate and how they can help raise the bicarbonate concentration of the blood.
3. Suppose a person has an arterial pH of 7.30, an abnormally low arterial P_{CO_2}, and an abnormally low bicarbonate concentration. Explain how these values might have been produced.

Table 16.12   Actions and effects of various types of diuretics.

Diuretic	Mechanisms of Action	Effect on K^+
Water, alcohol	Inhibit secretion of ADH so collecting ducts are less permeable to water	Slight or none
Furosemide, ethacrynic acid, thiazides	Inhibit active Cl^- transport in ascending limb so osmotic gradient in medulla is decreased	Increased K^+ excretion
Mannitol	Acts as osmotic diuretic, reducing water reabsorption in proximal tubule and loop of Henle (descending limb)	Slightly increased K^+ excretion
Acetazolamide	Inhibits carbonic anhydrase—has little effect in kidneys, but helps decrease eye pressure in treatment for glaucoma	None
Spironolactones	Inhibits action of aldosterone in distal tubule, reducing Na^+ reabsorption and K^+ secretion	Decreased K^+ excretion

Clinical Applications

The importance of kidney function in maintaining homeostasis, and the ease with which urine can be collected and used as a mirror of the plasma's chemical composition, make the clinical study of renal function and urine composition particularly significant. Further, the ability of the kidneys to regulate blood volume is exploited clinically in the management of high blood pressure.

Diuretics

There are a number of mechanisms by which substances can act as diuretics and increase urine volume. These mechanisms include (1) inhibition of ADH secretion (water, caffeine, and alcohol, for example); (2) inhibition of active Cl^- transport in the ascending loops (the thiazides and furosemide); (3) inhibition of aldosterone action; (4) inhibition of carbonic anhydrase; and (5) osmotic diuresis.

The thiazides and furosemide are the most potent and most commonly prescribed diuretics, as previously described. An increased concentration of solutes in the glomerular ultrafiltrate can act as osmotic diuretics. In diabetes mellitus, for example, urinary excretion of glucose and ketone bodies increase the obligatory water loss and cause polyuria. This effect is sometimes used clinically by administration of mannitol—a monosaccharide that cannot be reabsorbed at all by the tubules. Like the effect of thiazides, osmotic diuretics result in excessive secretion of K^+ into the distal tubule and may therefore cause hypokalemia.

Urinary loss of K^+ is sometimes controlled by combining *spironolactones* with other diuretics. Spironolactones inhibit aldosterone-stimulated Na^+ reabsorption and K^+ secretion in the distal tubules. They therefore reduce the loss of K^+ from the body and act as mild diuretics. The actions of different diuretics, and their effect on blood K^+, is summarized in table 16.12.

Renal Function Tests and Kidney Disease

Renal function can be tested by techniques that include the PAH clearance rate, which measures total blood flow to the kidneys, and the measurement of GFR by the inulin clearance rate. The plasma creatinine concentration, as previously described, also provides an index of renal function. These tests aid the diagnosis of kidney diseases such as glomerulonephritis and renal insufficiency.

Glomerulonephritis   Inflammation of the glomeruli, or glomerulonephritis, is currently believed to be an *autoimmune disease* that involves the person's own antibodies (as described in chapter 17). These antibodies may have been raised against the basement membrane of the glomerular capillaries, but more commonly, they appear to have been produced in response to streptococcus infections (such as strep throat). A variable number of glomeruli are destroyed in this condition, and the remaining glomeruli become more permeable to plasma proteins. Leakage of proteins into the urine results in decreased plasma colloid osmotic pressure and can therefore lead to edema (as discussed in chapter 12).

Renal Insufficiency   When nephrons are destroyed—as in chronic glomerulonephritis, infection of the renal pelvis *(pyelonephritis),* and loss of a kidney—or when kidney function is reduced by either arteriosclerosis or blockage by kidney stones, a condition of *renal insufficiency* may develop. This can cause hypertension, due primarily to retention of salt and water, and uremia (high plasma urea concentrations). The inability to excrete urea is accompanied by elevated plasma H^+ (acidosis) and elevated K^+, which are more immediately dangerous than the high urea. Uremic coma appears to result from these associated changes.

Patients with uremia, or the potential for developing uremia, are often placed on *dialysis* machines. The term *dialysis* refers to the separation of molecules on the basis of size by their ability to diffuse through an artificial semipermeable membrane. Urea and other wastes can easily pass through the membrane pores while plasma proteins are left behind (just as occurs across glomerular capillaries). The plasma is thus cleansed of these wastes as they pass from the blood into the dialyzing fluid. Unlike the tubules, however, the dialysis membrane cannot reabsorb Na^+, K^+, glucose, and other needed molecules. These substances are prevented from diffusing through the membrane by including them in the dialysis fluid. More recent techniques include the use of the patient's own peritoneal membranes (which line the body cavity) for dialysis.

The difficulties encountered in attempting to compensate for renal insufficiency, and the many dangers presented by this condition, are stark reminders of the importance of kidney function in maintaining homeostasis. The ability of the kidneys to regulate blood volume and chemical composition in a way that can be adjusted according to the body's needs requires great complexity of function. Homeostasis is maintained in large part by coordination of these functions with those of the cardiovascular and pulmonary systems, as described in the preceding six chapters.

Summary

The Nephron and Glomerular Filtration

I. The nephron consists of tubular and vascular components.
   A. The vascular components include capillary beds called glomeruli, which produce an ultrafiltrate that enters the tubules.
   B. This filtrate enters Bowman's capsule, located in the renal cortex.
      1. Fluid from Bowman's capsule travels through the proximal convoluted tubule, also located in the cortex.
      2. The tubule then forms a loop of Henle that enters the medulla and then reemerges in the cortex.
      3. Fluid from the loop then travels through the distal convoluted tubule, which empties into collecting ducts.
      4. Collecting ducts empty their content of urine into the calyces.
   C. Blood drained from the glomeruli in the efferent arterioles passes to peritubular capillaries that surround the tubules.
II. The glomerular filtrate is formed under the hydrostatic pressure of the blood.
   A. The hydrostatic pressure of the fluid in Bowman's capsule and the colloid osmotic pressure of the plasma oppose formation of filtrate.
   B. The glomerular ultrafiltrate is identical to plasma in composition and concentration, but is almost free of protein.
   C. The glomerular filtration rate (GFR) is regulated both extrinsically and intrinsically.
      1. Sympathetic nerves stimulate constriction of the afferent arterioles, and thus decrease GFR.
      2. The GFR remains relatively constant, in the absence of sympathetic nerve effects, as blood pressure changes within an "autoregulatory range."

Reabsorption of Salt and Water

I. Approximately 65 percent of the filtered salt and water is returned to the vascular system—or reabsorbed—in the proximal tubules.
   A. Since the filtrate is isosmotic with plasma, an osmotic gradient is created by the active extrusion of salt from the filtrate.
      1. The proximal tubule cells actively transport Na^+ ions into the peritubular blood, and Cl^- follows passively.
      2. The concentration difference created in this way draws water out of the filtrate by osmosis.
   B. Reabsorption of salt and water in the proximal tubule, together with reabsorption in the loop of Henle, occurs at a constant rate.
      1. Reabsorption in these "early regions" of the nephron accounts for 85 percent of the filtered volume.
      2. This reabsorption is not affected by hormones.
II. Salt and water transport in the loop of Henle form a countercurrent multiplier system that makes the renal medulla very hypertonic.
   A. The ascending limb actively extrudes Cl^- into the surrounding tissue fluid of the medulla, and Na^+ follows passively.
      1. Since the ascending limb is impermeable to water, the fluid that ascends to the distal tubule is hypotonic.
      2. Active salt transport from the ascending limb creates a hypertonic environment around the descending limb of the loop.
   B. Since the descending limb is passively permeable to water and salt, fluid in this region becomes increasingly hypertonic as it travels down into the medulla.
   C. The fluid that reaches the ascending limb is thus hypertonic—this increases the amount of salt that the ascending limb can extrude.
      1. Since salt is recycled between the ascending and descending limbs, the renal medulla remains very hypertonic.

2. The hypertonicity of the medullary tissue fluid is maintained by the vasa recta through countercurrent exchange, which minimizes the amount of salt carried away to the general circulation.

D. The hypertonicity of the medulla may also be increased by recycling of urea between the ascending limb and the collecting duct.

III. The collecting ducts are permeable to water but not to salt under the influence of antidiuretic hormone (ADH).

A. As hypotonic fluid in the collecting duct passes through the hypertonic medulla, water is drawn out of the ducts by osmosis and enters peritubular capillaries.

B. Antidiuretic hormone increases the permeability of the collecting ducts to water.

1. When ADH concentrations are low, less water is reabsorbed and a large volume of dilute urine is excreted—this occurs when a person is well hydrated and in diabetes insipidus.

2. When ADH concentrations are high, a small volume of hypertonic urine is excreted; at a maximum concentration of 1400 mOsm, about 400 ml of urine are excreted per day—this is the obligatory water loss.

Renal Clearance Rates

I. The kidneys excrete waste products by filtration into the tubules and by secretion.

A. The term *secretion* refers to active transport of a substance from the peritubular blood to the tubular fluid.

B. Waste products that are excreted in the urine are cleared from the blood to variable degrees.

C. When a substance is cleared from the blood, its concentration in the renal vein is lower than its concentration in the renal artery.

II. The glomerular filtration rate is measured by the inulin clearance rate.

A. Since inulin is filtered, but neither reabsorbed nor secreted, its clearance rate equals the filtration rate.

B. The clearance rate indicates the volume of plasma that contains the amount of a substance that is excreted per minute.

1. When a substance is reabsorbed, its clearance rate is less than the filtration rate.

2. When a substance is secreted, its clearance rate is greater than the filtration rate.

C. Since about 40 percent of the filtered urea is reabsorbed, only 60 percent of the plasma filtered per minute (60 percent $\times$ GFR) is cleared; this is about 75 ml per minute.

D. Almost all of the para-aminohippuric acid (PAH) contained in the renal blood is cleared because it is all secreted into the tubules; PAH clearance can thus be used to measure total renal blood flow.

E. Glucose and amino acids are not normally cleared from the blood because they are completely reabsorbed.

1. When the blood glucose is in the normal range, the carriers are not saturated.

2. At the renal plasma threshold for glucose, some glucose appears in the urine because carriers are in some nephrons saturated.

Renal Control of Electrolyte and Acid-Base Balance

I. In the distal tubules, Na^+ is reabsorbed and both K^+ and H^+ are secreted.

A. These processes are stimulated by aldosterone, a steroid hormone secreted by the adrenal cortex.

B. Aldosterone secretion increases in direct response to high K^+ concentration in the plasma, and as an indirect result of low plasma Na^+.

1. When the plasma Na^+ is low, blood volume falls.

2. A fall in blood volume indirectly affects the juxtaglomerular apparatus.

II. The juxtaglomerular apparatus is the region where the afferent arteriole is in contact with the distal tubule.

A. Granular cells in the afferent arteriole secrete renin in response to a fall in renal blood flow and in response to sympathetic stimulation.

B. Renin catalyzes the formation of angiotensin I, which is converted into angiotensin II.

1. Angiotensin II stimulates the adrenal cortex to secrete aldosterone.

2. Aldosterone promotes Na^+ reabsorption and thus helps to raise the blood volume.

C. The region of distal tubule in contact with the granular cells is called the macula densa; this region may inhibit renin secretion when there is an increase in filtered Na^+.

III. Secretion of H^+ indirectly promotes the reabsorption of bicarbonate (HCO_3^-).

A. Carbonic acid formed in the tubular fluid is converted to CO_2 and H_2O by carbonic anhydrase.

1. The CO_2 can diffuse into the blood, where it is used to produce another molecule of carbonic acid.

2. The tubule cells can also produce carbonic acid, and release the HCO_3^- from this into the blood while secreting the H^+ back into the filtrate.

3. The kidneys thus help to regulate the blood bicarbonate concentration and to excrete acid.

B. The acid remaining in the urine is buffered by ammonia and phosphates.

C. Regulation of plasma bicarbonate concentrations by the kidneys controls the metabolic component of acid base balance.

1. Under normal conditions, all of the filtered bicarbonate is reabsorbed.

2. Under acidotic conditions, HCO_3^- is released from the tubules into the blood while H^+ is excreted.

3. Under alkalotic conditions, bicarbonate is excreted in the urine and H^+ is retained in the blood.

D. Disturbances in acid-base balance can affect electrolyte balance.

1. Secretion of H^+ in place of K^+ may raise the plasma K^+ concentrations.

2. Under alkalotic conditions with less H^+ secretion, more K^+ is lost in the urine.

Self-Study Quiz

Match the following:
1. Active transport of Na^+; water follows passively
2. Active transport of Cl^-; impermeable to water
3. Passively permeable to salt and water
4. Passively permeable to water only

    (a) proximal tubule
    (b) descending limb
    (c) ascending limb
    (d) distal tubule
    (e) collecting duct

5. Antidiuretic hormone promotes the retention of water by stimulating:
   (a) active transport of water
   (b) active transport of Cl^-
   (c) active transport of Na^+
   (d) permeability of the collecting duct to water
6. Aldosterone stimulates Na^+ reabsorption and K^+ secretion in
   (a) the proximal tubule
   (b) the descending limb of loop of Henle
   (c) the ascending limb of loop
   (d) the distal tubule
   (e) the collecting duct
7. Reabsorption of water through the tubules occurs by:
   (a) osmosis
   (b) active transport
   (c) facilitated diffusion

8. Substance X has a clearance rate greater than zero but less than that of inulin. What can you conclude about substance X?
   (a) it is not filtered
   (b) it is filtered, but neither reabsorbed nor secreted
   (c) it is filtered and partially reabsorbed
   (d) it is filtered and secreted
9. Substance Y has a clearance rate greater than that of inulin. What can you conclude about Y?
   (a) it is not filtered
   (b) it is filtered, but neither reabsorbed nor secreted
   (c) it is filtered and partially reabsorbed
   (d) it is filtered and secreted
10. About 65 percent of the glomerular ultrafiltrate is reabsorbed in:
   (a) the proximal tubule
   (b) the loop of Henle
   (c) the distal tubule
   (d) collecting duct

11. Diuretics that inhibit active Cl^- transport would also have this effect:
   (a) the osmotic pressure of the blood would be decreased
   (b) the plasma K^+ concentration would be decreased
   (c) the urine volume would be decreased
12. Which of the following statements about glycosuria is TRUE?
   (a) Glycosuria occurs whenever a person has hyperglycemia.
   (b) Glycosuria occurs only when the plasma glucose concentrations exceeds a certain threshold value.
   (c) Glycosuria never occurs because glucose does not enter the glomerular ultrafiltrate.
   (d) Glucose is always found in normal urine.

The Immune System

Objectives

By studying this chapter, you should be able to:

1. Describe how a fever is produced, and explain the function of interferons in nonspecific immunity

2. Describe the characteristics of antigens, and define the terms *hapten* and *antigenic determinant site*

3. Distinguish between humoral and cell-mediated immunity, and describe the origin and functions of B cells and T cells

4. Describe the structure and origin of antibodies, and explain how antibodies promote destruction of invading pathogens

5. Describe the complement system, and explain how it functions together with antibodies to enhance phagocytosis

6. Describe the events that occur in a local inflammation

7. Describe the role of macrophages in the activation of lymphocytes by antigens

8. List the effects of the lymphokines, and describe how killer T cells promote the destruction of their victims

9. Describe how the ratio of helper T cells to suppressor T cells affects humoral immunity

10. Explain the functions of the thymus gland

11. Describe the process of active immunization, and describe how the clonal selection theory might explain this process

12. Define passive immunization, and give natural and clinical examples of this process

13. Define immunological tolerance, and explain the clonal deletion theory and immunological suppression theories of tolerance

14. Describe the histocompatibility antigens and explain their functional significance

15. Explain the ABO and Rh blood types and how agglutination occurs when the wrong blood types are mixed

16. Define autoimmunity and immune complex diseases, and explain how these are produced

17. Describe how immediate hypersensitization is produced and give examples of this condition

18. Describe delayed hypersensitization and give examples of this process

19. Describe the roles of natural killer cells and killer T cells in immunological surveillance against cancer; describe how tumor cells might escape this surveillance system

20. Discuss possible mechanisms that might explain the effect of aging and stress on the development of cancer

*T*he immune system includes all of the structures and processes that provide defense against potentially *pathogenic* (disease-causing) organisms such as viruses, bacteria, fungi, and worm parasites. These defenses can be grouped into two major categories: those that are **nonspecific** and those that are **specific.**

Table 17.1   Structures and defense mechanisms of nonspecific immunity.

	Structure	Mechanisms
External	Skin	Anatomic barrier to penetration by pathogens. Secretions have lysozyme (enzyme that destroys bacteria).
	Digestive tract	High acidity of stomach. Protection by normal bacterial population of colon.
	Respiratory tract	Secretion of mucus; movement of mucus by cilia; alveolar macrophages.
	Genitourinary tract	Acidity of urine. Vaginal lactic acid.
Internal	Phagocytic cells	Ingest and destroy bacteria, cellular debris, denatured proteins, and toxins.
	Interferons	Inhibit replication of viruses.
	Complement proteins	Promote destruction of bacteria and other effects of inflammation.
	Endogenous pyrogen	Secreted by leukocytes and other cells; produces fever.

The nonspecific defense mechanisms are inherited as part of the structure of each organism. They confer a "natural immunity" to some diseases—dogs and cats, for example, cannot catch the mumps—while providing first-line defenses against other diseases in a nonspecific manner. The epithelial membranes that cover the body surfaces (such as the epidermis and the lining of the digestive tract, lungs, and urogenital systems) restrict infection by most pathogens, for example. The strong acidity of gastric juice (pH 1–2) also helps to kill many microorganisms before they can invade the body. These external defenses are backed by internal defenses, such as phagocytosis, that function in both a specific and nonspecific manner (see table 17.1).

Each individual can acquire the ability to defend against specific pathogens by prior exposure to those pathogens. This specific immune response requires action by a type of leukocyte (white blood cell) known as *lymphocytes.* Internal nonspecific and specific defense mechanisms usually function together to combat infection, with lymphocytes interacting in a coordinated effort with other types of leukocytes.

Leukocytes

Unlike erythrocytes (red blood cells), which contain red hemoglobin pigment, leukocytes are almost invisible under the microscope unless they are stained. These cells are classified according to their staining properties. Those leukocytes that have well-stained granules in their cytoplasm are called **granular leukocytes,** while those that do not are **agranular** (or nongranular) **leukocytes.** The granular leukocytes are also identified by their oddly shaped nuclei, which in some cases are contorted into lobes separated by thin strands. The granular leukocytes are therefore also known as **polymorphonuclear leukocytes** (abbreviated **PMN**).

The stain used to identify leukocytes under a microscope is usually a mixture of a pink-to-red stain (eosin) and a blue-to-purple stain, which is known as a "basic" stain in the jargon of histology because it is positively charged. Granular leukocytes with pink-staining granules are therefore called *eosinophils,* while those with blue-staining granules are called *basophils.* Those with granules that have very little affinity for either stain are *neutrophils.* Neutrophils are the most abundant type of leukocyte, comprising 50 to 70 percent of the leukocytes in the blood.

There are two types of agranular leukocytes: *lymphocytes* and *monocytes.* Lymphocytes are usually the second most numerous type of leukocyte; they are small cells with round nuclei and little cytoplasm. Monocytes, in contrast, are the largest of the leukocytes and generally have kidney-to-horseshoe-shaped nuclei. In addition to these two cell types, there are a smaller number of large lymphocytes, or *plasma cells,* that are often difficult to distinguish from monocytes. The appearance and function of different types of leukocytes are summarized in table 17.2.

Table 17.2   Characteristics of the leukocytes (white blood cells).

Category	Leukocyte	Appearance	Percent of Total	Major Functions
Polymorphonuclear or granular	Neutrophils	Granules have weak affinity for stain.	55–75	Phagocytosis
	Eosinophils	Granules stain pink to red.	2–4	Phagocytose antigen-antibody complexes in allergic reactions
	Basophils	Granules stain blue to purple.	0.5–1.0	Release histamine during inflammation and allergic reactions
Agranular	Lymphocytes	No granules; round nucleus	20–40	Specific immune response— secrete antibodies (B cells) and produce cell-mediated immunity (T cells)
	Monocytes	No obvious granules; nucleus is kidney to horseshoe shaped.	3–8	Phagocytic; can be transformed into macrophages

Figure 17.1   Leukocytes, erythrocytes, and blood platelets.

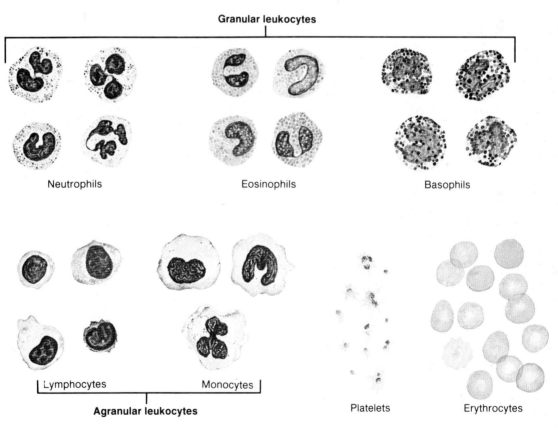

Granular leukocytes

Neutrophils   Eosinophils   Basophils

Lymphocytes   Monocytes

Agranular leukocytes

Platelets   Erythrocytes

Figure 17.2 Diapedesis. White blood cells squeeze through openings between capillary endothelial cells to enter underlying connective tissues.

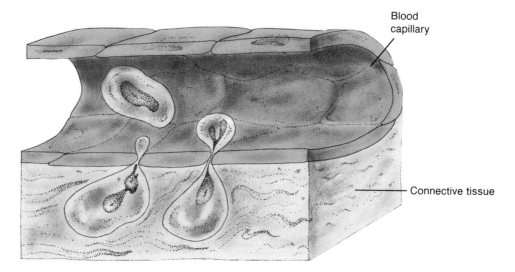

Blood capillary

Connective tissue

Table 17.3 Phagocytic cells and their location.

Phagocyte	Location
Neutrophils	Blood and all tissues
Monocytes	Blood and all tissues
Tissue macrophages (histiocytes)	All tissues (including spleen, lymph nodes, bone marrow)
Kupffer cells	Liver
Alveolar macrophages	Lungs
Microglia	Central nervous system

Nonspecific Immunity

The invading pathogens that have crossed the epithelial barriers enter connective tissues. These invaders—or chemicals, called *toxins,* secreted from them—may enter blood or lymphatic capillaries and be carried to other areas of the body. The invasion and spread of infection is fought in two stages: (1) first, nonspecific immunological defenses are employed; if these are sufficiently effective the pathogens may be destroyed without recourse to specifically acquired immunological defenses; (2) second, lymphocytes may be recruited and their specific actions used to reinforce the previously nonspecific immune defenses.

Phagocytosis

One of the most important nonspecific defense mechanisms is phagocytosis. There are two major groups of phagocytic cells: (1) neutrophils and (2) the cells of the *mononuclear phagocyte system*. This latter category includes monocytes in the blood, *macrophages* (literally, "big eaters," which are derived from monocytes) in the connective tissues, and organ-specific phagocytes in the liver, spleen, lymph nodes, lungs, and brain (see table 17.3).

Connective tissues contain a resident population of all leukocyte types. Neutrophils and monocytes in particular can be highly mobile within the connective tissue as they scavenge for invaders and for cellular debris. These leukocytes are recruited to the site of an infection by a process known as *chemotaxis*—movement towards chemical attractants. Neutrophils are the first to arrive at the site of an infection; monocytes arrive later, and can be transformed into macrophages as the battle progresses.

If the infection is sufficiently large, new phagocytic cells from the blood may join those already in the connective tissue. These new neutrophils and monocytes are able to squeeze through the tiny gaps between adjacent endothelial cells in the capillary wall and enter the connective tissue. This process is called **diapedesis** (shown in figure 17.2).

Figure 17.3 Phagocytosis by a neutrophil or macrophage. Phagocytic cell extends pseudopods around object to be engulfed (such as a bacterium). Dots represent lysosomal enzymes (L = lysosomes). If the pseudopods fuse to form a complete food vacuole (1), lysosomal enzymes are restricted to the organelle formed by the lysosome and food vacuole. If the lysosome fuses with the vacuole before fusion of the pseudopods is complete (2), lysosomal enzymes are released into the infected area of tissue.

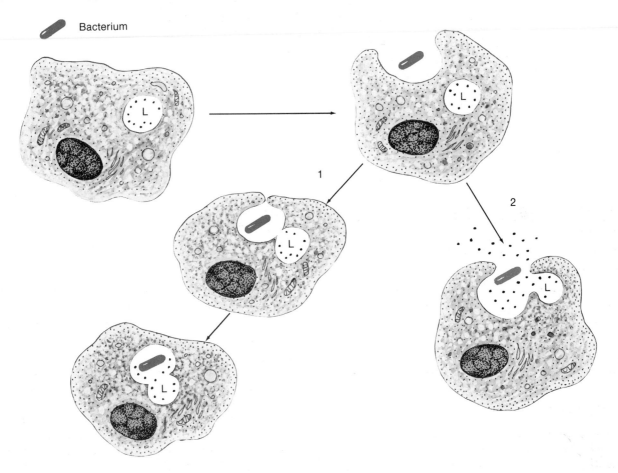

Bacterium

1

2

Phagocytic cells engulf particles in a manner similar to the way an amoeba eats. The particle becomes surrounded by cytoplasmic extensions called pseudopods, which ultimately fuse together. The particle, together with material from its immediate environment, thus becomes surrounded by a membrane derived from the cell membrane (see figure 17.3). The cytoplasmic movements required in phagocytosis are believed to be produced by actin and myosin proteins.

The ingested particle thus becomes contained within an organelle analogous to the "food vacuole" in an amoeba. This vacuole then fuses with lysosomes (organelles that contain digestive enzymes), so that the ingested particle and the digestive enzymes remain separated from the cytoplasm by a continuous membrane. Often, however, lysosomal enzymes are released before the food vacuole has been completely formed. When this occurs, free lysosomal enzymes may be released into the infected environment and contribute to inflammation.

The *Kupffer cells* in the liver, together with phagocytic cells in the spleen and lymph nodes, are **fixed phagocytes.** This term refers to the fact that these cells are immobile ("fixed") in the blood and lymph channels within these organs. As blood flows through the liver and spleen, and as lymph percolates through the lymph nodes, foreign chemicals and debris are removed by phagocytosis and chemically inactivated within the phagocytic cells. Invading pathogens are very effectively removed by these fixed phagocytes, so that blood is usually sterile after a few passes through the liver and spleen.

Fever

Fever may be a component of the nonspecific defense system. Body temperature is regulated by the hypothalamus, which contains a thermoregulatory control center (a "thermostat") that coordinates skeletal muscle shivering and activity of the sympathoadrenal system to maintain body temperature at about 37°C. This thermostat is reset upwards in response to a chemical called *endogenous pyrogen* (pyro = fire) secreted by leukocytes. Endogenous pyrogen secretion is stimulated by a chemical called *endotoxin,* which is released by certain bacteria.

Although high fevers are definitely dangerous, many believe that mild to moderate fevers may be a beneficial response that aids recovery from bacterial infections. There is some evidence to support this view, but the mechanisms that may be involved are not clearly understood. One interesting theory is that elevated body temperature may interfere with the nutritional requirements of some bacteria.

During a fever, the plasma iron concentration is reduced, while the bacterial requirements for iron are increased. The nutritional deficiency that results may weaken the bacteria or reduce their ability to multiply. These mechanisms, and indeed the entire question of whether a fever can be beneficial, are, however, controversial.

Interferons

In 1957, researchers demonstrated that cells infected with a virus produced polypeptides that "interfered with" the ability of a second, unrelated strain of virus to infect other cells in the same culture. These **interferons,** as they were called, thus produced a nonspecific (but transient) resistance to viral infection. This discovery produced a great deal of excitement, but further research in this area was hindered by the fact that human interferons could only be obtained in very small quantities, and animal interferons had little effect in humans. In 1980, however, various technological breakthroughs allowed researchers to introduce human interferon genes into bacteria—through a technique called *genetic recombination*—so that the bacteria could act as interferon "factories."

Figure 17.4 Sequence of events that can occur when human cells are infected with virus particles.

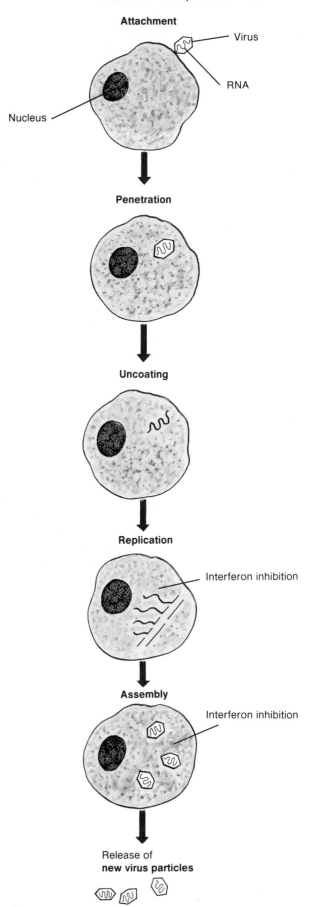

Table 17.4 Some of the proposed effects of interferons.

Stimulation	Inhibition
Macrophage phagocytosis	Cell division
Activity of cytotoxic ("killer") T cells	Tumor growth
Activity of natural killer cells	Maturation of adipose cells
Production of antibodies	Maturation of erythrocytes

Leukocytes, fibroblasts, and probably many other cells make their own characteristic type of interferons. These polypeptides act as messengers that protect other cells in the vicinity from viral infection. The viruses are still able to penetrate these other cells, but the ability of the viruses to replicate is reduced by interferons (see figure 17.4). This is apparently due to interferon-induced production of two host cell enzymes: one enzyme destroys viral mRNA, and the other enzyme prevents the initiation of viral protein synthesis.

Certain lymphocytes called *T cells* (discussed in more detail in a later section) release interferons in response to viral infections and perhaps as part of their immunological surveillance against cancer. Interferons may destroy cancer cells by both a direct effect on the cancer cells and by activation of other cells (including *natural killer cells,* discussed later in this chapter), which in turn attack the cancer cells. Some of the proposed effects of interferons are summarized in table 17.4.

Recent clinical trials have suggested that interferons may be effective in the treatment of such viral infections as viral hepatitis and herpes-induced "cold sores." While it is not currently believed that interferons will provide a "magic bullet" against cancer, it is hoped that they may significantly augment treatment of some forms of cancer. Notice that the emphasis is on treatment rather than prevention; while interferons could theoretically prevent the common cold and other viral infections, the costs involved would be prohibitive because of the short-term nature of their protection.

Acquired (Specific) Immunity

In 1890, von Behring demonstrated that a guinea pig which had been previously injected with a sublethal dose of diphtheria toxin could survive subsequent injections of otherwise lethal doses of that toxin. Further, von Behring demonstrated that this immunity could be transferred to a second, non-exposed animal by injections of serum from the immunized guinea pig. He concluded that the immunized animal had chemicals in its serum—which he called **antibodies**—that were responsible for the immunity. He also showed that these antibodies conferred immunity only to subsequent diphtheria infections; the antibodies were *specific* in their actions. It was later learned that antibodies are *proteins* produced by a particular type of lymphocyte.

Antigens

Antigens are molecules that stimulate antibody production, and which combine with specific antibodies. Most antigens are large molecules (such as proteins) with a molecular weight greater than about 10,000, and are foreign to the blood and other body fluids (although there are exceptions to both descriptions). The ability of a molecule to function as an antigen depends not only on its size but also on the complexity of its structure. Proteins are therefore more antigenic (are more powerful antigens) than large polysaccharides that have a simpler structure.

A large, complex foreign molecule can have a number of **antigenic determinant sites,** which are areas of the molecule that can specifically combine with different antibodies. Molecules that have more of these sites are better antigens, in terms of stimulating antibody production, than simpler molecules. Many plastics used for surgical tubing and artificial prostheses, for example, are not antigenic because of their simple, repeating structure.

Haptens   Many small organic molecules are not antigenic by themselves, but can become antigens if they bond to proteins (and thus become antigenic determinant sites on the proteins). This was discovered by Karl Landsteiner, the same man who discoverd the ABO blood groups (discussed later). By bonding these small molecules—which Landsteiner called *haptens*—to proteins in the laboratory, new antigens could be created for research or diagnostic purposes. Bonding of foreign haptens to the person's own proteins can also occur in the body; by this means, derivatives of penicillin, for example, that would otherwise be harmless can produce fatal allergic reactions in susceptible people.

Figure 17.5 Immunoassay using agglutination technique. Antibodies against a particular antigen are adsorbed to red blood cells or latex particles. When these particles are mixed with a solution that contains the appropriate antigen, the formation of antigen-antibody complexes produces clumping (agglutination) of the particles that can be seen with the unaided eye.

Antibodies (anti-X) attached to particles

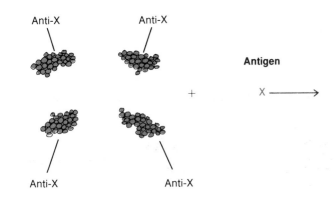

Agglutination (clumping) of particles

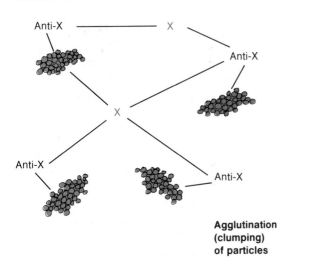

Immunoassays When the antigen or antibody is attached to the surface of a cell or to particles of latex rubber (in commercial diagnostic tests), the antigen-antibody reaction becomes visible because the cells or latex particles *agglutinate* (clump) as a result of antigen-antibody bonding. These agglutination reactions can be used to detect (assay) a variety of antigens, and so tests that utilize this procedure are called *immunoassays*. Blood typing and modern pregnancy testing are examples of such immunoassays.

Lymphocytes

Leukocytes, erythrocytes, and blood platelets are all ultimately derived *(stem)* from unspecialized cells in the bone marrow. These *stem cells* are said to be *pluripotent* because they can divide to produce different specialized cell types (they also replace themselves by cell division, so that the stem cell population is not exhausted). Lymphocytes derived in this manner "seed" the thymus, spleen, and lymph nodes after birth, producing self-replacing lymphocyte colonies in these organs.

The lymphocytes that become seeded in the thymus are called **T cells.** These cells have surface characteristics and an immunological function that is different from other lymphocytes. The thymus, in turn, seeds other organs; about 65 to 85 percent of the lymphocytes in blood, and most of the lymphocytes in lymph nodes, are T cells. Under appropriate stimulation (from antigens and thymus hormones) these cells divide to produce more T cells, resulting in self-replacing colonies. T cells therefore come from, or have an ancestor that came from, the thymus gland.

Lymphocytes that are not T cells are called **B cells.** The letter *B* is derived from immunological research performed in chickens. Chickens have an organ called the *bursa of Fabricius* at the hind end of their digestive tract that is the equivalent of the thymus located at the anterior end. The thymus processes T cells; the bursa processes B cells (in a chicken). Since mammals do not have a bursa, and since the site of processing of B cells in mammals is not known, the *B* is often translated as the "bursa equivalent" for humans and other mammals.

Both the B and T cells function in specific immunity. The B cells combat bacterial and some viral infections by secreting antibodies into blood and lymphatic fluid—they are therefore said to provide **humoral immunity** (the term *humor* is used in its ancient sense to mean fluid). The T cells attack host cells that have become infected with viruses or fungi, transplanted human cells, and cells that have become cancerous. The T cells do not secrete antibodies; they must come into close proximity or have actual physical contact with the victim cell in order to destroy it. T cells are therefore said to provide **cell-mediated immunity.**

Figure 17.6 Derivatives of "stem cells" in the bone marrow form leukocytes, erythrocytes, and macrophages.

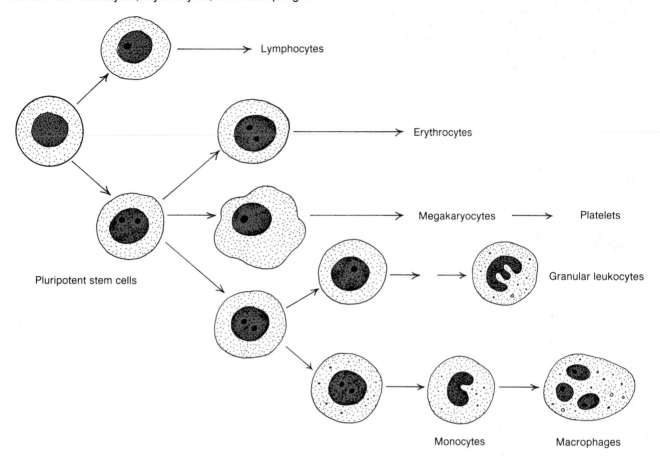

Table 17.5 Comparison of B and T lymphocytes.

Characteristic	B Lymphocyte	T Lymphocyte
Site where processed	Bursa of Fabricius or equivalent in mammals	Thymus
Type of immunity	Humoral (secretes antibodies)	Cell-mediated
Subpopulations	Memory cells and plasma cells	Cytotoxic (killer) T cells, helper cells, suppressor cells
Presence of surface antibodies	Yes—IgM or IgD	Not detectable
Receptors for antigens	Present—are surface antibodies	Present—nature unknown
Life span	Short	Long
Tissue distribution	High in spleen, low in blood	High in blood and lymph
Percent of blood lymphocytes	10–15%	75–80%
Transformed by antigens to:	Plasma cells	Small lymphocytes
Secretory product	Antibodies	Lymphokines
Immunity to viral infections	Enteroviruses, poliomyelitis	Most others
Immunity to bacterial infections	Streptococcus, staphylococcus, many others	Tuberculosis, leprosy
Immunity to fungal infections	None known	Many
Immunity to parasitic infections	Trypanosomiasis, maybe to malaria	Most others

Figure 17.7   B lymphocytes have antibodies on their surface that function as receptors for specific antigens. Interaction of antigens and antibodies on the surface stimulates cell division and the maturation of the B cell progeny into memory cells and plasma cells. Plasma cells produce and secrete large amounts of the antibody (note the extensive rough endoplasmic reticulum in these cells).

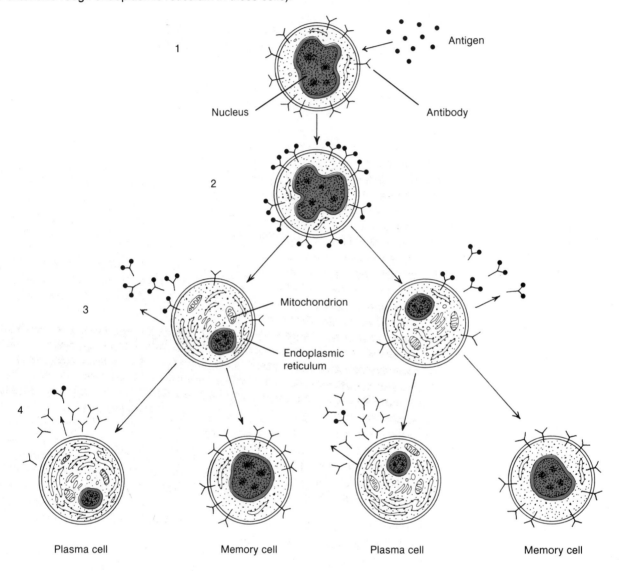

Figure 17.8 Separation of serum protein by electrophoresis. A = albumin, Alpha–1 (α_1), alpha–2 (α_2), beta (β), and gamma (γ) globulin.

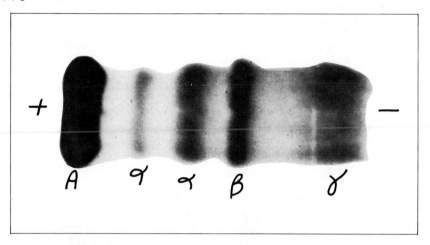

1. List the phagocytic cells of blood and connective tissues, and indicate which organs contain fixed phagocytes. Explain why phagocytosis is included as one of the nonspecific immune mechanisms.
2. Describe the actions of interferons, and explain why interferons are included in the nonspecific immune system.
3. Describe four properties that antigens usually have. Explain why proteins are more antigenic than polysaccharides.
4. Explain the meaning of the term *hapten*, using allergy to penicillin as an example.
5. Distinguish between B and T cells in terms of their origin and immune function.

Humoral Immunity

Exposure of a B cell to the appropriate antigen results in growth of the lymphocyte, followed by many cell divisions. Some of the progeny become **memory cells** that are indistinguishable from the original cell; others are transformed into **plasma cells.** Plasma cells are protein factories that produce about two thousand antibody proteins per second in their brief (five-to-seven day) life span.

The antibodies that are produced by plasma cells when B cells are exposed to a particular antigen react specifically with that antigen. Such antigens may be isolated molecules (toxins or haptens on isolated proteins) or may be molecules at the surface of an invading foreign cell. The specific bonding of antibodies to antigens serves to identify the enemy and to activate defense mechanisms that lead to the invader's destruction.

Antibodies

Antibodies—the proteins secreted by plasma cells—are also known as **immunoglobins.** These form the gamma globulin class of plasma proteins, as identified by a technique known as *electrophoresis* in which the classes of plasma proteins are separated by their rate of movement in an electric field. When this technique is employed, five distinct bands of proteins, corresponding to classes of plasma proteins, are seen: albumin, alpha-1-globulin, alpha-2-globulin, beta globulin, and gamma globulin (see figure 17.8).

The gamma globulin band is wide and diffuse because it represents a heterogenous class of molecules. Since antibodies are specific in their actions, it would be predicted that different types of antibodies would have different structures. An antibody against smallpox, for example, does not confer immunity on poliomyelitis and therefore must have a slightly different structure than an antibody against polio. Despite these differences, antibodies are structurally related and form only a few subclasses.

There are five subclasses of immunoglobins (abbreviated Ig): *IgG, IgA, IgM, IgD,* and *IgE*. Antibodies in each of these subclasses share similar biological functions but can have varied specificities for antigens. Most of the antibodies in serum, for example, are in the IgG subclass, while most antibodies in external secretions (saliva and milk, for example) are IgA. Antibodies in the IgE subclass are involved in allergic reactions, as will be described in a later section.

Figure 17.9   Antibodies are composed of four polypeptide chains—two are heavy (H) and two are light (L). In *(a)* these chains are shown as sticks; different regions within these chains are shown separately in *(b)*. The variable regions are abbreviated *V*, while the constant regions are abbreviated *C*. Antigens combine with the variable regions. Each antibody molecule is divided into a Fab (antigen-binding) fragment and a Fc (crystallizable) fragment.

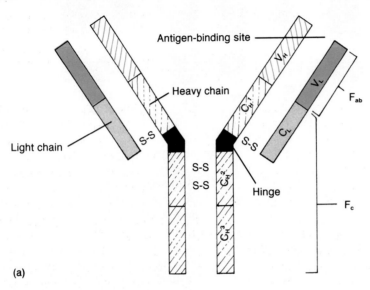

(a)

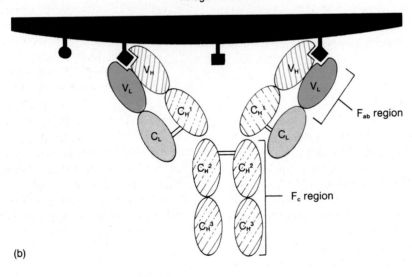

(b)

Table 17.6 The immunoglobins.

Immunoglobin	Heavy Chain	Light Chain	Examples of Functions
IgG	Gamma	Kappa or lambda	Main form of antibodies in circulation; production increased after immunization
IgA	Alpha	Same as others	Main antibody type in external secretions, such as saliva and mother's milk
IgE	Epsilon	Same as others	Responsible for allergic symptoms in immediate hypersensitivity reactions
IgM	Mu	Same as others	Function as antigen receptors on lymphocyte surface prior to immunization; secreted during primary response
IgD	Delta	Same as others	Function as antigen receptors on lymphocyte surface prior to immunization; other functions unknown

Antibody Structure All antibody molecules consist of four interconnected polypeptide chains; two longer, higher molecular weight chains (the *H chains*) are joined to two lighter, shorter *L chains*. Researchers have shown that these four chains are arranged in the form of a Y. The stalk of the Y has been called the "crystallizable fragment" (abbreviated F_c), while the top of the Y is the "antigen-binding fragment (F_{ab}).

The amino acid sequences of some antibodies have been determined using antibodies derived from people with multiple myelomas. These lymphocyte tumors are derived from division of a single B cell, forming a population of genetically identical cells called a *clone*. All of the antibodies secreted from each tumor are thus of a single type. Analysis of these *monoclonal antibodies* have shown that the F_c regions of different antibodies are the same (are *constant*), while the F_{ab} regions are *variable*. Variability of the antigen-binding regions is required for the specificity of antibodies for antigens.

Each immunoglobin subclass is characterized by its own unique constant region of the heavy chains. Substituting one of these chains for another, therefore, changes the antibody subclass. Before a B cell is stimulated by an antigen, for example, it has IgM antibodies on its surface that function as *receptors* for the antigen. After stimulation by antigens, the B cell becomes transformed and the IgM antibodies are replaced by IgG antibodies, which are secreted from the plasma cells. This replacement involves a change only in the constant part of the heavy chains; since the antigen-binding parts are unaltered, the antibody specificity remains the same when IgM is replaced by IgG.

Diversity of Antibodies It has been estimated that there are about 100 million trillion (10^{20}) antibody molecules in each individual, representing a few million different specificities for different antigens. Considering that antibodies to particular antigens can cross-react to some degree with closely related antigens, this tremendous antibody diversity usually insures that there are some antibodies that can combine with any antigen the person may encounter. These observations evoke a question that has long fascinated scientists—How can a few million different antibody proteins be produced? A person cannot possibly inherit a correspondingly large number of genes devoted to antibody production.

Two mechanisms have been proposed to explain antibody diversity. First, since different combinations of heavy and light chains can produce different antibody specificities, a person does not have to inherit a million different genes to code for a million different antibodies. If a few hundred genes code for different H chains and a few hundred code for different L chains, different combinations of these polypeptide chains could produce millions of different antibodies. Second, the diversity of antibodies could increase during development if, when some lymphocytes divided, the progeny received antibody genes that have been slightly altered by mutations. This is called *somatic mutation* since the mutation occurs in a body cell rather than in a sperm or ovum. The diversity of antibodies would thus increase as the lymphocyte population increases during development.

Figure 17.10 Fixation of complement proteins. The formation of an antibody-antigen complex causes complement protein C4 to be split into two subunits—C4a and C4b. The C4a subunit attaches (is fixed) to the membrane of the cell to be destroyed (such as a bacterium). This event triggers the activation of other complement proteins, some of which attach (are fixed) to C4a on the membrane surface.

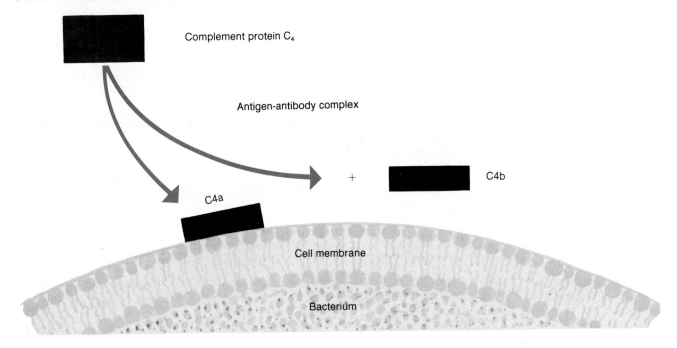

Complement protein C₄

Antigen-antibody complex

+   C4b

C4a

Cell membrane

Bacterium

Opsonization   Combination of antibodies with antigens does not itself produce destruction of the antigens or of the pathogenic organism that contains these antigens. Antibodies, rather, serve to identify the targets for immunological attack and to activate nonspecific immune processes that destroy the invader. Bacteria that are buttered with antibodies, for example, are better targets for phagocytosis by neutrophils and macrophages. The ability of antibodies to stimulate phagocytosis is termed *opsonization*. Immune destruction of bacteria is also promoted by antibody-induced activation of a system of proteins known as *complement*.

The Complement System

The complement system is an important nonspecific defense system that can be directed against specific invaders that have been "identified" by antibodies. The complement system consists of nine protein components, designated C1 through C9 (but since C1 is actually composed of three separate proteins there are actually a total of eleven proteins involved). These proteins are present in an inactive state within plasma and other body fluids and become activated by attachment of antibodies to antigens.

In terms of their relationship to the complement system, antibodies serve three functions: (1) *identification* of the invading pathogen; (2) *activation* of the complement proteins; and (3) *fixation of complement*—that is, attachment of activated complement to the membrane of invading cells. This latter function is very important because activated complement proteins can destroy the cells to which it is "fixed."

Figure 17.11 Complement proteins C5 through C9 (illustrated as a doughnut-shaped ring) puncture the membrane of the cell to which they are attached (fixed). This aids destruction of the cell.*

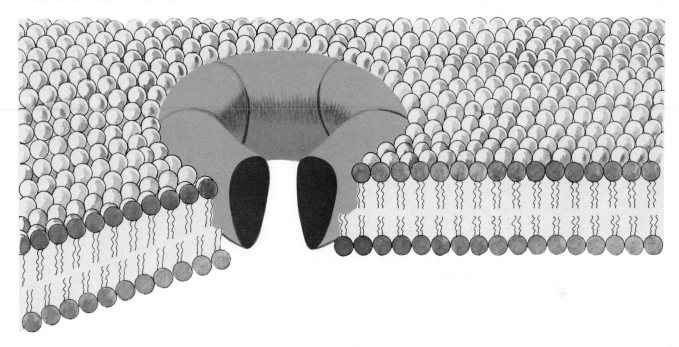

Antibodies of the IgG and IgM subclasses, which are attached to antigens on the invading cell's membrane, bond to the first complement protein complex (Cl), and by this means activate its enzyme activity. Activated Cl catalyzes hydrolysis of C4 into two fragments, designated C4a and C4b. The C4b fragment itself bonds to antigens in the cell membrane (is "fixed"), and becomes an active enzyme that splits C2 into two fragments—C2a and C2b. The C2a in turn becomes attached to C4b and becomes activated, cleaving C3 into C3a and C3b fragments. One of these fragments (C3a) is released, while the other (C3b) becomes attached to the growing complex of complement proteins. These processes are repeated for the remaining complement proteins in a "cascade" of reactions similar to those involved in the blood-clotting sequence (described in chapter 13).

Complement proteins C5 through C9 that become fixed to antigens in a cell membrane create large pores (see figure 17.11) in the membrane. These pores allow the osmotic influx of water, so that the victim cell (usually a bacterium) swells and bursts. Notice that the complement, not the antibodies, kill the bacteria—antibodies only serve as activators of this process. Other molecules can also activate the complement system in an alternate nonspecific pathway that bypasses the early phases of the specific pathway described here.

Complement fragments (particularly C3a and C5a) that are not fixed but are instead liberated into the surrounding area have a number of effects. These include (1) *chemotaxis*—C3a and C5a attract phagocytic cells (neutrophils and macrophages) to the site of complement system activation; (2) *opsonization*—like antibodies, these complement fragments enhance phagocytosis (macrophages have receptors for C3b, so that this fragment may form "bridges" between the macrophage and the pathogen); and (3) fragments C3a and C5a *stimulate mast cells* (a cell type found in connective tissues) to release various chemicals, including histamine. As a result of the release of histamine and other molecules from mast cells, there is increased blood flow to the infected area (due to vasodilation) and increased capillary permeability. The latter effect can result in leakage of plasma proteins into the surrounding connective tissues, producing local edema.

Figure 17.12   Entry of bacteria through a cut in the skin produces a local inflammatory reaction. In this reaction, antigens on the bacterial surface are coated with antibodies and ingested by phagocytic cells. Symptoms of inflammation are produced by release of lysosomal enzymes and by secretion of histamine and other chemicals from tissue mast cells (see text for more complete description).

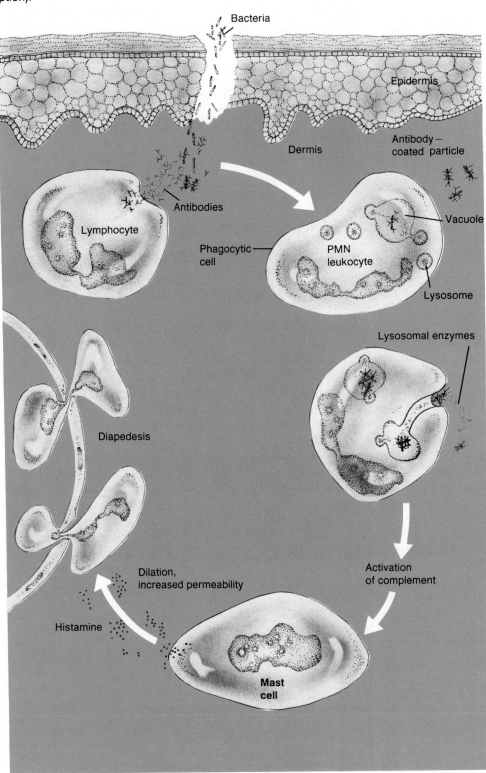

Table 17.7 Summary of events that occur in a local inflammation when a break in the skin permits entry of bacteria.

Category	Events
Nonspecific immunity	Bacteria enter through break in anatomic barrier of skin.
	Resident phagocytic cells—neutrophils and macrophages—engulf bacteria.
	Nonspecific activation of complement proteins occurs.
Specific immunity	B cells are stimulated to produce specific antibodies.
	Phagocytosis is enhanced by antibodies attached to bacterial surface antigens.
	Specific activation of complement proteins occurs, which stimulates phagocytosis, chemotaxis of new phagocytes to the infected area, and secretion of histamine from tissue mast cells.
	Diapedesis allows new phagocytic leukocytes (neutrophils and monocytes) to invade the infected area.
	Vasodilation and increased capillary permeability (as a result of histamine secretion) produce redness and edema.

Local Inflammation

Aspects of the nonspecific and specific immune responses and their interactions are well illustrated by the events that occur when bacteria enter a break in the skin and produce a local inflammation. The inflammatory reaction is initiated by nonspecific mechanisms of phagocytosis and complement system activation. Activated complement further increases this nonspecific response by attracting new phagocytes to the area (through chemotaxis) and by increasing the activity of these phagocytic cells (the process of opsonization).

After some time—a couple of hours to a few days—B lymphocytes are stimulated to produce antibodies against specific antigens that are part of the invading bacteria. Attachment of these antibodies to antigens in the bacteria greatly increases the previously nonspecific immune response. This occurs because of greater activation of complement, which directly destroys the bacteria and which—together with the antibodies themselves—promotes the phagocytic activity of neutrophils, macrophages, and monocytes.

As inflammation progresses, release of lysosomal enzymes from macrophages causes destruction of leukocytes and other tissue cells. These effects, together with those produced by release of histamine and other chemicals from mast cells in response to complement activation, produce the characteristic symptoms of a local inflammation: *redness* (due to vasodilation), *swelling* (edema), and *pus* (accumulation of dead leukocytes). If the infection continues, release of endogenous pyrogen from leukocytes and macrophages may produce a fever.

Cell-Mediated Immunity

The T lymphocytes provide specific immune protection without secreting antibodies. Unlike the B lymphocytes that can kill at a distance by secreting antibodies (like bombardiers—which also starts with the letter *B*), T lymphocytes must be near or in actual contact with their victims. This cell-mediated immunity is provided by **killer T cells,** which defend against viral and fungal infections, as well as some bacterial infections. Killer T cells are also responsible for rejection of transplanted tissue and for immunological surveillance against cancer. In addition to killer cells, other subpopulations of T cells function as **helper** and **suppressor cells** that modify the activity of B lymphocytes.

Figure 17.13 The clonal selection theory. In this example, particular antigens interact with specific membrane receptors on the T cell surface. As a result of this selection process, the lymphocyte that is specific for this antigen proliferates to produce a population of genetically identical cells (a clone).

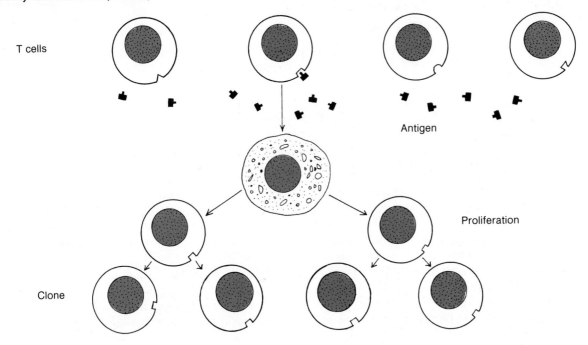

T cells

Antigen

Proliferation

Clone

While most bacterial infections are fought by B cells, as previously described, some are targets of cell-mediated attack. This is the case with the *tubercle bacili* that cause tuberculosis. If a person has become sensitized by a previous exposure to tuberculosis, injections of some of these bacteria under the skin produces inflammation after a 48-to-72-hour latent period. The ability of an unexposed guinea pig to produce this response can be transferred from an exposed animal by the latter animal's lymphocytes, but not by its serum. This demonstrates that the response to these bacteria—called a *delayed hypersensitivity reaction* (because of its long latent period) is mediated by T cells. Other examples of delayed hypersensitivity will be discussed later in the section on allergy.

Actions of T Cells

Unlike B cells, T cells do not have antibodies on their surface to act as receptors for specific antigens. The T cells do, however, have some type of receptor proteins on their surface that interact with specific antigens. This interaction stimulates the T cell to divide many times, creating a population of genetically identical cells—a clone. A similar process of clone selection occurs in B cells, as will be described in a later section dealing with immunizations.

Interaction with Macrophages In order for lymphocytes—both T cells and B cells—to be stimulated by many antigens, the antigens must first be processed by macrophages. The nature of this processing is not well understood, but since intimate contact is required between lymphocyte and macrophage, it is believed that the macrophage "presents" the antigen to the lymphocyte. Perhaps the macrophage orients the antigen in such a way as to maximize interaction with the surface receptors on the lymphocyte.

Figure 17.14 Interactions between macrophages and T cells. In this example, activation of helper T cells occurs after the foreign antigens have been "processed" by the macrophages. Interaction between T cells and macrophages requires that both cells have the same histocompatibility antigens. Once activated, the helper T cells can stimulate cytotoxic ("killer") T cells and B cells.

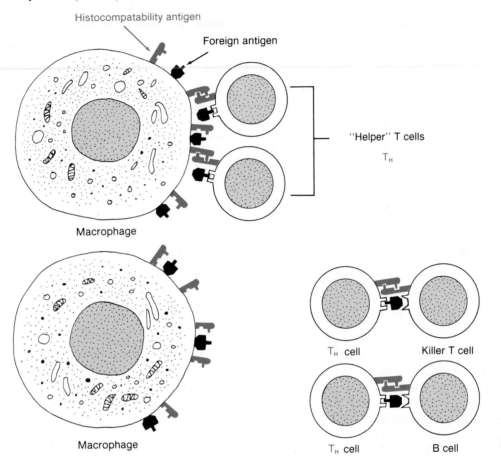

Histocompatability antigen

Foreign antigen

"Helper" T cells

T_H

Macrophage

Macrophage

T_H cell   Killer T cell

T_H cell   B cell

Macrophages from one person cannot activate lymphocytes from another. This is due to the fact that tissue cells from each person have a unique combination of surface proteins called *histocompatibility antigens,* which serve as genetic markers (and which can be used for tissue typing). Macrophage-lymphocyte interaction apparently requires proper fitting of surface receptors to these antigens (see figure 17.14).

Killer T cells and Lymphokines   Cell-mediated destruction of victim cells is due to the action of killer T cells, which secrete a family of low-molecular weight polypeptides called **lymphokines.** The lymphokines include (1) interferon; (2) lymphotoxin, which by unknown means destroys the victim cells; (3) macrophage-activating factor, which promotes increased phagocytic activity (makes macrophages more "hungry"); (4) chemicals that attract leukocytes to the area by chemotaxis; and (5) macrophage-migration inhibiting factor, which apparently prevents phagocytic cells from leaving the area. Killer T cells thus promote destruction of their victims both directly, by means of interferon and lymphotoxin, and indirectly, via activity of phagocytic cells.

Figure 17.15 A given antigen can stimulate production of both B and T cell clones. The ability to produce B cell clones, however, is also influenced by the relative effects of helper and suppressor T cells.

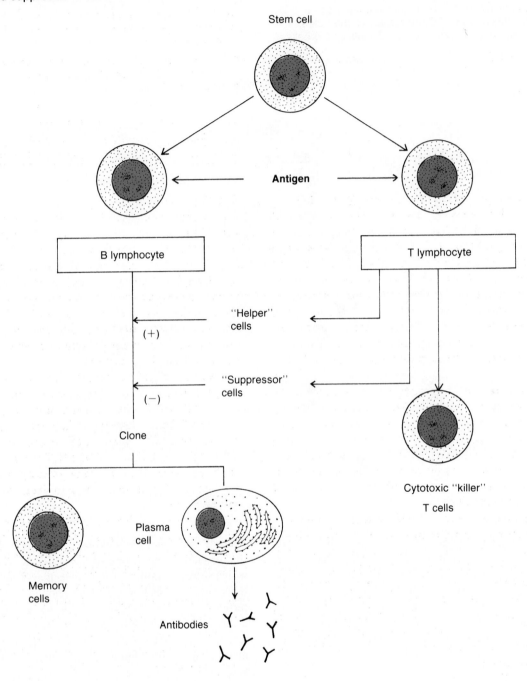

Table 17.8 Summary of different interactions that can occur between cells of the immune system.

Cell	Interacts with	Mechanism
Macrophage	T lymphocytes	Macrophages "process" or "present" antigens in such a manner that the antigen can more effectively activate T cells.
Helper T cells	B cells	Helper T cells augment the production of B cell clones in response to antigen stimulation, or in some unknown way aid the production of antibodies.
Cytotoxic (killer) T cells	Macrophages	Chemicals, called lymphokines, secreted by T cells promote macrophage activity and prevent macrophages from migrating away from the infected area.
Suppressor T cells	B cells	Suppressor T cells prevent the formation of B cell clones or in some unknown way inhibit production of antibodies.
B cells	T cells	Blocking antibodies secreted by plasma cells derived from B cells may prevent cytotoxic T cells from destroying their target cells.

Helper and Suppressor T Cells   There appears to be, in addition to killer T cells, two additional subpopulations of T cells that do not directly participate in immune destruction of invading pathogens. Instead, these subpopulations of T cells modulate the activity of B cells and thus affect humoral immunity (secretion of antibodies). Activity of B cells is increased by *helper T cells* and decreased by *suppressor T cells.* The amount of antibodies secreted by B cells in response to antigens is thus affected by the relative number of helper versus suppressor T cells that developed in response to that same antigen (see figure 17.15).

Effect of Blocking Antibodies on T Cells   Activity of the T cell system, as previously described, can affect activity of B cells and secretion of antibodies. Secretion of antibodies, in turn, can affect the activity of T cells and cell-mediated immunity. Antigens that are bound to antibodies may in some way be protected from cell-mediated attack by T cells—these *blocking antibodies* can thus inhibit cell-mediated immunity.

Proper function of the immune system, in summary, requires extremely complex and poorly understood interactions between macrophages, T cells, B cells, and antibodies. These interactions appear to provide a system of checks and balances that normally prevent immune defense mechanisms from committing offenses against the host. This intricate system of restraints (or "fail-safe" mechanisms, to use an analogy) does not always serve its function, however. Many diseases result when these safeguards have failed.

The Thymus Gland
The thymus gland extends from below the thyroid in the neck down into the thoracic cavity. This organ grows during childhood and reaches its maximum size at the time of puberty, after which time it gradually regresses with age. Lymphocytes from the fetal liver and spleen in prenatal life, and from the bone marrow postnatally, "seed" the thymus and become transformed into T cells. These cells in turn enter the blood and seed lymph nodes and other organs, where they divide to produce new T cells when stimulated by antigens.

Small T cells that have not yet been stimulated by antigens have very long life spans—months or perhaps years. Still, new T cells must be continuously produced to provide efficient cell-mediated immunity. Since the thymus atrophies after puberty, this organ may not be able to provide new T cells in later life. Colonies of T cells in the lymph nodes and other organs are apparently able to produce new T cells under the stimulation of various **thymus hormones.**

Two hormones that are believed to be secreted by the thymus—*thymopoietin I* and *thymopoietin II*—may promote the transformation of lymphocytes into T cells. This possibility is supported by evidence that lymphocytes from bone marrow that are incubated with these two hormones take on the surface antigens that are characteristic of T cells. Another thymus hormone—called *thymosin*—may promote the maturation of T lymphocytes.

Injections of thymosin into mice that have had their thymus surgically removed has been shown to restore cell-mediated immunity, and a similar effect was demonstrated in a child who had a cell-mediated immune deficiency. These observations support the view that thymus hormones are critical for T cell development. This association could have great clinical significance, because secretion of thymus hormones and cell-mediated immunity has been shown to decline with advancing age, while susceptibility to viral infections and cancer increases with age.

Figure 17.16 Active immunity to a pathogen can be gained by exposure to the fully virulent form, or by inoculation with a pathogen whose virulence (ability to cause disease) has been attenuated (reduced), but whose antigens are the same as in the virulent form.

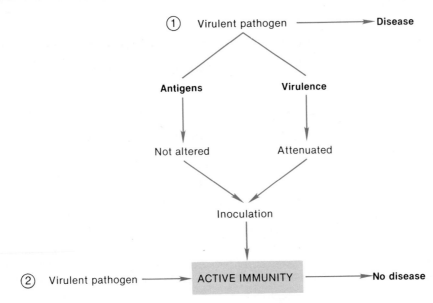

1. Illustrate the structure of an antibody molecule, labeling the constant and variable regions, the F_c and F_{ab} parts, and the heavy and light chains. Explain how the antibody specificity and immunoglobin class would be affected by variations in this structure.
2. Define opsonization, and name two types of molecules that promote this process.
3. Describe how the binding of antibodies to antigens can produce the symptoms of a local inflammation.
4. Describe the role of macrophages in the specific immune response.
5. Define the term *lymphokines*, indicating the type of cell that produces them, and list their actions.
6. Explain the function of helper and suppressor T cells.

Immunizations

It was first known in the mid-eighteenth century that the fatal effects of smallpox could be prevented by inducing mild cases of the disease. This was accomplished at that time by rubbing needles into the pustules of people who had mild forms of smallpox and injecting these needles into healthy people. Understandably, this method of immunization was extremely unpopular.

Acting on the observation that milkmaids who contracted cowpox—a disease similar to smallpox, but less dangerous (*virulent*)—were immune to smallpox, an English physician named Jenner inoculated a healthy boy with needles rubbed in the pustules of milkmaids with cowpox. When the boy recovered, Jenner inoculated him with an otherwise deadly amount of smallpox, from which he also recovered. This experiment, performed in 1796, began the first widespread (though controversial) immunization program.

A similar, but more sophisticated demonstration of the effectiveness of immunizations, was performed by Louis Pasteur almost 100 years later. Pasteur isolated the bacteria that caused anthrax and heated them until their ability to cause disease was greatly reduced (their virulence was *attenuated*). He then injected these attenuated bacteria into twenty-five sheep, leaving twenty-five unimmunized. Several weeks later, before a gathering of scientists, he injected all fifty sheep with the completely active anthrax bacteria. All twenty-five of the unimmunized sheep died—all twenty-five of the immunized sheep survived.

Figure 17.17 Comparison of antibody production in the primary response (upon first exposure to an antigen) to antibody production in the secondary response (upon subsequent exposure to the antigen). The greater secondary response is believed to be due to development of lymphocyte clones produced during the primary response.

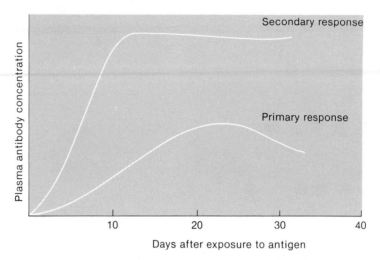

Figure 17.18 The greater antibody production in the secondary response compared to the primary response is due mainly to increased amounts of IgG antibodies. Notice that the amount of IgM antibodies is about the same in both primary and secondary responses.

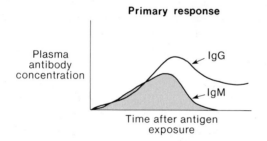

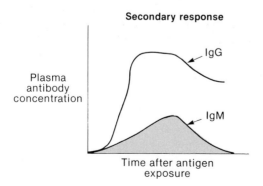

Active Immunity and the Clonal Selection Theory

When a person is exposed to a particular pathogen for the first time there is a latent period of five-to-ten days before measurable amounts of specific antibodies against the foreign antigens appear in the blood. This sluggish **primary response** may not be sufficient to protect the individual against the disease caused by the pathogen. Antibody concentrations in the blood during this primary response reach a plateau in a few days and decline after a few weeks.

Subsequent exposure of the same individual to the same antigens results in a **secondary response.** Compared to the primary response, antibody production during the secondary response is much more rapid. Maximum antibody concentrations in the blood are reached in less than two hours, and are maintained for a longer time than in the primary response. This rapid rise in antibody production to levels higher than that achieved during the primary response is usually sufficient to prevent the disease.

Before stimulation by a particular antigen, the B cells that are able to make the appropriate antibodies against that antigen produce mainly IgM antibodies. These IgM antibodies therefore account for a high proportion of the antibodies made during the primary response. In contrast, most antibodies made during the secondary response are in the IgG subclass (due to substitution of the constant heavy chains, as previously discussed). Since the improved performance of the B cells during a secondary response appears to represent a type of immune "memory," the IgG antibodies are sometimes called *memory antibodies.*

Figure 17.19 According to the clonal selection theory, exposure to an antigen stimulates the production of lymphocyte clones and the maturation of some members of B cell clones into antibody-secreting plasma cells.

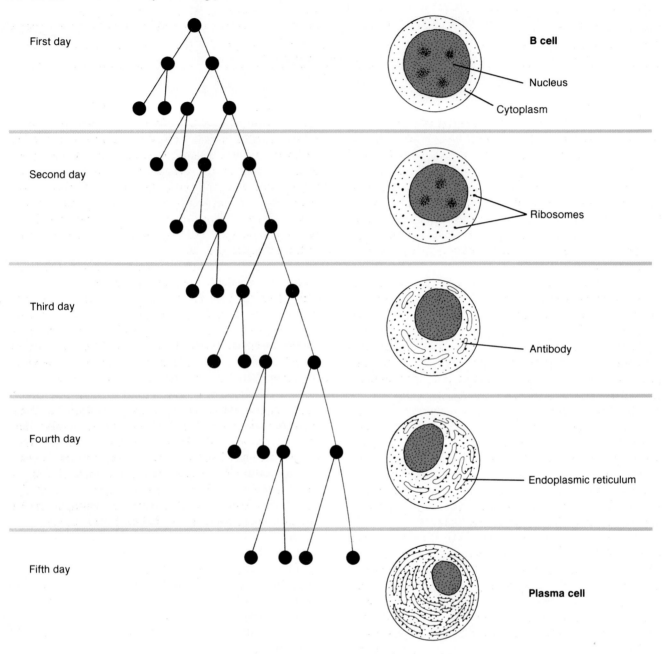

First day

Second day

Third day

Fourth day

Fifth day

B cell

Nucleus

Cytoplasm

Ribosomes

Antibody

Endoplasmic reticulum

Plasma cell

Table 17.9 Summary of the clonal selection theory (with regard to B cells).

Process	Results
Lymphocytes inherit the ability to produce specific antibodies.	Prior to antigen exposure, lymphocytes are already present in the body that can make the appropriate antibodies.
Antigens interact with antibody receptors on the lymphocyte surface.	Antigen-antibody interaction stimulates cell division and the development of lymphocyte clones containing memory cells and plasma cells that secrete antibodies.
Subsequent exposure to the specific antigens produce a more efficient response.	Exposure of lymphocyte clones to specific antigens results in greater and more rapid production of specific antibodies.

Clonal Selection Theory The immunization procedures of Jenner and Pasteur were effective because the people who were inoculated produced a secondary rather than a primary response when exposed to the virulent pathogens. This protection is not simply due to accumulations of antibodies in the blood, because secondary responses may occur even after antibodies produced by the primary response have disappeared. Immunizations, therefore, seem to produce a type of "learning," in which the ability of the immune system to combat a particular pathogen is improved by prior exposure.

The mechanisms by which secondary responses are produced are not well understood; the **clonal selection theory,** however, appears to account for most of the evidence. According to this theory, B lymphocytes *inherit* the ability to produce particular antibodies (and T cells inherit the ability to respond to particular antigens). Some lymphocytes, therefore, can respond to smallpox and produce antibodies against it even if the person has never been exposed to this disease.

Exposure to smallpox antigens stimulates these specific lymphocytes to divide many times until a large population of genetically identical cells—a clone—is produced. Some of the cells of this clone become plasma cells that secrete antibodies for the primary response; others become memory cells that can be stimulated to secrete antibodies during the secondary response.

Notice that, according to the clonal selection theory, antigens do not "cause" the production of appropriate antibodies. Rather, antigens "select" lymphocytes that are already able to make antibodies against that antigen. This is analogous to evolution by natural selection—the environmental agent (in this case, antigens) acts on the genetic diversity already present to cause increasing numbers of the individuals that are selected. Interaction between antigens and receptor molecules on the lymphocyte surface causes the lymphocyte to divide and produce a clone. In the process, the lymphocytes are "transformed"—IgM antibodies are changed into IgG antibodies.

Active Immunity The development of a secondary response, possibly by means of the clonal selection mechanism, provides *active immunity* against the specific pathogens. Development of active immunity requires prior exposure to the specific antigens, during which time a primary response is produced and the person is likely to get the disease. Some parents, for example, deliberately expose their children to others who have measles, chicken pox, and mumps so that their children will be immunized to these diseases in later life.

Clinical immunization programs induce primary responses by inoculating people with pathogens whose virulence has been attenuated or destroyed (such as Pasteur's heat-inactivated anthrax bacteria), or by use of strains that are less virulent but antigenically similar to the pathogens (such as Jenner's cowpox inoculations). These procedures provide the benefits of a secondary response to the virulent pathogens that may later be encountered.

The first successful polio vaccine (the Salk vaccine) was composed of viruses that had been inactivated by treatment with formaldehyde. These "killed" viruses were injected into the body, in contrast to the currently used oral (Sabin) vaccine. The oral vaccine contains "living" viruses that have attenuated virulence. These viruses invade the epithelial lining of the intestine and multiply, thereby sensitizing the immune system to antigens in the virulent viruses that are able to infect nerve tissue.

Table 17.10 Comparison of active and passive immunity.

Characteristic	Active Immunity	Passive Immunity
Injection of person with	Antigens	Antibodies
Source of antibodies	The person inoculated	An animal that is inoculated with the antigen; or the mother
Method	Injection with killed or attenuated pathogens or their toxins	Natural—transfer of antibodies across the placenta; artificial—injection with antibodies
Time to develop resistance	5 to 14 days	Immediately after injection
Duration of resistance	Long (perhaps years)	Short (days to weeks)
When used	Before exposure to pathogen	Before or after exposure to pathogen

Passive Immunity

Immunological competence—the ability to mount an immune response—does not develop until about a month after birth. The fetus, therefore, cannot develop active immunity and immunologically reject its mother. The immune system of the mother is fully competent but does not reject fetal antigens, for reasons that are incompletely understood (this will be discussed in chapter 20). Some IgG antibodies from the mother do cross the placenta and enter the fetal circulation, however, and these serve to confer **passive immunity** to the fetus.

The fetus and newborn baby are therefore immune to the same antigens as the mother. Since the baby did not itself produce the lymphocyte clones needed to form these antibodies, such passive immunity disappears when the infant is about one month old. If the infant is breast-fed it can receive additional antibodies of the IgA subclass in its mother's first milk (the *colostrum*).

Passive immunizations are used clinically to protect people who have been exposed to extremely virulent infections or toxins, such as snake venom. In these cases the affected person is injected with **antiserum** (serum containing antibodies), also called **antitoxin,** from an animal that has previously been exposed to the pathogen. The animal develops the lymphocyte clones and active immunity, which produce a high antibody concentration in its blood; the person who is injected with antiserum from these animals does not develop active immunity. If the person is again exposed to the same pathogen, therefore, he or she must again be passively immunized.

Continued treatment with antiserum from lower animals may produce *serum sickness*. This disease results from the fact that serum from animals (horses and sheep are commonly used) contains many proteins that are foreign and therefore antigenic to people. Although injections of antiserum may protect against a specific disease, the host may mount an immune response against the animal proteins, and this immune response itself may produce disease.

Table 17.11 Common uses of passive immunizations.

Measles	Hyperimmune globulin immediately after exposure or with weakly attenuated vaccines
Rubella	Pregnant women exposed to rubella
Infectious hepatitis	Immediately after exposure
Tetanus	Immediately after injury if not immunized or if immunization is outdated
Rh disease	Within 72 hours after delivery
Hypogammaglobulinemia	Pooled human gamma globulin

Source: From Barrett, James T.: Textbook of immunology, ed. 4, St. Louis, 1983, The C.V. Mosby Co.

1. Explain the characteristics of the primary and secondary immune responses, and draw graphs to illustrate your discussion.
2. Explain the clonal selection theory and how this theory accounts for the development of a secondary response.
3. Describe two methods used to induce active immunity. Explain how passive immunity is produced, and give natural and clinical examples of this type of immunization.

Tolerance and Non-Self Recognition

The immune system normally produces antibodies against foreign antigens but not against potential antigens that are the products of the individual's own genes. The ability to produce antibodies against **non-self**-antigens, while tolerating **self**-antigens, occurs early in life when immunological competence is established (the first month or so of postnatal life). If a fetal mouse of one strain receives transplanted antigens from a different strain, therefore, it will not recognize tissue transplanted later in life from the other strain as foreign, and as a result will not immunologically reject the transplant.

Figure 17.20 According to the clone deletion theory, tolerance to self-antigens may be due to destruction of lymphocytes that are able to make autoantibodies (antibodies against self-antigens).

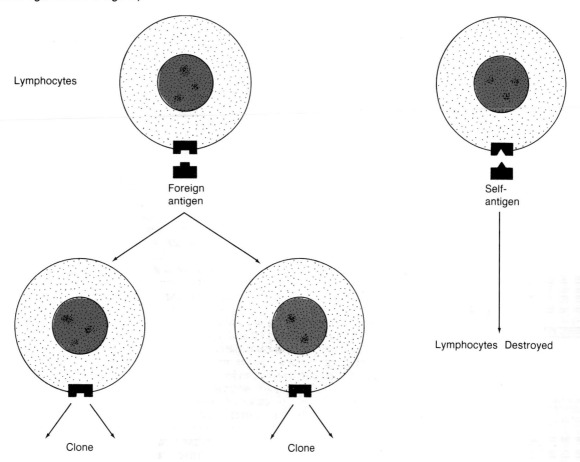

Lymphocytes

Foreign antigen

Self-antigen

Lymphocytes Destroyed

Clone

Clone

Tolerance

The ability of an individual's immune system to recognize and tolerate self-antigens requires continuous exposure of the immune system to those antigens. If this exposure occurs when the immune system is weak—such as in fetal and early postnatal life—tolerance is more complete and long lasting than when exposure occurs later in life. Self-antigens that are normally hidden from the blood, such as thyroglobulin within the thyroid gland and lens protein in the eye, are therefore not tolerated. Exposure of these self-proteins results in antibody production just as if these proteins were foreign. Antibodies made against self-proteins are called **autoantibodies.**

Two major theories have been exposed to account for immunological tolerance: (1) *clone deletion;* and (2) *immunological suppression.* According to the clone deletion theory, tolerance to self-antigens is achieved by destruction of the lymphocytes that inherit the ability to make autoantibodies (see figure 17.20). It is not known how interaction of self-antigens with these lymphocytes might lead to their destruction.

According to the immunological suppression theory, the lymphocytes that make autoantibodies are present throughout life but are normally inhibited from attacking self-antigens. This is presumably due to the effects of suppressor T cells. Alteration in the ratio of suppressor to helper T cells for self-antigens later in life, therefore, might result in the production of autoantibodies.

Figure 17.21 There are four human histocompatibility antigens (or human leukocyte antigens—HLA). Each of these antigens is coded by a gene located on chromosome number 6.

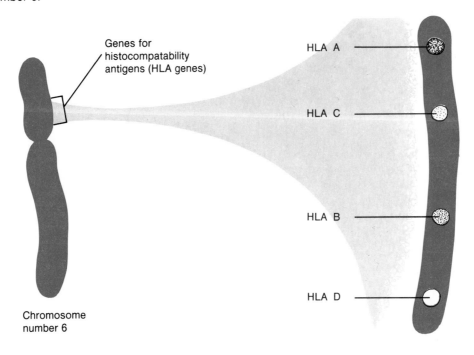

Genes for histocompatability antigens (HLA genes)

HLA A

HLA C

HLA B

HLA D

Chromosome number 6

Histocompatibility Antigens

Tissue that is transplanted from one person to another contains antigens that are foreign to the host (except for transplants between identical twins). This is because every tissue cell, with the exception of mature red blood cells, is genetically marked by a characteristic combination of **histocompatibility antigens** on its surface. The greater the difference in these antigens between donor and recipient in a transplant, the greater will be the chance of transplant rejection. Prior to organ transplants, therefore, the "tissue type" of the recipient is matched to that of potential donors. Since the person's white blood cells are used for this purpose, an alternate name for histocompatibility antigens in humans is **human leukocyte antigens,** abbreviated **HLA.**

The histocompatibility antigens (HLA) are proteins that are coded by four genes located on chromosome number 6. These four genes are labeled A, B, C, and D. Each of these genes can code for only one protein in a given individual, but this protein can be different in different people. Two people, for example, may both have antigen A3, but one might have antigen B17 while the other has antigen B21. The closer two people are related the more similar their histocompatibility antigens will be.

Interaction between T cells and macrophages, which is often required for activation of the T cells as previously discussed, requires that both cells have the same histocompatibility antigens (see figure 17.22). Likewise, the ability of killer T cells to destroy another human cell (which may have been infected with a virus, for example) requires that both the T cells and the victim cell have the same histocompatibility antigens. A culture of T cells

Figure 17.22 *(a)* is an electron micrograph showing contact between a macrophage *(left)* and a lymphocyte *(right)*. As illustrated in *(b),* such contact between a macrophage and a T cell requires that both cells share the same histocompatibility (or HLA) antigens. By presenting antigens to the T cell, the interaction of the macrophage with the T cell may help convert the T cell into a helper or a suppressor or a cytotoxic (killer) T cell.

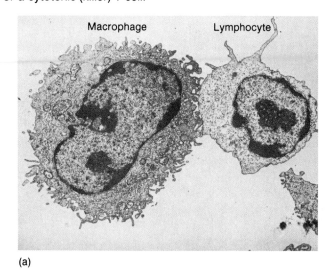

(a)

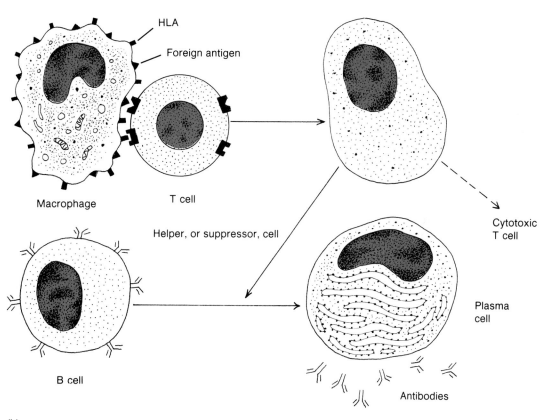

(b)

could thus not provide effective cell-mediated immunity against infected cells *in vitro* that are derived from an unrelated person.

Great clinical interest has been generated by the observation that certain diseases are much more common in people who have particular histocompatibility antigens. Ankylosing spondylitis (a type of rheumatoid arthritis), for example, is much more common in people who have antigen B27, and psoriasis (a skin disorder) is three times more common in people with antigen B17 than in the general population. Other diseases that have a high correlation with particular histocompatibility antigens include Hodgkin's disease (cancer of the lymph nodes), myasthenia gravis, Graves' disease, and juvenile-onset diabetes (discussed in chapter 20).

Red Blood Cell Antigens and Blood Typing

The histocompatibility antigens found on the surface of tissue cells and leukocytes are far more varied than the antigens on red blood cells. The major group of red blood cell antigens is known as the **ABO system,** in which an individual may have one of only four possible types: A, B, AB, or O. A person who is *type A* has A antigens on his or her red blood cells, while one who is *type B* has B antigens. If both A and B antigens are on the red blood cells, the person is *type AB,* while *type O* people have neither A nor B antigens on the red blood cells.

Each person inherits two genes (one from each parent) that control production of the ABO antigens. The genes for A or B antigens are dominant to the gene for O, since the latter simply indicates the absence of A or B. A person who is type A, therefore, may have inherited the A gene from both parents (may have the genotype AA), or may have inherited the A gene from one parent and the O gene from the other parent (and have the genotype AO). Likewise, a person who is type B may have either the BB or BO genotypes. It follows that a type O person must have inherited the O gene from both parents, and have the genotype OO. A person who is type AB has both the A and B antigens and has the genotype AB (notice that both genes express themselves—there is no dominance-recessive relationship between them).

The immune system is tolerant to its own red blood cell antigens as it is to its own histocompatibility antigens. A person who is type A, for example, does not normally produce anti-A antibodies. Surprisingly, however, type A people do have antibodies against the B antigen, and type B people have antibodies against the A antigen. These antibodies are not directed specifically against the opposite blood type—they can't be, because most people who are type A have never encountered type B antigens and most type B people have not been exposed to type A antigens.

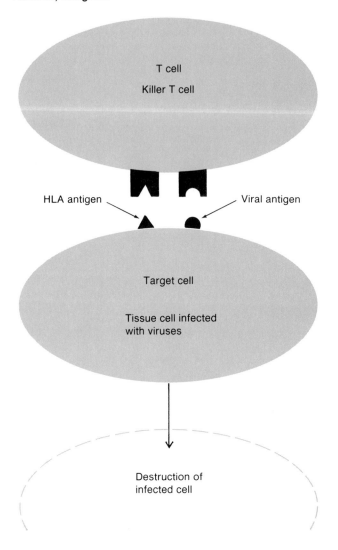

Figure 17.23   In order for a killer T cell to destroy a tissue cell infected with viruses, the T cell and the target cell must share the same histocompatibility (or HLA for humans) antigens.

Rather, antibodies directed against commonly encountered bacteria—such as *pneumonococcus* and *Escherichia coli* (normally found in the large intestine)—are believed to cross-react with the A or B antigens. A person who is type A therefore acquires antibodies that can react with B antigens by exposure to these bacteria, but does not develop antibodies that can react with A antigens because this is prevented by tolerance mechanisms.

People who are type AB develop tolerance to both of these antigens, and thus do not produce either anti-A nor anti-B antibodies. Those who are type O, in contrast, do not develop tolerance to either antigen and therefore have both anti-A and anti-B antibodies in their plasma (see table 17.13). People who are type O are known as *universal donors* and those who are type AB are *universal recipients* for blood transfusions.

Table 17.12 Association between particular human histocompatibility antigens and diseases.

Disease	HLA Antigen	Frequency in Patients (%)	Frequency in Controls (%)
Ankylosing spondylitis	B27	90	7
	B13	18	4
Psoriasis	B17	29	8
	B16	15	5
Graves' disease	B8	47	21
Coeliac disease	B8	78	24
Dermatitis herpetiformis	B8	62	27
Myasthenia gravis	B8	52	24
SLE	B15	33	8
Multiple sclerosis	A3	36	25
	B7	36	25
Acute lymphatic leukemia	A2	63	37
	B35	25	16
Hodgkin's disease	A1	39	32
	B8	26	22
Chronic hepatitis	B8	68	18
Ragweed hayfever	B7	50	19

Source: From H. McDervitt and W. Bodmer, Lancet *1*, 1269, 1974. Reprinted by permission.

Table 17.13 The ABO system of antigens on red blood cells.

Antigen on RBC's	Antibody in Plasma
A	anti-B
B	anti-A
O	anti-*A* and anti-*B*
AB	no antibody

Transfusion Reactions Before transfusions are performed, a *major cross-match* is made by mixing serum from the recipient with blood cells from the donors. If the types do not match—if the donor is type A, for example, and the recipient is type B—the recipient's antibodies attach to the donor's red blood cells and form "bridges" that cause the cells to clump together, or **agglutinate,** (see figure 17.24). Because of this agglutination reaction, the A and B antigens are sometimes called *agglutinogens* and the antibodies against them are called *agglutinins.* Transfusion errors that result in such agglutination in the blood can result in blockage of small blood vessels and organ damage. Also, hemoglobin released from damaged red blood cells can result in kidney disease.

Figure 17.24 Agglutination (clumping) of red blood cells occurs when cells with A-type antigens are mixed with anti-A antibodies, and when cells with B-type antigens are mixed with anti-B antibodies. No agglutination would occur with type O blood (not shown).

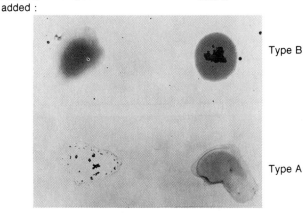
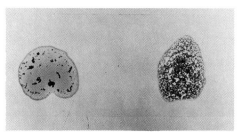

In emergencies, type O blood can be given to people who are type A, B, or AB (as well as type O). Since the type O cells lack A and B antigens, the recipient's antibodies cannot cause agglutination. The type O blood does contain anti-A and anti-B antibodies, but if the transfusion is not too large these antibodies will become too diluted in the recipient's blood to cause agglutination of the recipient's cells. This is why type O is called the universal donor. People who are type AB, and therefore lack anti-A and anti-B antibodies, can receive any blood type in emergencies (are universal recipients) because they cannot agglutinate the donor cells. Notice that, in determining blood transfusions, it is the effect of the *recipient's antibodies on the donor's red blood cells* that is of major importance. The effect of the donor's antibodies on the recipient's red blood cells assumes increasing importance as more blood is transfused.

Rh Factor   Another important antigen found on most red blood cells is the *Rh factor* (*Rh* stands for Rhesus monkey, where this antigen was first discovered). People who have this antigen are said to be *Rh positive,* while those who do not are *Rh negative.* There are fewer Rh negative people because this condition is recessive to Rh positive. The Rh factor is of particular significance when Rh negative mothers give birth to Rh positive babies.

At the time of birth, some of the Rh antigens from the baby may become exposed to the mother's immune system. Since the mother is Rh negative, this antigen is foreign to her and she can thus become sensitized to it (by developing lymphocyte clones that make antibodies against the Rh factor). This does not affect the first baby, but antibodies against the Rh factor can cross the placenta and attack the red blood cells of subsequent Rh positive fetuses. The second baby may thus be born anemic—with a condition called **erythroblastosis fetalis.** A third Rh positive fetus can die *in utero* from this condition.

Erythroblastosis fetalis can be prevented by injecting the Rh negative mother with *antibodies against the Rh factor* (one trade name for this is RhoGAM) within seventy-two hours after the birth of each Rh positive baby. These antibodies combine with the Rh antigens and by this means prevent the antigens from sensitizing the mother's immune system. Prevention of Rh disease is therefore achieved by use of specific "blocking antibodies." This program has been extremely successful, and is currently the only example of the use of specific immunological suppression in the prevention or treatment of any disease.

1. Define immunological tolerance, and explain the clone deletion and immunological suppression theories of tolerance.
2. Describe the nature, physiological functions, and clinical applications of the histocompatibility antigens.
3. Describe, verbally and with an illustration, the agglutination reaction that occurs when blood types of the ABO system are mixed together. Explain why agglutinins are present in type A, B, and O blood, but not in type AB blood.
4. Describe how erythroblastosis fetalis is produced and the technique used to prevent this condition.

The ability of the normal immune system to tolerate self-antigens while identifying and attacking foreign antigens provides specific defense against invading pathogens. In every individual, however, this system of defense against invaders commits domestic offenses at times. These offenses can cause diseases that range in severity from the sniffles to almost instant death.

Diseases caused by the immune system can be grouped into three interrelated but distinct categories: (1) *autoimmune diseases;* (2) *immune complex diseases;* and (3) *allergy,* or *hypersensitivity.* It is important to remember that these diseases are not caused by foreign pathogens but by the abnormal response of the immune system to particular antigens.

Autoimmunity

Autoimmune diseases are those produced by failure of the immune system to recognize and tolerate self-antigens. This failure results in the production of autoantibodies that cause inflammation and organ damage. Such autoimmune destruction can occur as a result of five known mechanisms.

1. **An antigen that does not normally circulate in the blood may become exposed to the immune system.** Thyroglobulin protein that is normally trapped within thyroid follicles (described in chapter 20), for example, can stimulate production of autoantibodies that cause destruction of the thyroid (in *Hashimoto's thyroiditis*). Similarly, autoantibodies produced to lens protein in a damaged eye may cause destruction of the healthy eye (in *sympathetic ophthalmia*).

Figure 17.25 Autoimmune thyroiditis in a rabbit, induced experimentally by injection with thyroglobulin. Compare the picture of a normal thyroid *(top)* with that of the diseased thyroid *(bottom)*. The grainy appearance of the diseased thyroid is due to infiltration with large numbers of lymphocytes and macrophages.

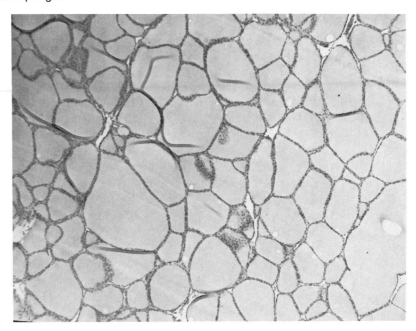

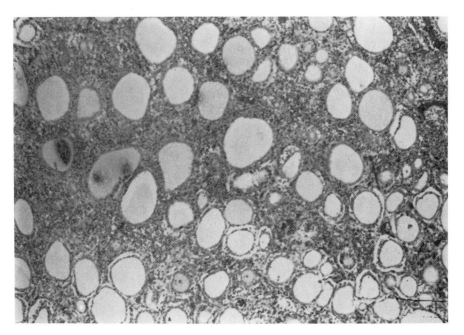

Figure 17.26   One proposed mechanism for the
development of rheumatoid arthritis. An initial joint
inflammation (1) results in the release of lysosomal
enzymes (2) that cause damage to IgG antibodies (3).
These damaged IgG antibodies act as antigens (4) to
stimulate B lymphocytes, resulting in the production of
IgM antibodies (5) against those damaged IgG antibodies/
antigens. This, in turn, produces further inflammation (6).

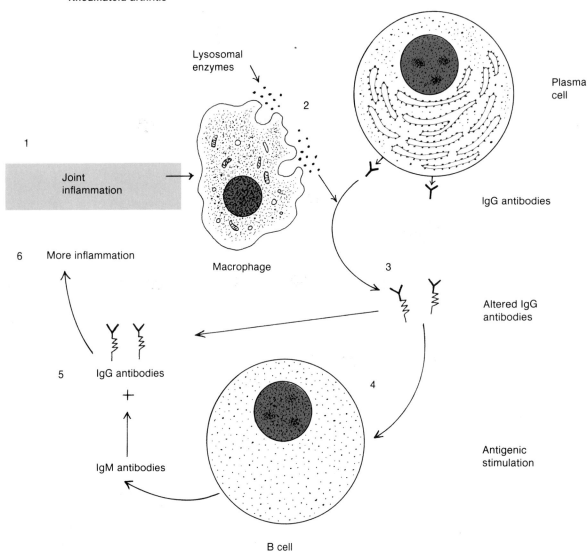

2. **A self-antigen, which is otherwise tolerated, may be
   altered by combining with a foreign hapten.** The
   disease *thrombocytopenia* (low platelet count), for
   example, can be caused by autoimmune destruction
   of platelets (thrombocytes). This occurs when drugs
   such as aspirin, sulfonamide, antihistamines,
   digoxin, and others combine with platelet proteins
   to produce new antigens. The symptoms of this
   disease usually stop when the person stops taking
   these drugs.

3. **A self-antigen may become damaged and as a result,
   may expose new antigenic sites.** The disease
   rheumatoid arthritis is believed to develop in this
   way. An initial inflammation of the joints may
   result in the release of digestive enzymes from
   lysosomes in phagocytic cells. When these enzymes
   digest part of IgG antibodies they expose new
   antigenic sites, which stimulate production of IgM
   antibodies directed against the altered IgG
   antibodies (which are the antigens in this process).

Figure 17.27 A model that has been proposed to explain why the frequency of autoimmune diseases increases with age. Early in life, those lymphocytes that produce autoantibodies may be destroyed. With time, however, some descendants of the remaining lymphocytes may mutate to produce autoantibodies. When this occurs later in life, these cells are not destroyed and may cause autoimmune diseases. Other models have also been proposed to explain autoimmunity (see text).

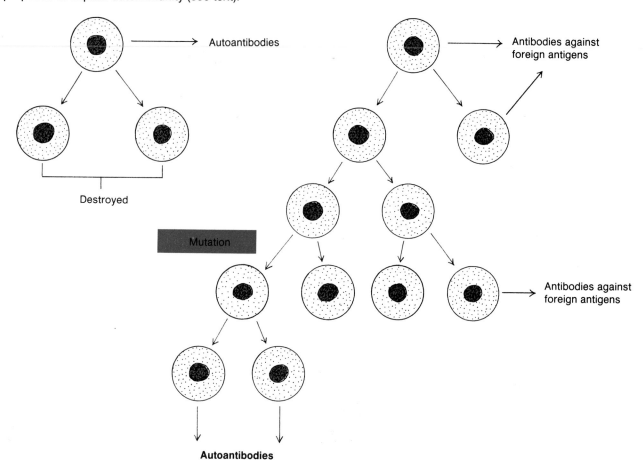

The IgM antibodies that a person develops against his own IgG antibodies are called *rheumatoid factor,* and can be used to diagnose this condition.

4. **Antibodies produced against foreign antigens may cross-react with self-antigens.** Autoimmune diseases of this sort can occur, for example, as a result of *streptococcus* bacterial infections. Antibodies produced in response to antigens in this bacterium may cross-react with self-antigens in the heart and kidneys. Inflammation induced by such autoantibodies can produce heart damage (including the valve defects characteristic of *rheumatic fever*) and damage to the glomerular capillaries in the kidneys *(glomerulonephritis).*

5. **Genes that code for non-self antibodies may mutate to produce autoantibodies.** This would not cause particular diseases, but would lead with age to an increased frequency of autoimmune diseases in general. An increased frequency of autoimmune diseases, and an increased amount of autoantibodies in the blood of apparently healthy people, does in fact occur with age, as predicted by this mutation theory.

Table 17.14   Examples of autoimmune diseases.

Disease	Antigen	Ig and/or T Cell Response
Postvaccinal and postinfectious encephalomyelitis	Myelin, cross-reactive	T cell
Aspermatogenesis	Sperm	T cell
Sympathetic ophthalmia	Uvea	T cell
Hashimoto's disease	Thyroglobulin	IgG and T cell
Graves' disease		Long-acting thyroid stimulator (LATS)
Autoimmune hemolytic disease	I, Rh, and others on surface of RBCs	IgM and IgG
Thrombocytopenic purpura	Hapten-platelet or hapten-adsorbed antigen complex	IgG
Myasthenia gravis	Myosin	IgG
Rheumatic fever	Streptococcal cross-reactive with heart	IgG and IgM
Glomerulonephritis	Streptococcal cross-reactive with kidney	IgG and IgM
Rheumatoid arthritis	IgG	IgM to $Fc(\gamma)$
Systemic lupus erythematosus	DNA, nucleoprotein, RNA, etc.	IgG

Source: From Barrett, James T.: Textbook of immunology, ed. 4, St. Louis, 1983, The C.V. Mosby Co.

It is not known why some people are more resistant than others to autoimmune diseases. One possibility is that those who succumb to autoimmune diseases may have an abnormally high ratio of helper T cells to suppressor T cells, which aids the development of the B cell clones that produce autoantibodies. If this is true, however, the causes of such imbalances in the subpopulations of T cells would still remain to be determined.

Immune Complex Diseases

The term *immune complexes* refers to combinations of antibodies with antigens that are free, rather than attached to bacterial or other cells. Formation of such complexes activates complement proteins and promotes inflammation. This inflammation is normally self-limiting because the immune complexes are removed by phagocytic cells. When large numbers of immune complexes are continuously formed, however, the inflammation may be prolonged. Also, dispersion of immune complexes to other sites can lead to widespread inflammations and organ damage. Damage produced by the inflammatory response to antigens—either foreign or self-antigens—is called **immune complex disease.**

Immune complex disease can result from infections by bacteria, parasites, and viruses. In viral hepatitis B, for example, an immune complex that consists of viral antigens and antibodies can cause widespread inflammation of arteries *(periarteritis)*. Note that the arterial damage is not caused by the hepatitis virus itself, but by the inflammatory process.

Immune complex diseases can also result from the formation of complexes between self-antigens and autoantibodies. In rheumatoid arthritis, for example, inflammation of the synovial joints is produced by the continuous generation of immune complexes consisting of IgG antibodies and the rheumatoid factor (IgM antibodies, as previously described). Activation of complement proteins, release of histamine and other chemicals from mast cells, and release of lysosomal enzymes from phagocytic cells produce the symptoms of inflammation characteristic of this disease.

Another immune complex disease that has an autoimmune basis is *systemic lupus erythematosus (SLE)*. People with SLE produce antibodies against their own DNA and nuclear protein. This can result in the formation of immune complexes throughout the body, including the glomerular capillaries where glomerulonephritis may be produced.

Table 17.15 Allergies; comparison between immediate and delayed hypersensitivity reactions.

Characteristic	Immediate Reaction	Delayed Reaction
Time for onset of symptoms	Within several minutes	Within one to three days
Lymphocytes involved	B cells	T cells
Immune effector	IgE antibodies	Cell-mediated immunity
Allergies most commonly produced	Hay fever, asthma, and most other allergic conditions	Contact dermatitis (such as to poison ivy and poison oak)
Therapy	Antihistamines and adrenergic drugs	Corticosteroids (such as cortisone)

Table 17.16 Some of the chemicals released by mast cells that are responsible for immediate hypersensitivity.

Chemical	Derivation	Action
Histamine	From histidine	Contracts smooth muscles in bronchioles; dilates blood vessels; increases capillary permeability
Serotonin	From tryptophane	Contracts smooth muscles
Slow-reacting substance of anaphylaxis	Fatty acids—leukotrienes and prostaglandins	Prolonged contraction of smooth muscle; increased capillary permeability
Eosinophil chemotactic factor	Polypeptides	Attracts eosinophils
Bradykinen and related compounds	Polypeptides	Slow smooth muscle contraction

Allergy

The term *allergy* is sometimes used to indicate any abnormal immune reaction; more commonly, however, allergy is used synonymously with *hypersensitivity*. In this sense, there are two major forms of allergy: (1) *immediate hypersensitivity;* and (2) *delayed hypersensitivity*. These terms emphasize the different periods of time required for the onset of symptoms in the two forms of allergy. The different latent periods reflect fundamental differences in the mechanisms involved; immediate hypersensitivity is the result of an abnormal B cell response, while delayed hypersensitivity is due to an abnormal T cell response (see table 17.15).

Immediate Hypersensitivity   Immediate hypersensitivity can produce such symptoms as allergic rhinitis (chronic runny or stuffy nose), conjunctivitis (red eyes), allergic asthma, atopic dermatitis (uticaria, or hives) and others. These symptoms result from the production of antibodies of the **IgE** subclass, instead of the normal IgG antibodies, in response to particular antigens. Antigens that provoke the production of IgE antibodies can be called *allergens*.

Unlike IgG antibodies, IgE antibodies do not circulate in the blood, but instead attach to tissue mast cells (their F_c "stalks" combine with membrane receptor proteins in these cells). When the person is again exposed to the same allergen, the allergen bonds to the F_{ab} portion of the antibodies on the mast cell surface. This stimulates the mast cells to secrete various chemicals, including *histamine*. In addition, mast cells and leukocytes release *prostaglandins* and related fatty acids called *leukotrienes*. These chemical mediators of allergy cause constriction of bronchioles, vasodilation of blood vessels, and increased capillary permeability, resulting in the symptoms of allergy.

Figure 17.28 Allergy (immediate hypersensitivity) is produced when antibodies of the IgE subclass attach to tissue mast cells. Combination of these antibodies with allergens (antigens that provoke an allergic reaction) causes the mast cell to secrete histamine and other chemicals that produce the symptoms of allergy.

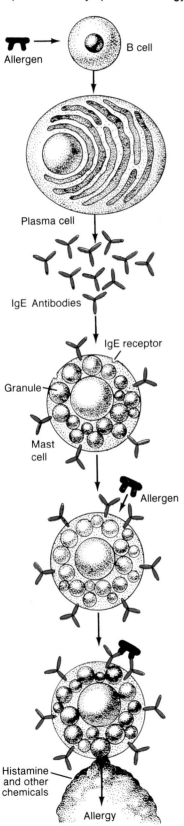

Allergen

B cell

Plasma cell

IgE Antibodies

IgE receptor

Granule

Mast cell

Allergen

Histamine and other chemicals

Allergy

Figure 17.29 Skin test for allergy. If an allergen is injected into the skin of a sensitive individual, a typical flare and wheal response occurs within several minutes.

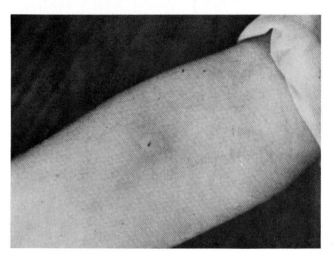

Immediate hypersensitivity to particular antigens is commonly tested by injecting various antigens under the skin (see figure 17.29). Within about twenty minutes (which is why the term *immediate* is used) a **flare-and-wheal reaction** is produced if the person is allergic to that antigen. This reaction is due to the release of histamine and other chemical mediators—the flare is due to vasodilation, and the wheal results from local edema.

Allergens that provoke immediate hypersensitivity include various foods, bee stings, and pollen grains. The most common allergy of this type is seasonal hay fever, which may be provoked by ragweed *(Ambrosia)* pollen grains. People with chronic allergic rhinitis due to an allergy to "dust" or "feathers" are actually allergic to a tiny mite (see figure 17.30) that lives in dust (particularly in pillows and bedding) and eats the scales of skin that are constantly shed from the body.

Figure 17.30 *(a)* Scanning electron micrograph of ragweed *(Ambrosia)*, which is responsible for hayfever. *(b)* Scanning electron micrograph of the house dust mite *(Dermatophagoides farinae)*, which lives in dust and is often responsible for year-long allergic rhinitis and asthma.

(a)

(b)

Anaphylaxis   When two researchers in 1902 attempted to immunize a dog to sea urchin poison, they were surprised to find that subsequent injection of this poison resulted in death. The attempted "immunization," in other words, produced an effect opposite to protection (prophylaxis). They therefore called this abnormal response *anaphylaxis.* The dog died from anaphylactic shock—a rapid drop in blood pressure caused by excessive IgE-stimulated release of histamine and other mediators of immediate hypersensitivity. It has been found that as little as one nanogram (1,000-millionth of a gram) of IgE can be fatal.

Desensitization   A common treatment for allergic conditions is "allergy shots," or desensitization. This treatment involves the regular injection of increasing amounts of the allergen. The effectiveness of this treatment appears to be variable, and the mechanisms by which protection against allergy is gained are poorly understood. One theory is that desensitization is produced by the development of blocking antibodies of the IgG subclass, which bond the allergen in the circulation and thus prevent it from attaching to IgE antibodies on the mast cell membrane.

Delayed Hypersensitivity   As the name implies, a longer time is required for the development of symptoms in delayed hypersensitivity (hours to days) than in immediate hypersensitivity. This may be due to the fact that immediate hypersensitivity is mediated by antibodies, while delayed hypersensitivity results from an abnormal T cell response.

One of the best known examples of delayed hypersensitivity is *contact dermatitis* caused by poison ivy *(Rhus radicans),* poison oak *(Rhus diversiloba),* and poison sumac *(Rhus toxicodendron).* The skin tests for tuberculosis—the tine test and the Mantoux test—also rely on delayed hypersensitivity reactions. If a person has been exposed to the tubercle bacillus and has, as a result, developed T cell clones, skin reactions appear within a few days after the tubercle antigens are rubbed into the skin with small needles (tine test) or are injected under the skin (Mantoux test).

Figure 17.31 A killer T cell *(a)* contacts a cancer cell (the larger cell), in a manner that requires specific interaction with antigens on the cancer cell. The killer T cell releases lymphokines, including toxins that cause death of the cancer cell *(b)*.

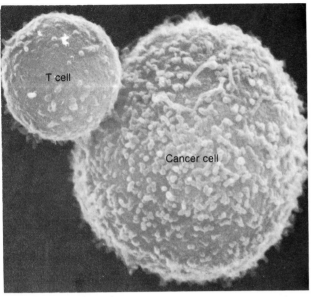

(a)   Andrejs Liepins

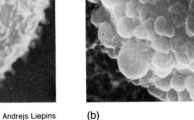

(b)   Andrejs Liepins

Tumor Immunology

The study of tumors, a field called *oncology* (from the Greek word *oncos* for *tumor*), has revealed that tumor biology is similar to, and interrelated with, the function of the immune system. Most tumors, or *neoplasms* (literally, "new growth") appear to be clones of single cells that have become transformed. This is similar to the development of lymphocyte clones in response to specific antibodies. Lymphocyte clones, however, are under complex inhibitory control systems—such as that exerted by suppressor cells and negative feedback control by antibodies. The division of tumor cells, in contrast, is not effectively controlled by normal inhibitory mechanisms. Tumor cells are also relatively unspecialized—they **dedifferentiate,** which means that they become similar to the less-specialized cells of an embryo.

Tumors are described as **benign** when they are relatively slow growing and limited to a specific location (warts, for example). Benign tumors do not undergo *metastasis,* a term that refers to the dispersion of tumor cells and resultant seeding of new tumors in different locations. **Malignant** tumors grow more rapidly and do metastasize. The term **cancer,** as it is usually used, refers to malignant, life-threatening tumors.

Immunological Surveillance

As tumors dedifferentiate, they reveal surface antigens that can stimulate immune destruction of the tumor cells. Consistent with the concept of dedifferentiation, some of these antigens are proteins produced in embryonic or fetal life that are not normally present in postnatal life after tolerance to self-antigens is established. These embryonic and fetal antigens can thus be treated as "foreign" and as fit subjects for immunological attack. Release of two such antigens into the blood has provided the basis for laboratory diagnosis of some cancers. *Carcinoembryonic antigen (CEA)* tests are useful in the diagnosis of colon cancer, for example, and tests for *alpha-fetoprotein* (normally produced only by the fetal liver) help in the diagnosis of liver cancer.

Tumor cells are attacked by the cell-mediated immune system, although humoral immunity (antibodies) may have a supportive role. Killer T cells destroy tumor cells by direct contact and release of lymphokines (see figure 17.31). Exposure of the immune system to tumor cells stimulates the development of T cell clones that specifically interact with the newly revealed antigens on the tumor cells.

Natural Killer Cells   Researchers observed that a strain of hairless mice, which genetically lacks a thymus and T lymphocytes, does not suffer from a particularly high incidence of tumor production. This surprising observation led to the discovery of *natural killer (NK) cells*. Unlike killer T cells, NK cells destroy tumors in a nonspecific fashion, and do not require prior exposure for sensitization to the tumor antigens. Activity of natural killer cells is stimulated by interferon (one of the lymphokines secreted by T cells). The identity of NK cells is currently under investigation—they appear to have some of the surface "markers" characteristic of T cells, but they differ from T cells in some other respects.

Both natural killer cells and killer T cells are believed to provide **immunological surveillance** against cancer—they normally destroy transformed cells before these produce a tumor. The fact that tumors do sometimes grow indicates that the tumor cells can in some way "escape" the surveillance system. Different mechanisms have been proposed to explain this escape: (1) the tumor cells may release antigens that combine with antibodies and T cell receptors in such a way as to block destruction of the tumor; (2) antibodies made by B cells against tumor antigens may "protect" the tumor against attack by killer cells (that is, blocking antibodies may be produced); and (3) helper T cells may aid the production of blocking antibodies, and suppressor T cells may inhibit the production of antibodies that could destroy the tumor.

Immunization against Cancer   Attempts to immunize people against cancer have thus far been unsuccessful. This is because (1) passive immunization with antibodies could block cell-mediated immunity rather than destroy the tumor; and (2) active immunization is difficult because each tumor that is produced by chemical carcinogens has different, unique antigens. Tumors produced by specific viruses have antigens characteristic of that virus, but so far only one type of rare cancer—Burkitt's lymphoma—has been demonstrated to be caused by a virus. This is the Epstein-Barr virus that is also responsible for infectious mononucleosis.

Nonspecific activation of the immune system to treat cancer has been attempted with limited success. In the early part of this century, William B. Coley observed that some cancer patients had spontaneous remissions after a bacterial infection. Coley therefore attempted to activate the immune system of cancer patients by injecting a mixture of killed bacteria *(Coley's toxins).* More recently, attempts to activate the immune system in a nonspecific fashion have been made using BCG (Bacille Calmette-Guérin), which has proven successful in activating cell-mediated immunity against tuberculosis and leprosy.

Effects of Stress and Aging

Susceptibility to cancer varies greatly. The Epstein-Barr virus that causes cancer in a few individuals in Africa, for example, can also be found in healthy people throughout the world. In most cases the virus is harmless; in some cases mononucleosis (involving limited proliferation of white blood cells) is produced. Only rarely does this virus cause the uncontrolled proliferation of leukocytes occurring in Burkitt's lymphoma. The reasons for these different responses to Epstein-Barr virus, and indeed for different susceptibilities of people to other forms of cancer, are very poorly understood.

It is known that cancer risk increases with age. According to one theory, this is due to the fact that aging lymphocytes accumulate genetic errors over the years that decrease their effectiveness. Secretion of thymus hormones also decreases with age in parallel with decreased cell-mediated immune competence. These effects may decrease the killer T cell population and interfere with the elaborate interactions between helper and suppressor T cells, B cells, antibodies, and perhaps natural killer cells.

Numerous experiments have demonstrated that tumors placed in stressed mice grow more rapidly than tumor placed in control mice. These effects of stress may perhaps be explained by observations that (1) steroid hormones secreted by the adrenal cortex (particularly hydrocortisone and cortisone) are known to suppress the immune system—these compounds are therefore often given to reduce chronic inflammation and to retard the immune rejection of organ transplants; and (2) stress stimulates secretion of ACTH from the anterior pituitary gland, and this hormone stimulates secretion of hydrocortisone from the adrenal cortex (as described in chapter 20).

Experimentally stressed mice do, in fact, have increased levels of adrenal steroid secretion, resulting in lowered white blood cell count, thymus involution, and decreased mass of lymph nodes and spleen. Such mice are more susceptible to viral infections and have enhanced rates of tumor growth. It is fascinating to consider that procedures that reduce psychological stress (such as "laugh therapy" and others) and injections of BCG, or *Coley's toxins,* may provide beneficial effects through the same mechanism—a general strengthening of the ability of the immune system to destroy tumor cells.

Summary

Nonspecific and Specific Immunity

I. Nonspecific immunity against infection is provided by various structures and processes.
   A. The skin and gastric acidity protect the body from invasion by pathogens.
   B. Fever, produced in response to secretion of endogenous pyrogen from leukocytes, may retard the growth or reproduction of bacteria.
   C. Interferons are molecules secreted by leukocytes and other cells that confer a transient and nonspecific immunity against viral infection.
   D. Phagocytosis by neutrophils and tissue macrophages, as well as by fixed phagocytes in the liver, spleen, and lymph nodes, provide nonspecific immunity.

II. Specific immunity against particular pathogens is acquired by exposure to their antigens.
   A. Antigens are molecules that stimulate production of antibody proteins by lymphocytes, and that combine in a specific way with those antibodies.
   B. Most molecules that are antigenic are large and foreign to the body.
      1. In order for a molecule to function effectively as an antigen, it must have a complex structure—this is why proteins are more antigenic than polysaccharides.
      2. Large complex proteins, with many antigenic determinant sites that can stimulate antibody production, are the most powerful antigens.
   C. Small foreign molecules that can become antigens by attaching to proteins are called haptens.

D. Specific immunity is provided by lymphocytes.
   1. Lymphocytes called B cells can be transformed into plasma cells by antigens and produce antibodies.
   2. Lymphocytes that are processed by the thymus, and are thus called T cells, provide cell-mediated immunity.

Humoral and Cell-Mediated Immunity

I. Humoral immunity refers to the secretion of antibodies into plasma and other body fluids by B cells and plasma cells.
   A. Antibodies are proteins that consist of four polypeptide chains—two H chains and two L chains.
   B. Part of the structure of an antibody has a constant amino acid sequence (the F_c fragment) and part has a variable amino acid sequence (the F_{ab} fragment).
      1. Each of these fragments consists of both H and L polypeptide chains.
      2. The variable or F_{ab} part of the antibody combines with antigens in a specific manner.
   C. There are several immunoglobin (antibody) subclasses.
      1. Each subclass—IgA, IgM, IgG, IgE, and IgA—has its characteristic constant heavy chains.
      2. IgM antibodies serve as receptors for antigens on B cells; IgG antibodies circulate in the blood; and IgE antibodies contribute to allergic reactions.

II. Combination of antibodies with antigens provokes an inflammation reaction.
   A. Antibodies promote opsonization—increased phagocytic activity by neutrophils and macrophages.

B. Antibody-antigen complexes activate the complement system of proteins.
   1. This produces a cascade of reactions in which some complement proteins are "fixed" to the victim cell and some are liberated.
   2. The fixed complement proteins produce pores in the victim cell membrane that cause destruction of the cell.
   3. The liberated complement proteins stimulate phagocytosis and attract phagocytes to the infected area.
   4. Complement also stimulates release of histamine and other chemicals from mast cells, which cause vasodilation and increased capillary permeability.

III. Cell-mediated immunity is provided by the T cells.
   A. Killer T cells must come into close proximity or have physical contact with their victim cells.
   B. Killer T cells fight viral infections, fungal infections, and some bacterial infections; cause rejection of transplanted tissue; and participate in immunological surveillance against cancer cells.
   C. Chemicals called *lymphokines* are secreted by killer T cells.
      1. The lymphokines include interferon and lymphotoxin, a chemical that directly kills the victim cell.
      2. Lymphokines also include chemicals that enhance phagocytosis by macrophages.
   D. Other subpopulations of T cells function indirectly, by modifying the activity of B cells and humoral immunity.
      1. Helper T cells and suppressor T cells can increase and decrease the secretion of antibodies by B cells.
      2. Activity of T lymphocytes, in turn, may be decreased by blocking antibodies secreted from B cells.

Immunization and Tolerance

I. Prior exposure to a particular pathogen can cause active immunization.

A. Upon first exposure, antibody production is sluggish and antibodies are of the IgM class—this is the primary response.

B. On subsequent exposures, antibody production occurs more rapidly and to a greater degree, and the antibodies are of the IgG subclass—this is the secondary response.

C. According to the clonal selection theory, lymphocytes that are already able to make the appropriate antibody are stimulated to produce a clone.

1. The ability to make a particular antibody is inherited by the lymphocyte.

2. The combination of antigen with antibody receptor proteins cause the lymphocyte to divide many times to produce a large population of cells able to make the appropriate antibody.

3. In the process of clonal development, IgM antibodies are replaced with IgG antibodies.

D. Clinical immunizations are performed using either attenuated pathogens or related organisms that are less virulent but antigenically similar to the pathogens.

II. Passive immunity occurs when an individual obtains antibodies produced by a different person or an animal.

A. Newborn babies and fetuses are passively immunized by antibodies from the mother.

B. People may be injected with antisera or antitoxins developed by horses or sheep.

III. The immune system normally distinguishes between self-antigens and non-self-antigens.

A. The person does not normally make autoantibodies against self-antigens.

1. According to the clonal deletion theory, lymphocytes able to make autoantibodies may be destroyed early in life.

2. According to the immunological suppression theory, production of autoantibodies is normally inhibited by the action of suppressor T cells.

B. The genetic identity of human cells, and the distinction between self-antigens and non-self-antigens, is based in part on histocompatibility antigens on the surface of every tissue cell.

1. There are four genes on chromosome number 6 that code for these antigens.

2. Immunological acceptance of transplants depends on the degree of similarity in histocompatibility antigens between donor and recipient.

3. The human histocompatibility antigens are usually detected on the surface of leukocytes and so are called human leukocyte antigens (HLA).

C. The major antigens on red blood cells are of the ABO system and the Rh antigen.

1. The blood type of a person indicates the type of antigens present on the erythrocytes.

2. People of one blood type produce antibodies against the opposite blood type.

3. Type O is the universal donor and type AB is the universal recipient in blood transfusions.

Immune Diseases and Tumor Immunology

I. Autoimmune diseases are caused by antibodies against self-antigens.

A. Attachment of autoantibodies to self-antigens activates complement proteins and stimulates an inflammatory reaction.

B. Immune complex diseases—involving complexes of antibodies to free antigens—can be produced by autoantibodies.

1. Rheumatoid arthritis and glomerulonephritis are examples of such diseases.

2. The symptoms of autoimmune diseases are produced by the immune system, not by a foreign pathogen.

II. Allergy includes immediate and delayed hypersensitivity.

A. Immediate hypersensitivity—such as hay fever and allergic rhinitis—is produced by IgE antibodies.

1. The IgE antibodies attach to mast cells.

2. Combination of IgE antibodies with antigens stimulates the mast cells to secrete histamine and other mediators of the allergic reaction.

3. Anaphylactic shock can be produced by excessive production of IgE antibodies.

B. Delayed hypersensitivity—such as contact dermatitis and the skin reaction to tuberculosis tests—is a cell-mediated (T cell) response.

III. Killer T cells and natural killer (NK) cells are believed to provide immunological surveillance against cancer.

A. Natural killer cells are somewhat like T cells, but they do not require prior sensitization to the tumor.

B. Immune protection against cancer is weakened in aging and with stress.

1. During aging, genetic errors may accumulate that decrease the effectiveness of immunological surveillance.

2. Stress causes increased secretion of adrenal steroids such as hydrocortisone, which may weaken the ability of the immune system to combat cancer.

Self-Study Quiz

Multiple Choice

Match the *cell type* with its *secretion:*

1. Killer T cells
2. Mast cells
3. Plasma cells
4. Macrophages

(a) antibodies
(b) lymphokines
(c) lysosomal enzymes
(d) histamine

5. Which of the following offers nonspecific defense against viral infections?
   (a) antibodies
   (b) complement system
   (c) interferon
   (d) histamine

6. Which of the following statements about the F_{ab} portion of antibodies is TRUE?
   (a) it bonds to antigens
   (b) its amino acid sequences are variable
   (c) it consists of both H and L chains
   (d) all of these

7. Which of the following statements about complement proteins C3a and C5a is FALSE?
   (a) they are released during the complement fixation process
   (b) they stimulate chemotaxis of phagocytic cells
   (c) they promote the activity of phagocytic cells
   (d) they produce pores in the victim cell membrane

8. Mast cell secretion during an immediate hypersensitivity reaction is stimulated when antigens combine with:
   (a) IgG antibodies
   (b) IgE antibodies
   (c) IgM antibodies
   (d) IgA antibodies

9. During a secondary immune response:
   (a) more antibodies are produced more quickly than in a primary response
   (b) antibody production lasts longer than in a primary response
   (c) antibodies of the IgG subclass are produced, while IgM antibodies are produced during the primary reaction
   (d) Lymphocyte clones are believed to develop
   (e) all of these

10. Which of the following cell types is believed to "present" antigens to lymphocytes?
    (a) macrophages
    (b) neutrophils
    (c) mast cells
    (d) natural killer cells

11. According to the clonal selection theory, the ability of each B lymphocyte to make antibodies is:
    (a) acquired by exposure to the appropriate antigen
    (b) inherited—either from the parents or from somatic mutation

12. Which of the following statements about T lymphocytes is FALSE?
    (a) some T cells promote the activity of B cells
    (b) some T cells suppress the activity of B cells
    (c) some T cells secrete interferons
    (d) some T cells produce antibodies

13. Delayed hypersensitivity—such as the skin reaction in a tuberculosis test—is mediated by:
    (a) T cells
    (b) B cells
    (c) plasma cells

14. The universal donor in a blood transfusion is:
    (a) type A
    (b) type B
    (c) type AB
    (d) type O

18

The Digestive System

Objectives

By studying this chapter, you should be able to:

1. Describe the general structure of the digestive tract and the specializations of the mucosa, including gastric pits and glands, villi, and crypts of Lieberkühn

2. Describe the secretions of the gastric mucosa, how pepsinogen is converted to pepsin, and the functions of pepsin and hydrochloric acid in gastric juice

3. Describe the location and functions of the brush border enzymes of the intestine

4. Describe the composition and functions of pancreatic juice and the composition and functions of bile

5. Describe the flow of blood and bile in the liver, and explain how the liver functions to modify the chemical composition of the blood

6. Describe the formation of free bilirubin, conjugated bilirubin, and urobilinogen, and explain the enterohepatic circulation of urobilinogen

7. Explain the steps involved in the digestion and absorption of carbohydrates, proteins, and lipids

8. Explain how gastric motility and secretion are regulated by the nervous and endocrine systems

9. Explain how secretion of pancreatic juice and bile are regulated by the nervous and endocrine systems

10. Discuss the caloric, mineral, and vitamin requirements of the human diet

Unlike plants, which can form organic molecules using inorganic compounds such as carbon dioxide, water, and ammonia, humans and other animals must ingest their organic molecules preformed. This is done by eating either the plants, or the animals that ate the plants (the herbivores), or by eating the animals that ate the herbivores. Some of these ingested molecules are needed for their caloric (energy) value—obtained by the reactions of cell respiration, and used in the production of ATP (as described in chapter 5)—while the balance is used to make additional tissue.

Figure 18.1 Digestion of food molecules occurs by means of hydrolysis reactions.

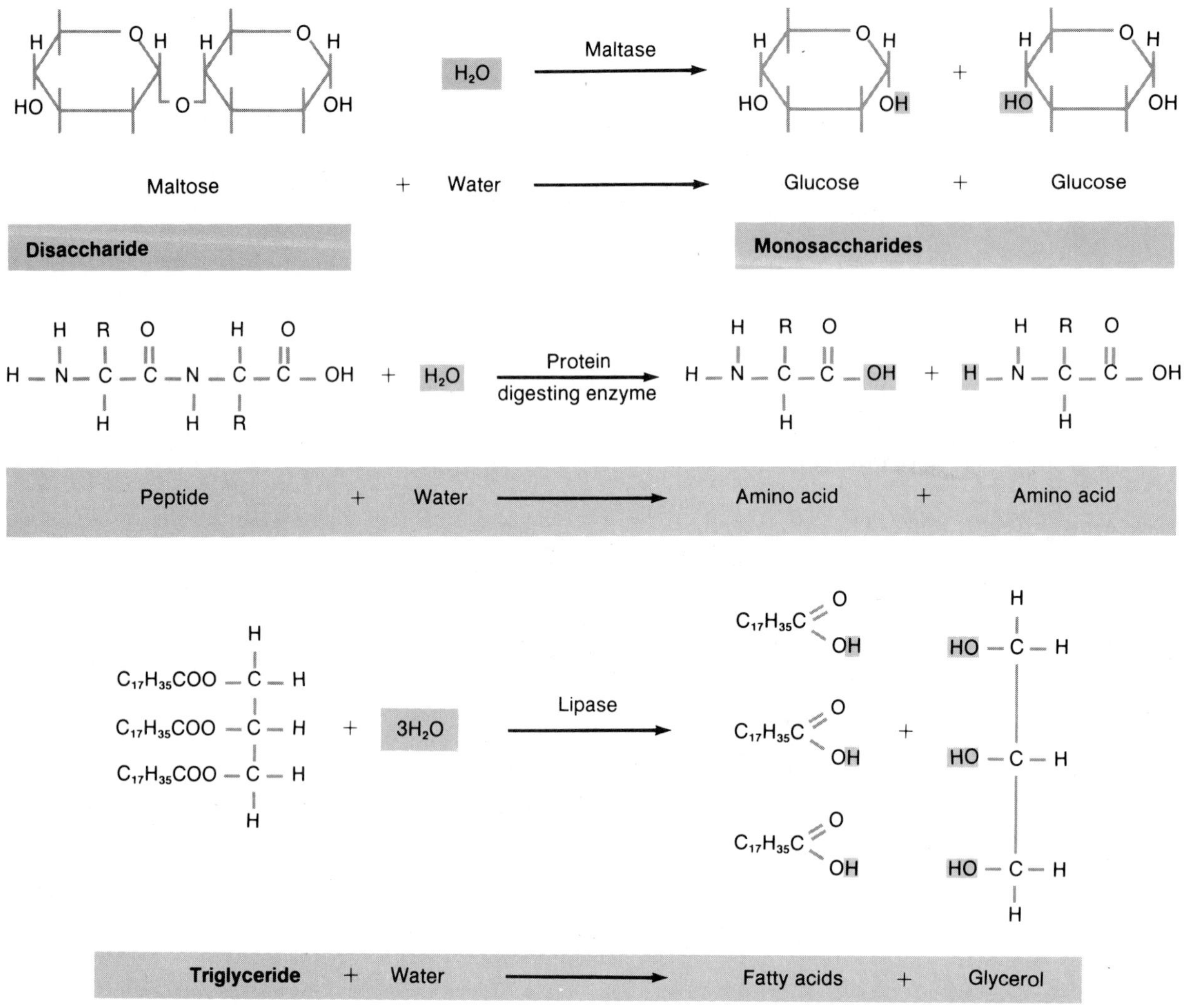

Most of the organic molecules that are ingested are similar to the molecules that form the structure of human tissues. These are generally large molecules *(polymers)* that are composed of subunits, or *monomers.* Within the digestive tract, these molecules are *digested* by hydrolysis reactions (described in chapter 2 and reviewed in figure 18.1) that cleave the polymers into monomers. The monomers thus formed are then absorbed across the wall of the intestine into the blood. The digestive tract is therefore said to have two primary functions: **digestion** of food and **absorption** of the digestion products.

Since the composition of food is similar to the composition of the body tissues, enzymes that digest food could also digest the person's own tissues. This does not normally occur because there are a variety of protective devices that inactivate digestive enzymes in the body and keep these enzymes separated from the cytoplasm of tissue cells. The fully active digestive enzymes are normally limited in their location to the lumen (the cavity) of the digestive tract.

The lumen of the digestive tract is continuous with the external environment because it is open at both ends (mouth and anus). Indigestible material, such as cellulose from plant-cell walls, pass from one end to the other without crossing the epithelial lining of the digestive tract (that is, without being absorbed). In this sense, these indigestible materials never enter the body, and the harsh conditions required for digestion thus occur *outside* the body.

Figure 18.2 Organs of the digestive system.

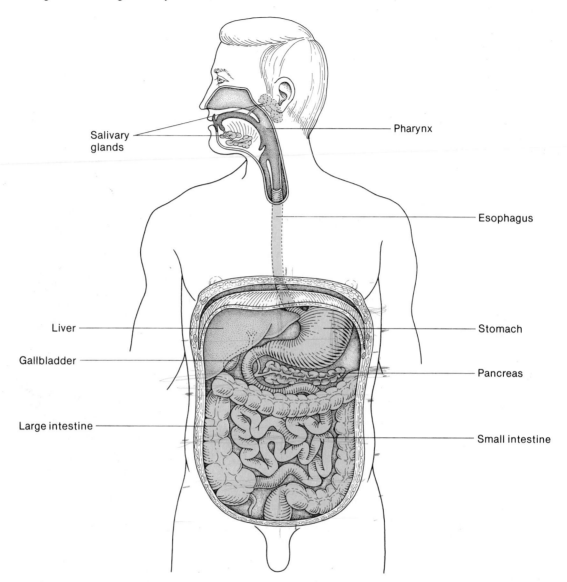

In *planaria* (a type of flatworm) the digestive tract has only one opening—the mouth is also the anus. Each cell that lines the digestive tract is thus exposed to food, absorbable digestion products, and waste products. The two-ended digestive tract of higher organisms, in contrast, allows one-way transport, which is insured by wavelike muscle contractions and by the action of valves. This one-way transport allows different regions of the digestive tract to be specialized for different functions, as a "dis-assembly line."

The digestive system, exclusive of the mouth and anal openings, is derived from a tube of embryonic tissue known as *endoderm*. In addition to forming the epithelial lining of the **digestive tract** (esophagus, stomach, small intestine, and large intestine), regions of endoderm form outpouchings that develop into the **liver, gallbladder,** and **pancreas.** In postnatal life these organs maintain continuity with the digestive tract by means of ducts that empty into the intestine and function to aid the digestion of molecules in the intestine.

Figure 18.3   Illustration of the major layers of the intestine. Insert shows how folds of mucosa form projections called *villi*.

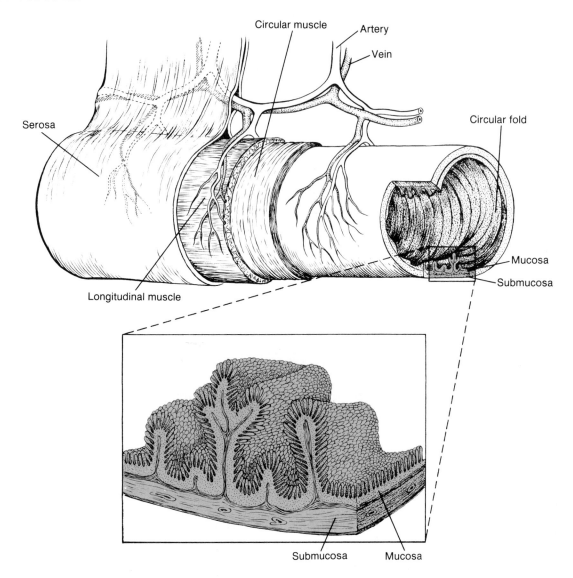

A cross section of the gastrointestinal tract reveals four major layers. From the innermost layer outwards, these include (1) *mucosa,* or *mucous membrane;* (2) *submucosa;* (3) *muscularis externa;* and (4) *serosa.* The serosa and the submucosa are connective tissue layers; the serosa covers the outer surface of the digestive tract, and the extracellular spaces of the submucosa provide space for blood vessels, nerves, and mucous-secreting exocrine glands. The muscularis externa, in most parts of the digestive tract, consists of an inner circular and an outer longitudinal layer of smooth muscle.

The innermost layer—the mucosa—is itself divided into three parts: (1) an *epithelium* that lines the tract—this is a stratified squamous epithelium in the esophagus and anal canal, and a simple columnar epithelium in the rest of the tract; (2) the *lamina propria*—a very thin layer of connective tissue containing lymphatic vessels, blood capillaries, and many lymphocytes; and (3) the *muscularis mucosa*—a thin ribbon of smooth muscle that marks the boundary between the mucosa and the submucosa (see figure 18.4).

Figure 18.4 Illustration of a cross section of the intestine showing layers and structures.

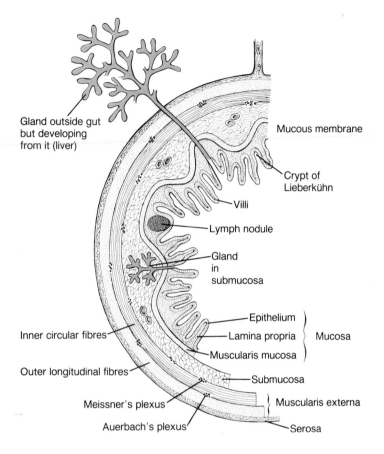

Gland outside gut but developing from it (liver)

Mucous membrane

Crypt of Lieberkühn

Villi

Lymph nodule

Gland in submucosa

Inner circular fibres

Epithelium

Lamina propria

Muscularis mucosa

Mucosa

Outer longitudinal fibres

Submucosa

Muscularis externa

Meissner's plexus

Auerbach's plexus

Serosa

Movements and secretions of the gastrointestinal tract are controlled in part by parasympathetic and sympathetic nerves. Preganglionic parasympathetic fibers enter the wall of the digestive tract and synapse with postganglionic fibers that originate in terminal ganglia within the tract (as described in chapter 9). The postganglionic fibers, in turn, innervate the smooth muscles and glands. Preganglionic and postganglionic fibers form a network, or "plexus," in the region of the terminal ganglia: those in the submucosa form *Meissner's plexus,* and those in the muscularis externa form *Auerbach's plexus* (see figure 18.4). Postganglionic sympathetic fibers, which arise outside of the digestive tract in collateral ganglia (described in chapter 9), pass through the plexus without synapsing to innervate the digestive tract.

Although different regions of the digestive tract have similar structures, their differences in structure and function provide the basis for the sequential processing of food. Each of these regions will thus be discussed in sequence before their coordinated activities—in terms of the digestion and absorption of specific food groups—will be considered.

Esophagus

Prior to swallowing, food is usually chewed and mixed with saliva. Saliva contains a starch-digesting enzyme—*salivary amylase,* or *ptyalin*—and mucus, which makes the masticated food more slippery. This aids the process of swallowing, by which food is passed through a muscular tube—the esophagus—to the stomach.

The esophagus is lined with a stratified squamous epithelium that is continuous with the epithelium of the pharynx. This is the last stratified squamous epithelium that the ingested material will encounter until it reaches the anal canal. The muscles of the first one-third of the esophagus, like those in the pharynx and mouth, are voluntary, striated muscles. These voluntary, striated muscles are needed for control of swallowing. The muscles of the last two-thirds of the esophagus, like those of the rest of the digestive tract, are involuntary smooth muscles. As in other parts of the digestive tract, the wall of the esophagus contains mucous-secreting exocrine glands and lymphatic nodules with lymphocytes (see figure 18.5).

Figure 18.5   Microscopic appearance of the esophagus.

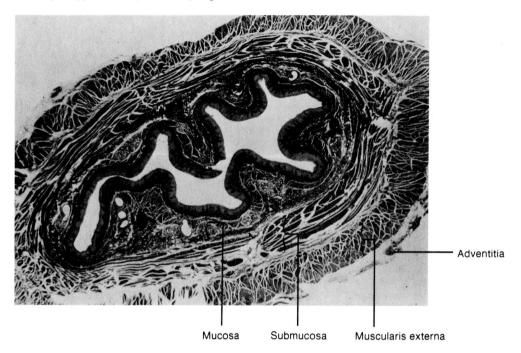

Adventitia

Mucosa   Submucosa   Muscularis externa

Peristalsis   Swallowed food is pushed from one end of the esophagus to the other by a wavelike muscular contraction called *peristalsis.* Peristaltic waves are produced by constriction of the lumen as a result of circular muscle contraction, followed by shortening of the tube by longitudinal muscle contraction. These contractions progress from the superior end of the esophagus to the *gastroesophageal junction* at a rate of about 2–4 cm per second, as they empty the contents of the esophagus into the cardiac region of the stomach.

Heartburn and Hiatal Hernia   The last 2–4 cm of the esophagus behaves like a sphincter muscle, even though it does not differ anatomically from other regions of the esophagus. This *lower esophageal sphincter,* as it is called, normally remains closed until swallowed food is pushed through it by peristalsis. When a person is not swallowing, the pressure in the abdominal cavity (where the stomach is located) is greater than the pressure in the esophagus (which is subject to the subatmospheric pressure of the thoracic cavity). Opening of the lower esophageal sphincter at inappropriate times could thus result in the movement of gastric juice into the esophagus. This causes an unpleasant sensation commonly referred to as *heartburn* (although the heart is not involved).

The esophagus passes from the thoracic to the abdominal cavities through an opening in the diaphragm— the *esophageal hiatus.* One of the most common disorders of the gastrointestinal tract is *hiatal hernia,* which occurs when part of the stomach protrudes through the esophageal hiatus into the thoracic cavity.

Stomach

Swallowed food is delivered from the esophagus to the *cardiac region* of the stomach (see figure 18.7). An imaginary horizontal line drawn through the cardiac region divides the stomach into an upper *fundus* and a lower *body,* which comprises about two-thirds of the stomach. The distal portion of the stomach is called the *antrum.* Contractions of the stomach push partially digested food from the antrum through the pyloric sphincter, and thus into the first part of the small intestine.

The inner surface of the stomach is thrown into long folds called *rugae* (shown in figure 18.7), which can be seen with the unaided eye. Microscopic examination of the gastric mucosa shows that it is likewise folded. The openings of these folds into the stomach lumen are called **gastric pits.** The cells that line the folds deeper in the mucosa secrete various products into the stomach; these form the exocrine **gastric glands** (see figure 18.8).

Figure 18.6   Peristalsis in the esophagus.

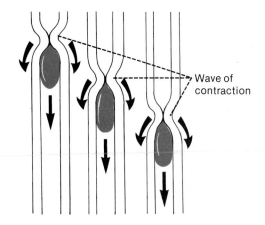

Wave of contraction

Figure 18.7   Structure of the stomach.

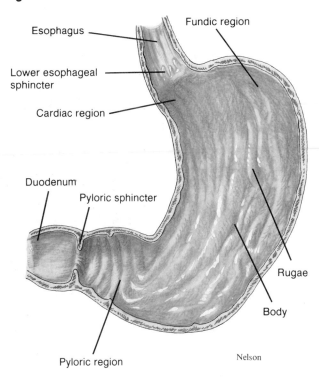

Esophagus

Fundic region

Lower esophageal sphincter

Cardiac region

Duodenum

Pyloric sphincter

Rugae

Body

Pyloric region

Nelson

Figure 18.8   Microscopic structure of the stomach.

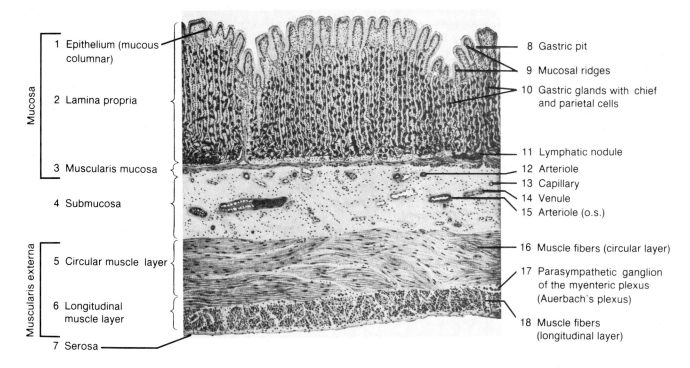

Mucosa

1 Epithelium (mucous columnar)

2 Lamina propria

3 Muscularis mucosa

4 Submucosa

Muscularis externa

5 Circular muscle layer

6 Longitudinal muscle layer

7 Serosa

8 Gastric pit

9 Mucosal ridges

10 Gastric glands with chief and parietal cells

11 Lymphatic nodule

12 Arteriole

13 Capillary

14 Venule

15 Arteriole (o.s.)

16 Muscle fibers (circular layer)

17 Parasympathetic ganglion of the myenteric plexus (Auerbach's plexus)

18 Muscle fibers (longitudinal layer)

Table 18.1 Secretions of the fundus and antrum regions of the stomach.

Stomach Region	Cell Type	Secretions
Fundus	Parietal cells	Hydrochloric acid; intrinsic factor
	Chief cells	Pepsinogen
	Goblet cells	Mucus
	Argentaffen cells	Histamine, serotonin
Antrum	G cells	Gastrin
	Chief cells	Pepsinogen
	Goblet cells	Mucus

There are several different types of cells in the gastric glands that secrete different products. These include (1) *goblet cells,* which secrete mucus; (2) *parietal cells,* which secrete hydrochloric acid (HCl); (3) *chief cells,* which secrete pepsinogen (an inactive form of the protein-digesting enzyme pepsin); (4) *Argentaffen cells,* which secrete serotonin and histamine; and (5) *G cells,* which secrete the hormone *gastrin* (discussed in a later section). In addition to these products, the gastric mucosa (probably the parietal cells) secretes a polypeptide called *intrinsic factor,* which is required for absorption of vitamin B_{12} in the intestine.

Pepsin and Hydrochloric Acid

Secretion of hydrochloric acid by parietal cells makes gastric juice very acidic, with a pH less than 2. This strong acidity serves three functions: (1) ingested proteins are denatured at low pH—that is, their tertiary structure (see chapter 2) is altered so that they become more digestible; (2) under acidic conditions, weak pepsinogen enzymes partially digest each other—this frees the active pepsin enzyme as small peptide fragments are removed; and (3) pepsin is most active under very acidic conditions—it has a pH optimum of about 2.0.

Figure 18.9 Section of the gastric mucosa showing gastric pits and glands.

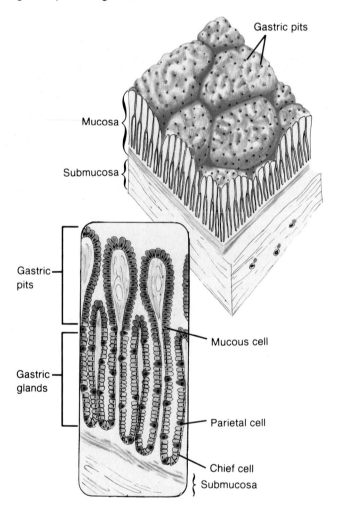

Figure 18.10 The gastric mucosa secretes the inactive enzyme pepsinogen and hydrochloric acid (HCl). Through a process of autocatalysis, the active enzyme pepsin is produced. Pepsin digests proteins into shorter polypeptides.

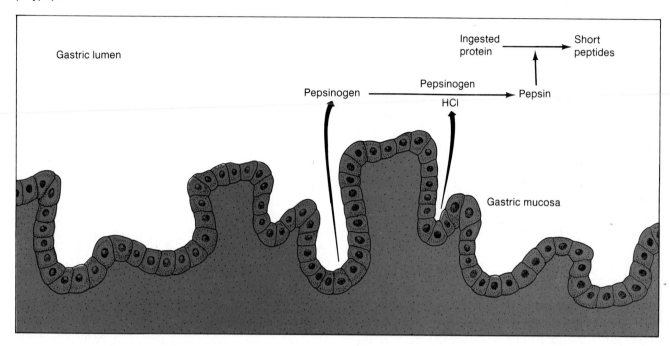

Hydrochloric acid breaks the weak bonds that stabilize the tertiary structure of ingested proteins—the acid does *not* hydrolyze the peptide bonds between amino acids. Hydrolysis of these peptide bonds—digestion of proteins—is catalyzed by pepsin. This digestive activity, together with the muscular contractions of the stomach that help mix the stomach contents with acid and mucus, converts "lumpy" food into a soft, homogenous mixture called **chyme.** The stomach serves as a reservoir for this chyme and regulates the rate at which chyme enters the small intestine.

Digestion and Absorption in the Stomach

Proteins are only partially digested in the stomach, as will be described in a later section; carbohydrates and fats are left essentially undigested until they enter the intestine. The partially digested proteins in chyme are completely digested, and absorbed, in the intestine. Patients with partial gastric resections, therefore, and even those with complete gastrectomies (removal of the stomach), can still adequately digest and absorb their food.

The only function of the stomach that appears to be essential for life is secretion of intrinsic factor. This polypeptide, as previously discussed, is needed for the intestinal absorption of vitamin B_{12}. Vitamin B_{12} is required for maturation of red blood cells in the bone marrow. A patient with a gastrectomy can therefore develop *pernicious anemia* (low red blood cell count due to vitamin B_{12} deficiency) unless they receive B_{12} injections.

Almost all of the products of digestion are absorbed (enter the blood) through the wall of the intestine. The only commonly ingested substances that can be absorbed across the stomach wall are *alcohol* and *aspirin.* This occurs as a result of the lipid solubility of these molecules. Under the acidic conditions of gastric juice, aspirin assumes a non-ionized form that can pass through the lipid cell membranes of the epithelial cells that line the stomach (see figure 18.11). This can damage the mucosal barrier and cause gastric bleeding. The significance of this bleeding depends on the amount of aspirin ingested and the susceptibility of the person.

Gastritis and Peptic Ulcers

The gastric mucosa is normally resistant to the harsh conditions of low pH and pepsin activity in gastric juice. The reasons for resistance to pepsin digestion are not well understood. Resistance to hydrochloric acid appears to be due to three interrelated mechanisms: (1) the stomach lining is covered by a thin layer of alkaline mucus that may offer some protection; (2) the epithelial cells of the mucosa are joined together by tight junctions that prevent acid from leaking into the submucosa; and (3) epithelial cells that are damaged are exfoliated (shed) and replaced by new cells. This latter process results in the loss of about one-half million cells a minute, so that the entire epithelial lining is replaced every three days.

These three mechanisms provide a barrier that prevents self-digestion of the stomach. Breakdown of the gastric mucosa barrier may occur at times, however, perhaps as a result of the detergent action of bile salts that are regurgitated from the small intestine through the pyloric sphincter. When this occurs, acid can leak through the mucosal barrier to the submucosa. This causes direct damage and stimulates inflammation. The histamine released from mast cells during inflammation (see chapter 17) may stimulate further acid secretion and result in further damage to the mucosa. The inflammation that occurs during these events is known as *acute gastritis.*

Once the mucosal barrier is broken, acid can produce craterlike holes and even complete perforation of the stomach wall. These are called *gastric ulcers,* and are usually located in the lesser curvature of the stomach. Ulcers of the first 3–4 cm of the duodenum (the first part of the small intestine), however, are about four times more common than gastric ulcers. The injuries caused by gastric acid in the stomach and duodenum are known as **peptic ulcers.** It has been estimated that about 10 percent of American men and 4 percent of American women will develop peptic ulcers during their lifetime.

The duodenum is normally protected from gastric acid by the buffering action of bicarbonate in alkaline pancreatic juice (as described in later sections). People who develop duodenal ulcers, however, apparently produce excessive amounts of gastric acid (as shown in table 18.2) that cannot be neutralized by pancreatic juice. Excess gastric acid secretion is apparently due to excessive stimulation of the gastric parietal cells by parasympathetic fibers.

Small ulcers may heal themselves, but large ulcers—particularly those that perforate major blood vessels—may require emergency surgery. People with gastritis and peptic ulcers must avoid substances that stimulate acid secretion (such as caffeine and alcohol), and they often require *antacid* therapy.

Common antacids include sodium bicarbonate, calcium carbonate, magnesium hydroxide (milk of magnesia), aluminum hydroxide (such as Amphojel), and combinations that include both aluminum and magnesium hydroxide (Maalox and Mylanta). While these antacids serve to neutralize acid, they do not make gastric juice neutral (pH 7.0). It should be recalled that pH is an inverse logarithmic function. If the pH of gastric juice is originally 1.3, neutralization of 90 percent of the acid would raise the pH to 2.3, and neutralization of 99 percent of the acid would raise the pH to 3.3.

Figure 18.11 In acidic gastric juice, aspirin is converted into a lipid-soluble, nonionized molecule that is able to pass through cell membranes into the gastric epithelial cells. This process is accompanied by small amounts of gastric bleeding.

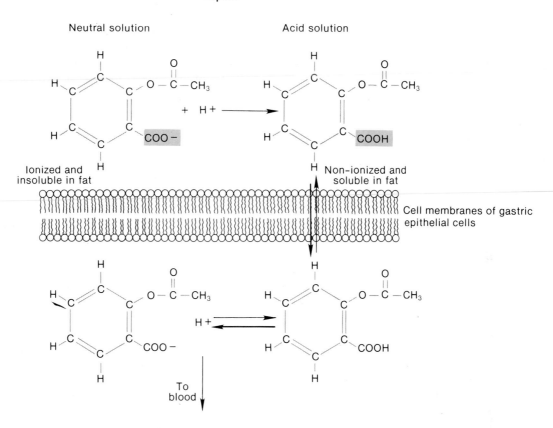

Table 18.2 Gastric secretion of hydrochloric acid, measured in milliequivalents (mEq). One milliequivalent is equal to one millimole of H^+.

Condition	Day (mEq/hr)	Night (mEq/12 hrs)	Maximal (mEq/hr)
Normal	2–3	18	16–20
Gastric ulcer	2–4	8	16–20
Duodenal ulcer	4–10	60	25–40
Zollinger-Ellison syndrome (G cell tumor)	30	120	45

Figure 18.12   The microscopic structure of the duodenum.

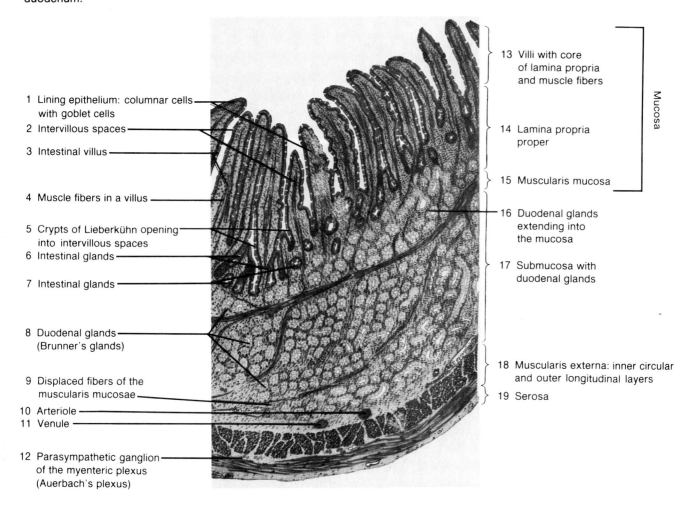

1 Lining epithelium: columnar cells with goblet cells

2 Intervillous spaces

3 Intestinal villus

4 Muscle fibers in a villus

5 Crypts of Lieberkühn opening into intervillous spaces

6 Intestinal glands

7 Intestinal glands

8 Duodenal glands (Brunner's glands)

9 Displaced fibers of the muscularis mucosae

10 Arteriole

11 Venule

12 Parasympathetic ganglion of the myenteric plexus (Auerbach's plexus)

13 Villi with core of lamina propria and muscle fibers

14 Lamina propria proper

15 Muscularis mucosa

16 Duodenal glands extending into the mucosa

17 Submucosa with duodenal glands

18 Muscularis externa: inner circular and outer longitudinal layers

19 Serosa

Mucosa

Intestine

The **small intestine** averages 6.5–7.0 m (twenty-one feet) in length. The first 20–30 cm (ten inches) from the pyloric sphincter is the *duodenum.* The next two-fifths of the small intestine is the *jejunum,* and the last three-fifths is the *ileum.* The ileum empties into the large intestine through the ileocecal valve.

The products of digestion are absorbed across the epithelial lining of the intestinal mucosa. This occurs primarily in the jejunum, although some absorption also occurs in the duodenum and ileum. Absorption occurs at a rapid rate as a result of the high mucosal surface area in the intestine. The mucosa and submucosa form large folds—the *plicae circulares,* which can be observed with the unaided eye; surface area is further increased by microscopic folds of mucosa called *villi* and by foldings of the epithelial cell membrane (which can only be seen with an electron microscope) called *microvilli.*

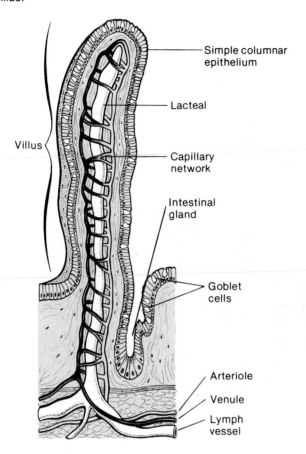

Figure 18.13 Diagram of the structure of an intestinal villus.

Villus

Simple columnar epithelium

Lacteal

Capillary network

Intestinal gland

Goblet cells

Arteriole

Venule

Lymph vessel

Villi and Microvilli

Each villus is a fingerlike projection of mucosa that projects into the intestinal lumen. The villi are covered with columnar epithelial cells, and interspersed among these are mucous-secreting *goblet cells*. The lamina propria forms a connective tissue core of each villus and contains numerous lymphocytes, blood capillaries, and a lymphatic vessel called the *central lacteal*. Absorbed monosaccharides and amino acids are secreted into the blood capillaries; absorbed fat enters the central lacteals.

Epithelial cells at the tips of the villi are continuously exfoliated and are replaced by cells that are pushed up from the bases of the villi. The epithelium at the base of the villi invaginates downwards at various points to form narrow pouches that open through pores to the intestinal lumen. These structures are called the *crypts of Lieberkühn*. Although the crypts look somewhat like gastric pits and glands, they do not secrete digestive enzymes. Instead, the crypts appear to function as "nurseries" where new epithelial cells are formed by mitotic division. New cells formed in the crypts are pushed upwards and replace those that exfoliate from the tips of the villi. By this means the entire *enteric* (a term that refers to intestine) epithelium is completely replaced every three to six days.

Figure 18.14 Intestinal villi and crypts of Lieberkühn. The crypts serve as sites for production of new epithelial cells. The time required for migration of these new cells to the tip of the villi is shown in *(b)*. Epithelial cells are exfoliated from the tips of the villi.

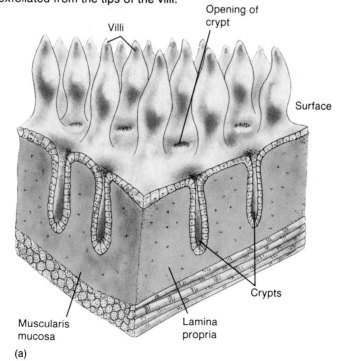

Villi

Opening of crypt

Surface

Crypts

Muscularis mucosa

Lamina propria

(a)

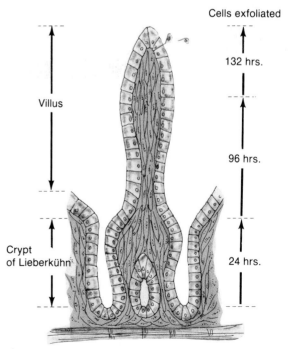

Cells exfoliated

132 hrs.

Villus

96 hrs.

Crypt of Lieberkühn

24 hrs.

(b)

Figure 18.15 The cell membrane of epithelial cells that line the small intestine are folded into microvilli. Glycoproteins within the membranes of the microvilli contain polysaccharides that extend out into the lumen and form the glycocalyx that covers the brush border epithelium *(b)*. A diagram of an intestinal epithelial cell is shown in *(c)*. Notice that the microvilli contain actin filaments attached to a terminal web of myosin filaments. This allows the microvilli to shorten.

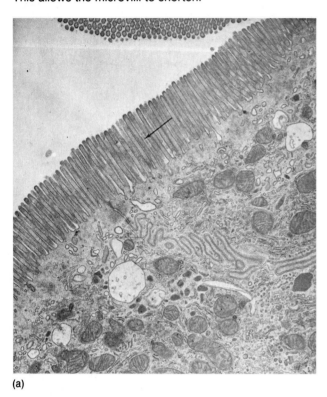

(a)

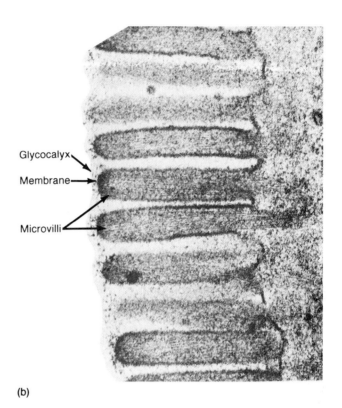

Glycocalyx

Membrane

Microvilli

(b)

The surface area of each villus is greatly increased by the fact that each columnar epithelial cell contains about 3,000–6,500 *microvilli* that project into the lumen. Microvilli are fingerlike projections formed by foldings of the cell membrane. These can only be clearly seen in an electron microscope. In a light microscope, the microvilli produce a somewhat vague **brush border** on the edges of the columnar epithelial cells. The term *brush border* is thus often used synonymously with microvilli in descriptions of the intestine.

Intestinal Enzymes

In addition to providing a large surface area for absorption, the cell membranes of the microvilli contain digestive enzymes. These enzymes are not secreted into the lumen, but instead remain attached to the cell membrane with their active sites exposed to the chyme. These **brush border enzymes** hydrolyze disaccharides, polypeptides, and other substrates (see table 18.3). One brush border enzyme, *enterokinase,* is required for activation of the protein-digesting enzyme *trypsin,* which enters the intestine in pancreatic juice.

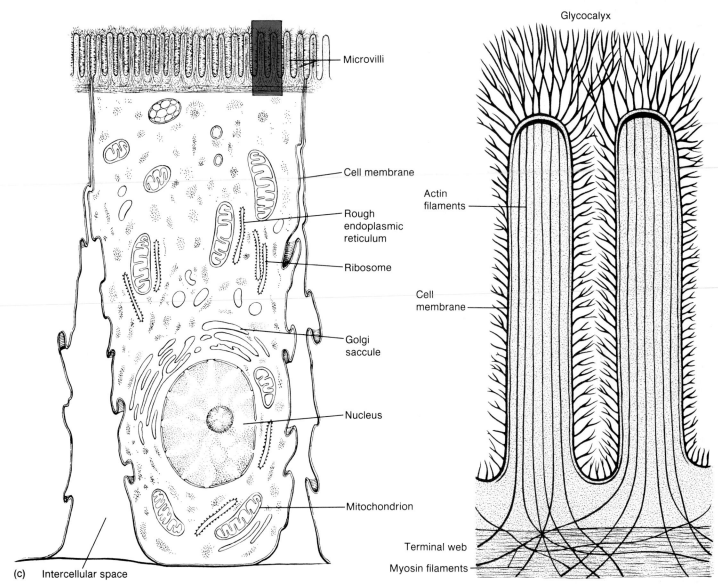

Glycocalyx

Microvilli

Cell membrane

Rough
endoplasmic
reticulum

Ribosome

Golgi
saccule

Nucleus

Mitochondrion

Actin
filaments

Cell
membrane

Terminal web

Myosin filaments

(c)   Intercellular space

Table 18.3   Brush border enzymes attached to the cell membrane of microvilli in the small intestine.

Category	Enzyme	Comments
Disaccharidase	Sucrase	Digests sucrose to glucose and fructose. Deficiency produces gastrointestinal disturbances.
	Maltase	Digests maltose to glucose.
	Lactose	Digests lactose to glucose and galactose. Deficiency produces gastrointestinal disturbances (lactose intolerance).
Peptidase	Aminopeptidase	Produces free amino acids, dipeptides, and tripeptides.
	Enterokinase	Activates trypsin (and indirectly other pancreatic juice enzymes). Deficiency results in protein malnutrition.
Phosphatase	Ca^{++}, Mg^{++}—ATPase	Needed for absorption of dietary calcium. Enzyme activity regulated by vitamin D.
	Alkaline phosphatase	Removes phosphate groups from organic molecules. Enzyme activity may be regulated by vitamin D.

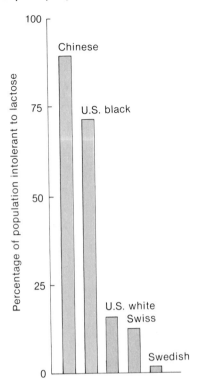

Figure 18.16 Frequency of lactose intolerance among different groups of people.

Lactose Intolerance   Milk sugar, or lactose, is a disaccharide of glucose and galactose. In order to utilize this carbohydrate as an energy source the disaccharide must be hydrolyzed into its component monosaccharides, which are absorbed through the small intestine. This is accomplished by the action of a brush border enzyme—*lactase.* This enzyme is active in almost all children under the age of four, but becomes inactive in most adults in the world, with the exception of whites in the United States and northern Europeans (see figure 18.16).

The inability to digest lactose is called *lactose intolerance.* When a person with this condition drinks milk or eats dairy products (with the exception of yogurt and cheese, in which lactose is already digested by bacteria), the undigested lactose accumulates in the intestine and increases the osmolality of chyme. This results in osmotic accumulation of water and diarrhea, as the excess water is excreted with waste material. The presence of undigested lactose in the large intestine also serves as an energy source for resident bacteria (which have lactase) that produce CO_2 and methane gas as a product of cell respiration. This gas can produce feelings of discomfort, cramps, and other unpleasant symptoms. In severe forms, the fluid loss from diarrhea may cause significant dehydration.

Intestinal Contractions and Motility
Like cardiac muscle, intestinal smooth muscle is capable of spontaneous electrical activity and automatic, rhythmic contractions. Spontaneous depolarization begins in the longitudinal smooth muscle and is conducted to the circular smooth muscle layer across *nexuses.* The term *nexus* is used here to indicate an electrical synapse between smooth muscle cells. The spontaneous depolarizations—called **pacesetter potentials**—decrease in amplitude as they are conducted from one muscle cell to another, much like excitatory postsynaptic potentials (EPSPs—described in chapter 7). Also like EPSPs, pacesetter potentials stimulate production of action potentials in the smooth muscle cells through which they are conducted.

The nexuses conduct the pacesetter potentials, not the action potentials. Action potentials are therefore limited to those smooth muscle cells that are depolarized to threshold by the spreading pacesetter potentials. When this occurs, the action potentials stimulate smooth muscle contraction. The rate at which this automatic activity occurs is influenced by autonomic nerves. Intestinal smooth muscle contraction is stimulated by parasympathetic (vagus nerve) innervation and reduced by sympathetic nerve activity; this is exactly opposite to the effects of these nerves on cardiac muscle.

The small intestine has two major types of contractions: (1) *peristalsis* and (2) *segmentation.* Peristalsis is much weaker in the small intestine than it is in the esophagus or stomach. **Intestinal motility**—movement of chyme through the intestine—is relatively slow (about 1 cm per minute), and is due primarily to the fact that the pressure at the pyloric end of the intestine is greater than the pressure at the distal end that empties into the large intestine.

The major contractile activity of the small intestine is *segmentation.* This term refers to muscular constrictions of the lumen that occur simultaneously at different intestinal segments (see figure 18.18). Segmentation slows the motility of the intestine (chyme is moved more slowly), but serves to mix the chyme more thoroughly.

Figure 18.17 The smooth muscle of the gastrointestinal tract produces and conducts spontaneous pacesetter potentials. As these potential changes reach a threshold level of depolarization they stimulate production of action potentials, which in turn stimulate smooth muscle contraction.

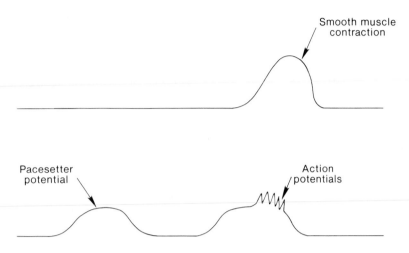

Smooth muscle contraction

Pacesetter potential

Action potentials

Figure 18.18 Segmentation in the small intestine. Simultaneous contractions of many segments of the intestine help to mix the chyme with digestive enzymes and mucus.

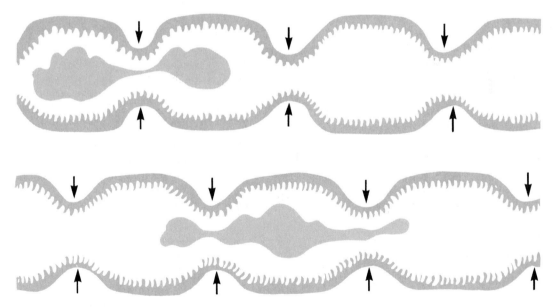

Figure 18.19   Different regions of the large intestine and their anatomical relationships to other structures of the digestive system. The relative size of the small intestine has been **shortened** in this illustration for the sake of clarity.

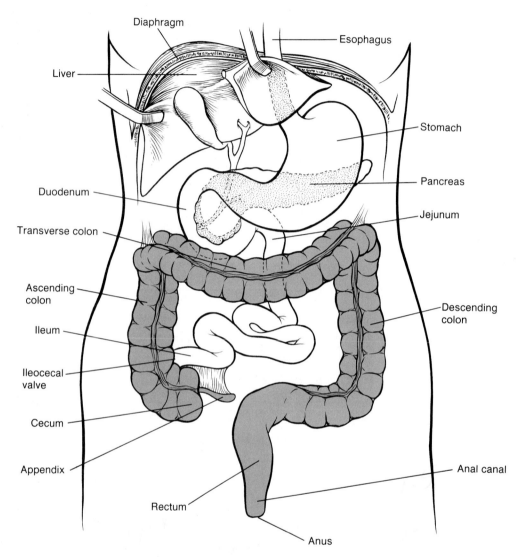

Large Intestine

Chyme from the ileum passes into the *cecum,* which is a blind-ending pouch at the beginning of the large intestine, or *colon.* Waste material then passes in sequence through the ascending colon, transverse colon, descending colon, sigmoid colon, rectum, and anal canal. Waste material (feces) is excreted through the anus.

As in the small intestine, the mucosa of the large intestine contains many scattered lymphocytes and lymphatic nodules, and is covered by columnar epithelial cells and mucous-secreting goblet cells. Although this epithelium does form crypts of Lieberkühn, there are no villi in the large intestine—the intestinal mucosa therefore presents a flat appearance (see figure 18.20). The outer surface of the colon bulges outwards to form pouches, or **haustra** (see figure 18.21). Occasionally, the muscularis externa of the haustra may become so weakened that the wall forms a more elongated outpouching, or diverticulum (divert = turned aside). Inflammation of these structures is called *diverticulitis.*

Figure 18.20   Surface view of the mucosa of the large intestine, as seen with the scanning electron microscope. Arrow points to the opening of a crypt of Lieberkühn.

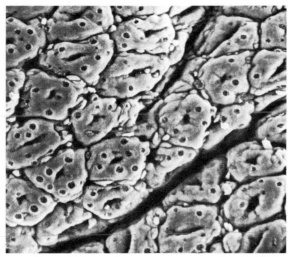

Kessel and Kardon

Figure 18.21   Radiograph after a barium enema showing the haustra of the large intestine.

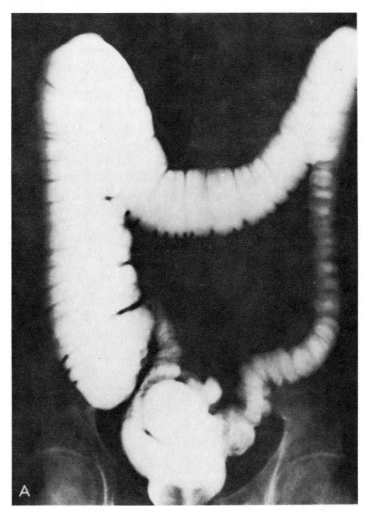

As material passes through the large intestine, Na⁺, K⁺, and water are absorbed. It has been estimated that as much as 300 ml of water per day can be absorbed across the 1.5 m (about five feet) of colon. The waste material that is left is then passed to the *rectum,* which is the final 15 cm of the colon. This causes an increase in rectal pressure and the urge to defecate. If defecation is to occur, the feces descend into the *anal canal,* which is the last 2–4 cm of the gastrointestinal tract.

Anal Sphincters and Defecation   If the urge to defecate is denied, feces are prevented from entering the anal canal by constriction of the *internal anal sphincter,* which is composed of smooth muscle. In this case the feces remain in the rectum, and may even back up into the sigmoid colon. Defecation normally occurs when the rectal pressure rises to a particular level that is determined, to a large degree, by habit. At this point the internal sphincter relaxes to admit feces into the anal canal.

During the act of defecation the *external anal sphincter,* composed of striated muscle, also relaxes and opens the anus. Excretion is accomplished by contractions of abdominal and pelvic skeletal muscles, which increase intra-abdominal pressure and raise the pelvic floor, and by contractions of the colon.

Disorders of the Colon   The **vermiform** ("wormlike") **appendix** is a short, thin outpouching of the cecum. It does not function in digestion, but like the tonsils, it contains numerous lymphatic nodules (see figure 18.23) and is subject to inflammation—a condition called *appendicitis.* This is commonly detected in its later stages by pain in the lower right quadrant of the abdomen. Rupture of the appendix can cause inflammation of the surrounding body cavity—*peritonitis.* This dangerous event may be prevented by surgical removal of the inflamed appendix (appendectomy).

Figure 18.22   Structure of the rectum, anal canal, and anus.

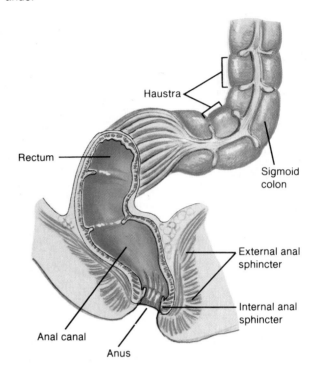

Constipation occurs when fecal material accumulates because of longer-than-normal periods between defecations. The slower rate of elimination allows more time for absorption of water, so that the waste products become "harder" in constipation. Though uncomfortable and sometimes painful, this condition is not usually dangerous. In fact, periods of up to one year between defecations have been recorded. *Diarrhea* is usually produced when waste materials pass too quickly through the colon, so that insufficient time is allowed for water absorption. In *irritable bowel syndrome,* for example, emotional disturbances result in increased colon motility and decreased water absorption. Excessive diarrhea may result in dangerous levels of dehydration and electrolyte imbalance.

Figure 18.23 Microscopic appearance of a cross section of the human appendix.

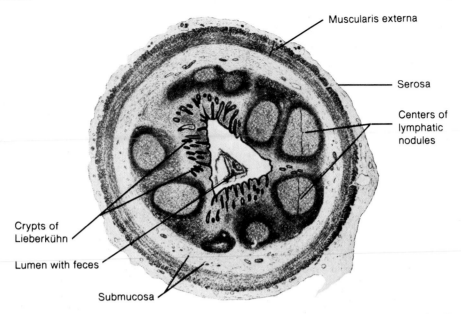

Muscularis externa

Serosa

Centers of lymphatic nodules

Crypts of Lieberkühn

Lumen with feces

Submucosa

1. List the secretory products of gastric parietal and chief cells, and describe how pepsinogen becomes activated. Describe the roles of HCl and of pepsin in protein digestion.
2. Describe the intestinal structures that increase surface area, and explain the function of the crypts of Lieberkühn.
3. Define the term *brush border enzymes* and list examples. Describe how the symptoms of lactose intolerance are produced.
4. Describe peristalsis and compare it with intestinal segmentation.
5. Explain how pacesetter potentials are produced and transmitted and how autonomic nerve activity affects electrical excitation and contraction of intestinal smooth muscle.

Pancreas, Liver, and Gallbladder

The liver, which is the largest internal organ, lies immediately beneath the diaphragm in the abdominal cavity. Attached to the inferior surface of the liver, between the right and quadrate lobes (see figure 18.24), is the pear-shaped gallbladder. This organ is approximately 10 cm long by 3.5 cm wide. The pancreas is located behind the stomach, as can be seen by retracting the lesser curvature of the stomach (see figure 18.24).

Bile is produced by the liver and drained into *hepatic ducts,* which merge to form a single *common hepatic duct.* The common bile duct is joined by the *cystic duct* that drains the gallbladder. The gallbladder serves to store and concentrate the bile. The single duct formed by merger of the common hepatic duct and cystic duct is called the *common bile duct,* and functions to transport bile to the duodenum.

Figure 18.24   The anatomical location of the pancreas in relation to the stomach, liver, and gallbladder.

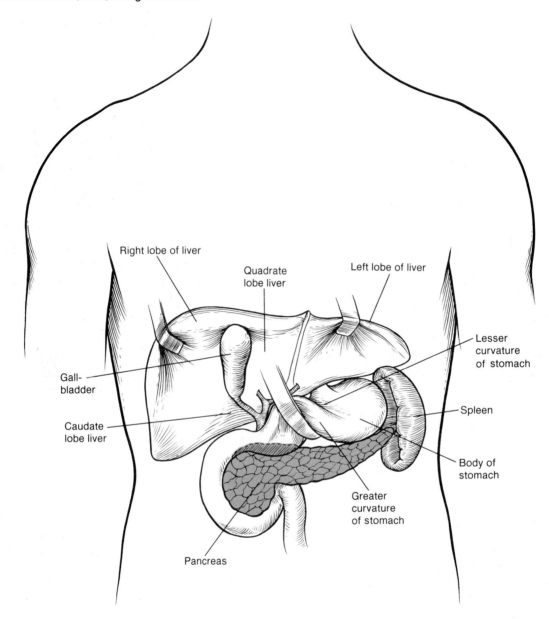

Right lobe of liver

Quadrate lobe liver

Left lobe of liver

Lesser curvature of stomach

Gall-bladder

Caudate lobe liver

Spleen

Body of stomach

Greater curvature of stomach

Pancreas

The exocrine secretion of the pancreas, known as **pancreatic juice,** is drained by the *pancreatic duct.* This duct runs the entire length of the pancreas, and eventually joins with the common bile duct to form a single opening into the duodenum. This single opening is the *greater duodenal papilla.* Excretion of pancreatic juice and bile through this opening is controlled by the *sphincter of Oddi.*

Pancreas

The pancreas is both an exocrine and an endocrine gland. The endocrine structures, known as the *islets of Langerhans,* are scattered clusters of cells that secrete the hormones *insulin* and *glucagon.* The actions of these hormones will be discussed in chapter 20. Most of the pancreas is composed of exocrine cells that are grouped into clusters called *pancreatic acini,* which secrete pancreatic juice into the pancreatic duct.

Figure 18.25 The pancreatic duct joins the common bile duct to empty its secretions through the duodenal papilla into the duodenum. Release of bile and pancreatic juice into the duodenum is controlled by the sphincter of Oddi.

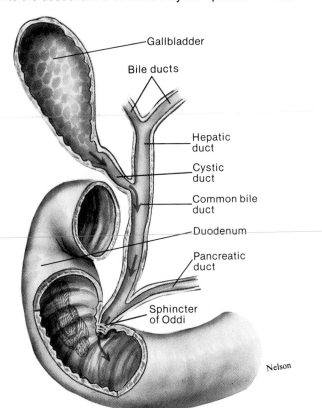

Gallbladder
Bile ducts
Hepatic duct
Cystic duct
Common bile duct
Duodenum
Pancreatic duct
Sphincter of Oddi
Nelson

Figure 18.26 The pancreas is both an exocrine and an endocrine gland. Pancreatic juice—the exocrine product—is secreted by acinar cells into the pancreatic duct. Scattered "islands" of cells, called the islets of Langerhans, secrete the hormones insulin and glucagon into the blood.

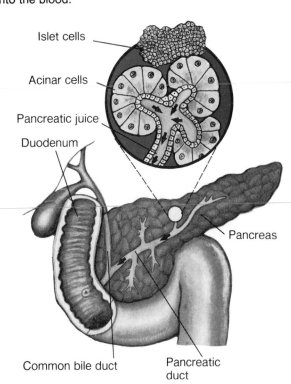

Islet cells
Acinar cells
Pancreatic juice
Duodenum
Pancreas
Common bile duct
Pancreatic duct

Pancreatic Juice Pancreatic juice contains water, bicarbonate, and a wide variety of digestive enzymes that are secreted into the duodenum. These enzymes include (1) *amylase,* which digests starch; (2) *trypsin,* which digests protein; and (3) *lipase,* which digests triglycerides; other pancreatic enzymes are indicated in table 18.4. It should be noted that complete digestion of protein and polysaccharides in the small intestine requires the hydrolytic activity of both pancreatic enzymes and brush border enzymes.

Most pancreatic enzymes are produced as inactive molecules, or *zymogens,* which helps to minimize the risk of self-digestion within the pancreas. The inactive form of typsin—called trypsinogen—is activated within the small intestine by the catalytic action of the brush border enzyme *enterokinase.* Enterokinase converts trypsinogen to active trypsin. Trypsin, in turn, activates the other zymogens of pancreatic juice by cleaving off polypeptide sequences that inhibit the activity of these enzymes (see figure 18.27).

Activation of trypsin is therefore the triggering event for activation of other pancreatic enzymes. Actually, the pancreas does produce small amounts of active trypsin, yet the other enzymes don't become active until pancreatic juice enters the duodenum. This is because pancreatic juice also contains a small protein called *pancreatic trypsin inhibitor,* which attaches to trypsin and inactivates it in the pancreatic acini and ducts.

Disorders of the Pancreas Inflammation of the pancreas may result when the various safeguards against self-digestion are insufficient. *Acute pancreatitis* is believed to be caused by reflux of pancreatic juice and bile from the duodenum into the pancreatic duct. In addition to the adverse effects that these two agents may directly have, trypsin can activate phospholipase that may destroy the phospholipid structure of cell membranes.

Table 18.4   Enzymes in pancreatic juice. Most are secreted in an inactive form (as zymogens) and are activated within the duodenum.

Enzyme	Zymogen	Activator	Action
Trypsin	Trypsinogen	Enterokinase	Cleaves internal peptide bonds
Chymotrypsin	Chymotrypsinogen	Trypsin	Cleaves internal peptide bonds
Elastase	Proelastase	Trypsin	Cleaves internal peptide bonds
Carboxypeptidase	Procarboxypeptidase	Trypsin	Cleaves last amino acid from carboxyl-terminal end of polypeptide
Phospholipase	Prophospholipase	Trypsin	Cleaves fatty acids from phospholipids such as lecithin
Lipase	None	None	Cleaves fatty acids from glycerol
Amylase	None	None	Digests starch to maltose and short chains of glucose molecules
Cholesterolesterase	None	None	Releases cholesterol from its bonds with other molecules
Ribonuclease	None	None	Cleaves RNA to form short chains
Deoxyribonuclease	None	None	Cleaves DNA to form short chains

Figure 18.27   The pancreatic protein-digesting enzyme *trypsin* is secreted in an inactive form known as trypsinogen. This inactive enzyme (zymogen) is activated by a brush border enzyme, enterokinase *(En)*, located in the cell membrane of microvilli. Active trypsin in turn activates other zymogens in pancreatic juice.

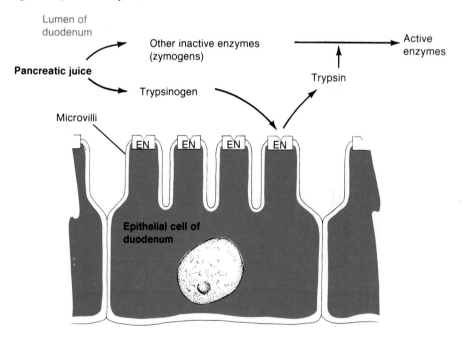

Figure 18.28 The liver receives two vascular supplies. The hepatic artery carries oxygen to the liver cells. The hepatic portal vein carries blood from the intestine to the liver. Blood that has passed through the liver enters the vena cava and returns to the general circulation.

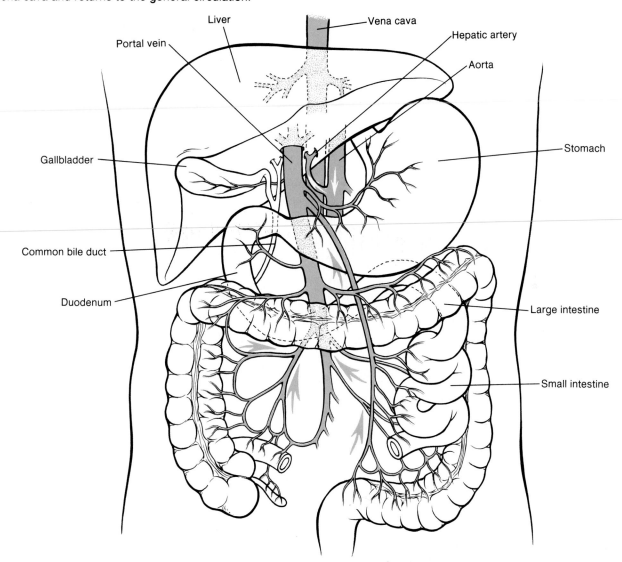

While pancreatic enzymes are exocrine secretions (because they are secreted into the pancreatic duct), small amounts of these enzymes do "leak" into the blood. Trypsin activity in blood, however, cannot normally be measured because this enzyme is inactivated by two plasma proteins—alpha-1-antitrypsin and alpha-2-macroglobulin. Pancreatic amylase activity in the blood, in contrast, can be measured (this enzyme is not inhibited, but it is not active in blood because blood does not contain glycogen). Measurements of pancreatic amylase activity in plasma are commonly performed to assess the health of the pancreas.

Structure and Function of the Liver

The products of digestion that are absorbed into blood capillaries in the intestine do not directly enter the general circulation. Instead, this blood is delivered first to the liver. Capillaries in the digestive tract drain into the *hepatic portal vein,* which carries this blood to capillaries in the liver; it is not until the blood has passed through this second capillary bed that it enters the general circulation through the *hepatic vein* that drains the liver. The term **portal system** is used to describe this unique pattern of circulation: capillaries → vein → capillaries → vein. In addition to receiving venous blood from the intestine, the liver also receives arterial blood via the *hepatic artery* (see figure 18.28).

Figure 18.29 The structure of the liver. Hepatocytes are arranged in plates so that blood that passes through sinusoids *(a)* will be in contact with each liver cell. *(b)* is a scanning electron micrograph of the liver.

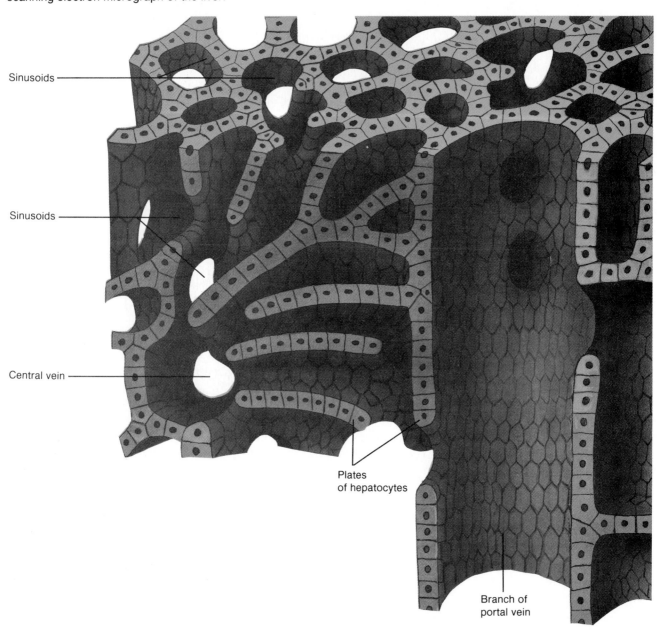

Sinusoids

Sinusoids

Central vein

Plates
of hepatocytes

Branch of
portal vein

(a)

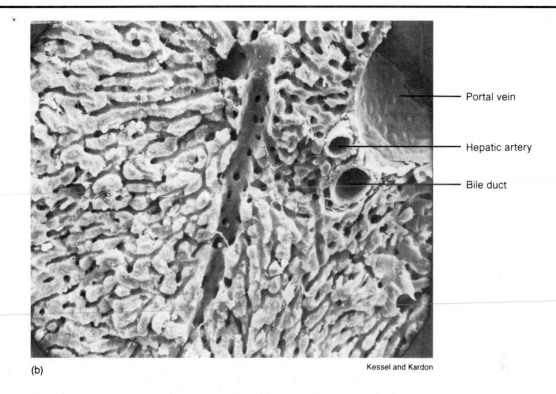

Portal vein

Hepatic artery

Bile duct

(b)

Kessel and Kardon

Although the liver is the largest internal organ, it is, in a sense, only one to two cells thick. This is because the liver cells, or *hepatocytes,* form one-to-two-cell thick **plates** that are separated from each other by large capillary spaces called **sinusoids** (see figure 18.29). The sinusoids are lined with phagocytic *Kupffer cells,* but the large intercellular gaps between adjacent Kupffer cells make these sinusoids more highly permeable than other capillaries. The plate structure of the liver and the high permeability of the sinusoids allows each hepatocyte to have direct contact with the blood.

Structure and Function of Liver Lobules

The hepatic plates are arranged into functional groups called *liver lobules.* In the middle of each lobule is a *central vein,* and at the periphery of each lobule are branches of the hepatic portal vein and of the hepatic artery, which open into the spaces *between* hepatic plates. Arterial blood and portal venous blood thus mix as the blood flows within the sinusoids from the periphery of the lobule to the central vein. The central veins of different liver lobules converge to form the hepatic vein, which carries blood from the liver to the inferior vena cava.

Bile is produced by hepatocytes and secreted into thin channels called *bile canaliculi* located *within* each hepatic plate (see figure 18.30). These bile canaliculi are drained at the periphery of each lobule by *bile ducts,* and these in turn drain into the hepatic duct, which carries the bile to the gallbladder (via the cystic duct) and to the intestine (via the common bile duct). Since blood travels in the sinusoids and bile travels in the canaliculi within the hepatic plates, blood and bile do not mix in the liver lobules.

Enterohepatic Circulation   In addition to the normal constituents of bile (see table 18.5), a wide variety of exogenous compounds, such as drugs, are secreted by the liver into the bile ducts. Analogous to the renal clearance—in which compounds are removed from blood and excreted in urine—the liver can "clear" the blood of particular compounds by excreting them into the intestine with the bile. The liver also can clear the blood by other mechanisms that will be described in the next section.

Figure 18.30   Flow of blood and bile in a liver lobule. Blood flows within sinusoids from portal vein to central vein (from the periphery to the center of a lobule). Bile flows within hepatic plates from the center to bile ducts at the periphery of a lobule.

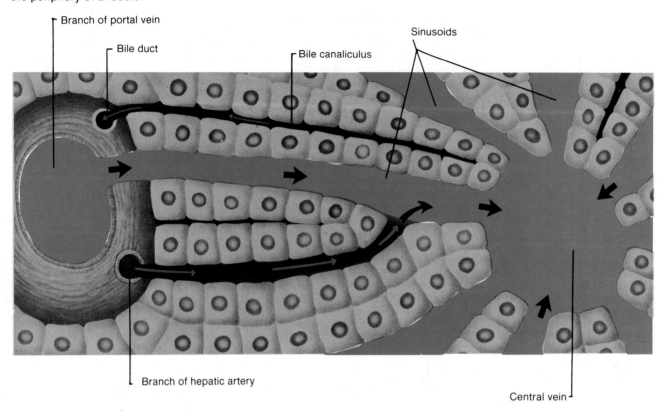

Branch of portal vein

Bile duct

Bile canaliculus

Sinusoids

Branch of hepatic artery

Central vein

Table 18.5   Compounds that the liver excretes into the bile. Many of these are absorbed in the intestine and return to the liver by way of the hepatic portal vein. Compounds that recycle in this way are said to have an enterohepatic circulation.

	Compound	Comments
Endogenous (naturally occurring)	Bile salts, urobilinogen, cholesterol	High percentage is absorbed and has an enterohepatic circulation
	Lecithin	Small percentage is absorbed and has an enterohepatic circulation
	Bilirubin	No enterohepatic circulation
Exogenous (drugs)	Ampicillin, streptomycin, tetracycline	High percentage is absorbed and has an enterohepatic circulation
	Sulfonamides, penicillin	Small percentage is absorbed and has an enterohepatic circulation

Figure 18.31 Enterohepatic circulation. Substances excreted in the bile may be absorbed by the intestinal epithelium and be recycled to the liver via the hepatic portal vein.

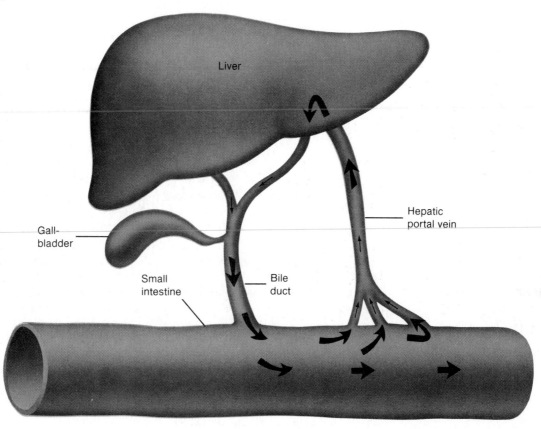

Many compounds that are released with the bile into the intestine are not excreted with the feces, however. Some of these can be absorbed through the small intestine and enter the hepatic portal blood. These absorbed molecules are thus carried back to the liver, where they can be again secreted by hepatocytes into the bile ducts. Compounds that recirculate between the liver and intestine in this way are said to have an *enterohepatic circulation* (see figure 18.31).

Detoxification of the Blood The liver can remove biologically active molecules such as hormones and drugs from the blood by (1) excreting these compounds in the bile, as previously discussed; (2) phagocytosis of these compounds by the Kupffer cells that line the sinusoids; and (3) chemical alteration of these molecules within the hepatocytes.

Ammonia, for example, is a very toxic molecule produced by the action of bacteria in the intestine. The observation that portal blood has an ammonia concentration that is four to fifty times greater than in the hepatic vein indicates that this compound is removed by the liver. The liver has the enzymes needed to convert ammonia into less toxic *urea* molecules, which are released into the blood and eliminated by the kidneys in the urine. Similarly, the liver converts toxic porphyrins into bile pigment, or *bilirubin,* and toxic purines into *uric acid.*

Steroid hormones and other nonpolar compounds (those that are not very water soluble), such as many drugs, are inactivated in their passage through the liver by modifications in their chemical structure. The liver has enzymes that convert these molecules into more polar (more water-soluble) forms by *hydroxylation* (addition of OH^- groups) and by *conjugation* with highly polar groups such as sulfate and glucuronic acid. Polar derivatives of steroid hormones and drugs are less biologically active and, because of their increased water solubility, are more easily excreted by the kidneys into the urine.

Table 18.6   Summary of the major categories of liver functions.

Functional Category	Actions
Detoxification of blood	Phagocytosis by Kupffer cells Chemical alteration of biologically active molecules (hormones and drugs) Production of urea, uric acid, and other molecules that are less toxic than parent compounds Excretion of molecules in bile
Carbohydrate metabolism	Conversion of blood glucose to glycogen and fat Production of glucose from liver glycogen and from other molecules (amino acids, lactic acid) by gluconeogenesis Secretion of glucose into the blood
Lipid metabolism	Synthesis of triglyceride and cholesterol Excretion of cholesterol in bile Production of ketone bodies from fatty acids
Protein synthesis	Production of albumin Production of plasma transport proteins Production of clotting factors (fibrinogen, prothrombin, and others)
Secretion of bile	Synthesis of bile salts Conjugation and excretion of bile pigment (bilirubin)

Secretion of Glucose, Triglycerides, and Ketone Bodies   The liver helps to regulate the blood glucose concentration by either removing glucose from the blood or adding glucose to the blood, according to the needs of the body. After a carbohydrate-rich meal, some of the glucose in the hepatic portal blood is removed by the liver and converted into glycogen and triglycerides (fat). During fasting, the liver adds glucose to the blood by hydrolysis of its stored glycogen (a process called *glycogenolysis*) and by conversion of noncarbohydrate molecules to glucose (a process called *gluconeogenesis*). The liver also contains the enzymes required to convert free fatty acids into ketone bodies, which are secreted into the blood in large amounts during prolonged fasting. These process, and their hormonal control, are discussed in detail in chapter 20.

Production of Plasma Proteins   Plasma albumin and most of the plasma globulins (with the exception of the immunoglobins) are produced in the liver. Albumin comprises about 70 percent of the total plasma protein and contributes most to the colloid osmotic pressure of the blood (see chapter 12). The globulins produced by the liver have a wide variety of functions, including transport of steroid and thyroid hormones, transport of cholesterol and triglycerides, inhibition of trypsin activity, and blood clotting. Clotting factors I (fibrinogen), II (prothrombin), III, V, VII, IX, X, and XI are all produced by the liver. A summary of liver function is presented in table 18.6.

Table 18.7   Composition of the bile.

Component	Concentration
pH	5.7–8.6
Bile salts	140–2230 mg/100 ml
Lecithin	140–810 mg/100 ml
Cholesterol	97–320 mg/100 ml
Bilirubin	12–70 mg/100 ml
Urobilinogen	5–45 mg/100 ml
Sodium	145–165 mEq/L
Potassium	2.7–4.9 mEq/L
Chloride	88–115 mEq/L
Bicarbonate	27–55 mEq/L

Bile Production and Excretion

The liver produces and secretes 250–1,500 ml of bile per day into the hepatic ducts. Most of this bile travels through the hepatic ducts to the cystic duct, and from there to the gallbladder. The gallbladder absorbs water and electrolytes, so that bile stored in the gallbladder becomes four to ten times more concentrated than the bile secreted by the liver. The major constituents of bile include bile salts, bile pigment (bilirubin), phospholipids (mainly lecithin), cholesterol, and inorganic ions (see table 18.7).

Figure 18.32 Enterohepatic circulation of urobilinogen. Bacteria in the intestine convert bile pigment (bilirubin) into urobilinogen. Some of this pigment goes out in the feces; some is absorbed by the intestine and is recycled through the bile. A portion of the urobilinogen that is absorbed goes into the general circulation and is therefore filtered by the kidneys into the urine.

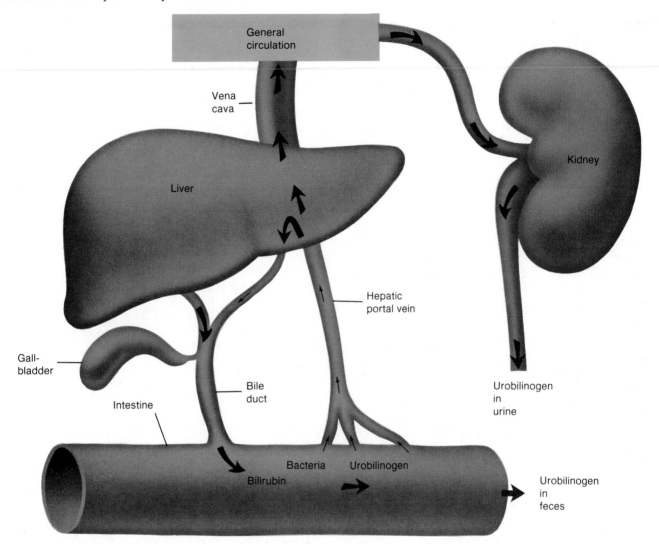

Bilirubin and Urobilinogen Bile pigment, or *bilirubin,* is produced in the spleen, liver, and bone marrow from heme groups (minus the iron) derived from hemoglobin. Without the protein part of hemoglobin, the **free bilirubin** is not very water soluble and thus must be carried in the blood attached to albumin proteins. This protein-bound bilirubin cannot be filtered by the kidneys into the urine, nor can it be directly excreted by the liver into the bile.

The liver can take some of the free bilirubin out of the blood and conjugate (combine) it with glucuronic acid. This **conjugated bilirubin** is water soluble, and is secreted into the bile (not into the blood). Once the conjugated bil-

irubin enters the intestine it is converted by bacteria into another pigment—**urobilinogen**—which is partially responsible for the color of the stools. About 30 to 50 percent of the urobilinogen, however, is absorbed by the intestine and enters the blood. Some of this absorbed urobilinogen is carried to the liver and reenters the bile (has an enterohepatic circulation); the rest enters the general circulation. The urobilinogen in plasma, unlike free bilirubin, is not attached to albumin and therefore is easily filtered into the urine, giving the urine its characteristic yellow color.

Figure 18.33 The two major bile acids (which form bile salts) in humans.

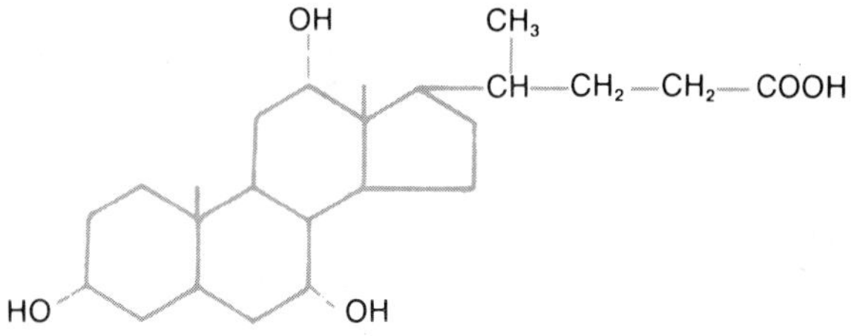

Cholic acid

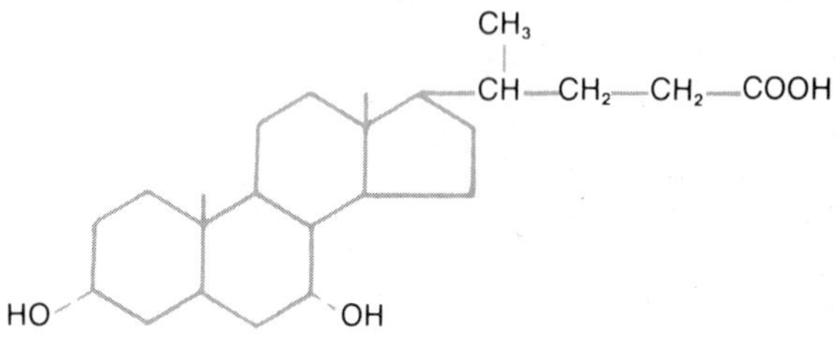

Chenodeoxycholic acid

Bile Salts The bile salts are derivatives of cholesterol (a non-water soluble lipid) that have two to four polar groups on each molecule. The principal bile salts in humans are *cholic acid* and *chenodeoxycholic acid* (see figure 18.33). In aqueous solutions these molecules "huddle" together to form loose aggregates known as **micelles;** the nonpolar parts are located in the central region of the micelle (away from water), while the polar groups face water around the periphery of the micelle. Cholesterol in the bile is absorbed into the micelles so that the micelles that enter the intestine are mixtures of bile salts and cholesterol. The bile salts, and the micellar structure that they form, aid the digestion and absorption of fats (as will be discussed in a later section).

Disorders of the Liver and Gallbladder
Inflammation of the liver, or *hepatitis,* may result from a variety of causes such as viral infections, alcohol abuse, allergy, and drugs (including tranquilizers and general anesthetics). This condition is usually reversible and most patients completely recover. As might be predicted, people with hepatitis have depressed levels of plasma albumin and increased concentrations of gamma globulin (antibodies).

In *cirrhosis,* large numbers of liver lobules are destroyed and replaced with permanent connective tissue and "regenerative nodules" of hepatocytes. These regenerative nodules don't have the platelike structure of normal liver tissue, and are therefore less functional. One indication of this decreased function is the entry of ammonia from the hepatic portal blood into the general circulation (see figure 18.34). Cirrhosis may be caused by chronic alcohol abuse, viral hepatitis, and other agents that attack liver cells.

Jaundice Jaundice is a yellow staining of the tissues produced by high blood concentrations of either free or conjugated bilirubin. Since free bilirubin is derived from heme, abnormally high concentrations of this pigment may result from an unusually high rate of red blood cell destruction. This can occur, for example, as a result of Rh disease (erythroblastosis fetalis, as described in chapter 17) in the second Rh positive baby born to an Rh negative mother. Jaundice also may occur in otherwise healthy premature or underweight infants because of the fact that their hemoglobin levels normally decrease from about 19 g per 100 ml to 14 g per 100 ml near the time of birth. This condition is called *physiological jaundice of the newborn.* Jaundice caused by free bilirubin under these circumstances is usually transient and does not indicate a disease state. Premature infants may also develop jaundice due to the fact that the hepatic enzymes that conjugate bilirubin (a reaction needed for excretion of bilirubin in the bile) don't normally mature until about the tenth month of fetal life. Jaundice due to high levels of conjugated bilirubin in the blood is commonly produced in adults when excretion of bile is blocked by gallstones.

Gallstones There are two types of gallstones: one with bilirubin as the major component, the other with cholesterol as the major component. About three-fourths of the gallstones found in the United States are cholesterol-containing gallstones. Formation of these gallstones apparently occurs when there are not enough bile salts to dissolve cholesterol in the micelles. Since cholesterol is very insoluble in water, the cholesterol crystalizes under these circumstances and forms gallstones. This condition is sometimes treated with the bile salt chenodeoxycholic acid, which helps to dissolve the stones slowly.

Figure 18.34 Hepatic processing of intestinal ammonia in the normal liver and in cirrhosis. The normal liver converts most of the ammonia produced by intestinal bacteria into urea, which is secreted into the vena cava. The liver in cirrhosis is unable to process the ammonia in this way, and therefore ammonia enters the systemic circulation.

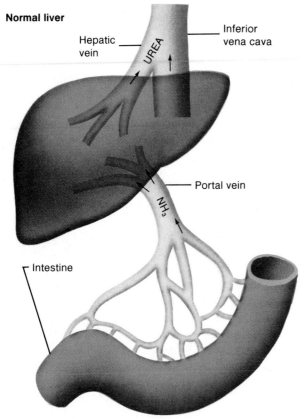

Figure 18.35 Cholesterol gallstones in the gallbladder, as seen in X ray.

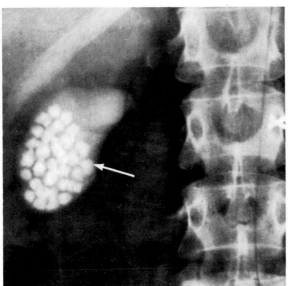

1. Use a flow chart to describe how pancreatic enzymes become activated in the intestine.
2. Describe the structure of the liver and the pathways for the flow of blood and bile.
3. Explain how the liver inactivates and excretes compounds such as hormones and drugs.
4. Define the term *enterohepatic circulation*, using urobilinigen as an example. Use a flow chart to describe the sequence of events between the formation of free bilirubin and the excretion of urobilinogen in the urine.
5. Explain why jaundice in hemolytic anemia is due to high levels of free bilirubin, and explain how high blood levels of conjugated bilirubin might be produced.

Table 18.8 Summary of the sources and activities of the major digestive enzymes.

Organ	Source	Substrate	Enzymes	Optimum pH	Products
Mouth	Saliva	Starch	Salivary amylase	6.7	Maltose
Stomach	Gastric glands	Protein	Pepsin	1.6–2.4	Shorter polypeptides
Duodenum	Pancreatic juice	Starch	Pancreatic amylase	6.7–7.0	Maltose, maltriose, and oligosaccharides
		Polypeptides	Trypsin, chymotrypsin, carboxypeptidase	8.0	Amino acids, dipeptides, and tripeptides
		Triglycerides	Pancreatic lipase	8.0	Fatty acids and monoglycerides
	Epithelial membranes	Maltose	Maltase	5.0–7.0	Glucose
		Sucrose	Sucrase	5.0–7.0	Glucose + fructose
		Lactose	Lactase	5.8–6.2	Glucose + galactose
		Polypeptides	Aminopeptidase	8.0	Amino acids, dipeptides, tripeptides

Digestion and Absorption of Carbohydrates, Proteins, and Lipids

The caloric value of food is found predominantly in its content of carbohydrates, lipids, and proteins. In the average American diet, carbohydrates account for approximately 50 percent of the total calories ingested, protein accounts for 11 to 14 percent, and lipids account for the balance of the total calories. These foods consist primarily of long combinations of subunits (monomers) that must usually be digested by hydrolysis reactions into the free monomers before absorption can occur. The characteristics of the major digestive enzymes are summarized in table 18.8.

Digestion and Absorption of Carbohydrates

Most of the ingested carbohydrates are in the form of starch, which is a long polysaccharide of glucose in the shape of straight chains with occasional branchings. The most commonly ingested sugars are the disaccharide sucrose (table sugar, consisting of glucose and fructose) and the disaccharide lactose (milk sugar, consisting of glucose and galactose). Digestion of starch begins in the mouth with the action of **salivary amylase,** or **ptyalin.** This enzyme cleaves some of the bonds between adjacent glucose molecules, but most people don't chew their food long enough for significant digestion to occur in the mouth. The digestive activity of salivary amylase stops when the food enters the stomach because this enzyme is inactivated at the low pH of gastric juice.

Digestion of starch therefore occurs mainly in the duodenum as a result of the action of **pancreatic amylase.** This enzyme cleaves the straight chains of starch to produce the disaccharide *maltose* and the trisaccharide *maltriose.* Pancreatic amylase, however, cannot hydrolyze the bond between glucose molecules at the branch points in the starch. As a result, short, branched chains of glucose molecules—called *oligosaccharides*—are released together with maltose and maltriose by activity of this enzyme (see figure 18.36).

Maltose, maltriose, and the oligosaccharides of glucose released from partially digested starch, together with the disaccharides sucrose and lactose, are hydrolyzed to their monosaccharides by brush border enzymes located on the microvilli of duodenal epithelial cells. Absorption of these monosaccharides across the membrane of the microvilli occurs by means of coupled transport (described in chapter 6). In this process, glucose binds to the same carrier as Na^+, and enters the duodenal cell as Na^+ diffuses down its electrochemical gradient. This is a type of active transport because energy from ATP is needed to maintain the Na^+ gradient. Glucose is then secreted from the epithelial cells into capillaries within the villi.

Figure 18.36 Pancreatic amylase digests starch into maltose, maltriose, and short oligosaccharides containing branch points in the chain of glucose molecules.

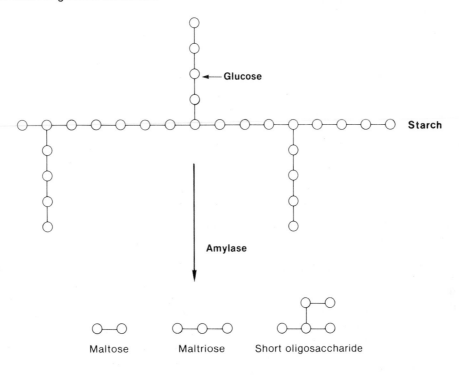

Digestion and Absorption of Protein

Protein digestion begins in the stomach with the action of pepsin. This results in the liberation of some amino acids, but the major products of pepsin digestion are short chain polypeptides. This activity helps to produce a more homogenous chyme, but is not essential for the complete digestion of protein that occurs—even in patients with total gastrectomies—in the small intestine.

Most protein digestion occurs in the duodenum and jejunum. The pancreatic juice enzymes *trypsin, chymotrypsin,* and *elastase* cleave peptide bonds within the interior of the polypeptide chains. These enzymes are thus grouped together as *endopeptidases*. Enzymes that remove amino acids from the ends of polypeptide chains, in contrast, are called *exopeptidases*. These include the pancreatic juice enzyme *carboxypeptidase,* which removes amino acids from the carboxyl-terminal end of polypeptide chains, and the brush border enzyme *aminopeptidase*. Aminopeptidase cleaves amino acids from the amino-terminal ends of polypeptide chains.

As a result of the action of these enzymes, polypeptide chains are digested into free amino acids, dipeptides, and tripeptides. The free amino acids are absorbed through the epithelial cells of the intestinal mucosa and secreted into blood capillaries. The dipeptides and tripeptides may enter the epithelial cells, but are then digested within these cells into amino acids, which are secreted into the blood. Newborn babies appear to be capable of absorbing a substantial amount of undigested protein; in adults, however, only the free amino acids enter the portal vein. Foreign food protein, which would be very antigenic, does not normally enter the blood.

Figure 18.37 Polypeptide chains are digested into free amino acids, dipeptides, and tripeptides by the action of pancreatic juice enzymes and brush border enzymes. The amino acids, dipeptides, and tripeptides enter duodenal epithelial cells. Dipeptides and tripeptides are hydrolyzed into free amino acids within the epithelial cells, and these products are secreted into capillaries that carry them to the hepatic portal vein.*

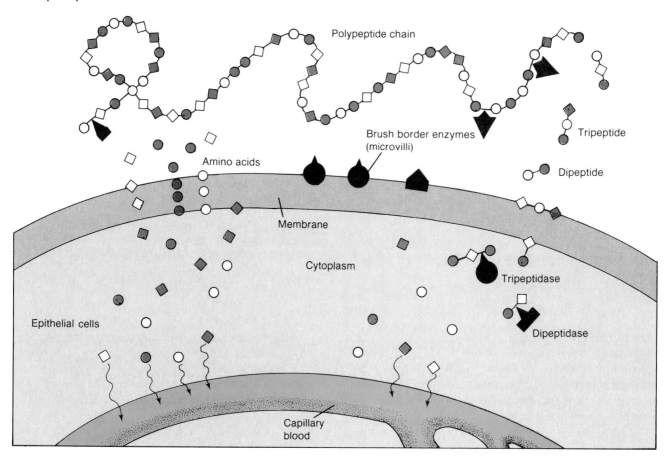

*From "The Lining of the Small Intestine," by Moog, Florence. Copyright 1981 by Scientific American, Inc. All rights reserved.

Digestion and Absorption of Lipids

Fat enters the intestine undigested in the form of small droplets (similar to those produced when oil and water are shaken, as in a salad dressing). Bile salts, which are polar on one side and nonpolar on the other, function as detergents and break up these droplets to form a suspension of much finer droplets. This process is called **emulsification.** The emulsified fat is not digested—the bonds joining the subunits together are not hydrolyzed—but the total surface area is increased by the emulsification process. This aids digestion because the fat-digesting action of pancreatic lipase, which is a water-soluble enzyme, can only occur at the surface of the fat droplets.

Digestion of Fat Almost all fat digestion occurs in the small intestine as a result of the action of pancreatic lipase and phospholipase A. Lipase removes two of the three fatty acids from each triglyceride molecule, and thus liberates free fatty acids and monoglycerides (glycerol attached to one fatty acid). Phospholipase A likewise digests phospholipids such as lecithin into fatty acids and lysolecithin (the remainder of the lecithin molecule after the two fatty acids are removed).

Absorption of Fat Free fatty acids, monoglycerides, and lysolecithin are more polar than the undigested lipids, and are able to be moved more easily into the micelles of bile salts, cholesterol, and lecithin that are secreted in the bile. Entry of the products of lipid digestion into the micelles produces "mixed micelles," which are smaller than the undigested lipid droplets. As a result of their small size, micelles are able to move rapidly into the brush border of the intestinal epithelium.

Figure 18.38 Pancreatic juice lipase digests fat (triglycerides) by cleaving off the first and third fatty acids. This produces free fatty acids and monoglycerides. Sawtooth structures indicate hydrocarbon chains in the fatty acids.

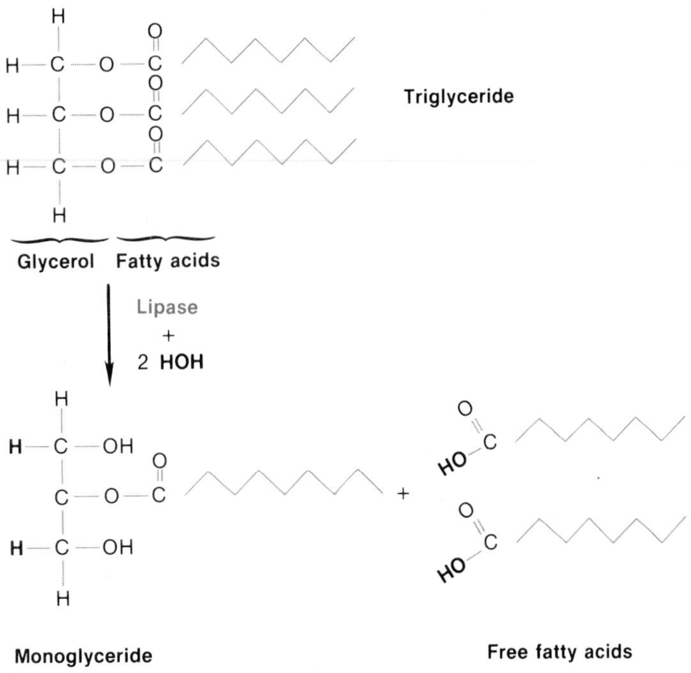

Glycerol   Fatty acids

Triglyceride

Lipase
+
2 HOH

Monoglyceride

Free fatty acids

Figure 18.39 Steps in the digestion of fat (triglycerides), and the entry of fat digestion products (fatty acids and monoglycerides) into micelles of bile salts secreted by the liver.

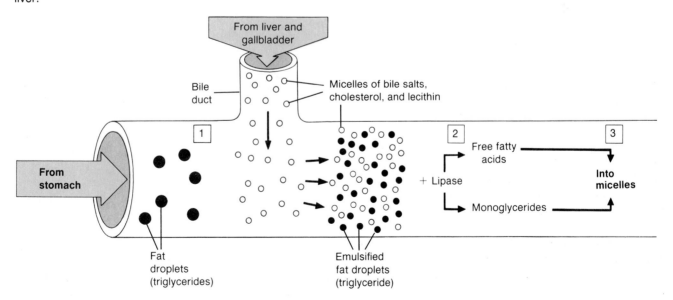

Step 1 Emulsification of fat droplets by bile salts

Step 2 Hydrolysis of triglycerides in emulsified fat droplets into fatty acid and monoglycerides

Step 3 Dissolving of fatty acids and monoglycerides into micelles to produce "mixed micells"

Figure 18.40 Fatty acids and monoglycerides from micelles within the small intestine are absorbed by epithelial cells and converted intracellularly into triglycerides. These are then combined with protein to form chylomicrons, which enter the lymphatic vessels (lacteals) of the villi. These lymphatic vessels transport the chylomicrons to the thoracic duct, which empties them into the venous blood.

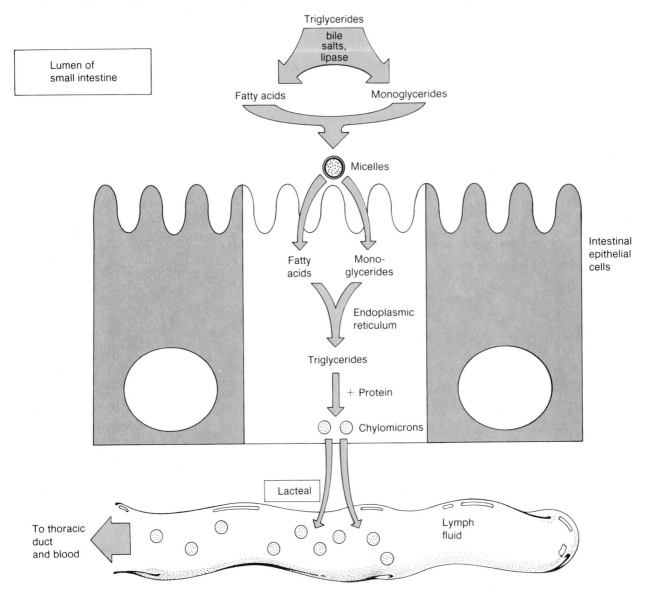

Free fatty acids, monoglycerides, and lysolecithin can leave the micelles and pass through the membrane of the microvilli to enter intestinal epithelial cells. There is also some evidence to suggest that the micelles may be transported intact into the epithelial cells, and that the lipid digestion products may be removed intracellularly from these micelles. In any event, these products are used to *resynthesize* triglycerides and phospholipids within the epithelial cells. This is different from the absorption of amino acids and monosaccharides, which pass through the intestinal epithelial cells without being altered.

Table 18.9 Summary of the physiological effects of gastrointestinal hormones.

Secreted by	Hormone	Effects
Stomach	Gastrin	Stimulates parietal cells to secrete HCl Stimulates chief cells to secrete pepsinogen Maintains structure of gastric mucosa
Small intestine	Secretin	Stimulates water and bicarbonate secretion in pancreatic juice Potentiates actions of cholecystokinen on pancreas
Small intestine	Cholecystokinin (CCK)	Stimulates contraction of the gallbladder Stimulates secretion of pancreatic juice enzymes Potentiates action of secretin on pancreas Maintains structure of exocrine pancreas (acini)
Small intestine	Gastric inhibitory peptide (GIP)	Inhibits gastric emptying Inhibits gastric acid secretion Stimulates secretion of insulin from endocrine pancreas (Islets of Langerhans)

Triglycerides, phospholipids, and cholesterol are then combined with protein inside the epithelial cells to form aggregates called **chylomicrons.** These tiny combinations of lipids and protein are secreted into the lymphatic capillaries of the intestinal villi. Absorbed lipids thus pass through the lymphatic system, eventually entering venous blood by way of the thoracic duct. Absorption of lipids is thus significantly different from that of amino acids and monosaccharides, which directly enter the portal blood.

Transport of Lipids in Blood Once the chylomicrons are in the blood, their triglyceride content is removed by the enzyme *lipoprotein lipase,* found in plasma. This enzyme hydrolyzes triglycerides, and thus provides free fatty acids and glycerol for use by the tissue cells. The remaining "remnant particle," containing cholesterol and protein, is passed to the liver where it is further processed to form *low-density lipoproteins.* Low-density lipoproteins help transport cholesterol to various organs, including the arteries (as described in chapter 13).

1. List the enzymes involved in carbohydrate digestion, indicating their origin, site of action, substrates, and products.
2. List the enzymes involved in protein digestion, indicating their origin, site of action, substrates, whether they are endopeptidases or exopeptidases, and their products. Compare the characteristics of pepsin and trypsin.
3. Describe how bile salts aid both the digestion and absorption of fats. Describe how absorption of fat differs from absorption of amino acids and monosaccharides.

Neural and Endocrine Regulation of the Digestive System

The motility and glandular secretions of the gastrointestinal tract are, to a large degree, automatic. Neural and endocrine control mechanisms, however, can stimulate or inhibit these automatic functions to help coordinate the different stages of digestion. The sight, smell, or taste of food, for example, can stimulate salivary and gastric secretion via activation of the vagus nerve; this helps to "prime" the digestive system in preparation for a meal. Stimulation of the vagus, in this case, originates in the brain and is a conditioned reflex (as Pavlov demonstrated by training dogs to salivate in response to a bell). The vagus nerve is also involved in reflex control of one part of the digestive system by another—these "short reflexes" (which don't involve the brain) will be described in later sections.

The gastrointestinal tract is both an endocrine gland and a target organ for the actions of various hormones. Indeed, the first hormones to be discovered were gastrointestinal hormones. In 1902, Bayliss and Starling discovered that the duodenum produced a chemical regulator, which they named **secretin:** in 1905, these scientists proposed that secretin was but one of many yet undiscovered chemical regulators produced by the body. They coined the term *hormones* for this new class of regulators. Other investigators in 1905 discovered that an extract from the stomach antrum stimulated gastric acid secretion. The hormone **gastrin** was thus the second hormone to be discovered.

The chemical structures of gastrin, secretin, and another hormone from the duodenum—**cholecystokinin (CCK)**—were determined in the 1960s. More recently, a fourth hormone produced by the small intestine, **gastric inhibitory peptide (GIP),** has been added to the list of proven gastrointestinal hormones. The effects of these hormones are summarized in table 18.9.

Figure 18.41 Stimulation of gastric acid secretion by the presence of proteins in the stomach lumen and by the hormone gastrin. Secretion of gastrin is inhibited by gastric acidity. This forms a negative feedback loop.

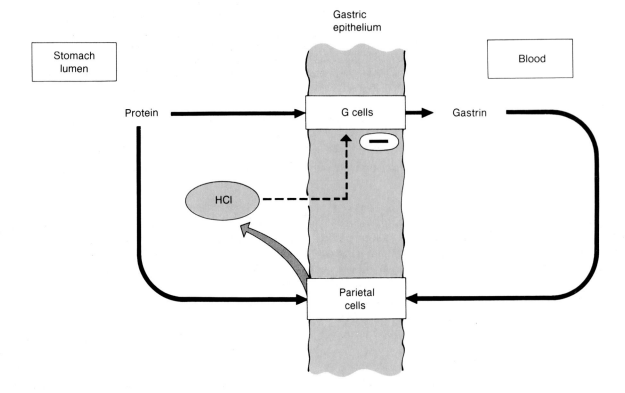

Regulation of Gastric Function

Gastric motility and secretion are, to some extent, automatic. Waves of contraction that serve to push chyme through the pyloric sphincter, for example, are initiated spontaneously by pacesetter cells in the greater curvature of the stomach. Secretion of hydrochloric acid (HCl) and pepsinogen, likewise, can be stimulated in the absence of neural and hormonal influences by the presence of cooked or partially digested protein in the stomach. The effects of autonomic nerves and hormones are superimposed on this automatic activity. Extrinsic control of gastric function is conveniently divided into three phases: (1) cephalic phase; (2) gastric phase; and (3) intestinal phase.

Cephalic Phase   The cephalic phase of gastric regulation refers to control by the brain via the vagus nerve. As previously discussed, various conditioned stimuli can evoke gastric secretion. Activation of the vagus nerve can stimulate HCl and pepsinogen secretion by three mechanisms: (1) direct vagal stimulation of gastric parietal and chief cells (this is the primary mechanism); (2) vagal stimulation of gastrin secretion by G cells, which in turn stimulates parietal cells to secrete HCl and stimulates chief cells to secrete pepsinogen; and (3) stimulation of secretion by increases in gastric blood flow.

Gastric Phase   The presence of short polypeptides and amino acids in the stomach stimulates G cells to secrete gastrin, and also stimulates the parietal and chief cells to secrete HCl and pepsinogen, respectively. This is called the gastric phase of regulation. Since gastrin also stimulates HCl and pepsinogen secretion, a *positive feedback mechanism* develops: as more HCl and pepsinogen are secreted, more short polypeptides and amino acids are released from the ingested protein, thus stimulating more secretion of gastrin and more secretion of HCl and pepsinogen (see figure 18.41).

Secretion of HCl during the gastric phase is also regulated by a *negative feedback mechanism*. As the pH of gastric juice drops, so does the secretion of gastrin—at a pH of 2.5, gastrin secretion is reduced, and at a pH of 1.0, gastrin secretion is totally abolished. Secretion of HCl thus declines accordingly. The presence of proteins and polypeptides in the stomach helps to buffer the acid and thus to prevent a rapid fall in gastric pH; more acid can thus be secreted when proteins and polypeptides are present than when they are absent. Arrival of protein into the stomach thus stimulates acid secretion two ways—by the positive feedback mechanism previously discussed, and by inhibition of the negative feedback control of acid secretion. The amount of acid secreted is, by means of these effects, closely matched to the amount of protein ingested.

Table 18.10 The cephalic, gastric, and intestinal phases in the regulation of gastric acid secretion.

Phase of Regulation	Description
Cephalic phase	1. Sight, smell, and taste of food cause stimulation of vagus nuclei in brain 2. Vagus stimulates acid secretion 　　*a)* Direct stimulation of parietal cells (major effect) 　　*b)* Stimulation of gastrin secretion; gastrin stimulates acid secretion (lesser effect)
Gastric phase	1. Distension of stomach stimulates vagus nerve; vagus stimulates acid secretion 2. Amino acids and peptides in stomach lumen stimulate acid secretion 　　*a)* Direct stimulation of parietal cells (lesser effect) 　　*b)* Stimulation of gastrin secretion; gastrin stimulates acid secretion (major effect) 3. Gastrin secretion inhibited when pH of gastric juice falls below 2.5
Intestinal phase	1. Neural inhibition of gastric emptying and acid secretion 　　*a)* Arrival of chyme in duodenum causes distension, increase in osmotic pressure 　　*b)* These stimuli activate a neural reflex that inhibits gastric activity 2. Gastric inhibitory peptide (GIP) secreted by duodenum in response to fat in chyme; GIP inhibits gastric acid secretion

Intestinal Phase The intestinal phase of gastric regulation refers to inhibition of gastric activity when chyme enters the small intestine. Investigators in 1886 demonstrated that the addition of olive oil to a meal inhibits gastric emptying, and in 1929 it was shown that the presence of fat inhibits gastric secretion. This inhibitory intestinal phase of gastric regulation is due to both a neural reflex originating from the duodenum and to a chemical hormone secreted by the duodenum.

Arrival of chyme into the duodenum increases its osmolality. This stimulus, together with stretch of the duodenum and possibly other stimuli, produces a neural reflex that results in inhibition of gastric motility and secretion. The presence of fat in the chyme also stimulates the duodenum to secrete a hormone that inhibits gastric function. This inhibitory hormone is believed to be gastric inhibitory peptide (GIP).

The inhibitory neural and endocrine mechanisms during the intestinal phase prevent further passage of chyme from the stomach to the duodenum. This gives the duodenum time to process the load of chyme that it has previously received. A breakfast of bacon and eggs therefore takes a longer time to pass through the stomach—and makes one feel "fuller" for a longer time—than does a breakfast of pancakes and syrup.

Regulation of Pancreatic Juice and Bile Secretion

Arrival of chyme into the duodenum stimulates the intestinal phase of gastric regulation and, at the same time, stimulates reflex secretion of pancreatic juice and bile. Entry of new chyme is thus retarded as the previous load is digested. Secretion of pancreatic juice and bile is stimulated both by neural reflexes initiated in the duodenum and by the secretion of the duodenal hormones cholecystokinin (CCK) and secretin.

Pancreatic Juice Secretion of pancreatic juice is stimulated by both secretin and CCK. Release of secretin occurs in response to a fall in duodenal pH below 4.5; this occurs for only a short time, however, because the acidic chyme is rapidly neutralized by alkaline pancreatic juice. Release of cholecystokinin occurs in response to the fat content of chyme in the duodenum.

Secretin stimulates production of water and bicarbonate by the pancreas. Since bicarbonate neutralizes the acidic chyme, and since secretin is secreted in response to the low pH of chyme, this completes a negative feedback loop where the effects of secretin inhibit its secretion. Cholecystokinin, in contrast, stimulates production of pancreatic enzymes such as trypsin, lipase, amylase, and others. Secretin and CCK can have different effects on the same cells (the exocrine pancreatic acinar cells) because their effects are mediated by different intracellular compounds or "messengers." The second messenger (see chapter 19) of secretin action is cyclic AMP (cAMP), while the second messenger for CCK is Ca^{++}.

Secretion of Bile The liver secretes bile continuously, but this secretion is greatly augmented following a meal. This augmented secretion is due to increased release of secretin and CCK. Secretin is the major stimulator of bile secretion, and CCK enhances this effect. The presence of chyme in the intestine does not only stimulate increased production of bile by the liver (through the effects of secretin and CCK), it also stimulates contraction of the gallbladder. Contraction of the gallbladder occurs in response to neural reflexes from the duodenum and in response to the effects of CCK.

Trophic Effects of Gastrointestinal Hormones

Patients with tumors of the stomach antrum have high acid secretion and hyperplasia (growth) of the gastric mucosa. Surgical removal of the antrum reduces gastric secretion by 60 percent and prevents hypertrophy of the

Table 18.11   Regulation of pancreatic juice and bile secretion by the hormones *secretin* and *cholecystokinin* (CCK), and by the neurotransmitter *acetylcholine* released from parasympathetic nerve endings.

Description	Secretin	CCK	Acetylcholine (vagus nerve)
Stimulus for release	Acidity of chyme decreases duodenal pH below 4.5	Fat and protein in chyme	Sight, smell of food; distension of stomach
Second messenger	Cyclic AMP	Ca^{++}	Ca^{++}
Effect on pancreatic juice	Stimulates water and bicarbonate secretion; potentiates action of CCK	Stimulates enzyme secretion; potentiates action of secretin	Stimulates enzyme secretion
Effect on bile	Stimulates secretion	Potentiates action of secretin; stimulates contraction of gallbladder	Stimulates contraction of gallbladder

gastric mucosa. Patients with peptic ulcers are sometimes treated by vagotomy—cutting of the vagus. This also reduces acid secretion by 60 percent, but has no effect on the gastric mucosa. These observations suggest that the hormone gastrin, secreted by the antrum, may exert a stimulatory, or "trophic," effect on the gastric mucosa. The structure of the gastric mucosa, in other words, is dependent on the effects of gastrin.

In a similar way, the structure of the acinar (exocrine) cells of the pancreas is dependent upon the trophic effect of CCK. Perhaps this explains why the stomach and pancreas atrophy during starvation. Since neural reflexes appear to be capable of regulating digestion, perhaps the primary function of the gastrointestinal hormones is trophic—that is, maintenance of the structure of their target organs.

1. Describe, by means of flow charts, the positive feedback and negative feedback mechanisms that operate during the gastric phase of HCl and pepsinogen secretion.
2. Describe the mechanisms involved in the intestinal phase of gastric regulation, and explain why a fatty meal takes longer to leave the stomach than a meal low in fat.
3. Explain the hormonal mechanisms involved in the production and release of pancreatic juice and bile into the small intestine.

Nutritional Requirements

Living tissue is an unstable, extremely complex state of matter that can only be maintained by the constant expenditure of energy. The energy needed for synthesis, active transport, cellular movements, and other activities essential for life is obtained directly from ATP, and indirectly from the cell respiration of glucose, fatty acids, amino acids, and other organic molecules. These molecules are obtained ultimately from food, but they can be obtained more immediately from stored glycogen, fat, and

Table 18.12   Weight in kilograms (kg) and energy content in kilocalories (kcal) of energy stores in the body of a normal 70 kg man and an obese 140 kg man. Notice that fat stores are greatly increased in obesity, while glycogen and protein stores are only slightly increased.

Fuel Supply	Normal Man (70 kg)		Obese Man (140 kg)	
Fat	15.0 kg	141,000 kcal	80.0 kg	752,000 kcal
Protein	6.0 kg	24,000	8.0 kg	32,000
Glycogen				
Muscle	0.12 kg	480	0.160 kg	640
Liver	0.07 kg	280	0.700 kg	280
Glucose	0.02 kg	80	0.025 kg	100
		165,840 kcal		785,020 kcal

Source: Adapted from Cahill, G. F. and Owen, O. E. in Rowland, C. V. Jr. (ed.) Anorexia and Obesity. Copyright © 1970, Little, Brown & Company, Boston, MA. Used with permission.

protein if food sources are temporarily inadequate. Examination of the data in table 18.12 reveals that most of the stored energy—measured in kilocalories, or "big calories"—is in the form of fat.

When the caloric intake is greater than the energy expenditure, excess calories are stored primarily as fat. This is true regardless of the source of the calories—carbohydrates, protein, or fat—because these molecules can be converted to fat by the metabolic pathways described in chapter 5. The endocrine control of total body metabolism will be described in detail in chapter 20. Since obesity greatly increases the risk of cardiovascular diseases and diabetes mellitus, people are advised to maintain an appropriate weight for their height and body frame (see table 18.13).

Food, in addition to its caloric value, also supplies the body with molecules that cannot be produced metabolically by humans. These include the *essential amino acids* and the *essential, polyunsaturated fatty acids*. The nine essential amino acids are lysine, methionine, valine, leucine, isoleucine, tryptophane, phenylalanine, threonine, and histidine. The essential fatty acids are linoleic acid and linolenic acid.

Table 18.13   1983 height and weight tables indicating body weights at which longevity is greater.

Men

Height Feet	Height Inches	Small Frame	Medium Frame	Large Frame
5	2	128–134	131–141	138–150
5	3	130–136	133–143	140–153
5	4	132–138	135–145	142–156
5	5	134–140	137–148	144–160
5	6	136–142	139–151	146–164
5	7	138–145	142–154	149–168
5	8	140–148	145–157	152–172
5	9	142–151	148–160	155–176
5	10	144–154	151–163	158–180
5	11	146–157	154–166	161–184
6	0	149–160	157–170	164–188
6	1	152–164	160–174	168–192
6	2	155–168	164–178	172–197
6	3	158–172	167–182	176–202
6	4	162–176	171–187	181–207

Weights at ages 25–59 based on lowest mortality. Weight in pounds according to frame (in indoor clothing weighing 5 lbs., shoes with 1″ heels).

Women

Height Feet	Height Inches	Small Frame	Medium Frame	Large Frame
4	10	102–111	109–121	118–131
4	11	103–113	111–123	120–134
5	0	104–115	113–126	122–137
5	1	106–118	115–129	125–140
5	2	108–121	118–132	128–143
5	3	111–124	121–135	131–147
5	4	114–127	124–138	134–151
5	5	117–130	127–141	137–155
5	6	120–133	130–144	140–159
5	7	123–136	133–147	143–163
5	8	126–139	136–150	146–167
5	9	129–142	139–153	149–170
5	10	132–145	142–156	152–173
5	11	135–148	145–159	155–176
6	0	138–151	148–162	158–179

Weights at ages 25–59 based on lowest mortality. Weight in pounds according to frame (in indoor clothing weighing 3 lbs., shoes with 1″ heels).
Source: Reprinted with permission of the Metropolitan Life Insurance Companies.

Mineral Requirements   Since human tissues are composed of the same types of molecules as the food, adequate intake of the elements carbon, hydrogen, oxygen, nitrogen, and sulfur—which form the bulk of food and human tissues—can usually be obtained from a diet with adequate caloric value. Other elements that are required in gram amounts per day are sodium, potassium, magnesium, calcium, chlorine, and phosphorous. In addition, the following *trace elements* are recognized as essential: iron, zinc, manganese, fluorine, copper, molybdenum, chromium, selenium, and iodine. These must be ingested in amounts ranging from fifty micrograms to eighteen milligrams per day (see table 18.14).

Table 18.14   Recommended daily intake of trace elements.

Element	Intake (mg/day)
Iron (males)	10
Iron (females)	18
Zinc	15
Manganese	2.5 to 5.0
Fluorine	1.5 to 4.0
Copper	2.0 to 3.0
Molybdenum	0.15 to 0.5
Chromium	0.05 to 0.2
Selenium	0.05 to 0.2
Iodine	0.15

Source: Courtesy of Food and Nutrition Board of the National Academy of Science, © 1980. Reprinted by permission.

Vitamins   Vitamins are small organic molecules that are essential for metabolism but that cannot be made by the body. There are two major groups of vitamins—the fat-soluble vitamins (A, D, E, and K) and the water-soluble vitamins. Water-soluble vitamins include thiamine (B_1), riboflavin (B_2), niacin (B_3), pyridoxine (B_6), pantothenic acid, biotin, folic acid, B_{12}, and vitamin C (ascorbic acid).

Many of the water-soluble vitamins serve as coenzymes. Thiamine, for example, is required for the activity of the enzyme that converts pyruvic acid to acetyl coenzyme A. Riboflavin and niacin are needed for production of FAD and NAD, respectively; these compounds serve as coenzymes that transfer hydrogens during cell respiration (see chapter 5). Pyridoxine is a cofactor for enzymes involved in amino acid metabolism. Deficiencies of the water-soluble vitamins can, for obvious reasons, have widespread effects in the body.

Fat-soluble vitamins generally have more specialized functions than do the water-soluble vitamins. Vitamin K, for example, is required for the production of prothrombin and for clotting factors VII, IX, and X. Vitamin D is converted into a hormone that participates in the regulation of calcium balance (see chapter 20). The visual pigments in the eye are produced from vitamin A. Vitamin E is believed to be an antioxidant, and claims have been made that this vitamin, together with vitamin C, may help protect against a wide variety of diseases. This is currently the subject of much research and controversy.

Table 18.15   Recommended daily allowances for vitamins and minerals.

	Infants	Children	Adolescents 15–18 yrs		Adults 23–50 yrs	
	0–6 mos	*4–6 yrs*	*Males*	*Females*	*Males*	*Females*
Weight, kg (lb)	6 (13)	20 (44)	66 (145)	55 (120)	70 (154)	55 (120)
Height, cm (in)	60 (24)	112 (44)	176 (69)	163 (64)	178 (70)	163 (64)
Protein, g	kg × 2.2	30	56	46	56	44
Fat-soluble vitamins†						
Vitamin A, μg	420	500	1000	800	1000	800
Vitamin D, μg	10	10	10	10	5	5
Vitamin E activity, mg	3	6	10	8	10	8
Water-soluble vitamins						
Ascorbic acid, mg	35	45	60	60	60	60
Folacin, μg	30	200	400	400	400	400
Niacin, mg	6	11	18	14	18	13
Riboflavin, mg	0.4	1.0	1.7	1.3	1.6	1.2
Thiamine, mg	0.3	0.9	1.4	1.1	1.4	1.0
Vitamin B_6, mg	0.3	1.3	2.0	2.0	2.2	2.0
Vitamin B_{12}, μg	0.5	2.5	3.0	3.0	3.0	3.0
Minerals						
Calcium, mg	360	800	1200	1200	800	800
Phosphorus, mg	240	800	1200	1200	800	800
Iodine, μg	40	90	150	150	150	150
Iron, mg	10	10	18	18	10	18
Magnesium, mg	50	200	400	300	350	300
Zinc, mg	3	10	15	15	15	15

Source: Courtesy of Food and Nutrition Board, *Recommended Dietary Allowances*, 9th ed., National Academy of Sciences © 1980. Reprinted by permission.
†Microgram

Summary

Structure of the Gastrointestinal Tract

I. The four layers of the tract include, from the lumen outwards, the mucosa, submucosa, muscularis externa, and the serosa.
  A. The submucosa and serosa are composed of connective tissue.
  B. The muscularis externa is composed of an inner circular and outer longitudinal layer of smooth muscle.
  C. Terminal parasympathetic ganglia are located in the submucosa (Meissner's plexus) and in the muscularis externa (Auerbach's plexus).
II. The mucosa consists of an inner epithelium of simple columnar cells, a connective tissue layer (the lamina propria), and a thin muscularis mucosa.
  A. The gastric mucosa is highly specialized, forming gastric pits and gastric glands.
    1. Parietal cells in the gastric glands secrete HCl, while chief cells secrete trypsinogen.
    2. G cells secrete the hormone *gastrin.*
  B. The highly folded structure of the small intestine provides a high surface area for absorption.
    1. The submucosa is folded to form large plicae circularis.
    2. The mucosa is folded to form projections called villi, with cores of lamina propria.
    3. The cell membrane of epithelial cells is folded to form microvilli.
  C. The intestinal epithelium invaginates below the villi to form crypts of Lieberkühn.
    1. The crypts do not secrete enzymes, but instead appear to produce new epithelial cells by mitotic division.
    2. Intestinal enzymes are located in the microvilli— these are called brush border enzymes.

Pancreas and Liver

I. The pancreas is both an endocrine and an exocrine gland.
  A. The endocrine cells of the pancreas are located in the islets of Langerhans, and secrete the hormones *insulin* and *glucagon.*
  B. Most of the pancreas is composed of exocrine cells that are grouped into clusters called acini.
    1. The acini produce bicarbonate and pancreatic enzymes.
    2. The products of the acini, called pancreatic juice, are secreted into the pancreatic duct.
  C. The enzymes of pancreatic juice are activated in the duodenum by trypsin.
    1. The pancreas produces trypsin in an inactive form called trypsinogen.
    2. Trypsinogen is converted to trypsin by activity of the brush border enzyme *enterokinase.*
    3. Active trypsin converts other pancreatic juice enzymes to their active forms.
II. The liver is composed of one-to-two-cell-thick plates of hepatocytes
  A. Wide blood channels called sinusoids separate the plates of hepatocytes.
    1. Blood from the hepatic portal vein (which drains the digestive tract) and from the hepatic artery empty into the periphery of liver lobules.
    2. This blood is drained into the central vein in each lobule, which carries the blood to the hepatic vein.
  B. Bile is secreted by hepatocytes into bile canaliculi located within the hepatic plates and is drained into bile ducts.
  C. Bile from the liver is stored in the gallbladder and is released together with pancreatic juice into the duodenum.

Digestion of Carbohydrates, Proteins, and Lipids

I. Starch digestion begins in the mouth with the action of salivary amylase.
  A. Starch, and the disaccharides sucrose and lactose, are digested in the intestine into their monosaccharides.
    1. Pancreatic amylase digests starch.
    2. Sucrose and lactose are digested by enzymes located in the intestinal brush border.
  B. Monosaccharides are absorbed through the intestinal mucosa into the hepatic portal blood.
II. Proteins are partially digested in the stomach.
  A. Chief cells secrete inactive pepsinogen, which is activated by HCl in the lumen of the stomach.
  B. Pepsin partially digests protein into short polypeptide chains.
  C. The partially digested protein and other ingested molecules are mixed with acid and mucus to form chyme.
  D. Digestion of proteins is completed in the small intestine.
    1. Trypsin and other protein-digesting enzymes in pancreatic juice are released into the duodenum.
    2. Aminopeptidase, located on the intestinal brush border, removes amino acids from the amino terminal end of polypeptide chains.
  E. Amino acids are absorbed into the portal blood.
III. Fat digestion occurs almost entirely in the small intestine.
  A. Ingested fat is emulsified into a fine suspension of droplets by the detergent action of bile salts.
  B. Ingested triglycerides and phospholipids are digested by pancreatic lipase and by phospholipase A, respectively.
  C. The products of lipid digestion enter micelles composed of bile salts, cholesterol, and lecithin.

D. Once the lipid digestion products enter intestinal epithelial cells, they are combined to form triglycerides and lecithin within these cells.

E. Triglycerides and lecithin are combined with protein within the intestinal epithelial cells to form chylomicrons.
  1. Chylomicrons are secreted into lymphatic capillaries within the intestinal villi.
  2. Chylomicrons are passed from the lymphatic system to venous blood through the thoracic duct.
  3. The triglyceride content of chylomicrons is removed in the blood by the action of lipoprotein lipase.

Regulation of Digestive Function

I. Motility and secretion of the digestive tract is, to a large degree, automatic.
  A. Spontaneous depolarizations called pacesetter potentials are transmitted across electrical synapses, called nexuses, between smooth muscle cells.
  B. Pacesetter potentials stimulate production of action potentials, which in turn produce spontaneous contraction of the muscles.
  C. The rate of this muscular contraction, and of glandular secretion, is increased by parasympathetic nerves and decreased by sympathetic nerves.

II. Neural and endocrine reflexes help to coordinate the activity of the digestive system.
  A. Salivary and gastric secretion is stimulated by the sight, smell, and taste of food.
    1. These are conditioned stimuli that evoke these responses through learning.
    2. Glandular secretion is stimulated by the vagus nerve that is activated, under these circumstances, by the brain.
  B. Regulation of gastric function occurs in three phases.
    1. In the cephalic phase, secretion of HCl and pepsinogen occurs in response to vagus nerve stimulation.
    2. In the gastric phase, HCl and pepsinogen secretion is stimulated by the presence of polypeptides in the stomach and by the secretion of the hormone gastrin.
    3. In the intestinal phase, gastric motility and secretion is inhibited by neural reflexes from the duodenum, and by secretion of gastric inhibitory peptide from the duodenum.

III. Arrival of chyme into the duodenum stimulates secretion of pancreatic juice and bile.
  A. This is promoted by both neural and endocrine reflexes.
  B. The acidity of chyme stimulates release of secretin from the duodenum.
    1. Secretin promotes the secretion of water and bicarbonate in pancreatic juice.
    2. Secretin also stimulates secretion of bile by the liver.
  C. The fat content of chyme stimulates secretion of cholecystokinin (CCK) from the duodenum.
    1. CCK stimulates production of pancreatic enzymes.
    2. CCK also stimulates contraction of the gallbladder and release of bile into the small intestine.
  D. Gastrointestinal hormones may be necessary for maintenance of the structure of the digestive system—this is called a trophic effect.

Self-Study Quiz

1. Which of the following statements about gastric secretion of hydrochloric acid is FALSE?
   (a) HCl is secreted by parietal cells
   (b) HCl hydrolyzes peptide bonds
   (c) HCl helps convert inactive pepsinogen to active pepsin
   (d) HCl produces a pH that allows pepsin to be maximally active

2. Intrinsic factor:
   (a) is secreted by the stomach
   (b) is a polypeptide
   (c) promotes absorption of vitamin B_{12} in the intestine
   (d) helps prevent pernicious anemia
   (e) all of these

3. Intestinal enzymes such as lactase are:
   (a) secreted by the intestine into the chyme
   (b) produced by the crypts of Lieberkühn
   (c) produced by the pancreas
   (d) attached to the cell membrane of microvilli in the epithelial cells of the mucosa

4. Most digestion occurs in:
   (a) the mouth
   (b) the stomach
   (c) the small intestine
   (d) the large intestine

5. Which of the following statements about trypsin is TRUE?
   (a) Trypsin is derived from trypsinogen in the duodenum by the digestive activity of pepsin
   (b) Active trypsin is secreted into the acini of the pancreas
   (c) Trypsin is produced in the crypts of Lieberkühn
   (d) Trypsin digests starch to maltose
   (e) The pH optimum of trypsin is less than 2
   (f) Trypsin is activated by the brush border enzyme *enterokinase*

6. Peptic ulcers:
   (a) are produced by the digestive action of pepsin
   (b) can result from excessive vagus stimulation of the pancreas
   (c) occur most often in the stomach
   (d) can occur when HCl leaks into the submucosa

7. During the gastric phase, secretion of HCl and pepsinogen is stimulated by:
   (a) vagal nerve stimulation that originates in the brain
   (b) polypeptides in the stomach lumen and gastrin secretion
   (c) secretin and cholecystokinin from the duodenum
   (d) all of these

8. Secretion of HCl by the stomach mucosa is inhibited by
   (a) neural reflexes from the duodenum
   (b) secretion of gastric inhibitory peptide from the duodenum
   (c) the lowering of gastric pH
   (d) all of these

9. The first organ to receive the blood-borne products of digestion is:
   (a) the pancreas
   (b) the liver
   (c) the heart
   (d) the brain

10. Which of the following statements about hepatic portal blood is TRUE?
    (a) it contains ingested fat
    (b) it contains ingested protein
    (c) it is mixed with bile in the liver
    (d) it drains directly into the inferior vena cava and is carried to the general circulation
    (e) all of these statements are false

The Endocrine System: *Secretion and Actions of Hormones*

Objectives

By studying this chapter, you should be able to:

1. Define the terms *hormone, prohormone,* and *prehormone*

2. Describe how hormone secretion is controlled by negative feedback inhibition

3. Describe the relationship between the hypothalamus and the posterior pituitary, and explain how secretions of the posterior pituitary are regulated

4. List the hormones of the anterior pituitary, and describe how the secretion of each is controlled by hypothalamic-releasing and -inhibiting hormones

5. Explain how secretion of hypothalamic-releasing hormones and anterior pituitary hormones is regulated by feedback loops from the target endocrine glands

6. Describe the two categories of hormones secreted by the adrenal cortex, and explain the involvement of the pituitary and adrenals in the general adaptation syndrome

7. Describe the formation, secretion, and transport of thyroid hormones, and explain how an iodine deficiency can lead to the formation of a goiter

8. Explain the mechanisms of action of steroid and thyroid hormones

9. Define the term *second messenger,* and describe how production of cAMP serves as a second messenger in the action of particular hormones

10. Describe how Ca^{++} can function as a second messenger in the action of some hormones

11. Explain how hormones can have permissive, synergistic, and antagonistic effects, and give examples of each type of hormone interaction

*T*he functions of tissue cells, and of the organs they compose, are regulated by many kinds of "biologically active" chemicals. These include ions, neurotransmitters released by axon endings, local chemicals produced within the organs, and hormones. **Hormones** are biologically active compounds secreted into the blood by endocrine glands. **Endocrine glands** are ductless glands (as compared to exocrine glands, which secrete their products into ducts) that secrete hormones into the blood.

Table 19.1 A partial list of the endocrine glands.

Endocrine Gland	Major Hormones	Primary Target Organs	Primary Effects
Adrenal cortex	Cortisol Aldosterone	Liver, muscles Kidneys	Glucose metabolism; N$^+$ retention, K$^+$ excretion.
Adrenal medulla	Epinephrine	Heart, bronchioles, blood vessels	Adrenergic stimulation.
Hypothalamus	Releasing and inhibiting hormones	Anterior pituitary	Regulates secretion of anterior pituitary hormones.
Intestine	Secretin and cholecystokinen	Stomach, liver, and pancreas	Inhibits gastric motility; stimulates bile and pancreatic juice secretion.
Islets of Langerhans (pancreas)	Insulin Glucagon	Many organs Liver and adipose tissue	Insulin promotes cellular uptake of glucose and formation of glycogen and fat; glucagon stimulates hydrolysis of glycogen and fat.
Ovaries	Estradiol-17β and progesterone	Female genital tract and mammary glands	Maintains structure of genital tract; promotes secondary sexual characteristics.
Parathyroids	Parathyroid hormone	Bone, intestine, and kidneys	Increases Ca^{++} concentration in blood.
Pineal	Melatonin	Hypothalamus and anterior pituitary	Affects secretion of gonadotrophic hormones.
Pituitary, anterior	Trophic hormones	Endocrine glands and other organs	Stimulates growth and development of target organs; stimulates secretion of other hormones.
Pituitary, posterior	Vasopressin Oxytocin	Kidneys, blood vessels Uterus, mammary glands	Vasopressin promotes water retention and vasoconstriction. Oxytocin stimulates contraction of uterus and mammary secretory units.
Stomach	Gastrin	Stomach	Stimulates acid secretion.
Testes	Testosterone	Prostate, seminal vesicles, other organs	Stimulates secondary sexual development.
Thymus	Thymosin	Lymph nodes	Stimulates white blood cell production.
Thyroid	Thyroxine (T$_4$) and triiodothyronine (T$_3$)	Most organs	Growth and development; stimulates basal rate of cell respiration (basal metabolic rate or BMR).

Recent evidence suggests that neurotransmitters, hormones, and other biologically active molecules may have a common evolutionary origin. This would explain, for example, why some chemicals that are well-established hormones may also function as neurotransmitters in the brain (as discussed in chapter 7). Whether or not the nervous and endocrine systems share a common evolutionary origin, they do share a common function—that of regulating the activities of other organ systems (and of each other) in the service of the total organism.

Hormones affect the metabolism of their *target cells* (those capable of responding to specific hormones), and by this means help regulate (1) total body metabolism; (2) growth; and (3) reproduction. The effects of hormones on body metabolism and growth are discussed in chapter 20; the regulation of reproductive function by hormones is discussed in chapter 21.

Organs such as the adrenals, thyroid, parathyroids, pituitary, and pineal glands appear to be exclusively endocrine in function. Other organs of the endocrine system secrete hormones while they serve additional functions. The stomach and small intestine, for examples, secrete hormones that help regulate their digestive functions (see chapter 18). The pancreas is both an exocrine and an endocrine gland—pancreatic juice is an exocrine secretion that enters the pancreatic duct, while the hormones insulin and glucagon are secreted into the blood by clusters of cells known as the islets of Langerhans (see figure 19.1). Likewise, the skin, thymus, liver, and kidneys secrete hormones and perform other functions described in previous chapters.

Figure 19.1 *(a)* Anatomy of some of the endocrine glands. *(b)* Islets of Langerhans within the pancreas.

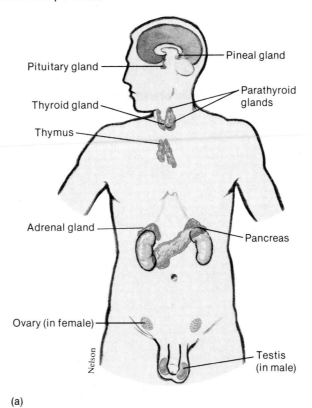

Pituitary gland

Pineal gland

Thyroid gland

Parathyroid glands

Thymus

Adrenal gland

Pancreas

Ovary (in female)

Nelson

Testis (in male)

(a)

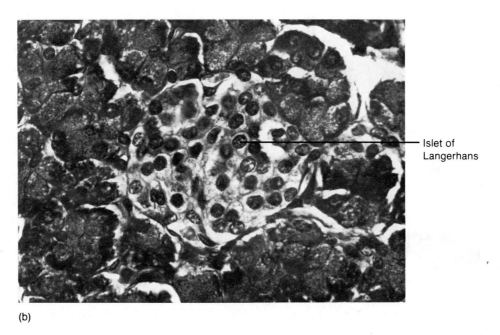

Islet of Langerhans

(b)

Figure 19.2   Simplified biosynthetic pathways for steroid hormones. Notice that progesterone (a hormone secreted by the ovaries) is a common precursor in the formation of all other steroid hormones, and that testosterone (the major androgen secreted by the testes) is a precursor in the formation of estradiol–17β, the major estrogen secreted by the ovaries.

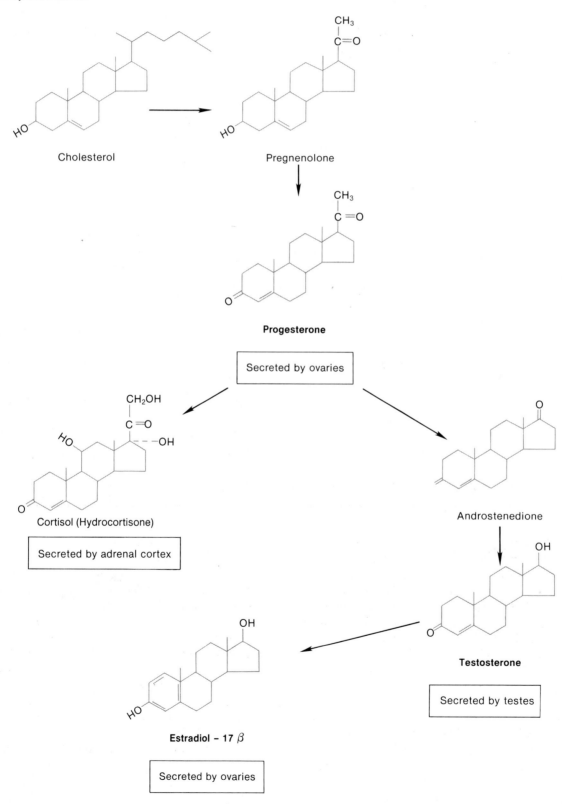

Cholesterol

Pregnenolone

Progesterone

Secreted by ovaries

Cortisol (Hydrocortisone)

Secreted by adrenal cortex

Androstenedione

Testosterone

Secreted by testes

Estradiol – 17 β

Secreted by ovaries

Figure 19.3 The thyroid hormones. Thyroxine (T₄) and triiodothyronine (T₃) are secreted in a ratio of 9:1.

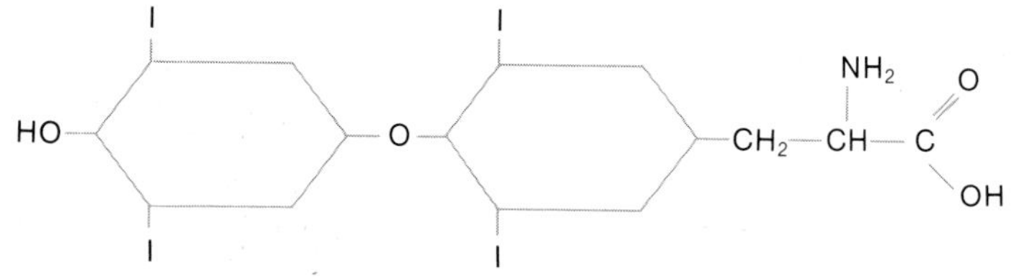

Thyroxine, or tetraiodothyronine (T₄)

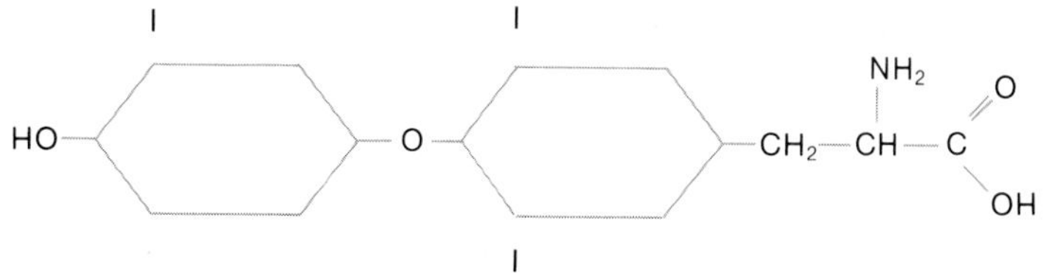

Triiodothyronine (T₃)

Table 19.2 Some examples of polypeptide and glycoprotein hormones.

Hormone	Structure	Gland	Primary Effects
Vasopressin	8 amino acids	Posterior pituitary	Water retention and vasoconstriction
Oxytocin	8 amino acids	Posterior pituitary	Uterine and mammary contraction
Insulin	21 and 30 amino acids (double chain)	β cells in islets of Langerhans	Cellular glucose uptake, lipogenesis, and glycogenesis
Glucagon	29 amino acids	α cells in islets of Langerhans	Hydrolysis of stored glycogen and fat
ACTH	39 amino acids	Anterior pituitary	Stimulation of adrenal cortex
Parathyroid hormone	84 amino acids	Parathyroid	Increases blood Ca^{++} concentration
FSH, LH, TSH	Glycoproteins	Anterior pituitary	Stimulates growth, development, and secretion of target glands

Types of Hormones

The hormones secreted by different endocrine glands include (1) **biogenic amines** such as *catecholamines* (epinephrine and norepinephrine) and other derivatives of amino acids; (2) **polypeptides** such as insulin, antidiuretic hormone, growth hormone, and others that are composed of chains of amino acids that are too short to be called proteins; (3) **glycoproteins** such as FSH, LH, and others that consist of large proteins combined with carbohydrates; and (4) **steroids.**

Steroid hormones, which are derived from cholesterol (see figure 19.2), are lipids and thus are not water soluble. The gonads—testes and ovaries—secrete *sex steroids;* the adrenal cortex secretes *corticosteroids* such as hydrocortisone, as well as small but significant amounts of sex steroids.

The major thyroid hormones are composed of two derivatives of the amino acid *tyrosine* that are bonded together. These hormones are unique because they contain iodine. When the hormone contains four iodine atoms it is called *tetraiodothyronine (T₄),* or *thyroxine.* When it contains three atoms of iodine it is called *triiodothyronine (T₃).* Although these hormones are not steroids, they are like steroids in that they are relatively small and nonpolar molecules. Steroid and thyroid hormones are active when taken orally (as a pill); sex steroids are contained in the contraceptive pill, and thyroid hormones are ingested by people whose own thyroid gland is deficient (who are hypothyroid). Other types of hormones cannot be taken orally because they would be digested into inactive fragments.

Figure 19.4 The negative feedback control of insulin
secretion by changes in the blood glucose concentration.
The effects of insulin are described in chapter 20.

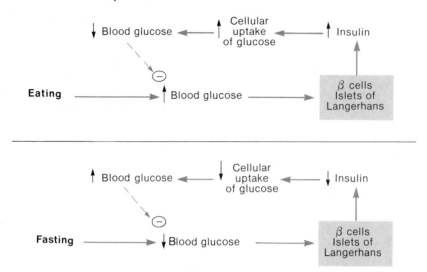

Control of Hormone Secretion

Hormones are secreted in response to specific chemical stimuli. The presence of protein in the stomach, for example, stimulates the secretion of gastrin (see chapter 18), and a rise in plasma glucose stimulates insulin secretion from the islets of Langerhans in the pancreas. Hormones are secreted in response to neurotransmitters released from central and autonomic axons; acetylcholine (ACh) from parasympathetic endings can stimulate secretion of gastrin, insulin, and other hormones, for example. Hormones can also stimulate the secretion of other hormones—the hypothalamus secretes hormones that stimulate the anterior pituitary, which in turn secretes hormones that stimulate some other endocrine glands.

Secretion of hormones is controlled by inhibitory as well as by stimulatory influences. The effect of a given hormone's action can inhibit its own secretion. Secretion of insulin—which helps to lower plasma glucose concentration—is stimulated by a rise in blood glucose, for example, and is inhibited by a fall in blood glucose levels. The lowering of blood glucose by insulin thus has an inhibitory feedback effect on further insulin secretion. This closed-loop control system is called **negative feedback inhibition** (see figure 19.4).

Inhibition by negative feedback loops helps to prevent excessive increases or decreases in hormone secretion, and thus helps to prevent excessive fluctuations in the effects of hormone action. In a **positive feedback** loop, in contrast, a hormone would have effects that promote its further secretion and result in an "explosive" surge of the hormone's concentration in the blood. Such a positive feedback system is seen in the hormonal control of ovulation, as described in chapter 21.

Activation and Inactivation of Hormones

Hormone molecules that affect metabolism of target cells are often derived from less active "parent" molecules, or precursors. When cells of the islets of Langerhans produce insulin and cells of the anterior pituitary produce ACTH (adrenocorticotrophic hormone), for examples, these molecules are derived from larger, less active peptides called **prohormones.** Under normal conditions prohormones are either not secreted at all or are only secreted in extremely small amounts.

In the case of insulin and ACTH, conversion of the less active prohormones to the fully active hormones occurs in the endocrine glands. Thyroxine (T_4), testosterone (a steroid secreted from the testes), and vitamin D_3 (secreted from the skin), however, are converted into more active forms by other organs (see table 19.3). Thyroxine is converted within its target cells into T_3, for example, and testosterone is converted into a variety of active products, as described in chapter 21, within its target organs.

Table 19.3 Conversion of some prehormones into biologically active derivatives.

Endocrine Gland	Prehormone	Active Products	Comments
Skin	Vitamin D_3	1,25-dihydroxy vitamin D_3	Hydroxylation reactions occur in the liver and kidneys.
Testes	Testosterone	Dihydrotestosterone (DHT)	DHT and other 5α-reduced androgens are formed in most androgen-dependent tissue.
		Estradiol-17β (E_2)	E_2 is formed in the brain from testosterone, where it is believed to affect both endocrine function and behavior; small amounts of E_2 are also produced in the testes.
Thyroid	Thyroxine (T_4)	Triiodothyronine (T_3)	Conversion of T_4 to T_3 occurs in almost all tissues.

Figure 19.5 The pituitary gland.

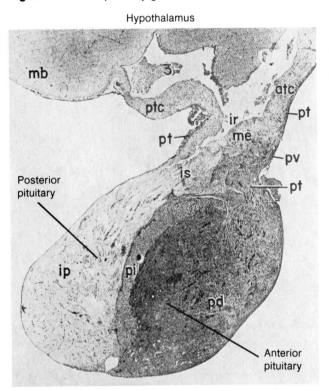

Since the form of the molecules that are secreted by some endocrine glands are different from the final form that is active in the target cells, these less active molecules that are secreted into the blood are sometimes called **prehormones.**

The concentration of hormones (or prehormones) in the blood primarily reflects the rate of secretion by the endocrine glands—hormones do not generally "accumulate" in the blood. The *half-life* of hormones in the blood is quite short, ranging from under two minutes to a few hours for most hormones. This is because hormones are rapidly removed from the blood by target organs and by the liver. Hormones removed from the blood by the liver are converted by enzymatic reactions to less active products; steroids, for example, are converted to more polar derivatives in the liver. These less active, more water-soluble derivatives are released into the blood and are excreted in the urine and bile.

Hypothalamus and Pituitary

The **pituitary gland,** or **hypophysis,** is actually two glands located together in one organ. The *posterior pituitary,* or *neurohypophysis,* is derived embryologically from a downgrowth of the brain and contains nerve fibers that originate in the hypothalamus. It is, in a sense, an extension of the brain (the connective tissues meninges that cover the brain also cover the pituitary gland). The *anterior pituitary,* or *adenohypophysis* (adeno = glandular), in contrast, is derived from epithelial tissue from the embryonic mouth. This epithelial tissue buds off from the epithelial membrane in the roof of the mouth and joins the neurohypophysis to form a single organ—the pituitary gland. The posterior part of the pituitary contains nerve tissue, therefore, while the anterior part is truly glandular (glands, remember, are derived from epithelial membranes).

Figure 19.6   The posterior pituitary, or neurohypophysis, stores and secretes hormones (vasopressin and oxytocin) produced in neuron cell bodies within the supraoptic and paraventricular nuclei of the hypothalamus. These hormones are transported to the posterior pituitary by nerve fibers of the hypothalamo-hypophyseal tract.

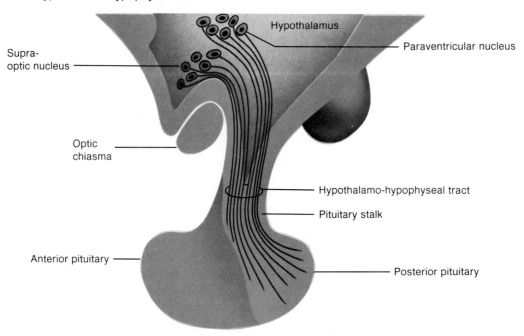

Hypothalamic Control of the Posterior Pituitary

The posterior pituitary secretes two hormones: **antidiuretic hormone (ADH;** also called **vasopressin)** and **oxytocin.** Vasopressin causes water retention by the kidneys and vasoconstriction; oxytocin stimulates contractions of the uterus and contraction of the secretory units within mammary glands (as described in chapter 21). Although these two hormones are secreted by the posterior pituitary, they are actually produced within neuron cell bodies in the *supraoptic* and *paraventricular nuclei* of the hypothalamus. These nuclei within the hypothalamus are thus endocrine glands; the hormones they produce are transported along axons of the *hypothalamo-hypophyseal tract* to the posterior pituitary that stores and later secretes these hormones. The posterior pituitary is thus more of a storage organ than a true gland.

Secretion of ADH and oxytocin from the posterior pituitary is controlled by *neuroendocrine reflexes.* In nursing mothers, for example, the stimulus of suckling acts via sensory nerve impulses to the hypothalamus to stimulate secretion of oxytocin (this is described in chapter 21). The secretion of ADH is stimulated by osmoreceptor neurons in the hypothalamus and is inhibited by sensory impulses from stretch receptors in the left atrium of the heart, as described in previous chapters.

Hypothalamic Control of the Anterior Pituitary

At one time the anterior pituitary was called the "master gland" because it secretes hormones that regulate some other endocrine glands. **Adrenocorticotrophic hormone (ACTH),** for example, stimulates the adrenal cortex to (1) secrete its steroid hormones and (2) maintain its proper structure. If ACTH secretion from the anterior pituitary is deficient, therefore, the adrenal cortex does not secrete its normal amount of steroid hormones and it atrophies.

Table 19.4 Hormones secreted by the anterior pituitary.

Hormone	Target Tissue	Stimulated by Hormone	Regulation of Secretion
ACTH (adrenocorticotrophic hormone)	Adrenal cortex	Secretion of glucocorticoids	Stimulated by CRH (corticotrophin-releasing hormone); inhibited by glucocorticoids
TSH (thyroid-stimulating hormone)	Thyroid gland	Secretion of thyroid hormones	Stimulated by TRH (thyrotrophin-releasing hormone); inhibited by thyroid hormones
GH (growth hormone)	Most tissue	Protein synthesis and growth; lipolysis and increased blood glucose	Inhibited by somatostatin; stimulated by growth hormone-releasing hormone
FSH (follicle-stimulating hormone) and LH (luteinizing hormone)	Gonads	Gamete production and sex steroid hormone secretion	Stimulated by GnRH (gonadotrophin-releasing hormone); inhibited by sex steroids
Prolactin	Mammary glands and other sex accessory organs	Milk production Controversial actions in other organs	Inhibited by PIH (prolactin-inhibiting hormone)
MSH (melanocyte-stimulating hormone)	Melanocytes (pigment cells)	Darkening of skin	Stimulated by CRH; may be inhibited by hypothalamic hormone

Excessive secretion of ACTH, in contrast, causes excessive secretion and hypertrophy (growth) of the adrenal cortex. The anterior pituitary hormone thus has a trophic effect (a "feeding" effect) on the adrenal cortex. Similarly, the anterior pituitary secretes **thyroid-stimulating hormone (TSH;** also called *thyrotrophic hormone*) and **gonadotrophic hormones** that stimulate the thyroid and gonads, respectively. These and other hormones secreted by the anterior pituitary are listed in table 19.4.

Releasing and Inhibiting Hormones   Contrary to earlier concepts, the anterior pituitary cannot be considered a "master gland" because its secretions are controlled by the hypothalamus. Since axons do not enter the anterior pituitary, hypothalamic control is chemical—that is, hormonal—rather than neural. Neurons in the hypothalamus produce releasing and inhibiting hormones, which are transported to axon endings in the basal portion of the hypothalamus. This region, known as the **median eminence,** contains blood capillaries that are drained by venules in the stalk of the pituitary.

The venules that drain the median eminence deliver blood to a second capillary bed in the anterior pituitary. Since this second capillary bed delivers venous rather than arterial blood, the vascular link between the median eminence and the anterior pituitary comprises a portal system. (This is analogous to the hepatic portal system that delivers venous blood from the intestine to the liver). The vascular connection between the hypothalamus and the anterior pituitary is thus called the **hypothalamo-hypophyseal portal system.**

Neurons of the hypothalamus secrete short polypeptides into this portal system that regulate secretion of anterior pituitary hormones. Thyrotrophin-releasing hormone **(TRH),** for example, stimulates secretion of TSH, and corticotrophin-releasing hormone **(CRH)** stimulates secretion of ACTH from the anterior pituitary. A single releasing hormone—gonadotrophin-releasing hormone, or **GnRH**—appears to stimulate secretion of the two gonadotrophic hormones (*FSH* and *LH*—see chapter 21) from the anterior pituitary. Secretion of prolactin and of growth hormone from the anterior pituitary is regulated by hypothalamic inhibitory hormones known as *PIH* (prolactin-inhibiting hormone) and *somatostatin,* respectively. Very recently, a specific hypothalamic-releasing hormone that stimulates growth-hormone secretion has also been identified. The secretion of growth hormone thus appears to be controlled by both inhibiting and releasing hormones from the hypothalamus.

Figure 19.7 Hormones secreted by the anterior pituitary, and the target organs for these hormones.

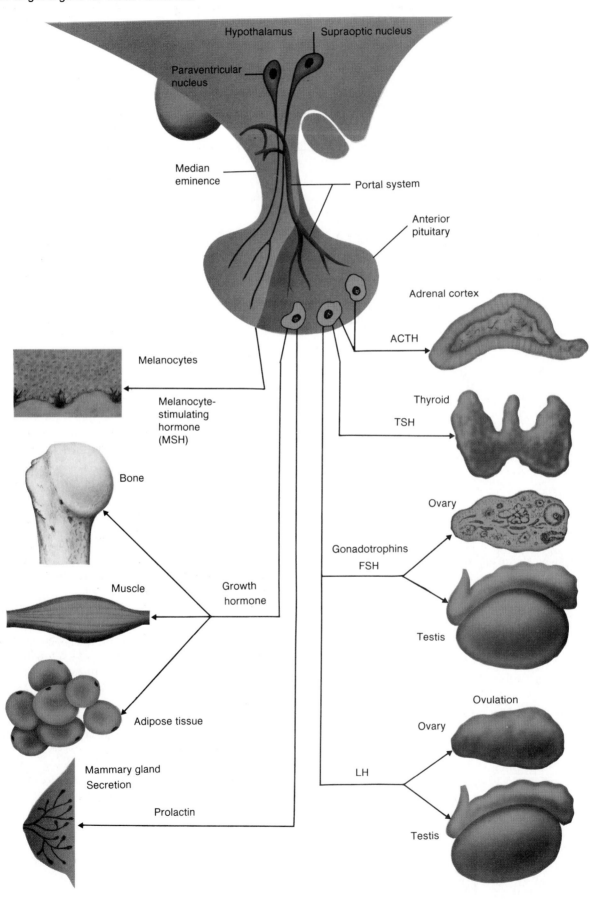

Table 19.5 Some hypothalamic hormones involved in the control of the anterior pituitary.

Hypothalamic Hormone	Structure	Effect on Anterior Pituitary	Action of Anterior Pituitary Hormone
Corticotrophin-releasing hormone (CRH)	Presently unknown	Stimulates secretion of adrenocorticotrophic hormone (ACTH)	Stimulates secretions of adrenal cortex
Gonadotrophin-releasing hormone (GnRH)	10 amino acids	Stimulates secretion of follicle-stimulating hormone (FSH) and luteinizing hormone (LH)	Stimulates gonads to produce gametes (sperm and ova) and secrete sex steroids
Prolactin-inhibiting hormone (PIH)	Presently unknown	Inhibits prolactin secretion	Stimulates production of milk in mammary glands
Somatostatin	14 amino acids	Inhibits secretion of growth hormone	Stimulates anabolism and growth in many organs
Thyrotrophin-releasing hormone (TRH)	3 amino acids	Stimulates secretion of thyroid-stimulating hormone (TSH)	Stimulates secretion of thyroid gland

Figure 19.8 Secretion of thyroxine from the thyroid is stimulated by thyroid-stimulating hormone (TSH) from the anterior pituitary. Secretion of TSH is stimulated by thyrotrophin-releasing hormone (TRH) secreted from the hypothalamus into the hypothalamo-hypophyseal portal system. This stimulation is balanced by negative feedback inhibition by thyroxine, which decreases the responsiveness of the anterior pituitary to stimulation by TRH.

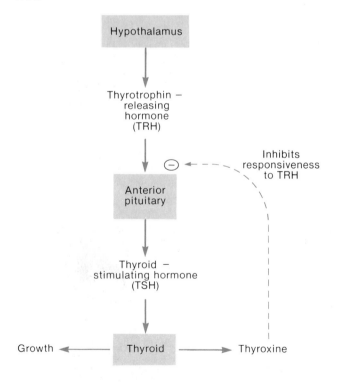

Feedback Control of Anterior Pituitary Secretion

In view of its secretion of releasing and inhibiting hormones, the hypothalamus might be considered the "master gland." The chain of command, however, is not linear; the hypothalamus and anterior pituitary are controlled by the effects of their own actions. In the endocrine system, to use an analogy, the general takes orders from the private. The hypothalamus and anterior pituitary are not master glands because their secretions are controlled by the target glands they regulate.

The anterior pituitary secretion of ACTH, TSH, and the gonadotrophins (FSH and LH) is controlled by both hypothalamic hormones and by *negative feedback inhibition* from target gland hormones. Secretion of ACTH is inhibited by a rise in corticosteroid secretion, for example, and TSH is inhibited by an increase in thyroxine secretion from the thyroid. These negative feedback relationships are easily demonstrated by removal of the target glands. Castration (surgical removal of the gonads), for example, produces a rise in FSH and LH secretion. In a similar manner, removal of the adrenals or the thyroid would result in an abnormal increase in ACTH or TSH secretion from the anterior pituitary.

These effects demonstrate that under normal conditions the target glands exert an inhibitory effect on the anterior pituitary. This inhibitory effect can occur at two levels: (1) the target gland hormones can inhibit the secretion of specific hypothalamic-releasing hormones, and (2) the target gland hormones can inhibit the sensitivity of the anterior pituitary to stimulation by releasing hormones. Thyroxine, for example, appears to inhibit the responsiveness of the anterior pituitary to TRH, and thus acts to reduce TSH secretion. Sex steroids reduce secretion of gonadotrophins by inhibiting both GnRH secretion and the ability of the anterior pituitary to respond to stimulation by GnRH (see figure 19.9).

Figure 19.9 Negative feedback control of gonadotrophin secretion. The possible short negative feedback loop involving inhibition of GnRH secretion by retrograde transport of gonadotrophins (see text) is shown by a dotted arrow.

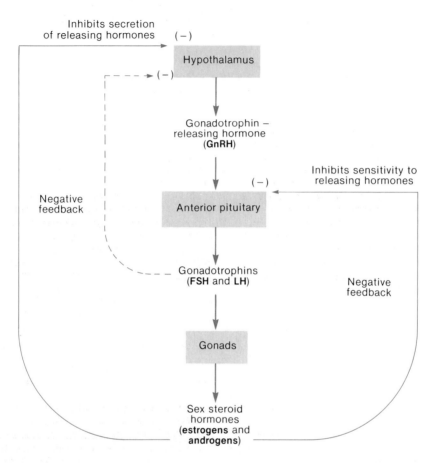

Recent evidence suggests that there may be retrograde transport of blood from the anterior pituitary to the hypothalamus. This may permit a *short feedback loop* where a particular trophic hormone inhibits the secretion of its releasing hormone from the hypothalamus. High secretion of TSH, for example, may inhibit further secretion of TRH by this means. The existence of retrograde transport and of short feedback loops is difficult to prove and is currently controversial.

In addition to negative feedback control of the anterior pituitary, there is one known example of *positive feedback* in which a target gland hormone stimulates secretion of a pituitary hormone. During the middle phase of the menstrual cycle the ovaries secrete increasing amount of a sex steroid called estradiol-17β, which stimulates a "surge" of LH secretion from the anterior pituitary. This positive feedback effect (discussed in conjunction with ovulation in chapter 21) is unique for this phase of the menstrual cycle; at other times this sex steroid inhibits LH secretion.

The positive feedback effect of estradiol-17β on LH secretion is believed to result from activation of neurons in the amygdaloid nucleus in the limbic system of the brain (see chapter 9). This illustrates the fact that activity of the hypothalamus, and thus of the anterior pituitary and its target glands, is influenced by *higher brain functions*. As a result of the effects of higher brain activity on the hypothalamus and pituitary, emotional disturbances can affect ovarian secretion and the menstrual cycle (see chapter 21). Stress, as described in a later section in this chapter, produces an increase in CRH secretion from the hypothalamus, which in turn results in elevated ACTH and corticosteroid secretion. Growth hormone is secreted in increased amounts during sleep, and secretion of LH likewise follows a *circadian* ("about a day") *rhythm* as a result of the influence of higher brain centers on the hypothalamus.

Figure 19.10 Structure of the adrenal gland showing the three zones of the adrenal cortex.

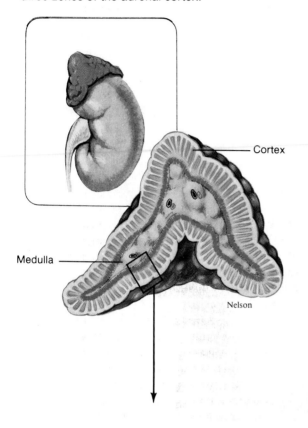

Cortex

Medulla

Nelson

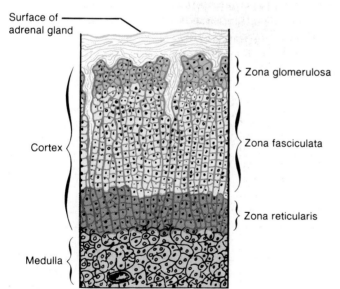

Surface of adrenal gland

Cortex
- Zona glomerulosa
- Zona fasciculata
- Zona reticularis

Medulla

1. List the endocrine glands that secrete steroid hormones. Describe the structure of thyroid hormones, and indicate the common chemical property of thyroid and steroid hormones that allows both to be active when taken orally.
2. Define the term *prehormone* and give three examples.
3. Define a "negative feedback loop," and illustrate this with a flow chart showing how insulin secretion is controlled by the plasma glucose concentration.
4. Distinguish between the anterior and posterior pituitary in terms of their differences in *(a)* embryonic origin and structure; *(b)* the origin of their hormones; and *(c)* the mechanisms by which the hypothalamus exerts control over their secretions.
5. Use a flow chart to illustrate the negative feedback control of CRH and ACTH secretion. Explain how this system would be affected by *(a)* an injection of ACTH; *(b)* surgical removal of the pituitary; *(c)* injection of corticosteroids; and *(d)* surgical removal of the adrenal glands.

Adrenal and Thyroid Glands

Through its secretion of trophic hormones, the anterior pituitary controls the function of the adrenal glands, thyroid gland, and gonads. A discussion of the endocrine function of the gonads will be included, together with their reproductive function, in chapter 21. Similarly, secretion of growth hormone and prolactin by the anterior pituitary will be discussed, together with their functions in regulating metabolism and reproduction, in chapters 20 and 21, respectively.

Structure and Function of the Adrenal Glands

Each adrenal gland, like the pituitary, is actually composed of two different glands that are located together in the same organ. The inner part, or **adrenal medulla,** is derived from modified postganglionic sympathetic neurons, and secretes catecholamine hormones (epinephrine and norepinephrine) into the blood upon stimulation by the sympathetic nervous system (see chapter 9). The **adrenal cortex,** or outer part of the adrenals, is derived from a different embryonic tissue, secretes different hormones, and is under a different control system than the adrenal medulla.

Figure 19.11 Simplified pathways for the synthesis of steroid hormones in the adrenal cortex. The adrenal cortex produces steroids that regulate Na$^+$ and K$^+$ balance (mineralocorticoids), steroids that regulate glucose balance (glucocorticoids), and small amounts of sex steroid hormones.

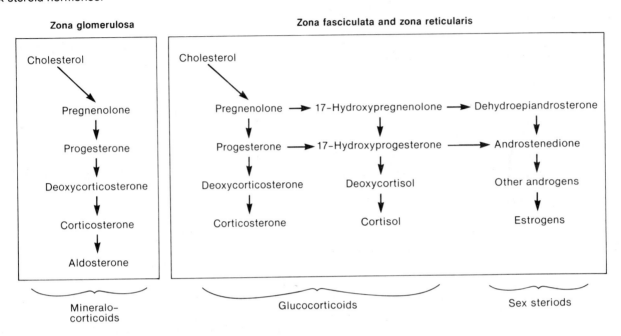

The adrenal cortex secretes steroid hormones called corticosteroids, or corticoids for short. There are two functional categories of corticosteroids: (1) **mineralocorticoids,** which regulate Na$^+$ and K$^+$ balance, and (2) **glucocorticoids,** which help regulate metabolism of glucose and other organic molecules. The major mineralocorticoid is *aldosterone,* and the major glucocorticoid is *hydrocortisone* (cortisol). The regulatory functions of mineralocorticoids were discussed, together with kidney function, in chapter 16; the effects of glucocorticoids on energy balance in the body are discussed in chapter 20. In addition to glucocorticoids and mineralocorticoids, the adrenal cortex secretes small amounts of sex steroids (see figure 19.11).

Based on its microscopic appearance, the adrenal cortex is divided into three zones. In the outer zone, or *zona glomerulosa,* the cells are grouped into clusters and are believed to secrete mineralocorticoids (such as aldosterone) in response to stimulation by angiotensin II and elevated blood K$^+$ concentration (as described in chapter 16). The middle zone and perhaps the inner zone, called the *zona fasciculata* and the *zona reticularis,* respectively, are believed to secrete glucocorticoids such as hydrocortisone under stimulation by ACTH. This difference in function of the zones is evidenced by the effects of surgical removal of the pituitary (hypophysectomy): the structure of the zona glomerulosa remains unaltered, while the inner zones atrophy when ACTH is withdrawn.

Addison's Disease and Cushing's Syndrome The effects of inadequate and excessive glucocorticoid secretion are evident in two disease states. *Addison's disease,* or chronic adrenal insufficiency, is characterized by abnormally low secretion of corticosteroids due to atrophy of the adrenal cortex. This produces a variety of seemingly unrelated symptoms—weight loss, gastrointestinal problems, generalized weakness, and other disturbances—which imply that corticosteroids have generalized effects on the body (described in chapter 20). Since corticosteroid secretion is low, ACTH secretion from the anterior pituitary is abnormally high. Melanocyte-stimulating hormone (MSH) is derived from the same parent molecule (prohormone) as ACTH, and is also secreted in increased amounts in Addison's disease. High secretion of ACTH and MSH results in the darkening of the skin in people with Addison's disease.

Cushing's syndrome is characterized by excessively high secretion of corticosteroids. High levels of corticosteroids cause muscle wasting and osteoporosis (described in chapter 20), as well as structural changes sometimes described as "buffalo hump" and "moon face." Similar effects are also seen when people with inflammatory diseases are treated with corticosteroids, which are given to reduce inflammation and the immune response.

Figure 19.12 Comparison of organs in a normal rat with those from a stressed ("alarmed") rat. In the stressed rat the adrenals are enlarged, the thymus and lymph nodes are atrophied, and the gastric mucosa shows blood-covered ulcers.

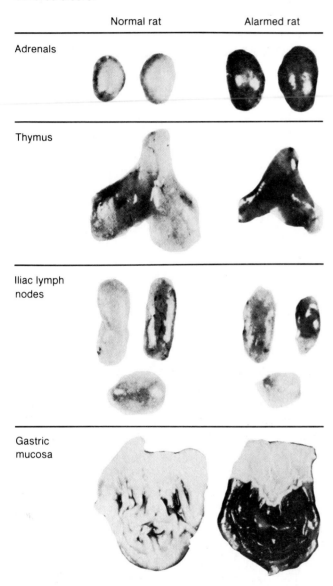

	Normal rat	Alarmed rat
Adrenals		
Thymus		
Iliac lymph nodes		
Gastric mucosa		

Figure 19.13 Activation of the pituitary-adrenal axis by nonspecific stress.

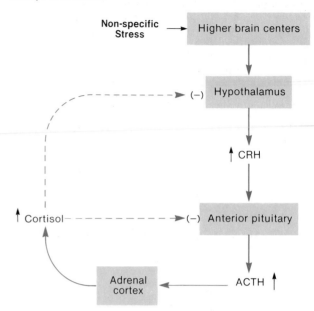

The General Adaption Syndrome   In 1936, Hans Selye discovered that injections of cattle ovaries into rats: (1) stimulated growth of the adrenal cortex; (2) caused atrophy of the lymphoid tissue of the spleen, lymph nodes, and thymus; and (3) produced bleeding peptic ulcers. At first he thought that these ovarian extracts contained a specific hormone that caused these effects. He later discovered that injections of a variety of substances, including foreign chemicals such as formaldehyde, could produce the same effects. Indeed, the same pattern of effects occurred when he placed rats in cold environments or when he dropped rats in water and made them swim until they were exhausted.

The specific pattern of effects produced by these procedures suggested that these effects were the result of something that the procedures shared in common. Selye reasoned that all of the procedures were *stressful,* and that the pattern of changes he observed represented a specific response to any stressful agent. He later discovered that all forms of stress produce these effects because all stressors *stimulate the pituitary-adrenal axis.* Under stressful conditions, there is increased secretion of ACTH and thus increased secretion of corticosteroids from the adrenal cortex.

Figure 19.14   Thyroid gland, showing numerous thyroid follicles. Each follicle consists of follicular cells surrounding fluid known as colloid. The gaps between the follicular cells and the colloid are artifacts made in the preparation of the slide.

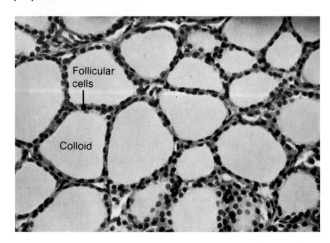

Figure 19.15   The radioactive iodine uptake test of thyroid function. The active accumulation of iodine is increased in hyperthyroidism and decreased in hypothyroidism compared to the normal (euthyroid) condition.

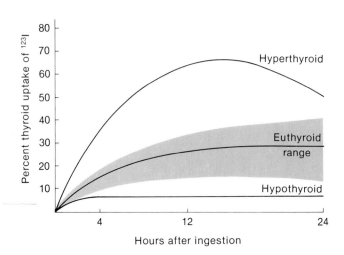

Since stress is so very difficult to define, many people prefer to define stress operationally as any stimulus that activates the pituitary-adrenal axis. Using this criterion, it has been found that pleasant changes in one's life—such as marriage, a recent promotion, and so on—can be as stressful as unpleasant changes. On this basis, Selye has attempted to define stress as "a nonspecific response of the body to readjust itself following any demand made upon it."

Selye termed this nonspecific response the **general adaptation syndrome (GAS).** Stress, in other words, produces GAS. There are three stages in the response to stress: (1) the *alarm reaction,* when the adrenal glands are activated; (2) the *stage of resistance,* in which readjustment occurs; and (3) if the readjustment is not complete, the *stage of exhaustion* may follow, leading to sickness and death.

Glucocorticoids such as hydrocortisone and cortisone are used medically to treat inflammations and reduce the symptoms of autoimmune diseases, because these steroids decrease inflammation and inhibit the immune response (see chapter 17). Similarly, high levels of glucocorticoid secretion during the phase of exhaustion may contribute to increased susceptibility to disease. Although this subject is controversial, it has served to focus attention on stress as one of the contributing factors in disease.

The Thyroid Gland

The thyroid gland is composed primarily of hollow structures called **thyroid follicles.** Each follicle consists of a protein-rich fluid known as *colloid,* surrounded by *follicular cells* (see figure 19.14). *Thyroglobulin* protein within the colloid serves as the precursor (parent compound) for the two hormones produced by the thyroid follicles—T_4 (tetraiodothyronine or thyroxine) and T_3 (triiodothyronine).

The thyroid follicles actively accumulate iodide (I^-) from the blood and secrete it into the colloid. This ability—through the use of radioactive iodide—can be measured and used clinically to assess thyroid function (see figure 19.14). Once the iodide is in the colloid it is oxidized to iodine (I) and attached to tyrosines (a type of amino acid) in the polypeptide chains of thyroglobulin. Attachment of one iodine to tyrosine produces *monoiodotyrosine (MIT);* attachment of two iodines produces *diiodotyrosine (DIT).*

Within the colloid, enzymes modify the structure of MIT and DIT and couple them together. When two DIT molecules, appropriately modified, are coupled together, a molecule of tetraiodothyronine (T_4, or thyroxine) is formed. Combination of one MIT with one DIT produces triiodothyronine (T_3). Note that within the colloid the T_4 and T_3 are still attached to thyroglobulin.

Figure 19.16 Stages in the formation and secretion of thyroid hormones. Iodide is actively accumulated by the follicular cells. In the colloid it is converted into iodine and attached to tyrosine amino acids within thyroglobulin protein. Pinocytosis of iodinated thyroglobulin, coupling of MIT and DIT, and release of thyroid hormones are stimulated by TSH from the anterior pituitary.

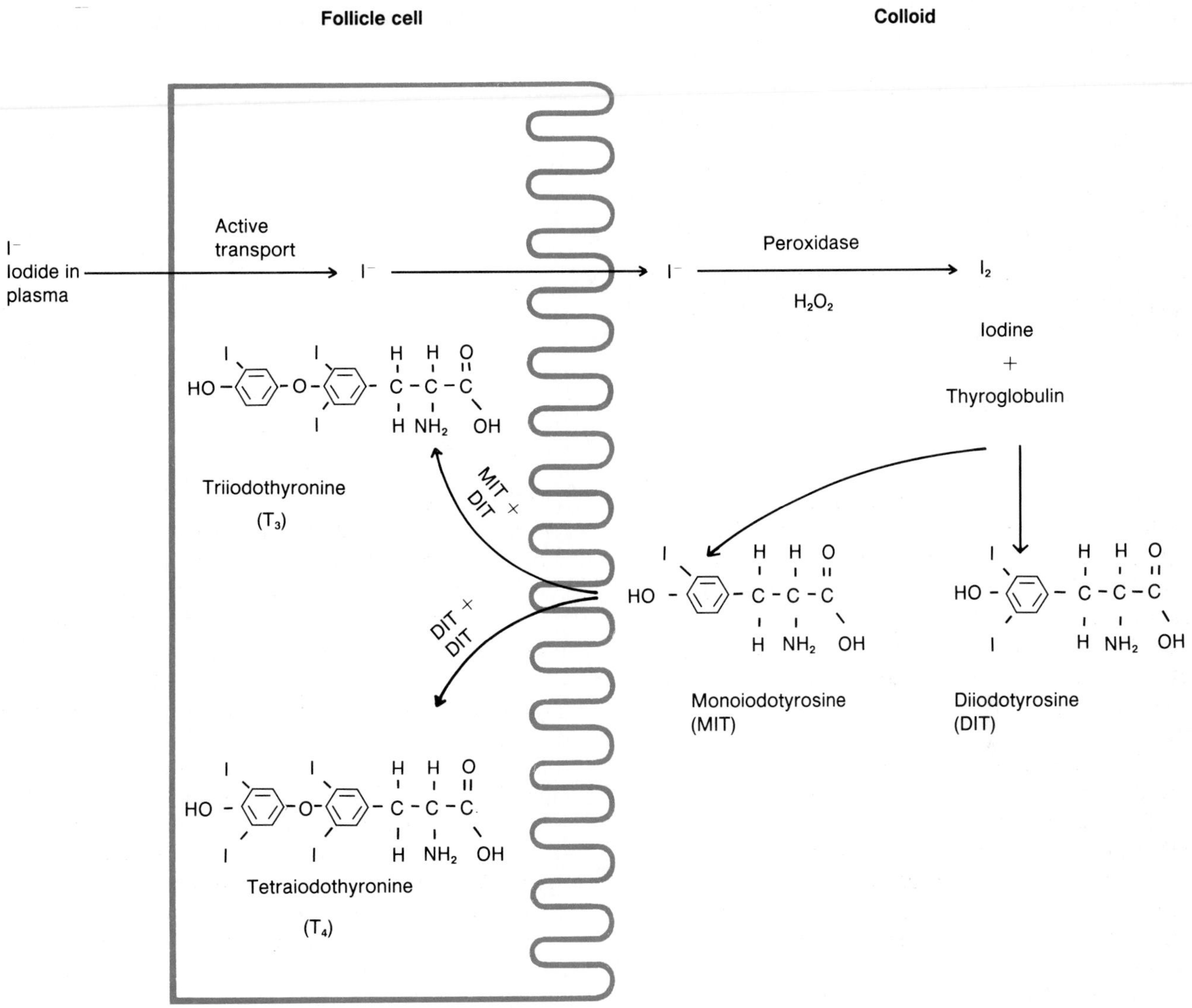

Under the influence of *thyroid-stimulating hormone (TSH)* secreted from the anterior pituitary, the colloid is taken up by pinocytosis into the follicular cells. Fusion of pinocytotic vesicles with lysosomes within these follicular cells allows digestion of thyroglobulin and the release of T_4 and T_3 (in a ratio of about 90 percent to 10 percent) into the blood.

Transport of T_4 and Conversion to T_3   Thyroxine and, to a lesser degree, T_3, like steroid hormones, are not very water soluble and so must travel in the blood attached to plasma proteins. These carrier proteins include *thyroid-binding globulin (TBG)*, an alpha-2-globulin, and a "pre-albumin." These carrier proteins have a higher affinity for T_4 than for T_3. As a result, although the thyroid secretes much more T_4 than T_3, the amount of unbound (or "free") T_3 is about ten times greater than the free T_4 in the plasma.

Approximately 99.96 percent of the thyroxine (T_4) in plasma is attached to carrier proteins; 0.04 percent is free. Only the free thyroxine can enter target cells; the protein-bound thyroxine serves as a "reservoir" that can contribute new free thyroxine when the free thyroxine concentration is diminished by entry into target cells. Thyroxine, as previously discussed, is a prehormone—the T_4 is converted in the target cells into triiodothyronine (T_3), which is the biologically active molecule responsible for the effects of thyroid hormones on body metabolism (see chapter 20) and on the negative feedback control of TSH secretion.

Iodine-deficiency Goiter   Secretion of TSH from the anterior pituitary is stimulated by TRH from the hypothalamus, as previously described, and is inhibited by negative feedback from thyroxine (which is converted to T_3 within the anterior pituitary). The anterior pituitary is a target organ for thyroxine, and responds to elevated thyroxine levels by decreasing its secretion of TSH. Conversely, a fall in plasma thyroxine concentration results in a rise in TSH secretion due to release from negative feedback inhibition.

The importance of negative feedback control of pituitary hormones is well illustrated by the effects of iodine deprivation. When inadequate amounts of iodine are ingested in the diet, the ability of the thyroid to produce T_4 and T_3 is reduced. This causes a fall in the secretion of thyroid hormones and, after awhile, results in a fall in circulating thyroid hormones (two weeks are required for a significant fall because of the large reservoir of protein-bound thyroxine). This fall in thyroxine decreases negative feedback inhibition of TSH. The anterior pituitary thus greatly increases its secretion of TSH, which in turn causes enlargement (hypertrophy) of the thyroid. This produces a bulge in the neck—a goiter.

Figure 19.17   Most thyroxine (tetraiodothyronine, or T_4), is transported in the blood attached to thyroid-binding globulin (TBG). The small amount of T_4 that dissociates from TBG can enter target cells and be converted into triiodothyronine (T_3), which is the active form of the hormone.

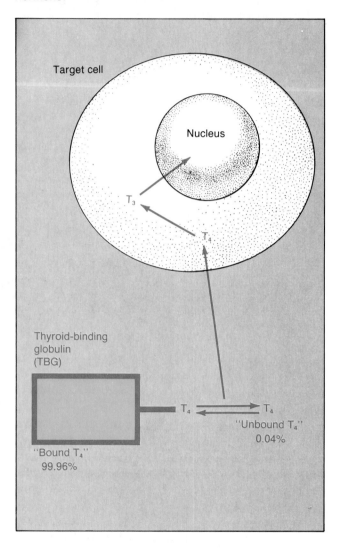

Radiotherapy of Thyroid Cancer   Thyroid cancer, particularly when it occurs in younger adults, has a more optimistic prognosis than most other forms of cancer. It is extremely slow growing, and usually spreads only into the lymph nodes of the neck (although other sites can be invaded). Surgical treatment involves removal of the thyroid and of the affected cervical lymph nodes. This surgical treatment is followed by swallowing solutions containing radioactive iodine (^{131}I), which is selectively transported into cells of the thyroid gland and into cancerous thyroid cells that have travelled ("metastasized") to other regions of the body. These cells are then killed by the radioactively labelled iodine.

Figure 19.18 Mechanism of goiter formation in iodine deficiency.

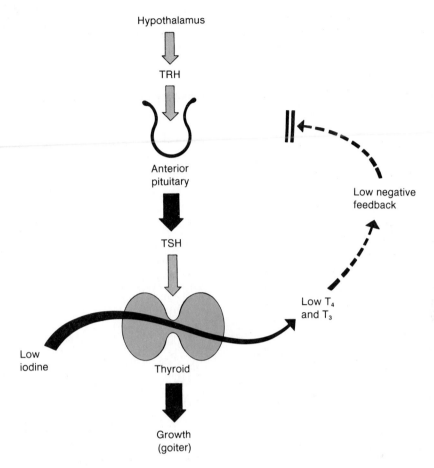

Thyroid-stimulating hormone (TSH), secreted from the anterior pituitary, stimulates the active accumulation of iodine into thyroid cells. Such active transport of iodine is obviously required for the thyroid cells to produce and secrete thyroid hormone (thyroxine). Therefore, in order for the ingested radioactive iodine to be effective in the treatment of thyroid cancer, the blood levels of TSH must be raised to high levels. This is accomplished by removing most of the patient's thyroid gland.

When the patient's thyroid gland is surgically removed the blood levels of thyroxine gradually decline, and—through the decrease in negative feedback inhibition that results—the secretion of TSH increases. A period of about four to five weeks is required after removal of the thyroid for the TSH levels to rise sufficiently. At this point the high secretion of TSH that results from low thyroxine levels can stimulate accumulation of radioactive iodine into thyroid cells and promote the destruction of both normal and metastasized cells.

1. Draw a flow diagram describing the control of corticosteroid secretion. Explain why people with Addison's disease may have darkening of the skin.
2. Describe how stress affects the endocrine system and how these effects may lead to increased susceptibility to disease.
3. Describe how thyroid hormones are transported in the blood. Explain why destruction of the thyroid (as might be produced by exposure to high levels of radiation) would not result in a significant fall in plasma thyroxine for a couple of weeks.
4. Using a flow diagram, explain how a diet deficient in iodine can result in the formation of a goiter.

Table 19.6   Functional categories of hormones, based on the location of their receptor proteins and the mechanisms of their action.

Types of Hormones	Secreted by	Location of Receptors	Effects of Hormone-Receptor Interaction
Catecholamines, polypeptides, glycoproteins	All glands except adrenal cortex, gonads, and thyroid	Outer surface of cell membrane	Stimulates production of intracellular "second messenger," which activates previously inactive enzymes
Steroids	Adrenal cortex, testes, ovaries	Cytoplasm of target cells	Stimulates translocation of hormone-receptor complex to nucleus and activation of specific genes
Thyroxine (T_4)	Thyroid	Nucleus of target cells	After conversion of triiodothyronine (T_3), activates specific genes

Mechanisms of Hormone Action

The specific effects of different hormones on body metabolism and reproductive function are discussed in chapters 20 and 21. The differences in the effects of hormones on their target tissues are due to differences in the characteristics of the target cells; a given hormone can thus have multiple effects in the body. Epinephrine, for example, can produce dilation of bronchiolar smooth muscles, constriction of vascular smooth muscles, and hydrolysis of liver glycogen. The manner in which these effects are produced—the *mechanism* of the hormone's action—however, is similar in each case.

Hormones are delivered by the blood to every cell in the body, but only the *target cells* are able to respond to each particular hormone. In order to respond to a hormone, a target cell must have a specific **receptor protein** for that hormone. While hormones can bond weakly and in a nonspecific fashion to many components of both target and non-target cells, bonding to a receptor protein is characterized by its *specificity, high affinity* (high bond strength), and *low capacity*. The latter characteristic refers to the fact that there are a limited number of receptor proteins per target cell (usually a few thousand), so that bonding to receptor proteins can reach a saturation value (see figure 19.19). Notice that these characteristics of receptor proteins are similar to those of enzyme and transport proteins discussed in previous chapters.

The location of a hormone's receptor protein in target cells depends on the chemical nature of the hormone. Based on the location of their receptor proteins, hormones can be grouped into three categories: (1) receptor proteins within the nucleus of target cells—*thyroid hormones;* (2) receptor proteins in the cytoplasm of target cells—*steroid hormones;* and (3) receptor proteins in the outer surface of the cell membrane of target cells—*catecholamine, polypeptide,* and *glycoprotein hormones.*

Figure 19.19   Nonspecific and specific binding of hormones to receptor proteins in target cells. Note that bonding of hormones to specific receptors occurs with a higher affinity than to nonspecific sites. The flatness of the specific bonding curve indicates that receptors can be saturated at relatively low hormone concentrations.

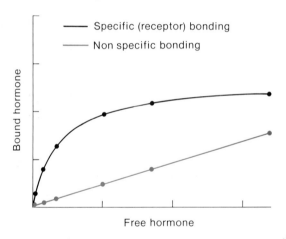

Mechanisms of Steroid and Thyroid Hormone Action

Steroids and thyroid hormones are similar in size and in the fact that they are nonpolar and thus are not very water soluble. Unlike other hormones, therefore, steroids and thyroid hormones (primarily thyroxine) do not travel dissolved in the aqueous plasma, but instead are transported to their target cells attached to carrier proteins. These hormones then dissociate from the carrier proteins in the blood and easily pass through the lipid component of the target cell's membrane.

Figure 19.20 Mechanism of steroid hormone (H) action on the target cells.

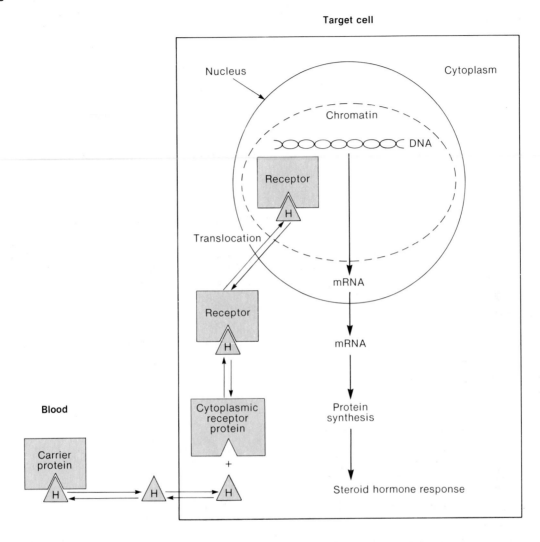

Steroid Hormones   Once through the cell membrane, steroid hormones attach to *cytoplasmic receptor proteins* in the target cells. The steroid hormone-receptor protein complex then *translocates* to the nucleus, and attaches by means of the receptor protein to the chromatin (see chapter 3). The sites of attachment are termed *acceptor sites,* and are specific for the target tissue. This specificity is believed to be determined by acidic (nonhistone) proteins in the chromatin. According to one theory, part of the receptor bonds to an acidic protein while a different part of the receptor bonds to DNA.

Attachment of the receptor protein-steroid complex to the acceptor site "turns on genes." Through mechanisms that are poorly understood, specific genes become activated by this process and produce nuclear RNA, which is then processed into messenger RNA (mRNA—see chapter 3). This new mRNA enters ribosomes and codes for the production of new proteins. Since some of these newly synthesized proteins may be enzymes, the metabolism of the target cell is changed in a specific manner. Steroid hormones, in short, affect their target cells by stimulating genetic transcription (RNA synthesis) followed by genetic translation (protein synthesis).

Figure 19.21 Mechanism of action of T₃ (triiodothyronine) on the target cells.

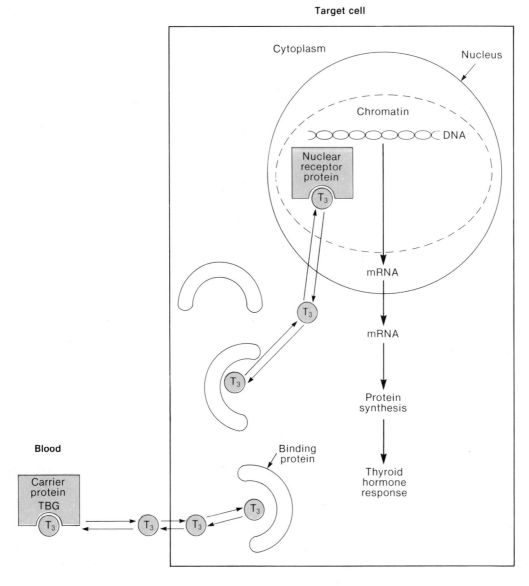

Thyroxine Once thyroxine, in the free form after dissociation from thyroid-binding globulin in the blood, passes through the target cell membrane, it is converted in the cytoplasm into T₃. Inactive T₃ receptor proteins are already in the nucleus attached to chromatin. These receptors are inactive until T₃ enters the nucleus from the cytoplasm. Attachment of T₃ to the chromatin-bound receptors activates previously inactive genes and results in the production of new mRNA and new proteins. This sequence of events is summarized in figure 19.21.

Membrane Receptor Proteins and Second Messengers
Catecholamine, polypeptide, and glycoprotein hormones cannot pass through the lipid barrier of the target cell membrane. While some of these hormones may enter the cell by pinocytosis, most of the effects of these hormones are believed to result from interaction with receptor proteins on the outer surface of the target cell membrane. Since these hormones do not have to enter the target cells to exert their effects, other molecules must mediate the effects of these hormones within the target cells. If one thinks of hormones as "messengers" from the endocrine glands, the intracellular mediators of the hormone's action can be called **second messengers.**

Cyclic AMP Cyclic AMP (abbreviated cAMP) was the first second messenger to be discovered and is the best understood. Catecholamines, including epinephrine and norepinephrine, use cAMP as their second messenger. (This was discussed in relation to adrenergic synaptic transmission in chapter 7.) It was later discovered that the effects of many (but not all) polypeptide and glycoprotein hormones are mediated by cAMP.

Figure 19.22   Cyclic AMP (cAMP) as a second messenger in the action of catecholamine, polypeptide, and glycoprotein hormones.

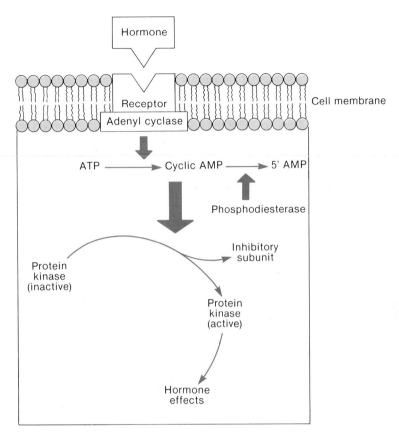

Table 19.7   Hormones that appear to use cAMP as a second messenger and hormones that appear to use other second messengers.

Hormones That Stimulate Adenyl Cyclase	Hormones That Do Not Stimulate Adenyl Cyclase
Adrenocorticotrophic hormone (ACTH)	Catecholamines (α-adrenergic)
Calcitonin	Growth hormone (GH)
Epinephrine (β-adrenergic)	Insulin
Follicle-stimulating hormone (FSH)	Oxytocin
Glucagon	Prolactin
Luteinizing hormone (LH)	Somatomedin
Parathyroid hormone	Somatostatin
Thyrotrophin-releasing hormone (TRH)	
Thyroid-stimulating hormone (TSH)	
Vasopressin	

Bonding of catecholamine and some other hormones (see table 19.7) to their membrane receptor proteins activates an enzyme called *adenyl cyclase*. This enzyme is built into the cell membrane and when activated, catalyzes the following reaction:

$$ATP \xrightarrow{\text{adenyl cyclase}} cAMP + PP_i$$

Adenosine triphosphate (ATP) is thus converted into cAMP plus two inorganic phosphates (*pyrophosphate*, abbreviated PP_i). As a result of interaction of the hormone with its receptor and activation of adenyl cyclase, therefore, the intracellular concentration of cAMP is increased. Cyclic AMP activates a previously inactive enzyme in the cytoplasm called *protein kinase*. This enzyme is produced in an inactive form, but becomes activated when attachment of cAMP to an inhibitory subunit of this enzyme causes the inhibitory subunit to dissociate from the active, catalytic subunit of the enzyme (see figure 19.22). The hormone, in summary—acting through an increase in cAMP production—causes an increase in protein kinase activity within its target cells.

Figure 19.23   Effects of cAMP as a second messenger in
the action of epinephrine and glucagon. These hormones
(acting through cAMP) promote hydrolysis of glycogen
(glycogenolysis) in the liver.

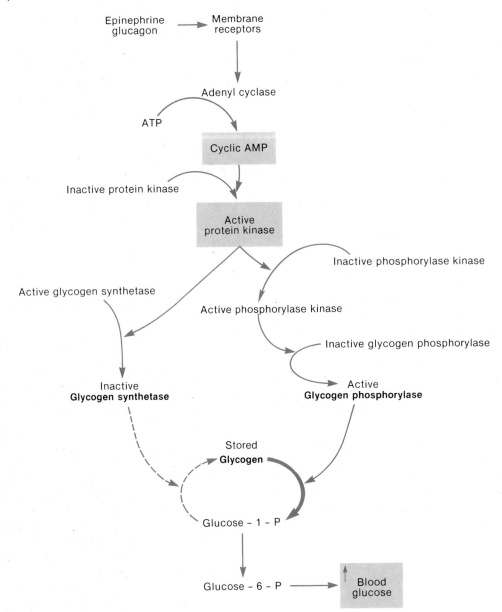

Table 19.8   Sequence of events that occur in the stimulation of target organs by catecholamine, polypeptide, and glycoprotein hormones.

1. The hormones combine with their receptors on the outer surface of target cell membranes.

2. Hormone-receptor interaction stimulates activation of adenyl cyclase on the cytoplasmic side of the membranes.

3. Activated adenyl cyclase catalyzes the conversion of ATP to cyclic AMP (cAMP) within the cytoplasm.

4. Cyclic AMP activates protein kinase enzymes that were already present in the cytoplasm in an inactive state.

5. Activated cAMP-dependent protein kinase transfers phosphate groups to (phosphorylates) other enzymes in the cytoplasm.

6. The activity of specific enzymes is either increased or inhibited by phosphorylation.

7. Altered enzyme activity mediates the target cell's response to the hormone.

Active protein kinase catalyzes the attachment of phosphate groups to different enzymes in the target cells. This causes some of these enzymes to become activated, and some to become inactivated. Cyclic AMP, acting through protein kinase, thus modulates the activity of enzymes that are already present in the target cell. This alters the metabolism of the target tissue in a characteristic way.

The response to a rise in cAMP and activation of protein kinase is characteristic of the target tissue. The testis, for example, responds to increased cAMP levels (stimulated by luteinizing hormone from the anterior pituitary—see chapter 21) by increasing its secretion of male sex hormone. The liver responds to increased cAMP (stimulated by epinephrine) by hydrolysis of stored glycogen and secretion of glucose into the blood. This results from protein kinase-induced stimulation of the enzyme that hydrolyzes glycogen and inhibition of the enzyme that catalyzes the formation of glycogen (see figure 19.23).

Like all biologically active molecules, cAMP must be rapidly inactivated for its regulation to be effective. This function is served by an enzyme called *phosphodiesterase* within the target cells, which hydrolyzes cAMP into inactive fragments. Drugs that inhibit the activity of phosphodiesterase thus prevent the breakdown of cAMP and result in increased concentrations of cAMP within the target cells. The drug *theophylline,* for example, is used clinically to raise cAMP levels within bronchiolar smooth muscle. This duplicates the effect of epinephrine on the bronchioles and produces dilation of the bronchioles in people who suffer from asthma.

Other Second Messengers   Liver cells respond to epinephrine by an increase in cAMP and a breakdown of glycogen; the action of insulin, in contrast, produces an increase in the production of glycogen. Clearly, since the action of epinephrine is mediated by cAMP, the action of insulin in liver cells must be mediated by a different second messenger.

Some of the hormones that do not use cAMP as a second messenger may act via production of **cyclic guanosine monophosphate (cGMP).** Indeed, cAMP and cGMP produced in response to the action of different hormones may act antagonistically within the target cells. Control of mitotic division and the cell cycle, for example, appears to be controlled by the ratio of cAMP to cGMP (as discussed in chapter 3). In addition to these "cyclic nucleotides," **calcium ion (Ca^{++})** may also function as a second messenger in the action of certain hormones.

Active transport carriers help maintain a very steep concentration gradient of Ca^{++}, which is present at much higher concentration in the extracellular fluid than it is within the cytoplasm of cells. Transient inhibition of this Ca^{++} pump produces a sudden inflow of Ca^{++} that can serve as a second messenger. The inflow of Ca^{++} that is produced by depolarization, for example, serves as a messenger that stimulates release of neurotransmitter chemicals from presynaptic nerve endings (see chapter 7), and serves to couple electrical excitation to striated muscle contraction (as described in chapter 8).

Contraction of smooth muscles, like that of striated muscles, is calcium dependent. Calcium ions in smooth muscles do not bond to troponin, as in striated muscles, but instead bond to a related protein known as **calmodulin.** In smooth muscles this protein serves as a subunit of an enzyme that activates myosin cross-bridges. In other tissues, calmodulin activation may mediate the effects of various hormones.

Figure 19.24   Hydrolysis of glycogen (glycogenolysis) in liver and skeletal muscles in response to cyclic AMP (cAMP) and Ca^{++} as second messengers. Both protein kinase and calmodulin are proteins activated by the second messengers.

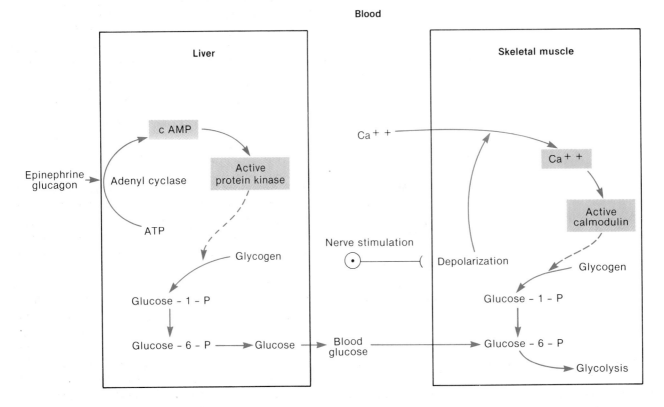

Activation of calmodulin by a sudden influx of Ca^{++} is analogous to activation of protein kinase by cAMP. Recent experiments, in fact, suggest that insulin may exert at least some of its effects by inhibiting active extrusion of Ca^{++}, and thus by promoting Ca^{++}-induced activation of calmodulin. Activated calmodulin, like activated protein kinase, affects various enzymes in the target cells. Glycogen breakdown in the liver in response to epinephrine, for example, is mediated by cAMP and protein kinase (as previously described); glycogen breakdown in skeletal muscles in response to nerve stimulation appears to be mediated by Ca^{++} inflow and activation of calmodulin (see figure 19.24).

A given cell may respond to a variety of different hormones that produce similar as well as antagonistic effects. The response of a target cell to these different hormones may depend upon interactions between different second messengers. Calmodulin, for example, has been found to stimulate adenyl cyclase—and thus to raise cAMP levels—under some circumstances, and to stimulate phosphodiesterase activity (and thus to lower cAMP) in other

instances. These effects, together with the antagonism between cAMP and cGMP, suggest that the effects of various hormones on a target cell may depend on complex and poorly understood interactions between different second messengers.

Effects of Hormone Concentrations on Tissue Response

The effects of hormones are very concentration dependent. Normal tissue responses are only produced when the hormones are present within their normal, or *physiological,* range of concentrations. When some hormones are present in abnormally high, or *pharmacological,* concentrations (as when they are taken as drugs), they may produce non-characteristic effects. This may be partly due to the fact that when a hormone is present at pharmacological levels it may bond significantly to receptors other than its own. Also, since some steroid hormones can be converted by target cells into other biologically active molecules, administration of pharmacological amounts of one steroid can result in production of abnormally high amounts of its conversion products, which may have different effects in the body.

Table 19.9 Comparison of the effects of cyclic AMP (cAMP) and Ca^{++} as second messengers in different tissues.

Second Messenger	Mechanism	Effects: Nerves	Effects: Muscles	Effects: Hormones
cAMP	Activation of cAMP-dependent protein kinase	Opens ion channels in postsynaptic membrane	Phosphorylation of myosin in skeletal muscle in response to depolarization (significance controversial)	Activation or inactivation of target cell enzymes in response to many polypeptide and glycoprotein hormones
Ca^{++}	Activation of Ca^{++}-dependent regulatory protein (calmodulin)	Mediates release of neurotransmitter in response to depolarization	Phosphorylation of myosin activates actomyosin ATPase in smooth muscles and stimulates contraction	May mediate the effects of insulin on target cells; may increase cAMP by stimulation of adenyl cyclase or decrease cAMP by stimulation of phosphodiesterase

Pharmacological doses of hormones—particularly steroids—can thus have widespread and often damaging "side effects." People with inflammatory diseases who are treated with high doses of cortisone over long periods of time, for example, may develop characteristic changes in bone and soft tissue structure. The contraceptive pill—which contains sex steroids—has a number of potential side effects (discussed in chapter 21) that could not have been predicted at the time the pill was first introduced.

Priming Effects   Variations in hormone concentration within the normal, physiological range can affect the responsiveness of target cells. This is due in part to the effects of polypeptide and glycoprotein hormones on the number of their receptor proteins in target cells. Small amounts of gonadotrophin-releasing hormone (GnRH), secreted into the portal blood, for example, increase the sensitivity of anterior pituitary cells to further GnRH stimulation. This is a priming effect; subsequent stimulation by GnRH thus causes greater secretion of gonadotrophic hormones (FSH and LH) from the anterior pituitary.

Desensitization and Downregulation   Prolonged exposure to high concentrations of polypeptide and glycoprotein hormones, in contrast to the priming effect of low concentrations, has been found to *desensitize* the target cells. Subsequent exposure to the same concentration of the same hormone thus produces less of a target tissue response. This desensitization may be due to the fact that high concentrations of these hormones cause a decrease in the number of receptor proteins in their target cells—a phenomenon called *downregulation*. Such desensitization and downregulation of receptors has been shown to occur, for example, in adipose cells exposed to high concentrations of insulin and in testicular cells exposed to high concentrations of LH.

However, only 1 percent of the membrane receptors for polypeptide and glycoprotein hormones are required to produce a maximum response by the target organ. The remaining 99 percent of the membrane receptors have thus been called *spare receptors*. Although they do not contribute to the maximum response, spare receptors increase the sensitivity of the target cells to low concentrations of hormones. Since only 1 percent of the membrane receptors need be occupied with hormones to produce a maximum response, it seems unlikely that downregulation of receptor numbers can entirely account for desensitization in response to high hormone concentrations. Indeed, there is evidence that desensitization may be due, at least in part, to interference with the response of the target cells at steps beyond the bonding of hormones to their membrane receptors.

Hormone Interactions

The responsiveness of a target tissue to a particular hormone is affected not only by the concentration of that hormone (as described in the previous section), but also by the effects of other hormones on that tissue. One hormone can modify the response of a target cell to another hormone by (1) changing the number of receptor proteins for the other hormone, or altering the sensitivity of the cell after the other hormone has combined with its receptor; (2) altering the activity of enzymes needed to activate the other hormone (if the other hormone functions as a prehormone); and (3) altering the activity of enzymes that inactivate the other hormone.

Table 19.10   Types of interactions between the effects of different hormones on target cells.

Interactions	Description	Example
Permissive effects	The effects of one hormone on a target organ require previous or simultaneous exposure of the organ to other hormones.	Estrogen stimulates the production of progesterone receptors in the uterus, thus having a permissive effect on the response of the uterus to progesterone.
Synergistic effects	The normal response of a target organ depends on the sum of the effects of different hormones.	Follicle-stimulating hormone (FSH) is required for the beginning and final stages of sperm production in testes; androgen is required for the intermediate stages of spermatogenesis.
Antagonistic effects	The effects of one hormone block or are opposite to the effects of another in a target organ.	Estrogen inhibits the formation of prolactin receptors in mammary glands, thus antagonizing the ability of prolactin to stimulate milk production.

Permissive and Synergistic Effects   A hormone is said to have a *permissive effect* when it enhances the responsiveness of a target organ to another hormone or when it increases the activity of the other hormone. Prior exposure of the uterus to estrogen (a female sex steroid), for example, induces the formation of progesterone (another female sex steroid) receptors, and thus increases the response of the uterus when it is subsequently exposed to progesterone. Estrogen thus has a permissive effect on the responsiveness of the uterus to progesterone. Prior exposure of the testes to FSH, as another example, may stimulate production of LH receptors and enhance the responsiveness of the testes to LH stimulation during puberty (the effects of these hormones are discussed in chapter 21).

Vitamin D_3 is a prehormone that must first be converted by enzymes in the kidney and liver into the active hormone—1,25-dihydroxy vitamin D_3—which helps to raise blood calcium levels. Parathyroid hormone (PTH) has a permissive effect on the actions of 1,25-dihydroxy vitamin D_3 because it stimulates the activity of these hydroxylating enzymes. Conversely, 1,25-dihydroxy vitamin D_3 also has a permissive effect on the ability of PTH to stimulate intestinal absorption of Ca^{++}.

Two or more hormones are said to exert *synergistic effects* when their effects are complementary, so that the sum of their effects cannot be reproduced by any one hormone separately. Development of sperm, for example, requires the synergistic action of both FSH and androgens during puberty (see table 19.10). Likewise, the ability of the mammary glands to produce and excrete milk requires the synergistic action of many hormones—estrogen, cortisol, prolactin, oxytocin, and others.

Antagonistic Effects   In some situations the actions of one hormone antagonize the effects of another. Lactation during pregnancy, for example, is prevented because the high concentration of estrogen in the blood inhibits the ability of the breasts to respond to prolactin. High levels of progesterone, as another example, inhibit the response of the uterus to estrogen.

Summary   The effects of a given hormone in the body are influenced by the concentration of that hormone, and by the concentration of other hormones as well. Alterations in the concentration of a particular hormone—in disease states, by surgical removal of a gland, or by administration of hormones as drugs—can thus produce many different and seemingly unrelated effects. Since hormones exert permissive, synergistic, and antagonistic effects with other hormones, and since the "characteristic" effects of the hormone are concentration dependent, the effects of hormone administration or withdrawal are widespread and often unpredictable.

1. Using diagrams, explain how steroid hormones and thyroxine exert their effects on their target cells.
2. Describe, using a diagram, how cyclic AMP is produced within a target cell in response to hormone stimulation. List three hormones that use cAMP as a second messenger.
3. Explain how cAMP functions as a second messenger to mediate the effects of some hormones. Describe how Ca^{++} might also serve as a second messenger in the action of some hormones.
4. Describe how the concentration of a particular polypeptide or glycoprotein hormone can affect the responsiveness of its target cells to further stimulation by the same hormone.
5. Describe three different ways in which the effects of one hormone can influence the response of a target cell to another hormone; use the terms *permissive, synergistic,* and *antagonistic* in your answer.

Summary

Introduction

I. There are several different classes of hormones.
   A. Catecholamines include epinephrine and norepinephrine.
   B. Polypeptide hormones include insulin, ADH, and others.
   C. Glycoprotein hormones, such as growth hormone, LH, FSH, and others, consist of large proteins combined with carbohydrates.
   D. Steroid hormones, which are secreted by the adrenal cortex (corticosteroids) and the gonads (sex steroids), are derived from cholesterol.

II. Hormones are secreted by endocrine glands.
   A. Endocrine glands lack ducts and therefore secrete their hormones into the vascular system.
   B. Exocrine glands, in contrast, have ducts that carry their secretions to the outside of epithelial membranes (and thus outside the body).

III. Active hormones are produced from less active precursors.
   A. When the active hormones are produced from precursors within the endocrine gland, these precursor molecules are called prohormones (e.g., proinsulin).
   B. When the active molecules are produced within the cells of the target organ, the precursor secreted by the endocrine gland is called a prehormone.
      1. Thyroxine (T_4) is converted to T_3 within the target organs, and testosterone is converted to other active products (such as dihydrotestosterone) within target organs.
      2. Vitamin D_3 is converted into 1,25-dihydroxy vitamin D_3 (the active form) in the liver and kidneys.

Hypothalamus and Pituitary

I. The pituitary gland is actually one organ composed of two separate structures.
   A. The posterior pituitary is derived from a downgrowth of the brain, and is thus continuous with the hypothalamus.
   B. The anterior pituitary is derived from the roof of the embryonic mouth.

II. The hormones of the posterior pituitary are actually produced by neurons in the hypothalamus.
   A. These hormones are then transported along axons to be stored in the posterior pituitary (neurohypophysis).
   B. The posterior pituitary is thus a storage organ, which secretes ADH and oxytocin in response to neural input from sensory nerves that stimulate neuroendocrine reflexes.

III. The anterior pituitary produces "trophic hormones."
   A. The term *trophic* means *feeding*—these hormones stimulate other tissues to grow and stimulate other endocrine glands to secrete their hormones.
   B. Adrenocorticotrophic hormone (ACTH), thyroid-stimulating hormone (TSH), and the gonadotrophic hormones (FSH and LH) stimulate the adrenal cortex, thyroid, and gonads, respectively.
   C. The anterior pituitary also secretes growth hormone and prolactin, whose functions are discussed in chapter 20 and 21, respectively.

IV. Hormone secretion by the anterior pituitary is controlled by the hypothalamus.
   A. The hypothalamus secretes releasing and inhibiting hormones.
      1. Corticotrophin-releasing hormone (CRH) stimulates the anterior pituitary to secrete ACTH.
      2. Thyrotrophin-releasing hormone stimulates the thyroid to secrete TSH.
      3. Gonadotrophin-releasing hormone (GnRH) stimulates the anterior pituitary to secrete FSH and LH.
      4. Somatostatin inhibits the secretion of growth hormone from the anterior pituitary.
      5. Prolactin-inhibiting hormone (PIH) inhibits the secretion of prolactin.
   B. The releasing and inhibiting hormones are secreted by the hypothalamus into the hypothalamo-hypophyseal portal system.

V. Secretion of anterior pituitary hormones is inhibited by negative feedback and, in one case, is stimulated by positive feedback mechanisms.
   A. A rise in target gland hormones can inhibit the secretion of hypothalamic-releasing hormones and/or inhibit the response of the anterior pituitary to the releasing hormones.
   B. Corticosteroids and sex steroids inhibit the secretion of ACTH and gonadotrophic hormones (FSH and LH), respectively.
      1. This is due to inhibition of releasing hormones and to inhibition of the response of the anterior pituitary to the releasing hormones.
      2. Thyroxine inhibits the response of the anterior pituitary to stimulation by TRH from the hypothalamus.
   C. Positive feedback occurs during the menstrual cycle, when the sex steroid estradiol-17β from the ovaries stimulates LH secretion from the anterior pituitary (described in chapter 21).

Adrenal and Thyroid Glands

I. The adrenal glands consist of two parts.
   A. The adrenal medulla is a modified sympathetic ganglion and secretes epinephrine in response to stimulation by preganglionic sympathetic nerve fibers.
   B. The adrenal cortex secretes two categories of steroid hormones.
      1. The outer zona glomerulosa secretes mineralocorticoids in response to stimulation by angiotensin II or high plasma K^+ concentration.

2. The middle and inner zones (zona fasciculata and zona reticularis) secrete glucocorticoids in response to stimulation by ACTH.

C. High levels of glucocorticoids exert negative feedback inhibition on CRH and ACTH secretion.

1. In Addison's disease, in which the adrenal cortex fails to secrete its hormones, ACTH secretion is abnormally high.

2. Excess secretion of corticosteroids, as in Cushings' syndrome, inhibits ACTH secretion.

II. Stress stimulates CRH, ACTH, and corticosteroid secretion.

A. This response to stress is part of the general adaptation syndrome (GAS).

B. Activation of the pituitary-adrenal axis occurs in response to all forms of stress.

C. High levels of glucocorticoids inhibit inflammation and suppress the immune response.

1. This can occur when corticoids such as cortisone are taken as drugs (in pharmacological doses).

2. Chronic high levels of corticoids may make the individual more susceptible to disease.

III. The thyroid gland is composed of hollow structures called follicles that accumulate iodide from the blood.

A. Iodine is attached to amino acids (tyrosines) in thyroglobulin protein within the colloid of the follicles.

B. Upon stimulation by TSH, the follicle cells pinocytose the colloid and digest the thyroglobulin, thus releasing the thyroid hormones.

1. Thyroxine—tetraiodothyronine (T_4)—is formed from two diiodotyrosine molecules; this is the major thyroid hormone secreted.

2. Triiodothyronine (T_3) is formed from one molecule of diiodotyrosine and one monoiodotyrosine.

C. The two thyroid hormones (T_4 and T_3) are transported in the blood attached to carrier proteins.

D. Thyroxine exerts negative feedback control of TSH secretion from the anterior pituitary.

1. In the absence of iodine in the diet, secretion of thyroxine declines and thus secretion of TSH is increased.

2. High secretion of TSH in this state causes growth of the thyroid and the production of a goiter.

Mechanisms of Hormone Action

I. In order to respond to a hormone, the target cells must contain specific receptor proteins for the hormone.

A. Receptors are characterized by their specificity for a hormone, and by the fact that they bond to the hormone with a high affinity but low capacity.

B. The location of the receptor proteins is variable, depending on the chemical nature of the hormone.

1. Receptors for catecholamine, polypeptide, and glycoprotein hormones are in the cell membrane.

2. Receptors for steroid hormones are in the cytoplasm.

3. Receptors for thyroid hormones are in the nucleus.

II. Once steroid hormones combine with cytoplasmic receptor proteins, the complex translocates into the nucleus.

A. The hormone-receptor complex attaches to specific acceptor sites in the chromatin and by this means, activates specific genes.

B. Once the genes are activated, they produce mRNA and the proteins that are coded by this mRNA.

C. Enzymatic activity of the proteins produced in this way affects the target cells and mediates the actions of the hormone.

D. Thyroxine is converted within the target cells to triiodothyronine, which then enters the nucleus.

1. T_3 combines with receptors that are already attached to the chromatin.

2. Combination of T_3 with the nuclear receptors activates genes and stimulates new production of mRNA and proteins.

III. Catecholamines, polypeptides, and glycoprotein hormones bind to receptors located in the outer surface of the cell membrane of the target cells.

A. This stimulates a rise in the intracellular concentration of a molecule or ion that acts as a second messenger to mediate the effects of the hormone.

B. In cells that use cyclic AMP (cAMP) as a second messenger, combination of a hormone with its receptor activates adenyl cyclase.

1. This enzyme is also part of the cell membrane and catalyzes the conversion of ATP within the cell to cAMP and two inorganic phosphates.

2. A rise in cAMP, in turn, activates a previously inactive enzyme in the cell called protein kinase.

3. Activated protein kinase transfers phosphates to various enzyme proteins, which activates some and inactivates others.

4. The effects of this enzyme activation and inactivation mediates the effects of the hormone in the target cell.

5. cAMP is inactivated within the target cell by the enzyme phosphodiesterase.

C. In some target cells, some hormones appear to use cyclic guanosine monophosphate (cGMP) as a second messenger.
D. Calcium ion (Ca^{++}) may function as a second messenger in the action of some hormones in some target cells.
   1. A steep concentration gradient for Ca^{++} is maintained by the active extrusion of Ca^{++} from the cell.
   2. A sudden rise in the intracellular Ca^{++} concentration can activate a protein called calmodulin, which serves to initiate changes characteristic of the hormone.
IV. Hormones can interact in a permissive, synergistic, or antagonistic way.
   A. A hormone has a permissive effect if it increases the sensitivity of a target cell to the effects of another hormone.
   B. Hormones interact synergistically when their effects complement each other.
   C. Some hormones antagonize the effects of others by a variety of mechanisms.
   D. Because of hormonal interactions, the effects of hormones when they are given in large doses are difficult to predict.

Self-Study Quiz

Multiple Choice

Match the hormone with the primary agent that stimulates its secretion:

1. Epinephrine
2. Thyroxine
3. Corticosteroids
4. ACTH
5. Insulin

   (a) TSH
   (b) ACTH
   (c) glucose
   (d) CRH
   (e) sympathetic nerves

Match the hormone to its mechanism of action:

6. Insulin
7. Epinephrine
8. Steroid hormones
9. Triiodothyronine (T_3)

   (a) uses cAMP as a second messenger
   (b) uses second messengers other than cAMP
   (c) attaches to cytoplasmic receptors
   (d) attaches to nuclear receptors

10. Steroid hormones are secreted by:
   (a) the adrenal cortex
   (b) the gonads
   (c) the thyroid
   (d) a and b
   (e) b and c
11. The hormones ADH and oxytocin are PRODUCED by:
   (a) the anterior pituitary
   (b) the posterior pituitary
   (c) the hypothalamus
   (d) the islets of Langerhans

12. Secretion of which of the following hormones would be INCREASED in an iodine-deficiency goiter?
   (a) TSH
   (b) thyroxine
   (c) T_3
   (d) all of these
13. What happens to the responsiveness of a target cell when it is exposed to high concentrations of a polypeptide hormone (such as insulin)?
   (a) responsiveness increases
   (b) responsiveness decreases
   (c) responsiveness either increases or decreases
   (d) there is no change in responsiveness

14. If one hormone stimulates production of receptors for a second hormone, these effects of the first hormone are said to be:
   (a) synergistic
   (b) permissive
   (c) antagonistic
   (d) complementary

Endocrine Control of Metabolism

*T*he term *metabolism* refers to all of the chemical changes that occur within the cells of the body. On the basis of energy flow, these reactions comprise two categories: **anabolism** and **catabolism.** Anabolic reactions generally require the input of energy (from hydrolysis of ATP), and result in the formation of large, energy-rich molecules such as triglycerides, glycogen, and protein. Catabolism refers to reactions in which these large molecules are hydrolyzed to yield their subunits and to the reactions of cell respiration in which these subunits are oxidized to provide energy for ATP production.

Figure 20.1 The balance of metabolism can be tilted towards anabolism (synthesis of energy reserves) or catabolism (utilization of energy reserves) by the combined actions of various hormones. Growth hormone and thyroxine have both anabolic and catabolic effects.

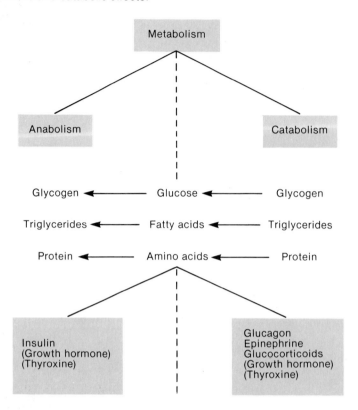

The molecules that can be oxidized for energy by the processes of cell respiration may be derived from the **energy reserves** of glycogen, fat, or protein in the body, or can be derived from the products of digestion that are absorbed through the small intestine. Since these molecules—glucose, amino acids, fatty acids, and others—are carried by the blood and serve as energy substrates for cell respiration, they can be called **energy carriers.**

Absorption of energy carriers from the intestine is not continuous; it rises to high levels following meals and tapers towards zero between meals. Despite this, the plasma concentration of glucose and other energy carriers does not remain high during periods of absorption and does not normally fall below a certain level during periods of fasting. During the phase of absorption, energy carriers are removed from the blood and deposited as energy reserves from which withdrawals can be made during times of fasting. This assures that there will be an adequate plasma concentration of energy carriers to sustain tissue metabolism at all times.

Because of differences in cellular enzyme content, different organs have different *preferred energy sources.* The brain has an almost absolute requirement for blood glucose as its energy source, for example. A fall in plasma concentration of glucose below about 50 mg per dl (deciliter) can thus "starve" the brain and have disastrous consequences. Resting skeletal muscles, in contrast, use fatty acids as their preferred energy source. Similarly, ketone bodies (derived from fatty acids), lactic acid, and amino acids can be used to different degrees as the energy sources of various organs (see chapter 5). The plasma must thus contain sufficient concentrations of all of these energy carriers at all times.

The rate of deposit and withdrawal of energy carriers into and from the energy reserves and the conversion of one type of energy carrier into another are regulated by the actions of hormones. The balance between anabolism and catabolism is determined by the antagonistic effects of hormones such as insulin, glucagon, growth hormone, thyroxine, and others, as summarized in figure 20.1 and in table 20.1.

Table 20.1 Summary of the endocrine regulation of metabolism.

Hormone	Blood Glucose	Carbohydrate Metabolism	Protein Metabolism	Lipid Metabolism
Insulin	Decreased	↑Glycogen formation ↓Glycogenolysis ↓Gluconeogenesis	↑Amino acid transport	↑Lipogenesis ↓Lipolysis ↓Ketogenesis
Glucagon	Increased	↓Glycogen formation ↑Glycogenolysis ↑Gluconeogenesis	No direct effect	↑Lipolysis ↑Ketogenesis
Growth hormone	Increased	↑Glycogen formation ↑Gluconeogenesis ↓Glucose utilization	↑Protein synthesis	↓Lipogenesis ↑Lipolysis ↑Ketogenesis
Glucocorticoids	Increased	↑Glycogen formation ↑Gluconeogenesis	↓Protein synthesis	↓Lipogenesis ↑Lipolysis ↑Ketogenesis
Epinephrine	Increased	↓Glycogen formation ↑Glycogenolysis ↑Gluconeogenesis	No direct effect	↑Lipolysis ↑Ketogenesis
Thyroxine	No effect	↑Glucose utilization	↑Protein synthesis	No direct effect

Insulin and Glucagon

Scattered within a "sea" of pancreatic exocrine tissue (the acini—see chapter 18) are islands of hormone-secreting cells. These **islets of Langerhans** contain three distinct cell types that secrete three different hormones. The most numerous are **beta cells;** these comprise 60 percent of each islet and they are the cells that secrete the hormone *insulin.* The **alpha cells** comprise about 25 percent of each islet and secrete the hormone *glucagon.* The least numerous cell type, **delta cells,** produce *somatostatin* (identical to the inhibitor of growth hormone secretion that is produced by the hypothalamus).

All three hormones are polypeptides. Insulin consists of two polypeptide chains—one twenty-one amino acids long and another that is thirty amino acids long—joined together by disulfide bonds. Glucagon is a twenty-one-amino acid polypeptide, and somatostatin contains fourteen amino acids. Insulin was the first of these hormones to be discovered (in 1921). The importance of insulin in the prevention of diabetes was immediately appreciated, and clinical use of insulin in the treatment of diabetes began almost immediately after its discovery. The physiological role of glucagon was discovered later, and the importance of glucagon in the development of diabetes has only recently been suspected, as will be described later in this chapter. The physiological significance of islet-secreted somatostatin is not currently known.

Figure 20.2 The cellular composition of a normal islet of Langerhans.

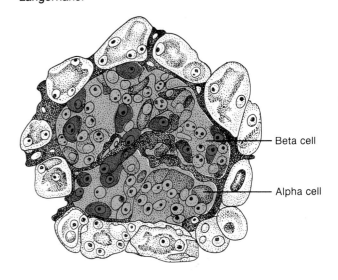

— Beta cell

— Alpha cell

Figure 20.3 The secretion of insulin from the B (beta) cells and A (alpha) cells of the islets of Langerhans is regulated to a large degree by the blood glucose concentration. A high blood glucose concentration stimulates insulin and inhibits glucagon secretion. Low blood glucose, conversely, stimulates glucagon and inhibits insulin secretion. The actions of insulin and glucagon provide negative feedback control of the blood glucose concentration.

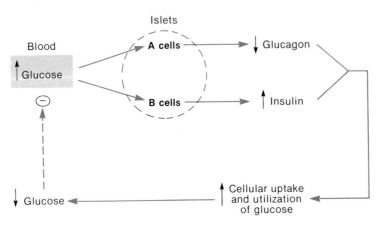

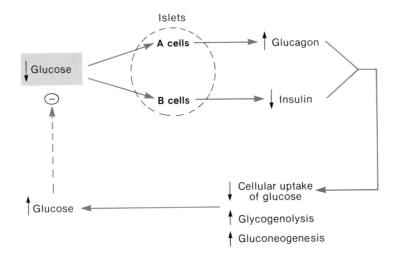

Regulation of Insulin and Glucagon Secretion

Insulin and glucagon secretion is largely regulated by the plasma concentrations of glucose and, to a lesser degree, of amino acids. The alpha and beta cells, therefore, act as both the sensors and effectors (see chapter 19) in this control system. Since the plasma concentration of glucose and/or amino acids rises during absorption of a meal and falls during fasting, secretion of insulin and glucagon likewise changes between periods of absorption and periods of fasting. These changes in insulin and glucagon secretion, in turn, cause changes in plasma glucose and amino acid concentrations, and thus help to maintain homeostasis via negative feedback loops.

Effect of Glucose During absorption of a carbohydrate meal the plasma glucose concentration rises. This rise in plasma glucose (1) stimulates the beta cells to secrete insulin, and (2) inhibits the secretion of glucagon from the alpha cells. Since *insulin stimulates the cellular uptake of plasma glucose* into many organs (as described in later sections), a rise in insulin secretion promotes the lowering of plasma glucose concentrations. Glucagon, conversely, antagonizes the effects of insulin on plasma glucose (it acts to raise the plasma glucose concentration, as will be described later). Inhibition of glucagon secretion thus complements the effects of increased insulin secretion during the absorption of a carbohydrate meal; both changes help to lower the plasma concentration of glucose.

Figure 20.4 Changes in blood glucose and plasma insulin concentrations after ingestion of one hundred grams of glucose in an oral glucose tolerance test.

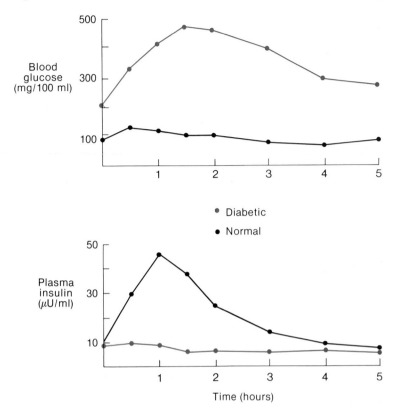

Diabetic

Normal

Time (hours)

During fasting, the plasma glucose concentration falls. During fasting, therefore: (1) insulin secretion decreases, and (2) glucagon secretion increases. These changes in hormone secretion prevent the cellular uptake of glucose into organs such as muscles, liver, and adipose tissue, and promote the release of glucose from the liver (through the actions of glucagon). A negative feedback loop is therefore completed (see figure 20.3), which helps to retard the fall in plasma glucose concentration that occurs during fasting.

The ability of the beta cells to secrete insulin, and the ability of insulin to lower blood glucose, is measured clinically by the oral **glucose tolerance test** (see figure 20.4). In this procedure a person drinks a glucose containing solution, and blood samples are periodically taken for plasma glucose measurements. In a normal person the rise in blood glucose produced by drinking this solution is reversed to normal levels within two hours following glucose ingestion. People with *diabetes mellitus*—due to inadequate insulin secretion or action—in contrast, maintain a state of high plasma glucose concentration during this time. People who have *hypoglycemia* (low plasma glucose concentrations due to excessive insulin secretion) have lower-than-normal plasma glucose concentrations five hours following glucose ingestion.

Effects of Ingested Protein   While ingestion of carbohydrates causes a rise in insulin and a fall in glucagon secretion, ingestion of proteins stimulates secretion of both hormones. The rise in insulin secretion helps to complete a negative feedback loop because insulin promotes the uptake of amino acids from blood into muscles and the incorporation of these amino acids into muscle protein. The importance of a simultaneous rise in glucagon secretion following a protein meal can be understood in terms of the need to maintain a constant plasma glucose concentration. A rise in insulin secretion, if not accompanied by a simultaneous rise in glucagon secretion, would cause hypoglycemia in the absence of ingested carbohydrates. A rise in both insulin and glucagon secretions thus promotes the lowering of the amino acid concentration in the plasma, while the glucose concentration remains relatively constant due to the antagonistic effects of insulin and glucagon.

Table 20.2 Regulation of insulin and glucagon secretion.

Regulator	Effect on Insulin Secretion	Effect on Glucagon Secretion
Hyperglycemia	Stimulates	Inhibits
Hypoglycemia	Inhibits	Stimulates
Gastric inhibitory peptide	Stimulates	Stimulates (?)
Sympathetic nerves	Inhibits	Stimulates
Parasympathetic nerves	Stimulates	Inhibits
Amino acids	Stimulates	Stimulates
Somatostatin	Inhibits*	Inhibits*

*Inhibitory effects of somatostatin on insulin and glucagon secretion may not occur physiologically.

Effects of Autonomic Nerves The islets of Langerhans receive both parasympathetic and sympathetic innervation. Activation of the parasympathetic system during meals stimulates insulin secretion at the same time that gastrointestinal function is stimulated. Activation of the sympathetic system, in contrast, stimulates glucagon secretion and inhibits insulin secretion. The effects of glucagon, together with those of epinephrine (discussed in a later section), produce a "stress hyperglycemia" when the sympathoadrenal system is activated.

Effects of Gastric Inhibitory Peptide (GIP) Surprisingly, insulin secretion increases more rapidly following glucose ingestion than it does following an intravenous injection of glucose. This is due to the fact that the intestine, in response to glucose ingestion, secretes a hormone that stimulates insulin secretion before the glucose is absorbed. Insulin secretion thus begins to rise in anticipation of a rise in blood glucose. The intestinal hormone that mediates this effect is believed to be gastric inhibitory peptide (GIP).

The mechanisms that regulate insulin and glucagon secretion, and the actions of these two hormones, normally prevent the plasma glucose concentrations from rising above 170 mg per 100 ml after a meal, or from falling below about 50 mg per 100 ml between meals. This regulation is important because abnormally high blood glucose can damage tissue cells (as may occur in diabetes mellitus, described in a later section), and abnormally low blood glucose can damage the brain. This latter effect results from the fact that glucose enters the brain by facilitated diffusion; when the rate of this diffusion is too low, due to low plasma glucose concentration, the supply of metabolic energy for the brain may become inadequate. It is important to remember that the brain relies almost entirely on blood glucose as an energy source.

Effects of Insulin and Glucagon: Period of Absorption

The lowering of plasma glucose by insulin is, in a sense, a side effect of the primary actions of this hormone. Insulin is the major hormone that promotes anabolism in the body. During absorption of the products of digestion, and the subsequent rise in the plasma concentrations of energy carriers, insulin promotes the cellular uptake of glucose and its incorporation into glycogen and fat. As previously described, insulin also promotes the cellular uptake of amino acids and its incorporation into muscle proteins. The stores of large, energy-reserve molecules are thus increased while the plasma concentrations of glucose and amino acids are reduced.

Synthesis of triglycerides (fat) within adipose cells depends upon insulin-stimulated glucose uptake from plasma. Once inside the adipose cells, glucose can be converted into α-glycerol phosphate and acetyl CoA (acetyl coenzyme A). Condensation of three fatty acids (derived from acetyl CoA—see chapter 5) with glycerol yields triglycerides. While the adipose cells can also utilize free fatty acids from the blood, they cannot incorporate glycerol from the blood into triglycerides. This is because adipose cells lack the enzyme needed to convert glycerol to α-glycerol phosphate, which is the precursor needed in triglyceride synthesis. Entry of blood glucose into adipose cells—which is directly dependent on insulin—thus determines the rate at which fat is produced.

A non-obese 70 kg man has approximately 10 kg of stored fat. Since 250 g of fat can supply the energy requirements for one day, this reserve fuel is sufficient for about forty days. Glycogen is less efficient as an energy reserve, and less is stored in the body; there is about 100 g of glycogen in the liver and about 200 g in skeletal muscles. Insulin promotes the cellular uptake of glucose into liver and muscles and the conversion of glucose into glucose-6-phosphate. In liver and muscles, this molecule is changed to glucose-1-phosphate, which is used as the precursor of glycogen. Once the stores of glycogen in liver and muscles are filled, continued ingestion of excess calories results in continued production of fat rather than of glycogen.

Figure 20.5 The synthesis of triglycerides (fat) within adipose cells. Notice that fat can be produced from glucose and from fatty acids released by hydrolysis of plasma lipoproteins. See text for more complete description.

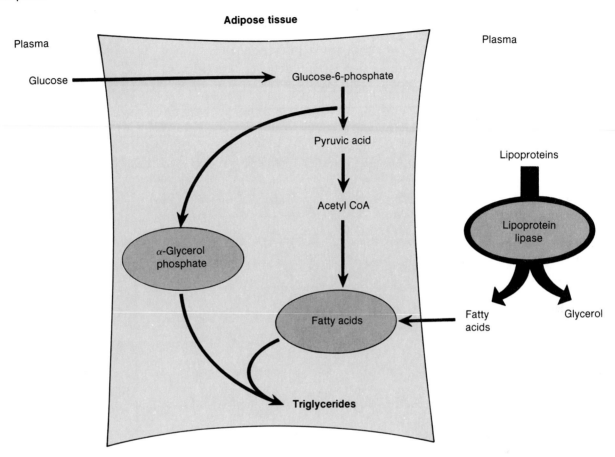

Effects of Insulin and Glucagon: Period of Fasting

Glucagon stimulates, and insulin suppresses, the hydrolysis of liver glycogen (**glycogenolysis**). During times of fasting, when glucagon secretion is high and insulin secretion is low, therefore, liver glycogen is used as a source of additional blood glucose. This process is essentially the reverse of that which formed glycogen, and results in the liberation of free glucose from glucose-6-phosphate by the action of an enzyme called *glucose-6-phosphatase*. Only the liver has this enzyme, and therefore only the liver can use its stored glycogen as a source of additional blood glucose. Since muscles lack glucose-6-phosphatase, their glycogen can only be used for glycolysis by the muscles themselves.

Since there is only about 100 g of stored liver glycogen, adequate blood glucose levels could not be maintained for very long during fasting using this source alone. The low levels of insulin secretion that occur during fasting, however, together with elevated glucagon secretion, promote **gluconeogenesis:** the formation of glucose from noncarbohydrate molecules. Low insulin secretion allows the release of amino acids from skeletal muscles; glucagon stimulates production of enzymes in the liver that convert amino acids to pyruvic acid (transamination and deamination), and that convert pyruvic acid to glucose. During prolonged fasting, gluconeogenesis in the liver, using amino acids from muscles, may be the only source of blood glucose.

Figure 20.6 Increased glucagon secretion and decreased insulin secretion during fasting favors catabolism. These hormonal changes result in elevated release of glucose, fatty acids, ketone bodies, and amino acids into the blood. Notice that the liver secretes glucose that is derived both from breakdown of liver glycogen and from conversion of amino acids in gluconeogenesis.

Fasting (↓insulin, ↑glucagon)

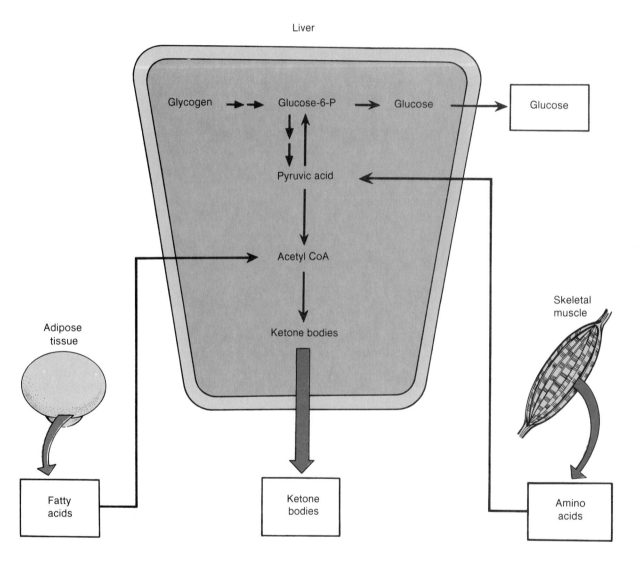

Secretion of glucose from the liver during fasting thus compensates for the low plasma concentrations of glucose and helps to provide glucose that is needed by the brain. In the presence of low insulin secretion, however, other organs cannot utilize blood glucose as an energy source. This helps to spare glucose for the brain, but alternative energy sources are needed for these other organs. The major alternate energy sources that are used by the liver and skeletal muscles are free fatty acids and ketone bodies.

Glucagon, in the presence of low insulin levels, stimulates in adipose cells an enzyme called hormone-sensitive lipase. This enzyme catalyzes the hydrolysis of stored triglycerides and the release of free fatty acids (as well as glycerol) into the blood. Glucagon, in the presence of low insulin levels, also stimulates enzymes in the liver that convert some of these fatty acids into ketone bodies. These derivatives of fatty acids—including acetoacetic acid, β-hydroxybutyric acid, and acetone—can also be released into the blood. Several organs in the body can use ketone bodies, like fatty acids, as a source of acetyl CoA in aerobic respiration.

Figure 20.7 Inverse relationship between insulin and glucagon secretion during absorption of a meal and during fasting. Changes in the insulin:glucagon ratio tilts metabolism toward anabolism during absorption of food and toward catabolism during fasting.

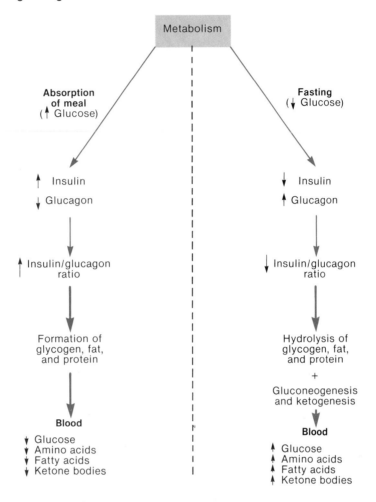

Through stimulation of **lipolysis** (breakdown of fat) and **ketogenesis** (formation of ketone bodies), the high glucagon and low insulin levels that are found during fasting provide energy carriers in the blood for use by the liver, muscles, and other organs. Through liver glycogenolysis and gluconeogenesis, these hormonal changes help to provide adequate levels of blood glucose to sustain the metabolism of the brain. The antagonistic actions of insulin and glucagon (see table 20.3) thus produce appropriate metabolic responses during periods of fasting and periods of absorption (see figure 20.7).

Table 20.3 Comparison of the metabolic effects of insulin and glucagon.

Actions	Insulin	Glucagon
Cellular glucose transport	↑	0
Glycogen synthesis	↑	↓
Glycogenolysis in liver	↓	↑
Gluconeogenesis	↓	↑
Amino acid uptake; protein synthesis	↑	0
Inhibition of amino acid release; protein degradation	↓	0
Lipogenesis	↑	0
Lipolysis	↓	↑
Ketogenesis	↓	↑

Code: ↑ = Increased
　　　 ↓ = Decreased
　　　 0 = No direct effect

Table 20.4   Comparison of juvenile-onset and maturity-onset diabetes mellitus.

Characteristics	Juvenile-onset (Type I)	Maturity-onset (Type II)
Usual age at onset	Under 20 years	Over 40 years
Development of symptoms	Rapid	Slow
Percent of diabetics	About 10%	About 90%
Development of ketoacidosis	Common	Rare
Association with obesity	Rare	Common
Beta cells of islets	Destroyed	Usually not destroyed
Insulin secretion	Decreased	Normal or increased
Autoantibodies to islet cells	Present	Absent
Associated with particular HLA antigens	Yes	No
Usual treatment	Insulin injections	Diet; oral stimulators of insulin secretion

1. Explain how the secretions of insulin and glucagon are affected by *(a)* absorption of a carbohydrate meal; *(b)* absorption of a protein meal; and *(c)* fasting.
2. Describe how the synthesis of fat in adipose cells, and the breakdown of fat, is regulated by insulin and glucagon during periods of absorption and periods of fasting.
3. Define the following terms and indicate how the processes they describe are affected by insulin and glucagon: *(a) glycogenesis; (b) gluconeogenesis;* and *(c) ketogenesis.*
4. Explain why the liver, but not skeletal muscles, can secrete glucose into the blood. Describe two pathways that contribute to hepatic secretion of glucose.
5. Using a flow diagram, describe how changes in insulin and glucagon secretion affect the plasma concentrations of glucose, amino acids, fatty acids, and ketone bodies during fasting.

Diabetes Mellitus

Chronic high blood glucose—hyperglycemia—is the hallmark of the disease **diabetes mellitus.** The name of this disease is derived from the fact that glucose is present in urine when the blood glucose concentration exceeds the ability of the kidneys to reabsorb filtered glucose (see chapter 16). The hyperglycemia of diabetes mellitus—which is seen even during fasting—results from either insufficient secretion of insulin by the beta cells of the islets of Langerhans, or from inability of the secreted insulin to stimulate cellular uptake of glucose from the blood. Diabetes mellitus, in short, results from inadequate insulin secretion or action.

There are two forms of diabetes mellitus. In **juvenile-onset (type I,** or **insulin-dependent)** diabetes, the beta cells are destroyed and secrete little or no insulin. This form of the disease accounts for only 10 percent of the cases of diabetes mellitus in the country. Most—about 90 percent—of the people who have diabetes have **maturity-onset (type II,** or **non-insulin dependent)** diabetes mellitus. As the name implies, the onset of symptoms in this disease normally occurs after the age of forty. Maturity-onset diabetes is usually slow to develop, occurs most commonly in people who are overweight, and is hereditary. Also, unlike type I diabetes mellitus, people with type II diabetes can have normal or even elevated levels of insulin in their blood (see table 20.4).

Juvenile-onset (Type I) Diabetes Mellitus

Juvenile-onset, or type I, diabetes mellitus results when the beta cells of the islets of Langerhans are destroyed by a virus or other environmental agent (although susceptibility to these agents may be inherited). Similar effects can be produced by surgical removal of the pancreas from an experimental animal. Removal of insulin in this way results in hyperglycemia and the appearance of glucose in the urine. Without insulin, glucose cannot enter adipose cells; the rate of fat synthesis thus lags behind the rate of fat breakdown and large amounts of free fatty acids are liberated from the adipose cells.

Figure 20.8 Sequence of events in which insulin deficiency (and perhaps glucagon excess) in diabetes mellitus may result in ketoacidosis, electrolyte disturbances, and dehydration. These changes may, in turn, lead to diabetic coma and death.

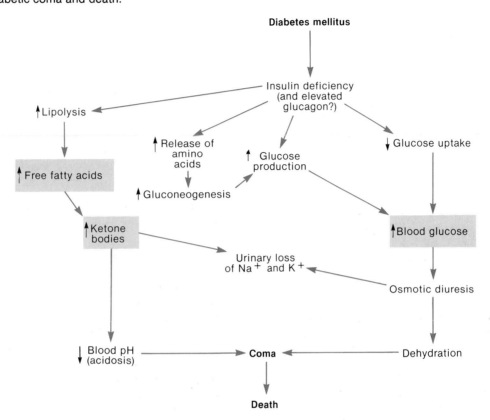

In a person with uncontrolled type I diabetes, many of the fatty acids released from adipose cells are converted into ketone bodies in the liver. This may result in elevated ketone body concentration in the blood (ketosis) and, if the buffer reserve of bicarbonate is neutralized, may result in *ketoacidosis*. During this time the glucose and excess ketone bodies that are excreted in the urine act as osmotic diuretics, as described in chapter 16, and cause excessive excretion of water in the urine (*polyuria*). This can result in severe *dehydration* that, together with ketoacidosis, may lead to coma and death (see figure 20.8).

Notice that the lack of insulin in type I diabetes produces effects that are an exaggeration of those normally produced when insulin levels are low in fasting—glucose is unavailable to the tissue cells, and energy carriers such as fatty acids and ketone bodies are released into the blood. In a sense, the tissues of a person with diabetes mellitus *are* fasting, because without insulin the blood glucose cannot enter the tissue cells. The alpha cells of the islets also respond in type I diabetes similarly to the way they do in fasting—glucagon is secreted in abnormally high amounts.

Role of Glucagon in Diabetes Mellitus When a non-diabetic person eats a carbohydrate-rich meal, the high glucose concentration inhibits glucagon secretion. In a diabetic, however, hyperglycemia does not inhibit glucagon secretion. This suggests that inhibition of glucagon secretion by high blood glucose requires the action of insulin on the alpha cells; since diabetics have decreased insulin action, they have high glucagon secretion despite high blood glucose.

Diabetics therefore have *low insulin and high glucagon secretion.* Glucagon, it may be recalled, stimulates glycogenolysis in the liver and thus helps to raise blood glucose concentrations. Glucagon also stimulates production of enzymes in the liver that convert fatty acids to ketone bodies, and thus acts to promote ketosis. Some researchers believe that the full symptoms of diabetes result from both low insulin and from excessive glucagon secretion. Low insulin levels may cause hyperglycemia and increased release of fatty acids and amino acids into the blood. High glucagon secretion may contribute to the hyperglycemia of diabetes and may be largely responsible for the development of ketoacidosis.

Somatostatin   The hyperglycemia and ketoacidosis of some people with type I diabetes cannot be controlled by insulin injections alone. In some clinical experiments, injections of somatostatin together with insulin appear to be more effective at controlling these symptoms of diabetes. Somatostatin—a hormone produced by the delta cells of the islets (as well as by the hypothalamus)—suppresses both glucagon and insulin secretion. A person injected with both insulin and somatostatin would thus have high blood levels of insulin and low levels of glucagon. These results lend further support to the idea that glucagon contributes to the symptoms of diabetes. Use of somatostatin in the treatment of this disease, however, is hindered by the fact that somatostatin also inhibits growth hormone secretion.

Long-term Complications of Diabetes Mellitus   Diabetes is the second leading cause of blindness in the country, and people with diabetes frequently have circulatory problems that increase the tendency to get gangrene and increase the risk of atherosclerosis. The causes of damage to the retina and lens of the eyes and to blood vessels are not well understood. It is believed, however, that these problems may result from long-term exposure to high blood glucose.

Two mechanisms have been proposed to explain how hyperglycemia may contribute to these complications. First, hyperglycemia has been found to "glycosylate" blood proteins, which may in some way lead to organ damage. Second, hyperglycemia appears to cause cells to accumulate glucose derivatives such as sorbitol, which draw water and may cause osmotic damage to the cells.

Insulin Resistance: Obesity and Maturity-onset (Type II) Diabetes

The effects produced by insulin, or any hormone, depend on the concentration of that hormone in the blood and on the sensitivity of the target tissue to given amounts of the hormone. The sensitivity of the target tissue to a hormone could decrease due to decreased numbers or affinity of the receptor proteins in the target cells or to other changes that occur at steps after hormone-receptor protein interaction. A state of resistance to the effects of the hormone would be produced in the target tissue as a result of these mechanisms.

Effects of Obesity and Exercise   Tissue responsiveness to insulin varies under normal conditions. *Exercise* appears to increase the sensitivity of target tissue to insulin. This is seen by the faster return of blood glucose to normal levels in a glucose tolerance test in persons who are physically trained, and by the observation that normal blood glucose concentrations are maintained by a lower level of insulin secretion.

Non-diabetic people who are *obese,* in contrast, require higher levels of circulating insulin to maintain normal blood glucose concentrations. This may be due to the fact that adipose cells get larger in obesity, so that their number of surface receptor proteins is "diluted" over a larger area. Since insulin promotes fat deposition, and thus makes fat cells "fatter," the decreased insulin sensitivity of large fat cells provides a negative feedback mechanism for slowing the rate of fat formation in people who are obese. Non-diabetic obese people, in summary, have high insulin levels and correspondingly decreased insulin sensitivity, so that their blood glucose concentrations are in the normal range.

Maturity-onset (Type II) Diabetes   Maturity-onset, or type II, diabetes mellitus appears to result from a genetic predisposition for insulin resistance. Like type I diabetes, type II diabetes is characterized by hyperglycemia and an elevated glucagon secretion. Unlike people with type I diabetes, however, most people with maturity-onset diabetes have normal—or even elevated—levels of insulin secretion. The beta cells in a person with type II diabetes, therefore, are usually healthy.

Since obesity decreases insulin sensitivity, as previously described, people who are genetically predisposed to insulin resistance may develop symptoms of diabetes when they become obese. Conversely, people who inherit the tendency towards developing maturity-onset diabetes may not develop hyperglycemia if they lose weight and exercise because these effects increase insulin sensitivity. Indeed, most people with maturity-onset diabetes can control their hyperglycemia with diet and exercise. If diet and exercise are not sufficient, these people can take drugs orally that (1) stimulate increased insulin secretion, and (2) stimulate increased tissue responsiveness to insulin. Insulin, of course, cannot be taken orally, because it is a polypeptide that would be digested before being absorbed.

Hypoglycemia

People with type I diabetes mellitus are dependent upon insulin injections to prevent hyperglycemia and ketoacidosis. If inadequate insulin is injected, the person may enter a coma as a result of the dehydration and ketoacidosis that develop. If an overdose of insulin is injected, however, a coma may also be produced as blood glucose levels fall to abnormally low levels in response to the insulin overdose. The hypoglycemia (low blood glucose) that results from insulin overdose and the hyperglycemia that accompanies insufficient insulin in diabetes represent two different mechanisms that can each lead to coma. The physical signs and symptoms in diabetic and hypoglycemic coma, however, are sufficiently different (see table 20.5) to allow hospital personnel to distinguish between these two types.

Table 20.5 Comparison of coma due to diabetic ketoacidosis and to hypoglycemia.

	Diabetic Ketoacidosis	Hypoglycemia
Onset	Hours to days	Minutes
Causes	Insufficient insulin; other diseases	Excess insulin; insufficient food; excessive exercise
Symptoms	Excessive urination and thirst; headache, nausea, and vomiting	Hunger, headache, confusion, stupor
Physical Findings	Deep, labored breathing; breath has acetone odor; blood pressure decreased, pulse weak; skin is dry	Pulse, blood pressure, and respiration are normal; skin is pale and moist
Laboratory Findings	Urine: glucose present, ketone bodies increased Plasma: glucose and ketone bodies increased, bicarbonate decreased	Urine: no glucose; ketone bodies at normal concentration Plasma: glucose concentration low, bicarbonate normal

Figure 20.9 Idealized oral glucose tolerance test in a person with reactive hypoglycemia. The blood glucose concentration falls below the normal range within five hours of glucose ingestion as a result of excessive insulin secretion.

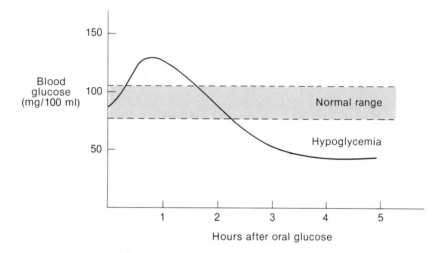

Hypoglycemic coma is most commonly produced by an overdose of injected insulin. Less severe symptoms of hypoglycemia, however, are most frequently produced by oversecretion of insulin from the islets after a carbohydrate meal. This **reactive hypoglycemia** is caused by an exaggerated response of the beta cells to a rise in blood glucose, and is most commonly seen in adults who are prediabetic. When people with reactive hypoglycemia ingest a carbohydrate meal the blood glucose levels fall below normal within five hours after the meal as a result of excessive insulin secretion. People with this condition must therefore limit their intake of carbohydrates and eat small meals at frequent intervals, rather than two or three large meals per day.

The symptoms of reactive hypoglycemia include tremor, hunger, weakness, perspiration, blurred vision, and impaired mental ability. The appearance of some of these symptoms, however, does not necessarily indicate hypoglycemia, nor does a given level of low blood glucose always produce these symptoms (some people have these symptoms at normal blood glucose levels, and other people do not have these symptoms even though their blood glucose is below normal levels). For these reasons the glucose tolerance test (see figure 20.9) and other tests are needed to confirm the diagnosis of reactive hypoglycemia.

Figure 20.10 Cyclic AMP (cAMP) serves as a second messenger in the actions of epinephrine and glucagon on liver and adipose tissue metabolism.

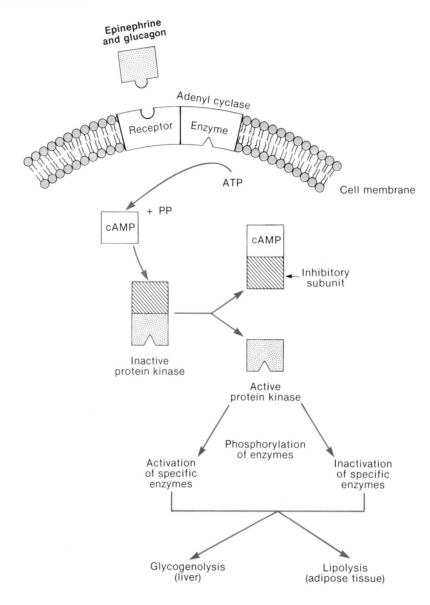

1. Describe the immediate dangers of uncontrolled type I diabetes mellitus. Explain how these conditions relate to those that occur during normal fasting and to the dangers of certain high protein-zero carbohydrate diets.
2. Compare type I and type II diabetes in terms of (a) etiology (cause); (b) insulin and glucagon secretion; and (c) cause of the hyperglycemia.
3. Explain both graphically and verbally the events that occur in reactive hypoglycemia. Describe which people are at risk of developing this condition and why they should eat frequent small meals rather than a few large meals.

Adrenal Hormones

The adrenal gland consists of two different parts that have different embryonic origins, that secrete different hormones, and that are regulated by different control systems. The **adrenal medulla** secretes catecholamine hormones—epinephrine and lesser amounts of norepinephrine—in response to sympathetic nerve stimulation. The **adrenal cortex** secretes corticosteroid hormones. These are grouped into two functional categories: *mineralocorticoids* such as aldosterone, which regulate Na^+ and K^+ balance (described in chapter 16), and *glucocorticoids* such as hydrocortisone, which participate in metabolic regulation. Secretion of the glucocorticoids is stimulated by ACTH, as described in chapter 19.

Figure 20.11 Catabolic actions of glucocorticoids help raise the blood concentration of glucose and other energy-carrier molecules.

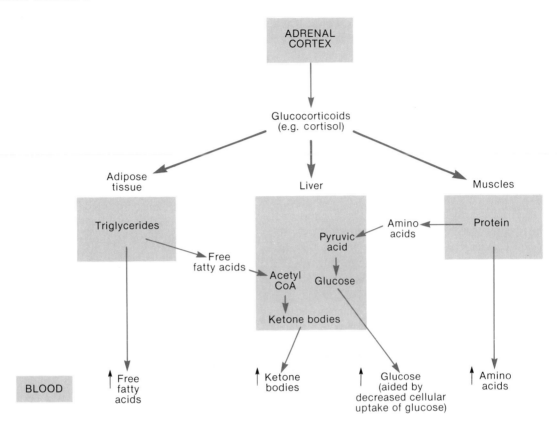

Metabolic Effects of Epinephrine

The metabolic effects of epinephrine are similar to those of glucagon. Glucagon, it may be recalled, promotes glycogenolysis and secretion of glucose from the liver, and stimulates lipolysis and the release of fatty acids from adipose tissue during periods of fasting. Since the metabolic effects of epinephrine are similar, these help to raise the blood concentrations of glucose and fatty acids when the sympathoadrenal system is activated during the "fight or flight" reaction. The effects of epinephrine and glucagon in liver and adipose cells are mediated by cyclic AMP as a second messenger (see figure 20.10).

Metabolic Effects of Glucocorticoids

Hydrocortisone (also called cortisol) and other glucocorticoids are secreted by the adrenal cortex in response to ACTH stimulation. Secretion of ACTH from the anterior pituitary occurs as a characteristic response to stress (as described in chapter 19). Since prolonged fasting or prolonged exercise certainly qualify as stressors, ACTH, and

thus glucocorticoid secretion, is stimulated under these conditions. Increased secretion of glucocorticoids during prolonged fasting or exercise supports the effects of increased glucagon secretion and decreased insulin secretion from the islets.

Like glucagon, hydrocortisone and other glucocorticoids promote lipolysis and ketogenesis and stimulate synthesis of hepatic enzymes involved in gluconeogenesis. The most distinctive metabolic effect of hydrocortisone is its promotion of gluconeogenesis, because (unlike glucagon) hydrocortisone stimulates the breakdown of muscle proteins and subsequent liberation of amino acids into the blood. This provides the substrates that enzymes in the liver need for production of glucose. The release of energy carriers—amino acids, glucose, fatty acids, and ketone bodies—into the blood in response to hydrocortisone helps to compensate for a state of prolonged fasting or prolonged exercise. Whether these metabolic responses are beneficial in other stressful states is open to question.

Thyroxine

The thyroid follicles secrete thyroxine, also called tetraiodothyronine (T_4), in response to stimulation by TSH from the anterior pituitary. Almost all organs in the body are targets of thyroxine action. Thyroxine itself, however, is not the active form of the hormone within the target cells; as described in chapter 19, thyroxine is a prehormone that must first be converted to triiodothyronine (T_3) within the target cells to be active. Acting via its conversion to T_3 within target cells, thyroxine (1) regulates the rate of cell respiration; and (2) contributes to proper growth and development, particularly during early childhood.

Thyroxine and Cell Respiration

Thyroxine stimulates the rate of cell respiration in almost all cells in the body. This effect is believed to be due to thyroxine-induced lowering of cellular ATP concentrations. As described in chapter 5, ATP exerts an end-product inhibition of certain key respiratory enzymes so that when ATP concentrations increase, the rate of cell respiration decreases in a negative feedback fashion. Conversely, a lowering of cellular ATP concentrations, as may occur in response to thyroxine, stimulates an increase in cell respiration through a lowering of this end-product inhibition.

The mechanisms of thyroxine action are not entirely understood. One effect of thyroxine, however—the stimulation of Na^+/K^+ pump activity in cell membranes—could account for the thyroxine-induced lowering of cellular ATP concentrations. Active transport of Na^+ and K^+ across the membrane represents a significant energy "sink" in the cell, accounting for about 12 percent of the calories consumed by the body at rest. Through stimulation of Na^+/K^+ pumps, therefore, thyroxine could significantly decrease cellular ATP concentrations and, by this means, stimulate the rate of cell respiration (see figure 20.12).

Effect on Basal Metabolic Rate

When a person wakes up in the morning—before stress, muscular activity, and digestion of food contribute to energy expenditures—the total energy consumption of the body is at its lowest or basal (baseline) level. This is the *basal metabolic rate (BMR)*. Most commonly measured by the rate of oxygen consumption, the BMR indicates the "idling speed" of the body. Since activity of the Na^+/K^+ pumps contributes significantly to the energy consumed in this state, and since the activity of these pumps is set by the level of thyroxine secretion, the BMR can be used as an index of thyroid function. Indeed, such measurements were used clinically to evaluate thyroid function prior to development of direct chemical measurements of T_4 and T_3 in blood.

Effect on Heat Production

The coupling of energy-releasing reactions to energy-requiring reactions can never be 100 percent efficient; a proportion of the energy is always lost as heat. Much of the energy liberated during cell respiration, and much of the energy liberated by the hydrolysis of ATP, escapes as heat. Since thyroxine stimulates both ATP consumption and cell respiration, the actions of thyroxine result in the production of metabolic heat.

The heat-producing, or *calorigenic* (calor = heat), *effects* of thyroxine are required for cold adaptation. This does not mean that people who are cold adapted have higher than normal levels of thyroxine secretion. Rather, thyroxine levels in the normal range coupled with increased activity of the sympathoadrenal system are responsible for cold adaptation. Thyroxine exerts a permissive effect (see chapter 19) on the ability of the sympathoadrenal system to increase heat production in response to cold stress.

Hypothyroidism and Hyperthyroidism

As might be predicted from the effects of thyroxine, people who are hypothyroid have an abnormally low BMR and experience weight gain and lethargy. In the absence of normal levels of thyroid secretion there is decreased ability to adapt to cold stress. Another symptom of hypothyroidism is *myxedema*—accumulation of mucoproteins in subcutaneous connective tissues. Hypothyroidism may result from a variety of causes, including insufficient TRH secretion from the hypothalamus, insufficient TSH secretion from the anterior pituitary, or insufficient iodine in the diet. Hypothyroidism due to lack of iodine is accompanied by excessive TSH secretion (as described in chapter 19), which causes excessive growth of the thyroid and the development of a goiter.

Figure 20.12 The mechanism that has been proposed to explain the effects of thyroid hormones on basal metabolic rate. Through activation of genes and the stimulation of protein synthesis, thyroid hormones increase activity of the Na^+/K^+ pump. This active transport carrier accounts for a large percentage of the energy expenditures of the cell. The concentration of ATP therefore declines as a result of increased energy usage, and the decreased ATP concentrations stimulate increased cellular respiration.

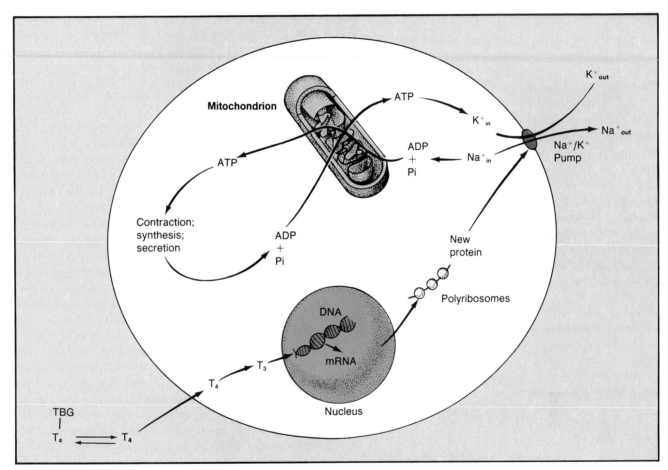

Goiters due to iodine deficiency are no longer very common in the United States because of the availability of iodized salt and other sources of iodine. Goiters, however, can be produced by another mechanism. In *Graves' Disease,* autoantibodies are believed to be produced that have TSH-like effects on the thyroid. These autoantibodies—called *LATS* (long-acting thyroid stimulator)—stimulate both secretion and growth of the thyroid. In this case, therefore, a goiter is found associated with a hyperthyroid condition.

Hyperthyroidism, as might be predicted from the actions of thyroxine, produces a high BMR associated with weight loss, nervousness, irritability, and intolerance to heat. These and other symptoms of hyperthyroidism are compared with the symptoms of hypothyroidism in table 20.6.

Table 20.6 Comparison of hypothyroidism and hyperthyroidism.

	Hypothyroid	Hyperthyroid
Growth and Development	Impaired growth	Accelerated growth
Activity and Sleep	Decreased activity; increased sleep	Increased activity; decreased sleep
Temperature Tolerance	Intolerance to cold	Intolerance to heat
Skin Characteristics	Coarse, dry skin	Smooth skin
Perspiration	Absent	Excessive
Pulse	Slow	Rapid
Gastrointestinal Symptoms	Constipation; decreased appetite; increased weight	Frequent bowel movements; increased appetite; decreased weight
Reflexes	Slow	Rapid
Psychological Aspects	Depression and apathy	Nervous, "emotional"
Plasma T_4 Levels	Decreased	Increased

Thyroxine in Growth and Development

Through its stimulation of cell respiration, thyroxine promotes increased consumption of energy carriers such as glucose, fatty acids, and other molecules. These effects, however, are mediated at least in part through activation of genes; thyroxine thus stimulates both RNA and protein synthesis (described in chapter 19). As a result of its stimulation of protein synthesis throughout the body, thyroxine is considered to be an anabolic hormone like insulin and growth hormone.

Because of its stimulation of protein synthesis, thyroxine is needed for growth of the skeleton and—most importantly—for proper development of the central nervous system. This latter effect is particularly significant during prenatal development and the first two years of postnatal life. Hypothyroidism during this time may result in *cretinism*. Unlike dwarfs, who have normal thyroid activity but low secretion of growth hormone (described in the next section), cretins suffer from severe mental retardation.

Growth Hormone

The anterior pituitary secretes **growth hormone (GH),** also called **somatotrophic hormone,** in larger amounts (about 500 µg per day) than any other of its hormones. As its name implies, growth hormone stimulates growth in children and adolescents. The continued high secretion of growth hormones in adults, particularly under conditions of fasting and other forms of stress, implies that this hormone can have important metabolic effects even after the growing years have ended.

Regulation of Growth Hormone Secretion

Secretion of growth hormone is inhibited by somatostatin, which is produced by the hypothalamus and secreted into the hypothalamo-hypophyseal portal system. A newly discovered hypothalamic-releasing hormone appears to stimulate growth hormone secretion; growth hormone thus seems to be unique among the anterior pituitary hormones in that its secretion is controlled by both a releasing and an inhibiting hormone from the hypothalamus. These hypothalamic hormones are apparently released in a circadian ("about a day") pattern, because growth hormone secretion increases during sleep and decreases during periods of wakefulness.

Growth hormone secretion is stimulated by an increase in the plasma concentrations of amino acids and by a decrease in the plasma glucose concentrations. Secretion of growth hormone is therefore stimulated during absorption of a high protein meal, when amino acids are absorbed through the intestine. Secretion of growth hormone is also stimulated during prolonged fasting, when plasma glucose is low and the plasma amino acid concentration is high due to breakdown of skeletal muscle proteins.

Effects of Growth Hormone on Metabolism

The fact that growth hormone secretion is increased during fasting and also during absorption of a protein meal reflects the complex nature of this hormone's actions. Growth hormone has both anabolic and catabolic effects; it promotes protein synthesis (anabolism) while it stimulates the release of fatty acids from adipose tissue. In the former effect, growth hormone is similar to insulin; its latter effect is similar to that of glucagon.

Effect on Lipid and Carbohydrate Metabolism Increased secretion of growth hormone reinforces the effects of glucagon, which is also secreted in increased amounts at this time. Growth hormone stimulates lipolysis and the release of free fatty acids from adipose tissue. A rise in the free fatty acid concentration of plasma provides an alternate energy source for tissues, as previously described, so that glucose can be spared for the brain.

A rise in plasma fatty acid concentration results in decreased rates of glycolysis within many organs. This inhibition of glycolysis by fatty acids, perhaps together with more direct "anti-insulin" effect of growth hormone, causes decreased glucose utilization by the tissues. Experimental animals that have had their pituitary removed (and thus lack growth hormone) as a consequence experience hypoglycemia and are more sensitive to the effects of insulin than intact animals. The anti-insulin actions of growth hormone are called its *diabetogenic effect*.

Effect on Protein Metabolism Growth hormone stimulates the cellular uptake of amino acids and protein synthesis in many organs of the body. These actions are useful during a protein-rich meal; amino acids are removed from the blood and used to form proteins, while the plasma concentrations of glucose and fatty acids are increased to provide alternate energy sources. The anabolic effect of growth hormone on protein synthesis is particularly important during the growing years, when it contributes to growth hormone-stimulated increases in bone length and increased mass of many soft tissues.

Effects of Growth Hormone on Body Growth

The anabolic effects of growth hormones and their resultant stimulation of skeletal growth are mediated indirectly. This was demonstrated by the observation that growth hormone by itself cannot stimulate bone growth *in vitro* (outside the body); plasma from growth hormone-stimulated animals must be present. This is because the growth-promoting effects are stimulated directly by polypeptides called **somatomedins,** which are produced by the liver and secreted into the blood in response to growth hormone stimulation. The somatomedins therefore are responsible for, or mediate, the stimulation of skeletal growth induced by growth hormone (see figure 20.13).

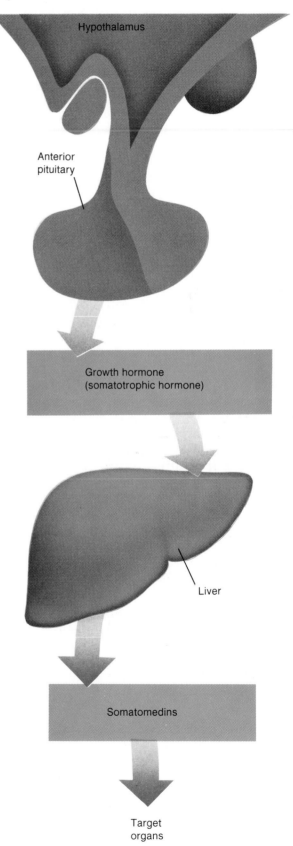

Figure 20.13 Hepatic production of polypeptides called *somatomedins* is stimulated by growth hormone. These compounds in turn produce the anabolic effects that are characteristic of the actions of growth hormone in the body.

Figure 20.14 Effects of growth hormone. The growth-promoting, or anabolic, effects of growth hormone are mediated indirectly via stimulation of somatomedin production by the liver.

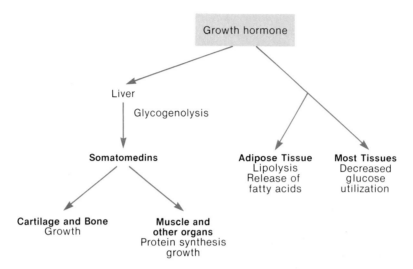

Growth of Cartilage and Bone Bone can only grow in width by the addition of new bone on already existing surfaces. This is accomplished by *osteoblasts* located in the connective tissue (periosteum) that covers the bone. The *osteocytes* that are entombed within the calcified bone (see chapter 1) remain alive because of blood supplied to them through small channels called canaliculi, but these cells are prevented from producing new bone by their calcified surroundings. Living cartilage, in contrast, remains uncalcified; cells called *chondrocytes* within the cartilage can secrete new organic material that makes the cartilage grow thicker from the inside.

The long bones of a child can grow in length because they contain a layer of cartilage at each end, between the heads (epiphyses) and the shaft (diaphysis) of the bone. These layers of cartilage are called the **epiphyseal discs.** Growth hormone—acting through somatomedins—stimulates cell division of the chondrocytes and secretion of new organic material so that the epiphyseal discs grow in thickness.

When the epiphyseal discs increase in thickness, the cartilage on the diaphysis side becomes calcified. Since cartilage lacks canaliculi to nourish the chondrocytes, the calcified cartilage disintegrates and osteoblasts produce bone around the disintegrating cartilage (see figure 20.15). Only the cartilage on the diaphysis side of the epiphyseal discs are replaced by bone. The epiphyseal discs thus increase in thickness in response to growth hormone (via somatomedin) stimulation, and then return to their original thickness when part of the discs become new bone. In this way, growth hormone promotes an increase in bone length throughout the growing years.

Influence of Nutrition and Androgens on Growth Adequate diet—particularly of protein—is required for the production of somatomedins. This helps to explain the common observation that many children are significantly taller than their immigrant parents, who may not have had an adequate diet in their youth. Children with protein malnutrition (*kwashiorkor*) have low growth rates and low somatomedin levels, despite the fact that growth hormone secretion is abnormally elevated. When these children eat an adequate diet, somatomedin levels and growth rates increase while growth hormone secretion decreases to the normal range.

Figure 20.15 Long bones grow in length by means of a disc of cartilage—the epiphyseal disc—located between the epiphysis and diaphysis of the bone. Since the cartilage matrix is not calcified toward the epiphyseal side, cartilage cells (chondrocytes) can secrete new matrix, allowing the cartilage to grow "from within." Calcification causes the cartilage on the diaphyseal side to degenerate, and osteoblasts secrete bone over these degenerating spicules of cartilage.

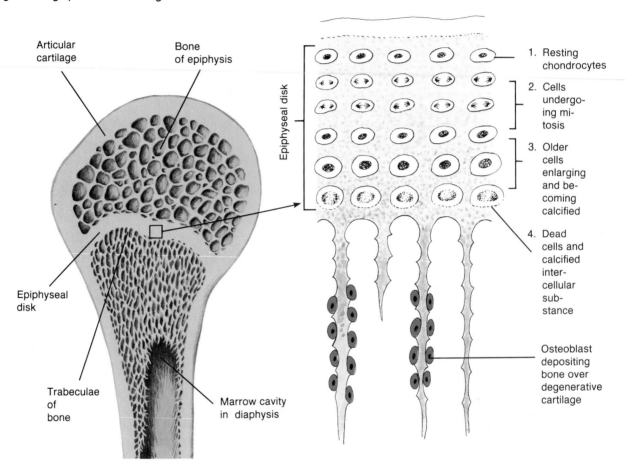

Somatomedin concentrations in plasma normally reach adult levels by about the age of eight years. The growth spurt that occurs at puberty is therefore not due to growth hormone-stimulated somatomedin production. This appears rather to be due to increased secretion of androgens in both sexes. Ultimately, however, high androgen secretion stops growth by "sealing" the epiphyseal discs—that is, by stimulating the conversion of these layers of cartilage to bone. Growth therefore stops, despite the fact that growth hormone secretion in adults is not significantly lower than that in children.

Gigantism, Acromegaly, and Dwarfism

Excessive secretion of growth hormone in children can produce **gigantism,** in which people may grow up to eight feet tall. Excessive growth hormone secretion that occurs after the epiphyseal discs have sealed cannot, of course, produce increases in height. Oversecretion of growth hormone in the adult results in elongation of the jaw and deformities in the bones of the face, hands, and feet. This is accompanied by growth of soft tissues and coarsening of the skin to give the characteristic appearance of **acromegaly** (acro = highest or outermost; mega = big—see figure 20.16).

Figure 20.16 The progression of acromegaly in one individual, from age nine *(a)*, sixteen *(b)*, thirty-three *(c)*, and fifty-two *(d)* years. The coursening of features and disfigurement are evident by age thirty-three and severe at age fifty-two.

(a)

(b)

(c)

(d)

Inadequate secretion of growth hormone during the growing years results in **dwarfism.** An interesting variant of this is *Laron dwarfism,* in which there is genetic insensitivity to the effects of growth hormone. In this case, growth hormone secretion is normal, but somatomedin production is abnormally low. It is also believed that the short stature of African pygmies may be due to a genetically low sensitivity to the effects of growth hormone.

---

1. Use a flow diagram to describe the effects of glucocorticoids on the metabolism of the liver and skeletal muscles. Explain how epinephrine contributes to the plasma concentration of energy carriers during the "fight or flight" reaction.
2. Explain how thyroid hormones may set the basal metabolic rate, and why people who are hypothyroid have a tendency to gain weight and are less resistant to cold stress.
3. Describe the effects of growth hormone on the metabolism of lipids, glucose, and amino acids.
4. Explain how growth hormone stimulates growth of long bones. Describe the hormonal defects that cause acromegaly, dwarfism, and cretinism.

Calcium and Phosphate Balance

In addition to regulating the blood concentrations of energy carriers, as part of the regulation of total energy balance in the body, hormones also serve to regulate the blood concentrations of inorganic ions. Plasma concentrations of sodium and potassium, for example, are regulated by the effects of mineralocorticoids such as aldosterone (see chapter 16). Plasma calcium and phosphate concentrations are regulated by three hormones: **parathyroid hormone (PTH), 1,25-dihydroxy vitamin D₃, and calcitonin.**

The skeleton, in addition to providing support for the body, serves as a large store of calcium and phosphate in the form of hydroxyapatite crystals. The relative amount of calcium and phosphate in the blood and in the bones in the form of these crystals is determined by the relative activity of two types of cells. These are *osteoblasts,* which produce bone as previously described, and *osteoclasts,* which dissolve calcium phosphate crystals and promote bone resorption (dissolution of bone).

Formation and resorption of bone occur constantly at rates determined by the hormonal balance. Body growth during the first two decades of life occurs because bone formation proceeds at a faster rate than bone resorption. By age fifty or sixty, the rate of bone resorption often exceeds the rate of bone formation. The constant activity of osteoblasts and osteoclasts allows bone to be remodeled throughout life. The position of the teeth, for example, can be changed by orthodontic appliances—braces—that cause bone resorption on the pressure-bearing side and bone formation on the opposite side of the tooth socket. The bony tooth socket, as a result, moves.

Figure 20.17   Resorption of bone by osteoclasts *(a)* and formation of new bone by osteoblasts *(b)*. Both resorption and deposition (formation) occur simultaneously throughout the body.

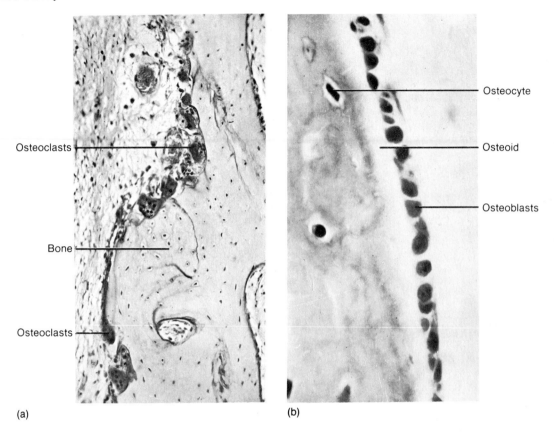

Osteoclasts

Bone

Osteoclasts

(a)

Osteocyte

Osteoid

Osteoblasts

(b)

The plasma concentration of calcium and phosphate is maintained, despite changing rates of bone formation and resorption, by hormonal control of intestinal absorption and urinary excretion of these ions. These hormonal mechanisms are very effective and maintain the plasma calcium and phosphate concentrations within narrow limits. Plasma calcium, for example, is normally maintained at about 2.5 millimolar or 5 milliequivalents per liter (a milliequivalent equals a millimole times the valence of the ion—in this case, times two).

Maintenance of normal plasma calcium concentrations is important because of the wide variety of effects that calcium has in the body. In addition to its role in bone formation, excitation-contraction coupling in muscles, and second messenger function in the action of some hormones (discussed previously), calcium is needed to maintain proper membrane permeability. An abnormally high plasma calcium concentration decreases membrane permeability, while an abnormally low calcium level increases the permeability of cell membranes to Na^+ and other substances. Hypocalcemia (low-plasma calcium) therefore enhances the excitability of nerves and muscles, and can result in muscle spasm (tetany).

Parathyroid Hormone and 1,25-Dihydroxy Vitamin D₃

Parathyroid hormone, an eighty-four-amino acid polypeptide secreted by the parathyroid glands, acts to help raise the plasma concentration of calcium. The effects of parathyroid hormone include (1) stimulation of bone resorption; (2) stimulation of calcium reabsorption in the kidneys, so that less calcium is excreted in the urine; and (3) inhibition of renal phosphate reabsorption, so that more phosphate is excreted in the urine. Parathyroid hormone also promotes formation of 1,25-dihydroxy vitamin D₃, and so indirectly helps to raise plasma calcium levels via the effects of this other hormone.

Production of 1,25-dihydroxy vitamin D₃ begins in the skin, where vitamin D₃ is produced from its precursor molecule (7-dehydrocholesterol) under the influence of sunlight. When the skin does not make sufficient vitamin D₃ because of insufficient exposure to sunlight, this compound must be ingested in the diet—this is why it is called a vitamin. Whether this compound is secreted into the blood from the skin or is secreted into the blood from the intestine (after ingestion), vitamin D₃ functions as a *prehormone,* which must be chemically changed in order to be biologically active.

Figure 20.18   Pathway for the production of the hormone 1,25–dihydroxy vitamin D_3. This hormone is produced in the kidneys from the inactive precursor, 25–hydroxy vitamin D_3 (formed in the liver). This latter molecule is produced from vitamin D_3 secreted by the skin.

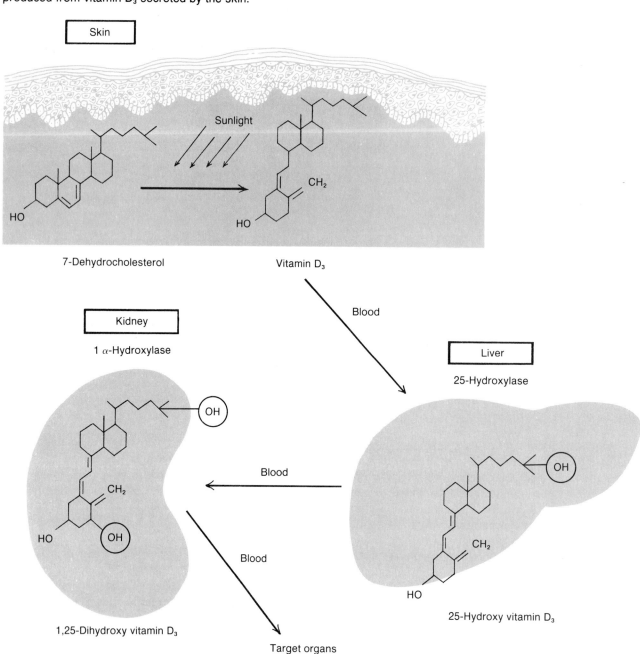

An enzyme in the liver adds a hydroxyl (OH) group to carbon number 25, which converts vitamin D_3 into 25-hydroxy vitamin D_3. In order to be active, however, another hydroxyl group must be added to the first carbon. Hydroxylation of the first carbon is accomplished by an enzyme—$l\alpha$-hydroxylase—in the kidneys, which converts the molecule to 1,25-dihydroxy vitamin D_3. Activity of this enzyme is stimulated by parathyroid hormone.

The hormone 1,25-dihydroxy vitamin D_3 helps to raise the plasma concentration of calcium and phosphate by stimulating (1) resorption of bone; (2) intestinal absorption of calcium and phosphate; and (3) renal reabsorption of phosphate. Notice that unlike parathyroid hormone, 1,25-dihydroxy vitamin D_3 directly stimulates intestinal absorption of calcium and phosphate. Also unlike parathyroid hormone, vitamin D_3 promotes the reabsorption of phosphate in the kidneys.

Negative Feedback Control of Calcium and Phosphate Balance

Secretion of parathyroid hormone is controlled by the plasma calcium concentrations. Secretion of parathyroid hormone is stimulated by low calcium and is inhibited by high calcium concentrations. Low concentrations of plasma calcium are thus corrected by the effects of increased secretion of parathyroid hormone and by the effects of increased production of 1,25-dihydroxy vitamin D_3 that result from PTH stimulation.

It is possible that plasma calcium levels might fall while phosphate levels are normal. In this case, the increased secretion of parathyroid hormone and production of 1,25-dihydroxy vitamin D_3 that result could abnormally raise phosphate levels while acting to restore normal calcium levels. This is prevented by the effects of parathyroid hormone on the kidneys: PTH inhibits phosphate reabsorption so that more phosphate is excreted in the urine. In this way blood calcium levels can be raised to normal without excessively raising blood phosphate concentrations.

Figure 20.19 A decrease in plasma Ca^{++} directly stimulates secretion of parathyroid hormone (PTH). Production of 1,25-dihydroxy vitamin D_3 also rises when Ca^{++} is low because PTH stimulates the final hydroxylation step in the formation of this compound in the kidneys.

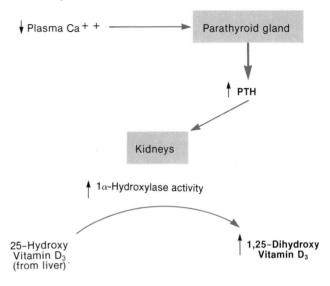

Figure 20.20 Negative feedback loop that returns low blood Ca^{++} concentrations to normal without simultaneously raising blood phosphate levels above normal.

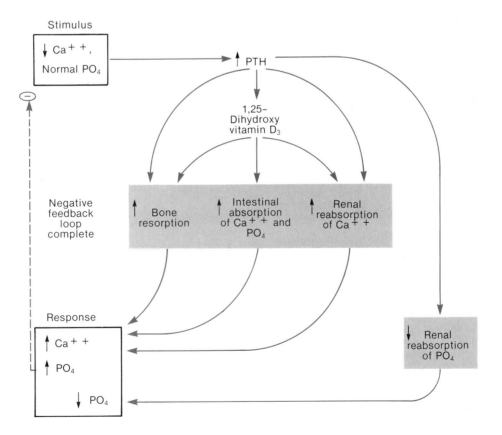

Figure 20.21 Negative feedback control of calcitonin secretion.

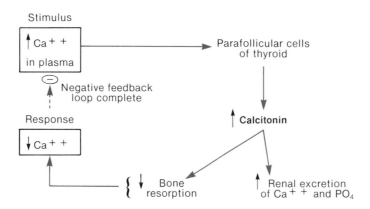

Calcitonin

Experiments in the 1960s revealed that high blood calcium in dogs may be lowered by a hormone secreted from the thyroid gland. Lowering of blood calcium, in other words, could be achieved by secretion of this hormone as well as by inhibition of PTH secretion and inhibition of 1,25-dihydroxy vitamin D_3 formation. The calcium-lowering hormone was found to be a thirty-two-amino acid polypeptide secreted by *parafollicular,* or *C cells,* in the thyroid that were distinct from the follicular cells that secrete thyroxine. This thyroid hormone, whose action was antagonistic to that of PTH and to 1,25-dihydroxy vitamin D_3, was called **calcitonin.**

Secretion of calcitonin is stimulated by high plasma calcium levels, and acts to lower calcium levels by (1) inhibiting activity of osteoclasts, thus reducing bone resorption and (2) stimulating renal clearance of calcium and phosphate by decreasing their reabsorption. Regulation of calcitonin secretion and the effects of calcitonin action are summarized in figure 20.21.

While it is attractive to think that calcium balance is regulated by the effects of antagonistic hormones, the significance of calcitonin in human physiology remains unclear. Patients with their thyroid glands surgically removed (as for Graves' disease) are *not* hypercalcemic, as would be expected if calcitonin were needed to lower blood calcium levels. Similarly, patients who receive injections of many times the normal amount of calcitonin do not become hypocalcemic. The ability of very large, pharmacological doses of calcitonin to inhibit osteoclast activity and bone resorption, however, is clinically useful in the treatment of *Paget's disease* (in which osteoclast activity causes softening of bone).

Osteitis Fibrosa Cystica, Osteomalacia, and Osteoporosis

Hypersecretion of parathyroid hormone can produce a disease called **osteitis fibrosa cystica.** As might be predicted from the effects of PTH, this disease is characterized by increased bone resorption. As a result, areas of bone are produced that have a low content of calcium phosphate, and the blood concentration of calcium is increased. The latter effect helps to promote calcification of soft tissues and is associated with kidney stone production.

Osteomalacia   Softening or demineralization of bone that occurs when plasma calcium and phosphate concentrations are low is called *osteomalacia.* The most common form of this condition—known as **rickets**—is caused by a deficiency in 1,25-dihydroxy vitamin D_3. Insufficient production of this compound is usually due to sun deprivation or to inadequate dietary intake of vitamin D. Rickets can also be produced, however, in kidney disease or as a result of inadequate PTH secretion. In both of these cases, low activity of the kidney enzyme 1α-hydroxylase interferes with the last step in the formation of 1,25-dihydroxy vitamin D_3.

Osteoporosis   A decrease in bone mass that occurs due to an imbalance between the rates of bone resorption and formation, while plasma calcium and phosphate levels remain normal, is called *osteoporosis.* This condition is most common in women over sixty and in men over seventy years old. While the causes of osteoporosis are not well understood, the fall in estrogen (female sex hormone) that occurs in women after the menopause (see chapter 21) is believed to be a contributing factor. Women who develop osteoporosis after their menopause are thus sometimes treated by estrogen therapy.

Table 20.7 Comparison of some major bone diseases.

Disease	Cause	Plasma Calcium	Plasma Phosphate
Osteitis fibrosa cystica	Excessive PTH secretion	High	Low or normal
Osteomalacia	Deficient 1,25-dihydroxy vitamin D_3	Low or normal	Low
Osteoporosis	Variable causes, not well understood	Normal	Normal

People with osteoporosis can also be helped by increasing the amount of ingested calcium. This increased need for dietary calcium may be due to the fact that people with osteoporosis have lower blood levels of 1,25-dihydroxy vitamin D_3, and thus suffer from decreased intestinal absorption of calcium.

The development of osteoporosis may be caused, at least in part, by an age-related decrease in the ability of kidney 1α-hydroxylase to respond to PTH stimulation. In one study, for example, infusion of PTH into young subjects produced an increase in 1,25-dihydroxy vitamin D_3, while no such increase was observed when elderly subjects were infused with PTH. The reason for this age-related change is not known.

1. Construct a flow diagram to show the negative feedback control of plasma Ca^{++} levels through secretion of parathyroid hormone.
2. List the steps in the formation of 1,25-dihydroxy vitamin D_3, and indicate how this formation is influenced by parathyroid hormone.
3. Compare the effects of parathyroid hormone, 1,25-dihydroxy vitamin D_3, and calcitonin on (a) plasma Ca^{++} levels; (b) plasma PO_4^{-3} levels; (c) intestinal absorption of calcium; (d) bone resorption and formation; and (e) renal reabsorption of calcium and phosphate.

Summary

Insulin and Glucagon

I. Insulin is the major anabolic hormone in the body.
   A. Insulin promotes the cellular uptake of glucose from blood.
      1. Glucose is converted to glycogen in the liver and muscles, and into fat in adipose tissue.
      2. The blood glucose concentration, as a result, is lowered by the action of insulin.
   B. Insulin also promotes uptake of amino acids from blood into skeletal muscles, and synergizes with growth hormone to promote protein synthesis.

II. Glucagon promotes catabolism.
   A. Glucagon, like epinephrine, stimulates hydrolysis of stored glycogen and fat.
   B. Glucagon, like the glucocorticoids (such as hydrocortisone), stimulates the hepatic enzymes that convert fatty acids to ketone bodies.
   C. The actions of glucagon thus help to raise the blood concentration of glucose and other energy carriers.

III. Secretions of insulin and glucagon are regulated by the blood concentrations of glucose and amino acids.
   A. During absorption of a carbohydrate meal, when blood glucose is rising, insulin secretion is stimulated and glucagon secretion is inhibited.
      1. The ratio of insulin to glucagon thus increases.
      2. Metabolism is tilted towards hydrolysis of energy reserves.
   B. During absorption of a protein meal, when the blood concentration of amino acids is rising, secretions of both insulin and glucagon are stimulated.
   C. During fasting, when the blood concentrations of all energy carriers decreases, glucagon secretion rises, while insulin secretion declines.

IV. Diabetes mellitus and reactive hypoglycemia are caused by defects in the secretion or action of islet of Langerhans hormones.
   A. Juvenile-onset (type I) diabetes is caused by destruction of beta cells and consequent lack of insulin secretion.

      1. This is accompanied by abnormally high glucagon secretion.
      2. Lack of insulin and high glucagon cause hyperglycemia, increased lipolysis, and increased ketogenesis.
      3. The combination of ketoacidosis and dehydration (which results from urinary glucose and ketone bodies acting as osmotic diuretics) can produce a diabetic coma and death.
   B. Maturity-onset (type II) diabetes is usually due to lack of tissue sensitivity to insulin.
      1. People with this condition usually have normal or elevated secretion of insulin, together with high glucagon secretion.
      2. Low tissue sensitivity to insulin is aggravated by obesity and improved by exercise.
   C. Reactive hypoglycemia occurs when excessive insulin is secreted in response to a rise in blood glucose.

Effects of Growth Hormone, Thyroxine, and Glucocorticoids

I. Growth hormone has an anabolic effect on protein synthesis, but a catabolic effect on glycogen and fat synthesis.
   A. The anabolic effect on protein synthesis in skeletal muscles is mediated by polypeptides called somatomedins.
      1. Somatomedins are produced by the liver under stimulation from growth hormone.
      2. Somatomedin production also requires the permissive effects of adequate insulin and adequate protein in the diet.
   B. Growth hormone exerts an "anti-insulin" effect on utilization of blood glucose.
   C. Through the action of somatomedins, growth hormone stimulates growth of cartilage and thus stimulates growth of bones.
      1. Long bones grow through a thickening of cartilage in the epiphyseal discs.
      2. After the epiphyseal discs thicken, they become calcified and degenerate on the diaphyseal (shaft) side; the degenerating cartilage is replaced by bone.
   D. Gigantism is caused by excessive growth hormone secretion during the growing years.
      1. Excessive growth hormone secretion in later life produces acromegaly.
      2. Inadequate growth hormone secretion during the growing years produces dwarfism.

II. Thyroxine, acting via its conversion to triiodothyronine (T_3) in the target cells, stimulates cell respiration in most tissues of the body.
   A. Thyroxine is believed to increase activity of the Na^+/K^+ pumps, and thus to decrease cellular concentrations of ATP.
   B. Decreased concentrations of ATP, in turn, stimulates enzymes involved in cell respiration.
   C. Since the Na^+/K^+ pumps contribute significantly to the total energy expenditure at rest, thyroxine levels set the basal metabolic rate.
      1. The basal metabolic rate (BMR) is the oxygen consumption (due to energy consumption) of the entire body under resting conditions.
      2. Hypothyroidism causes a low BMR, while hyperthyroidism causes a high BMR.
      3. Since metabolic heat is produced by cell respiration, people with hypothyroidism have decreased resistance to cold, while those with hyperthyroidism have decreased resistance to heat stress.

III. Thyroxine also stimulates the synthesis of many types of proteins in many organs.
   A. Thyroxine, therefore, is considered to be an anabolic hormone.
   B. Adequate amounts of thyroxine are required for normal growth and development; hypothyroidism in small children can result in cretinism.

IV. Glucocorticoids such as hydrocortisone (cortisol) are secreted in response to ACTH stimulation during times of stress.
   A. Hydrocortisone acts to increase the plasma concentration of all energy carriers.
   B. Unlike other hormones, glucocorticoids stimulate the breakdown of muscle proteins that helps to increase the plasma concentration of amino acids.
   C. In the liver, glucocorticoids stimulate synthesis of enzymes that convert amino acids to glucose.
      1. Stimulation of gluconeogenesis from amino acids helps to maintain the blood glucose concentrations during the stresses of prolonged fasting or exercise.
      2. Glucocorticoids also stimulate the hepatic enzymes involved in converting fatty acids to ketone bodies.

Regulation of Calcium and Phosphate Balance

I. A fall in blood calcium levels stimulates secretion of parathyroid hormone (PTH).
   A. Parathyroid hormone helps to raise blood calcium levels in a variety of ways.
      1. PTH stimulates bone resorption, which raises blood calcium and phosphate levels.
      2. PTH stimulates the kidney enzyme 1α-hydroxylase, which completes formation of 1,25-dihydroxy vitamin D_3 (a molecule that stimulates intestinal absorption of calcium and phosphate).
      3. PTH stimulates renal reabsorption of calcium without stimulating renal reabsorption of phosphate.
   B. Together with 1,25-dihydroxy vitamin D_3, PTH helps to maintain high enough calcium and phosphate concentrations to permit proper bone growth.

II. Vitamin D_3 is a prehormone formed in the skin or ingested in the diet.
   A. Vitamin D_3 is converted into 25-hydroxy vitamin D_3 in the liver, and then into the active hormone, 1,25-dihydroxy vitamin D_3, in the kidneys.
   B. The final hydroxylation reaction in the kidneys is catalyzed by the enzyme 1α-hydroxylase.
      1. This enzyme is stimulated by the effects of parathyroid hormone.
      2. Deficiency of this enzyme, or of the kidneys to respond to PTH, or of the precursor vitamin D_3, produces a softening of bones known as rickets.

C. 1,25-dihydroxy vitamin D_3 promotes increased bone resorption, increased intestinal absorption of calcium and phosphate, and increased renal reabsorption of both calcium and phosphate.

III. Parafollicular cells of the thyroid secrete calcitonin in response to a rise in blood calcium levels.
   A. Calcitonin promotes bone deposition, and thus helps to lower blood calcium levels.
   B. The significance of calcitonin in human physiology, however, has not been established.

IV. Osteoporosis is a decrease in bone density that occurs as a result of age-related changes.
   A. Women after their menopause, and men after about the age of sixty, experience an increased rate of bone resorption compared to bone deposition.
   B. Osteoporosis may result from inadequate intestinal absorption of calcium and phosphate due to insufficient production of 1,25-dihydroxy vitamin D_3.
   C. Some experiments suggest that there is a decrease in the ability of PTH to stimulate kidney 1α-hydroxylase enzyme in people with osteoporosis.

Self-Study Quiz

Multiple Choice
Match the following:

1. Absorption of carbohydrate meal
2. Absorption of protein meal
3. Fasting

(a) rise in insulin, rise in glucagon
(b) fall in insulin, rise in glucagon
(c) rise in insulin, fall in glucagon
(d) fall in insulin, fall in glucagon

Match the following:

4. Growth hormone
5. Thyroxine
6. Hydrocortisone

(a) increased protein synthesis, increased cell respiration
(b) protein catabolism in muscles, gluconeogenesis in liver
(c) protein synthesis in muscles, decreased glucose utilization
(d) fall in blood glucose, increased fat synthesis

7. A lowering of blood glucose concentration promotes:
   (a) decreased lipogenesis
   (b) increased lipolysis
   (c) glycogenolysis
   (d) all of these
8. Glucose can be secreted into the blood by:
   (a) liver
   (b) muscles
   (c) brain
   (d) all of these
9. Maturity-onset (type II) diabetes is caused by:
   (a) destruction of beta cells
   (b) lack of insulin
   (c) decreased tissue sensitivity to insulin
   (d) hyperglycemia

10. The basal metabolic rate is determined primarily by:
    (a) hydrocortisone
    (b) insulin
    (c) growth hormone
    (d) thyroxine
11. Somatomedins are required for the anabolic effects of:
    (a) hydrocortisone
    (b) insulin
    (c) growth hormone
    (d) thyroxine
12. Increased intestinal absorption of calcium is stimulated directly by:
    (a) parathyroid hormone
    (b) 1,25-dihydroxy vitamin D_3
    (c) calcitonin
    (d) all of these

13. A rise in blood calcium levels directly stimulates:
    (a) parathyroid hormone secretion
    (b) calcitonin secretion
    (c) 1,25-dihydroxy vitamin D_3 formation
    (d) all of these

Reproduction

Objectives

By studying this chapter, you should be able to:

1. Explain how the development of gonads and accessory sex organs occurs in male and female embryos

2. Describe the hormonal changes that occur during puberty and the changes in body structure that result

3. Describe the effects of pituitary gonadotrophins on the testes and the effects of testicular hormones on the secretion of gonadotrophins

4. Describe the structure and functions of the seminiferous tubules and of the interstitial Leydig cells, and explain how these two compartments interact

5. Describe the hormonal control of spermatogenesis and the possible functions served by Sertoli cells in this process

6. Describe oogenesis and the stages of ovarian follicle development through ovulation and the formation of a corpus luteum

7. Describe the mechanisms that control ovulation in terms of the hormonal interactions between the ovaries and the pituitary

8. Describe the formation, function, and fate of the corpus luteum during a nonfertile cycle

9. Explain how cyclic variation in hormone secretion affects the structure of the endometrium during the course of the menstrual cycle

10. Describe the events that occur between the time of fertilization and the implantation of the blastocyst into the endometrium

11. Explain how menstruation and further ovulation are normally prevented during pregnancy

12. Explain how the placenta is formed, and describe the functions of the placenta

13. Explain the factors involved in labor and parturition

14. Describe the hormonal interactions involved in breast development during pregnancy and in lactation following parturition

"*A* chicken is an egg's way of making another egg." Phrased in more modern terms, genes are "selfish." Genes, according to this view, do not exist in order to make a well-functioning chicken (or other organism). The organism, rather, exists and functions so that the genes can survive beyond the mortal life of individual members of a species. Whether or not one accepts this rather cynical view, it is clear that reproduction is an essential function of life. The incredible complexity of structure and function in living organisms could not be produced in successive generations by chance; mechanisms must exist to transmit the blueprint (genetic code) from one generation to the next. *Sexual reproduction,* in which genes from two individuals are combined in random and novel ways with each new generation, offers the further advantage of introducing great variability into a population. This variability of genetic constitution helps to insure that some members of a population will survive changes in the environment over evolutionary time.

Figure 21.1 The human life cycle. Numbers in parentheses indicate haploid state (23 chromosomes) and diploid state (46 chromosomes).

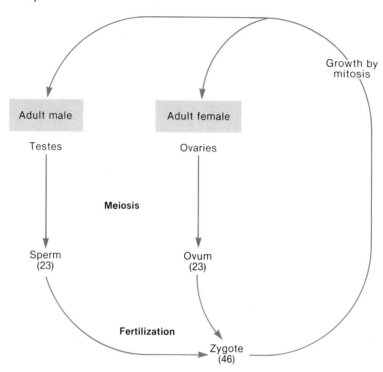

Sex Determination

In sexual reproduction, **germ cells,** or **gametes,** (sperm and ova) are formed within the *gonads* (testes and ovaries) by a process of reduction division, or *meiosis*. During this type of cell division, the normal number of chromosomes in most human cells—forty-six—is halved, so that each gamete receives twenty-three chromosomes. Fusion of a sperm and egg cell (ovum) in the act of **fertilization** results in restoration of the original chromosome number of forty-six in the fertilized egg (the *zygote*). Growth of the zygote into an adult member of the next generation occurs by means of mitotic cell divisions, as described in chapter 3. When this individual reaches puberty, mature sperm or ova will be formed by meiosis within the gonads so that the life cycle can be continued (see figure 21.1).

Each zygote inherits twenty-three chromosomes from its mother and twenty-three chromosomes from its father. This does not produce forty-six different chromosomes, but rather, twenty-three pairs of *homologous chromosomes*. Each member of a homologous pair, with the important exception of the sex chromosomes, looks like the other and contains similar genes (such as those coding for eye color, height, and so on). These homologous pairs of chromosomes can be photographed and numbered (as shown in figure 21.2). Each cell that contains forty-six chromosomes (that is *diploid*) has two chromosomes number 1, two chromosomes number 2, and so on through chromosomes number 22. The first twenty-two pairs of chromosomes are called **autosomal chromosomes.**

The twenty-third pair of chromosomes are the **sex chromosomes.** In a female these consist of two X chromosomes, while in a male there is one X chromosome and one Y chromosome. The X and Y chromosomes look different and contain different genes. This is the exceptional pair of homologous chromosomes mentioned earlier.

Figure 21.2 Homologous pairs of chromosomes obtained from a human diploid cell. The first twenty-two pairs of chromosomes are called the autosomal chromosomes. The sex chromosomes are *(a)* XY for a male and *(b)* XX for a female.

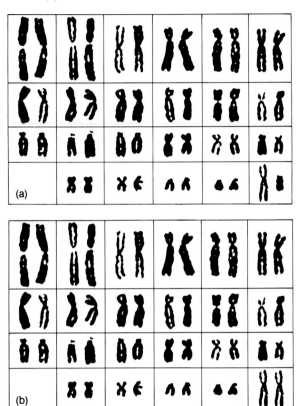

Figure 21.3 The nuclei of cheek cells from females *(A and B)* have Barr bodies *(arrows)*. These are formed from one of the X chromosomes, which is inactive. No Barr body is present in the cell obtained from a male *(C)* because males have only one X chromosome, which remains active. Some white blood cells from females have a "drumsticklike" appendage that is not found in white blood cells from males *(D)*.

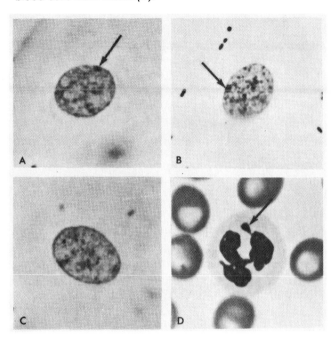

When a diploid cell (with forty-six chromosomes) undergoes meiotic division, its daughter cells receive only one chromosome from each homologous pair of chromosomes. The gametes are therefore said to be *haploid* (they contain only half the number of chromosomes in the diploid parent cell). Each sperm, for example, will receive only one chromosome of homologous pair number 5—either the one originally contributed by the organism's mother, or the one originally contributed by the father. Which of the two chromosomes—maternal or paternal—ends up in a given sperm is completely random. This is also true for the sex chromosomes, so that approximately half of the sperm produced will contain an X and approximately half will contain a Y chromosome.

A similar random assortment of maternal and paternal chromosomes into ova will occur in a woman's ovary. Since females have two X chromosomes, however, all of the ova will normally contain one X chromosome. Since all ova contain one X chromosome, while some sperm are X bearing and others are Y bearing, the *chromosomal sex of the zygote is determined by the sperm.* If a Y-bearing sperm fertilizes the ovum, the zygote will be XY and male; if an X-bearing sperm fertilizes the ovum, the zygote will be XX and female.

While each diploid cell in a woman's body inherits two X chromosomes, it appears that only one of each pair of X chromosomes remains active. The other X chromosome forms a clump of inactive "heterochromatin" that can often be seen as a dark spot called a *Barr body* at the edge of the nucleus of cheek cells (see figure 21.3). This provides a convenient test for chromosomal sex in cases in which there is suspicion that the chromosomal sex may differ from the apparent ("phenotypic") sex of the individual. Also, some of the nuclei in polymorphonuclear leukocytes (a type of white blood cell—see chapter 17) in females have a "drumstick" appendage not seen in neutrophils from males.

Figure 21.4 Chromosomes from a person with Turner's syndrome *(a)* and a variant of Klinefelter's syndrome *(b)*.

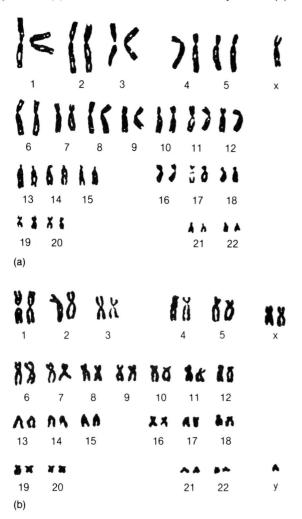

(a)

(b)

Figure 21.5 Formation of chromosomal sex of the embryo and development of the gonads. The very early embryo has "indifferent gonads" that can develop into either testes or ovaries. If the embryonic cells have Y chromosomes, which code for H-Y antigens, the gonads become testes. If no Y chromosomes, and therefore no H-Y antigens, are present the gonads become ovaries. Embryonic testes develop quickly, forming seminiferous tubules (which will produce sperm later in life) and Leydig cells (which secrete testosterone during embryonic development).

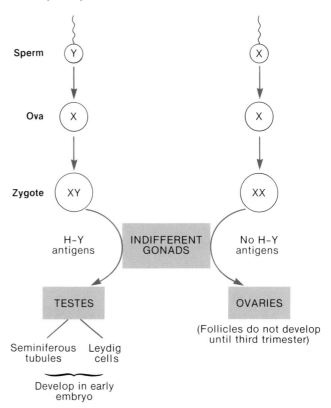

Formation of Testes and Ovaries

The gonads of males and females appear similar for the first forty or so days of development following conception. During this time, cells that will give rise to sperm (called *spermatogonia*) and cells that will give rise to ova (called *oogonia*) migrate from the yolk sac to the embryonic gonads. At this stage in development the gonads have the potential to become either testes or ovaries.

It is currently believed that the Y chromosome in XY embryos masculinize the gonads (cause them to develop into testes) through production of cell-surface proteins coded by genes in the Y chromosome. These male "markers" are a type of histocompatibility antigen (as described in chapter 17), and are therefore known as **H-Y antigens.**

Embryos that are chromosomal females (XX) lack Y chromosomes and therefore lack H-Y antigens. In the absence of these antigens the gonads develop into ovaries.

Notice that it is the presence or absence of the Y chromosome that determines whether the embryo will have testes or ovaries. This point is well illustrated by two genetic abnormalities. In *Klinefelter's syndrome* the affected person has forty-seven instead of forty-six chromosomes because of the presence of an extra X chromosome. These people, with XXY genotypes, develop testes despite the presence of two X chromosomes. Patients with *Turner's syndrome,* who have the genotype XO (and therefore have only forty-five chromosomes) develop ovaries.

The structures that will eventually produce sperm within the testes—the **seminiferous tubules**—appear very early in embryonic development (between forty-three and fifty days following conception). Although spermatogenesis begins during embryonic life, it is arrested until the onset of puberty (this will be described in a later section). At about day sixty-five the **Leydig cells** appear in the embryonic testes. These cells are located in the interstitial

Figure 21.6 Embryonic development of male and female sex accessory organs and external genitalia. In the presence of testosterone and Müllerian inhibition factor (MIF) secreted by the testes, male structures develop. In the absence of these secretions, female structures develop.

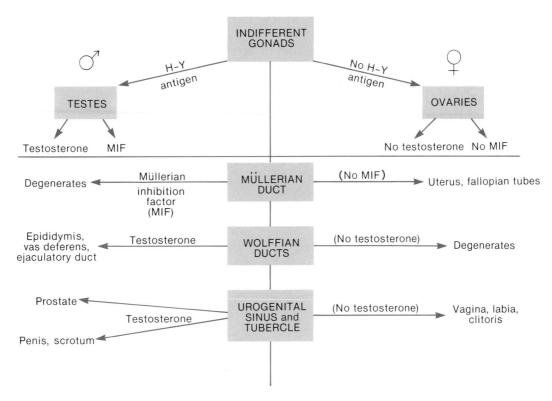

tissue between adjacent convolutions of the seminiferous tubules, and constitute the endocrine tissue of the testes. In contrast to the rapid development of the testes, the functional units of the ovaries—called the **ovarian follicles**—do not appear until the third trimester of pregnancy (at about day 105).

The early-appearing Leydig cells in the embryonic testes secrete large amounts of male sex hormones, or *androgens,* (andro = man; gen = forming). The major androgen secreted by these cells is **testosterone.** Testosterone secretion begins as early as eight to ten weeks after conception, reaches a peak at twelve to fourteen weeks, and thereafter declines to very low levels by the end of the second trimester (at about twenty-one weeks). Testosterone secretion during embryonic development in the male serves a very important function (described in the next section); similarly high levels of testosterone will not appear again in the life of the individual until the time of puberty.

As the testes develop they move within the abdominal cavity and gradually descend into the *scrotum.* Complete descent of the testes is sometimes not complete until shortly after birth. The cooler temperature of the scrotum, which is usually three to four degrees centrigrade lower than the temperature of the body cavity, is needed for spermatogenesis. This requirement is illustrated by the fact that spermatogenesis does not occur in males with undescended testes—a condition called *cryptorchidism* (crypt = hidden; orch = testes).

Development of Sexual Accessory Organs and External Genitalia

In addition to testes and ovaries, various internal accessory sexual organs are needed for reproductive function. Most of these sex accessory organs are derived from two systems of embryonic ducts. Male accessory organs are derived from the **Wolffian ducts,** and female accessory organs are derived from the **Müllerian ducts** (see figure 21.6). Interestingly, both male and female embryos between about day twenty-five and day fifty have both duct systems, and therefore have the potential to form the accessory organs characteristic of either sex.

Figure 21.7   Development of male and female external genitalia from common embryonic structures. Development of male structures requires stimulation by dihydrotestosterone (DHT), derived from testosterone secreted from the Leydig cells of the embryonic testes.

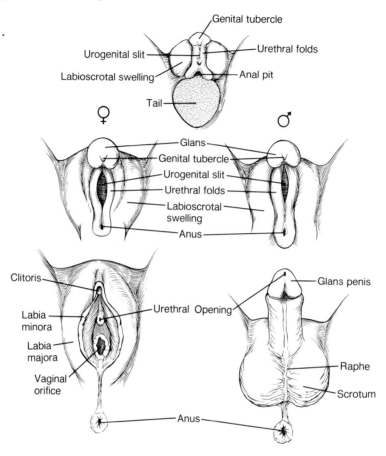

Experimental removal of the testes (castration) from male embryonic animals results in regression of the Wolffian ducts and development of the Müllerian ducts into female accessory organs: the **uterus** and **uterine (or fallopian tubes).** *Female sex accessory organs, therefore, develop as a result of the absence of testes rather than as a result of the presence of ovaries.*

The developing seminiferous tubules within the testes secrete a polypeptide called *Müllerian inhibition factor,* which causes regression of the Müllerian ducts beginning about day sixty. Secretion of testosterone by the Leydig cells of the testes subsequently causes growth and development of the Wolffian ducts into *male sex accessory organs:* the **epididymis, vas deferens, seminal vesicles,** and **ejaculatory duct.** The structure and function of the sex accessory organs will be described in later sections.

Two other embryonic structures, the *urogenital sinus* and the *urogenital tubercle,* are also masculinized by secretions of the testes. The urogenital sinus becomes another sex accessory organ—the **prostate.** The urogenital tubercle is converted into the male external genitalia— the *penis* and the *scrotum.* In the absence of testicular secretions, the urogenital sinus and tubercle form the female external genitalia: *vagina, labia majora* and *labia minora,* and *clitoris.* The labia are homologous to the scrotum, and the clitoris is homologous to the penis (see figure 21.7).

Masculinization of the urogenital sinus and tubercle into the prostate and male external genitalia occurs as a result of testosterone secreted by the embryonic testes. Testosterone itself, however, is not the active agent within these target organs. Once inside the target cells, testosterone is converted by means of an enzyme called *5α-reductase* into the active hormone known as **dihydrotestosterone (DHT),** which directly mediates the androgen effect in these organs.

Figure 21.8 Conversion of testosterone, secreted by the Leydig cells of the testes, into dihydrotestosterone (DHT) within the target cells. This reaction involves the addition of a hydrogen (and removal of the double carbon bond) in the first (A) ring of the steroid.

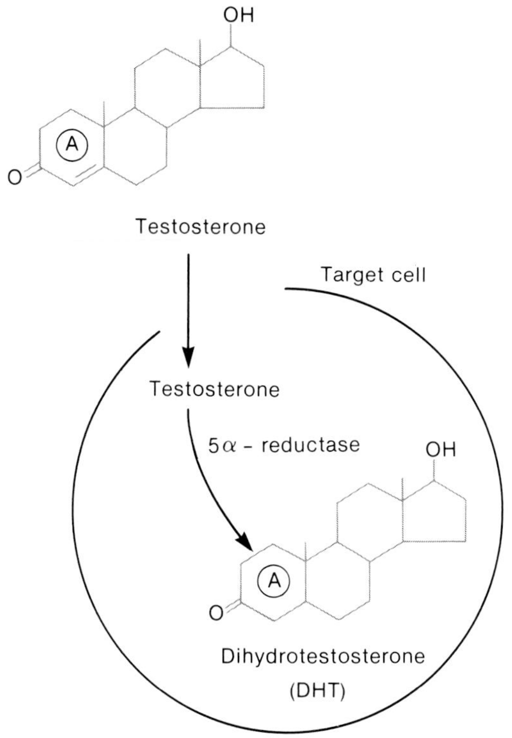

In summary, the genetic sex is determined by whether a Y-bearing or an X-bearing sperm fertilizes the ovum; the presence or absence of a Y chromosome in turn determines whether the gonads of the embryo will be testes or ovaries; the presence or absence of testes, finally, determines whether the sex accessory organs and external genitalia will be male or female. This regulatory pattern of sex determination makes sense in light of the fact that both male and female embryos develop within an environment high in estrogen, which is secreted by the mother's ovaries and the placenta. If estrogen determined the sex, all embryos would become feminized.

Disorders of Sex Determination

A *hermaphrodite* is an individual with both testes and ovaries, or with testicular and ovarian tissue located together in the same gonad. This condition is extremely rare, and appears to be caused by the fact that some embryonic cells have H-Y antigens on their surface while others do not. More common (though still rare) disorders of sex determination involve individuals with either testes or ovaries, but not both, who have sex accessory organs and external genitalia that are incompletely developed or are inappropriate for their chromosomal sex. These individuals are called *pseudohermaphrodites* (pseudo = false).

Table 21.1 Timetable for developmental changes of the gonads and genital tracts.

Approximate Time after Fertilization			Developmental Changes	
Days	Trimester	Indifferent	Male	Female
19	First	Germ cells migrate from yolk sac.		
25–30		Wolffian ducts begin development.		
44–48		Müllerian duct begins development.		
50–52		Urogenital sinus and tubercle develop.		
43–60			Tubules and Sertoli cells appear. Müllerian duct begins to regress.	
60–75			Leydig cells appear and begin testosterone production. Wolffian ducts grow.	Formation of vagina begins. Regression of Wolffian ducts begin.
105	Second			Development of ovarian follicles begins.
120				Uterus is formed.
160–260	Third		Testes descend into scrotum. Growth of external genitalia occurs.	Formation of vagina complete.

Source: From J. D. Wilson, Annual Review of Physiology, 40, (1978), p. 279. Reprinted by permission.

The most common cause of female pseudohermaphroditism is *congenital adrenal hyperplasia.* This condition, which is inherited as a recessive trait, is caused by excessive secretion of androgens from the adrenal cortex. A female with this condition would have Müllerian duct derivatives (uterus and fallopian tubes), because the adrenal doesn't secrete Müllerian inhibition factor, but would have partially masculinized external genitalia and Wolffian duct derivatives.

An interesting cause of male pseudohermaphroditism is known as *testicular feminization syndrome.* These individuals have normally functioning testes but lack receptors for testosterone. Thus, although large amounts of testosterone are secreted, the embryonic tissues cannot respond to this hormone. Female genitalia therefore develop, but the vagina ends blindly (uterus and fallopian tubes don't develop because of secretion of Müllerian inhibition factor). Male sex accessory organs likewise cannot develop because the Wolffian ducts lack testosterone receptors. A person with this condition appears externally to be a normal prepubertal girl, but she has testes in her body cavity and no sex accessory organs.

Some male pseudohermaphrodites have normally functioning testes and normal testosterone receptors, but genetically lack the ability to produce the enzyme 5α-reductase. Individuals with *5α-reductase deficiency* have normal Wolffian duct derivatives (epididymis, vas deferens, seminal vesicles, and ejaculatory duct) because development of these structures is stimulated directly by testosterone. These male accessory organs terminate in a vagina, however, because development of male external genitalia requires the action of DHT, which cannot be produced from testosterone in the absence of 5α-reductase.

---

1. Explain the meaning of the terms *diploid* and *haploid,* and describe how the chromosomal sex of an individual is determined.
2. Explain how the chromosomal sex determines whether testes or ovaries will be formed.
3. List the male and female sex accessory organs, and explain how the development of one or the other sets of these organs is determined.
4. Describe the abnormalities in testicular feminization syndrome and in 5α-reductase deficiency, and explain how these abnormalities are produced.

Puberty

The gonads are formed and receive spermatogonia (sperm-forming cells) or oogonia (egg-forming cells) early in embryonic development, as described in the previous section. The embryonic testes during the first trimester of pregnancy are active endocrine glands, secreting the high amounts of testosterone needed to masculinize the male embryo's external genitalia and sex accessory organs. Ovaries, in contrast, don't mature until the third trimester of pregnancy, and in the absence of testosterone secretion (and Müllerian inhibition factor secretion) female external genitalia and sex accessory organs develop. During the second trimester of pregnancy testosterone secretion in the male declines, so that, at the time the baby is born, the gonads of both sexes are relatively inactive.

There is no difference in the blood concentrations of *sex steroids*—androgens and estrogens—in prepubertal boys and girls. This does not appear to be due to deficiencies in the ability of the gonads to produce these hormones, but rather to lack of sufficient stimulation. During *puberty,* the gonads secrete increased amounts of sex steroid hormones as a result of increased stimulation by **gonadotrophic hormones** from the anterior pituitary gland.

Interactions between Hypothalamus, Pituitary, and Gonads

The anterior pituitary gland produces and secretes two gonadotrophic hormones—**FSH (follicle-stimulating hormone)** and **LH (luteinizing hormone).** Although these two hormones are named according to their actions in the female, the same hormones are secreted by the male's pituitary. The gonadotrophic hormones of both sexes have three primary effects on the gonads: (1) stimulation of spermatogenesis or oogenesis (formation of sperm or ova); (2) stimulation of gonadal hormone secretion; and (3) maintenance of the structure of the gonads (the gonads atrophy if the pituitary is removed).

Secretion of both LH and FSH from the anterior pituitary is stimulated by a hormone produced by the hypothalamus and secreted into the hypothalamo-hypophyseal portal vessels (see chapter 10). This releasing hormone is called *LHRH (luteinizing hormone-releasing hormone).* Since attempts to find a separate FSH-releasing hormone have thus far failed, and since LHRH stimulates FSH as well as LH secretion, LHRH is often referred to as **gonadotrophin-releasing hormone** (and accordingly abbreviated **GnRH**).

Figure 21.9 Simplified biosynthetic pathway for the steroid hormones. Also indicated are the major sources of sex hormones in the blood.

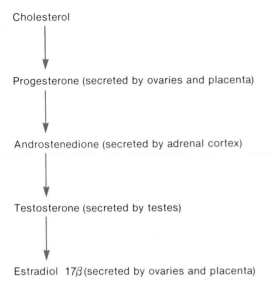

Cholesterol

↓

Progesterone (secreted by ovaries and placenta)

↓

Androstenedione (secreted by adrenal cortex)

↓

Testosterone (secreted by testes)

↓

Estradiol 17β (secreted by ovaries and placenta)

If a male or female animal is castrated (has its gonads surgically removed), secretion of FSH and LH increases to much higher levels than those measured in the intact animal. This demonstrates that the gonads secrete products that have a **negative feedback inhibition** of gonadotrophin secretion. This negative feedback is exerted in large part by sex steroids: estrogen and progesterone in the female, and testosterone in the male. The structure and biosynthetic pathways of these steroids are shown in figure 21.9.

The negative feedback effects of steroid hormones are believed to occur by means of two mechanisms: (1) inhibition of GnRH secretion, and (2) inhibition of the pituitary's response to a given amount of GnRH secreted from the hypothalamus. In addition to steroid hormones, there is evidence that the testes (and perhaps the ovaries as well) may secrete a polypeptide hormone called **inhibin** that specifically inhibits FSH secretion in response to GnRH, without affecting secretion of LH. Inhibin will be discussed more fully in later sections.

Figure 21.10 illustrates the similarities in gonadal regulation of males and females. Important differences in hypothalamus-pituitary-gonad interactions exist, however, between males and females. Secretion of gonadotrophins and sex steroids are more or less constant in adult males. Secretion of gonadotrophins and sex steroids in adult females, in contrast, shows cyclic variations (during the menstrual cycle). Also, during the normal female cycle, estrogen exerts a positive feedback effect on LH secretion.

Figure 21.10 Interactions between the hypothalamus, anterior pituitary, and gonads. Sex steroids secreted by the gonads have a negative feedback effect on secretion of GnRH (gonadotrophin-releasing hormone) and on secretion of gonadotrophins. The gonads may also secrete a polypeptide hormone called *inhibin* that exerts negative feedback control of FSH secretion.

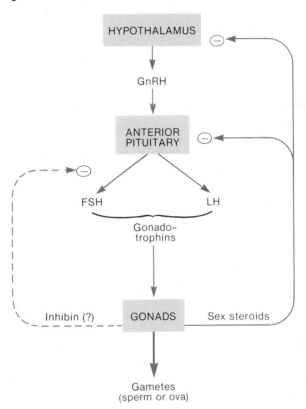

Such positive feedback does not normally occur in men. The reasons for these sexual differences in the patterns of gonadotrophin secretion will be described in a later section.

The Onset of Puberty

Secretion of FSH and LH is high in the newborn, but falls to very low levels a few weeks after birth. Gonadotrophin secretion remains low until the beginning of puberty, which is marked by rising levels of FSH followed by LH secretion. Experimental evidence suggests that this rise in gonadotrophin secretion is a result of two processes: (1) maturational changes in the brain that result in increased GnRH secretion by the hypothalamus, and (2) decreased sensitivity of gonadotrophin secretion to the negative feedback effects of sex steroid hormones.

Table 21.2   Development of secondary sexual characteristics and other changes that occur during puberty in girls.

Characteristic	Age of First Appearance	Hormonal Stimulation
Appearance of breast bud	8–13	Estrogen, progesterone, growth hormone, thyroxine, insulin, cortisol
Pubic hair	8–14	Adrenal androgens
Menarche (first menstrual flow)	10–16	Estrogen and progesterone
Axillary (underarm) hair	About two years after appearance of pubic hair	Adrenal androgens
Eccrine sweat glands and sebaceous glands; acne (from blocked sebaceous glands)	About same time as axillary hair growth	Adrenal androgens

Table 21.3   Development of secondary sexual characteristics and other changes that occur during puberty in boys.

Characteristic	Age of First Appearance	Hormonal Stimulation
Growth of testes	10–14	Testosterone, FSH, growth hormone
Pubic hair	10–15	Testosterone
Body growth	11–16	Testosterone, growth hormone
Growth of penis	11–15	Testosterone
Growth of larynx (voice lowers)	Same time as growth of penis	Testosterone
Facial and axillary (underarm) hair	About two years after appearance of pubic hair	Testosterone
Eccrine sweat glands and sebaceous glands; acne (from blocked sebaceous glands)	About same time as facial and axillary hair growth	Testosterone

The maturation of the hypothalamus or other regions of the brain that leads to increased GnRH secretion at the time of puberty appears to be programmed—children without gonads show increased FSH secretion at the normal time. Also during this period of time, a given amount of sex steroids has less of a suppressive effect on gonadotrophin secretion than the same dose would have if administered prior to puberty. This suggests that the sensitivity of the hypothalamus and the pituitary to negative feedback effects decreases at puberty, which would also help account for rising gonadotrophin secretion at this time.

During late puberty there is a "pulsatile" secretion of gonadotrophins—FSH and LH secretion *increase during periods of sleep,* and decrease during periods of wakefulness. These pulses of increased gonadotrophin secretion during puberty stimulate a rise in sex steroid secretion from the gonads. Increased secretion of testosterone from the testes and of **estradiol-17β** (estradiol is the major *estrogen,* or female sex steroid) from the ovaries during puberty in turn produces changes in body appearance characteristic of the two sexes. Such **secondary sexual characteristics** (see tables 21.2 and 21.3) are the physical manifestations of the hormonal changes occurring during puberty.

It has been observed that the age of onset of puberty is related to the amount of body fat and level of physical activity of the child. Ballerinas, for example, reach *menarche* (the age when menstrual flow first occurs) at a greater average age than less active girls with more body fat. The mechanisms by which body fat and physical activity may influence pituitary-gonadal function are not currently understood.

The Pineal Gland

The role of the pineal gland in human physiology is poorly understood. It is known that the pineal, a gland located deep within the brain, secretes the hormone **melatonin** as a derivative of the amino acid tryptophane (see figure 21.11) and that production of this hormone is influenced by light-dark cycles.

The pineal glands of some lower vertebrates have photoreceptors that are directly sensitive to environmental light. While no such photoreceptors are present in the pineal glands of mammals, secretion of melatonin has been shown to increase at night and decrease during daylight (as illustrated in figure 21.12). The inhibitory effect of light on melatonin secretion in mammals is indirect. Pineal secretion is stimulated by postganglionic sympathetic neurons that originate in the superior cervical ganglion; activity of these neurons is inhibited by nerve tracts that are activated by light striking the retina.

Figure 21.11 Simplified biosynthetic pathway for the pineal gland hormone *melatonin*.

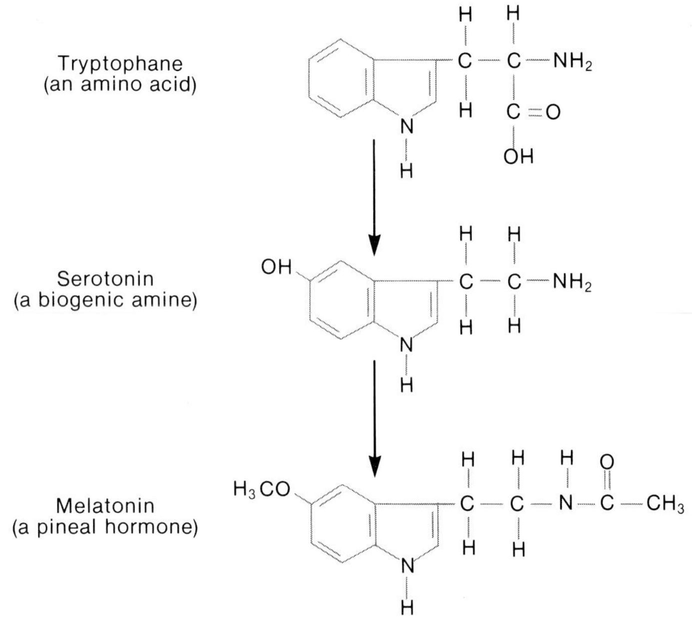

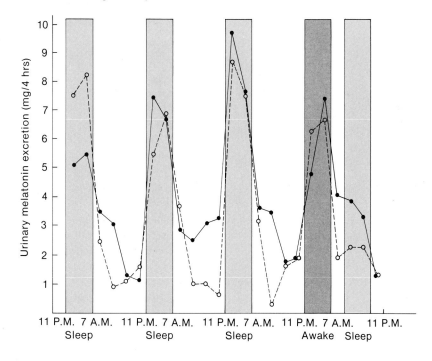

Tryptophane (an amino acid)

Serotonin (a biogenic amine)

Melatonin (a pineal hormone)

Figure 21.12 Melatonin secretion from the pineal increases at night and decreases during the day.

Figure 21.13   Negative feedback relationships between
the anterior pituitary and testes. The seminiferous tubules
are the targets of FSH action; the interstitial Leydig cells
are targets of LH action. Testosterone secreted by the
Leydig cells inhibits LH secretion; inhibin secreted by the
tubules may inhibit FSH secretion.

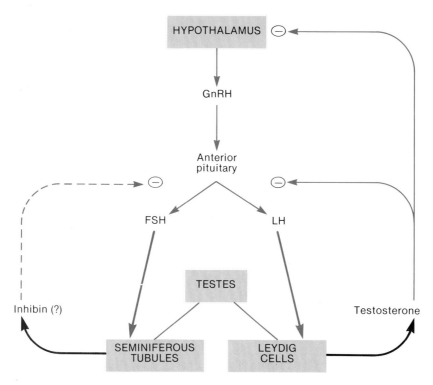

There is abundant experimental evidence that, in rats
and other lower animals, melatonin inhibits gonadotrophin secretion and thus has an "anti-gonad" effect. While
a similar effect has been demonstrated in humans, the significance of pineal function in the normal regulation of
reproductive activity is currently unclear.

1. Using a flow diagram, show the negative feedback
   control that the gonads exert on GnRH and
   gonadotrophin secretion. Explain the effects of
   castration on FSH and LH secretion and the effects of
   removal of the pituitary on the structure of the gonads
   and accessory sexual organs.
2. Describe the two mechanisms that have been proposed
   to explain the rise in sex steroid secretion that occurs at
   puberty.
3. Describe the effect of light on the pineal secretion of
   melatonin and the neural pathways that mediate this
   effect. Describe the proposed function of melatonin in
   regulation of the reproductive system.

Male Reproductive System

The testes consist of two parts, or "compartments"—the
seminiferous tubules, where spermatogenesis occurs, and
the interstitial tissue, which contains androgen-secreting
Leydig cells. About 90 percent of the weight of an adult
testis (which averages 20 g) is comprised of seminiferous
tubules. The interstitial tissue is a thin web of connective
tissue (containing Leydig cells) between convolutions of
the tubules.

There is a strict compartmentation in the testes with
regards to gonadotrophin action. Cellular receptor proteins for FSH are located exclusively in the seminiferous
tubules, while LH receptor proteins are confined exclusively to the interstitial tissue. Secretion of testosterone by
the Leydig cells is stimulated by LH but not by FSH.
Spermatogenesis in the tubules is stimulated by FSH. The
apparent simplicity of this compartmentation, however, is
an illusion because the two compartments can interact with
each other in complex ways.

Figure 21.14 Testosterone secreted by the Leydig cells of the testes can be converted into active metabolites in the brain and other target organs. These active metabolites include DHT, and other 5a-reduced androgens, and estradiol.

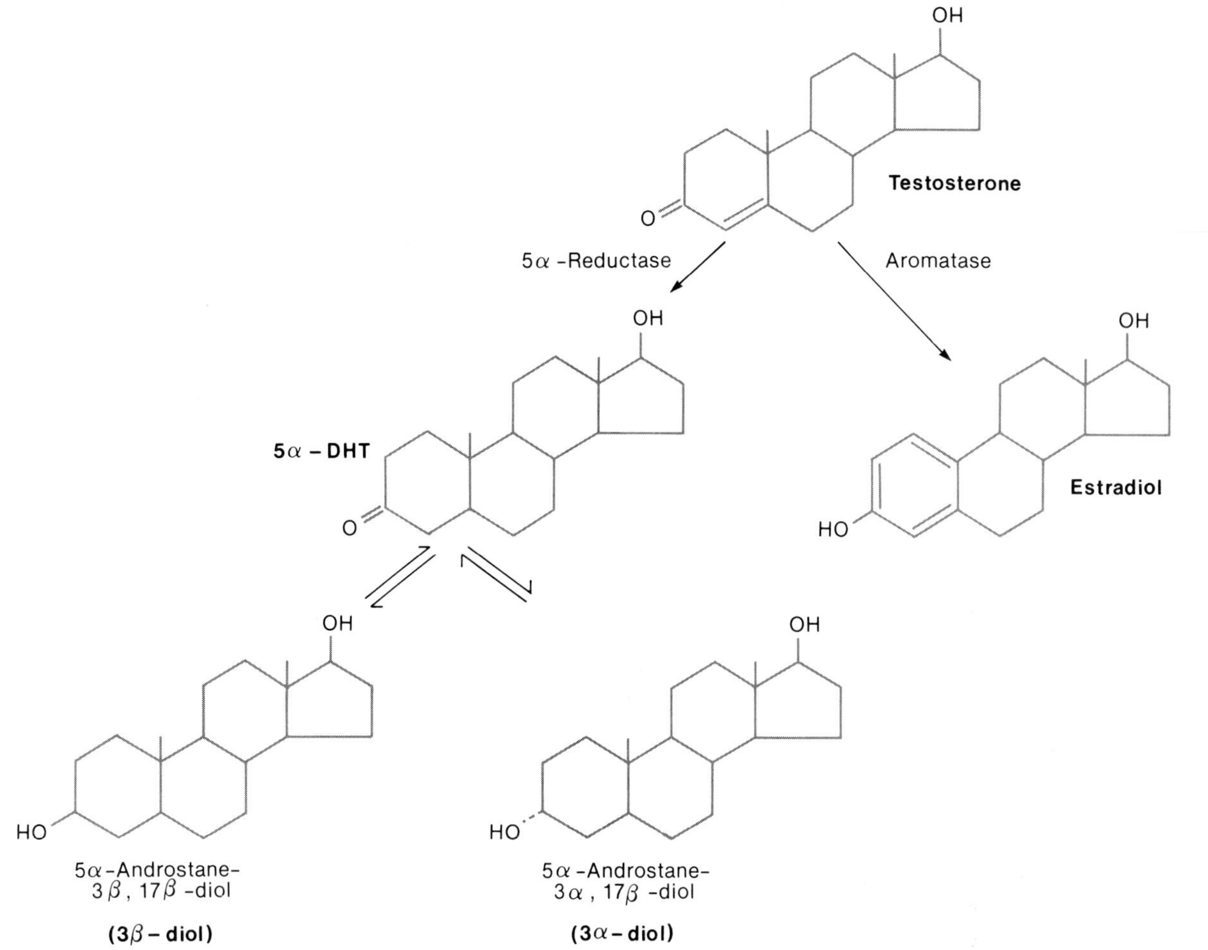

Control of Gonadotrophin Secretion

Castration of a male animal results in an immediate rise in FSH and LH secretion. This demonstrates that hormones secreted by the testes exert negative feedback inhibition of gonadotrophin secretion. If testosterone is injected into the castrated animal, secretion of LH can be returned to the previous (pre-castration) levels. This provides a classical example of negative feedback—LH stimulates testosterone secretion by the Leydig cells, and testosterone inhibits pituitary secretion of LH.

The amount of testosterone that is sufficient to suppress LH, however, is not sufficient to suppress the postcastration rise in FSH secretion in most experimental animals. In rams and bulls, a water-soluble (and, therefore, peptide rather than steroid) product of the seminiferous tubules specifically suppresses FSH secretion. This hormone is called **inhibin.** Though controversial, there is evidence that the seminiferous tubules of the human testes also produce inhibin.

Testosterone Derivatives in the Brain The brain contains testosterone receptors and is a target organ for this hormone. The effects of testosterone on the brain, such as suppression of LH secretion, are not mediated directly by testosterone, however, but rather by its derivatives that are produced within the brain cells. Testosterone may be converted by the enzyme 5α-reductase to dihydrotestosterone (DHT), as previously described. The DHT, in turn, can be changed by other enzymes into other 5α-reduced androgens—abbreviated 3α-diol and 3β-diol (see figure 21.14). Alternatively, testosterone can be converted within the brain to estradiol-17β. Though usually regarded as a female sex steroid, estradiol is therefore an active compound in normal male physiology! Estradiol is formed from testosterone by an enzyme called *aromatase,* in a reaction known as *aromatization* (this term refers to the presence of an aromatic carbon ring—see chapter 1—not to an odor).

Table 21.4   Summary of some of the actions of androgens in the male.

Category	Action
Sex determination	Growth and development of Wolffian ducts into epididymis, vas deferens, seminal vesicles, and ejaculatory ducts Development of urogenital sinus and tubercle into prostate Development of male external genitalia (penis and scrotum)
Spermatogenesis	At puberty: completion of meiotic division and early maturation of spermatids After puberty: maintenance of spermatogenesis
Secondary sexual characteristics	Growth and maintenance of accessory sexual organs Growth of penis Growth of facial and axillary hair Body growth
Anabolic effects	Protein synthesis and muscle growth Growth of bones Growth of other organs (including larynx) Erythropoiesis (red blood cell formation)

Testosterone Secretion and Age   The negative feedback effects of testosterone, and perhaps inhibin, help to maintain a constant secretion of gonadotrophins in men, resulting in relatively constant levels of androgen secretion from the testes. This contrasts with the cyclic secretion of gonadotrophins and ovarian steroids in women. Also unlike in women, who experience an abrupt cessation in sex steroid secretion during menopause (as described in a later section), secretion of androgens declines only gradually and to varying degrees in men over fifty years of age. This decline in testosterone secretion, when it occurs, is not due in humans to decreasing gonadotrophin secretion because these hormones are in fact elevated at that time (as a result of less negative feedback). The causes of these age-related changes in testicular function are not currently known.

Endocrine Functions of the Testes

Testosterone is by far the major androgen secreted by the adult testis. This hormone, or derivatives of it (the 5α-reduced androgens and estradiol), is responsible for initiation and maintenance of the body changes associated with puberty in men. Androgens are sometimes called *anabolic steroids* because they stimulate growth of muscles and other structures (see table 21.4). Increased testosterone secretion during puberty is also required for growth of the sex accessory organs—primarily the seminal vesicles and prostate. Removal of androgens by castration results in atrophy of these organs.

Androgens stimulate growth of the larynx (causing lowering of the voice), increased hemoglobin synthesis (males have higher hemoglobin levels than females), and bone growth. The effect of androgens on bone growth is self limiting, however, because androgens ultimately cause conversion of cartilage to bone in the epiphyseal discs, thus "sealing" the discs and preventing further lengthening of the bones (as described in chapter 20).

It was once thought that the Leydig cells of the interstitial tissue were the only source of testicular androgens. While the Leydig cells are the major source of testosterone, it is now known that the seminiferous tubules are also able to produce small amounts of testosterone if they are supplied with appropriate precursors (such as progesterone). Probably of more importance, the tubules in adult testes are able to convert testosterone, supplied to them by the Leydig cells, into 5α-reduced androgens (DHT and 3α-diol). The physiological significance of the endocrine function of the tubules is not currently understood.

While androgens are by far the major secretory product of the testes, the testes do produce and secrete small amounts of estradiol. The source of these estrogens is not known—some evidence suggests that the tubules produce estradiol as a result of FSH stimulation, while other evidence suggests that estradiol comes from the Leydig cells as a result of LH stimulation. The physiological significance of estrogens secreted into the blood by the testes is unclear.

The testes contain estrogen receptors, which are confined to the interstitial tissue. Receptors for androgens, in contrast, are confined to the seminiferous tubules. This suggests the possibility that androgens and estrogens might have regulatory functions within the testes themselves. This possibility has proven correct for androgens (they are needed for spermatogenesis, as described in the next section). The physiological role of estrogens in the testes remains to be demonstrated.

The two compartments of the testes interact with each other. Testosterone from the Leydig cells is metabolized by the tubules into other active androgens and serves to regulate tubular functions. The tubules also secrete products—5α-reduced androgens and possibly other hormones—which could conceivably influence Leydig cell function. Such interactions are suggested by evidence that, in the pubertal male rat, exposure to FSH augments the

Figure 21.15 Interactions between the two compartments of the testes. Testosterone secreted by the interstitial Leydig cells stimulates spermatogenesis in the tubules. Secretions of the tubules may affect the sensitivity of the Leydig cells to LH stimulation.

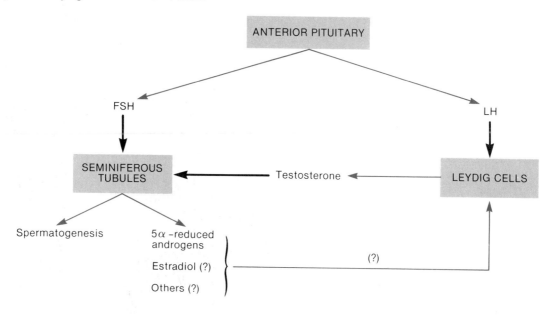

Table 21.5 Stages of meiosis.

Stage	Events
First meiotic division	
Prophase I	Chromosomes appear double-stranded. Each strand, called a chromatid, contains duplicate DNA joined together by a structure known as a centromere.
	Homologous chromosomes pair up side by side.
Metaphase I	Homologous chromosome pairs line up at equator.
	Spindle apparatus complete.
Anaphase I	Homologous chromosomes are separated; each member of a homologous pair moves to opposite poles.
Telophase I	Cytoplasm divides to produce two haploid cells.
Second meiotic division	
Prophase II	Chromosomes appear, each containing two chromatids.
Metaphase II	Chromosomes line up single file along equator as spindle formation is completed.
Anaphase II	Centromeres split and chromatids move to opposite poles.
Telophase II	Cytoplasm divides to produce two haploid cells from each of the haploid cells formed at telophase I.

responsiveness of the Leydig cells to LH. Since FSH can only directly affect the tubules, the FSH-induced enhancement of LH responsiveness must be mediated by products secreted from the tubules. The nature of these regulators is not currently known.

Spermatogenesis

The germ cells that migrate from the yolk sac to the testes during early embryonic development become "stem cells" called **spermatogonia** within the outer region of the seminiferous tubules. Spermatogonia are diploid cells (with forty-six chromosomes) that ultimately give rise to mature haploid gametes by a process of cell division called meiosis.

Meiosis, or reduction division, occurs in two parts. In the first part of this process the DNA duplicates, and homologous chromosomes are separated (during anaphase I) into two daughter cells at telophase I. Since each daughter cell contains only one of each homologous pair of chromosomes, the cells formed at the end of this *first meiotic division* contain twenty-three chromosomes each and are haploid. Each of the twenty-three chromosomes at this stage, however, consists of two strands (called *chromatids*) of identical DNA. During the *second meiotic division,* these duplicate chromatids are separated (at anaphase II) into daughter cells at telophase II. Meiosis of one diploid spermatogonia cell therefore produces four haploid daughter cells.

Figure 21.16   Meiosis, or reduction division. In the first meiotic division the homologous chromosomes of a diploid parent cell are separated into two haploid daughter cells. Each of these chromosomes contain duplicate strands, or chromatids. In the second meiotic division these chromatids are distributed to two new haploid daughter cells.

First meiotic division

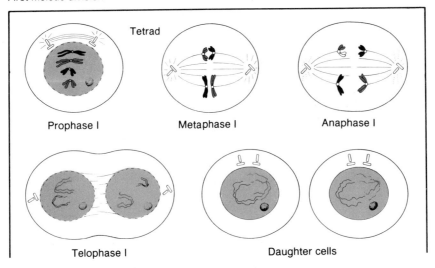

Second meiotic division

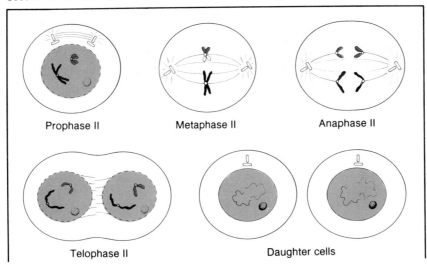

Figure 21.17 Spermatogonia undergo mitotic division to replace themselves and produce a daughter cell that will undergo meiotic division. This cell is called a primary spermatocyte. Upon completion of the first meiotic division the daughter cells are called secondary spermatocytes. Each of these completes a second meiotic division to form spermatids. Notice that the four spermatids produced by meiosis of a primary spermatocyte are interconnected. Each spermatid forms a mature spermatozoan.

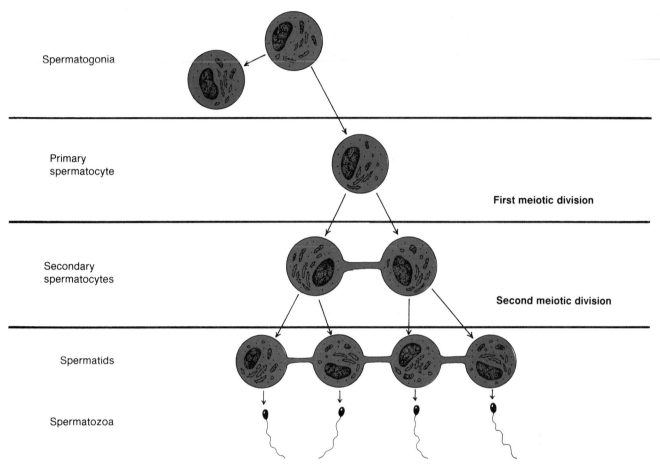

Actually, only about 1,000–2,000 stem cells migrate from the yolk sac into the embryonic testes. In order to produce many millions of sperm throughout adult life, these spermatogonia cells duplicate themselves by mitotic division, and only one of the two cells—now called a **primary spermatocyte**—undergoes meiotic division. In this way, spermatogenesis can occur continuously without exhausting the number of spermatogonia.

When a diploid primary spermatocyte completes the first meiotic division (at telophase I), the two haploid daughter cells thus produced are called **secondary spermatocytes.** At the end of the second meiotic division, each of the two secondary spermatocytes produce two haploid **spermatids.** One primary spermatocyte therefore produces four spermatids.

The stages of spermatogenesis are arranged sequentially in the wall of the seminiferous tubule. The epithelial wall of the tubule—called the *germinal epithelium*—is indeed composed of germ cells in different stages of spermatogenesis. The spermatogonia and primary spermatocytes are located toward the outer side of the tubule, while spermatids and mature spermatozoa are located on the side of the tubule facing the lumen.

At the end of the second meiotic division, the four spermatids produced by meiosis of one primary spermatocyte are interconnected with each other—their cytoplasm does not completely pinch off at the end of each division. Development of these interconnected spermatids into separate, mature spermatozoa (a process called **spermiogenesis**) requires the participation of another type of cell in the tubules, the **Sertoli cells.**

Figure 21.18 Seminiferous tubules are illustrated in cross section with surrounding interstitial tissue in *(a)*. The stages of spermatogenesis within the germinal epithelium of a seminiferous tubule is illustrated in *(b)*.

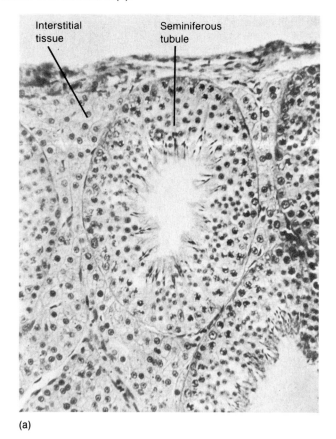

Interstitial tissue

Seminiferous tubule

(a)

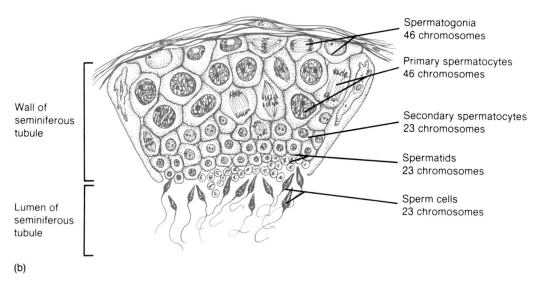

Wall of seminiferous tubule

Lumen of seminiferous tubule

Spermatogonia
46 chromosomes

Primary spermatocytes
46 chromosomes

Secondary spermatocytes
23 chromosomes

Spermatids
23 chromosomes

Sperm cells
23 chromosomes

(b)

Figure 21.19   Relationship between a Sertoli cell and spermatids within the wall of a seminiferous tubule.

Lumen of tubule

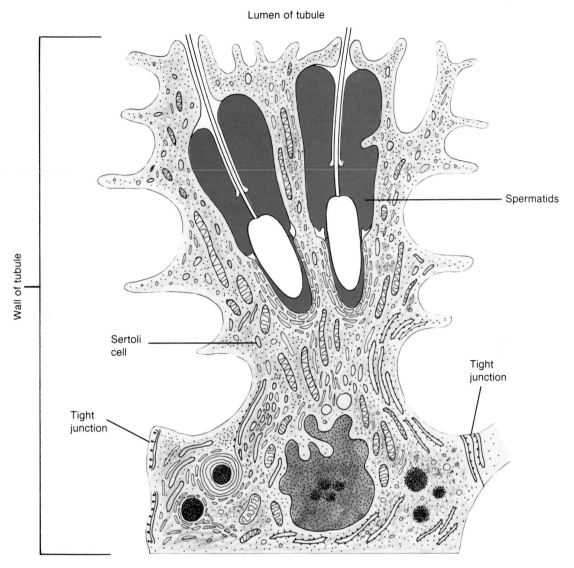

Wall of tubule

Spermatids

Sertoli cell

Tight junction

Tight junction

Outside of tubule

Sertoli Cells   The Sertoli cells are the only non-germinal cell type in the tubules. They form a continuous layer, connected by tight junctions, around the circumference of each tubule. In this way the Sertoli cells comprise a *blood-testis barrier,* because molecules from the blood must pass through the cytoplasm of Sertoli cells before entering germinal cells. The cytoplasm of the Sertoli cells extends through the width of the tubule and envelops the developing germ cells, so that it is often difficult to tell where the cytoplasm of Sertoli cells and germ cells are separated.

In the process of *spermiogenesis* (conversion of spermatids to spermatozoa), most of the spermatid cytoplasm is eliminated. This occurs through phagocytosis by Sertoli cells of the "residual bodies" of cytoplasm from the spermatids. Many believe that phagocytosis of residual bodies may transmit informational molecules from germ cells to Sertoli cells. The Sertoli cells, in turn, may provide many molecules needed by the germ cells. It is known, for example, that the X chromosome of germ cells is inactive during meiosis. Since this chromosome contains genes needed to produce many essential molecules, it is believed that these molecules are provided by the Sertoli cells during this time.

Figure 21.20 Processing of spermatids into
spermatozoa (spermiogenesis). As the spermatids
develop into spermatozoa most of their cytoplasm is
pinched off as residual bodies and ingested by the
surrounding Sertoli cell cytoplasm.

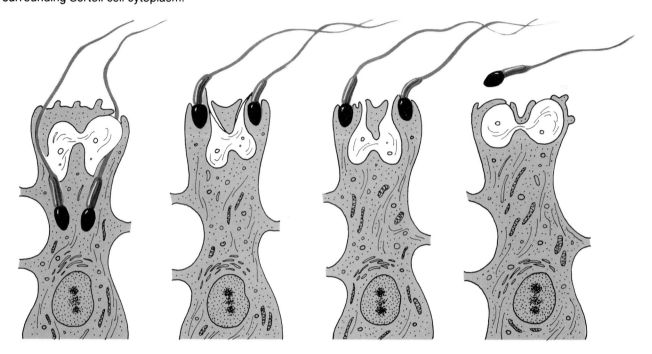

The importance of Sertoli cells in tubular function is
further evidenced by the fact that *FSH receptors are confined to the Sertoli cells.* Any effect of FSH on the tubules, therefore, must be mediated through the action of
Sertoli cells. The mechanisms by which these effects are
accomplished, and indeed, the nature of all chemical interactions between Sertoli and germ cells, are currently
unknown.

Hormonal Control of Spermatogenesis   The very beginning of spermatogenesis—formation of primary spermatocytes and entry into early prophase I—is apparently
somewhat independent of hormonal control and, in fact,
starts during embryonic development. Spermatogenesis is
arrested at this stage, however, until puberty, when testosterone secretion rises. Testosterone is required for completion of meiotic division and for the early stages of
spermatid maturation. This effect is probably not mediated directly by testosterone but, rather, by some of the
molecules produced from testosterone by the tubules.

The later stages of spermatid maturation during puberty appears to require stimulation by FSH. This FSH
effect is, as previously mentioned, mediated by Sertoli cells
that contain FSH receptors and that surround and interact with the spermatids. During puberty, therefore, both
FSH and androgens are needed for initiation of spermatogenesis.

Experiments in rats, and more recently in humans,
have revealed that spermatogenesis within the adult testis
can be maintained by androgens alone, in the absence of
FSH. It appears, in other words, that FSH is needed to
initiate spermatogenesis at puberty, but that once spermatogenesis has begun, FSH may no longer be required.
The physiological importance of FSH in the adult male
is, as a result of these experiments, currently controversial.

At the conclusion of spermiogenesis, spermatozoa are
released into the lumen of the seminiferous tubules. The
spermatozoa contain a *head* (with DNA inside), a *body,*
and a *tail* (see figure 21.22). Although the tail will ultimately be capable of flagellar movement, the sperm at this
stage are non-motile. Motility and other maturational
changes occur outside the testes in the epididymis.

Figure 21.21 Endocrine control of spermatogenesis. During puberty both testosterone and FSH are required to initiate spermatogenesis. In the adult, however, testosterone alone can maintain spermatogenesis.

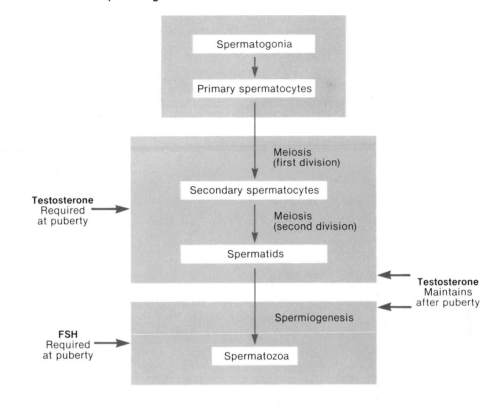

Figure 21.22 (a) Diagram of the parts of spermatozoan. (b) Scanning electron micrograph of a spermatozoan.

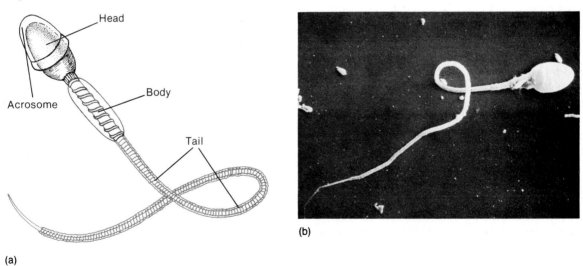

Male Sex Accessory Organs

The seminiferous tubules are connected at both ends to the *rete testis* (see figure 21.23). Spermatozoa and tubular secretions are moved to this area of the testis and are drained via the *ductuli efferentes* into the **epididymis.** The epididymis is a single-coiled tube, four to five meters long if stretched out, that receives the tubular products. Spermatozoa enter at the "head" of the epididymis and are drained from its "tail" by a single tube, the **vas deferens.**

Spermatozoa that enter the head of the epididymis are non-motile. During their passage through the epididymis, spermatozoa gain motility and undergo other maturational changes—sperm that leave the epididymis are more resistant to changes in pH and temperature, are motile, and are able to fertilize an ovum. Sperm obtained from the seminiferous tubules, in contrast, cannot fertilize an ovum. The epididymis serves as a site for sperm maturation and for storage of sperm between ejaculations.

The vas deferens carries sperm from the epididymis out of the scrotum and into the body cavity. In its passage, the vas deferens obtains fluid secretions of the **seminal vesicles** and **prostate** gland. This fluid, now called *semen,* is carried by the ejaculatory duct to the *urethra* (see figure 21.24). Sperm can be stored within a widened area of the vas deferens known as the *ampulla.*

Figure 21.23   Spermatozoa travel from seminiferous tubules to rete testis, to ductuli efferentes, to the epididymis. Spermatozoa leave the scrotum and enter the body cavity in a tube called the vas deferens.

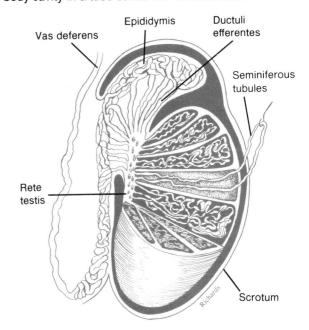

Figure 21.24   The male reproductive system.

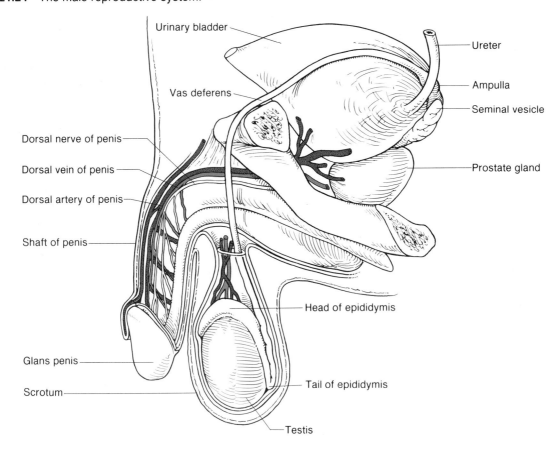

Table 21.6 Control of erection and ejaculation.

Regulation	Effect	Result
Parasympathetic nerves	Vasodilation produces increased blood flow into erectile tissues—corpus cavernosum and corpus spongiosum. As tissues become turgid, venous outflow is partially occluded, further increasing accumulation of blood.	Erection
Sympathetic nerves	Peristaltic waves of contraction in tubular system—primarily epididymis, vas deferens, and ejaculatory ducts, and contraction of prostate and seminal vesicles.	Ejaculation

Figure 21.25 A cross section of the penis, showing the erectile tissues—corpus cavernosum and corpus spongiosum.

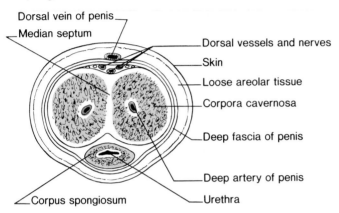

Dorsal vein of penis
Median septum
Dorsal vessels and nerves
Skin
Loose areolar tissue
Corpora cavernosa
Deep fascia of penis
Deep artery of penis
Corpus spongiosum
Urethra

Table 21.7 Some characteristics and reference values used in clinical examination of semen.

Characteristic	Reference Value
Volume of ejaculate	1.5–5.0 ml
Sperm count	40–250 million/ml
Sperm motility	
Percent of motile forms:	
1 hour after ejaculation	70% or more
3 hours after ejaculation	60% or more
Leukocyte count	0–2000/ml
pH	7.2–7.8
Fructose concentration	150–600 mg/100 ml

Source: Modified from Glasser, L. Seminal fluid and subfertility. Diagnostic Medicine, July/August 1981, p. 28. By permission.

The seminal vesicles and prostate are androgen-dependent accessory sexual organs—they atrophy if androgen is withdrawn by castration. The seminal vesicles secrete fluid containing fructose (which serves as an energy source for the spermatozoa), citric acid, coagulation proteins, and prostaglandins. The prostate secretes a liquefying agent and the enzyme *acid phosphatase,* which is often measured clinically to assess prostate function. Abnormal growth of the prostate often occurs in older men—a condition called *benign prostatic hyperplasia (BPH).* The causes of this condition are not well understood.

Erection, Emission, and Ejaculation

Erection, accompanied by increases in length and width of the penis, is achieved as a result of blood flow into the "erectile tissues" of the penis. These erectile tissues include two paired structures—the *corpora cavernosa*—located on the dorsal side of the penis, and one unpaired *corpus spongiosum* on the ventral side. The urethra runs through the center of the corpus spongiosum. The erectile tissue forms columns extending the length of the penis, although the corpus cavernosum does not extend all the way to the tip.

Erection is achieved as a result of parasympathetic nerve-induced vasodilation of arterioles that allows blood to flow into the penis. As the erectile tissues become engorged with blood and the penis becomes turgid, venous outflow of blood is partially occluded, thus aiding erection. The term *emission* refers to the movement of semen into the urethra, and *ejaculation* refers to the forcible expulsion of semen from the urethra out of the penis. Emission and ejaculation are stimulated by sympathetic nerves, which cause peristaltic contractions of the tubular system, contractions of the seminal vesicles and prostate, and contractions of muscles at the base of the penis. Sexual function in the male thus requires the synergistic action (rather than antagonistic action) of the parasympathetic and sympathetic systems.

Male Fertility

The male ejaculates about 1.5–5.0 ml of semen. The bulk of this fluid (45 to 80 percent) is produced by the seminal vesicles, while about 15 to 30 percent is contributed by the prostate. The sperm content in human males averages 60–150 million per milliliter in the ejaculated semen. Some of the values of normal human semen are summarized in table 21.7.

Figure 21.26 Simplified illustration of a vasectomy in which a segment of the vas deferens is removed through an incision in the scrotum.

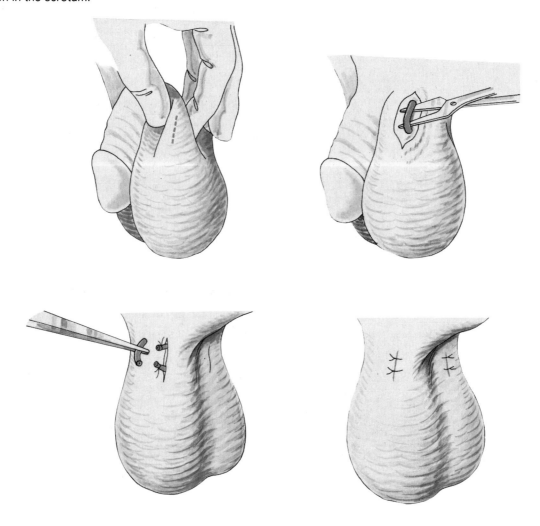

A sperm concentration below about 20 million per milliliter is termed *oligospermia,* and is associated with decreased fertility. A total sperm count below about 50 million per ejaculation is clinically significant in male infertility. In addition to low sperm counts as a cause of infertility, some men have antibodies against their own sperm (this is very common in men with vasectomies). Some women also produce antibodies against sperm.

Vasectomy is commonly performed as a contraceptive method. In this procedure the vas deferens is cut and tied or, in some cases, a valve or similar device is inserted.

This procedure interferes with sperm transport but does not directly affect secretion of androgens from Leydig cells in the interstitial tissue. Since spermatogenesis continues, the sperm produced cannot be drained from the testes and instead accumulates in "crypts" that form in the seminiferous tubules and vas deferens. These crypts present sites of inflammatory reactions in which spermatozoa are phagocytosed and destroyed by the immune system.

While androgen secretion from the Leydig cells is not directly affected by a vasectomy, it is clear that tubular function may be affected by the inflammatory reactions that result. Since the tubules have been shown to secrete 5α-reduced androgens and perhaps other hormones, and since the physiological importance of this endocrine function is not presently understood, it is clear that more research is needed to establish the long-term safety of this contraceptive procedure.

Figure 21.27   Anatomical relationships between ovaries, fallopian tubes, and uterus.

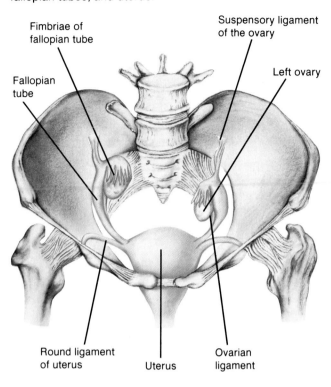

Fimbriae of fallopian tube

Suspensory ligament of the ovary

Fallopian tube

Left ovary

Round ligament of uterus

Uterus

Ovarian ligament

Figure 21.28   The tissue layers of the uterus.

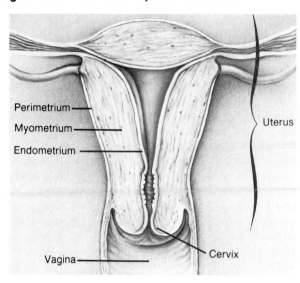

Perimetrium

Myometrium

Endometrium

Uterus

Vagina

Cervix

1. Describe the effects that castration has in the male on FSH and LH secretion, and explain the experimental evidence that suggests that the testes produce a polypeptide that specifically inhibits FSH secretion.
2. Describe the two compartments of the testes with respect to their *(a)* structure; *(b)* function; and *(c)* response to gonadotrophin stimulation. Describe two ways that these compartments appear to interact.
3. Using a diagram, describe the stages of spermatogenesis. Explain why spermatogenesis can continue throughout life without using up all of the spermatogonia.
4. Describe the structure and proposed functions of Sertoli cells in the seminiferous tubules.
5. Explain how FSH and androgens synergize to stimulate sperm production at puberty. Describe the hormonal requirements for spermatogenesis after puberty.

Female Reproductive System

The two ovaries are located within the body cavity, suspended by means of ligaments from the pelvic girdle. Extensions, or *fimbriae,* of the **fallopian tubes** partially cover each ovary. Ova that are released from the ovary—in a process called *ovulation*—are normally drawn into the fallopian tubes by the action of the ciliated epithelial lining of the tubes. The lumen of each fallopian tube is continuous with the **uterus** (or womb), a pear-shaped muscular organ also suspended within the pelvic girdle by ligaments (see figure 21.27).

The uterus narrows to form the *cervix* (cervi = neck), which opens to the **vagina.** The only physical barrier between the vagina and uterus is a plug of *cervical mucus.* These structures—the vagina, uterus, and fallopian tubes—constitute the sex accessory organs of the female. Like the sex accessory organs of the male, the female genital tract is affected by gonadal steroid hormones. Cyclic changes in ovarian secretion, as will be described in the next section, cause cyclic changes in the epithelial lining of the genital tract. The epithelial lining of the uterus, known as the *endometrium* (see figure 21.28), is most dramatically altered during the ovarian cycle.

The vaginal opening is located immediately posterior to the opening of the urethra. Both openings are covered by an inner *labia minora* and an outer *labia majora.* The *clitoris,* which is homologous to the penis as previously described, is located at the anterior margin of the labia minora.

Figure 21.29   The external female genitalia.

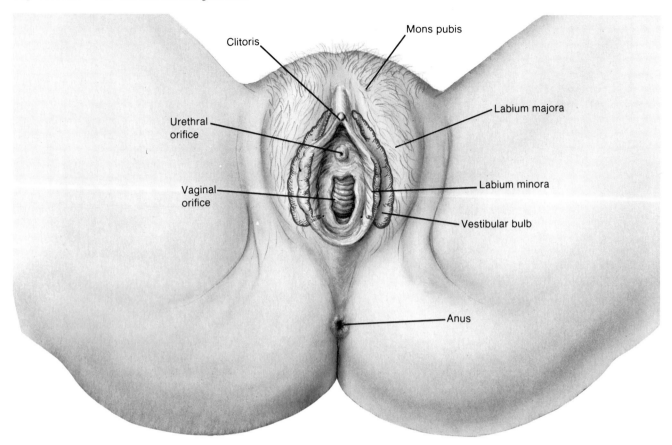

The Ovaries

The germ cells that migrate into the ovaries during early embryonic development multiply, so that by about five months of gestation (prenatal life) the ovaries contain approximately six to seven million oogonia. Production of new oogonia stops at this point and never resumes again. Towards the end of gestation the oogonia begin meiosis—at which time they are called **primary oocytes** (analogous to primary spermatocytes). Like spermatogenesis in the prenatal male, oogenesis is arrested at prophase I of the first meiotic division until puberty.

Unlike spermatogenesis, which continues throughout life because the spermatogonia duplicate themselves before becoming primary spermatocytes, oogenesis cannot continue indefinitely because no new primary oocytes can be formed. The number of primary oocytes, in fact, declines quite rapidly throughout a woman's reproductive years. The ovaries of a newborn girl contain about 2 million oocytes, but this number is reduced to about 300,000–400,000 by the time the girl enters puberty. Oogenesis ceases entirely at menopause (the time menstruation stops).

Oocytes are contained within hollow structures called *ovarian follicles*. Primary oocytes that are not stimulated to complete the first meiotic division are contained within tiny follicles called **primordial follicles.** In response to gonadotrophin stimulation, some of these oocytes and follicles get larger, and the follicular cells divide to produce numerous small *granulosa cells* that surround the oocyte and fill the follicle. A follicle at this stage in development is called a **primary follicle.**

Some primary follicles will be stimulated to grow still bigger and develop a fluid-filled cavity—the *antrum*—at which time they are called **secondary follicles.** The granulosa cells of secondary follicles form a ring around the circumference of the follicle and form a mound that supports the ovum. This mound is called the *cumulus oophorous* (literally, an egg-bearing hill). Some granulosa cells also encircle the ovum, forming a *corona radiata* ("radiant crown"). Between the oocyte and the corona radiata is a thin gel-like layer of proteins and polysaccharides called the *zona pellucida* ("transparent zone").

Figure 21.30 *(a)* Photomicrograph of primordial and secondary follicles. *(b)* Diagram of the parts of a secondary follicle.

Primordial follicles

Secondary follicle

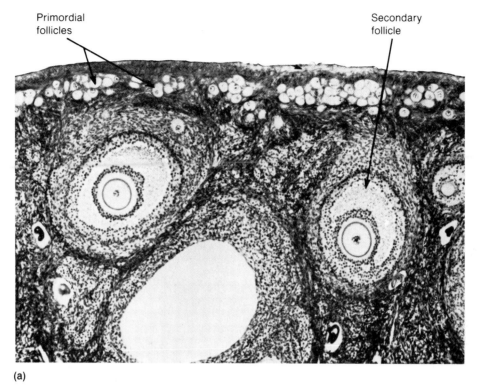

(a)

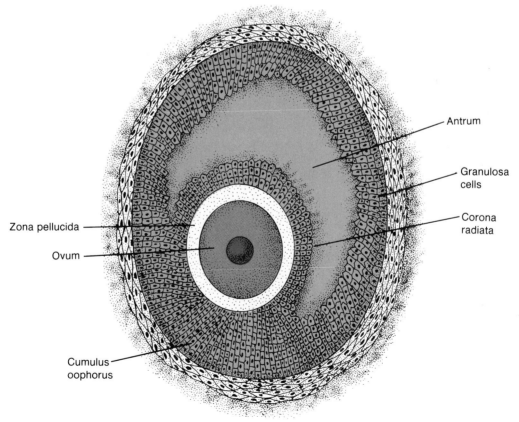

Antrum

Granulosa cells

Corona radiata

Zona pellucida

Ovum

Cumulus oophorus

(b)

Figure 21.31 *(a)* shows a primary oocyte at metaphase I of meiosis. Note the alignment of chromosomes *(arrow).* *(b)* shows a secondary oocyte formed at the end of the first meiotic division and the first polar body *(arrow).*

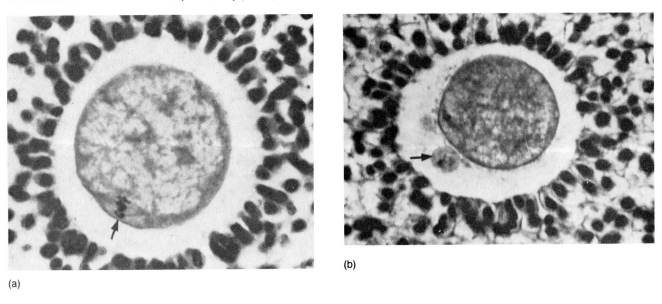

(a)

(b)

Figure 21.32 A Graafian follicle within the ovary of a monkey.

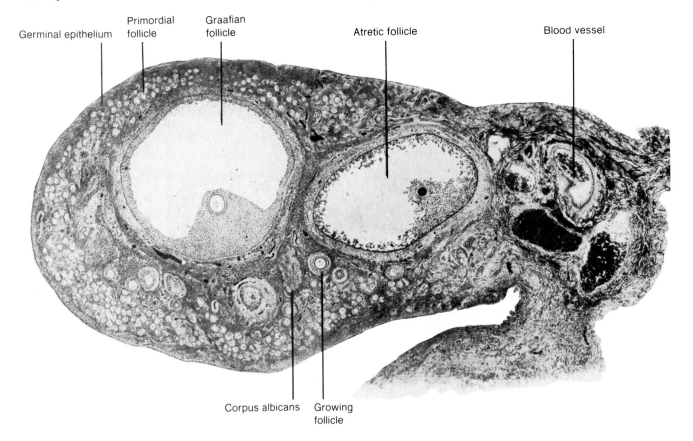

Germinal epithelium

Primordial follicle

Graafian follicle

Atretic follicle

Blood vessel

Corpus albicans

Growing follicle

Figure 21.33 Ovulation from a rabbit ovary.

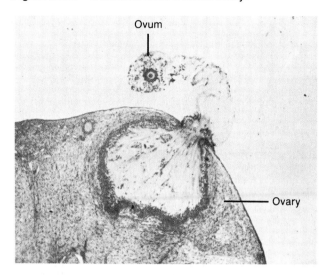

As the follicles develop, the primary oocyte completes its first meiotic division. This does not form two complete cells, however, because only one cell—the **secondary oocyte**—gets all of the cytoplasm. The other cell formed at this time becomes a small *polar body* (see figure 21.31), which eventually fragments and disappears. The secondary oocyte enters the second meiotic division, but meiosis is arrested at metaphase II and is never completed unless fertilization occurs.

Ovulation

Usually, by about ten to fourteen days after the first day of menstruation, only one follicle has continued its growth to become a mature **Graafian follicle** (see figure 21.32). The Graafian follicle is, in fact, so large that it forms a bulge on the surface of the ovary. Under proper hormonal stimulation, this follicle will rupture—much like the popping of a blister—and extrude its ovum into the fallopian tube in the process of **ovulation.** The ovarian ends of the fallopian tubes are flared out into fimbria, which catch the ovulated egg and funnel it into the fallopian tubes. The uptake and movement of the egg through the fallopian tubes are accomplished by the beating of a ciliated epithelium and by peristaltic waves of muscle contraction.

During each monthly cycle only one follicle in one ovary usually ovulates. The other follicles regress and are said to become *atretic*. The follicle that ovulates then undergoes structural and biochemical changes to become a **corpus luteum** (yellow body). Unless fertilization occurs the corpus luteum regresses at the end of each monthly cycle when menstruation begins.

The Menstrual Cycle

Humans, apes, and old-world monkeys have cycles of ovarian activity that repeat at approximately one-month intervals—hence the term *menstrual cycles* (menstru = monthly). The term *menstruation* is used to indicate the periodic shedding of the endometrium (inner epithelial lining of the uterus), which becomes thick and developed prior to the menstrual period under stimulation by ovarian steroid hormones. In primates (other than New World monkeys) this shedding of the endometrium is accompanied by bleeding. Other mammals, in contrast, do not bleed when they shed their endometrium, and therefore are not considered to have menstrual cycles.

Humans and other primates that have menstrual cycles may permit copulation at any time of the cycle. Non-primate mammals, in contrast, are sexually receptive only at a particular time in their cycles (shortly before or shortly after ovulation). These animals are therefore said to have *estrus cycles* (estrus = frenzy). Bleeding occurs in some animals (such as dogs and cows) that have estrus cycles shortly before they permit copulation. This bleeding is a result of high estrogen secretion from the ovaries and is not associated with shedding of the endometrium.

The bleeding that accompanies menstruation, in contrast, can be experimentally induced by removing the ovaries. This demonstrates that menstrual bleeding is caused by withdrawal of ovarian hormones. During the normal menstrual cycle, secretion of ovarian hormones rises and falls in a regular fashion, causing cyclic changes in the endometrium and other sex steroid-dependent tissues.

Phases of the Menstrual Cycle: Pituitary and Ovary

The average menstrual cycle has a duration of about twenty-eight days. Since it is a cycle, there is no beginning or end and the changes that occur are, for the most part, gradual. It is convenient, however, to call the first day of menstruation "day one" of each cycle, because menstrual flow is the most apparent of the changes that occur. It is also convenient to divide the cycle into four phases based on changes that occur in the ovary. These phases are (1) follicular; (2) ovulatory; (3) luteal; and (4) menstrual.

Figure 21.34 A corpus luteum in a human ovary.

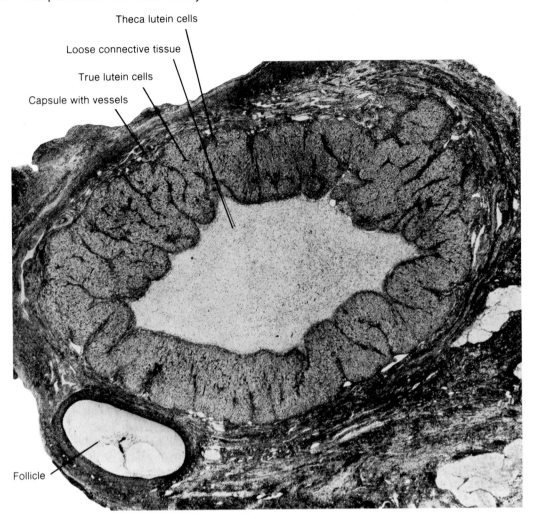

Theca lutein cells
Loose connective tissue
True lutein cells
Capsule with vessels
Follicle

Menstrual and Follicular Phases The **menstrual phase** lasts from day one to day four or five of the average cycle. During this time, secretion of ovarian steroid hormones are at their lowest ebb and the ovaries contain only primordial and primary follicles. During the **follicular phase,** which lasts from day five or six to day thirteen of the cycle, some of the primary follicles grow, form an antrum, and become secondary follicles. Towards the end of the follicular phase, one follicle reaches maturity and becomes a Graafian follicle. As the follicles grow, the granulosa cells secrete increasing amounts of estradiol, which reaches its highest concentration in the blood at about day twelve of the cycle.

Growth and development of the follicles, and secretion of estradiol, are stimulated by FSH (follicle-stimulating hormone). Although it has often been stated that FSH levels rise during the follicular phase, most studies fail to show significant changes in FSH concentration during this time of the cycle. There are, however, increasing numbers of FSH receptors in the granulosa cells at this time, so perhaps the increased growth and secretion of follicles is due to increased sensitivity of the follicles to FSH.

The rapid rise in estradiol secretion from the follicles during the follicular phase stimulates a "surge" of LH secretion from the anterior pituitary. This **LH surge** occurs at the last day of the follicular phase as a result of **positive feedback** of estradiol on the pituitary and hypothalamus.

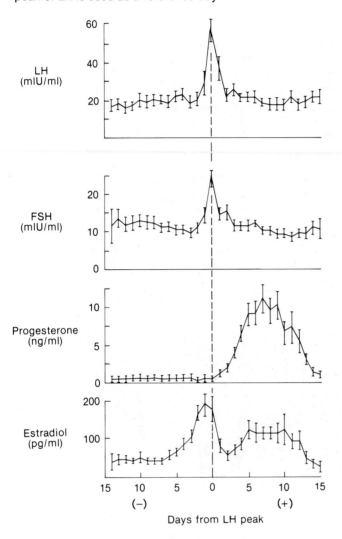

Figure 21.35 Average values for LH, FSH, progesterone, and estradiol during the menstrual cycle. The midcycle peak of LH is used as a reference day.

As estradiol concentrations rise in the blood, the pituitary first becomes more sensitive to GnRH (secreted from the hypothalamus) and then increased amounts of GnRH are released into the portal vessels. Since GnRH itself increases the pituitary's sensitivity to further GnRH (has a "priming effect"), small increases in releasing hormone can greatly stimulate secretion of gonadotrophins from the pituitary.

The LH surge begins about twenty-four hours before ovulation and reaches its peak about sixteen hours before ovulation. Since GnRH stimulates FSH as well as LH secretion, there is a simultaneous, though smaller, "surge" in FSH secretion. Since this is the only time in the cycle that FSH secretion significantly increases, many investigators believe that this mid-cycle peak in FSH acts as a stimulus for the development of new follicles for the next month's cycle.

Ovulation   Under the influence of FSH stimulation, the Graafian follicle grows so large that it becomes a thin-walled "blister" on the surface of the ovary. Growth of the follicle is accompanied by a rapid rate of increase in estradiol secretion. This rapid increase in estradiol, in turn, triggers the LH surge. Finally, the surge in LH secretion causes the wall of the Graafian follicle to rupture. Ovulation occurs, therefore, as a result of the sequential effects of FSH followed by LH on the ovarian follicles.

By means of the positive feedback effect of estradiol on LH secretion, the follicle in a sense sets the time for its own ovulation. This is because ovulation is triggered by the LH surge, and the LH surge is triggered by increased estradiol secretion, which occurs while the follicle grows. In this way the Graafian follicle does not normally ovulate until it has reached the proper size and degree of maturation.

In ovulation a secondary oocyte, arrested at metaphase II of meiosis, is released into the fallopian tube. This oocyte is still surrounded by a zona pellucida and corona radiata of granulosa cells as it begins its journey through the fallopian tube to the uterus. Normally only one ovary ovulates per cycle, with the right and left ovary alternating in successive cycles. Interestingly, if one ovary is removed the remaining ovary does not skip cycles, but ovulates every month. The mechanisms by which this regulation is achieved are not understood.

Luteal Phase   After ovulation, the empty Graafian follicle is stimulated by LH (luteinizing hormone) to become a new structure, the **corpus luteum.** This change in structure is accompanied by a change in function. Whereas the developing follicles secrete only estradiol, the corpus luteum secretes both estradiol and **progesterone.** Progesterone levels in the blood are negligible before ovulation but rise rapidly to reach a peak during the luteal phase approximately one week after ovulation (refer to figure 21.35).

The combined high levels of estradiol and progesterone secretion during the luteal phase exert a **negative feedback inhibition** of FSH and LH secretion. This serves to retard development of new follicles, so that further ovulation does not normally occur during that cycle. While this may seem like locking the barn door after the horse (ovum) has escaped, it does prevent more horses (ova) from escaping. In this way multiple ovulations (and possible pregnancies) on succeeding days of the cycle are normally prevented.

Figure 21.36 Phases of the menstrual cycle in relation to ovarian changes and hormone secretion.

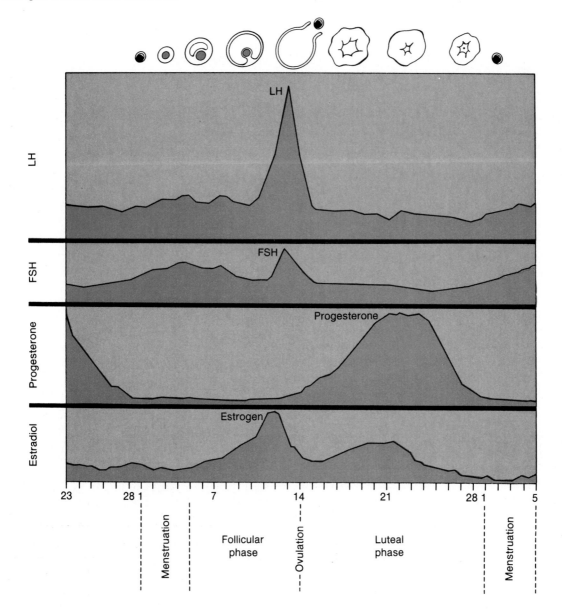

High levels of estrogen and progesterone secretion during the normal cycle do not persist for very long, however, and new follicles do start to develop towards the end of one cycle, in preparation for the next cycle. Estrogen and progesterone levels fall during the late luteal phase (starting about day twenty-two), because the corpus luteum regresses and stops functioning. In lower mammals, the decline in corpus luteum function is caused by a hormone secreted from the uterus known as *luteolysin.* A similar hormone has not yet been identified in humans, and the causes of corpus luteum regression in humans are not well understood. Luteolysis (breakdown of the corpus luteum) can be prevented by high LH secretion, but LH levels remain low during the luteal phase as a result of negative feedback inhibition by ovarian steroids.

With the declining function of the corpus luteum, estrogen and progesterone fall to very low levels by day twenty-eight of the cycle. Withdrawal of ovarian steroids causes menstruation (as described in the next section) and permits a new cycle of ovarian follicle development to progress.

Effects of Ovarian Hormones on the Endometrium

In addition to describing the female cycle in terms of phases of ovarian function, the cycle can also be described in terms of the changes that occur in the endometrium. Three phases can be identified on this basis: (1) proliferative phase; (2) secretory phase; and (3) menstrual phase.

Figure 21.37 Relationship between changes in ovaries and endometrium of the uterus during different phases of the menstrual cycle.

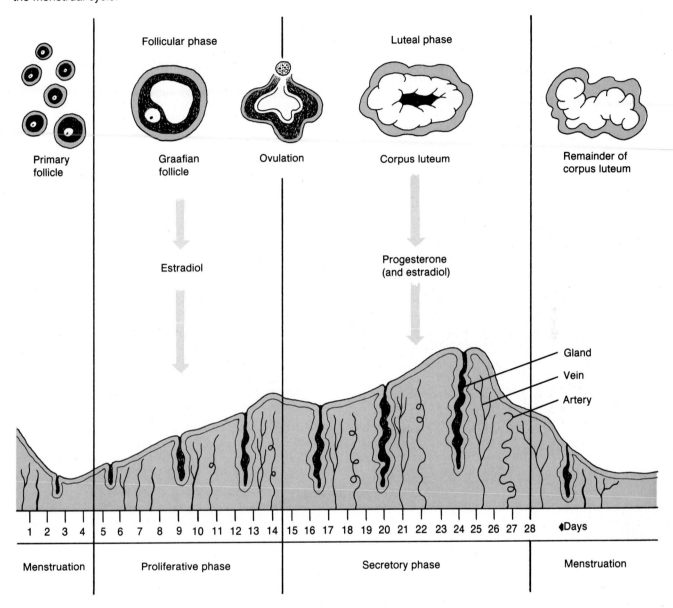

The **proliferative phase** of the endometrium occurs while the ovary is in its follicular phase. The increasing amounts of estradiol secreted by the developing follicles stimulate growth (proliferation) of the endothelial lining of the uterus (the endometrium). In humans and other primates, *spiral arteries* develop in the endometrium during this phase (see figure 21.37). Estradiol also stimulates the production of progesterone receptors in the uterus at this time, in preparation for the next phase of the cycle.

The **secretory phase** of the endometrium occurs when the ovary is in its luteal phase. This occurs as a result of increased progesterone secretion, which stimulates development of mucous glands. As a result of the combined actions of estradiol and progesterone, the uterus becomes thick, highly vascular, and "spongy" in appearance during the time of the cycle following ovulation. It is therefore well prepared to accept and nourish an embryo if fertilization has occurred.

Table 21.8 Phases of the menstrual cycle.

Phase of Cycle		Hormonal Changes		Tissue Changes	
Ovarian	Endometrial	Pituitary	Ovary	Ovarian	Endometrial
Follicular (days 5–13)	Proliferative	No apparent change	Estradiol secretion rises (due to FSH stimulation of follicles).	Follicles grow; Graafian follicle develops (due to FSH stimulation).	Mitotic division increases thickness of endometrium; spiral arteries develop (due to estradiol stimulation).
Ovulatory (day 14)	Proliferative	LH surge (and increased FSH) stimulated by positive feedback from estradiol	Slight fall in estradiol secretion.	Graafian follicle is ruptured and ovum is extruded into fallopian tube.	No change.
Luteal (days 15–28)	Secretory	LH and FSH decrease (due to negative feedback of steroids)	Progesterone and estrogen secretion increase.	Development of corpus luteum (due to LH stimulation).	Glandular development in endometrium (due to progesterone stimulation).
Menstrual (days 1–4)	Menstrual	FSH and LH remain low.	Estradiol and progesterone fall.	Corpus luteum regresses; primary follicles grow.	Inner two-thirds of endometrium is shed with accompanying bleeding.

The **menstrual phase,** as previously discussed, occurs as a result of the fall in ovarian sex steroid secretion during the late luteal phase. Necrosis (cellular death) and sloughing of the inner two-thirds of the endometrium may be produced by constriction of the spiral arteries. The spiral arteries appear to be responsible for bleeding during menstruation because lower mammals lack these arteries and don't bleed when they shed their endometrium.

The cyclic changes in ovarian secretion also cause other cyclic changes in the female genital tract. High levels of estrogen secretion, for example, cause cornification of the vaginal epithelium (the upper cells die and become filled with keratin). During the luteal phase of the cycle, the high levels of progesterone secreted from the ovary cause a change in the consistency of the cervical mucous plug. This allows sperm to penetrate the cervix and enter the uterus more easily.

Methods of Contraception

The *rhythm method* of birth control is used by many people, but popularity of this technique has declined as more successful methods of contraception have been introduced. In the rhythm method of birth control the woman attempts to predict the day of her ovulation and restricts intercourse to "safe" times of the cycle when fertilization cannot occur. The day of ovulation can be determined by a drop in basal body temperature or by a change in mucous discharge from the vagina. This technique is subject to great error, however, because most women don't ovulate on precisely the same day of the cycle in different months.

More popular methods of birth control include (1) sterilization; (2) oral contraceptives; (3) intrauterine devices (IUDs); and (3) barrier methods—including condoms for the male and diaphragms for the female. All of these techniques are effective, but the safety, side effects, and efficacy of each technique are different.

Sterilization techniques include *vasectomy* for the male and *tubal ligation* for the female. In the latter technique (which currently accounts for over 60 percent of sterilization procedures performed in the United States), the fallopian tubes are cut and tied. This is analogous to the procedure performed on the vas deferens in a vasectomy and prevents fertilization of the ovulated ovum. The long-term safety of vasectomy, as previously discussed, has not been well established, and evidence in experiments with Rhesus monkeys suggests that increased tendency towards atherosclerosis may be a side effect of this procedure. Symptoms of pelvic pain and other problems are found in about 15 percent of women who have tubal ligations.

About ten million women in the United States and sixty million women in the world are currently using **oral steroid contraceptives.** These contraceptives usually consist of a synthetic estrogen combined with a synthetic progesterone in the form of pills that are taken once each day for three weeks after the last day of a menstrual period. This procedure causes an immediate increase in blood levels of ovarian steroids (from the pill) that is maintained for the normal duration of a monthly cycle. As a result of *negative feedback inhibition* of gonadotrophin secretion, *ovulation never occurs.* The entire cycle is like a false luteal phase, with high levels of progesterone and estrogen and low levels of gonadotrophins.

Table 21.9 Effectiveness, as measured by failure rate (number of pregnancies per 100 woman-years of use), and side effects of various contraceptive methods.

Method	Failure Rate	Minor Side Effects	Major Side Effects
Female sterilization (tubal ligation)	1 in 200 to 1 in 1,000	None	Post-tubal symptoms
Male sterilization (vasectomy)	1 in 1,000	None	Possible atherosclerosis
Oral contraceptives	0.5	Pseudopregnancy	Thrombophlebitis, pulmonary embolus, cerebral thrombosis
Intrauterine device	2–3	Uterine bleeding and cramping	Uterine perforations, pelvic infections
Barrier methods (condom, diaphragm)	3–20	Allergic reactions; nuisance to use	None

Source: From W. Droegemueller and R. Bressler, *Annual Review of Medicine* (1980) Vol. *31*, p. 329. Reprinted by permission.

Figure 21.38 Insertion of an intrauterine device (IUD).

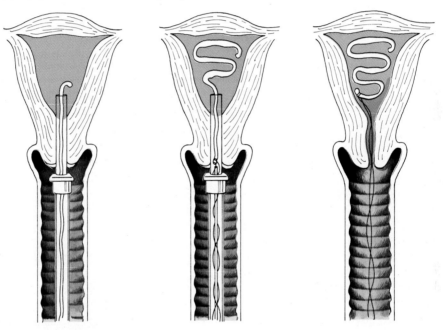

Since the contraceptive pills contain ovarian steroid hormones, the endometrium proliferates and becomes secretory just as it does during a normal cycle. In order to prevent abnormal growth of the endometrium women stop taking the pill after three weeks. This causes estrogen and progesterone levels to fall, and permits menstruation to occur. The contraceptive pill is an extremely effective method of birth control, but it does have potentially serious side effects—including increased incidence of thromboembolism, cardiovascular disorders, and increased incidence of endometrial and breast cancer. It has been pointed out, however, that the mortality risk of contraceptive pills is still much lower than the risk of death from complications of pregnancy—or from automobile accidents.

Intrauterine devices (IUDs) don't prevent ovulation, but instead prevent implantation of the embryo into the uterus in the event fertilization occurs. The mechanisms by which these contraceptive effects are produced are not well understood, but the efficiency of different IUDs appears to be related to their ability to cause inflammatory reactions in the uterus. Uterine perforations are the foremost complications of the use of IUDs.

Barrier contraceptives—condoms and diaphragms—are slightly less effective than the other methods of contraception, but they do not cause serious side effects. The failure rate of condoms—one of the oldest methods of contraception—is 12 to 20 pregnancies per 100 woman years of use, while the failure rate of the diaphragm is 12 to 18 pregnancies per 100 woman years.

The Menopause

The term *menopause* means literally "pause in the menses," and refers to the cessation of ovarian activity that occurs at about the age of fifty. During the postmenopausal years, which account for about a third of a woman's life span, no new ovarian follicles develop and the ovaries cease secreting estradiol. Like prepubertal boys and girls, the only estrogen found in the blood of postmenopausal women is that formed by aromatization of the weak androgen androstenedione, secreted principally by the adrenal cortex, into a weak estrogen called estrone.

It is the withdrawal of estradiol secretion from the ovaries that is most responsible for the many debilitating symptoms of menopause. These include vasomotor disturbances (which produce "hot flashes"), urogenital atrophy, and increased development of osteoporosis (see chapter 20). Estrogen replacement therapy, often in combination with progesterone, helps to alleviate these symptoms, although this treatment is given cautiously because of the increased risks associated with taking estrogen pills.

1. Describe the stages of oogenesis. Explain why oogenesis is only completed if fertilization occurs and why only one mature ovum is produced at the end of meiosis.
2. Describe the hormonal mechanisms that insure that ovulation occurs only when a mature Graafian follicle is produced.
3. Explain how the contraceptive pill works and why the effects of the pill can be described as producing a false luteal phase.
4. Using a diagram, show how the blood concentrations of estradiol and progesterone vary during the menstrual cycle. Describe the formation, function, and fate of the corpus luteum.
5. Describe the cyclic changes in the endometrium, indicating which ovarian steroid hormones cause these changes.

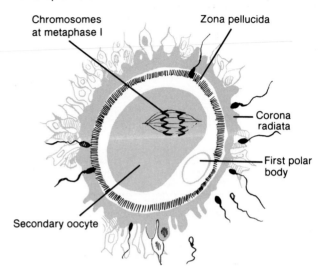

Figure 21.39 Ovulation releases a secondary oocyte, surrounded by its zona pellucida and corona radiata, into the fallopian tube.

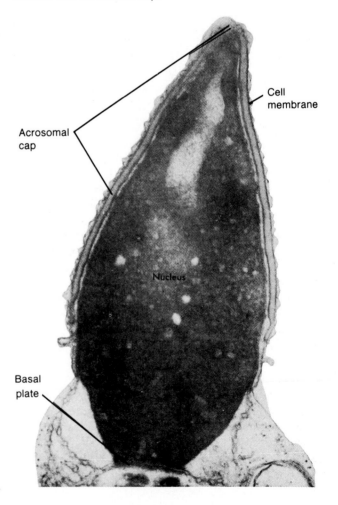

Figure 21.40 The head of a human sperm, showing the nucleus and acrosomal cap.

Figure 21.41 A secondary oocyte, arrested at metaphase II of meiosis, is released at ovulation. If this cell is fertilized it completes its second meiotic division and produces a second polar body.

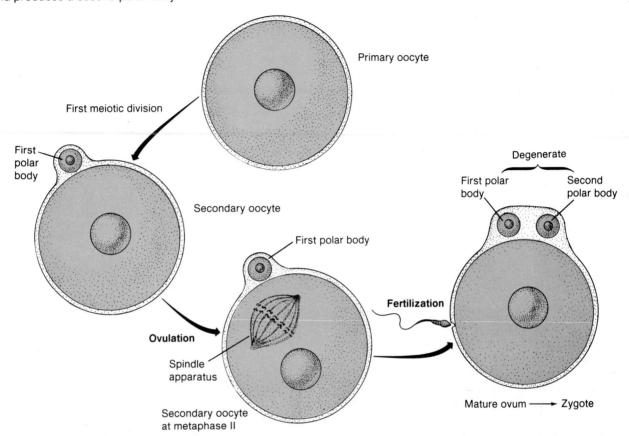

Fertilization, Pregnancy, and Parturition

During the act of sexual intercourse a man ejaculates an average of 300 million sperm into the vagina. This tremendous number is needed because of the high sperm fatality rate—only about 100 survive to enter each fallopian tube. During their passage through the female reproductive tract the sperm gain the ability to fertilize an ovum. This process is called **capacitation.** The changes that occur in capacitation are not known. Experiments have shown, however, that freshly ejaculated sperm are infertile; they must be present in the female tract for at least seven hours before they can fertilize an ovum.

A woman usually ovulates only one ovum a month, for a total number of less than 450 during her reproductive years. Each ovulation releases a secondary oocyte arrested at metaphase of the second meiotic division. The secondary oocyte, as previously described, enters the fallopian tube surrounded by its zona pellucida (a thin transparent layer of protein and polysaccharides) and corona radiata of granulosa cells (see figure 21.39).

The head of each spermatozoan is capped by an organelle called an *acrosome* (see figure 21.40). The acrosome contains digestive enzymes—a trypsinlike protein-digesting enzyme, and hyaluronidase (which digests hyaluronic acid, an important constituent of connective tissues). Fertilization normally occurs in the fallopian tubes. When sperm meets ovum, an **acrosomal reaction** occurs that exposes the acrosome's digestive enzymes and allows the sperm to penetrate the corona radiata and the zona pellucida. The acrosomal enzymes are not released in this process; rather, the sperm tunnels its way through these barriers by digestion reactions that are localized to its acrosomal cap.

As the first sperm tunnels its way through the zona pellucida, a chemical change in the zona occurs that prevents other sperm from entering. Only one sperm, therefore, is allowed to fertilize one ovum. As fertilization occurs, the secondary oocyte is stimulated to complete its second meiotic division. Like the first meiotic division, the second produces one cell that contains all of the cytoplasm—the mature ovum or egg cell—and one polar body. The second polar body, like the first, ultimately fragments and disintegrates.

Figure 21.42 Fertilization and union of chromosomes from the sperm and ovum to form the zygote.

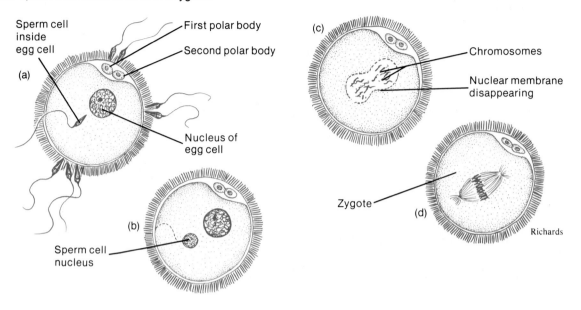

Figure 21.43 Cleavage of a human ovum fertilized in a laboratory *(in vitro)*. Cleavage produces embryos at 4-cell stage *(a)*, 8-cell stage *(b)*, 16–32-cell stage—a morula— *(c)*, and blastocyst stage *(d)*.

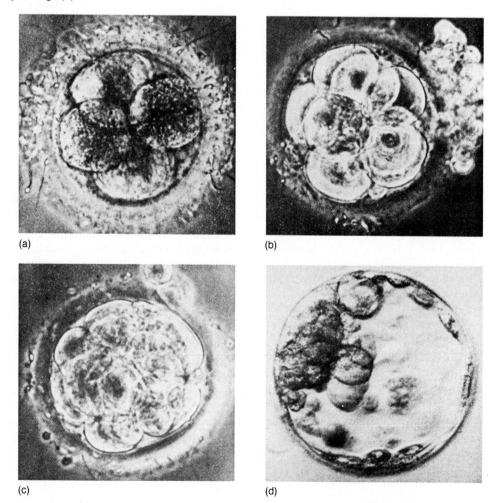

Figure 21.44 The blastocyst adheres to the endometrium as implantation begins, at about the sixth day following fertilization.

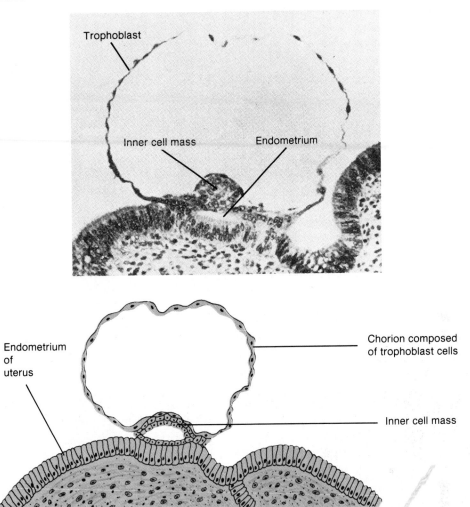

At fertilization, the entire sperm enters the cytoplasm of the much larger egg cell. Within twelve hours the nuclear membrane in the ovum disappears, and the haploid number of chromosomes (twenty-three) in the ovum is joined by the haploid number of chromosomes from the sperm. A fertilized egg, or *zygote,* containing the diploid number of chromosomes (forty-six) is thus formed.

A secondary oocyte that is ovulated but not fertilized does not complete its second meiotic division, but instead disintegrates twelve to twenty-four hours after ovulation. Fertilization therefore cannot occur if intercourse takes place beyond one day following ovulation. Sperm, in contrast, can survive up to three days in the female reproductive tract. Fertilization therefore can occur if intercourse is performed within three days prior to the day of ovulation.

Cleavage and Formation of a Blastocyst

At about thirty to thirty-six hours after fertilization the zygote divides by mitosis—a process called *cleavage*—into two smaller cells. The rate of cleavage is thereafter accelerated. A second cleavage, performed about forty hours after fertilization, produces four cells. By about fifty to sixty hours after fertilization a third cleavage occurs, producing a ball of eight cells called a **morula** (= mulberry). This very early embryo enters the uterus three days after ovulation had occurred.

Cleavage continues so that a morula consisting of thirty-two to sixty-four cells is produced by the fourth day after fertilization. The embryo remains unattached to the uterine wall for the next two days, during which time it undergoes changes that convert it into a hollow structure called a **blastocyst.** The blastocyst consists of two parts: (1) an *inner cell mass,* which will become the body of the fetus; and (2) a surrounding *chorion,* which will become part of the placenta. The cells that form the chorion are called *trophoblast cells.*

Figure 21.45 The egg is normally fertilized in the distal third of the fallopian tube and begins cleavage and early embryonic development during its three-day journey to the uterus. By the sixth day the embryo, now at the blastocyst stage, implants into the endometrium of the uterus.

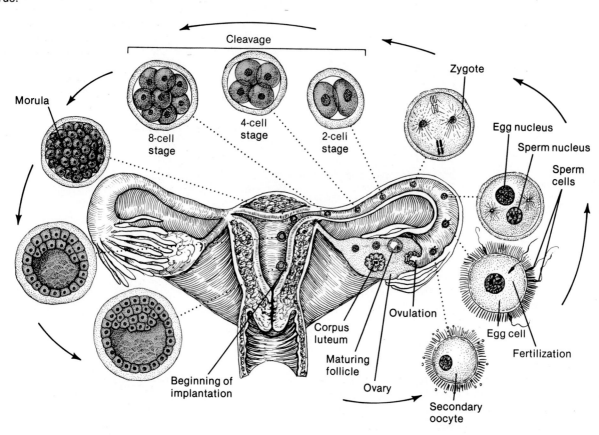

On the sixth day following fertilization the blastocyst attaches to the uterine wall, with the side containing the inner cell mass against the endometrium. This begins the process of **implantation,** or **nidation,** and by the seventh day the blastocyst is usually completely buried in the endometrium.

Implantation and Formation of the Placenta

In nonfertile cycles the corpus luteum begins to decrease its secretion of steroids about ten days after ovulation. This withdrawal of steroids, as previously described, causes necrosis and sloughing of the endometrium following day twenty-eight of the nonfertile cycle. If fertilization and implantation have occurred, however, these events must obviously be prevented to maintain the pregnancy.

Chorionic Gonadotrophin   The blastocyst saves itself from being eliminated with the endometrium by secreting a hormone that indirectly prevents menstruation. Even before the sixth day when implantation occurs, the trophoblast cells of the chorion secrete **chorionic gonadotrophin** (**hCG**—the *h* stands for *human*). This hormone is identical to LH in its effects and therefore is able to maintain the corpus luteum past the time when it would otherwise regress. Secretion of estradiol and progesterone is thus maintained and menstruation is normally prevented.

Secretion of hCG declines by the tenth week of pregnancy. Actually, this hormone is only required for the first five to six weeks of pregnancy, because the placenta itself becomes an active steroid hormone secreting gland. At the fifth to sixth week the mother's corpus luteum begins to regress (even in the presence of hCG), but the placenta secretes more than sufficient amounts of steroids to maintain the endometrium and prevent menstruation.

Figure 21.46 Human chorionic gonadotrophin (hCG) is secreted by trophoblast cells during the first trimester of pregnancy. This hormone maintains the mother's corpus luteum for the first 5½ weeks. After that time the placenta becomes the major sex hormone-producing gland, secreting increasing amounts of estrogen and progesterone throughout pregnancy.

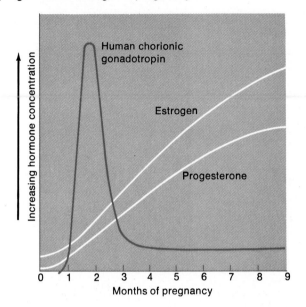

All *pregnancy tests* assay for the presence of hCG in blood or urine, because this is the only hormone that is secreted by the blastocyst but not by the mother's endocrine glands. Modern pregnancy tests detect the presence of hCG by use of antibodies against hCG (see chapter 17) or by use of cellular receptor proteins for hCG.

Chorionic Membranes Between days seven and twelve, as the blastocyst becomes completely embedded in the endometrium, the chorion becomes a two-cell-thick structure, consisting of an inner **cytotrophoblast** layer and an outer **syncytiotrophoblast** layer. The inner cell mass (which will become the body of the fetus), meanwhile, also develops two cell layers. These are the *ectoderm* (which will form such organs as the nervous system and skin) and the endoderm (which will eventually form the gut and its derivatives). A third, middle embryonic layer—the *mesoderm*—is not yet seen at this stage. The embryo at this stage is a two-layer-thick disc separated from the cytotrophoblast of the chorion by an **amniotic cavity** (see figure 21.47).

Figure 21.47 After the blastocyst *(a)* implants into the endometrium on day six, it is transformed by day eight into the structures shown in *(b)* and *(c)*. As the embryo develops into a disc of two cell layers (ectoderm and endoderm) the trophoblast likewise forms two layers. The outer syncytiotrophoblast extends into the endometrium and, through the action of digestive enzymes, creates blood-filled cavities in the endometrium around the embryo. The inner cytotrophoblast layer forms an enveloping membrane, the chorion, around the embryo.

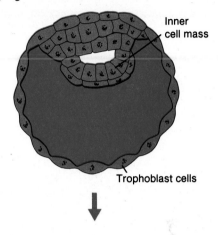

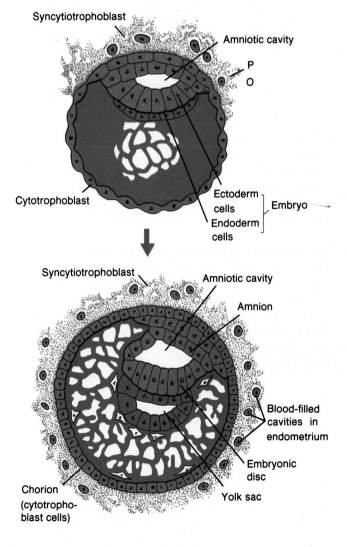

Figure 21.48   After the syncytiotrophoblast has created blood-filled cavities in the endometrium, these cavities are invaded by extensions of the cytotrophoblast *(a)*. These extensions, called villi, branch extensively to produce the chorion frondosum *(b)*. The developing embryo is surrounded by a membrane known as the amnion.

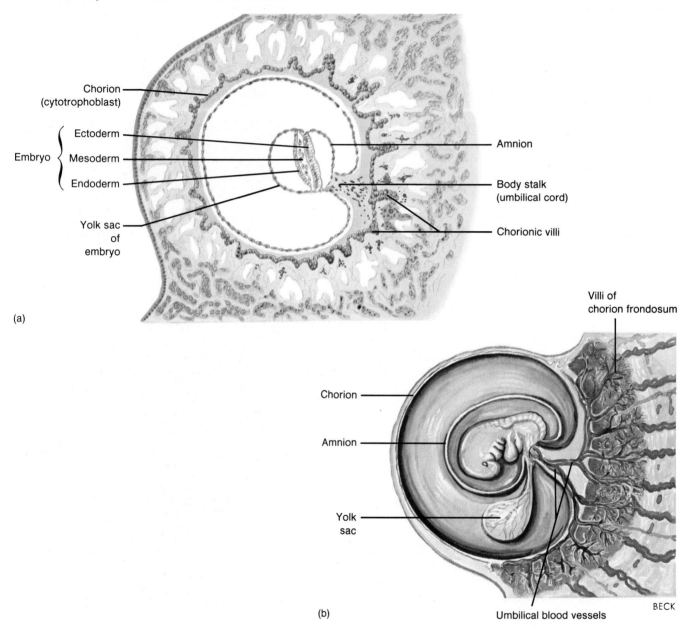

(a)

(b)

As the syncytiotrophoblast invades the endometrium, it secretes protein-digesting enzymes that create many small, blood-filled cavities in the maternal tissue. The cytotrophoblast then forms projections, or *villi,* that grow into these pools of venous blood, producing a leafy appearing structure called the *chorion frondosum* (frond = leaf). This occurs only on the side of the chorion that faces the uterine wall. As the embryonic structures grow, the other side of the chorion bulges into the cavity of the uterus, loses its villi, and becomes smooth in appearance.

Formation of Placenta and Amniotic Sac   As the blastocyst is implanted in the endometrium and the chorion develops, the cells of the endometrium also undergo changes. These changes, including cellular growth and accumulation of glycogen, are called the **decidual reaction.** The maternal tissue in contact with the chorion frondosum is called the *decidua basalis.* These two structures—chorion frondosum (fetal tissue) and decidua basalis (maternal tissue)—together form the functional unit known as the **placenta.**

Figure 21.49 Blood from the fetus is carried to and from the chorion frondosum by umbilical arteries and veins. The maternal tissue between the chorionic villi is known as the decidua basalis, and this tissue, together with the villi, form the functioning placenta. The space between chorion and amnion is obliterated, and the fetus lies within the fluid-filled amniotic sac.

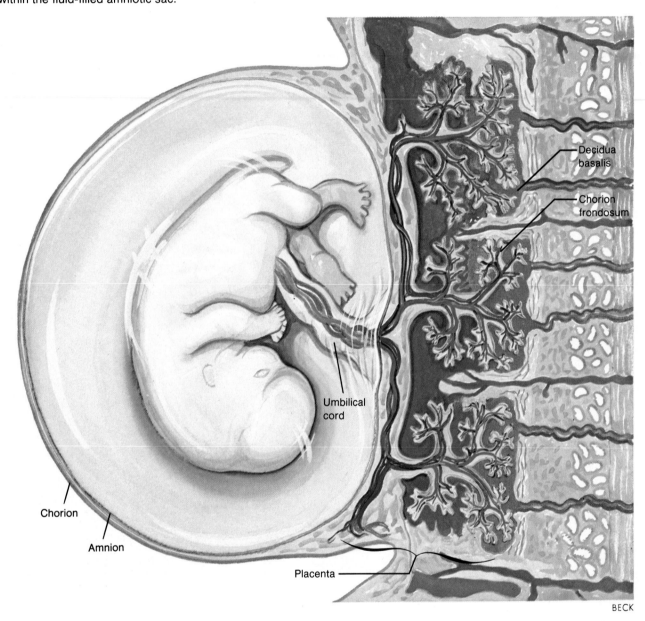

Decidua basalis

Chorion frondosum

Umbilical cord

Chorion

Amnion

Placenta

BECK

The human placenta is a disc-shaped structure that is continuous at its outer surface with the smooth part of the chorion, which bulges into the uterine cavity. Immediately beneath the chorionic membrane is the amnion, which has grown to envelop the entire fetus. The fetus, together with its umbilical cord, is therefore located within the fluid-filled **amniotic sac.**

Amniotic fluid is formed initially as an isotonic secretion that is later increased in volume and decreased in concentration by urine from the fetus. Amniotic fluid also contains cells that are sloughed off from the fetus, placenta, and amniotic sac. Since all of these cells are derived from the same fertilized egg, all have the same genetic composition. Many genetic abnormalities can be detected by aspiration of this fluid and examination of the cells thus obtained. This procedure is called **amniocentesis.**

Figure 21.50 Amniocentesis. In this procedure amniotic fluid, together with suspended cells, is withdrawn for examination. Various genetic diseases can be detected prenatally by this means.

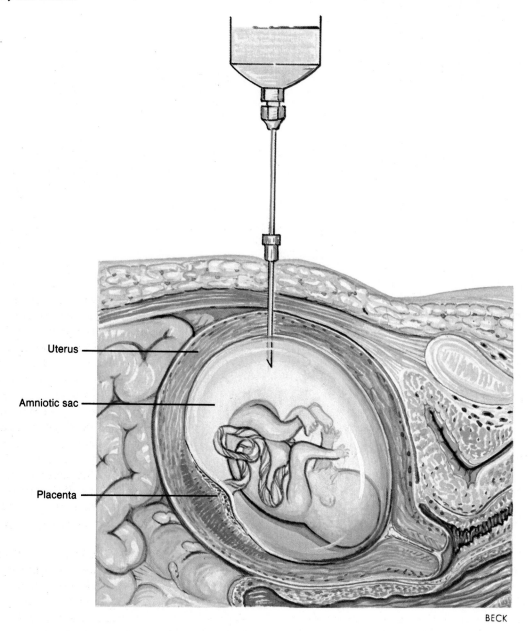

Uterus

Amniotic sac

Placenta

BECK

Amniocentesis is usually performed at the fourteenth or fifteenth weeks of pregnancy, when the amniotic sac contains 175–225 ml of fluid. Genetic diseases such as Down's syndrome (where there are three instead of two chromosomes number 21) can be detected by examining chromosomes; diseases such as Tay-Sach's disease, in which there is a defective enzyme involved in formation of myelin sheaths, can be detected by biochemical techniques.

Major structural abnormalities, which may not be predictable from genetic analysis, can often be detected by *ultrasound*. Sound wave vibrations are reflected from the interface of tissues with different densities—such as the interface between the fetus and amniotic fluid—and used to produce an image. This technique is so sensitive that it can be used to detect a fetal heartbeat several weeks before it can be detected by a stethoscope.

Figure 21.51 Chromosomes obtained by amniocentesis. The fetus is female, as shown by the presence of two X chromosomes. This fetus, however, has three number 13 chromosomes—a genetic disease that causes severe head and brain deformity, leading to early death of the fetus if it is not aborted.

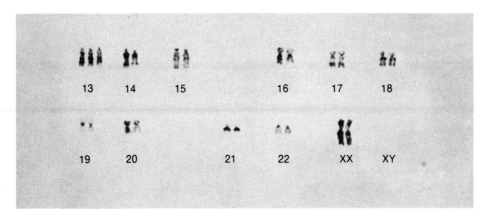

Exchange of Molecules across the Placenta

The *umbilical artery* delivers fetal blood to vessels within the villi of the chorion frondosum of the placenta. This blood circulates within the villi and returns to the fetus via the *umbilical vein*. Maternal blood is delivered to and drained from the cavities within the decidua basalis that are located between the chorionic villi. In this way, maternal and fetal blood are brought close together but never mix within the placenta.

The placenta serves as a site for the exchange of gases and other molecules between the maternal and fetal blood. Oxygen diffuses from mother to fetus, and carbon dioxide diffuses in the opposite direction. Nutrient molecules and waste products likewise pass between maternal and fetal blood; the placenta is, after all, the only link between the fetus and the outside world.

The placenta is not merely a passive conduit for exchange between maternal and fetal blood, however. It has a very high metabolic rate, utilizing about a third of all the oxygen and glucose supplied by the maternal blood. The rate of protein synthesis is, in fact, higher in the placenta than it is in the liver. Like the liver, the placenta produces a great variety of enzymes capable of converting biologically active molecules (such as hormones and drugs) into less active, more water-soluble forms. In this way, potentially dangerous molecules in the maternal blood are often prevented from harming the fetus.

Endocrine Functions of the Placenta

The placenta secretes both steroid hormones and glycoprotein hormones that have actions similar to those of some anterior pituitary hormones. This latter category of hormones includes **chorionic gonadotrophin (hCG)**, **placental lactogen (hPL)**, and **chorionic thyrotrophin (hCT)**.

Glycoprotein Hormones from the Placenta   The importance of chorionic gonadotrophin in maintaining the mother's corpus luteum for the first 5½ weeks of pregnancy has been previously discussed. There is also some evidence that hCG may in some way help to prevent immunological rejection of the implanting embryo. Placental lactogen acts together (synergizes with) growth hormone from the mother's pituitary to produce a "diabetic-like" effect in the pregnant woman. The effects of these two hormones stimulate (1) lipolysis and increased plasma fatty acid concentration; (2) decreased maternal utilization of glucose and, therefore, increased blood glucose concentrations; and (3) polyuria (excretion of large volumes of urine), thereby producing a degree of dehydration and thirst. This "diabetic-like" effect in the mother helps to spare glucose for the placenta and fetus that, like the brain, use glucose as their primary energy source.

Steroid Hormones from Placenta   After the first 5½ weeks of pregnancy, when the corpus luteum regresses, the placenta becomes the major sex steroid-producing gland. The blood concentration of estrogens, as a result of placental secretion, rises to levels more than 100 times greater than those existing at the beginning of pregnancy. The placenta also secretes large amounts of progesterone, changing the estrogen/progesterone ratio in the blood from 1:100 at the beginning of pregnancy to a ratio of close to 1:1 towards full term.

Table 21.10   Hormones secreted by placenta.

Hormones	Effects
Pituitary-like hormones	
Chorionic gonadotrophin (hCG)	Similar to LH; maintains mother's corpus luteum for first 5½ weeks of pregnancy; may be involved in suppressing immunological rejection of embryo
Placental lactogen (hPL)	Similar to prolactin; synergizes with growth hormone to promote, in the mother, increased fat breakdown and fatty acid release from adipose tissue, and to promote the sparing of glucose use by maternal tissues ("diabetic-like" effects)
Chorionic thyrotrophin (hCT)	Similar to TSH; physiological significance uncertain
Sex steroids	
Progesterone	Helps maintain endometrium during pregnancy; helps suppress gonadotrophin secretion; promotes uterine sensitivity to oxytocin; helps stimulate mammary gland development
Estrogens	Helps maintain endometrium during pregnancy; helps suppress gonadotrophin secretion; helps stimulate mammary gland development; inhibits prolactin secretion

Figure 21.52   Circulation of blood within the placenta. Maternal blood is delivered to and drained from the spaces between the chorionic villi. Fetal blood is brought to blood vessels within the villi by branches of the umbilical artery, and is drained by branches of the umbilical vein.

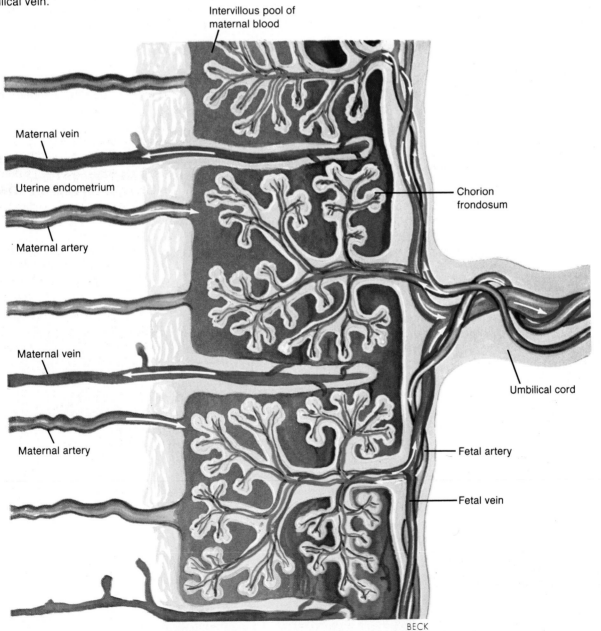

BECK

Figure 21.53 Secretion of progesterone and estrogen from the placenta requires a supply of cholesterol from the mother's blood and cooperation with fetal enzymes that convert progesterone to androgens.

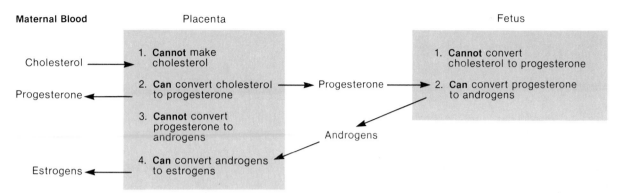

The placenta, however, is an "incomplete endocrine gland" because it cannot produce estrogen and progesterone without the aid of precursors supplied to it by both the mother and the fetus. The placenta, for example, cannot produce cholesterol from acetate and so must be supplied with cholesterol from the mother's circulation. Cholesterol, which is a steroid containing twenty-seven carbons, can then be converted by enzymes in the placenta into steroids that contain twenty-one carbons—such as progesterone. The placenta, however, lacks the enzymes needed to convert progesterone into androgens (which have nineteen carbons) and estrogens (which have eighteen carbons).

In order for the placenta to produce estrogens it needs to cooperate with steroid-producing tissues in the fetus. Fetus and placenta, therefore, form a single functioning system in terms of steroid hormone production. This system has been called the **fetal-placental unit** (see figure 21.53).

The ability of the placenta to convert androgens into estrogen helps to protect the female embryo from becoming masculinized by the androgens secreted from the mother's adrenal glands. In addition to producing estradiol, the placenta secretes large amounts of a weak estrogen called *estriol*. Production of estriol increases tenfold during pregnancy, so that by the third trimester estriol accounts for about 90 percent of the estrogens excreted in the mother's urine. Since almost all of this estriol comes from the placenta (rather than from maternal tissues), measurements of urinary estriol can be used clinically to assess the health of the placenta.

Labor and Parturition

Powerful contractions of the uterus in *labor* are needed for childbirth *(parturition)*. These uterine contractions are known to be stimulated by two agents: (1) *oxytocin,* a polypeptide hormone produced in the hypothalamus and secreted by the posterior pituitary; and (2) *prostaglandins,*

a class of cyclic fatty acids produced within the uterus itself. Labor can indeed be induced artificially by injections of oxytocin or by insertion of prostaglandins into the vagina as a suppository.

While the mechanisms operating during labor are well understood, the factors that lead to the initiation of parturition in humans remain unclear. In some lower mammals, the concentrations of estrogen and progesterone drop prior to parturition; this does not appear to be true in humans. In sheep and goats, labor seems to be initiated by corticosteroids from the fetal adrenal cortex, which stimulates production of prostaglandins in the uterus. The uterine contractions that result, in turn, stimulate increased secretion of oxytocin. The timing of parturition in humans, however, does not appear to be similarly dependent on the fetal adrenal cortex.

It appears that both oxytocin and prostaglandins are required in humans; oxytocin-induced contractions do not lead to dilation of the cervix and progressive labor in the absence of prostaglandins. Indeed, both alcohol—which inhibits oxytocin secretion—and indomethacin (which inhibits prostaglandin production) can be used to prevent pre-term births. A mechanism has recently been proposed to explain how oxytocin and prostaglandins might cooperate to initiate labor in humans.

It has been demonstrated that the concentration of oxytocin receptors increases dramatically in the uterus during gestation. While the blood concentrations of oxytocin remain constant, the sensitivity of the uterus to oxytocin increases as a result of the increased receptors. The threshold amount of oxytocin required to induce uterine contractions in a woman at term is almost one-hundredth the amount required in a non-pregnant woman. Increased oxytocin receptors and sensitivity may result from the effects of the rising levels of estrogens that occur during gestation.

Table 21.11 Possible sequence of events leading to the onset of labor in humans.

Step	Event
1	High estrogen secretion from the placenta stimulates production of oxytocin receptors in the uterus
2	Uterine muscle (myometrium) becomes increasingly sensitive to effects of oxytocin during pregnancy
3	Oxtocin may stimulate production of prostaglandins in the uterus
4	Prostaglandins may stimulate uterine contractions
5	Contractions of the uterus stimulate oxytocin secretion from the posterior pituitary
6	Increased oxytocin secretion stimulates increased uterine contractions, creating a positive feedback loop and resulting in labor.

In addition to finding increased oxytocin receptors in the myometrium (muscular layer of the uterus), increased oxytocin receptors were also found in the nonmuscular decidua. It has been proposed that oxytocin (possibly coming from the fetus as well as from the mother) may stimulate prostaglandin production in the decidua. Prostaglandins diffusing into the myometrium from the decidua may then act together with the increased sensitivity of the myometrium to oxytocin and stimulate the onset of labor. These events are summarized in table 21.11.

Lactation

The changes that occur in a pregnant woman's mammary glands and the regulation of lactation provide excellent examples of hormonal interactions and neuroendocrine regulation. Growth and development of the mammary glands during pregnancy, for example, require the permissive actions (see chapter 19) of insulin, cortisol, and thyroid hormones. In the presence of adequate amounts of these hormones, high levels of estradiol stimulate development of the milk secreting units *(alveoli)* of these glands, and progesterone stimulates proliferation of the mammary ducts (see figure 21.54).

Production of milk proteins, such as casein and lactalbumin, is stimulated after parturition by **prolactin,** a hormone secreted by the anterior pituitary. Secretion of prolactin is controlled primarily by *prolactin-inhibiting hormone (PIH)* which is produced by the hypothalamus and secreted into the portal vessels. Secretion of PIH is stimulated by high levels of estrogen; during pregnancy, consequently, the high levels of estrogen prepare the breast for lactation but prevent prolactin secretion.

Figure 21.54 Structure of the mammary glands.

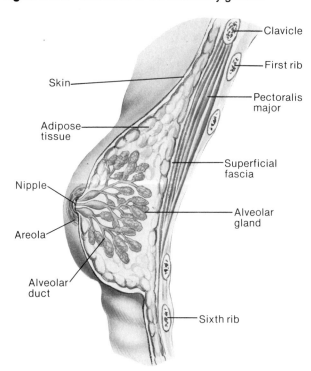

After parturition, when the placenta is eliminated, declining levels of estrogen are accompanied by increasing secretion of prolactin. Lactation therefore commences. If a woman does not wish to breast-feed her baby she may be injected with a powerful synthetic estrogen (diethylstilbestrol, or DES) which inhibits further prolactin secretion.

The act of nursing helps to maintain high levels of prolactin secretion via a *neuroendocrine reflex.* Sensory neurons in the breast, activated by the stimulus of suckling, are relayed to the hypothalamus. Continued secretion of high levels of prolactin results in secretion of milk from the alveoli into the ducts. In order for the baby to get the milk, however, action of another hormone is needed.

The stimulus of suckling also results in the reflex secretion of *oxytocin* from the posterior pituitary. This hormone is produced in the hypothalamus and stored in the posterior pituitary (as described in chapter 19); its secretion results in the **milk-ejection reflex.** Oxytocin stimulates contractions of the mammary ducts, as well as of the uterus (this is why women who breast-feed regain uterine muscle tone faster than women who bottle-feed). For reasons that are poorly understood, breast-feeding often (but not always) inhibits ovulation for the first two to three months—and sometimes for the first year or so—after parturition. This produces a natural spacing of births.

Figure 21.55 Hormonal control of mammary gland development during pregnancy and lactation. Note that milk production is prevented during pregnancy by estrogen inhibition of prolactin secretion. This inhibition is accomplished by stimulation of PIH (prolactin-inhibiting hormone) secretion from the hypothalamus.

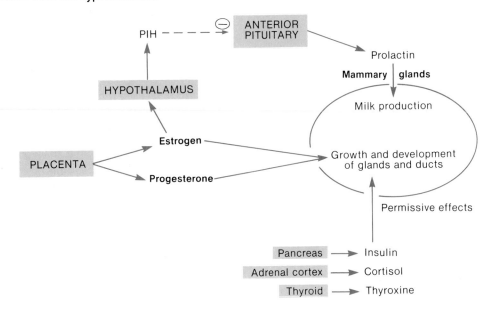

Figure 21.56 Lactation occurs in two stages: milk production (stimulated by prolactin) and milk ejection (stimulated by oxytocin). The stimulus of suckling triggers a neuroendocrine reflex that results in increased secretion of oxytocin and prolactin.

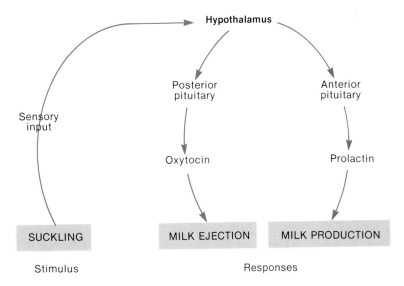

Table 21.12 Hormonal factors affecting lactation.

Hormones	Major Source	Effects
Insulin, cortisol, thyroid hormones	Pancreas, adrenal cortex, and thyroid	Permissive effects—adequate amounts of these must be present for other hormones to exert their effects on mammary glands
Estrogen and progesterone	Placenta	Growth and development of secretory units (alveoli) and ducts in mammary glands
Prolactin	Anterior pituitary	Production of milk proteins, including casein and lactalbumin
Oxytocin	Posterior pituitary	Stimulates milk-ejection reflex

Concluding Remarks

It may seem strange to end a textbook of physiology with the topics of pregnancy and parturition. This is done in part for practical reasons; these topics are complex, and require much prior knowledge of subjects covered earlier in the text. Also, it seems appropriate to end at the beginning, at the start of a new life. While generations of researchers have accumulated an impressive body of knowledge, the study of physiology is still young and rapidly growing. I hope that this introductory textbook will serve students' immediate practical needs, as a foundation for understanding current applications, and will provide a good basis for a lifetime of further study.

1. Describe the changes that occur in the sperm and ovum during fertilization.
2. Describe the origin of hCG, and explain why it is needed to maintain pregnancy for the first ten weeks.
3. Name the fetal and maternal components of the placenta, and describe the circulation in these two components. Explain how fetal and maternal gas exchange occurs.
4. List the names and major functions of the glycoprotein hormones and sex steroids secreted by the placenta.
5. Name the two factors that stimulate uterine contraction during labor, and describe the proposed mechanisms that may initiate labor in humans.
6. Describe the hormonal interactions required for breast development during pregnancy and for lactation after delivery.

Summary

Sex Determination and Puberty

I. Gametes contain only one of each pair of homologous chromosomes.
   A. Through reduction division, or meiosis, haploid gametes with twenty-three chromosomes are formed from diploid cells with twenty-three pairs of homologous chromosomes.
   B. All ova contain one X chromosome; half of the sperm contain X chromosomes and half contain Y chromosomes.
      1. A zygote (fertilized egg) with two X chromosomes becomes female.
      2. A zygote with the genotype XY becomes male.
II. If an embryo has a Y chromosome, the gonads become testes.
   A. This appears to be due to a particular cell-surface protein called the H-Y antigen.
   B. If the embryo is XX, an ovary develops instead of a testis.
   C. The chromosomal sex thus determines the type of gonads the embryo will have.

III. Embryonic testes secrete large amounts of the strong androgen, testosterone.
   A. Testosterone, and its intracellular derivatives (such as dihydrotestosterone) stimulate development of male sex accessory organs and male genitalia.
   B. Testes also secrete Müllerian inhibition factor, which prevents development of female sex accessory organs.
   C. In the absence of testosterone and the Müllerian inhibition factor, female sex accessory organs and external genitalia develop.
IV. Gonadal secretion of sex steroids increases during puberty.
   A. Testosterone from the testes of males and androgens from the adrenal cortex of females stimulates a growth spurt at puberty.

   B. Androgens stimulate increased muscle mass and other male secondary sexual characteristics; estrogens from the ovaries stimulate female secondary sexual characteristics.
   C. Increased secretion of sex steroids results from increased stimulation of the gonads by gonadotrophic hormones.
      1. Increased secretion of gonadotrophins may occur in part as a result of maturational changes in the central nervous system.
      2. The sensitivity of the hypothalamus and pituitary to negative feedback may decrease at puberty and thus contribute to the rise in gonadotrophin secretion.
   D. The pineal gland secretes melatonin, which may have an anti-gonad effect that delays puberty.

Male Reproductive System

I. Each testis consists of two parts or compartments.
  A. The seminiferous tubules produce sperm and contain FSH receptors.
  B. The interstitial tissue contains Leydig cells, which have LH receptors.
  C. LH stimulates the Leydig cells to secrete androgens; FSH stimulates the seminiferous tubules to produce sperm.
  D. Testosterone from the Leydig cells exerts negative feedback inhibition on LH secretion.
  E. A polypeptide called inhibin secreted from the tubules is believed to exert negative feedback inhibition on FSH secretion.

II. Spermatogenesis occurs by meiotic division of the cells that line the seminiferous tubules.
  A. Diploid spermatogonia divide by mitosis to replace themselves so that spermatogenesis can continue throughout life.
  B. When a spermatogonium undergoes meiotic division it becomes a diploid primary spermatocyte.
    1. At the end of the first meiotic division, two secondary spermatocytes are produced.
    2. At the end of the second meiotic division, four haploid spermatids are produced.
  C. Conversion of spermatids to spermatozoa is called spermiogenesis.
  D. Sertoli cells in the seminiferous tubules surround the germinal cells and appear necessary in the conversion of spermatids to spermatozoa.
    1. Sertoli cells phagocytose the cytoplasm of spermatids and appear to interact metabolically with the spermatids.
    2. Sertoli cells contain FSH receptors, so that stimulation of spermatogenesis by FSH must act via the Sertoli cells.
  E. The initiation of sperm production at puberty requires stimulation by both FSH and androgens.

The Menstrual Cycle

I. Menstruation refers to the shedding of the inner lining (endometrium) of the uterus during the first four days of the menstrual cycle.
  A. Shedding of the endometrium in a menstrual cycle is accompanied by bleeding—this is not true in lower mammals.
  B. Menstruation results from a decrease in the blood levels of ovarian steroid hormones.

II. During the follicular phase of the cycle (day five to thirteen), ovarian follicles develop under stimulation from FSH.
  A. Primordial follicles with diploid primary oocytes are converted to primary follicles that contain haploid secondary oocytes.
    1. Meiosis is arrested at metaphase II.
    2. Only one of the two cells formed at the end of the first meiotic division becomes a secondary oocyte—the other becomes a small polar body that eventually disintegrates.
  B. By day thirteen, one follicle—the Graafian follicle—has become a thin-walled blister on the ovary.
  C. As the follicle grows during the follicular phase, granulosa cells in the follicle secrete the strong estrogen known as estradiol—17β.
  D. Rising levels of estradiol at this time stimulate growth of the endometrium—this is called the proliferative phase of the endometrium.

III. Ovulation occurs at about the fourteenth day of the cycle.
  A. Towards the end of the follicular phase, the rising levels of estradiol in the blood stimulate the anterior pituitary to secrete LH.
    1. This is a positive feedback effect of estrogen on the pituitary.
    2. LH is secreted in a rapid "surge."
  B. Rupture of the Graafian follicle and extrusion of the ovum is caused by the LH surge.
    1. The LH surge requires the positive feedback effects of estradiol.
    2. The rise in estradiol occurs as a result of growth of the follicle; in this way the time of ovulation coincides with the full development of the follicle.
  C. The ovulated follicle is normally drawn into the fallopian tubes by the action of cilia.

IV. Under LH stimulation, the empty follicle becomes a corpus luteum.
  A. The luteal phase of the cycle lasts from day fourteen to day twenty-eight.
  B. The corpus luteum secretes progesterone as well as estradiol.
  C. Progesterone stimulates development of glands and vessels in the endometrium—this is the secretory phase of the endometrium.
  D. The high levels of estradiol and progesterone during the luteal phase of the cycle inhibit secretion of gonadotrophins from the pituitary.
    1. This prevents further growth and ovulation of follicles during that cycle and represents a negative feedback control mechanism.
    2. Estradiol thus has a positive feedback effect during the follicular phase and a negative feedback effect during the luteal phase.
  E. Lack of LH stimulation at the end of the luteal phase, together with the possible release of a luteolysin hormone from the uterus, causes the corpus luteum to regress at the end of the cycle.
    1. As the corpus luteum regresses, blood levels of estradiol and progesterone decline.
    2. As estradiol and progesterone decline, the developed endometrium is shed in menstruation.

Fertilization, Pregnancy, and Parturition

I. If the ovum (a secondary oocyte) becomes fertilized in the fallopian tubes, it completes meiotic development and becomes a mature ovum.
  A. Digestive enzymes in the acrosome cap of a sperm allow it to tunnel through the surrounding layers of granulosa cells and zona pellucida to the ovum.
  B. A diploid zygote is formed when the haploid numbers of chromosomes in the sperm and ovum join.

C. Mitotic division, or cleavage, then produces a ball of cells called a morula.

D. By the time the embryo enters the uterus it has reached the blastocyst stage of development.

1. The blastocyst consists of an inner cell mass that will become the placenta, and a surrounding layer of trophoblast cells that forms the chorionic membrane.

2. The trophoblast cells secrete human chorionic gonadotrophin (hCG), which maintains the mother's corpus luteum of the first ten weeks of pregnancy.

3. The corpus luteum thus continues to secrete ovarian steroids, which maintain the endometrium until the placenta secretes steroid hormones.

II. On the sixth day the blastocyst is implanted into the endometrium.

A. Implantation is complete by the eighth day, at which time the chorion consists of two layers.

1. The outer syncytiotrophoblast layer secretes digestive enzymes that create blood-filled cavities.

2. The inner cytotrophoblast layer invades these cavities to form a highly branched structure called the chorion frondosum.

B. The fetal chorion frondosum, together with associated maternal tissues (the decidua), form the placenta.

C. Fetal and maternal blood do not mix across the placenta; nutrients and wastes instead diffuse from one circulation to the other.

D. The placenta is very active metabolically.

1. The placenta, like the liver, can convert drugs and hormones to less active forms.

2. The placenta also secretes a wide variety of hormones.

E. The placenta secretes sex steroids and glycoprotein hormones that are similar to pituitary hormones.

III. Labor and birth (parturition) require the action of oxytocin from the posterior pituitary and prostaglandins produced in the uterus.

A. The mechanisms that initiate labor in humans are not well understood.

B. It is known, however, that there is an increase in oxytocin receptors in the uterus during gestation.

1. The uterus thus becomes increasingly sensitive to oxytocin.

2. An increase in sensitivity to oxytocin could also stimulate production of prostaglandins.

IV. Lactation requires the synergistic action of many hormones.

A. Insulin, cortisol, and estradiol act synergistically during pregnancy to stimulate mammary gland development.

B. Estradiol, however, also stimulates production of prolactin-inhibiting hormone in the hypothalamus.

1. PIH inhibits secretion of prolactin—an anterior pituitary hormone that stimulates milk production.

2. Milk production, therefore, cannot commence until after parturition, when estradiol levels have fallen.

C. A neuroendocrine reflex ejects milk from the nipple.

1. The stimulus of suckling results in secretion of oxytocin from the posterior pituitary gland.

2. Oxytocin stimulates contraction of the milk glands and ducts.

Self-Study Quiz

Multiple Choice
Match the following:

1. Menstrual phase
2. Follicular phase
3. Luteal phase
4. Ovulation

    (a) high estrogen and progesterone; low FSH and LH
    (b) low estrogen and progesterone
    (c) LH surge
    (d) increasing estrogen; low LH and low progesterone

5. A person with the genotype XXY has:
   (a) ovaries
   (b) testes
   (c) both ovaries and testes
   (d) neither ovaries nor testes
6. An embryo with the genotype XX develops female sex accessory organs because of:
   (a) androgens
   (b) estrogens
   (c) absence of androgens
   (d) absence of estrogens
7. In the male:
   (a) FSH is not secreted by the pituitary
   (b) FSH receptors are located in the Leydig cells
   (c) FSH receptors are located in the spermatogonia
   (d) FSH receptors are located in the Sertoli cells

8. Secretion of FSH in a male is inhibited by negative feedback effects of:
   (a) inhibin secreted from the tubules
   (b) inhibin secreted from the Leydig cells
   (c) testosterone secreted from the tubules
   (d) testosterone secreted from the Leydig cells
9. Which of the following statements is TRUE?
   (a) sperm are not motile until they pass through the epididymis
   (b) sperm require capacitation in the female genital tract before they can fertilize an ovum
   (c) an ovum does not complete meiotic division until it has been fertilized
   (d) all of these

10. The corpus luteum is maintained for the first ten weeks of pregnancy by:
    (a) hCG
    (b) LH
    (c) estrogen
    (d) progesterone
11. Fertilization normally occurs in:
    (a) the ovaries
    (b) the fallopian tubes
    (c) the uterus
    (d) the vagina
12. The placenta is formed from:
    (a) the fetal chorion frondosum
    (b) the maternal decidua basalis
    (c) both of these
    (d) neither of these
13. Uterine contractions are stimulated by:
    (a) oxytocin
    (b) prostaglandins
    (c) prolactin
    (d) both *(a)* and *(b)*;
    (d) both *(b)* and *(c)*
14. Contraction of the mammary glands and ducts during the milk ejection reflex is stimulated by:
    (a) prolactin
    (b) oxytocin
    (c) estrogen
    (d) progesterone

Appendixes

Appendix A
Answers to Self-Study Quizzes

Chapter 1

1. (d) 4. (b) 7. (b)
2. (d) 5. (a) 8. (b)
3. (b) 6. (c) 9. (a)

Chapter 2

1. (c) 5. (c) 9. (d)
2. (b) 6. (b) 10. (b)
3. (a) 7. (c) 11. (a)
4. (d) 8. (d) 12. (d)

Chapter 3

1. (d) 5. (d) 9. (a)
2. (b) 6. (b) 10. (b)
3. (a) 7. (a)
4. (c) 8. (c)

Chapter 4

1. (b) 4. (a) 7. (e)
2. (d) 5. (d) 8. (d)
3. (d) 6. (e)

Chapter 5

1. (b) 5. (d) 9. (a)
2. (a) 6. (c) 10. (d)
3. (c) 7. (a) 11. (b)
4. (e) 8. (c) 12. (d)

Chapter 6

1. (c) 5. (b) 9. (b)
2. (b) 6. (d) 10. (d)
3. (a) 7. (a)
4. (c) 8. (a)

Chapter 7

1. (c) 5. (d) 9. (b)
2. (b) 6. (d) 10. (c)
3. (a) 7. (e)
4. (c) 8. (c)

Chapter 8

1. (b) 5. (d) 9. (a)
2. (b) 6. (b) 10. (c)
3. (d) 7. (d) 11. (b)
4. (b) 8. (c)

Chapter 9

1. (d) 5. (c) 9. (b)
2. (a) 6. (c) 10. (b)
3. (c) 7. (d) 11. (d)
4. (b) 8. (d) 12. (c)

Chapter 10

1. (b) 5. (d) 9. (b)
2. (a) 6. (a) 10. (b)
3. (c) 7. (c) 11. (b)
4. (c) 8. (d) 12. (c)

Chapter 11

1. (c) 5. (b) 9. (d)
2. (b) 6. (c) 10. (b)
3. (e) 7. (c)
4. (a) 8. (a)

Chapter 12

1. (a) 5. (e) 9. (c)
2. (d) 6. (b) 10. (a)
3. (e) 7. (b) 11. (c)
4. (c) 8. (e) 12. (d)

Chapter 13

1. (d) 5. (d) 9. (e)
2. (b) 6. (c) 10. (a)
3. (a) 7. (e)
4. (c) 8. (b)

Chapter 14

1. (a) 5. (b) 9. (a)
2. (c) 6. (c) 10. (c)
3. (d) 7. (a)
4. (d) 8. (b)

Chapter 15

1. (c) 5. (c) 9. (d)
2. (b) 6. (a) 10. (b)
3. (a) 7. (c)
4. (e) 8. (b)

Chapter 16

1. (a) 5. (d) 9. (c)
2. (c) 6. (d) 10. (b)
3. (b) 7. (a) 11. (d)
4. (e) 8. (c) 12. (b)

Chapter 17

1. (b) 6. (d) 11. (b)
2. (d) 7. (d) 12. (d)
3. (a) 8. (b) 13. (a)
4. (c) 9. (e) 14. (d)
5. (c) 10. (a)

Chapter 18

1. (b) 5. (f) 9. (b)
2. (e) 6. (d) 10. (e)
3. (d) 7. (b)
4. (c) 8. (d)

Chapter 19

1. (e) 6. (b) 11. (c)
2. (a) 7. (a) 12. (a)
3. (b) 8. (c) 13. (b)
4. (d) 9. (d) 14. (b)
5. (c) 10. (d)

Chapter 20

1. (c) 6. (b) 11. (c)
2. (a) 7. (d) 12. (b)
3. (b) 8. (a) 13. (b)
4. (c) 9. (b)
5. (a) 10. (d)

Chapter 21

1. (b) 6. (c) 11. (b)
2. (d) 7. (d) 12. (c)
3. (a) 8. (a) 13. (d)
4. (c) 9. (d) 14. (b)
5. (b) 10. (a)

Appendix B
Supplementary Readings

A deeper understanding of topics covered in this introductory textbook can be gained by study of more *specialized texts* and by careful reading of *review articles* published in scientific and medical journals. These review articles summarize the results that are reported in recent research papers. Review articles are more current than textbooks, but generally assume that the reader has a thorough background in the biology and biochemistry of the field under review. Study of the pertinent chapters of this textbook should make most of the material in these articles accessible to the beginning student. A basic understanding of the experimental techniques used to study many of the topics covered in this textbook and in the review articles suggested can be gained by reading relevant articles in *Scientific American*. These articles are written at the level of the intelligent layman and therefore provide—together with this textbook—a good foundation for the study of more advanced review articles and texts.

Cell Physiology

Allfrey, V. G., and Mirsky, A. E. September 1961. How cells make molecules. *Scientific American.*

Brown, D. D. 1981. Gene expression in eurokaryotes. *Science* **211:** 667.

Chambon, P. May 1981. Split genes. *Scientific American.*

Crick, F. October 1962. The genetic code. *Scientific American.*

Crick, F. 1979. Split genes and RNA splicing. *Science* **204:** 264.

Davidson, E. H., and Britten, R. J. 1979. Regulation of gene expression: Possible role of repetitive sequences. *Science* **204:** 1052.

De Duve, C. May 1963. The lysosome. *Scientific American.*

De Duve, C. May 1983. Microbodies in the living cell. *Scientific American.*

Dustin, P. August 1980. Microtubules. *Scientific American.*

Dyson, R. D. 1978. *Essentials of cell biology.* 2d ed. Boston: Allyn & Bacon.

Fox, C. F. February 1972. The structure of cell membranes. *Scientific American.*

Hayflick, H. January 1980. The cell biology of human aging. *Scientific American.*

Hinkle, P., and McCarty, R. E. March 1978. How cells make ATP. *Scientific American.*

Holter, N. September 1961. How things get into cells. *Scientific American.*

Jackson, R. L., and Gotto, A. M. 1974. Phospholipids in biology and medicine. *New England Journal of Medicine* **290:** 24.

Lake, J. A. August 1981. The ribosome. *Scientific American.*

Lazarides, E., and Revel, J. P. May 1979. The molecular basis of cell movement. *Scientific American.*

Lehninger, A. L. September 1961. How cells transform energy. *Scientific American.*

Lodish, H. F., and Rothman, J. E. January 1979. The assembly of cell membranes. *Scientific American.*

McKusick, V. A. 1981. The anatomy of the human genome. *Hospital Practice* **16:** 82.

Mazia, D. September 1961. How cells divide. *Scientific American.*

Mazia, D. January 1974. The cell cycle. *Scientific American.*

Miller, O. L., Jr. March 1973. The visualization of genes in action. *Scientific American.*

Neutra, M., and Lelond, C. P. February 1969. The Golgi apparatus. *Scientific American.*

Nirenberg, M. W. March 1963. The genetic code II. *Scientific American.*

Nomura, M. October 1969. Ribosomes. *Scientific American.*

Palade, G. 1975. Intracellular aspects of the process of protein synthesis. *Science* **189:** 347.

Porter, K. R., and Tucker, J. B. March 1981. The ground substance of the living cell. *Scientific American.*

Prockop, D. J., and Guzman, N. A. 1977. Collagen disease and the biosynthesis of collagen. *Hospital Practice* **12:** 61.

Rothman, J. E. 1981. The Golgi apparatus: Two organelles in tandem. *Science* **213:** 1212.

Rothman, S. S. 1975. Protein transport by the pancreas. *Science* **190:** 747.

Schwartz, J. H. April 1980. The transport of substances in nerve cells. *Scientific American.*

Sharon, N. May 1974. Glycoproteins. *Scientific American.*

———. November 1980. Carbohydrates. *Scientific American.*

Sheeler, P., and Bianchi, D. E. 1980. *Cell biology: Structure, biochemistry, and function.* New York: John Wiley & Sons.

Singer, S. J., and Nicolson, G. L. 1972. The fluid mosaic model of the structure of cell membranes. *Science* **175:** 720.

Sloboda, R. D. 1980. The role of microtubules in cell structure and cell division. *American Scientist* **68:** 290.

Stein, G.; Stein, J. S.; and Kleinsmith, L. J. February 1975. Chromosomal proteins and gene regulation. *Scientific American.*

Nervous System

———. 1981. A *Nature* survey of the neurosciences. *Nature* **293:** 515.

Axelrod, J. June 1974. Neurotransmitters. *Scientific American.*

Barchas, J. D.; Akil, H.; Elliot, G. R.; Holman, R. B.; and Watson, S. J. 1978. Behavioral neurochemistry: Neuroregulators and behavioral states. *Science* **200:** 964.

Barde, Y. A.; Edgar, D.; and Thoen, H. 1983. New neurotrophic factors. *Annual Review of Physiology* **45:** 601.

Bartus, R. T.; Dean, R. L., III; Beer, B.; and Lippa, A. S. 1982. The cholinergic hypothesis of geriatric memory dysfunction. *Science* **217:** 408.

Bloom, F. E. May 1981. Neuropeptides. *Scientific American.*

Coyle, J. T.; Price, D. L.; and DeLong, M. R. 1983. Alzheimer's disease: A disorder of cortical cholinergic innervation. *Science* **219:** 1184.

Evarts, E. V. July 1973. Brain mechanisms in movement. *Scientific American.*

French, J. D. May 1957. The reticular formation. *Scientific American.*

Geschwind, N. April 1972. Language and the brain. *Scientific American.*

Gordon, B. December 1972. The superior colliculus of the brain. *Scientific American.*

Hoffman, B. B., and Lefkowitz, R. J. 1980. Alpha-adrenergic receptor subtypes. *New England Journal of Medicine* 302: 1390.

Hubel, D. H. September 1979. The brain. *Scientific American.*

Hubel, D. H. 1979. The visual cortex of normal and deprived monkeys. *American Scientist* 67: 532.

Hubel, D. H., and Wiesel, T. September 1979. Brain mechanisms of vision. *Scientific American.*

Hudspeth, A. J. February 1983. The hair cells of the inner ear. *Scientific American.*

Iverson, L. I. September 1979. The chemistry of the brain. *Scientific American.*

Jacobs, B. L., and Trulson, M. E. 1979. Mechanisms of action of LSD. *American Scientist* 67: 396.

Kandel, E., and Schwartz, J. H., eds. 1981. *Principles of neural science.* New York: Elsevier-North Holland Pub. Co.

Kandel, E. R. 1979. Psychotherapy and the single synapse. *New England Journal of Medicine* 301: 1028.

Kandel, E. R. September 1979. Small systems of neurons. *Scientific American.*

Katz, B. November 1952. The nerve impulse. *Scientific American.*

Kety, S. S. September 1979. Disorders of the human brain. *Scientific American.*

Keynes, R. D. March 1979. Ion channels in the nerve cell membrane. *Scientific American.*

Krieger, D. T., and Martin, J. B. 1981. Brain peptides. *New England Journal of Medicine* 304: first part, p. 876; second part, p. 944.

Kuffler, S. W., and Nicholls, J. G. 1976. *From neuron to brain: A cellular approach to the function of the nervous system.* Sunderland, Mass: Sinauer Associates, Inc. Publishers.

Lefkowitz, R. B. 1976. Beta-adrenergic receptors: Recognition and regulation. *New England Journal of Medicine* 295: 323.

Lester, H. A. February 1977. The response of acetylcholine. *Scientific American.*

Llinas, R. R. January 1975. The cortex of the cerebellum. *Scientific American.*

Luria, A. R. March 1970. The functional organization of the brain. *Scientific American.*

MacNichol, E. F., Jr. December 1964. Three-pigment color vision. *Scientific American.*

Morrell, P., and Norton, W. May 1980. Myelin. *Scientific American.*

Nathanson, J. A., and Greengard, P. August 1977. "Second messengers" in the brain. *Scientific American.*

Noback, C. E., and Demerest, R. J. 1975. *The human nervous system: Basic principles of neurobiology.* 2d ed. New York: McGraw-Hill Book Company.

O'Brien, D. F. 1982. The chemistry of vision. *Science* 218: 961.

Parker, D. E. November 1980. The vestibular apparatus. *Scientific American.*

Patterson, P. H., Potter, D., and Furshpan, E. J. July 1978. The chemical differentiation of nerve cells. *Scientific American.*

Purves, D., and Lichtman, J. W. 1983. Specific connections between nerve cells. *Annual Review of Physiology* 45: 553.

Routtenberg, A. November 1978. The reward system of the brain. *Scientific American.*

Rushton, W. A. H. March 1975. Visual pigments and color blindness. *Scientific American.*

Schwartz, J. H. April 1980. The transport of substances in nerve cells. *Scientific American.*

Snyder, S. H. March 1977. Opiate receptors and internal opiates. *Scientific American.*

Snyder, S. H. 1980. Brain peptides as neurotransmitters. *Science* 209: 976.

Sperry, R. W. January 1964. The great cerebral commissure. *Scientific American.*

Stevens, C. F. September 1979. The neuron. *Scientific American.*

Thai, L. J. March 1983. Neurotransmitters and receptors in neurologic diseases. *Medical Times.*

von Bekesy, G. August 1957. The ear. *Scientific American.*

Wurtman, R. J. April 1982. Nutrients that modify brain function. *Scientific American.*

Muscle Physiology

Bourne, G. H., ed. 1973. *The structure and function of muscle.* 2d ed. 4 vols. New York: Academic Press.

Cohen, C. November 1975. The protein switch of muscle contraction. *Scientific American.*

Drachman, D. B. 1978. Myasthenia gravis. *New England Journal of Medicine* 298: 136.

Eisenberg, E., and Greene, L. E. 1980. The relation of muscle biochemistry to muscle physiology. *Annual Review of Physiology* 42: 293.

Grinnell, A. D., and Brazier, M. A. B., eds. 1981. *Regulation of muscle contraction: Excitation-contraction coupling.* New York: Academic Press.

Hoyle, G. April 1970. How is muscle turned on and off? *Scientific American.*

Huxley, H. E. December 1965. The mechanism of muscular contraction. *Scientific American.*

Huxley, H. E. 1969. The mechanism of muscular contraction. *Science* 164: 1356.

Junge, D. 1981. *Nerve and muscle excitation.* 2d ed. Sunderland, Mass.: Sinauer Associates, Inc.

Margaria, R. March 1972. The sources of muscular energy. *Scientific American.*

Merton, P. A. May 1972. How we control the contraction of our muscles. *Scientific American.*

Murray, J. H., and Weber, A. February 1974. The cooperative action of muscle proteins. *Scientific American.*

Peterson, B. W. 1979. Reticulospinal projections to spinal motor nuclei. *Annual Review of Physiology* 41: 127.

Porter, K. R., and Franzini-Armstrong, C. March 1965. The sarcoplasmic reticulum. *Scientific American.*

Stefani, E., and Chiarandini, D. J. 1982. Ion channels in skeletal muscles. *Annual Review of Physiology* 44: 357.

Stein, R. B., ed. 1980. *Nerve and muscle: Membranes, cells, and systems.* New York: Plenum Publishing Corp.

Bishop, M. December 1982. Oncogenes. *Scientific American.*

Broder, S., and Waldmann, T. A. 1978. The suppressor cell network in cancer. *New England Journal of Medicine* **299:** first part, p. 1281; second part, p. 1335.

Buisseret, P. D. August 1982. Allergy. *Scientific American.*

Burke, D. C. March 1977. The status of interferon. *Scientific American.*

Burnet, F. M. 1960. *Immunology, aging, and cancer: Medical aspects of mutation and selection.* San Francisco: W. H. Freeman & Co.

Burnet, F. M. 1976. *Immunology: Readings from* Scientific American. San Francisco: W. H. Freeman & Co.

Butterworth, A. E., and David, J. R. 1981. Eosinophil function. *New England Journal of Medicine* **304:** 154.

Capra, J. D., and Edmunson, A. B. January 1977. The antibody combining site. *Scientific American.*

Cunningham, B. A. October 1977. The structure and function of histocompatibility antigens. *Scientific American.*

Dannenberg, A. M. 1975. Macrophages in inflammation and infection. *New England Journal of Medicine* **293:** 489.

Dausset, J. 1981. The major histocompatibility complex in man: Past, present, and future concepts. *Science* **213:** 1469.

Hamburger, R. N. 1976. Allergy and the immune system. *American Scientist* **64:** 157.

Herberman, R. B., and Ortaldo, J. R. 1981. Natural killer cells: Their role in defense against disease. *Science* **214:** 24.

Koffler, D. July 1980. Systemic lupus erythematosus. *Scientific American.*

Leder, P. November 1982. The genetics of antibody diversity. *Scientific American.*

Lerner, R. A. February 1983. Synthetic vaccines. *Scientific American.*

Lisak, R. P. 1983. Myasthenia gravis: Mechanisms and management. *Hospital Practice* **18:** 101.

Marsh, D. G.; Meyers, D. A.; and Bias, W. B. 1981. The epidemiology and genetics of atopic allergy. *New England Journal of Medicine* **305:** 1551.

Milstein, C. October 1980. Monoclonal antibodies. *Scientific American.*

Oettgen, H. F. 1981. Immunological aspects of cancer. *Hospital Practice* **16:** 93.

Old, L. J. May 1977. Cancer immunology. *Scientific American.*

Reif, A. E. 1981. The causes of cancer. *American Scientist* **69:** 437.

Riley, V. 1981. Psychoneuroendocrine influences on immunocompetence and neoplasia. *Science* **212:** 1100.

Roit, I. 1980. *Essential immunology.* Boston: Blackwell Scientific Publications.

Rose, N. R. February 1981. Autoimmune diseases. *Scientific American.*

Rowlands, D. T., and Danielle, R. P. 1975. Surface receptors in the immune response. *New England Journal of Medicine* **293:** 26.

Russel, P. S., and Cosimi, A. B. 1979. Transplantation. *New England Journal of Medicine* **301:** 470.

Samuelsson, B. 1983. Leukotrines: Mediators of immediate hypersensitivity and inflammation. *Science* **220:** 568.

Schwartz, R. S. 1981. Therapeutic uses of immune suppression and enhancement. *Hospital Practice* **16:** 93.

Sell, S. 1980. *Immunology, immunopathology, and immunity.* 3d ed. New York: Harper & Row.

Stewart, E. E. 1981. Clinical status of the interferons: Will their promise be kept? *Hospital Practice* **16:** 212.

Stobo, J. D. 1981. Autoimmune antireceptor disease. *Hospital Practice* **16:** 49.

Strakosch, C. R.; Wenzel, B. E.; Row, V. V.; and Volpe, R. 1982. Immunology of autoimmune thyroid diseases. *New England Journal of Medicine* **307:** 1499.

Tannock, I. F. 1983. Biology of tumor growth. *Hospital Practice* **18:** 81.

Vitetta, E. S., and Uhr, J. W. 1975. Immunoglobin receptors revisited. *Science* **189:** 964.

Weissmann, G. 1978. Leukocytes as secretory organs of inflammation. *Hospital Practice* **13:** 53.

Weksler, M. E. The senescence of the immune system. *Hospital Practice* **16:** 53.

Yelton, D. E., and Scharff, M. D. 1980. Monoclonal antibodies. *American Scientist* **63:** 510.

Zinkernagel, R. M. 1978. Major transplantation antigens in host responses to infection. *Hospital Practice* **13:** 83.

Digestive System

Bleich, H. L., and Boro, E. S. 1979. Protein digestion and absorption. *New England Journal of Medicine* **300:** 659.

Carey, M. C.; Small, D. M.; and Bliss, C. M. 1983. Lipid digestion and absorption. *Annual Review of Physiology* **45:** 651.

Chou, C. C. 1982. Relationship between intestinal blood flow and motility. *Annual Review of Physiology* **44:** 29.

Cohen, S. 1983. Neuromuscular disorders of the gastrointestinal tract. *Hospital Practice* **18:** 121.

Davenport, H. W. January 1972. Why the stomach does not digest itself. *Scientific American.*

Davenport, H. W. 1982. *Physiology of the digestive tract.* 5th ed. Chicago: Year Book Medical Publishers.

Dockray, G. J. 1979. Comparative biochemistry and physiology of gut hormones. *Annual Review of Physiology* **41:** 83.

Forte, J. G.; Machen, T. E.; and Obrink, K. J. 1980. Mechanisms of gastric H^+ and Cl^- transport. *Annual Review of Physiology* **42:** 111.

Freeman, H. J., and Kim, Y. S. 1978. Digestion and absorption of proteins. *Annual Review of Medicine* **29:** 99.

Gray, G. M. 1975. Carbohydrate digestion and absorption: Role of the small intestine. *New England Journal of Medicine* **292:** 1225.

Grossman, M. I. 1979. Neural and hormonal regulation of gastrointestinal function: An overview. *Annual Review of Physiology* **41:** 27.

Guth, P. H. 1982. Stomach blood flow and acid secretion. *Annual Review of Physiology* **44:** 3.

Johnson, L. R. 1977. Gastrointestinal hormones and their functions. *Annual Review of Physiology* **39:** 135.

Kappas, A., and Alvarez, A. P. June 1975. How the liver metabolizes foreign substances. *Scientific American.*

Kretchmer, N. October 1972. Lactose and lactase. *Scientific American.*

McGuigan, J. E. 1978. Gastrointestinal hormones. *Annual Review of Medicine* **29:** 99.

MacKowiak, P. A. 1982. The normal microbial flora. *New England Journal of Medicine* **307:** 83.

Moog, F. November 1981. The lining of the small intestine. *Scientific American.*

Richardon, I., and Withrington, P. G. 1982. Physiological regulation of the hepatic circulation. *Annual Review of Physiology* **44:** 57.

Rose, R. C. 1980. Water-soluble vitamin absorption. *Annual Review of Physiology* **42:** 157.

Salen, G., and Shefer, S. 1983. Bile acid synthesis. *Annual Review of Physiology* **45:** 679.

Schulz, I., and Stolze, H. H. 1980. The exocrine pancreas: The role of secretogogues, cyclic nucleotides, and calcium in enzyme secretion. *Annual Review of Physiology* **42:** 127.

Soll, A., and Walsh, J. H. 1979. Regulation of gastric acid secretion. *Annual Review of Physiology* **41:** 35.

Walsh, J. H., and Grossman, M. I. 1975. Gastrin. *New England Journal of Medicine* **292:** first part, p. 1324; second part, p. 1377.

Weems, W. A. 1981. The intestine as a fluid-propelling system. *Annual Review of Physiology* **43:** 9.

Weisbrodt, N. W. 1981. Patterns of intestinal motility. *Annual Review of Physiology* **43:** 21.

Williamson, R. C. 1978. Intestinal adaptation. *New England Journal of Medicine* **298:** first part, p. 1393; second part, p. 1444.

Winick, M., ed. 1980. *Nutrition and gastroenterology.* New York: John Wiley & Sons.

Wood, J. D. 1981. Intrinsic neural control of intestinal motility. *Annual Review of Physiology* **43:** 33.

Endocrinology and Metabolism

Austin, L. A., and Heath, H., III. 1981. Calcitonin: Physiology and pathophysiology. *New England Journal of Medicine* **304:** 269.

Baxter, J. D., and Funder, J. W. 1979. Hormone receptors. *New England Journal of Medicine* **300:** 117.

Cahill, G. F., and McDevitt, H. O. 1981. Insulin-dependent diabetes mellitus: The initial lesion. *New England Journal of Medicine* **304:** 454.

Catt, K. J., and Dufau, M. L. 1976. Basic concepts of the mechanism of action of peptide hormones. *Biology of Reproduction* **14:** 1.

DeLuca, H. F. 1980. The vitamin D hormonal system: Implications for bone disease. *Hospital Practice* **15:** 57.

Edelman, I. 1974. Thyroid thermogenesis. *New England Journal of Medicine* **290:** 1303.

Flier, J. S.; Kahn, C. R.; and Roth, R. 1979. Receptors, antireceptor antibodies, and mechanisms of insulin resistance. *New England Journal of Medicine* **300:** 413.

Ganong, W. F.; Alpert, L. C.; and Lee, T. C. 1974. ACTH and the regulation of adrenocortical secretion. *New England Journal of Medicine* **290:** 1006.

Gardner, L. I. July 1972. Deprivation dwarfism. *Scientific American.*

Guillemin, R., and Burgus, R. November 1972. The hormones of the hypothalamus. *Scientific American.*

Habener, J. F., and Mahaffey, J. E. 1978. Osteomalacia and disorders of vitamin D metabolism. *Annual Review of Medicine* **29:** 327.

Haussler, M. R., and McCain, T. A. 1977. Basic and clinical concepts related to vitamin D metabolism and action. *New England Journal of Medicine* **297:** first part, p. 974; second part, p. 1041.

Jacobs, S., and Cuatrecasas, P. 1977. Cell receptors in disease. *New England Journal of Medicine* **297:** 1383.

Katzenellenbogen, B. S. 1980. Dynamics of steroid hormone receptor action. *Annual Review of Physiology* **42:** 17.

McEwen, B. S. July 1976. Interactions between hormones and nerve tissue. *Scientific American.*

Morley, J. E. 1983. The aging endocrine system. *Postgraduate Medicine* **73:** 107.

Notkins, A. L. November 1979. The causes of diabetes. *Scientific American.*

O'Malley, B., and Schrader, W. T. February 1976. The receptors of steroid hormones. *Scientific American.*

Oppenheimer, J. H. 1979. Thyroid hormone action at the cellular level. *Science* **203:** 971.

Pastan, R. August 1972. Cyclic AMP. *Scientific American.*

Pfaff, D. W., and McEwen, B. S. 1983. Actions of estrogens and progestogens on nerve cells. *Science* **219:** 808.

Phillips, L. S., and Vassilopoulou-Sellin, R. 1980. Somatomedins. *New England Journal of Medicine* **302:** first part, p. 371; second part, p. 438.

Pike, J. E. November 1971. Prostaglandins. *Scientific American.*

Raisz, L. G., and Kream, B. E. 1981. Hormonal control of skeletal growth. *Annual Review of Physiology* **43:** 225.

Rivlin, R. S. 1975. Therapy of obesity with hormones. *New England Journal of Medicine* **292:** 26.

Roth, J. 1980. Insulin receptors in diabetes. *Hospital Practice* **15:** 98.

Roth, J., and Taylor, S. I. 1982. Receptors for peptide hormones: Alterations in diseases of humans. *Annual Review of Physiology* **44:** 639.

Schally, A. V. 1978. Aspects of the hypothalamic regulation of the pituitary gland. *Science* **202:** 18.

Selye, H. 1973. The evolution of the stress concept. *American Scientist* **61:** 692.

Sterling, S. 1979. Thyroid hormone action at the cell level. *New England Journal of Medicine* **300:** first part, p. 117; second part, p. 173.

Tata, J. R. 1975. Hormonal regulation of hormonal receptors. *Nature* **257:** 740.

Tepperman, J. 1980. *Metabolic and endocrine physiology.* 4th ed. Chicago: Year Book Medical Publishers.

Unger, R. H.; Dobbs, R. E.; and Orci, L. 1979. Insulin, glucagon, and somatostatin secretion in the regulation of metabolism. *Annual Review of Physiology* **40:** 307.

Unger, R. H., and Orci, L. 1977. The role of glucagon in the endogenous hyperglycemia of diabetes mellitus. *Annual Review of Medicine* **28:** 119.

Unger, R. H., and Orci, L. 1981. Glucagon and the A cell: Physiology and pathophysiology. *New England Journal of Medicine* **304:** first part, p. 1518; second part, p. 1575.

Van Wyk, J., and Underwood, L. E. 1978. Growth hormone, somatomedins, and growth failure. *Hospital Practice* **13:** 57.

Villee, D. B. 1975. *Human endocrinology: A developmental approach.* Philadelphia: W. B. Saunders Co.

Reproduction

Axelrod, J. 1974. The pineal gland: A neuroendocrine transducer. *Science* **184:** 1341.

Bardin, C. W. 1979. The neuroendocrinology of male reproduction. *Hospital Practice* **14:** 65.

Bartke, A.; Hafiez, A. A; Bex, F. J.; and Dalterio, S. 1978. Hormonal interaction in the regulation of androgen secretion. *Biology of Reproduction* **18:** 44.

Beaconsfield, P.; Birdwood, G.; and Beaconsfield, R. July 1980. The placenta. *Scientific American.*

Biggers, J. D. 1981. *In vitro* fertilization and embryo transfer in human beings. *New England Journal of Medicine* **304:** 336.

Binkley, S. April 1979. A timekeeping enzyme in the pineal gland. *Scientific American.*

Boyar, R. M. 1978. Control of the onset of puberty. *Annual Review of Medicine* **29:** 509.

Droegemueller, W., and Bressler, R. 1980. Effectiveness and risks of contraception. *Annual Review of Medicine* **31:** 329.

Everett, R. B., and MacDonald, P. C. 1979. Endocrinology of the placenta. *Annual Review of Medicine* **30:** 473.

Fink, G. 1979. Feedback action of target hormones on hypothalamus and pituitary with special reference to gonadal steroids. *Annual Review of Physiology* **41:** 571.

Frantz, A. G. 1978. Prolactin. *New England Journal of Medicine* **298:** 112.

Grobstein, C. March 1979. External human fertilization. *Scientific American.*

Grumbach, M. M. 1979. The neuroendocrinology of puberty. *Hospital Practice* **14:** 65.

Lipsett, M. B. 1980. Physiology and pathology of the Leydig cell. *New England Journal of Medicine* **303:** 682.

McGuire, W. L. 1980. Steroid receptors and breast cancer. *Hospital Practice* **15:** 83.

McKerns, K. W., ed. 1969. *The gonads.* New York: Appleton-Century-Crofts.

Means, A. R.; Dedman, J. R.; Tash, J. S.; Tindall, D. J.; van Sickle, M.; and Walsh, M. J. 1980. Regulation of the testis Sertoli cell by follicle-stimulating hormone. *Annual Review of Physiology* **42:** 59.

Naftolin, F., and Butz, E., eds. 1981. Sexual dimorphism. *Science* **211:** entire issue.

Ramaley, J. A. 1979. Development of gonadotropin regulation in the prepubertal mammal. *Biology of Reproduction* **20:** 1.

Reiter, E. O., and Grumbach, M. M. 1982. Neuroendocrine control mechanisms and the onset of puberty. *Annual Review of Physiology* **44:** 595.

Riddick, D. H. 1982. All about the human breast. *Contemporary OB/GYN* **19:** 101.

Simpson, E. R., and MacDonald, P. C. 1981. Endocrine physiology of the placenta. *Annual Review of Physiology* **43:** 163.

Stadel, B. V. 1981. Oral contraceptives and cardiovascular disease. *New England Journal of Medicine* **305:** 612.

Steinberger, E. 1971. Hormonal control of mammalian spermatogenesis. *Physiological Reviews* **51:** 1.

Troen, P., and Nankin, H. R., eds. 1977. *The testes in normal and infertile men.* New York: Raven Press.

Wachtel, S. S. 1977. H–Y antigen and the genetics of sex determination. *Science* **198:** 797.

Wilson, J. D. 1978. Sexual differentiation. *Annual Review of Physiology* **40:** 279.

Wurtman, R. J. 1980. The pineal as a neuroendocrine transducer. *Hospital Practice* **15:** 82.

Wurtman, R. J., and Moskowitz, M. A. 1977. The pineal organ. *New England Journal of Medicine* **296:** first part, p. 1329; second part, p. 1383.

Yen, S. S. C. 1977. Regulation of the hypothalamic-pituitary-ovarian axis in women. *Journal of Reproduction and Fertility* **51:** 181.

Yen, S. S. C. 1979. Neuroendocrine regulation of the menstrual cycle. *Hospital Practice* **14:** 83.

Glossary

a-, an- (Gr.) Not, without, lacking.

ab- (La.) Off, away from.

abdomen Body cavity between the thorax and pelvis.

abductor A muscle that moves the skeleton away from the median plane of the body or away from the axial line of a limb.

ABO system The most common system of classification for red blood cell antigens. On the basis of antigens on the red blood cell surface, individuals can be type A, type B, type AB, or type O.

absorption The transport of molecules across epithelial membranes into the body fluids.

accommodation To fit or adjust; specifically, the ability of the eyes to adjust their curvature so that an image of an object is focused on the retina at different distances.

acetylcholine A molecule—which is an acetic acid ester of choline—that functions as a neurotransmitter chemical at somatic motor nerve and parasympathetic nerve fibers.

acetylcholinesterase An enzyme in the membrane of postsynaptic cells that catalyzes the conversion of ACh into choline and acetic acid. This enzymatic reaction inactivates the neurotransmitter.

acidosis An abnormal increase in the H^+ concentration of the blood that lowers arterial pH below 7.35.

acromegaly A condition caused by hypersecretion of growth hormone from the pituitary after maturity and characterized by enlargement of the extremities, such as the nose, jaws, fingers, and toes.

actin A structural protein of muscle that, along with myosin, is responsible for muscle contraction.

action potential An all-or-none electrical event in an axon or muscle fiber in which the polarity of the membrane potential is rapidly reversed and reestablished.

active immunity Immunity involving sensitization, in which antibody production is stimulated by prior exposure to an antigen.

active transport Movement of molecules or ions across the cell membranes of epithelial cells by membrane carriers; expenditure of cellular energy (ATP) is required.

ad- (La.) Towards, next to.

adductor Muscle that moves the skeleton towards the midline of the body or towards the axial plane of a limb.

adenyl cyclase Enzyme found in cell membranes that catalyzes the conversion of ATP to cyclic AMP and pyrophosphate (PP_1). This enzyme is activated by interaction between a specific hormone and its membrane receptor protein.

ADH Antidiuretic hormone, also known as vasopressin. A hormone produced by the hypothalamus and secreted by the posterior pituitary gland, it acts on the kidneys to promote water reabsorption, thus decreasing the urine volume.

adipose tissue Fatty tissue; a type of connective tissue consisting of fat cells in a loose connective tissue matrix.

ADP Adenosine diphosphate; a molecule that, together with inorganic phosphate, is used to make ATP (adenosine triphosphate).

adrenal cortex The outer part of the adrenal gland. Derived from embryonic mesoderm, the adrenal cortex secretes corticosteroid hormones (such as aldosterone and hydrocortisone).

adrenal medulla The inner part of the adrenal gland. Derived from embryonic postganglionic sympathetic neurons, the adrenal medulla secretes catecholamine hormones—epinephrine and (to a lesser degree) norepinephrine.

adrenergic Adjective describing the actions of epinephrine, norepinephrine, or other molecules with similar activity (as in adrenergic receptor and adrenergic stimulation).

aerobic capacity The ability of an organ to utilize oxygen, and respire aerobically, to meet its energy needs.

agglutinate Clumping of cells (usually erythrocytes) due to specific chemical interaction between surface antigens and antibodies.

agranular leukocytes White blood cells (leukocytes) that do not contain cytoplasmic granules; specifically, lymphocytes and monocytes.

albumin A water-soluble protein that is the major component of the plasma proteins; produced in the liver.

aldosterone The principal corticosteroid hormone involved in regulation of electrolyte balance (mineralocorticoid).

allergens Antigens that evoke an allergic response rather than a normal immune response.

allergy A state of hypersensitivity caused by exposure to allergens; results in the liberation of histamine and other molecules with histamine-like effects.

allosteric The alteration of an enzyme's activity by combination with a regulator molecule; allosteric inhibition by an end product represents negative feedback control of an enzyme's activity.

alveoli Plural of alveolus; anatomical term for small, saclike dilations (as in lung alveoli).

amniocentesis Procedure to obtain amniotic fluid and fetal cells in this fluid through transabdominal perforation of the uterus.

amnion The inner fetal membrane that contains the fetus in amniotic fluid; commonly called the "bag of waters."

amphoteric Pertaining to having opposite characteristics; a molecule that can be positively or negatively charged, depending on the pH of its environment.

an- (Gr.) Without, not.

anabolic steroids Steroids with androgen-like stimulatory effects on protein synthesis.

anabolism Chemical reactions within cells that result in the production of larger molecules from smaller ones; specifically, the synthesis of protein, glycogen, and fat.

anaerobic respiration Form of cell respiration involving the conversion of glucose to lactic acid in which energy is obtained without the use of molecular oxygen.

anaphylaxis Unusually severe allergic reaction, which can result in cardiovascular shock and death.

androgens Steroids, containing eighteen carbons, that have masculinizing effects; primarily those hormones (such as testosterone) secreted by the testes, although weaker androgens are also secreted by the adrenal cortex.

anemia An abnormal reduction in the red blood cell count, hemoglobin concentration, or hematocrit, or any combination of these measurements. This condition is associated with a decreased ability of the blood to carry oxygen.

angina pectoris   A thoracic pain, often referred to the left pectoral and arm area, caused by myocardial ischemia.

angiotensin II   An eight-amino acid polypeptide formed from angiotensin I (a ten-amino acid precursor), which in turn is formed by cleavage of a protein (angiotensinogen) by the action of renin, an enzyme secreted by the kidneys. Angiotensin II is a powerful vasoconstrictor and a stimulator of aldosterone secretion from the adrenal cortex.

anions   Ions that are negatively charged, such as chloride, bicarbonate, and phosphate.

anterior   Anatomical term referring to a forward position of a structure.

anterior pituitary   Also called the adenohypophysis; secretes FSH (follicle-stimulating hormone), LH (luteinizing hormone), ACTH (adrenocorticotrophic hormone), TSH (thyroid-stimulating hormone), GH (growth hormone), and prolactin. Secretions of the anterior pituitary are controlled by hormones secreted by the hypothalamus.

antibodies   Immunoglobin proteins secreted by B lymphocytes that have transformed into plasma cells. Antibodies are responsible for humoral immunity. Their synthesis is induced by specific antigens, and they combine with these specific antigens but not with unrelated antigens.

anticodon   A base triplet provided by three nucleotides within a loop of transfer RNA, which is complementary in its base pairing properties to a triplet (the codon) in mRNA; matching of codon to anticodon provides the mechanism for translation of the genetic code into a specific sequence of amino acids.

antigen   A molecule able to induce the production of antibodies and able to react in a specific manner with antibodies.

antigenic determinant site   The region of an antigen molecule that specifically reacts with particular antibodies. A large antigen molecule may have a number of such sites.

antiserum   A serum that contains specific antibodies.

apneustic center   A collection of neurons in the brain stem that participates in the rhythmic control of breathing.

aqueous humor   A fluid produced by the ciliary body that fills the anterior and posterior chambers of the eye.

arteriosclerosis   A group of diseases characterized by thickening and hardening of the artery wall and in the narrowing of its lumen.

arteriovenous anastomoses   Direct connections between arteries and veins that bypass capillary beds.

artery   A vessel that carries blood away from the heart.

astigmatism   Unequal curvature of the refractive surfaces of the eye (cornea and/or lens), so that light that enters the eye along certain meridians does not focus on the retina.

atherosclerosis   A common type of arteriosclerosis found in medium and large arteries in which raised areas, or plaque, within the tunica intima are formed from smooth muscle cells, cholesterol, and other lipids. These plaques occlude arteries and serve as sites for the formation of thrombi.

atomic number   A whole number representing the number of positively charged protons in the nucleus of an atom.

atopic dermatitis   Allergic skin reaction to agents such as poison ivy and poison oak; a type of delayed hypersensitivity.

ATP   Adenosine triphosphate; the universal energy donor of the cell.

atretic   Without an opening; atretic ovarian follicles are those that fail to ovulate.

atropine   An alkaloid drug obtained from a plant of the species *Belladonna* that acts as an anticholinergic agent. Used medically to inhibit parasympathetic nerve effects and dilate the pupils of the eye, increase the heart rate, and inhibit movements of the intestine.

auto-   (Gr.) Self, same.

autoantibodies   Antibodies that are formed in response to, and which react with, molecules that are part of one's own body.

autosomal chromosomes   The paired chromosomes; those other than the sex chromosomes.

axon   The process of a nerve cell that conducts impulses away from the cell body.

Barr body   Microscopic structure in the cell nucleus produced from an inactive X chromosome in females.

basal ganglia   Gray matter, or nuclei, within the cerebral hemispheres, comprising the corpus striatum, amygdaloid nucleus, and claustrum.

basal metabolic rate (BMR)   The rate of metabolism (expressed as oxygen consumption or heat production) under resting or basal conditions fourteen to eighteen hours after eating.

basophil   The rarest type of leukocyte; a granular leukocyte with an affinity for blue stain in the standard staining procedure.

B cell lymphocyte   Lymphocytes that can be transformed by antigens into plasma cells that secrete antibodies (and are thus responsible for humoral immunity). The *B* stands for *bursa equivalent*.

benign   Not malignant or life threatening.

bi-   (La.) Two, twice.

bile   Fluid produced by the liver and stored in the gallbladder that contains bile salts, bile pigments, cholesterol, and other molecules. The bile is secreted into the small intestine.

bilirubin   Bile pigment derived from the breakdown of the heme portion of hemoglobin.

blocking antibodies   An antibody that is specific for antigens attacked by other antibodies or by T cells, and which, therefore, may block this attack.

blood-brain barrier   The structures and cells that selectively prevent particular molecules in the plasma from entering the central nervous system.

bradycardia   A slow cardiac rate; less than sixty beats per minute.

bradykinins   Short polypeptides that stimulate vasodilation and other cardiovascular changes.

bronchioles   The smallest air passages in the lungs, which contain smooth muscle and cuboidal epithelial cells.

buffer   A molecule that serves to prevent large changes in pH by either combining with H^+ or by releasing H^+ into solution.

bundle of His   A band of rapidly conducting cardiac fibers that originate in the A-V node and extend down the atrioventricular septum to the apex of the heart. This tissue conducts action potentials from the atria into the ventricles.

calcitonin   Also called thyrocalcitonin. A polypeptide hormone produced by the parafollicular cells of the thyroid and secreted in response to hypercalcemia. Acts to lower blood calcium and phosphate concentrations and may serve as an antagonist of parathyroid hormone.

calmodulin   A receptor protein for Ca^{++} located within the cytoplasm of target cells. Appears to mediate the effects of this ion on cellular activities.

calorie   A unit of heat equal to the amount of heat needed to raise the temperature of one gram of water by 1°C.

cAMP   Cyclic adenosine monophosphate; a second messenger in the action of many hormones, such as catecholamines, polypeptides, and glycoprotein hormones. Serves to mediate the effects of these hormones on their target cells.

cancer   A tumor characterized by abnormally rapid cell division and loss of specialized tissue characteristics. Usually refers to malignant tumors.

capacitation   Changes that occur within spermatozoa in the female reproductive tract that enable the sperm to fertilize ova; sperm that have not been capacitated in the female tract cannot fertilize ova.

capillary   The smallest vessels in the vascular system. Capillary walls are only one cell thick, and all exchanges of molecules between the blood and tissue fluid occur across the capillary wall.

carbonic anhydrase   An enzyme that catalyzes the formation or breakdown of carbonic acid. When carbon dioxide concentrations are relatively high, this enzyme catalyzes the formation of carbonic acid from CO_2 and H_2O. When carbon dioxide concentrations are low, the breakdown of carbonic acid to CO_2 and H_2O is catalyzed. These reactions aid the transport of carbon dioxide from tissues to alveolar air.

cardiac muscle   Muscle of the heart, consisting of striated muscle cells. These cells are interconnected into a mass called the myocardium.

cardiac output   The volume of blood pumped by either right or left ventricle per minute.

cardiogenic shock   Shock that results from low cardiac output in heart disease.

carrier-mediated transport   The transport of molecules or ions across a cell membrane by means of specific protein carriers. Includes both facilitated diffusion and active transport.

casts   Accumulation of proteins that produce molds of kidney tubules and that appear in urine sediment.

catabolism   Chemical reactions in a cell whereby larger, more complex molecules are converted into smaller molecules.

catecholamines   Group of molecules including epinephrine, norepinephrine, L-dopa, and related molecules that have effects that are similar to those produced by activation of the sympathetic nervous system.

cations   Positively charged ions, such as sodium, potassium calcium, and magnesium.

cell-mediated immunity   Immunological defense provided by T cell lymphocytes, which come into close proximity to their victim cells (as opposed to humoral immunity provided by secretion of antibodies by plasma cells).

cellular respiration   The energy-releasing metabolic pathways in a cell that oxidize organic molecules such as glucose, fatty acids, and others.

centri-   (La.) Center.

centrioles   Cell organelles that form the spindle apparatus during cell division.

centromere   The central region of a chromosome to which the chromosomal arms are attached.

chemoreceptors   Neural receptors sensitive to chemical changes in blood and other body fluids.

chemotaxis   Movement of an organism or a cell, such as a leukocyte, towards a chemical stimulus.

Cheyne-Stokes respiration   Breathing characterized by rhythmic waxing and waning of the depth of respiration, with regularly occurring periods of apnea (lack of breathing).

cholesterol   A twenty-seven-carbon steroid that serves as the precursor of steroid hormones.

cholinergic   Nerve endings that liberate acetylcholine as a neurotransmitter, such as those of the parasympathetic system.

chondrocytes   Cartilage-forming cells.

chorea   Occurrence of a wide variety of rapid, complex, jerky movements that appear to be well coordinated but that are performed involuntarily.

chromatids   Duplicated chromosomes that are joined together at the centromere and that separate during cell division.

chromatin   Threadlike structures in the cell nucleus consisting primarily of DNA and protein. Represents the extended form of chromosomes during interphase.

chyme   Mixture of partially digested food and digestive juices within the stomach and small intestine.

cilia   Plural of cilium; tiny hairlike processes that extend from the cell surface and beat in a coordinated fashion.

circadian rhythms   Physiological changes that repeat at about a twenty-four-hour period, and which are often synchronized to changes in the external environment, such as the day-night cycles.

cirrhosis   Liver disease characterized by loss of normal microscopic structure replaced by fibrosis and nodular regeneration.

clone   A group of cells derived from a single parent cell by mitotic cell division; since reproduction is asexual, the descendants of the parent cell are genetically identical. Term used when cells are separate individuals (as in white blood cells) rather than part of a growing organ.

CNS   Central nervous system; part of the nervous system consisting of the brain and spinal cord.

cochlea   The organ of hearing in the inner ear where nerve impulses are generated in response to sound waves.

codon   Sequence of three nucleotide bases in mRNA that specifies a given amino acid, and which determines the position of that amino acid in a polypeptide chain through complementary base pairing with an anticodon in transfer RNA.

coenzyme   An organic molecule, usually derived from a water-soluble vitamin, that combines with and activates specific enzyme proteins.

cofactor   A substance needed for the catalytic action of an enzyme; usually refers to inorganic ions such as Ca^{++} and Mg^{++}.

colloid osmotic pressure   Osmotic pressure exerted by plasma proteins, which are present as a colloidal suspension. Also called oncotic pressure.

com-, con-   (La.) With, together.

compliance   A measure of the ease with which a structure such as the lungs expands under pressure; a measure of the change in volume as a function of pressure changes.

cones   Photoreceptors in the retina of the eye that provide color vision and high visual acuity.

congestive heart failure   Inability of the heart to deliver an adequate blood flow, due to heart disease or hypertension. Associated with breathlessness, salt and water retention, and edema.

conjunctivitis   Inflammation of the conjunctiva of the eye; sometimes called "pink eye."

connective tissue   One of the four primary tissues, characterized by an abundance of extracellular material.

Conn's syndrome   Primary hyperaldosteronism; excessive secretion of aldosterone produces electrolyte imbalances.

contralateral   Affecting the opposite side of the body.

cornea   The transparent structure forming the anterior part of the connective tissue covering of the eye.

corpus callosum   A large, transverse tract of nerve fibers connecting the cerebral hemispheres.

cortex   The outer covering or layer of an organ.

corticosteroids   Steroid hormones of the adrenal cortex, consisting of glucocorticoids (such as hydrocortisone) and mineralocorticoids (such as aldosterone).

creatine phosphate   Organic phosphate molecule in muscle cells that serves as a source of high-energy phosphate for the synthesis of ATP. Also called phosphocreatine.

crenation   A notched or scalloped appearance of the red blood cell membrane caused by the osmotic loss of water from these cells.

cretinism   A condition caused by insufficient thyroid secretion during prenatal development or the years of early childhood. Results in stunted growth and inadequate mental development.

crypt-   (Gr.) Hidden, concealed.

cryptorchidism   A developmental defect in which the testes fail to descend into the scrotum and, instead, remain in the body cavity.

curare   Chemical derived from plant sources that causes flaccid paralysis by blocking ACh receptor proteins in muscle cell membranes.

Cushing's syndrome   Symptoms caused by hypersecretion of adrenal steroid hormones, due to tumors of the adrenal cortex or to ACTH-secreting tumors of the anterior pituitary.

cyanosis   A blue color given to the skin or mucous membranes by deoxyhemoglobin; indicates inadequate oxygen concentration in the blood.

cyto-   (Gr.) Cell.

cytochrome   Pigment in mitochondria that transports electrons in the process of aerobic respiration.

cytokinesis   Division of the cytoplasm, that occurs in mitosis and meiosis when a parent cell divides to produce two daughter cells.

cytoplasm   The semifluid part of the cell between the cell membrane and the nucleus, exclusive of membrane-bound organelles. Contains many enzymes and structural proteins.

cytoskeleton   A latticework of structural proteins in the cytoplasm arranged in the form of microfilaments and microtubules.

delayed hypersensitivity   An allergic response in which the onset of symptoms takes as long as two to three days after exposure to an antigen. Produced by T cells, it is a type of cell-mediated immunity.

dendrite   A relatively short, highly branched neural process that carries electrical activity to the cell body.

dentin   One of the hard tissues of the teeth; covers the pulp cavity, and is itself covered on its exposed surface by enamel and on the root surface by cementum.

depolarization   Loss of membrane polarity in which the inside of the cell membrane becomes less negative in comparison to the outside of the membrane. The term is also used to indicate the reversal of membrane polarity that occurs during the production of action potentials in nerve and muscle cells.

diabetes insipidus   Condition in which inadequate amounts of antidiuretic hormone (ADH) are secreted by the posterior pituitary. Results in inadequate reabsorption of water by the kidney tubules and, thus, in the excretion of a large volume of dilute urine.

diabetes mellitus   Appearance of glucose in the urine due to the presence of high plasma glucose concentrations, even in the fasting state. Disease caused by either lack of sufficient insulin secretion or by inadequate responsiveness of the target tissues to the effects of insulin.

diapedesis   Migration of white blood cells through the endothelial walls of blood capillaries into the surrounding connective tissues.

diarrhea   Abnormal frequence of defecation accompanied by abnormal liquidity of the feces.

diastole   Phase of relaxation in which the heart fills with blood. Unless accompanied by the modifier term *atrial*, diastole usually refers to the resting phase of the ventricles.

diffusion   Net movement of molecules or ions from regions of higher to regions of lower concentration.

digestion   The process of converting food into molecules that can be absorbed through the intestine into the blood.

diploid   Cells having two of each chromosome, or twice the number of chromosomes that are present in sperm or ova.

disaccharide   Class of double sugars; carbohydrates that yield two simple sugars, or monosaccharides, upon hydrolysis.

DNA   Deoxyribonucleic acid; composed of nucleotide bases and deoxyribose sugar; contains the genetic code.

dopa   A derivative of the amino acid tyrosine, L-dopa serves as the precursor for the neurotransmitter molecule dopamine. L-dopa is given to patients with Parkinson's disease to stimulate dopamine production.

dopamine   Serves as a neurotransmitter in the central nervous system; also is the precursor of norepinephrine, another neurotransmitter molecule.

dorsal   Anatomical term pertaining to a backward location.

dorsal root ganglion   Collections of cell bodies of sensory neurons that form swellings in the dorsal roots of spinal nerves.

ductus arteriosus   A fetal blood vessel connecting the pulmonary artery directly to the aorta.

dwarfism   Condition in which a person is undersized due to inadequate secretion of growth hormone.

dyspnea   Subjective difficulty in breathing.

ECG   Electrocardiogram (also abbreviated EKG); a recording of electrical currents produced by the heart.

E. coli   Species of bacteria normally found in the human intestine; full name is Escherichia coli.

ecto-   (Gr.) Outside, outer.

-ectomy   (Gr.) Surgical removal of a structure.

ectopic   Foreign, out of place.

ectopic focus   An area of the heart other than the S-A node that assumes pacemaker activity.

ectopic pregnancy   Embryonic development that occurs anywhere other than in the uterus (as in the fallopian tubes or body cavity).

edema   Swelling due to an increase in tissue fluid.

EEG   Electroencephalogram; a recording of the electrical activity of the brain from electrodes placed on the scalp.

effector organs   Collective term for muscles and glands that are activated by motor neurons.

electrolytes   Ions and molecules that are able to ionize and thus carry an electric current. The most common electrolytes in the plasma are Na^+, HCO_3^-, and K^+.

electrophoresis   Biochemical technique in which different molecules can be separated and identified by their rate of movement in an electric field.

elephantiasis   Disease caused by infection with a nematode worm in which the larvae block lymphatic drainage and produce edema; the lower areas of the body can become enormously swollen as a result.

EMG   Electromyogram; electrical recordings of the activity of skeletal muscles using surface electrodes.

emmetropia   A condition of normal vision in which the image of objects is focused on the retina, as opposed to nearsightedness (myopia) or farsightedness (hypermetropia).

emphysema   Lung disease in which alveoli are destroyed and the remaining alveoli become larger. Results in decreased vital capacity and increased airways resistance.

emulsification   The process of producing an emulsion or fine suspension; in the intestine, fat globules are emulsified by the detergent action of bile.

endergonic   Chemical reaction that requires the input of energy from an external source in order to proceed.

endo-   (Gr.) Within, inner.

endocrine glands   Glands that secrete hormones into the circulation rather than into a duct; also called ductless glands.

endocytosis   The cellular uptake of particles that are too large to cross the cell membrane; occurs by invagination of the cell membrane until a membrane-enclosed vesicle is inched off within the cytoplasm.

endoderm   The innermost of the three primary germ layers of an embryo; gives rise to the digestive tract and associated structures, respiratory tract, bladder, and urethra.

endogenous   A product or process arising from within the body; as opposed to exogenous products or influences from external sources.

endolymph   The fluid contained within the membranous labyrinth of the inner ear.

endometrium   The mucous membrane of the uterus, the thickness and structure of which vary with the phase of the menstrual cycle.

endoplasmic reticulum   An extensive system of membrane-enclosed cavities within the cytoplasm of the cell; those with ribosomes on their surface are called rough endoplasmic reticulum and participate in protein synthesis.

endorphins   A group of endogenous opioid molecules that may act as a natural analgesic.

endothelium   The simple squamous epithelium that lines blood vessels and the heart.

endotoxin   A toxin contained within certain types of bacteria that is able to stimulate the release of endogenous pyrogen and produce a fever.

enkephalins   Short polypeptides, containing five amino acids, that have analgesic effects; may function as neurotransmitters in the brain. The two enkephalins (which differ in only one amino acid) are endorphins.

enteric   A term referring to the intestine.

entropy   The energy of a system that is not available to perform work; a measure of the degree of disorder in a system, entropy increases whenever energy is transformed.

enzyme   A protein catalyst that increases the rate of specific chemical reactions.

epi-   (Gr.) Upon, over, outer.

epidermis   The stratified squamous epithelium of the skin, the outer layer of which is dead and filled with keratin.

epididymis   A tubelike structure outside the testes; sperm pass from the seminiferous tubules into the head of the epididymis, and pass from the tail of the epididymis to the vas deferens. The sperm mature, becoming motile, as they pass through the epididymis.

epinephrine   Also known as adrenalin; a catecholamine hormone secreted by the adrenal medulla in response to sympathetic nerve stimulation; acts together with norepinephrine released from sympathetic nerve endings to prepare the organism for "fight or flight."

epithelium   One of the four primary tissues, forming membranes that cover and line the body surfaces and forming exocrine and endocrine glands.

EPSP   Excitatory postsynaptic potential; a graded depolarization of a postsynaptic membrane in response to stimulation by a neurotransmitter chemical. EPSPs can be summated, but can only be transmitted short distances; EPSPs can stimulate the production of action potentials when a threshold level of depolarization is attained.

erythroblastosis fetalis   Hemolytic anemia in a newborn Rh positive baby caused by maternal antibodies against the Rh factor that have crossed the placenta.

erythrocytes   Red blood cells; the formed elements of blood that contain hemoglobin and transport oxygen.

essential amino acids   Those eight amino acids in adults or nine amino acids in children that cannot be made by the human body and therefore must be obtained in the diet.

estrus cycle   Cyclic changes in the structure and function of the ovaries and female reproductive tract, accompanied by periods of "heat" (estrus) or sexual receptivity; the lower mammalian equivalent of the menstrual cycle, but differing from the menstrual cycle in that the endometrium is not shed with accompanying bleeding.

ex-   (La.) Out, off, from.

exergonic   Chemical reactions that liberate energy.

exo-   (Gr.) Outside or outward.

exocrine gland   A gland that discharges its secretion through a duct to the outside of an epithelial membrane.

exocytosis   Process of cellular secretion in which the secretory products are contained within a membrane-enclosed vesicle; the vesicle fuses with the cell membrane so that the lumen of the vesicle is open to the extracellular environment.

extensor   A muscle that, upon contraction, increases the angle of a joint.

exteroreceptors   Sensory receptors that are sensitive to changes in the external environment (as opposed to interoreceptors).

extra-   (La.) Outside, beyond.

extraocular muscles   The muscles that insert into the sclera of the eye and that act to change the position of the eye in its orbit. As opposed to the intraocular muscles such as those of the iris and ciliary body within the eye.

facilitated diffusion   Carrier-mediated transport of molecules through the cell membrane along the direction of their concentration gradients; does not require the expenditure of metabolic energy.

FAD   Flavin adenine dinucleotide; a coenzyme derived from riboflavin that participates in electron transport within the mitochondria.

feces   The excrement discharged from the large intestine.

fertilization   Fusion of an ovum and sperm.

fibrillation   Condition of cardiac muscle characterized electrically by random and continuously changing patterns of electrical activity and resulting in the inability of the myocardium to contract as a unit and pump blood. Can be fatal if it occurs in the ventricles.

fibrin   The insoluble protein formed from fibrinogen by the enzymatic action of thrombin during the process of blood clot formation.

fibrinogen   Also called factor I; a soluble plasma protein that serves as the precursor of fibrin.

flagellum   A whiplike structure that provides motility for sperm.

flare-and-wheal reaction   Cutaneous reaction to skin injury or administration of antigens, produced by release of histamine and related molecules and characterized by local dema and a red flare.

flavoprotein   A conjugated protein that contains a flavin pigment; involved in electron transport within the mitochondria.

flexor   A muscle that, when it contracts, decreases the angle of a joint.

foramen ovale   An opening that is normally present in the atrial septum of a fetal heart that allows direct communication between the right and left atria.

fovea centralis   A tiny pit in the macula lutea of the retina that contains slim, elongated cones and that provides the highest visual acuity (clearest vision).

FSH   Follicle-stimulating hormone; one of the two gonadotrophic hormones secreted from the anterior pituitary. In females FSH stimulates development of the ovarian follicles, while in males it stimulates production of sperm in the seminiferous tubules.

GABA   Gamma-aminobutyric acid; believed to function as an inhibitory neurotransmitter in the central nervous system.

gametes   Collective term for haploid germ cells: sperm and ova.

ganglion   A grouping of nerve cell bodies located outside the brain and spinal cord.

gastric intrinsic factor   A glycoprotein secreted by the stomach and needed for the absorption of vitamin B_{12}.

gastrin   A hormone secreted by the stomach that stimulates the gastric secretion of hydrochloric acid and pepsin.

gates   A term used to describe structures within the cell membrane that regulate the passage of ions through membrane channels. Such gates may be chemically regulated (by neurotransmitters) or voltage regulated (in which they open in response to a threshold level of depolarization).

gen-   (Gr.) Producing.

genetic recombination   The formation of new combinations of genes, as by crossing-over between homologous chromosomes.

genetic transcription   Process by which RNA is produced, with a sequence of nucleotide bases that is complementary to a region of DNA.

genetic translation   Process by which proteins are produced, with amino acid sequences specified by the sequence of codons in messenger RNA.

gigantism   Abnormal body growth due to excessive secretion of growth hormone.

glomerular ultrafiltrate   Fluid filtered through the glomerular capillaries into Bowman's capsule of kidney tubules.

glomeruli   General term for a tuft or cluster; most often used to describe the tufts of capillaries in the kidneys that filter fluid into the kidney tubules.

glomerulonephritis   Inflammation of the renal glomeruli, associated with fluid retention, edema, hypertension, and appearance of protein in the urine.

glucagon   Polypeptide hormone secreted by the alpha cells of the islets of Langerhans in the pancreas that acts to promote glycogenolysis and raise the blood glucose levels.

glucocorticoids   Steroid hormones secreted by the adrenal cortex (corticosteroids) that affect the metabolism of glucose, protein, and fat. These hormones also have anti-inflammatory and immunosuppressive effects; the major glucocorticoid in humans is hydrocortisone (cortisol).

gluconeogenesis   The formation of glucose from non-carbohydrate molecules such as amino acids and lactic acid.

glycogen   A polysaccharide of glucose—also called *animal starch*— produced primarily in the liver and skeletal muscles. Similar to plant starch in composition, glycogen contains more highly branched chains of glucose subunits than does plant starch.

glycogenesis   Formation of glycogen from glucose.

glycogenolysis   Hydrolysis of glycogen to glucose-1-phosphate, which can be converted to glucose-6-phosphate, which then may be oxidized via glycolysis or (in the liver) converted to free glucose.

glycolysis   Metabolic pathway that converts glucose to pyruvic acid and that yields a net production of two ATP molecules and two molecules of reduced NAD. In anaerobic respiration the reduced NAD is oxidized by conversion of pyruvic acid to lactic acid. In aerobic respiration pyruvic acid enters the Krebs cycle in mitochondria, and reduced NAD is ultimately oxidized by oxygen to yield water.

glycosuria   Excretion of an abnormal amount of glucose in the urine (urine normally only contains trace amounts of glucose).

Golgi apparatus   Stacks of flattened membranous sacs within the cytoplasm of cells that are believed to bud off vesicles containing secretory proteins.

Golgi tendon organ   A tension receptor in the tendons of muscles that becomes activated by the pull exerted by a muscle on its tendons.

gonadotrophin hormones   Hormones of the anterior pituitary gland that stimulate gonadal function—formation of gametes and secretion of sex steroids. The two gonadotrophins are FSH (follicle-stimulating hormone) and LH (luteinizing hormone), which are essentially the same in males and females.

gonads   Collective term for testes and ovaries.

Graafian follicle   A mature ovarian follicle containing a single fluid-filled cavity, with the ovum located towards one side of the follicle perched on top of a hill of granulosa cells.

granular leukocytes   Leukocytes with granules in the cytoplasm; on the basis of the staining properties of the granules, these cells are of three types: neutrophils, eosinophils, and basophils.

Graves' disease   A hyperthyroid condition believed to be caused by excessive stimulation of the thyroid gland by autoantibodies; associated with exophthalmos (bulging eyes), high pulse rate, high metabolic rate, and other symptoms of hyperthyroidism.

growth hormone   A hormone secreted by the anterior pituitary that stimulates growth of the skeleton and soft tissues during the growing years and that influences the metabolism of protein, carbohydrate, and fat throughout life.

gyrus   A fold or convolution in the cerebrum.

haploid   A cell that has one of each chromosome type and therefore half the number of chromosomes present in most other body cells; only the gametes (sperm and ova) are haploid.

haptens   Small molecules that are not antigenic by themselves, but which—when combined with proteins—become antigenic and thus able to stimulate production of specific antibodies.

Haversian system   A Haversian canal and its concentrically arranged layers, or lamellae, of bone; constitutes the basic structural unit of compact bone.

hay fever   A seasonal type of allergic rhinitis caused by pollen; characterized by itching and tearing of the eyes, swelling of the nasal mucosa, attacks of sneezing, and often by asthma.

heart murmur   Abnormal heart sounds caused by an abnormal flow of blood in the heart due to structural defects, usually of the valves or septum.

helper T cells   A subpopulation of T cells (lymphocytes) that assist the stimulation of antibody production of B lymphocytes by antigens.

hematocrit   The ratio of packed red blood cells to total blood volume in a centrifuged sample of blood, expressed as a percentage.

heme   The iron-containing red pigment that, together with the protein globin, forms hemoglobin.

hemoglobin   The combination of heme pigment and protein within red blood cells that acts to transport oxygen and (to a lesser degree) carbon dioxide. Hemoglobin also serves as a weak buffer within red blood cells.

heparin   A mucopolysaccharide in many tissues, but most abundant in lungs and liver, that is used medically as an anticoagulant.

hepatic   Pertaining to the liver.

hepatitis   Inflammation of the liver.

Hering-Breuer reflex   A reflex in which distension of the lungs stimulates stretch receptors, which in turn act to inhibit further distension of the lungs.

hermaphrodite   An organism with both testes and ovaries.

hetero-   (Gr.) Different, other.

heterochromatin   A condensed, inactive form of chromatin.

hiatal hernia   A protrusion of an abdominal structure through the esophageal hiatus of the diaphragm into the thoracic cavity.

high-density lipoproteins   Combinations of lipids and proteins that migrate rapidly to the bottom of a test tube during centrifugation; carrier proteins for lipids such as cholesterol that appear to offer some protection from atherosclerosis.

histamine   A compound secreted by tissue mast cells and other connective tissue cells that stimulates vasodilation and increases capillary permeability; responsible for many of the symptoms of inflammation and allergy.

histones   A basic protein associated with DNA that is believed to repress genetic expression.

homeo-   (Gr.) Same.

homeostasis   The dynamic constancy of the internal environment, which serves as the principal function of physiological regulatory mechanisms. The concept of homeostasis provides a framework for the understanding of most physiological processes.

homologous chromosomes   The matching pairs of chromosomes in a diploid cell.

hormones   Regulatory chemicals secreted into the blood by endocrine cells and carried by the blood to target cells that respond to the hormones by an alteration in their metabolism.

humoral immunity   The form of acquired immunity in which antibody molecules are secreted in response to antigenic stimulation (as opposed to cell-mediated immunity).

hyaline membrane disease   A disease of some premature infants that lack pulmonary surfactant, and characterized by collapse of the alveoli (atelectasis) and pulmonary edema. Also called respiratory distress syndrome.

hydrocortisone   Also called cortisol; the principal corticosteroid hormone secreted by the adrenal cortex, with glucocorticoid action.

hydrophilic   A substance that readily absorbs water; "water loving."

hydrophobic   A substance that repels, and that is repelled by, water; "water fearing."

hyper-   (Gr.) Over, above, excessive.

hyperbaric oxygen   Oxygen gas present at greater than atmospheric pressure.

hypercapnia   Excessive concentration of carbon dioxide in the blood.

hyperglycemia   Abnormally increased concentration of glucose in the blood.

hyperkalemia   Abnormally high concentration of potassium in the blood.

hyperopia   Also called farsightedness; a refractive disorder in which rays of light are brought to a focus behind the retina as a result of the eyeball being too short.

hyperplasia   Increase in organ size due to an increase in cell numbers as a result of mitotic cell division.

hyperpolarization   An increase in the negativity of the inside of a cell membrane with respect to the resting membrane potential.

hypersensitivity   Another name for allergy; abnormal immune response that may be immediate (due to antibodies of the IgE class) or delayed (due to cell-mediated immunity).

hypertension   High blood pressure. Divided into primary or essential hypertension of unknown cause and secondary hypertension that develops as a result of other, known disease processes.

hypertonic   A solution with a greater solute concentration, and thus a greater osmotic pressure, than plasma.

hypertrophy   Growth of an organ due to an increase in the size of its cells.

hyperventilation   A high rate and depth of breathing that results in a decrease in the blood carbon dioxide concentration below normal.

hypo-   (Gr.) Under, below, less.

hypodermis   A layer of fat beneath the dermis of the skin.

hypothalamus   An area of the brain below the thalamus and above the pituitary gland. The hypothalamus regulates the pituitary gland and contributes to the regulation of the autonomic nervous system, among many other functions.

hypothalamic hormones   Hormones produced by the hypothalamus; these include antidiuretic hormone and oxytocin, which are secreted by the posterior pituitary gland; and both releasing and inhibiting hormones that regulate the secretion of the anterior pituitary.

hypothalamo-hypophyseal portal system   Vascular system that transports releasing and inhibiting hormones from the hypothalamus to the anterior pituitary.

hypothalamo-hypophyseal tract   Tract of nerve fibers (axons) that transport antidiuretic hormone and oxytocin from the hypothalamus to the posterior pituitary gland.

hypovolemic shock   A rapid fall in blood pressure due to diminished blood volume.

hypoxemia   A low oxygen concentration of the arterial blood.

immediate hypersensitivity   Hypersensitivity (allergy) that is mediated by antibodies of the IgE class and that results in the release of histamine and related compounds from tissue cells.

immunization   The process of increasing one's resistance to pathogens. In active immunity a person is injected with antigens that stimulate development of clones of specific B or T lymphocytes; in passive immunity a person is injected with antibodies made by another organism.

immunoassay   Detection or measurement of a molecule that acts as an antigen by reaction with specific antibodies. The antigen-antibody reaction may be followed in a variety of ways, such as agglutination (if the antigen is attached to visible cells or particles), fluorescence, or radioactivity (a radioimmunoassay, or RIA).

immunoglobins   Subclasses of the gamma globulin fraction of plasma proteins that have antibody functions, providing humoral immunity.

immunosurveillance   The function of the immune system to recognize and attack malignant cells that produce antigens that are not recognized as "self." This function is believed to be cell mediated rather than humoral.

implantation   The process by which a blastocyst attaches itself to and penetrates into the endometrium of the uterus.

inhibin   Believed to be a water-soluble hormone secreted by the seminiferous tubules of the testes that specifically exerts negative feedback inhibition of FSH secretion from the anterior pituitary gland.

insulin   A polypeptide hormone secreted by the beta cells of the islets of Langerhans in the pancreas that promotes anabolism of carbohydrates, fat, and protein. Insulin acts to promote the cellular uptake of blood glucose and, therefore, to lower the blood glucose concentration; insulin deficiency produces hyperglycemia and diabetes mellitus.

inter-   (La.) Between, among.

interferons   A group of small proteins that inhibit the multiplication of viruses inside host cells and that also have antitumor properties.

interneurons   Also called association neurons; those neurons within the central nervous system that do not extend into the peripheral nervous system; interposed between sensory (afferent) and motor (efferent) neurons.

interoreceptors   Sensory receptors that respond to changes in the internal environment (as opposed to exteroreceptors).

interphase   The interval between successive cell divisions; during this time the chromosomes are in an extended state and are active in directing RNA synthesis.

intra   (La.) Within, inside.

intrafusal fibers   Modified muscle fibers that are encapsulated to form muscle spindle organs, which are muscle stretch receptors.

intrapleural space   An actual or potential space between the visceral pleural membrane covering the lungs and the somatic pleural membrane lining the thoracic wall.

intrapulmonary space   The space within the air sacs and airways of the lungs.

inulin   A polysaccharide of fructose produced by certain plants that is filtered by the human kidneys, but neither reabsorbed nor secreted. The clearance rate of injected inulin is thus used to measure the glomerular filtration rate.

in vitro   Occurring outside the body, in a test tube or other artificial environment.

in vivo   Occurring within the body.

ion   An atom or group of atoms that has either lost or gained electrons, and that thus has a net positive or a net negative charge.

ionization   The dissociation of a solute to form ions.

ipsilateral   On the same side (as opposed to contralateral).

IPSP   Inhibitory postsynaptic potential; hyperpolarization of the postsynaptic membrane in response to a particular neurotransmitter chemical, which makes it more difficult for the postsynaptic cell to attain a threshold level of depolarization required to produce action potentials; responsible for postsynaptic inhibition.

ischemia   A rate of blood flow to an organ that is inadequate to supply sufficient oxygen and maintain aerobic respiration in that organ.

islets of Langerhans   Encapsulated groupings of endocrine cells within the exocrine tissue of the pancreas. The islets contain alpha cells that secrete glucagon and beta cells that secrete insulin.

iso-   (Gr.) Equal, same.

isoenzymes   Enzymes, usually produced by different organs, that catalyze the same reaction but that differ from each other in amino acid composition.

isometric contraction   Muscle contraction in which there is no appreciable shortening of the muscle.

isotonic contraction   Muscle contraction in which the muscle shortens in length and maintains approximately the same amount of tension throughout the shortening process.

isotonic solution   A solution having the same total solute concentration, osmolality, and osmotic pressure as the solution with which it is compared; a solution with the same solute concentration and osmotic pressure as plasma.

jaundice   A condition characterized by high blood bilirubin levels and staining of the tissues with bilirubin, which gives skin and mucous membranes a yellow color.

junctional complexes   The structures that join adjacent epithelial cells together, including the zonula occludens, zonula adherens, and macula adherens (desmosome).

juxta-   (La.) Near to, next to.

keratin   A protein that forms the principal component of the outer layer of the epidermis, and of hair and nails.

ketoacidosis   A type of metabolic acidosis produced by excessive production of ketone bodies, as in diabetes mellitus.

ketogenesis   The production of ketone bodies.

ketone bodies   Includes acetone, acetoacetic acid, and β-hydroxybutyric acid; derived from fatty acids via acetyl coenzyme A in the liver. Ketone bodies are oxidized by skeletal muscles for energy.

ketosis   An abnormal elevation in the blood concentration of ketone bodies that does not necessarily produce acidosis.

kilocalorie   Equal to 1,000 calories, which are units of heat (a kilocalorie is the amount of heat required to raise the temperature of 1 kilogram of water by 1°C). In nutrition the kilocalorie is called a big calorie (Calorie).

Klinefelter's syndrome   The syndrome produced in a male by the presence of an extra X chromosome (genotype XXY).

Krebs cycle   A cyclic metabolic pathway in the matrix of mitochondria by which the acetic acid part of acetyl CoA is oxidized and substrates are provided for reactions that are coupled to the formation of ATP.

Kupffer cells   Phagocytic cells that line the sinusoids of the liver, and which are part of the reticuloendothelial system.

lactose   Milk sugar; a disaccharide of glucose and galactose.

lactose intolerance   The inability of many adults to digest lactose due to loss of the ability of the intestine to produce lactase enzyme.

larynx   A structure consisting of epithelial tissue, muscle, and cartilage that serves as a sphincter guarding the entrance of the trachea and as the organ responsible for voice.

lesion   A wounded or damaged area.

leukocytes   White blood cells.

Leydig cells   The interstitial cells of the testes that serve an endocrine function by secreting testosterone and other androgenic hormones.

ligament   Dense regular connective tissue, containing many parallel arrangements of collagen fibers, that connects bones or cartilages and serves to strengthen joints.

lipogenesis   The formation of fat or triglycerides.

lipolysis   The hydrolysis of triglycerides into free fatty acids and glycerol.

low-density lipoproteins   Plasma proteins that transport triglycerides and cholesterol, and which are believed to contribute to arteriosclerosis.

lumen The cavity of a tube or hollow organ.

lung surfactant A mixture of lipoproteins (containing phospholipids) secreted by type II alveolar cells into the alveoli of the lungs; lowers surface tension and prevents collapse of the lungs as occurs in hyaline membrane disease, in which surfactant is absent.

luteinizing hormone (LH) A gonadotrophic hormone secreted by the anterior pituitary that, in a female, stimulates ovulation and the development of a corpus luteum. In a male, LH stimulates the Leydig cells to secrete androgens.

lymph The fluid in lymphatic vessels that is derived from tissue fluid.

lymphatic system The lymphatic vessels and lymph nodes.

lymphocytes A type of mononuclear leukocyte; the cells responsible for humoral and cell-mediated immunity.

lymphokines A group of chemicals released from T cells that contribute to cell-mediated immunity.

-lysis (Gr.) Breakage, disintegration.

lysosome Organelles containing digestive enzymes and responsible for intracellular digestion.

macro- (G.) Large.

macromolecules Large molecules; a term that usually refers to protein, RNA, and DNA.

macrophage A large phagocytic cell in connective tissue that contributes to both specific and nonspecific immunity.

malignant A structure or process that is life threatening.

mast cells A type of connective tissue cells that produce and secrete histamine and heparin.

maximal oxygen uptake The maximum amount of oxygen that can be consumed by the body per unit time during heavy exercise.

medulla oblongata A part of the brain stem; contains neural centers for the control of breathing and for regulation of the cardiovascular system via autonomic nerves.

mega- Large, great.

megakaryocyte A bone marrow cell that gives rise to blood platelets.

meiosis A type of cell division in which a diploid parent cell gives rise to haploid daughter cells; occurs in the process of gamete production in the gonads.

melanin A dark pigment found in the skin, hair, choroid layer of the eye, and substantia nigra of the brain; may also be present in certain tumors (melanomas).

melatonin A hormone secreted by the pineal gland that produces lightening of the skin in lower animals and that may contribute to regulation of gonadal function in mammals.

membrane potential The potential difference or voltage that exists between the inner and outer sides of a cell membrane; existing in all cells, but capable of being changed by excitable cells (neurons and muscle cells).

membranous labyrinth Communicating sacs and ducts within the bony labyrinth of the inner ear.

menarche The age at which menstruation begins.

Meniere's disease Deafness, tinnitus, and vertigo resulting from disease of the labyrinth.

menopause Cessation of menstruation, usually occurring at about age forty-eight to fifty.

menstrual cycle The cyclic changes in the ovaries and endometrium of the uterus that lasts about a month and that is accompanied by shedding of the endometrium, with bleeding. Occurs only in humans and the higher primates.

menstruation Shedding of the inner two-thirds of the endometrium with accompanying bleeding, due to lowering of estrogen secretion by the ovaries at the end of the monthly cycle. The first day of menstruation is taken as day one of the menstrual cycle.

meso- (Gr.) Middle.

mesoderm The middle embryonic tissue layer that gives rise to connective tissue (including blood, bone, and cartilage), blood vessels, muscles, the adrenal cortex, and other organs.

messenger RNA (mRNA) A type of RNA that contains a base sequence complementary to a part of the DNA that specifies the synthesis of a particular protein.

meta- (Gr.) Change.

metabolism All of the chemical reactions in the body; includes those that result in energy storage (anabolism) and those that result in the liberation of energy (catabolism).

metastasis A process whereby cells of a malignant tumor can separate from the tumor, travel to a different site, and divide to produce a new tumor.

micelles Colloidal particles formed by the aggregation of many molecules.

micro- (La.) Small; also, one-millionth.

microvilli Tiny fingerlike projections of a cell membrane; occurs on the apical (lumenal) surface of the cells of the small intestine and in the renal tubules.

micturition Urination.

mineralocorticoids Steroid hormones of the adrenal cortex (corticosteroids) that regulate electrolyte balance.

mitosis Cell division in which the two daughter cells receive the same number of chromosomes as the parent cell (both daughters and parent are diploid).

molal The number of moles of solute per kilogram of solvent.

molar The number of moles of solute per liter of solution.

mole The number of grams of a chemical that is equal to its formula weight (atomic weight for an element or molecular weight for a compound).

mono- (Gr.) One, single.

monoclonal antibodies Identical antibodies derived from a clone of genetically identical plasma cells.

monocyte A mononuclear, nongranular leukocyte that is phagocytic and able to be transformed into a macrophage.

monomers A single molecular unit of a longer, more complex molecule; monomers are joined together to form dimers, trimers, and polymers; hydrolysis of polymers eventually yields separate monomers.

monosaccharide Also called simple sugars; the monomers of more complex carbohydrates. Examples include glucose, fructose, and galactose.

-morph, morpho- (Gr.) Form, shape.

motile Capable of self-propelled movement.

motor neuron An efferent neuron that conducts action potentials away from the central nervous system and innervates effector organs (muscles and glands). Forms the ventral roots of spinal nerves.

mucous membrane The layers of visceral organs that include the lining epithelium, submucosal connective tissue, and (in some cases) a thin layer of smooth muscle (the muscularis mucosa).

muscle spindles Sensory organs within skeletal muscles that are composed of intrafusal fibers and are sensitive to muscle stretch; provide a length detector within muscles.

myelin sheath A sheath surrounding axons that is formed by successive wrappings of a neuroglial cell membrane. Myelin sheaths are formed by Schwann cells in the peripheral nervous system and by oligodendrocytes within the central nervous system.

myocardial infarction An area of necrotic tissue in the myocardium that is filled in by scar (connective) tissue.

myofibrils Subunits of striated muscle fibers that consist of successive sarcomeres; myofibrils run parallel to the long axis of the muscle fiber, and the pattern of their filaments provide the striations characteristic of striated muscle cells.

myogenic Originating within muscle cells; used to describe self-excitation by cardiac and smooth muscle cells.

myoglobin A molecule composed of globin protein and heme pigment, related to hemoglobin but containing only one subunit (instead of the four in hemoglobin) and found in striated muscles. Myoglobin serves to store oxygen in skeletal and cardiac muscle cells.

myoneural junction Also called the neuromuscular junction; a synapse between a motor neuron and the muscle cell that it innervates.

myosin The protein that forms the A bands of striated muscle cells; together with the protein actin, myosin provides the basis for muscle contraction.

myxedema A type of edema associated with hypothyroidism; characterized by accumulation of mucoproteins in tissue fluid.

NAD Nicotinamide adenine dinucleotide; a coenzyme derived from niacin that functions to transport electrons in oxidation-reduction reactions; helps to transport electrons to the electron transport chain within mitochondria.

naloxone A drug that antagonizes the effects of morphine and endorphins.

necrosis Cellular death within tissues and organs.

negative feedback Mechanisms in the body that act to maintain a state of internal constancy or homeostasis; effectors are activated by changes in the internal environment, and the actions of the effectors serve to counteract these changes and maintain a state of balance.

neoplasm A new, abnormal growth of tissue, as in a tumor.

nephron The functional unit of the kidneys, consisting of a system of renal tubules and a vascular component that includes capillaries of the glomerulus and the peritubular capillaries.

neuroglia The supporting tissue of the nervous system, consisting of neuroglial or glial cells. In addition to providing support, the neuroglial cells participate in the metabolic and bioelectrical processes of the nervous system.

neurons Nerve cells, consisting of a cell body that contains the nucleus, short branching processes called dendrites that carry electrical charges to the cell body, and a single fiber or axon that conducts nerve impulses away from the cell body.

neurotransmitter A chemical contained in synaptic vesicles in nerve endings that is released into the synaptic cleft, and that stimulates the production of either excitatory or inhibitory postsynaptic potentials.

neutron Electrically neutral particles that exist together with positively charged protons in the nucleus of atoms.

nexus A bond between members of a group; the type of bonds present in single-unit smooth muscles.

nidation Implantation of the blastocyst into the endometrium of the uterus.

Nissl bodies Granular-appearing structures in the cell bodies of neurons that have an affinity for basic stain; corresponds to ribonucleoprotein.

nodes of Ranvier Gaps in the myelin sheath of myelated axons, located approximately 1 mm apart.

norepinephrine A catecholamine released as a neurotransmitter from postganglionic sympathetic nerve endings and as a hormone (together with epinephrine) by the adrenal medulla.

nucleolus Dark-staining area within a cell nucleus; site where ribosomal RNA is produced.

nucleoplasm The protoplasm of a nucleus.

nucleotide The subunits of DNA and RNA macromolecules; each nucleotide is composed of a nitrogenous base (adenine, guanine, cytosine, and thymine or uracil), a sugar (deoxyribose or ribose), and a phosphate group.

nucleus The organelle, surrounded by a double saclike membrane called the nuclear envelope, that contains the DNA and genetic information of the cell.

nystagmus Involuntary, oscillatory movements of the eye.

obese Excessively fat.

oligo- (Gr.) Few, small.

oligodendrocytes A type of neuroglial cell; forms myelin sheaths around axons in the central nervous system.

oncology The study of tumors.

oncotic pressure The colloid osmotic pressure of solutions produced by proteins; in plasma, serves to counterbalance the outward filtration of fluid from capillaries due to hydrostatic pressure.

oo- (Gr.) Pertaining to an egg.

oogenesis Formation of ova in the ovaries.

opsonization The role of antibodies in enhancing the ability of phagocytic cells to attack bacteria.

optic disc The area of the retina where axons from ganglion cells gather to form the optic nerve and where blood vessels end and leave the eye; corresponds to the blind spot in the visual field due to the absence of photoreceptors.

organ A structure in the body composed of a number of primary tissues that perform particular functions.

organelle Membrane-enclosed structures within cells that perform specialized tasks; includes mitochondria, Golgi apparatus, endoplasmic reticulum, nuclei, and lysosomes; term also used for some structures not enclosed by membrane, such as ribosomes and centrioles.

organ of Corti The structure within the cochlea responsible for hearing. Consists of hair cells and supporting cells on the basilar membrane that help transduce sound waves into nerve impulses.

osmolality A measure of the total concentration of a solution; the number of moles of solute per kilogram of solvent.

osmoreceptors Sensory neurons that respond to changes in the osmotic pressure of the surrounding fluid.

osmosis The passage of solvent (water) from a more dilute to a more concentrated solution through a membrane that is more permeable to water than to the solute.

osmotic pressure A measure of the tendency for a solution to gain water by osmosis when separated by a membrane from pure water; directly related to the osmolality of the solution, it is the pressure required to just prevent osmosis.

osteo- (G.) Pertaining to bone.

osteoblast Cells that produce bone.

osteocyte Bone-forming cells that have become entrapped within a matrix of bone; these cells remain alive due to nourishment supplied by canaliculi within the extracellular material of bone.

osteomalacia Softening of bones due to deficiency of vitamin D and calcium.

osteoporosis Demineralization of bone, seen most commonly in the elderly. It may be accompanied by pain, loss of stature, and other deformities and fractures.

ovaries The gonads of a female that produce ova and secrete female sex steroids.

ovi- (La.) Pertaining to egg.

oviduct The part of the female reproductive tract that transports ova from the ovaries to the uterus. Also called the uterine or fallopian tube.

ovulation The extrusion of a secondary oocyte out of the ovary.

oxidative phosphorylation The formation of ATP using energy derived from electron transport to oxygen; occurs in the mitochondria.

oxidizing agent An atom that accepts electrons in an oxidation-reduction reaction.

oxyhemoglobin A compound formed by the bonding of molecular oxygen to hemoglobin.

oxyhemoglobin saturation The ratio, expressed as a percentage, of the amount of oxyhemoglobin compared to the total amount of hemoglobin in blood.

oxytocin One of the two hormones produced in the hypothalamus and secreted by the posterior pituitary (the other hormone is vasopressin); oxytocin stimulates contraction of uterine smooth muscles and promotes milk ejection in females.

pacemaker A group of cells that has the fastest spontaneous rate of depolarization and contraction in a mass of electrically coupled cells; in the heart, this is the sinoatrial, or S-A, node.

PAH Para-aminohippuric acid; a substance used to measure total renal plasma flow because its clearance rate is equal to the total rate of plasma flow to the kidneys; PAH is filtered and secreted, but not reabsorbed by the renal nephrons.

parathyroid hormone (PTH) A polypeptide hormone secreted by the parathyroid glands, PTH acts to raise the blood Ca^{++} levels primarily by stimulating reabsorption of bone.

Parkinson's disease Tremor of the resting muscles and other symptoms caused by inadequate dopamine-producing neurons in the basal ganglia of the cerebrum. Also called paralysis agitans.

parturition Birth.

passive immunity Specific immunity granted by administration of antibodies made by another organism.

Pasteur effect Decrease in the rate of glucose utilization and lactic acid production by exposure to oxygen.

pathogen Any disease-producing microorganism or substance.

pepsin The protein-digesting enzyme secreted in gastric juice.

peptic ulcer An injury to the mucosa of the esophagus, stomach, or small intestine caused by acidic gastric juice.

peri- (Gr.) Around, surrounding.

perilymph The fluid between the membranous and bony labyrinth of the inner ear.

perimysium The connective tissue surrounding a fascicle of skeletal muscle fibers.

periosteum Connective tissue covering bones; contains osteoblasts and is therefore capable of forming new bone.

peristalsis Waves of smooth muscle contraction in smooth muscles of tubular digestive tract, involving circular and longitudinal muscle fibers at successive locations along the tract; serves to propel contents of tract in one direction.

permease A term used to indicate membrane transport carriers, and to emphasize the similarity of specificity and other properties that transport carriers have with enzymes.

pH The pH of a solution is equal to 1 over the logarithm of the hydrogen ion concentration. The pH scale goes from zero to 14; a pH of 7.0 is neutral, while solutions with lower pH are acidic and solutions with higher pH are basic.

phagocytosis Cellular eating; the ability of some cells (such as white blood cells) to engulf large particles such as bacteria and digest these particles by merging the food vacuole containing these particles with a lysosome containing digestive enzymes.

phonocardiogram A visual display of the heart sounds.

photoreceptors Sensory cells (rods and cones) that respond electrically to light; located in the retinas of the eyes.

pineal gland A gland within the brain that secretes the hormone melatonin and that is affected by sensory input from the photoreceptors of the eyes.

pinocytosis Cell drinking; invagination of the cell membrane that forms narrow channels that pinch off into vacuoles; provides cellular intake of extracellular fluid and dissolved molecules.

plasma The fluid portion of the blood. Unlike serum (which lacks fibrinogen), plasma is capable of forming insoluble fibrin threads when in contact with test tubes.

plasma cells Cells derived from B lymphocytes that produce and secrete large amounts of antibodies; are responsible for humoral immunity.

platelets Disc-shaped structures, 2 to 4 micrometers in diameter, that are derived from bone marrow cells called megakaryocytes. Platelets circulate in the blood and participate (together with fibrin) in forming blood clots.

pluripotent A property of early embryonic cells; able to specialize in a number of ways to produce tissues characteristic of different organs.

pneumotaxic center A neural center in the pons that rhythmically inhibits inspiration in a manner independent of sensory input.

-pod, -podium (Gr.) Foot, leg, extension.

polar body A small daughter cell formed by meiosis that degenerates in the process of oocyte production.

polar molecule A molecule in which the shared electrons are not evenly distributed, so that one side of the molecule is relatively negatively (or positively) charged in comparison with the other side; polar molecules are soluble in polar solvents such as water.

poly- (Gr.) Many.

polydipsea Excessive thirst.

polymer A large molecule formed by the combination of smaller subunits, or monomers.

polymorphonuclear leukocyte A granular leukocyte containing a nucleus with a number of lobes connected by thin, cytoplasmic strands; includes neutrophils, eosinophils, and basophils.

polypeptide A chain of amino acids connected by covalent bonds called peptide bonds. A very large polypeptide is called a protein.

polyphagia Excessive eating.

polysaccharide A carbohydrate formed by covalent bonding of numerous monosaccharides; examples include glycogen and starch.

polyuria Excretion of an excessively large volume of urine in a given period.

posterior Anatomical term denoting a backside position.

posterior pituitary The part of the pituitary gland that is derived from the brain; secretes vasopressin (ADH) and oxytocin produced in the hypothalamus. Also called the neurohypophysis.

postsynaptic inhibition Inhibition of a postsynaptic neuron by axon endings that release a neurotransmitter that induces hyperpolarization (inhibitory postsynaptic potentials).

presynaptic inhibition Neural inhibition in which axoaxonic synapses inhibit the release of neurotransmitter chemicals from the presynaptic axon.

pro- (Gr.) Before, in front of, forward.

prolactin A hormone secreted by the anterior pituitary that stimulates lactation (acting together with other hormones) in the postpartum female. May also participate (along with the gonadotrophins) in regulating gonadal function in some mammals.

prophylaxis Prevention or protection.

proprioceptor Sensory receptor that provides information about body position and movement; includes receptors in muscles, tendons, and joints as well as the sense of equilibrium provided by the semicircular canals of the inner ear.

proto- (Gr.) First, original.

proton A unit of positive charge in the nucleus of atoms.

protoplasm General term that includes cytoplasm and nucleoplasm.

pseudo- (Gr.) False.

pseudohermaphrodite An individual with some of the physical characteristics of both sexes, but who lacks functioning gonads of both sexes; a true hermaphrodite has both testes and ovaries.

pseudopods Footlike extensions of the cytoplasm that enable some cells (with amoeboid motion) to move across a substrate; pseudopods also are used to surround food particles in the process of phagocytosis.

ptyalin Also called salivary amylase; an enzyme in saliva that catalyzes the hydrolysis of starch into smaller molecules.

puberty The period of time in an individual's life span when secondary sexual characteristics develop and fertility develops.

pulmonary circulation The part of the vascular system which includes the pulmonary artery and pulmonary veins; transports blood from the right ventricle of the heart, through the lungs, and back to the left atrium of the heart.

pupil The opening at the center of the iris of the eye.

pyrogen A fever-producing substance.

QRS complex The part of an electrocardiogram that is produced by depolarization of the ventricles.

reduced hemoglobin Hemoglobin with iron in the reduced ferrous state that is able to bond with oxygen but is not combined with oxygen. Also called deoxyhemoglobin.

reducing agent An electron donor in a coupled oxidation-reduction reaction.

REM sleep The stage of sleep in which dreaming occurs; associated with rapid eye movements (REM). REM sleep occurs three to four times each night and lasts from a few minutes to over an hour.

renal Pertaining to the kidneys.

renal plasma clearance rate The milliliters of plasma that is cleared of a particular solute per minute by excretion of that solute in the urine; if there is no reabsorption or secretion of that solute by the nephron tubules, this is equal to the glomerular filtration rate.

renal pyramids The medulla of the human kidney.

repolarization The reestablishment of the resting membrane potential after depolarization has occurred.

respiratory acidosis A lowering of the blood pH below 7.35 due to accumulation of CO_2 as a result of hypoventilation.

respiratory alkalosis A rise in blood pH above 7.45 due to excessive elimination of blood CO_2 as a result of hyperventilation.

respiratory distress syndrome Also called hyaline membrane disease; most frequently occurring in premature infants, this syndrome is caused by abnormally high alveolar surface tension as a result of a deficiency in lung surfactant.

retina The layer of the eye that contains neurons and photoreceptors (rods and cones).

rhodopsin   Visual purple; a pigment in rod cells that undergoes a photochemical dissociation in response to light, and in so doing stimulates electrical activity in the photoreceptors.

ribosomes   Particles of protein and ribosomal RNA that form the organelles responsible for translation of messenger RNA and protein synthesis.

rickets   A condition caused by deficiency of vitamin D, and associated with interference of normal ossification of bone.

rigor mortis   The stiffening of a dead body, due to depletion of ATP and the production of rigor complexes between actin and myosin in muscles.

RNA   Ribonucleic acid; a nucleic acid consisting of the nitrogenous bases adenine, guanine, cytosine, and uracil, the sugar ribose, and phosphate groups. There are three types of RNA found in cytoplasm: messenger RNA (mRNA), transfer RNA (tRNA), and ribosomal RNA (rRNA).

rods   One of the two categories of photoreceptors (along with cones) in the retina of the eye; rods are responsible for black-and-white vision under low illumination.

saccadic eye movements   Very rapid eye movements that occur constantly and that change the focus on the retina from one point to another.

saltatory conduction   The rapid passage of action potentials from one node of Ranvier to another in myelinated axons.

sarcomere   The structural subunit of a myofibril in a striated muscle, equal to the distance between two successive Z lines.

sarcoplasm   The cytoplasm of striated muscle cells.

sarcoplasmic reticulum   The smooth or agranular endoplasmic reticulum of skeletal muscle cells; surrounds each myofibril and serves to store Ca^{++} when the muscle is at rest.

Schwann cell   A neuroglial cell of the peripheral nervous system that forms sheaths around peripheral nerve fibers. Schwann cells also direct regeneration of peripheral nerve fibers to their target cells.

sclera   The tough white outer coat of the eyeball that is continuous anteriorly with the clear cornea.

second messenger   A molecule or ion whose concentration within a target cell is increased by the action of a regulator compound (e.g. hormone or neurotransmitter), and which stimulates the metabolism of that target cell in a way characteristic of the actions of that regulator molecule—that is, in a way that mediates the intracellular effects of that regulatory compound.

secretin   A polypeptide hormone secreted by the small intestine in response to acidity of the intestinal lumen; along with cholecystokinin, secretin stimulates secretion of pancreatic juice into the small intestine.

semicircular canals   Three canals of the bony labyrinth that contain endolymph that is continuous with the endolymph of the membranous labyrinth of the cochlea; the semicircular canals provide a sense of equilibrium.

semilunar valve   The valve flaps of the aorta and pulmonary artery at their juncture with the ventricles.

seminal vesicles   The paired organs that are located on the posterior border of the urinary bladder that empty their contents into the vas deferens and thus contribute to the semen.

seminiferous tubules   The tubules within the testes that produce spermatozoa by meiotic division of their germinal epithelium.

semipermeable membrane   A membrane with pores of a size that permits the passage of solvent and some solute molecules, but that restricts the passage of other solute molecules.

sensory neuron   An afferent neuron that conducts impulses from peripheral sensory organs into the central nervous system.

serosa   An outer epithelial membrane that covers the surface of a visceral organ.

Sertoli cells   Nongerminal, supporting cells in the seminiferous tubules. Sertoli cells envelop spermatids and appear to participate in the transformation of spermatids into spermatozoa.

serum   The fluid squeezed out of a clot as it retracts; the supernatant when a sample of blood clots in a test tube and is centrifuged; serum is plasma without fibrinogen (which has been converted to fibrin in clot formation).

sex chromosomes   The X and Y chromosomes; the unequal pairs of chromosomes involved in sex determination (which is due to the presence or absence of a Y chromosome). Females lack a Y chromosome and normally have the genotype XX; males have a Y chromosome and normally have the genotype XY.

shock   As it relates to the cardiovascular system, refers to a rapid, uncontrolled fall in blood pressure, which in some cases becomes irreversible and leads to death.

sickle-cell anemia   A hereditary, autosomal recessive trait that occurs primarily in people of African ancestry in which it evolved apparently as a protection (in the carrier state) against malaria. In the homozygous state, hemoglobin S is made instead of hemoglobin A; this leads to the characteristic sickling of red blood cells, hemolytic anemia, and organ damage.

sinus   A cavity.

sinusoids   Blood channels that appear as cavitylike in the surrounding tissue; function as a type of capillary space that is relatively large and lined (in the liver sinusoids) by phagocytic cells of the reticuloendothelial system.

sleep apnea   A temporary cessation of breathing during sleep, usually lasting for several seconds.

sliding filament theory   The theory that the thick and thin filaments of a myofibril slide past each other, while maintaining their initial length, during muscle contraction.

smooth muscle   Nonstriated, spindle-shaped muscle cells with a single nucleus in the center; involuntary muscle in visceral organs that is innervated by autonomic nerve fibers.

sodium/potassium pump   An active transport carrier, with ATPase enzymatic activity, that acts to accumulate K^+ within cells and extrude Na^+ from cells, thus maintaining gradients for these ions across the cell membrane.

soma-, somato-, -some   (Gr.) Body, unit.

somatomedins   A group of small polypeptides that are believed to be produced in the liver in response to growth hormone stimulation and are believed to mediate the actions of growth hormone on the skeleton and other tissues.

somatostatin   A polypeptide produced in the hypothalamus that acts to inhibit the secretion of growth hormone from the anterior pituitary; somatostatin is also produced in the islets of Langerhans of the pancreas, but its function there has not been established.

somatotrophic hormone   Growth hormone; an anabolic hormone secreted by the anterior pituitary that stimulates skeletal growth and protein synthesis in many organs.

sounds of Korotkoff   The sounds heard when blood pressure measurements are taken. These sounds are produced by the turbulent flow of blood through an artery that has been partially constricted by a pressure cuff.

spermatogenesis   The formation of spermatozoa, including meiosis and maturational processes in the seminiferous tubules.

spermiogenesis   The maturational changes that transform spermatids into spermatozoa.

sphygmo-   (Gr.) The pulse.

sphygmomanometer   A manometer (pressure transducer) used to measure the blood pressure.

spindle fibers   Filaments that extend from the poles of a cell to its equator and attach to chromosomes during the metaphase stage of cell division. Contraction of the spindle fibers pulls the chromosomes to opposite poles of the cell.

spironolocatones   Diuretic drugs that act as an aldosterone antagonist.

steroid   A lipid, derived from cholesterol, that has three six-sided carbon rings and one five-sided carbon ring. These form the steroid hormones of the adrenal cortex and gonads.

striated muscle   Skeletal and cardiac muscle, the cells of which exhibit cross-banding, or striations, due to arrangement of thin and thick filaments into sarcomeres.

stroke volume   The amount of blood ejected from each ventricle at each heartbeat.

sub-   (La.) Under, below.

substrate In enzymatic reactions, the molecules that combine with the active sites of an enzyme and that are converted to products by catalysis of the enzyme.

sulcus A groove or furrow; a depression in the cerebrum that separates folds, or gyri, of the cerebral cortex.

super-, supra- (La.) Above, over.

suppressor T cells A subpopulation of T lymphocytes that acts to inhibit the production of antibodies against specific antigens by B lymphocytes.

surfactant In the lungs, a mixture of phospholipids and proteins produced by alveolar cells that reduces the surface tension of the alveoli and contributes to the elastic properties of the lungs.

sym-, syn- (Gr.) With, together.

synapse A region where a nerve fiber comes into close or actual contact with another cell, and across which nerve impulses are transmitted either directly or indirectly (via release of chemical neurotransmitters).

synergistic Regulatory processes or molecules (such as hormones) that have complementary or additive effects.

systemic circulation The circulation that carries oxygenated blood from the left ventricle in arteries to the tissue cells, and that carries blood depleted in oxygen via veins to the right atrium; the general circulation, as compared to the pulmonary circulation.

systole The phase of contraction in the cardiac cycle. When unmodified, refers to contraction of the ventricles; the term *atrial systole* refers to contraction of the atria.

tachycardia Excessively rapid heart rate, usually applied to rates in excess of 100 beats per minute. In contrast to an excessively slow heart rate (below 60 beats per minute), which is termed bradycardia.

target organ The organ that is specifically affected by the action of a hormone or other regulatory process.

T cell A type of lymphocyte that provides cell-mediated immunity, in contrast to B lymphocytes that provide humoral immunity through secretion of antibodies. There are three subpopulations of T cells: cytotoxic, helper, and suppressor.

telo- (Gr.) An end; complete; final.

tendon The dense regular connective tissue that attaches a muscle to the bones of its origin and insertion.

testosterone The major androgenic steroid secreted by the Leydig cells of the testes after puberty.

tetanus Term used to mean either a smooth contraction of a muscle (as opposed to muscle twitching), or to a state of maintained contracture of high tension.

thalassemia A group of hemolytic anemias caused by hereditary inability to produce either the alpha or beta chain of hemoglobin. Found primarily among Mediterranean people.

thorax The part of the body cavity above the diaphragm; the chest.

threshold The minimum stimulus that just produces a response.

thrombocytes Blood platelets; disc-shaped structures in blood that participate in clot formation.

thrombus A blood clot, formed by formation of fibrin threads around a platelet plug.

thyroxine Also called tetraiodothyronine, or T_4. The major hormone secreted by the thyroid gland that regulates basal metabolic rate and stimulates protein synthesis in many organs; deficiency in early childhood produces cretinism.

tinnitus A ringing sound or other noise that is heard but is not related to external sounds.

toxin A poison.

toxoid A modified bacterial endotoxin that has lost toxicity but retains its ability to act as an antigen and stimulate antibody production.

trans- (La.) Across, through.

transamination The transfer of an amino group from an amino acid to an alpha-keto acid, forming a new keto acid and a new amino acid, without the appearance of free ammonia.

transpulmonary pressure The pressure difference across the wall of the lung, equal to the difference between intrapulmonary pressure and intrapleural pressure.

triiodothyronine Abbreviated T_3; a hormone secreted in small amounts by the thyroid; the active hormone in target cells formed from thyroxine.

tropomyosin A filamentous protein that attaches to actin in the thin filaments and that acts, together with another protein called troponin, to inhibit and regulate attachment of myosin cross-bridges to actin.

trypsin A protein-digesting enzyme in pancreatic juice that is released into the small intestine.

tympanic membrane The eardrum; a membrane separating the external from the middle ear that transduces sound waves into movements of the middle ear ossicles.

universal donor A person with blood type O, who is able to donate blood to people with other blood types in emergency blood transfusions.

universal recipient A person with blood type AB, who can receive blood of any type in emergency transfusions.

urea The chief nitrogenous waste product of protein catabolism in the urine, formed in the liver from amino acids.

uremia Retention of urea and other products of protein catabolism due to inadequate kidney function.

urobilinogen A compound formed from bilirubin in the intestine; some is excreted in the feces while some is absorbed and enters the enterohepatic circulation, where it may be excreted either in the bile or in the urine.

vasa-, vaso- (La.) Pertaining to blood vessels.

vasa vasora Blood vessels that supply blood to the walls of large blood vessels.

vasectomy Surgical removal of a portion of the vas (ductus) deferens to induce infertility.

vasoconstriction Narrowing of the lumen of blood vessels due to contraction of the smooth muscles in their walls.

vasodilation Widening of the lumen of blood vessels due to relaxation of the smooth muscles in their walls.

vein A blood vessel that returns blood to the heart.

ventilation Breathing; the process of moving air into and out of the lungs.

vertigo A feeling of movement or loss of equilibrium.

virulent Pathogenic, or able to cause disease.

zygote A fertilized ovum.

zymogens Inactive enzymes that become active when part of their structure is removed by another enzyme or by some other means.

Credits

Chapter 1

Ed Reschke: 1.1, 1.2, 1.3, 1.7b, 1.9, 1.10. Carolina Biological Supply Company: 1.4. John Walters and Associates: 1.5, 1.6, 1.8, 1.11, 1.14b, 1.15b. Bloom, William and Fawcett, Don W. *A Textbook of Histology,* 10th ed., W. B. Saunders Company, 1975: 1.7a, 1.15a. Carroll Weiss/ Camera M. D. Studios, Inc.: 1.12. From Hole, John W., Jr. *Human Anatomy and Physiology,* 2nd ed. © 1978, 1981 Wm. C. Brown Publishers, Dubuque, Iowa. All Rights Reserved. Reprinted by permission: 1.13, 1.14a. After T. Kuwabara, in Greep, Roy and Weiss, Leon. *Histology,* 2nd ed., McGraw-Hill Book Company, 1966: 1.15c. Ayres Associates: 1.16, 1.17, 1.18, 1.19, 1.20.

Chapter 2

John Walters and Associates: 2.1, 2.2, 2.3, 2.4, 2.6, 2.7, 2.8a–b, 2.20, 2.26b, 2.27b, 2.32. From Hole, John W., Jr. *Human Anatomy and Physiology,* 2nd ed. © 1978, 1981 Wm. C. Brown Publishers, Dubuque, Iowa. All Rights Reserved. Reprinted by permission: 2.5. Ayres Associates: 2.9a–c, 2.10, 2.11, 2.12, 2.13a–c, 2.14, 2.15a–b, 2.16a–b, 2.17a–b, 2.18, 2.19, 2.21, 2.22, 2.23, 2.24, 2.25, 2.26a, 2.29, 2.30, 2.31, 2.33. Ed Reschke: 2.28. From Fox, Stuart I., *A Laboratory Guide to Human Physiology,* 2nd ed. © 1976, 1980 Wm. C. Brown Publishers, Dubuque, Iowa. All Rights Reserved. Reprinted by permission: table 2.2.

Chapter 3

John Walters and Associates: 3.1, 3.15b, 3.21. Orci, L. and Perrelet, A. *Science* 181(1973)868–869, © 1973 by American Association for the Advancement of Science: 3.2. Richard Chao: 3.3, 3.7b, 3.9, 3.23. Kwang W. Jeon: 3.4. Sandra L. Wolin: 3.5. R. H. Albertin, M.A.: 3.6. Keith R. Porter: 3.7a, 3.20, 3.24. E. G. Pollack: 3.8. From Mader, Sylvia S., *Inquiry Into Life,* 3d. ed. © 1976, 1979, 1982 Wm. C. Brown Publishers, Dubuque, Iowa. All Rights Reserved. Reprinted by permission: 3.10. Ayres Associates: 3.11, 3.14, 3.17, 3.18, 3.19. Courtesy of the Upjohn Company: 3.12. Carolina Biological Supply Company (photos): 3.13a–e. From Mader, Sylvia S., *Inquiry Into Life,* 2d ed. © 1976, 1979 Wm. C. Brown Publishers, Dubuque, Iowa. All Rights Reserved. Reprinted by permission (line drawings): 3.13a–e. Oscar L. Miller, Jr. *Journal of Cell Physiology,* 74(1969)225–232: 3.15. Alexander Rich: 3.16. Daniel S. Friend: 3.22.

Chapter 4

Ayres Associates: 4.1, 4.2, 4.3, 4.4, 4.6, 4.7, 4.8, 4.9, 4.10, 4.11, 4.12, 4.13, 4.14, 4.15, 4.16, 4.17a–b, 4.18. John Walters and Associates: 4.5.

Chapter 5

Ayres Associates: 5.1, 5.2, 5.3, 5.4, 5.5, 5.6, 5.7, 5.8, 5.9, 5.11, 5.12, 5.13, 5.14, 5.15, 5.16, 5.17, 5.18, 5.19. H. Fernandes-Morán V., M.D., Ph.D.: 5.10.

Chapter 6

Ayres Associates: 6.1, 6.2, 6.4, 6.5, 6.6a–b, 6.7, 6.8, 6.9, 6.10, 6.12, 6.13, 6.14, 6.15, 6.16, 6.17, 6.18, 6.19, 6.20, 6.21. Dr. Carolyn Chambers: 6.3a. Kessel, R. G. and Kardon, R. H. *Tissues and Organs: A Text-Atlas of Scanning Electron Microscopy.* © 1979 by W. H. Freeman and Company: 6.3b. Richard Chao: 6.11.

Chapter 7

From Hole, John W., Jr. *Human Anatomy and Physiology,* 2nd ed. © 1978, 1981 Wm. C. Brown Publishers, Dubuque, Iowa. All Rights Reserved. Reprinted by permission: 7.1, 7.9a. John Walters and Associates: 7.2, 7.5, 7.7, 7.9b, 7.18, 7.21, 7.22, 7.23, 7.25, 7.26, 7.27. Ayres Associates: 7.4, 7.11, 7.12, 7.13, 7.15, 7.16, 7.17, 7.24, 7.28, 7.29. H. Webster, from Hubbard, John. *The Vertebrate Peripheral Nervous System.* Plenum Publishing Corporation, 1974: 7.6. Porter and Bonneville. *An Introduction to Fine Structures of Cells and Tissues,* 4th ed. © Lea and Febiger 1973: 7.8. Ed Reschke: 7.10. Bell et al. *Textbook of Physiology and Biochemistry,* 8th ed., Churchill Livingstone, Inc., 1972: 7.14. Gilula, Reeves, and Steinbach. *Nature,* 235: 262–265. © Macmillan Journals Limited: 7.19a and c. Randy Perkins, from Heuser and Reese. *Handbook of Physiology:* 7.20.

Chapter 8

From Hole, John W., Jr. *Human Anatomy and Physiology,* 2nd ed. © 1978, 1981 Wm. C. Brown Publishers, Dubuque, Iowa. All Rights Reserved. Reprinted by permission: 8.1. From Van De Graaff, Kent M., *Human Anatomy.* © 1984 Wm. C. Brown Publishers, Dubuque, Iowa. All Rights Reserved. Reprinted by permission: 8.2. Ed Reschke: 8.3a, 8.7. Kessel, R. G. and Kardon, R. H. *Tissues and Organs: A Text-Atlas of Scanning Electron Microscopy.* © 1979 by W. H. Freeman and Company: 8.3b, 8.21c. From Fox, Stuart I., *A Laboratory Guide to Human Physiology,* 2nd ed. © 1976, 1980 Wm. C. Brown Publishers, Dubuque, Iowa. All Rights Reserved. Reprinted by permission: 8.4. Courtesy HEALTHDYNE, Marietta, Georgia: 8.5. John Walters and Associates: 8.8, 8.9a–b, 8.10, 8.12, 8.13, 8.14, 8.15, 8.16a, 8.17, 8.23, 8.24, 8.25b, 8.26, 8.27. R. H. Albertin, M.A.: 8.11, 8.28. From Peele, T. L. *The Neuroanatomic Basis for Clinical Neurology,* 2nd ed. Copyright © 1961, McGraw-Hill Book Company, New York. Used with permission: 8.16b. Ayres Associates: 8.18, 8.25a, 8.29, 8.30, 8.32. Geraldine F. Gauthier: 8.19. Dr. H. E. Huxley: 8.20, 8.21a–b, 8.22a. Dr. V. Dubowitz, from Bourne, G. H. *The Structure and Function of Muscle,* vol. III, 2nd ed., Academic Press, 1973: 8.31. Costill, David L. *The Physician and Sports-medicine,* 2:36–41, 1974: 8.33a–b.

Chapter 9

Ed Reschke: 9.1. From Hole, John W., Jr. *Human Anatomy and Physiology,* 2nd ed. © 1978, 1981 Wm. C. Brown Publishers, Dubuque, Iowa. All Rights Reserved. Reprinted by permission: 9.2, 9.7. Avril Somylo, Ph.D.: 9.3. A. Ábrahám: 9.4. From Van De Graaff, Kent M., *Human Anatomy.* © 1984 Wm. C. Brown Publishers, Dubuque, Iowa. All Rights Reserved. Reprinted by permission: 9.5, 9.9, 9.12, 9.17a. John Walters and Associates: 9.6, 9.13, 9.15. R. H. Albertin, M.A.: 9.8, 9.10, 9.11 (9.10 and 9.11 are adapted from original paintings by Frank H. Netter, M.D. from *The CIBA Collection of Medical Illustrations,* copyright by CIBA Pharmaceutical Company, Division of CIBA-GEIGY Corporation), 9.16b, 9.17b (9.16b and 9.17b from Noback, Charles R., and Robert J. Demarest, *The Human Nervous System: Basic Principles of Neurobiology,* 2d ed. © Copyright 1975 McGraw-Hill Book Company. Used with permission). Ayres Associates: 9.14. Yokochi, C. and Rohen, J. W. *Photographic Anatomy of the Human Body,* 2nd ed. © Igaku-Shoin, Ltd., 1978: 9.16a.

Chapter 10

R. H. Albertin, M.A.: 10.1, 10.8 (10.8 is adapted from an original painting by Frank H. Netter, M.D. from *The CIBA Collection of Medical Illustrations,* copyright by CIBA Pharmaceutical Company, Division of CIBA-GEIGY Corporation), 10.15, 10.16, 10.23, 10.24, 10.26 (10.23, 10.24, and 10.26 are adapted from original paintings by Frank H. Netter, M.D. from *The CIBA Collection of Medical Illustrations,* copyright by CIBA Pharmaceutical Company, Division of CIBA-GEIGY Corporation). From Hole, John W., Jr. *Human Anatomy and Physiology,* 2nd ed. © 1978, 1981 Wm. C. Brown Publishers, Dubuque, Iowa. All Rights Reserved. Reprinted by permission: 10.2, 10.13, 10.14, 10.20b, 10.39, 10.46. John Walters and Associates: 10.3a–c, 10.5, 10.7,

10.10, 10.17a–c, 10.19, 10.25a–b, 10.27, 10.30, 10.35a–b, 10.37, 10.38, 10.45, 10.47, 10.48, 10.52. Ayres Associates: 10.4a–b, 10.6, 10.11a–d, 10.21, 10.32, 10.43, 10.44, 10.50, 10.51a–b. From Peele, T. L. *The Neuroanatomic Basis for Clinical Neurology*, 2nd ed. Copyright © 1961 McGraw-Hill Book Company, New York. Used with permission: 10.9. From Fox, Stuart I., *A Laboratory Guide to Human Physiology*, 2nd ed. © 1976, 1980 Wm. C. Brown Publishers, Dubuque, Iowa. All Rights Reserved. Reprinted by permission: 10.12, 10.20a, 10.22, 10.33, 10.42. Dean E. Hillman: 10.18. Kessel, R. G. and Kardon, R. H. *Tissues and Organs: A Text-Atlas of Scanning Electron Microscopy.* © 1979 by W. H. Freeman and Company: 10.28. Redrawn from J. Robert McClintic: *Basic Anatomy and Physiology*, John Wiley & Sons, Inc., Publishers, New York, 1975: 10.29a. Ernest W. Beck: 10.29b (From Bloom and Fawcett, *Textbook of Histology*, 10th edition. Copyright © 1975 W. B. Saunders, Philadelphia, PA. Reprinted by permission), 10.31. Vaughan, D. and Asbury, T. *General Ophthalmology*, 7th ed., Lange Medical Publishers, 1974: 10.36a–b. From Mader, Sylvia S. *Inquiry Into Life*, 3d ed. © 1976, 1979, 1982 Wm. C. Brown Publishers, Dubuque, Iowa. All Rights Reserved. Reprinted by permission: 10.40a–c. From Van De Graaff, Kent M. *Human Anatomy* © 1984 Wm. C. Brown Publishers, Dubuque, Iowa. All Rights Reserved. Reprinted by permission: 10.41. David H. Hubel: 10.49.

Chapter 11

From Hole, John W., Jr. *Human Anatomy and Physiology*, 2nd ed. © 1978, 1981 Wm. C. Brown Publishers, Dubuque, Iowa. All Rights Reserved. Reprinted by permission: 11.1, 11.4a, 11.21b. From Mader, Sylvia S. *Inquiry Into Life*, 3d ed. © 1976, 1979, 1982, Wm. C. Brown Publishers, Dubuque, Iowa. All Rights Reserved. Reprinted by permission: 11.2. From Van De Graaff, Kent M., *Human Anatomy* © 1984, Wm. C. Brown Publishers, Dubuque, Iowa. All Rights Reserved. Reprinted by permission: 11.3a, 11.15, 11.19. Yokochi, C. and Rohen, J. W. *Photographic Anatomy of the Human Body*, 2nd ed. © Igaku-Shoin, Ltd., 1978: 11.3b. Modified Schaffer, from Bloom, William and Fawcett, Don W. *A Textbook of Histology*, 10th ed., W. B. Saunders Company, 1975: 11.4b. John Walters and Associates: 11.5a, 11.6, 11.9, 11.11, 11.21a, 11.22a–b, 11.23 (*Emergency Medicine*, Volume 11, Issue #2. February 15, 1979), 11.29. Kessel, R. G. and Kardon, R. H. *Tissues and Organs: A Text-Atlas of Scanning*

Electron Microscopy. © 1979 by W. H. Freeman and Company: 11.5b. Historical Pictures Service, Chicago: 11.7. After Kölliker and von Ebner, from Bloom, William and Fawcett, Don W. *A Textbook of Histology*, 10th ed., W. B. Saunders Company, 1975: 11.8. Don Fawcett: 11.10. Ayres Associates: 11.12, 11.13, 11.14, 11.18, 11.20, 11.24, 11.30. Ernest W. Beck: 11.16, 11.17. Richard Menard: 11.25a–b. James Shaffer: 11.26a–c. Ed Reschke: 11.27. American Heart Association: 11.28a–b. From Kannel, W. B., W. P. Castell and T. Gordon. *Annals of Internal Medicine*, Vol. 90, page 85. Copyright © 1979 American College of Physicians. Philadelphia, PA. Used with permission: table 11.4.

Chapter 12

Ayres Associates: 12.1, 12.2, 12.3, 12.4, 12.5, 12.6 (From *Circulation*, by Bjorn Folkow and Eric Neil. Copyright © 1971 by Oxford University Press, Inc. Reprinted by permission), 12.10, 12.11, 12.12, 12.13, 12.16, 12.20, 12.23, 12.24 (Redrawn and reproduced with permission from Geddes, L. A. *The Direct and Indirect Measurement of Blood Pressure*. Copyright © 1970 by Year Book Medical Publishers, Inc., Chicago), 12.25. John Walters and Associates: 12.7, 12.8, 12.14, 12.15a and b, 12.17 (From Feigel, E. O., "Physics in the cardiovascular system," in *Physiology and Biophysics*, Volume II, T. C. Ruch and H. D. Patton, editors. Copyright © 1974 W. B. Saunders, Philadelphia, PA. Reprinted by permission), 12.19, 12.22. Markell, E. K. and Voge, M. *Medical Parasitology*, 5th ed., W. B. Saunders Company, 1981: 12.9. From Fox, Stuart I., *A Laboratory Guide to Human Physiology*, 2nd ed. © 1976, 1980 Wm. C. Brown Publishers, Dubuque, Iowa. All Rights Reserved. Reprinted by permission: 12.21. From Diem, K., and Lentner, C., Eds., *Documenta Geigy Scientific Tables*, 7th ed., J. R. Geigy S. A., Basle, Switzerland, 1970. With permission: table 12.6.

Chapter 13

Ayres Associates: 13.1 (From *Circulation*, by Bjorn Folkow and Eric Neil. Copyright © 1971 by Oxford University Press, Inc. Reprinted by permission), 13.2 (From Mellander, S., and B. Johansson in *Pharmacological Reviews*. Copyright © 1968. American Society for Pharmacology & Experimental Therapeutics. Williams & Wilkins Company. Baltimore, MD. Used with permission), 13.7, 13.11, 13.14, 13.15, 13.16. R. H. Albertin, M.A.: 13.3a (Reprinted with the permission of Leon Schlossberg), 13.5 (Reproduced by permission of *Practical Cardiology*), 13.8a. Carroll

Weiss/Camera M.D. Studios, Inc.: 13.3b. Donald S. Baim, from Hurst et al. *The Heart*, 5th ed., McGraw-Hill Book Company, 1982: 13.4a–b. John Walters and Associates: 13.6 (Modified from Astrand, P., and K. Rodahl. *Textbook of Work Physiology*. © Copyright 1977 McGraw-Hill Book Company. Used by permission), 13.10, 13.12. Fein, J. M. *Scientific American*, 238(1978)66: 13.8b. Niels, A. Lassen: 13.9a–b. Emil Bernstien and Eila Kairinen, Gillette Research Institute: 13.13. From Fox, Stuart I., *A Laboratory Guide to Human Physiology*, 2nd ed. © 1976, 1980 Wm. C. Brown Publishers, Dubuque, Iowa. All Rights Reserved. Reprinted by permission: table 13.3. From Wilson, R. F., editor, *Principles and Techniques of Critical Care*, Vol. 1. Copyright © 1977. F. A. Davis Company, Philadelphia, PA. Used with permission: table 13.5. From Mason, D. T., *Modern Concepts in Cardiovascular Disease*. Copyright © 1967, Volume 36. Reprinted by permission of the American Heart Association, Inc.: table 13.7. Abstracted by permission of *The New England Journal of Medicine*, Volume 302, pages 37–48, 1980.

Chapter 14

John Walters and Associates: 14.1, 14.3 (From Weibel, E. R. *Morphometry of the Human Lung*, Springer-Verlag OHG, Heidelberg, 1963. Reprinted by permission), 14.13, 14.15, 14.17, 14.18, 14.27, 14.28. American Lung Association: 14.2a–b, 14.6, 14.7. West, J. B. *Respiratory Physiology: The Essentials*, © 1979, Williams and Wilkins Company, Baltimore: 14.4a. From Hole, John W., Jr. *Human Anatomy and Physiology*, 2nd ed. © 1978, 1981 Wm. C. Brown Publishers, Dubuque, Iowa. All Rights Reserved. Reprinted by permission: 14.4b, 14.8, 14.21. Yokochi, C. and Rohen, J. W. *Photographic Anatomy of the Human Body*, 2nd ed. © Igaku-Shoin, Ltd., 1978: 14.5. Edward C. Vasquez, R. T., C.R.T., Department of Radiologic Technology, Los Angeles City College: 14.9a–b, 14.12. Ayres Associates: 14.10, 14.11, 14.14, 14.22, 14.23, 14.24, 14.26a–b. Comroe, J. H., Jr. *Physiology of Respiration*, © 1974 Year Book Medical Publishers, Inc., Chicago: 14.16a–b. Oscar Auerbach: 14.19. Courtesy of Warren E. Collins, Braintree, MA.: 14.20. Kessel, R. G. and Kardon, R. H. *Tissues and Organs: A Text-Atlas of Scanning Electron Microscopy*, © 1979 by W. H. Freeman and Company: 14.25a–b.

Chapter 15

John Walters and Associates: 15.1, 15.2a, 15.10. Ayres Associates: 15.2b, 15.3, 15.4, 15.5, 15.7, 15.13, 15.14, 15.15, 15.16 (From Wasserman, K., A. L. Van Kessel and G. G. Burton. *Journal of Applied Physiology*. Copyright © 1967 American Physiological Society. Bethesda, MD. Used with permission.) McCuray, P. R. *Sickle-Cell Disease*. © Medcom Inc. 1973, Reprinted with permission: 15.6a–d. From Hole, John W., Jr. *Human Anatomy and Physiology*, 2nd ed. © 1978, 1981 Wm. C. Brown Publishers, Dubuque, Iowa. All Rights Reserved. Reprinted by permission: 15.8, 15.9, 15.11, 15.12.

Chapter 16

From Mader, Sylvia S. *Inquiry Into Life*, 3d ed. © 1976, 1979, 1982 Wm. C. Brown Publishers, Dubuque, Iowa. All Rights Reserved. Reprinted by permission: 16.1, 16.2. From Van De Graaff, Kent M., *Human Anatomy*, © 1984 Wm. C. Brown Publishers, Dubuque, Iowa. All Rights Reserved. Reprinted by permission: 16.3a. Kessel, R. G. and Kardon, R. H. *Tissues and Organs: A Text-Atlas of Scanning Electron Microscopy.* © 1979 by W. H. Freeman and Company: 16.3b. John Walters and Associates: 16.4, 16.5, 16.7, 16.9a (Brenner, B. M., and R. Beeuwkes, III, *Hospital Practice*, July 1978, p. 40 [Artist: Nancy Lou Gahan]), 16.10, 16.12, 16.13, 16.14, 16.15a–b, 16.18, 16.19, 16.20a–c, 16.21a–d, 16.22, 16.23, 16.24a–b, 16.26, 16.27. From Hole, John W., Jr. *Human Anatomy and Physiology*, 2nd ed. © 1978, 1981 Wm. C. Brown Publishers, Dubuque, Iowa. All Rights Reserved. Reprinted by permission: 16.6, 16.8a. Ayres Associates: 16.11, 16.16, 16.17 (J. P. Kokko, *Hospital Practice*, Feb. 1979, p. 110 [Artist: Alan Iselin]), 16.25. Gordon F. Leedale/Biophoto Associates: 16.8b. Daniel Friend, from Bloom, William and Fawcett, Don W. *A Textbook of Histology*, 10th ed., W. B. Saunders Company, 1975: 16.9b.

Chapter 17

John Walters and Associates: 17.2, 17.3, 17.4, 17.5, 17.6, 17.7, 17.9a–b, 17.10, 17.11, 17.12, 17.13, 17.14, 17.15, 17.19, 17.20, 17.21, 17.22b, 17.23, 17.26, 17.27, 17.28. Ayres Associates: 17.16, 17.17, 17.18. Rosenthal, Alan S. *New England Journal of Medicine*, 303(1980)1153: 17.22a. Noel R. Rose, M.D.: 17.25a–b. Philip S. Norman, from Middleton, E., Reed, C. E. and Ellis, E. F. *Allergy: Principles and Practice*, C. V. Mosby Company, 1978: 17.29a–b. Kessel, R. G. and Shih,